Lecture Notes on Data Engineering and Communications Technologies

242

Series Editor

Fatos Xhafa, *Technical University of Catalonia, Barcelona, Spain*

The aim of the book series is to present cutting edge engineering approaches to data technologies and communications. It will publish latest advances on the engineering task of building and deploying distributed, scalable and reliable data infrastructures and communication systems.

The series will have a prominent applied focus on data technologies and communications with aim to promote the bridging from fundamental research on data science and networking to data engineering and communications that lead to industry products, business knowledge and standardisation.

Indexed by SCOPUS, INSPEC, EI Compendex.

All books published in the series are submitted for consideration in Web of Science.

Zhengbing Hu · Felix Yanovsky · Ivan Dychka ·
Matthew He
Editors

Advances in Computer Science for Engineering and Education VII

Springer

Editors
Zhengbing Hu
International Center of Informatics and Computer Science, Faculty of Applied Mathematics
National Technical University of Ukraine "Kyiv Polytechnic Institute"
Kyiv, Ukraine

Ivan Dychka
Faculty of Applied Mathematics
National Technical University of Ukraine
Kyiv, Ukraine

Felix Yanovsky
Faculty of Civil Engineering and Geoscience
Delft University of Technology (TU Delft)
Delft, The Netherlands

Faculty of Air Navigation, Electronics and Telecommunications
State University "Kyiv Aviation Institute"
Kyiv, Ukraine

Matthew He
Halmos College of Arts and Sciences
Nova Southeastern University
Florida, USA

ISSN 2367-4512 ISSN 2367-4520 (electronic)
Lecture Notes on Data Engineering and Communications Technologies
ISBN 978-3-031-84227-6 ISBN 978-3-031-84228-3 (eBook)
https://doi.org/10.1007/978-3-031-84228-3

This Springer imprint is published by the registered company Springer Nature Switzerland AG
The registered company address is: Gewerbestrasse 11, 6330 Cham, Switzerland

Preface

Computer science is playing an increasingly significant role in various engineering and education fields, including natural and social sciences. Computer science is constantly opening up new opportunities in artificial intelligence, information retrieval, various engineering fields, education, and talent cultivation, resulting in increasing understanding and attention from the scientific and industrial communities. Cultivating the next generation of professionals to correctly use and promote computer science has become a challenging task for higher education institutions around the world. Strengthening international cooperation and communication is undoubtedly momentous for solving the key issue of talent training, especially for cultivating talents with advanced computer skills and comprehensive qualities.

Driven by the aforementioned factors, the 7th International Conference on Computer Science, Engineering, and Educational Applications (ICCSEEA2024) received active responses from numerous schools and research institutions around the world. It has been jointly organized by the National Technical University of Ukraine "Igor Sikorsky Kyiv Polytechnic Institute", National Aviation University, Lviv Polytechnic National University, Kharkiv National University of Radio Electronics, Wuhan University of Technology, Polish Society of Operations and Systems, and International Association for Modern Education and Computer Science Research and was held online in Kyiv, Ukraine, during April 27–28, 2024. ICCSEEA2024 brings together top scholars from around the world to share their findings and discuss outstanding challenges in computer science, engineering, and educational applications.

Out of all the submissions, the best contributions to the conference were selected by the program committee and expert group for inclusion in this book.

May 2024

Zhengbing Hu
Felix Yanovsky
Ivan Dychka
Matthew He

Preface

Organization

General Chairs

Felix Yanovsky	Delft University of Technology, Netherlands
Ivan Dychka	National Technical University of Ukraine "Igor Sikorsky Kyiv Polytechnic Institute", Ukraine
Anatoliy Sachenko	Kazimierz Pułaski University of Technology and Humanities in Radom, Poland
Oleksandr Pavlov	National Technical University of Ukraine "Igor Sikorsky Kyiv Polytechnic Institute", Kyiv, Ukraine

Online Conference Organizing Chairs

Vadym Mukhin	National Technical University of Ukraine "Igor Sikorsky Kyiv Polytechnic Institute", Ukraine
A. V. Petrashenko	National Technical University of Ukraine "Igor Sikorsky Kyiv Polytechnic Institute", Ukraine
Orest Kochan	Lviv Polytechnic National University, Ukraine
Artem Volokyta	National Technical University of Ukraine "Igor Sikorsky Kyiv Polytechnic Institute", Kyiv, Ukraine

Program Chairs

Yaroslav Kornaga	National Technical University of Ukraine "Igor Sikorsky Kyiv Polytechnic Institute", Kyiv, Ukraine
Yevgeniya Sulema	National Technical University of Ukraine "Igor Sikorsky Kyiv Polytechnic Institute", Kyiv, Ukraine
Q. Y. Zhang	Wuhan University of Technology, China

Publication Chairs

Z. B. Hu	National Technical University of Ukraine "Igor Sikorsky Kyiv Polytechnic Institute", Ukraine
Felix Yanovsky	Delft University of Technology, Delft, Netherlands
Ivan Dychka	National Technical University of Ukraine "Igor Sikorsky Kyiv Polytechnic Institute", Kyiv, Ukraine
Matthew He	Nova Southeastern University, USA

Publicity Chairs

Roman Kochan	University of Bielsko-Biała, Bielsko-Biala, Poland
Solomiia Fedushko	Lviv Polytechnic National University, Ukraine
Oleksandra Yeremenko	Kharkiv National University of Radio Electronics, Kharkiv, Ukraine

Program Committee Members

Oleg Chertov	National Technical University of Ukraine "Igor Sikorsky Kyiv Polytechnic Institute", Kyiv, Ukraine
G. E. Zhang	Nanning University, China
Taras V. Panchenko	Kyiv National Taras Shevchenko University, Ukraine
M. Y. Zhang	Wuhan University of Technology, China
Volodymyr Sokolov	Borys Grinchenko Kyiv University, Ukraine
Yuriy Ushenko	Chernivtsi National University, Chernivtsi, Ukraine
M. Wang	Wuhan Technology and Business University, China
Ivan Izonin	Lviv Polytechnic National University, Lviv, Ukraine
Oleksii K. Tyshchenko	University of Ostrava, Czech Republic
Rabah Shboul	Al-albayt University, Jordan
L. J. Du	Wuhan University of Technology, China
Ihor Tereikovskyi	National Technical University of Ukraine "Igor Sikorsky Kyiv Polytechnic Institute", Ukraine
Q. L. Zhou	Shizuoka University, Japan

Conference Organizers and Supporters

National Technical University of Ukraine "Igor Sikorsky Kyiv Polytechnic Institute", Ukraine
National Aviation University, Ukraine
Lviv Polytechnic National University, Ukraine
Kharkiv National University of Radio Electronics, Kharkiv, Ukraine
Wuhan University of Technology, China
Polish Operational and Systems Society, Poland
International Research Association of Modern Education and Computer Science, Hong Kong

Contents

Computer Science and Software Engineering

Experimental Study and Value Analysis of Green Jadeite Image Acquisition Based on HSL Chromaticity Parameters

Lili Zhang[1](✉) and Xinqiang Yuan[2]

[1] Science and Technology Department, Wuhan Technology and Business University, Wuhan 430065, China
52844133@qq.com

[2] Gemological Institute, China University of Geoscience, Wuhan 430074, China

Abstract. The current jadeite market has not yet established a unified standard for evaluating the color of green jadeite. Based on the theory and technique of green jadeite color grading by HSL chromaticity, machine vision method is proposed to be used in this study. Based on the chromaticity principle, this study quantifies the parametric index of green jadeite color by capturing the color of green jadeite samples with a color digital camera under a white light source of color temperature 5500 K and analyzing the HSL value for each sample image color based on the HSL color space model as the theoretical basis. In addition, the study also cast the obtained HSL values onto a two-dimensional SL image by different hue (H) to obtain the HSL distribution intervals of emerald green color and its chromatic characteristics, and to determine the HSL value boundaries of various color types of green emerald. Based on the above information, we can distinguish and determine the hue of green jadeite, classify the color intensity and brightness of green jadeite, and quantify the color type of green jadeite.

Keywords: HSL value analysis · Chromaticity · HSL color space · Color threshold · Jadeite color

1 Introduction

Experimental research has been increasingly widely applied in various disciplinary fields, such as engineering, education, and even agriculture and commerce [1–3]. In the business of jewelry research, this is no exception. The assessment of jadeite jewellery usually includes both quality and value [4]. The traditional assessment of the quality and value of jadeite is based on an analysis of the various quality factors affecting the value of jadeite, combined with one's own experience, to make an estimate and judgement of the market value of jadeite [5–7]. However, due to the specificity of jadeite's own structure and the differences in the cultural backgrounds of appraisers, very different prices can occur for the same type of jadeite, leading to uncertainty about the value of jadeite and causing confusion and instability in the market [8]. Despite the promulgation of national jadeite grading standards and scholars' research into all aspects of jadeite, various levels

Z. Hu et al. (Eds.): ICCSEEA 2024, LNDECT 242, pp. 3–13, 2025.
https://doi.org/10.1007/978-3-031-84228-3_1

of quality assessment systems are being established, but there is a significant lag and inoperability in the commercial assessment and market value of jadeite research results, making it very difficult to assess the quality and value of jadeite jewellery. As a result, the market urgently needs the easiest and most accurate method of grading the colour of jadeite [9].

Research by scholars in recent years into the evaluation methods of jadeite colour has been divided into two broad categories, using quantitative colour measurement methods and semi-quantitative colour-card comparison methods, supported by the theory of chromaticity to evaluate the colour of jadeite [10].

The quantitative colour measurement method refers to the use of physical instruments to measure colour on behalf of the human eye, using the principles of chromaticity to provide an accurate digital representation of the physical characteristics of jadeite colour, which can more accurately reflect its visual characteristics of colour and is of great significance to the evaluation of jadeite grades. In the paper "Feasibility study on the chromaticity measurement of jadeite colour", the author Wang Rong et al. used the main wavelength values derived from the CIE 1931 chromaticity system to determine the dominant colour of jadeite, demonstrating the feasibility of using chromaticity in the measurement and evaluation of jadeite colour. That is, chromatic measurements can distinguish and determine the green tones of jadeite, they can delineate the concentration of jadeite, they can identify and rate the black tones of jadeite, and finally they can delineate the green range of saturated jadeite [11]. However, this research paper does not examine jadeite of the same hue, different transparency and texture, nor does it establish a system for evaluating jadeite colour. In 'The influence of colour rendering of light sources on the evaluation of the green colour of jadeite', the author, Wang Jinhuan, builds on previous research into the quantitative characterisation of green jadeite colour, quantifying the colour numerically and distinguishing the differences between green jadeite. This article selects the current international advanced colour difference formula and compares and analyses its calculation results, concluding that the CIE DE 2000 colour difference formula is more suitable for green jadeite grading [12]. In the article 'Green Jadeite Colour Index Characteristics', the author, Ziwei Tang, also uses the principles of chromaticity to quantitatively describe the colour of green jadeite by measuring the green jadeite reflection curve and chromaticity parameters using the USB2000 instrument, using the International Commission on Illumination CIE1931-XYZ chromaticity coordinates and colour index analysis method. She compared the colour of the visual samples with the instrumental data, and the experimental results were more consistent, indicating that the chromatic measurements were more accurate in characterising the colour of green jadeite, and that the colour of the different varieties of green jadeite samples was regularly distributed [13]. The scientific and accurate method of evaluating the colour of jadeite from quantitative colourimetric data has evolved to become more and more objective and reliable, but has the disadvantage that the instrument is not easy to carry around and has certain testing requirements. The theory is too complex and requires extensive measurements and calculations and is not suitable for widespread market dissemination.

In addition, Liu Yan proposed in his article "Green Jade Color Grading Method" that in jade trade, the color grading of jade is usually based on the inner eye evaluation, sometimes referring to the comparison of color stones, and sometimes not using color stones as a reference. Such visual color grading is not accurate. This article proposes a jade color grading method for color calibration in the CIE uniform color space through the measurement of double integrating spheres. Through this method, gemstone laboratory inspectors, jewelers and traders in the jade market can accurately and quickly grade the color of jade. In addition, the optical origin of "imperial green" and "apple green" in jadeite was determined through this method. The color appearance of imperial green is attributed to the high transmittance of dry light, while the color appearance of apple green is attributed to the high diffuse reflectance of light [14]. Xu Rupeng's article "Quantitative Description of Jade Color" uses the International Commission on Illumination CIE-L*a*b* (1976) chromaticity coordinates to quantitatively describe the color of jade [15]. Compare the color of the visual sample with the data of the Swiss Elrepho 2000 rapid colorimeter for testing jade, and quantitatively represent the color of the jade. This method starts from the perspective of chromaticity and first determines the color measurement conditions for measuring green jadeite samples. After comparing the conditions of observing jade in practice, it was determined that D65 light source was used as the illumination light source, and the color measurement condition was 0/45 (or 45/0). Based on this, a device for measuring the color of jade was designed. The device is based on an RS-650 color brightness meter, measuring the reflection spectrum of jade under 0/45 lighting and observation conditions, and using this to calculate the color coordinate position of the sample. In order to visually read the color of jadeite (yellow in green or blue in green), the CIE-Lab chromaticity coordinates were used in this project. The measurement uncertainty of the measured sample (with good color uniformity) was 3NBS, which was within the visual "distinguishable" range specified in chromaticity. The actual comparison was consistent with the visual results.

Another method is the study for the color card comparison method. The color card comparison method is represented by Munsell system and GemDialogue system, which is generally to label the colors in a three-dimensional model system, and to calibrate the hue, chroma and brightness of emerald color elements with the help of color cards or color plates. Scholars such as Hui Zhang and Peili Zhang measured the reflectance spectra of the samples using a spectrophotometer, calculated the reflectance and transmission spectra of the jadeite samples, calculated their chromaticity parameters, and then described the colors of the samples by two color characterization systems, GemDialogue and Munsell, for comparison, and concluded that the GemDialogue color card comparison method reflected the visual characteristics of jadeite samples to some extent [16]. However, the article did not quantify the parametric indicators of the color of the jadeite samples. In the paper "Green jadeite color grading study", Zhang Xiaowei determined the distribution range and distribution law of green jadeite in three-dimensional space by quantifying the color with three values of L^* a^* b^* in the CIE LAB color space [17]. In addition, he also analyzed the advantages and disadvantages of the existing color card, combined with the previous research on the distribution range and distribution law of jadeite green, and made a new green jadeite special color card. Therefore, the current gemstone color card used in the specific jadeite color is not detailed enough, the color division does not follow the characteristics of jadeite color, there are certain constraints on the evaluation of jadeite color.

In response to the shortcomings of the previously conducted studies, the innovations of this study are as follows:

The entry point of this study is relatively new, as HSL color space is used for the first time in green jadeite color grading. It is closer to people's understanding of color than RGB color space, and can present the hue, saturation and brightness of each green jadeite more visually and graphically through HSL two-dimensional coordinate cast point diagram.

This study uses a quick and convenient method to quantify the color of jadeite, and quantifies the color of all kinds of green jadeite, and gives a digital analysis of green jadeite whether it is uniform or uneven, deep or light in concentration, and high or low in brightness. In other words, one can have objective and comprehensive information about the color type, color index and uniformity of green jadeite through digital analysis.

2 Definition of Concept

The RGB parameter is derived from the visual trichromatic theory that all the colors that exist in nature can be synthesized from three colors: red *(R)*, green *(G)* and blue *(B)*. The formula of this principle is $C\ (C) = R\ (R) + G\ (G) + B\ (B)$, where *R (R), G (G), B (B)* indicate the number of red light, green and blue light, and the so-called RGB parameters, and *C (C)* indicates the color and brightness of these quantities of red, green and blue light mixture. The color CCD obtains RGB data directly. Although RGB parameters are simple and clear, and can completely express the diversified colors of jadeite, its disadvantage is that the expression is more abstract and different from people's understanding of color, which is not convenient to summarize the laws of jadeite color change and distribution, and is not conducive to the promotion of this research result. The chromaticity parameter of HSL is used in this research project. The three letters are HUE, which represent the range of colors perceived by the human eye, and the values are taken from *0°* to *360°* of the circle center angle, and each angle represents one color. As shown in Figs. 1 and 2, the colors are distributed on the hue ring with red, yellow, green, cyan, blue, and magenta at *0°, 60°, 120°, 180°, 240°*, and *300°*, respectively. In terms of the green color of jadeite, *120°* to *139°* is the more pure green color, the value less than *120°* is the yellowish green color, more than *150°* is the blueish green color. S (Saturation) is the saturation of the color, in percent, describing the variation of color purity in the same hue and lightness. L (Lightness) is the lightness of the color, also in percent, indicating the change of lightness and darkness of the color. The S value and L value determine the vividness of the color. In fact, HSL parameters are obtained by converting the RGB parameters and parsing them to be closer to our perception of colour.

3 Experimental Equipment and Method

3.1 Experimental Equipment

A professional photography lamp with a color temperature of *5500K* was used for the experiment, and the image acquisition tool was the model VMD-P500B industrial camera (Fig. 3). To ensure the consistency of each acquisition environment, the researchers

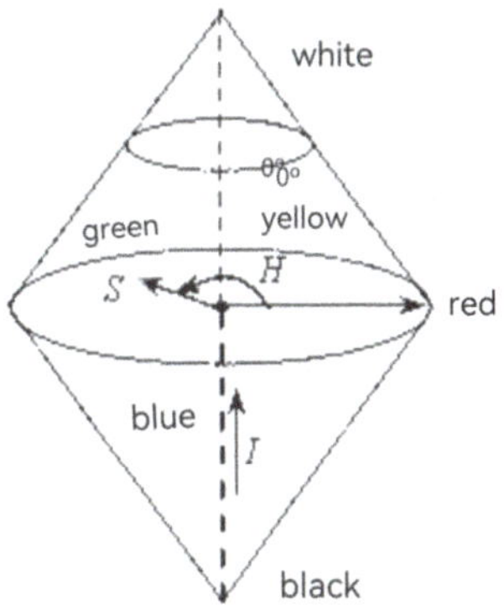

Fig. 1. HSL Three-dimensional Model

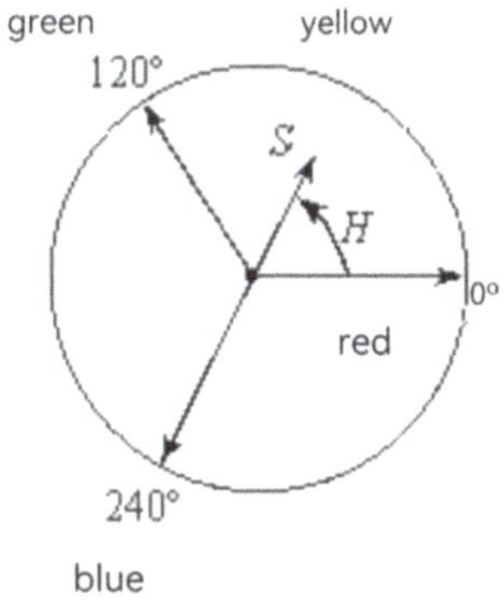

Fig. 2. HSL Color Wheel

strictly checked the index coefficients before the start of the acquisition, tested the color temperature of the lights with a color thermometer and used a *24*-color color card (Fig. 4) for color calibration, and the values should be in the range of 5500 K ± 100 K. In addition, the deviation of the HSL values of the green color card is guaranteed to be within ± 5, ensuring the consistency of the test conditions and the reproducibility of the experiment.

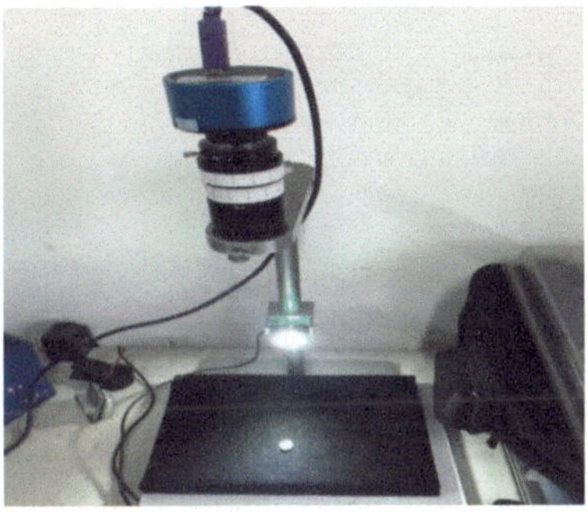

Fig. 3. The model VMD-P500B industrial camera

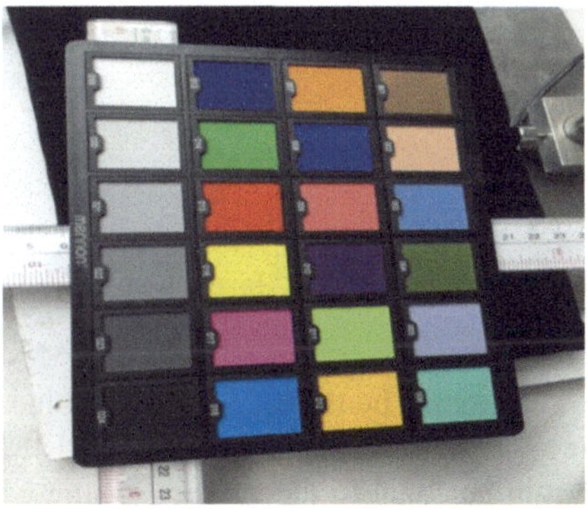

Fig. 4. 24-color color card

3.2 Experimental Method

This study proposes to use the machine vision method, which can also be referred to as the colour digital image analysis method. Colour digital capture devices use colour CCD chips, which are covered with light-sensitive photodiodes, each tiny photodiode constituting a so-called pixel. Colour CCDs have each pixel covered with a red, green or blue filter, making it possible for a given pixel to sense only red or green or blue light. When the jadeite is imaged by the digital lens onto the colour CCD chip, the pixels of the chip are sensitised to form an electrical signal, the intensity of which is expressed in 0–255 and is stored in the memory. These signals can be read by a computer and can be presented in colour on the display. If the read signal is analysed chromatically, quantitative data describing the colour of the image, such as R, G and B data, can be obtained. While RGB data (also known as tri-stimulus values) is the most basic chromatic parameter, i.e. any colour can be reproduced by mixing the three colours red (R), green (G) and blue (B), RGB data is the number of red, green and blue primary colours contained in this colour (as shown in Fig. 5). However, this description is different from what we are used to describing colours in everyday life and is not convenient to use. For this reason, RGB values can be converted by means of a conversion formula into hue *(H)*, saturation *(S)* and luminance *(L)* values (see formulae (1), (2) and (3).

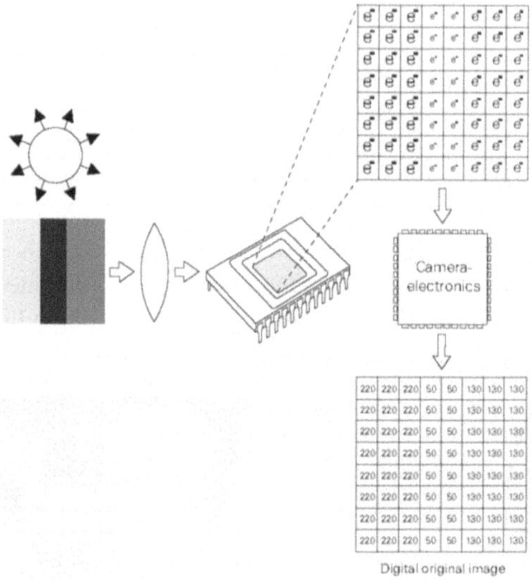

Fig. 5. Imaging process of color digital image equipment

Let (r, g, b) be the red, green and blue coordinates of a colour, respectively, whose values are real numbers between 0 and 1. Let max be equal to the largest of r, g and b. Let min be equal to the smallest of these values. To find the value of (h, s, l) in HSL space, here h ∈ [0, 360] is the angular hue angle and s, l ∈ [0, 1] are the saturation and luminance. The conversion Eqs. (1), (2), (3 are as follows.

$$\mathrm{H} = \begin{cases} 0^\circ, & if \max = \min \\ 60^\circ \times \frac{g-b}{\max - \min} + 0^\circ, & if \max = \mathrm{r\ and\ g} \geq \mathrm{b} \\ 60^\circ \times \frac{g-b}{\max - \min} + 360^\circ, & \mathrm{r\ and\ g} < \mathrm{b} \\ 60^\circ \times \frac{b-r}{\max - \min} + 120^\circ, & if \max = g \\ 60^\circ \times \frac{r-g}{\max - \min} + 240^\circ, & if \max = b \end{cases} \tag{1}$$

$$L = \frac{1}{2}(\max + \min) \tag{2}$$

$$S = \begin{cases} 0, & L = 0 \text{ or } \max = \min \\ \frac{\max - \min}{\max + \min} = \frac{\max - \min}{2l}, & if\ 0 < L \le \frac{1}{2} \\ \frac{\max - \min}{2-(\max + \min)} = \frac{\max - \min}{2-2l}, & if\ L > \frac{1}{2} \end{cases} \tag{3}$$

The RGB-HSL conversion formula is used to convert the RGB data obtained for each green emerald image colour into HSL data, and the cast point analysis is carried out based on the HSL data obtained. The advantages of the machine vision method of jadeite colour evaluation are:

(1) The technical conditions are relatively simple, and there are many colour digital photographic devices on the market now, so it is easy to find suitable equipment;
(2) He colour of the sample can be evaluated comprehensively, and any position of the sample being photographed can be evaluated;
(3) The evaluation is fast, and the results can be obtained within a few minutes after the sample has been captured by the image.

3.3 Setting the Sample Environment for Image Acquisition

The experiments were conducted using a professional photography lamp with a colour temperature of 5500K. The image acquisition tool was a model VMD-P500B industrial digital camera, which has 14 million pixels and an 8x zoom function, and has met the requirements of this experiment in terms of functionality. In this study, 4006 sample images were collected with the support of Shenzhen Jianxingli Emerald Company, of which 2379 valid sample images were used for testing. In order to ensure consistency in each collection environment, the index coefficients are strictly checked before the collection begins. The colour temperature of the lights is tested with a colour thermometer and the values in the range of 5500K $\pm$ 100 are checked with a 24-colour colour card, and the HSL values of the green colour card are deviated within the range of $\pm$ 5 to ensure the consistency of the test conditions and the repeatability of the experiment.

Before collecting the green jadeite colour images, 10 observers who had been working in the jadeite industry for many years on the subjective (naked-eye) colour classification of all physical samples were invited. To ensure the objectivity of the jadeite colour classification, the author provided a green jadeite colourimetric card as a reference when classifying the colours, with the colour classification criteria being as follows:

Emerald green: intense green colour in reflected light, emerald green in transmitted light.

Brilliant green: intense green colour in reflected light, less saturated and brighter than emerald green, yellow-green in transmitted light.

Yellow green: vibrant green colour in reflected light, yellow-green in transmitted light.

Bean green: bright green in reflected light, yellow-green, or bright green or blue-green in transmitted light.

Light bean green: brighter green colour in reflected light, less intense, green or blue-green tones in transmitted light.

Melon green: dark green colour in reflected light with a more pronounced grey tinge, green with a grey tinge in transmitted light.

Light green I: not a strong green colour in reflected light, green, or blue-green hue in transmitted light.

Light green II: the green colour is noticeably less intense in reflected light and light green in transmitted light.

Grey-green (oil-green): the green colour of the oil-green variety is distinctly impure, containing grey and blue elements, and is dull and not sufficiently vivid.

4 Experimental Results

In order to facilitate and intuitively summarize the data of point casting images, ORGIN software was used to comprehensively cast the HSL values of all green jadeite sample images tested in the experiment according to different hue value intervals (H values were divided into three intervals: *100° –119°, 120° –139°, 140° –159°*) on the two-dimensional map of the HSL color space model (the casting results are shown in Figs. 6, 7, and 8). The two-dimensional image of the HSL color space model is filled with H, S, and L values one by one using the HSL color mode in the drawing software. Produce an HSL two-dimensional image based on the three previously divided hue (H) value intervals: the brightness of the L value is on the horizontal axis, and the saturation of the S value is on the vertical axis (H is in degrees, S and L are in percentages).

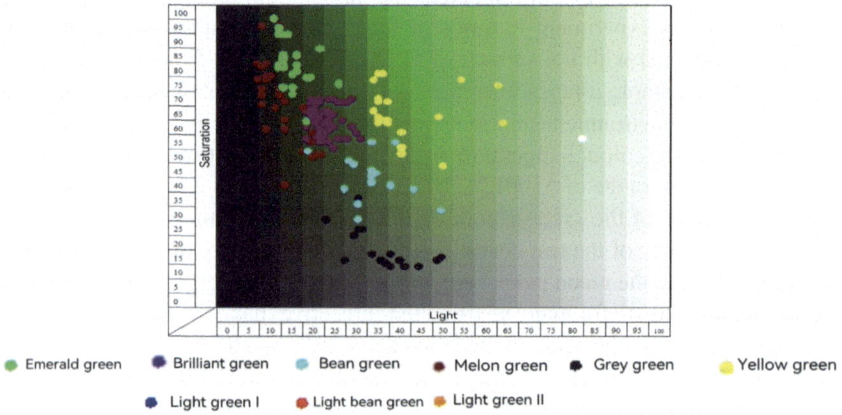

Fig. 6. Color distribution of green jadeite with H = *100°–119°*

5 Experimental Data Results and Analysis

From the above research experiments it can be concluded that, HSL values for green jadeite can be numerically described in terms of hue (hue), saturation and brightness, and can be used to differentiate between subtle colour differences. For example, if the jadeite is bean green, the range of cast points is very different for jadeite with yellow tones and jadeite with blue tones; it is also possible to distinguish between green jadeite with yellow, blue or grey tones.

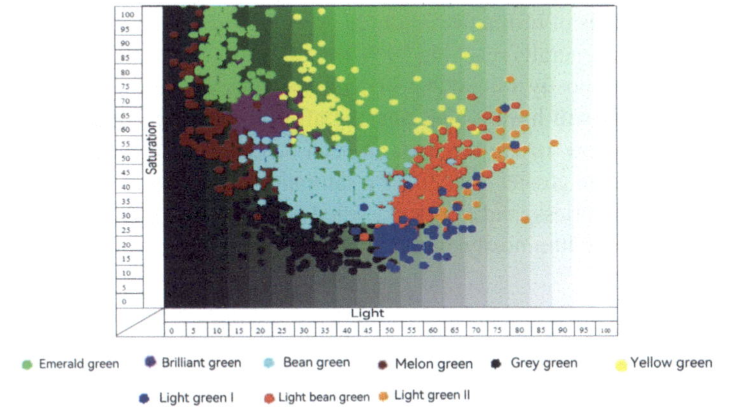

Fig. 7. Color distribution of green jadeite with H = *120°–139°*

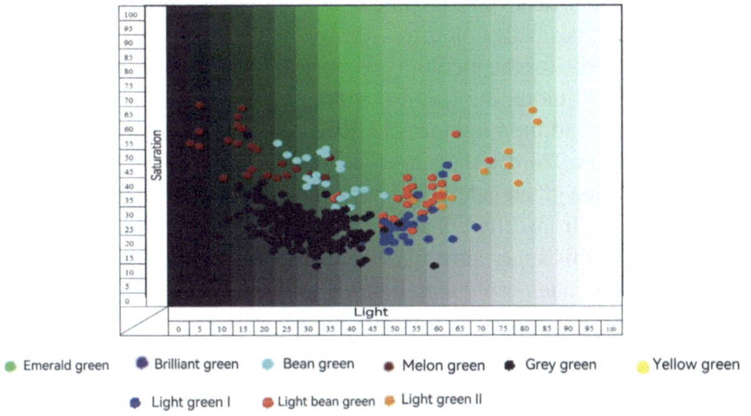

Fig. 8. Color distribution of green jadeite with H = *140°–159°*

The distribution of the HSL values of the green jadeite is generally characterized by a yellow tint in all nine colours, with each colour having a cast point between $H = 120°–139°$; the jadeite samples of emerald green, brilliant green and sun green have no cast point values between $H = 140°–159°$, proving that these three colours do not have a blue-green tint or a blue tint; while the jadeite samples of bean green, light green and light green have no cast point values between $H = 100°–119°$, proving that these three colours have a light or no yellow tint. The jadeite samples of pale green, light green and light green have no cast point value between $H = 100°–119°$, proving that the yellow tones of these three colours are light or absent; while the jadeite samples of melon green, bean green and grey-green (oil green) have HSL values covering all three colour ranges, proving that all three colours contain yellow tones, blue tones, yellow-green tones and grey tones, etc., which are the most common colours in the market. These are consistent with previous theoretical descriptions of green jadeite colour.

In general, samples of brilliantly coloured green jadeite are characterised by a high degree of saturation, generally reaching *70%–80%* and up to *90%–100%,* and a moderate degree of brightness, mostly around 20–30, with hue values of *120–149* all being green, and greater than *150* will have a blue tint that affects the beauty of the jadeite green. Brightness values below *10* are too dark and can be classified as black jadeite. Those with a luminosity value greater than *80*, or a saturation of less than *20*, are too pale and close to white (or colourless) and can be classified as white jadeite. There is a correlation between saturation and luminosity, with hue playing a secondary role.

6 Conclusion

Through this study, we can first distinguish and recognize the color tone of green jadeite. The H value represents the hue angle, when $H = 120°–140°$, Jadeite has a more pure green colour; when $H < 120°$, Jadeite has a yellowish green hue; when $H > 150°$, Jadeite has a bluish green hue. Colour intensity and brilliance of green jadeite.Secondly, the saturation S value can determine the intensity of green jadeite, when the saturation S value is less than 10%, basically colourless, grey or black jadeite; when the saturation *S* value is about *20%–40%,* for the medium green varieties of jadeite; when the saturation S value reaches *50%* or more, the green is more intense; when the saturation S value reaches *70%* or more, and the brightness L value is moderate, at about *20%–30%,* the jadeite colour is brilliant. Thirdly, the ranges and threshold boundaries to which the *Z-*, *S-* and *L-* values of the nine green jadeite species belong were derived from the analysis of H-value cast point plots from different angles. The green jadeite colours were refined and digitised. And finally, this study aims to describe the color of green jadeite objectively and explores applicable jadeite colour grading techniques. The experimental results show that this method has the advantages of multi-channel, fast and convenient compared to spectral measurement, and can achieve rapid testing of overall sample colour, which is conducive to the promotion and popularization in the market.

References

1. Tejera Hernandez, D.C.: An experimental study of K* algorithm. Int. J. Inform. Eng. Elec. Bus. (IJIEEB), **7**(2), 14–19 (2015)
2. Mahrous, A.F.: Experimental study of airlift pump performance with s-shaped riser tube bend. Int. J. Eng. Manufact. (IJEM), **3**(1), 1–12 (2013)
3. Chuntao, D.: An experimental study on college teacher's adoption of instructional technology. Int. J. Mod. Educ. Comput. Sci. (IJMECS) **3**(3), 47–54 (2011)
4. Shelekhov, I.Y., Shishelova, T.I., Smirnov, E.I.: Physical methods for studying jades of different colors. Eng. Technol. Syst. **30**(1), 149–161 (2020)
5. Kylie, J.: Jade and silver chain maille necklace. Lapidary J. Jewel. Artist: **Jewellery** artist **61**(3, 8, 2007), **71**(5), 26–33 (2017)
6. Pendergast, D.M.: Dressed to kill: jade beads and pendants in the maya lowlands. BEADS: J. Soc. Bead Res. **10**(1), 3 (1998)
7. Liggins, K., Douglas, J., Edwards, V., et al.: White-light emission characteristics of terbium - doped crystals. In: SPIE Conference on Optical Components and Materials (2018)

8. Jadeja, A.D., et al.: Mycoremediation of coomassie brilliant blue by aspergillus spp. Biochem. Ind. J. 8(5), 139–143 (2014)
9. Wilkins, C.J., Tennant, W.C., Williamson, B.E., et al.: Spectroscopic and related evidence on the coloring and constitution of New Zealand jade. Am. Miner. **88**(8–9), 1336–1344 (2003)
10. Pay, D.: Investigations into corundum and beryl color, jade artistry, the chivor emerald mine's history, and Swiss rhodonite. Gems. Gemol. **56**(1), 22–24 (2020)
11. Rong, W., Xinqiang, Y.: The feasibility study of jade colour measuring heart strongly. J. Gems Gemmol. 20–24 (2007). (in Chinese)
12. Jinhuan, W.: The Influence of the Color Rendering of Light Source on the Evaluation of Emerald Green. China University of Geosciences, Beijing (2010). (in Chinese)
13. Ziwei, T.: Characteristics of green jade color index. Guangxi Light Industry, pp. 21–23 (2011). (in Chinese)
14. Yan, L., Taijin, L., Hui, Z.: Color grading method for green jadeite. In: Proceedings of the China Jewelry and Jewelry Academic Exchange Conference, pp. 109–110 (2019)
15. Rupeng, X.: Quantitative description of jadeite color. J. Gemol. Gemol. **4**, 45–48 (2016)
16. Hui, Z., Beili, Z., Manjun, W.: Method of colour measurement in emerald color evaluation and application. J. Gems Gemmol. **1**, 16–20 (2006)
17. Xiaowei, Z.: Research on color classification of green jadeite. J. China Univ. Geosci. (Beijing) **2**, 22–23 (2019)

Fingerprint Localization Technology of WiFi Signal for Intelligent Parking Space Navigation

Xiaohua Mo and Xuanhan Huang(✉)

Guangxi Key Laboratory of International Join for China-ASEAN Comprehensive Transportation, Nanning University, Nanning 530000, Guangxi, China
2816633270@qq.com

Abstract. Combined with the current application of intelligent transportation technology in the parking lot designed a smart parking space navigation system based on WIFI signal fingerprint positioning technology, the system adopts the WiFi indoor positioning method of location fingerprinting, which has the advantages of higher precision, more terminals, faster speed, lower cost and other functions more complete. Intelligent parking space navigation system combines the WiFi method of extracting fingerprints based on indoor location, intelligent parking management system and cell phone application to accurately locate the objects in the parking lot, and through the real-time provided information and prompted routes, it judges the existence of empty parking spaces in the parking lot to carry out intelligent parking space navigation.

Keywords: AI technology · Intelligent parking space navigation · WiFi signal fingerprint positioning technology

1 Introduction

The background technologies of the intelligent parking space navigation system mainly involve wireless communication technology, mobile terminal technology, GPS positioning technology and GIS technology. These technologies are comprehensively applied to the collection, management, query, booking and navigation services of urban parking spaces in order to realize the integration of real-time updating, query, booking and navigation services of parking space resources. In addition, the parking space guidance system is also an important part of the intelligent parking space navigation system, which can guide the vehicle to enter the destination parking space smoothly and provide the information of empty parking space so that the driver can park easily [1]. In recent years, the application of intelligent parking space navigation system has become more popular in many public parking lots, commercial parking lots, residential communities and other places, which requires more intelligent parking space navigation system technology [2, 3].

At the same time, there have been a number of parking lots began to try to introduce AI technology, such as empty space reservation, empty space navigation these applications to a certain extent to improve the efficiency of the use of parking lots and management

Z. Hu et al. (Eds.): ICCSEEA 2024, LNDECT 242, pp. 14–24, 2025.
https://doi.org/10.1007/978-3-031-84228-3_2

results, such as the United States, Japan, Germany, France, the United Kingdom and other countries to gradually improve the parking policy and regulations to encourage the promotion of the development of intelligent parking industry through the means of new technologies and applications popularization [4–6]. Tesla of the United States has launched an intelligent parking system based on autopilot technology, which can realize the functions of automatic vehicle searching for parking space and automatic parking, etc. [7]. The government has stepped up its efforts to promote the development of intelligent parking industry in recent years. In recent years, the government has increased its policy support for the development of the smart parking industry, whether through capital guidance, encouraging cooperation between government and enterprises, or simplifying administration and decentralization, and accelerating the intelligent transformation of parking lots [8].

However, the GPS technology of the currently existing smart parking navigation system can track the user's location in real time, which has raised some concerns about privacy, GPS technology is mainly used underground, resulting in easy missing signals and lower accuracy problems, especially in some specific cases, GPS navigation system affects the parking efficiency, and the installation cost is higher [9].

2 Application of Intelligent Transportation Technology in Parking Lots

2.1 Intelligent Transportation Monitoring

For road traffic, the monitoring system is very important. Modern intelligent traffic monitoring system can not only use computer and camera equipment to detect the corresponding images, but also use recognition technology to analyze the traffic condition and traffic flow of the relevant road sections [10]. Intelligent transportation systems can play a role in the hub location of major urban roadways, and through the monitoring of major entrances and exits, they can effectively prevent the occurrence of accidental traffic accidents. On the other hand, the camera can also be used to collect data on the traffic flow and vehicle saturation of a specific road section, and then intelligently adjust the traffic signals at the intersection, so as to better control the operation of vehicles, ease traffic pressure and improve road safety. Traffic monitoring system can also be applied to parking lots to record the situation of parking spaces, deal with non-compliant vehicles in a timely manner, and improve the utilization rate of parking resources. Among them, sensor monitoring and license plate recognition technology are more mature parts.

The application of sensors monitoring in parking lots is mainly divided into two parts: parking space monitoring and vehicle management. The parking space detection mainly adopts ultrasonic parking space detector, which is installed above the front of each parking space to collect real-time parking space data in the parking lot and transmit it to the node controller through the network; in addition, video detection, trigger radar or trigger coil are also commonly used for monitoring the status of the parking space. Next is the vehicle management part. When a vehicle enters the parking lot, it will capture the license plate image through the license plate recognition system, obtain the license plate number from the image, and make a comparison. After the sensor detects that the parking space is empty, it transmits the information to the backend server through the communication module, so as to collect the parking space status of the whole parking lot in real time; at the same time, the driver can select the empty parking space online independently, and go to that parking space directly, in this way, to reduce the unnecessary traffic flow. When a vehicle enters the parking lot, it triggers the capturing machine to capture the license plate image through video detection, triggering radar or triggering coil, etc., and recognizes the license plate number through the license plate recognition system and inputs it into the database for comparison.

License plate recognition technology is the mainstream equipment of intelligent transportation and plays an important role in the parking system. It has the function of recognizing license plate and color and recording, and automatically confirming the time of vehicle entry and exit. When the car enters the entrance of the parking lot, it can be re-parked and enter directly without swiping the card. Intelligent license plate recognition system realizes electronic intelligent management such as fast vehicle entry, fast fee payment and fast boarding, which greatly improves the parking efficiency. In the unmanned parking system, as long as there is a license plate recognition and a car door, you can realize vehicle access and fee payment processing in a parking lot entrance or exit. In addition, the license plate recognition technology can establish unmanned fast channel for vehicles that do not charge parking fees (such as monthly trucks, internal free access vehicles), realizing no parking card, no parking in and out of the experience, and further improving the efficiency of vehicular traffic in and out of the entrances and exits. The application of license plate recognition technology not only improves the efficiency of the parking lot and reduces operating costs, but also provides a more convenient and faster parking experience for vehicle owners.

2.2 Car Parking Navigation System

The parking space navigation system is a comprehensive application innovation system, whose main purpose is to collect, manage, query, book and navigate services for urban parking spaces [11]. The system can update parking space information in real time, maximize the use of parking space resources, and provide users with convenient parking navigation services.

In the actual application process, the parking navigation system usually first collects real-time data in the parking lot through various sensors and devices, including the use of parking spaces and the location of vehicles. Then, these data are transmitted to the central processing system and analyzed and processed by indoor navigation and video detection algorithms. Finally, the processed parking space information will be displayed in the navigation system in real time, and users can view and select the vacant parking spaces through mobile terminals or in-vehicle devices, and the system will automatically plan the best route to guide users to find the target parking space quickly and accurately. In addition, some intelligent parking space navigation systems can also realize online booking function, users can book parking spaces in advance, to avoid the trouble of arriving at the parking lot and not being able to find a parking space.

3 Intelligent Parking Space Navigation WiFi Signal Fingerprint Positioning Technology Module

WiFi signal fingerprinting is a technique that associates a location in the physical environment with a certain "fingerprint", where a location corresponds to a unique fingerprint. This fingerprint can be one-dimensional or multi-dimensional, such as extracting the fingerprint from the signal of a specific location, then the fingerprint can be one or more features of the signal, the basic principle of fingerprint-based localization method is to use the spatial variability of wireless signals in different environments, and take the wireless signal features of a specific location as the fingerprint of the location, to build a database of the relationship between the location and fingerprints, and then estimate the user's location through the fingerprint matching method. Fingerprint matching to realize the estimation of the user's location [12]. The general framework of the system is shown in Fig. 1

Fig. 1. Diagram of the general framework of the system

3.1 Wi-Fi Signal Acquisition Module

This module is mainly responsible for collecting the signal strength RSSI (Received Signal Strength Indicator) and AP (Access Point) address sent by each AP in the surrounding area. The core function of the Wi-Fi signal acquisition module can be divided into two phases: the training phase and the online positioning phase. The core function of Wi-Fi signal acquisition module can be divided into two stages: training stage and online positioning stage.

During the training phase, the module samples locations in the localization area and collects wireless signal features at each sampled location to be deposited in the location-fingerprint database. These features may include signal strength, channel information, etc. Ideally, each geographic location in the target environment possesses a unique and distinguishable wireless signal feature, which serves as the fingerprint information of that location.

In the online localization phase, the Wi-Fi signal acquisition module senses the Wi-Fi signals received by the device in real time and matches these signal features with the features in the database to estimate its own location by pattern recognition.

3.2 Feature Extraction Module

The main task of the feature extraction module is to extract the features that can reflect the location differences from the collected Wi-Fi signals. Commonly used features include signal strength, signal-to-noise ratio, propagation delay and so on. These features need to be distinguishable enough, i.e., each geographic location has a unique distinguishable wireless signal feature, which serves as the "fingerprint information" of the location. By extracting and analyzing the features of Wi-Fi signals, accurate positioning and navigation of parking spaces can be achieved.

First, Wi-Fi signal acquisition devices need to be installed inside the parking lot, which can actively scan the surrounding Wi-Fi signals and record the Wi-Fi signal strength and other characteristics at each location point. The collected Wi-Fi signal data needs to be preprocessed, including data cleaning, noise removal, data alignment and other operations to ensure the accuracy and consistency of the data. After preprocessing, feature extraction needs to be performed on the Wi-Fi signal data. Feature extraction refers to extracting feature information from the original Wi-Fi signal data that can represent the location point, and common features include signal strength, signal frequency, channel and other features.

Then, the extracted features need to be appropriately represented, and common representation methods include vector representation, feature code representation, etc., for subsequent data analysis and comparison. The extracted feature information is stored in a feature library, which contains the Wi-Fi signal feature information at each location point, as the reference data for subsequent positioning and navigation.

Finally, when a vehicle enters a parking lot, it can locate the nearest available parking space and provide navigation guidance for the vehicle by using the pre-constructed feature library for localization and navigation with the collected Wi-Fi signals.

3.3 Model Training Module

The model training module of the intelligent parking space navigation Wi-Fi signal fingerprinting localization technology is a crucial part of the technology, which mainly uses the extracted features and the corresponding location information to train a classification or regression model, so that the model can learn how to predict the location of the parking space based on the signal features. The accuracy of this model directly determines the final localization result.

First, a large amount of Wi-Fi signal fingerprinting data needs to be prepared, including the Wi-Fi signal strength, frequency, channel and other characteristics at each location point. These data are usually obtained by Wi-Fi signal acquisition devices inside the parking lot. The collected data are labeled, i.e., each data point is marked with its corresponding location information, such as the coordinates of the parking space. These labeled data will be used as supervised learning data for training the model. Before model training, feature engineering is required to perform feature extraction and dimensionality reduction operations on the collected Wi-Fi signal data to provide more effective input data for the model.

Then, according to the specific localization requirements and data characteristics, appropriate machine learning or deep learning models are selected, and common models include SVM (support vector machines), neural networks, etc. Using the prepared labeled data, the selected model is trained. During the training process, the model will gradually adjust its own parameters to maximally fit the input data and labeled information. After the training is completed, the model needs to be evaluated, usually using methods such as cross-validation and test-set validation, to assess the model's performance in terms of localization accuracy, generalization ability, etc. Based on the evaluation results, the model is optimized.

Eventually, the trained model can be deployed into the intelligent parking space navigation system for real-time parking space localization and navigation services.

3.4 Positioning Decision Module

Users need to use their mobile devices for real-time positioning. When a user enters a parking lot, their device measures the signal strength of the surrounding APs and sends this data to the server. The server receives the data and inputs it into the trained model to get the predicted parking location. Finally, the server returns the predicted results to the user's device, guiding the user to find an available parking space.

4 Implementation of WiFi Signal Fingerprinting Localization Technology for Intelligent Parking Space Navigation

4.1 Technical Program

The implementation scheme of intelligent parking space navigation WiFi signal fingerprinting and localization technology includes five steps: data acquisition, feature extraction, model training, real-time localization, and system optimization.

First, a certain number of WiFi access points need to be deployed in the parking lot and the location information of each AP is recorded. Then, multiple measurements are taken between these APs by mobile devices (e.g., cell phones, tablets, etc.) to collect signal strength data at different locations. These data will be used as the basis for training and testing the model.

The collected signal strength data are then processed to extract features that reflect the differences in location. These features can include signal strength, signal-to-noise ratio, propagation delay, etc. The purpose of feature extraction is to reduce the dimensionality of the data and improve the computational efficiency of the model.

The extracted features and the corresponding location information are then used to train a classification or regression model. This model will learn how to predict the location of parking spaces based on signal features. Commonly used models include support vector machines, decision trees, neural networks, etc.

In practical applications, users need to use mobile devices for real-time localization. When a user enters a parking lot, their device measures the signal strength of the surrounding APs and sends this data to the server. After receiving the data, the server inputs it into the trained model to get the predicted parking space location. Finally, the server returns the predicted results to the user's device and guides the user to find an available parking space.

Finally, in order to improve the accuracy and stability of positioning, some optimization measures can be taken. For example, regularly update and maintain the location information of the AP; introduce multiple sensors (e.g., Bluetooth, infrared, etc.) for multi-source fusion localization; and use machine learning algorithms to automatically adjust the model parameters. The specific process is shown in Fig. 2.

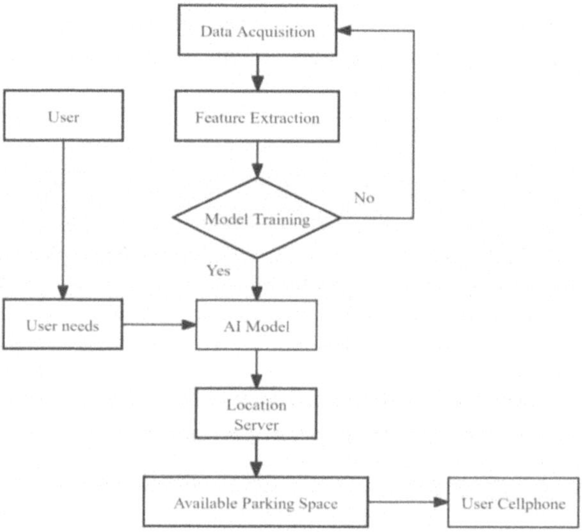

Fig. 2. Flowchart of the system implementation

Intelligent parking space navigation WiFi signal fingerprint localization technology is based on the "fingerprint" localization method, by collecting each sample point in the localization space for training, that is, collecting the signal characteristics of each sample point, constituting a fingerprint database. If a point to be located is to be localized, the signal characteristics of the point are collected and compared with the fingerprint database, and the training sample point in the database that is closest to the signal characteristics of the point to be localized is taken as the localization result of the localization point.

Specifically, WiFi signals propagate through the air at a known rate, and the moments between being announced by one AP or STA (station) and being accepted by another AP or perhaps STA can be converted into intervals by multiplying the speed of light. Then by means of four moment stamps (the difference between two moments), one device should send the moment stamps to another device before the second device is able to perform certain calculations.

Meanwhile, the RSSI of signal strength and AP address sent by each AP around can be obtained on the smartphone through the system, and the RSSI is used to locate the device. The main algorithms include triangulation algorithm and fingerprinting algorithm. The triangulation algorithm needs to know the location of the AP in advance, and is not suitable for use in occasions where the environment changes quickly. The fingerprint algorithm is similar to machine learning algorithm, which is divided into offline training phase and online positioning phase.

4.2 Analysis of Simulation Results

The underground parking lot of a farmers' market in a city is selected as the object, which has 237,000 permanent residents in the area, and the underground smart parking lot, which is divided into Zone A and Zone B, has a total of 269 parking spaces, as shown in the Fig. 3.

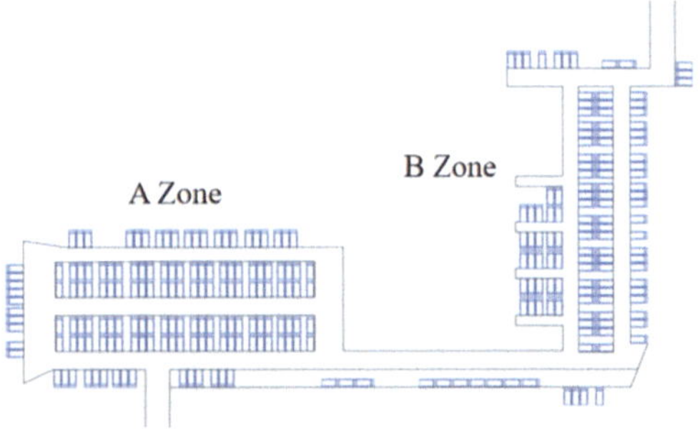

Fig. 3. Diagram of the parking lot

The simulation tested the parking time of 16 parking spaces in different locations, the free parking time in this parking lot is 30 min, the test results show that the ratio of the number of spaces with free parking time to the total number of spaces is 87.5%, the results are shown in the Table 1.

Also, the turnover rate of the parking lot was tested by simulation under different hourly traffic volumes and the results are shown in the Table 2.

From the simulation results, it can be seen that the WiFi signal fingerprint positioning technology has significant advantages in guiding the driver into the parking space, and can be applied to scenarios such as indoor navigation, shopping mall guiding, and smart home in addition to parking lot navigation. By adjusting and optimizing the model for the characteristics of different scenarios, a wider range of applications can be achieved.

Table 1. Vehicle parking time

Distance between parking space and entrance				Length (m)	Capacity (m)	Parking time(min)
No	Start section	End position	Distance(m)			
1	North entrance	51.25	47.26	12.00	2	9.45
2	North entrance	24.00	19.72	5.30	1	3.94
3	North entrance	32.91	28.06	5.30	1	5.61
4	North entrance	36.59	31.63	5.30	1	6.33
5	North entrance	44.16	39.31	5.30	1	7.86
6	North entrance	107.45	102.38	5.30	1	20.48
7	North entrance	144.77	139.57	5.30	1	27.91
8	North entrance	233.32	227.81	5.30	1	45.56
9	South entrance	72.98	67.46	5.30	1	13.49
10	South entrance	236.91	231.15	5.30	1	46.23
11	South entrance	53.66	50.80	5.30	1	10.16
12	South entrance	57.85	54.98	5.30	1	11.00
13	South entrance	82.71	80.17	5.30	1	16.03
14	South entrance	115.71	115.07	5.30	1	23.01
15	South entrance	120.44	119.56	5.30	1	23.91
16	South entrance	167.13	164.8	5.30	1	16.48

Table 2. Turnover rate of the parking lot

Test section	Hourly traffic volume (pcu/h)	Total parking spaces	Turnover rate
North and south entrance	300	269	1.11
North and south entrance	480	269	1.78

5 Conclusion

In order to alleviate the problem of imbalance between supply and demand of parking spaces in parking lots and to solve the problem that it is not easy for users to find parking spaces in large parking lots, this paper designs an intelligent parking space navigation system through WiFi location fingerprinting scheme on the basis of the application of intelligent transportation technology in parking lots, and the intelligent parking space navigation WiFi signal fingerprinting positioning technology module has the advantages of high precision, multiple terminals, fast speed, low cost, and other more perfect advantages to effectively realize the correct position of the vehicle.

It is reasonable to improve the intelligence of the parking lot vehicle navigation system through AI technology, and provide the urban planning department with reference opinions about parking lot layout, management and service, which helps to optimize the allocation of urban parking resources and improve the level of urban management.

Acknowledgment. This project is supported by Transportation Operation Subsidy Project of Guangxi Key Laboratory of International Join for China-ASEAN Comprehensive Transportation in 2021 (No. 21-220-21) and 2024 Young and Middle-aged Teachers in Guangxi Universities Scientific Research Basic Ability Enhancement Program "The Application of Artificial Intelligence in Nanning Parking Management" (No. 2024KY1890) and Nanning University Model Courses (No. 2022XJYYX09) and Nanning University Model Courses(No. 2023XJYYZ03).

References

1. Chiara, G.D., Cheah, L., Azevedo, C.L., et al.: A policy-sensitive model of parking choice for commercial vehicles in urban areas. Transp. Sci. **54**(3), 606–630 (2020)
2. Satpalkar, et al.: Smart city parking: a QR code-based approach. Int. J. Eng. Res. Technol. (IJERT) **5**(02), 1–9 (2016)
3. Laroca, R., et al.: A robust real-time automatic license plate recognition based on the yolo detector. In: International Joint Conference on Neural Networks (IJCNN), pp. 1–10 (2018)
4. Kaspi, M., Raviv, T., Tzur, M., et al.: Regulating vehicle sharing systems through parking reservation policies: analysis and performance bounds. Eur. J. Oper. Res. **251**(3), 969–987 (2016)
5. Mangiaracina, R., Tumino, A., Miragliotta, G., et al.: Smart parking management in a smart city: costs and benefits. In: 2017 IEEE International Conference on Service Operations and Logistics, and Informatics (SOLI). IEEE (2017)

6. Huseynova, A., Mazanova, O.: The leading role of digital technologies in the development of the smart city concept in Azerbaijan. Int. J. Inform. Technol. Comput. Sci. (IJITCS) **15**(3), 55–64 (2023)
7. Wang, W., Song, Y., Zhang, J., Deng, H.: Automatic parking of vehicles: a review of literatures. Int. J. Automot. Technol. **15**(6), 967–978 (2014)
8. Naveen, N., Nagaraj, G.C.: Enhancing early Alzheimer's disease detection: leveraging pre-trained networks and transfer learning. Int. J. Intell. Syst. Appl. (IJISA) **16**(1), 52–69 (2024)
9. Suleiman, G., Umar, A.O., Adekunle S.E.: Development and implementation of vehicle movement notification and location using GSM, GPS and Web based information system. Int. J. Comput. Netw. Inform. Secur. (IJCNIS), **10**(8), 36–43 (2018)
10. Mahboob, K., Ur Rehman, I., Ali, M., et al.: Development of Android controlled Arduino based intelligent car parking system. Int. J. Wirel. Microwave Technol. (IJWMT) **10**(1), 48–61 (2020)
11. Luciano, R.G., Cunanan, A.I., Mariano, R.P. et al.: A web-based parking lot management system designed and developed using multi-paradigm programming languages. Int. J. Inform. Eng. Elec. Bus. (IJIEEB) **15**(3), 57–67 (2023)
12. Bastwesy, M.R.M., ElShennawy, N.M., Faheem Saidahmed, M.T.: Deep learning sign language recognition system based on Wi-Fi CSI. Int. J. Intell. Syst. Appl. (IJISA) **12**(6), 33–45 (2020)

Design of an All-Terrain Unmanned Tracking Model Vehicle

Ran Qing, Wugang Li, Lanxiang Wei, Guanda An, and Fengmei Chen(✉)

School of Intelligent Manufacturing, Nanning University, Nanning, China
C16677041143@163.com

Abstract. The aim of this paper is to design an all-terrain unmanned tracking model vehicle to explore its transport and travelling capabilities on complex road conditions. The model vehicle, with STM32 microcontroller as the main control board, adopts a suspended six-wheeled body structure, equipped with lug-nut wheels and six independently powered direct current (DC) motors to ensure its driving function. In addition, it is equipped with a steep hill-climbing aid to make it more stable when climbing steep gradients. Two 360° rotating circumferential rudders and gyroscopes are used to realize cargo transport and maintain body stability. Eight infrared sensors are also provided to ensure stable tracking and correct turning of the model vehicle. Finally, the test results show that the design scheme has the characteristics of high reliability and stability, and can achieve effective transport and travelling functions on complex road conditions.

Keyword: All-Terrain unmanned tracking vehicle · Microcontroller · Infrared sensor

1 Introduction

With the increasing maturity of unmanned control systems and artificial intelligence technology, mechanical and electronic technology has achieved the integration of mechatronics [1]. In recent years, the continuous rise in labour costs has made the phenomenon of "machine for man" in the industrial field more and more common, which also brings a broad development prospect for the market of industrial robots [2]. The intelligent transformation of industrial robots has become a major development trend, and this trend is particularly obvious in the logistics industry [3, 4]. Although in the logistics industry, many times still use manual handling, but in some complex terrain road conditions, manual handling often can't meet the rapid development of market demand, and the cost is high, handling efficiency is also relatively low. Therefore, the use of machines instead of manual labour can not only improve the handling efficiency of goods, but also reduce costs in the long run [5, 6].

All-terrain vehicle (ATV) is a type of vehicle designed for non-road, all-weather, all-terrain and other complex road conditions and environments [7, 8]. Compared to ordinary vehicles, ATVs have better passability as well as better manoeuvrability and flexibility, and they can operate in a wide range of complex terrains and poor road conditions.

Z. Hu et al. (Eds.): ICCSEEA 2024, LNDECT 242, pp. 25–34, 2025.
https://doi.org/10.1007/978-3-031-84228-3_3

At present, although many designs of all-terrain unmanned vehicles have appeared [9–12], there are not many designs that can simultaneously satisfy the safe transport of goods and drive autonomously on a set trajectory. This is because the design of this all-terrain unmanned trajectory trolley combines these two requirements functionally and structurally, which is more challenging [13, 14]. The design of this all-terrain unmanned trajectory trolley is very necessary in order to achieve the safe transport of goods in an unmanned situation.

In this paper, an all-terrain unmanned tracking model vehicle is designed for practical transport and walking needs. The vehicle is able to travel autonomously in complex working conditions and roads, and has the ability to pass through various obstacles, such as pipelines, narrow bridges, steep slopes, and so on. After starting, it can travel autonomously. In the workplace, by drawing a line, the vehicles similar to the model vehicle are able to travel autonomously and transport goods according to the desired route. At the same time, autonomous travelling is not affected by uneven road surfaces or other special road conditions, effectively avoiding the problem of falling goods. This paper provides a reference sample for practical manufacturers to build an all-terrain vehicle for goods transport.

2 Overall Design Structure and Hardware Design

2.1 Overall Design Structure

In order to achieve the desired functionality, a reasonable and optimal design solution is required. After a comprehensive and detailed analysis and study of the all-terrain model vehicles, the following overall design scheme was determined, as shown in Fig. 1. The overall design scheme of this vehicle is divided into two major parts: hardware and software. The hardware part includes the vehicle frame structure, driving structure and transport structure; the software part includes the control system, recognition system and balance system.

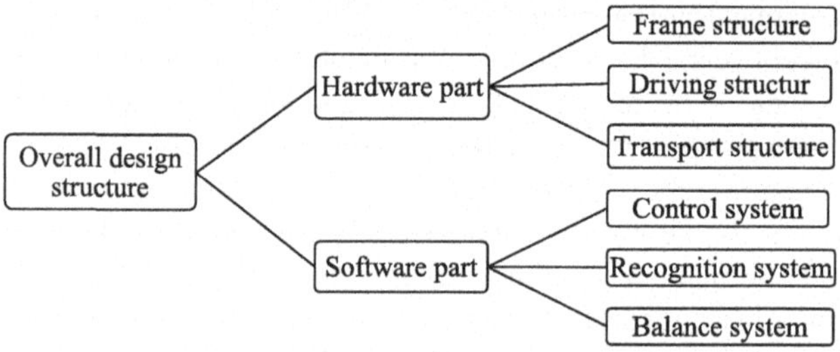

Fig. 1. Framework diagram of the overall design program

2.2 Hardware Design

The hardware part includes the vehicle frame structure, driving structure and transport structure. This design uses self-designed parts, some self-purchased parts and electronic modules. It contains a variety of parts such as tracks, spacers, servo double bending plates, silicone tyres, M01 servos, M*06* twin-axis motors, etc.

2.2.1 Frame Structure Design

According to the driving requirements and the analysis of obstacles, several frame structure schemes have been pre-designed. In order to ensure that the model vehicle can achieve the best performance, after in-depth analysis and discussion, there are two options that were finally determined: the first option is to use a four-wheeled structural frame; the second option is to use a suspended six-wheeled frame.

Scheme 1, four-wheel structure frame structure design: this scheme mainly uses four silicone tyres, two pieces of one-word connecting pieces, two servo double bending plates, two pieces of hole flat plates and other connecting parts to construct a four-wheel structure frame. The frame mainly fixes the four wheels on two *110 mm* × *70 mm* hole plates, and divides the four wheels into two parts, and uses the one-word connecting piece to connect these two parts to the front and rear ends of the two plates, thus constituting the four-wheeled structural frame. Scheme 2, Suspended six-wheeled frame structure design: this scheme mainly uses six silicone tyres, ten L-shaped plates, a hole plate and other connecting parts to produce a suspended six-wheeled frame. The adopted suspended six-wheeled body structure is to divide the body into two parts, the front four wheels as part of the suspension structure, which is mainly composed of L-shaped strip, bending T-shaped plate and a word connecting piece, and the middle end of the suspension structure is connected to the bottom of a *110 mm* × *70 mm* hole plate and fixed in the front of the hole plate; the rear two wheels are the other part of the hole plate is fixed to the bottom of the hole plate, and the two parts form a movable structure.

After completing the design of the above two scenarios, the two scenarios were iterated, and tested through different obstacles with the following results:

The four-wheeled frame in Scheme 1 can pass through different obstacles faster, but its stability is not high enough, and it will shake violently and deviate from the steps of steep slopes in the process of climbing steep slopes, which will easily lead to the overturning phenomenon. The suspension six-wheeled frame in Scheme 2 can pass through different obstacles smoothly and slowly, and has high stability when travelling. Climbing steep slopes with the movable structure is more stable and does not result in rollover. In addition, this frame has the advantages of high load carrying capacity, simple structure and control, flexible steering, and high efficiency [15].

After comparison and analysis, the suspension six-wheeled frame design was finally identified as the best solution. In the subsequent tests and practical applications, the suspended six-wheeled frame has shown good performance and stability, and is suitable for a variety of different fields and complex environments. The structural design sketch of the suspended six-wheeled frame is shown in Fig. 2.

2.2.2 Travelling Structure Design

The driving structure of this vehicle consists of six motors, six silicone tyres and one drive module.The detailed design is as follows: fix the six motors on the suspendable six-wheeled frame, fix the six silicone tyres on the six motors with screws, and use the

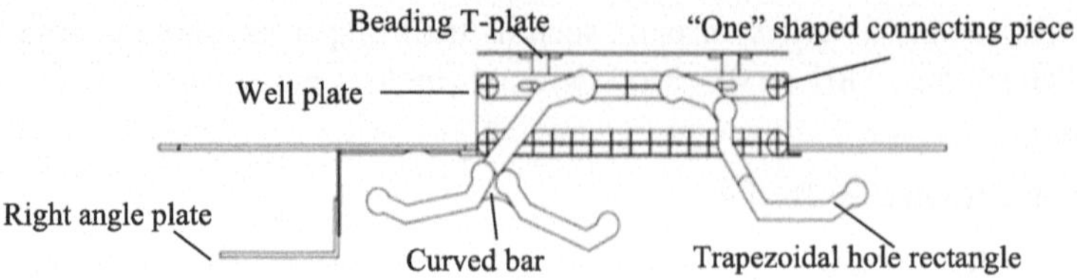

Fig. 2. Sketch of the structural design of the suspended six-wheeled frame

spare position on the upper part of the hole plate for placing a drive module, and the whole driving structure is completed. The design sketch of the driving structure is shown in Fig. 3.

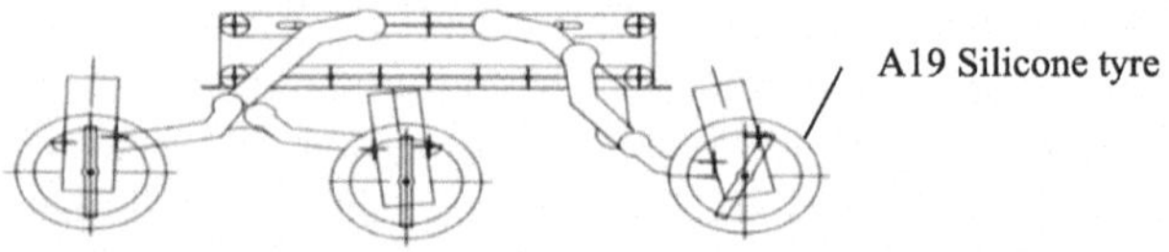

Fig. 3. Design sketch of driving structure

2.2.3 Transport Structure Design

The transport structure is designed as follows: firstly, a piece of *110 mm* × *70 mm* hole flat plate is used as the bottom support plate, which is mounted on the first helm. Then, two *110 mm* × *70 mm* hole plates as well as a number of bent T-plates and one-word connecting pieces are used as the carrier discs for transporting the goods. A servo was added to the bottom of the two holey flat plates to control the tipping angle of the carrier discs. In order to be able to transport the goods more accurately to the corresponding location, the best transport location parameters are derived through practical calculations, which can effectively avoid the problem that the goods cannot reach the specified location due to inaccurate transport. The sketch of the transport design structure is shown in Fig. 4.

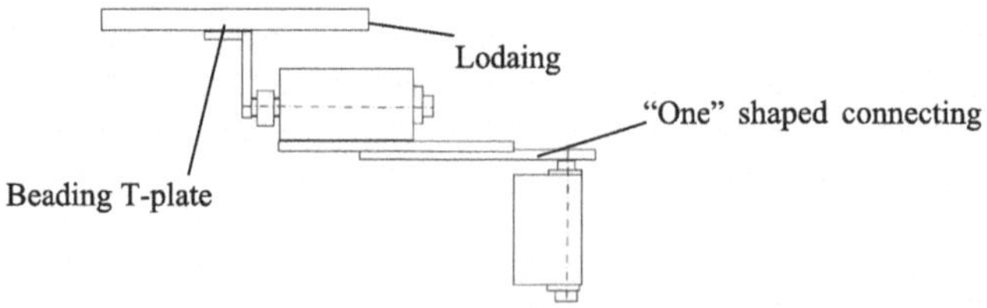

Fig. 4. Sketch of transportation structure

3 Software Design

3.1 Software Design of Identification System

This vehicle uses a homemade 8-channel infrared sensor to achieve the function of autonomous driving. The initial solution used a homemade grey scale sensor, but during the test it was found that the sensor was susceptible to external interference and it was difficult to accurately detect the black line situation under the reflective road surface. In order to solve this problem, the final solution uses a homemade 8-channel infrared sensor. In the running state of the model car, the detected black line conditions are converted into high and low level signals and delivered to the microcontroller. The microcontroller program processes the input signals and outputs four high and low level signals to the motor drive circuit to control the steering of the tracking model cart. At the same time, the output PWM (Pulse Width Modulation) signals are used to control the rotational speed of the trajectory model cart [16], as shown in Fig. 5.

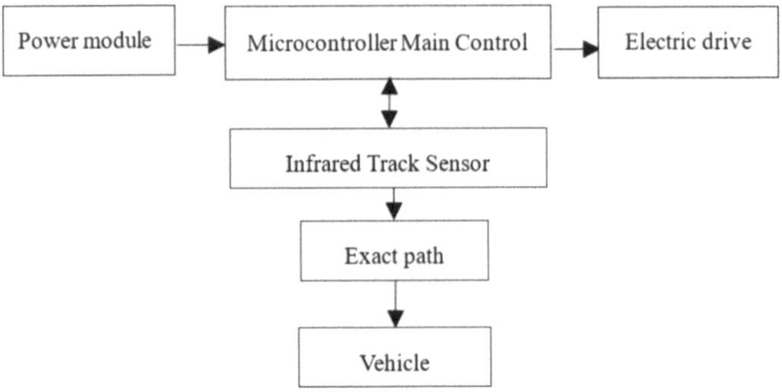

Fig. 5. Main design structure diagram

It uses an 8-way infrared sensor to realize the function of trajectory travel. This sensor can recognize the black area in the yellow background or the edge of the cliff and transmit the recognition signal processing to the controller. The output signal of the infrared sensor is very stable, which makes the tracking more stable and accurate. The model vehicle relies on these control signals to automatically commute and realize autonomous control when it is tracing forward. This enables the model vehicle to track forward more accurately and is easy to regulate.

It is possible for us to divide the 8-channel infrared sensors into left and right parts for the recognition of black and yellow lines. According to the offset of the left and right parts of the sensors in different situations, the following strategy can be adopted to control the travelling direction of the model vehicle: When both the left and right parts of the sensor are on the black line, the model car makes a straight forward movement. When the left part of the sensor deviates from the black line to the left, the model car makes a right turn. When the left and right parts of the sensor deviate from the black line to the left, the model car makes a big right turn. When the right part of the sensor deviates from the black line to the right, the model car makes a left turn. When the left and right parts of the sensor deviate from the black line to the right, the model cart makes a large left turn [17]. About the state where the left and right part of the infrared sensor is on the black line, a reference can be made to Fig. 6. to identify and control the direction. It should be noted that Fig. 6 is only an example, and may need to be adjusted and optimized according to the specific situation in the actual application.

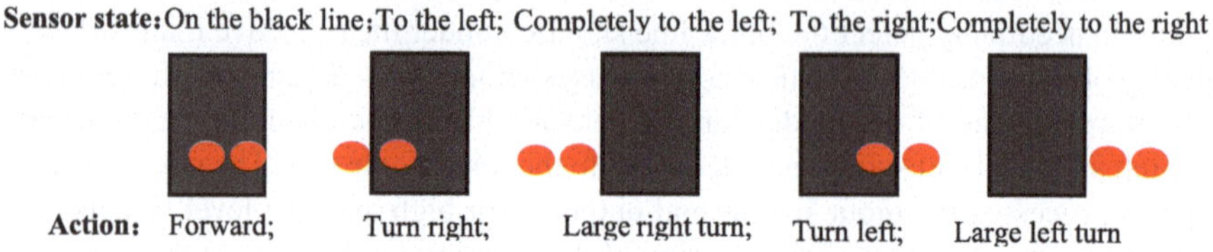

Fig. 6. Infrared sensor tracking situation model

3.2 Balance System Software Design

In order to make the all-terrain model trolley drive automatically and pass through various obstacles smoothly, *8*-channel infrared sensors are used for path tracking; at the same time, in order to avoid dropping the goods due to bumps on the road surface in the process of climbing steep slopes or driving automatically, a gyroscope is added to detect the stability of the model trolley, and the signals from the gyroscope are transmitted to the STM*32* MCU, and then the servo is used to control the load tray of its stability. The design idea of the overall control part is shown in Fig. 7.

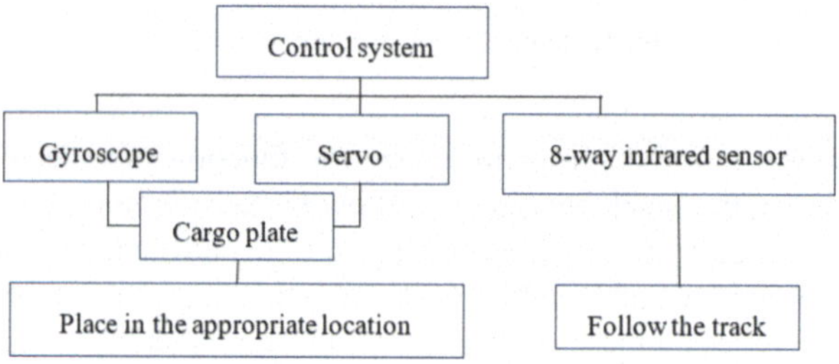

Fig. 7. Overall control part of the design idea

The control part is the core of the whole model car, which is able to perform control functions such as trajectory, steering, and dropping. When the main control board is powered on, it will initialize each module, and the 8-channel infrared sensors will start working after starting the DC motor to ensure that the model car can travel along the predetermined path. Under the system timer, the sensors read the path and external information at a frequency of *10* Hz. When the 8-channel infrared sensors detect a black line, they immediately generate a high level signal and pass it to the main controller; while when the sensors do not detect a black line, they immediately generate a low level signal and pass it to the main controller. The main controller comprehensively analyses these sensor signals and generates signals to control the actuator. In addition, a colour recognition module has been created. When this module is close to a colour card, the main controller receives a high level signal, while the other colour cards have a low level signal [18]. When the module detects a high level signal again, the model trolley will stop moving forward and place the goods in the appropriate position by rotating the servo.

In order to make the microcontroller output voltage and current to drive six DC motors, a DC motor driver chip is needed. And the microcontroller changes the speed of the DC motors by adjusting the duty cycle (i.e. PMK speed regulation). However, the microcontroller driving capability is limited, so PWM speed regulation using a relay is needed for more efficient driving. In order to better control the travelling direction of the model trolley, a motor driver module is used in this design. The motor's forward and reverse rotation is controlled by an internally contained 4-channel logic driver circuit and a high-voltage, high-current dual H-bridge driver L*298* [19, 20].

4 Working Principles and Processes of Autonomous Driving

4.1 Working Principles and Processes of Autonomous Driving

In order to realize the autonomous driving of the model vehicle, *8*-channel infrared sensors are divided into left and right parts. The working principle is: after the start of the main program run, the model vehicle is first initialized, and then it starts to read the values of the sensors. Let's assign a value to each sensor, when the sum of the sensor assignments of the left and right parts is negative, the microcontroller receives the signal and controls the motor to rotate, the wheel on the left side of the model trolley will reduce its speed, the middle wheel will rotate positively, and the model trolley will turn to the left. When the sum of the sensors of the left and right parts is positive, the microcontroller receives the signal and controls the motor to rotate, the wheel on the right side of the model car will reduce the rotational speed, the middle wheel reverses, and the model car turns to the right [21]. If the sum of the sensor assignments of the left and right parts is 0, the model cart goes straight. The flowchart is shown in Fig. 8.

4.2 Driving Working Principles and Processes for Passing Obstacles

In order to make the model car pass all kinds of obstacles smoothly, a suspended six-wheeled frame was adopted. When the cart travels to a narrow road, the sensor is triggered and sends a signal to the microcontroller, and the motor drives forward, using the power of the front two wheels and the middle wheel to shift the center of gravity to the rear wheel, and passes through the narrow road smoothly at a slow speed. When passing through a tunnel, the sensor sends a signal to the microcontroller, executes the turning instruction, the middle wheel turns forward and reverse, and passes through the pipe smoothly using the round hole piece as an aid. When climbing steep slopes, the sensor sends signals to the microcontroller, the motor drives forward, the front wheels are responsible for climbing the slope, the middle wheels are used as auxiliary wheels, and

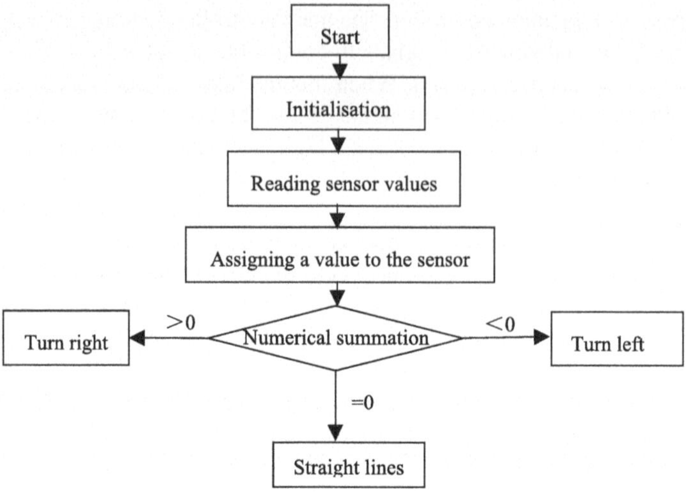

Fig. 8. Sensor operation flowchart

the rear wheels are driven to climb up slowly. After reaching the highest point, the motor drives the front wheel to make the model cart climb downward slowly, successfully completing the action of climbing steep slopes. The middle wheel plays an important role in avoiding the large up and down road bumps when the model car passes through the steep slope.

5 Conclusion

The final design of the vehicle model is shown in Fig. 9, which was done with respect to the design objectives and by using the hardware and software techniques mentioned in this paper. After testing and powering up the vehicle, the vehicle is able to follow the specified route accurately and remains stable while transporting the goods. The vehicle was also able to maintain good stability under various complex road conditions. After several tests, the results show that the performance of the vehicle is reliable and can be transformed into an actual vehicle that can be widely used in the transport industry.

With the continuous development and application of unmanned technology, the unmanned transport industry continues to grow. All-terrain vehicles can adapt to various complex terrains with their powerful passing ability and suspension system, thus enabling them to transport goods to their destinations smoothly. Therefore, all-terrain vehicles play an important role in the unmanned transport industry, improving the efficiency and flexibility of logistics and transport.

The model vehicle in this design not only has the ability to drive autonomously, but also can transport goods. In the case of bumpy road, the vehicle can maintain a stable driving state to ensure that the goods will not fall. Before using the model vehicle, it is necessary to plan the route of the vehicle in advance. Then, the vehicle learns about the road conditions and determines where to drop the goods. Finally, the vehicle is fine-tuned so that it can transport the goods according to the pre-designed route. This design

Fig. 9. Pictures of model vehicle

has great potential for application when road conditions are complex, and is particularly suitable for transporting goods. The unmanned transport system can achieve automated and intelligent transport, reduce the use of human resources and improve the efficiency of logistics and transport. At the same time, it has a broad development prospect due to its innovation and practicality.

In the future, all-terrain unmanned vehicles are expected to be integrated into the agricultural field, bringing new possibilities for automated agricultural operations. In the vast farmland, all-terrain vehicles can be equipped with various sensors and control systems to achieve automated agricultural operations such as sowing, fertilizing, spraying and harvesting. Its strong passing ability and manoeuvrability enable it to move freely in a variety of terrain conditions, improving agricultural productivity and reducing reliance on human resources, bringing greater convenience and benefits to the development of modern agriculture.

Acknowledgment. This project is supported by Nanning University Professor Cultivation Engineering Project. (Project No.: 2020JSGC01).

References

1. Wenchangjia, M., Jianhong, C.: Design and implementation of all-terrain vehicle based on ATMEGA328 controller. Autom. Technol. Appl. **39**(10), 14–18 (2020). (in Chinese)
2. Tsagaris, A., Polychroniadis, C., Tzotzis, A., Kyratsis, P.: Cost-effective robotic arm simulation and system verification. Int. J. Intell. Syst. Appl. (IJISA) **16**(2), 1–12 (2024)
3. Xiyuan, H., Zhichao, Y., Yuan, F., et al.: Software design and experiment of industrial handling robot control system. Indust. Control Comput. **34**(11), 77–78+80 (2021). (in Chinese)
4. Tanira, A.H., AbuHadrous, I.M.I.: An improved sampling dijkstra approach for robot navigation and path planning. Ternational J. Intell. Syst. Appl. (IJISA) **15**(6), 51–64 (2023)
5. Fang, R.: New intelligent storage robot and its application. Log. Technol. Appl. **23**(8), 80–83 (2018). (in Chinese)
6. Eljinini, M.A.H., Tayyar, A.: Collision-free random paths between two points. Int. J. Intell. Syst. Appl. (IJISA) **12**(3), 27–34 (2020)

7. Bickler, D.B.: Articulated suspension system, USA, 4840394A1, 20 Jun 1989
8. Chen, B.C., Wang, R.B., Lu, Y., et al.: Design and simulation research on a new type of suspension for lunar rover. In: IEEE International Conference on Computational Intelligence in Robotics and Automation, pp. 173–177. IEEE, Piscataway (2007)
9. Batelaan, J.: Development of an all-terrain vehicle suspension with an efficient, oval track. J. Terrramech. **35**(4), 209–223 (2015)
10. Mckinney, R.A., Zapata, M.J., Conrad, J.M., et al.: Components of an autonomous all-terrain vehicle. IEEE Southeastcon Proc. **6**(1), 416–419 (2010)
11. Dai, L., Wu, J.: Stability and vibrations of an all-terrain vehicle subjected to nonlinear structural deformation and resistance. Commun. Nonlinear Sci. Numer. Simul. **12**(1), 72–82 (2007)
12. Chugo, D., Kawabata, K., Kaetsu, H., Asama, H., Mishima, T.: Mechanical design of step-climbing vehicle with passive linkages. Bioinspiration Robot. Climb. Walk. Robots, pp. 429–440 (2007)
13. Tanira, A.H., AbuHadrous, I.M.I.: An improved sampling Dijkstra approach for robot navigation and path planning. Int. J. Intell. Syst. Appl. (IJISA), **15**(6), 51–64 (2023)
14. Yousif, M., Salim, A., Jummar, W.K.: A robotic path planning by using crow swarm optimization algorithm. Int. J. Math. Sci. Comput. (IJMSC) **7**(8), 20–25 (2021)
15. Fude, W.: Design and Research of a New Six-Wheeled All-Terrain Mobile Robot. Shenyang University of Technology, Shenyang (2012). (in Chinese)
16. Luo Qiang, X., Wencheng, L.Y.: Design of intelligent vehicle path recognition system based on laser sensor. Elec. Autom. **34**(5), 88–90 (2012). (in Chinese)
17. Bo, L., Wei, Y., Wendong, Z., et al.: Design of an autonomous seeking/ tracking system for an intelligent vehicle. Comput. Measure. Control **20**(10), 2798–2801 (2012). (in Chinese)
18. Yong, Z., Hailang, G., Qian, T.: Structural design and obstacle crossing analysis of a new all-terrain robot. Mech. Inform. **27**, 96–97 (2017). (in Chinese)
19. Lamon, P.: 3D position tracking for All-Terrain Robots. Ecole Polytechnique Federale De Lausanne, pp. 29–30 (2005)
20. Chang, Y., Tan, D.L., Wang, H.G.: Intelligent robots and systems. In: IEEE/RSJ International Conference, pp. 4169–4174 (2006)
21. Siegwart, R., Lamon, P., Estier, T., Lauria, M., Piguet, R.: Innovative design for wheeled locomotion in rough terrain. J. Robot. Autonom. Syst. **40**(2–3), 151–162. Elsevier

Accuracy Improvement of Spoken Language Identification System for Close-Related Languages

Oleh Romanovskyi[1], Ievgen Iosifov[1,2], Olena Iosifova[1], Volodymyr Sokolov[2(✉)], Pavlo Skladannyi[2], Igor Sukaylo[2], and Oleksii Tsarenok[1]

[1] Ender Turing OÜ, 11912 Tallinn, Estonia
[2] Borys Grinchenko Kyiv Metropolitan University, Kyiv 04053, Ukraine
v.sokolov@kubg.edu.ua

Abstract. Spoken Language Identification (SLI) systems have witnessed tremendous progress in adopting deep learning architectures. However, differentiating between acoustically and phonetically similar languages remains a significant challenge. This paper investigates the challenges of building SLI systems for similar languages and presents an accuracy improvement strategy. We provide experimental results on a set of closely related languages (Swedish-Norwegian, Czech-Slovak, Ukrainian-Russian, and Spanish-Catalan-Galician) and discuss the implications of our findings. This paper aims to research data impact on SLI models' accuracy, primarily focusing on comparing models trained on different datasets, data quality versus quantity comparison, and data balance demand to achieve the best possible SLI model. Experiments lead us to conclusions that the most impactful criteria on additional accuracy are region-specific SLI models (outperform heavy multilingual by 2.47% on average and much faster to train), adding new domain dataset (improvements for a new domain without significant risk to the original domain accuracy), balancing training data for the same amount of data per each language (provides stability in trained pairs/triples), and adding less but more quality data (30% high-quality data gives more accuracy than 100% of mid-quality data).

Keywords: Spoken Language Identification · SLI · Spoken Language Recognition · SLR · Language identification · LID · VoxLingua107 · Common Voice

1 Introduction

The ability to identify spoken languages accurately forms the bedrock of numerous applications, ranging from automated call routing systems to multilingual voice assistants to multilingual agent call center analytics software [1]. With the rapid advances in deep learning, SLI systems have achieved remarkable accuracy on diverse datasets [2].

Z. Hu et al. (Eds.): ICCSEEA 2024, LNDECT 242, pp. 35–52, 2025.
https://doi.org/10.1007/978-3-031-84228-3_4

However, as with many technologies that promise universal applicability, a caveat exists. Not all languages are identified with the same accuracy. While differentiating between acoustically distinct languages like English and Mandarin [3] may be straightforward for most models, the real challenge emerges when the task involves languages with many phonetic, historical, and acoustic characteristics. Consider, for instance, the Scandinavian languages Norwegian and Swedish or the South Asian languages Urdu [4] and Hindi [5]. Due to their intertwined histories and shared linguistic roots, these languages present significant challenges for conventional SLI systems. Misidentifications can be frequent, and the implications can range from minor inconveniences in consumer applications to significant misunderstandings in more critical scenarios [6].

It is crucial to emphasize that while many SLI models boast high accuracy rates, these figures can often mask the nuances. When dissected further, we notice that while these models perform exceptionally well on distinct languages, their performance can deteriorate considerably when faced with acoustically similar languages [7]. Such revelations necessitate a deeper dive into the challenges of similar languages and potential solutions.

This paper aims to research data impact on SLI models' accuracy, primarily focusing on comparing models trained on different datasets, data quality versus quantity comparison, and data balance demand to achieve the best possible SLI model.

By delineating the challenges in differentiating between closely related languages, we set the stage for introducing a novel fine-tuning strategy. Through this approach, we aim to augment the discriminative power of SLI systems, allowing them to tease apart even the most subtle nuances between similar languages, thereby pushing the boundaries of what's achievable in spoken language identification [8].

This work presents the results of experiments and methods to improve the accuracy of SLI models for close-related languages. In Sect. 2, we analyze the existing research, modern model architectures for the SLI problem, and dataset availability. Section 3 explain the experiment setup and which model and characteristics were used to train comparable models. In Sect. 4 Datasets selection and exploration were made for further experiments. Section 5 describes the approach and selection of language pairs/triples for training and evaluation experiments. The assessment of the experiment results and the authors' recommendations for the training and evaluation of SLI models are discussed in Sect. 6. The final section outlines potential directions for future research in this field.

2 Related Works

Shen et al. [9] conducted an extensive review of the current literature on SLI, with a particular focus on speech features and architectural designs. As for architecture, older SVM and HMM models for such tasks are losing heavily to newer models. For example, [10, 11] only achieved accuracy to an average of 66.55% for each language (trained just for four languages), much lower than any modern architecture can achieve even trained for 107 languages.

Draghici et al. [12] explored former designs of SLI systems, specifically those employing Convolutional Neural Networks (CNN) and Convolutional Recurrent Neural Network (CRNN) architectures. Their research also involved a set of seven languages.

Despite the growing intricacy of these systems, they achieved notable success, with the CNN architecture reaching an accuracy of 71% and the CRNN architecture attaining 83%.

That is why modern architectures are used for SLI problems Emphasized Channel Attention, Propagation, and Aggregation in Time Delay Neural Networks (ECAPA-TDNN) [13–15], CNN [16–18], and Cross-Lingual Speech Representations (XLSR) [19, 20], that significantly outperforms old architectures based on Long Short-Term Memory (LSTM). In [21] was evaluated two new model architectures TC-ResNet10 and LECAPAT focused on speed and achieving close to bigger architectures accuracy results but almost two order-of-magnitude (OOM) faster and four OOM smaller. While the speed is incredible, trade-offs by accuracy are still not acceptable for business-related tasks.

As for available datasets, the most widely recognized standards for appraising novel SLI/SLR models and techniques are based on the NIST LRE evaluation datasets [22]. These datasets predominantly consist of narrowband conversational telephone speech. Substantial volumes of conversational telephone data from specific languages are utilized to develop competitive systems for NIST LREs. The Linguistic Data Consortium (LDC) often provides such data, but it can be costly, amounting to thousands of dollars. For instance, the established Kaldi [23] framework for LRE07 requires 18 different LDC SLR datasets, which total $15,400 for non-members of the LDC3. This high cost presents a significant barrier to new research teams seeking to contribute to the SLR academic field.

From open datasets, the most known is VoxLingua107 [24], which is excellent and opened the door to training heavily multilingual SLI models instead of 5–10 languages, as it consists of 107 languages. On the other hand, limitations of the current dataset are that the development set, with manually verified language labels, is very limited for VoxLingua107 and just has labels for 33 languages, and because of the dataset automated collection process—the accuracy of training dataset for closely related languages is low for some languages (e.g., a lot of Spanish dialogues in the Galician train set or a lot of Russian in Ukrainian train set).

The Common Voice (CV) dataset [25] also has a very limited number of languages. SLI model improvement is described in [26], which focuses on the impact of foreign accents, and [27], which focuses on model efficiency and classifying unseen languages and different acoustic environments without additional training.

3 Experiments Setup Architecture

We utilized an SLI system built on a deep CNN NVIDIA TitaNet architecture [16, 17], which is part of the Nemo toolkit [28] displayed in Fig. 1. TitaNet-LID-BxRxC encoder is based on ContextNet architecture where *B* is the number of blocks, *R* is the number of repeated "basic" blocks, and filters in the convolution layers of each block.

The TitaNet-LID-BxRxC, as depicted in Fig. 1, features an encoder with a 1D depth-wise channel separable convolutional design, mirroring the ContextNet architecture, and a decoder. This encoder is composed of *B* blocks, where each block contains *R* "basic" blocks repeated and *C* convolutional filters in each block. The encoder of TitaNet-LID

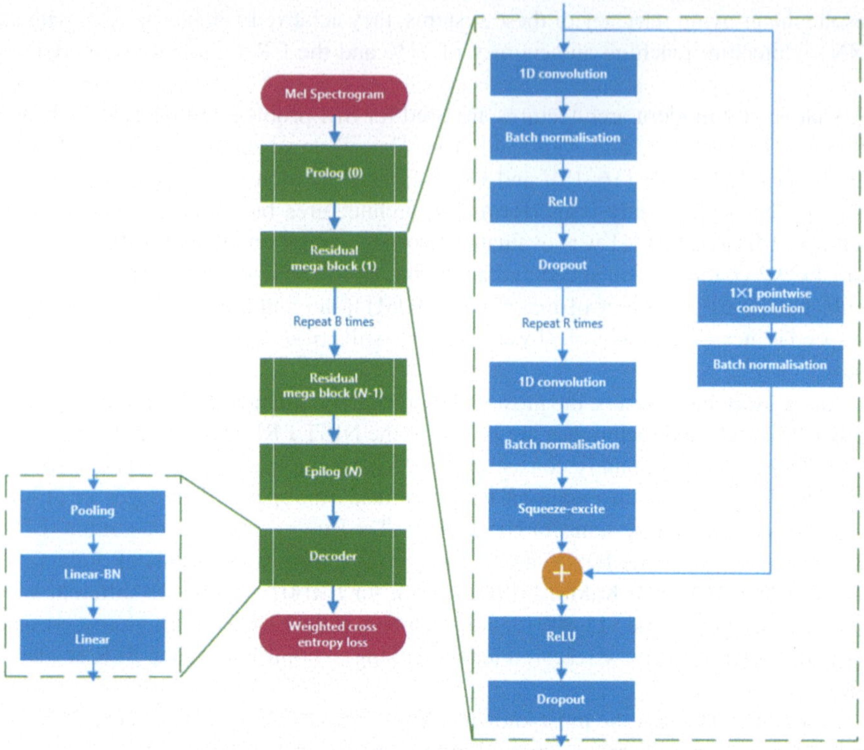

Fig. 1. Spoken Language Identification Model Architecture

begins with an initial block B_0 and then proceeds with a series of residual mega blocks from B_1 to $B_N - 1$.

Each mega block is composed of R "basic" blocks, concluding with a squeeze-excite module. A basic block includes a 1D time-channel separable convolutional module (referred to as 1D convolution in Fig. 1), featuring a kernel K. This is followed in sequence by BatchNorm, ReLU, and Dropout components. Each 1D convolution module is made up of two components: a depth-wise convolutional layer and a 1×1 pointwise convolutional layer. These basic blocks are reiterated R times and are interconnected residually with "Squeeze-and-Excitation" (SE) layers that include global average pooling.

The kernel K in repeated "basic" blocks are 7, 11, and 15. The encoder concludes with a final block, B_N, which produces intermediary audio features. These encoded audio features are then transferred to the decoder for classification. The decoder comprises two sections: a statistics pooling layer (calculating mean and standard deviation) that converts variable-length input audio features into a fixed-length feature representation, and two linear layers. The first linear layer has an output size of 512, while the second linear layer performs a transformation from 512 to the final class count N.

All models were trained for 40 epochs on one node with two GPUs with a batch size of 32 per GPU and precision fp16. We utilized the Adam optimizer and Cosine

Annealing learning rate scheduler with a warm-up ratio of 10%. A maximum learning rate of 0.001 and a minimum learning rate of 0.0001 were used. Random seed 42 was used for all pieces of training. All experiment models were trained using the same training procedure.

4 Dataset Selection and Comparison

As were already mentioned in related works, the most widely recognized standards for appraising novel SLI/SLR models and techniques are based on the NIST LRE evaluation datasets but their limitations are they are quite expensive. That is why we set two criteria for datasets for the current paper:

1. Selected datasets should be open.
2. Selected datasets should cover as many languages as it is in the world.

After the evaluation of available datasets next datasets were selected: to train the general SLI model we used a large multilingual dataset VoxLingua107 [24] and for model accuracy improvement, we used CV [25].

VoxLingua107 train dataset is composed of 6,628 h of speech, which have been sourced from YouTube videos through automatic scraping, as referenced in [24]. The language identified in the video's title and description categorizes these speech segments into different languages. A data-driven post-filtering method was employed to enhance language labeling accuracy. This approach helped to eliminate segments that were probably not in the specified language. As a result, the accuracy of correctly labeled segments in the dataset increased to 98%, as verified by crowd-sourced evaluation. VoxLingua107 features speech recordings in 107 distinct languages, with approximately 2.54 mln individual speech segments. The average duration of these segments is 9.4 s, and each language contributes about 62 h of data. The development subset of VoxLingua107 includes 4.5 h of spontaneous speech extracted from YouTube videos covering 33 languages. In this subset, the linguistic classification of each speech segment has been confirmed by a minimum of two native or highly proficient speakers sourced from the crowd. This development set comprises a total of 1,608 speech segments.

VoxLingua107 test/evaluation dataset. As a development set with manually verified language labels, it is constrained for VoxLingua107 and just has labels for 33 languages. It was not sufficient for experiments of the current paper, so we decided to divide 10% of the training dataset for test/evaluation purposes.

CV dataset [25] is an extensive, multilingual compilation of transcribed vocal recordings aimed at research and development in speech technology. This dataset encompasses 9,283 h of recorded audio. Additionally, it features demographic metadata, containing information such as age, gender, and accent. Of the total collection, 7,335 h across 60 languages have been validated for use.

CV-Clean dataset is a subset of the CV dataset with verified labels. Verification was made using an automated pipeline [29] in two steps: (1) speech recognition using open source Whisper v2 model and (2) filtering out all segments where the label of CV is not equal to what is predicted by the speech recognizer label.

CV-Balanced dataset is a subset of CV / CV-Clean datasets truncated/balanced by the smallest dataset in experiment pair/triple languages for a specific region. For example, if Language A has 500 h and Language B only has 100 h of audio, the balanced dataset will consist of only 100 h for 'anguages.

5 Languages Selection for Experiments

5.1 Approach to Getting Low-Accuracy Languages

We have conducted a series of experiments with the following steps to get low-accuracy languages:

1. Train the General SLI model for 107 languages using the VoxLingua107 dataset.
2. Identify languages with Low accuracy.
3. Select language pairs to improve the quality of SLI. Based on the confusion matrix between languages and the availability of CV datasets.
4. For each selected pair:

- Train SLI using the subsection of the VoxLingua107 dataset for selected pairs to improve the accuracy of both languages in pairs.
- Train SLI using the subsection of {VoxLingua107 + CV} datasets for selected pairs to improve the accuracy of both languages in pairs.
- Clean CV dataset.
- Train SLI using the subsection of {VoxLingua107 + CV-Clean} datasets for selected pairs to improve the accuracy of both languages in pairs.

5.2 Language-Region Selection for Experiments

Step 1. Selection of languages for experiments by low accuracy of the model trained for 107 languages on the VoxLingua107 dataset. We have trained a baseline model for all 107 languages present in VoxLingua107 to identify which languages can not be trained for high accuracy. As mentioned before, we divided the training dataset and reserved 10% of the dataset for evaluation purposes. The F1 score of all 107 languages is presented in Fig. 2. For further experiments, we were interested only in languages with an accuracy of less than 95%. Forty languages resulted from the 95% threshold.

Step 2. Select languages for experiments based on the availability of the CV dataset. At this step, we narrow the list of languages for experiments by evaluating available CV datasets for languages from Step 1. We were interested in pairs with open datasets available to train on. Hence, we assessed the availability of CV datasets for these 40 languages, presented in Table 1. CV datasets are absent for some languages (see Table 2).

Step 3. Selection languages for experiments by a confusion matrix. We evaluated the confusion matrix for languages with available datasets in Table 2. A confusion matrix is a table used to evaluate the performance of classification models. It is organized to show the comparison between the actual and predicted classifications. The matrix provides a visual and numerical way to understand a model's accuracy, indicating how many predictions were right or wrong and the nature of the errors. We build a confusion matrix using the Python *sklearn* library.

We were interested in significant numbers in non-diagonal positions on a confusion matrix. These points showed when the model prediction was incorrect and what language

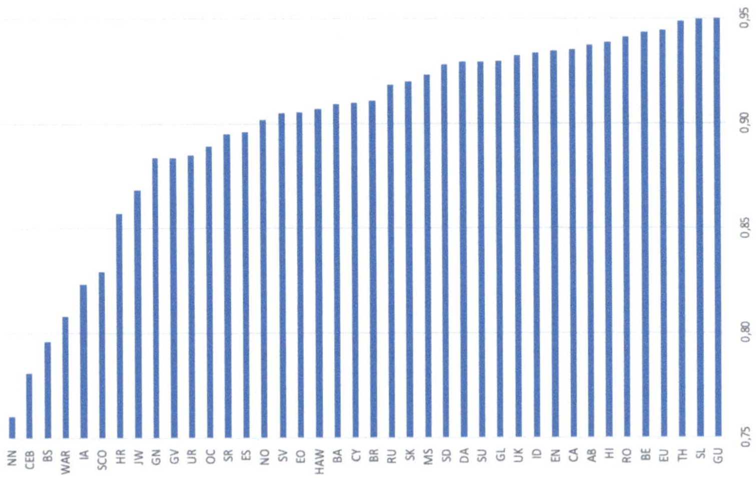

Fig. 2. List of low-accuracy language of the model trained on VoxLingua107

they predicted instead of the correct ones. This allows us to understand model confusion for languages to get pairs/triples most often confused with each other. The criteria by which we selected languages for experiments was that the model should make more than 20 mistakes for incorrect language (see Fig. 3).

After evaluating low accuracy and high confusion, we devised language-region pairs/triples to experiment with, presented in Table 3.

The 'Confused with' column represents a relative number of mistakes the model mistakenly uses incorrect language. For example, for Spanish audio samples, the model 4.3% times incorrectly predicted Galician instead of expected Spanish.

5.3 Dataset Training

Training datasets for experiments are presented in Table 4 with relevant sizes in minutes, hours, and number of samples to understand the disbalance of available data for region languages. Five datasets were used in the experiments: VoxLingua107 (Vox107), CV, CV-Clean (CV-Cl), CV-Balanced (CV-Bal), and CV-Clean-Balanced (CV-Cl-Bal).

As can be seen from Table there are some limitations of train and test datasets. CV and CV-Clean for Norwegian and as a consequence—CV-Clean-Balanced for Norwegian and Swedish are minimal (less than one minute), which we use in the experiment, just for consistency of charts results but not reliable for such a small amount of data.

6 Experiment Results

6.1 Model to Accuracy Impact Classes

Firstly, we decided to evaluate the impact on the accuracy of heavily multilingual (107 languages model) vs. regional models (per region model for 2–3 languages). This is important because (1) often, businesses do not need a whole 107-language model

Table 1. F1 score comparison for low-accuracy languages of the model trained on VoxLingua107.

Code	Language	F1 score, %		Code	Language	F1 score, %
GU	Gujarati	95.0		CY	Welsh	91.0
SL	Slovenian	95.0		BA	Bashkir	90.9
TH	Thai	94.9		HAW	Hawaiian	90.7
EU	Basque	94.4		EO	Esperanto	90.5
BE	Belarusian	94.3		SV	Swedish	90.5
RO	Romanian	94.1		NO	Norwegian	90.2
HI	Hindi	93.9		ES	Spanish	89.6
AB	Abkhazian	93.7		SR	Serbian	89.5
CA	Catalan	93.5		OC	Occitan	88.9
EN	English	93.5		UR	Urdu	88.5
ID	Indonesian	93.4		GV	Manx	88.4
UK	Ukrainian	93.3		GN	Guarani	88.4
GL	Galician	93.0		JW	Javanese	86.8
SU	Sundanese	92.9		HR	Croatian	85.7
DA	Danish	92.9		SCO	Scots	82.9
SD	Sindhi	92.8		IA	Interlingua	82.4
MS	Malay	92.3		WAR	Waray	80.8
SK	Slovak	92.0		BS	Bosnian	79.6
RU	Russian	91.8		CEB	Cebuano	78.1
BR	Breton	91.1		NN	Norwegian Nynorsk	76.0

but need the most accurate region-specific model and (2) most confusion was found within regional languages (like Spanish-Catalan-Galician), and it is the main focus of the current paper.

Both models were trained and tested using the VoxLingua107 dataset:

- Model-voxlingua-107 is 90% of VoxLingua107 for training and 10% for testing.
- Model-voxlingua-regional (per region model for 2–3 languages) is 90% of VoxLingua107 for region languages (see Table 4) for training and 10% for testing.

As shown in Fig. 4 below, models explicitly trained for 2–3 languages show significantly higher accuracy than those heavily multilingual (107 languages model). The average F1 score improvement is 2.47%, while the Median is 1.65% in absolute numbers.

Model-voxlingua-107 is a unified SLI model trained for 107 languages and model-voxlingua-region is a region-specific SLI model trained for specific 2–3 languages of the selected region.

Table 2. Common Voice dataset size.

Code	Language	Recorded, hours	Validated, hours	Number of voices
SL	Slovenian	14	11	146
TH	Thai	420	171	8
EU	Basque	159	105	1
BE	Belarusian	1 632	1 586	8 205
RO	Romanian	44	19	408
HI	Hindi	20	14	396
AB	Abkhazian	85	60	400
CA	Catalan	3 328	2 554	35 062
EN	English	3 347	2 532	88 904
ID	Indonesian	64	29	516
UK	Ukrainian	105	94	1
GL	Galician	61	38	1
DA	Danish	13	12	243
MS	Malay	10	3	128
SK	Slovak	27	22	216
RU	Russian	260	228	3
BR	Breton	26	12	196
CY	Welsh	155	122	1 800
BA	Bashkir	268	258	912
EO	Esperanto	1 897	1 432	1 682
SV	Swedish	54	45	808
ES	Spanish	2 188	526	25 338
SR	Serbian	6	4	144
OC	Occitan	13	2	141
UR	Urdu	221	63	316
GN	Guarani	25	4	140
IA	Interlingua	17	14	66
NN	Norwegian Nynorsk	2	2	32

6.2 Accuracy Degradation by Evaluation of External Dataset

To understand how accurately the model performs on different data, we evaluated the accuracy of region-trained models on VoxLingua107 datasets using CV datasets.

We evaluated only region-trained models because they are the most accurate (see Fig. 5). As we can see from the accuracy Fig. 2, all languages perform much worse on CV dataset data. Let's see if we can improve the accuracy of models.

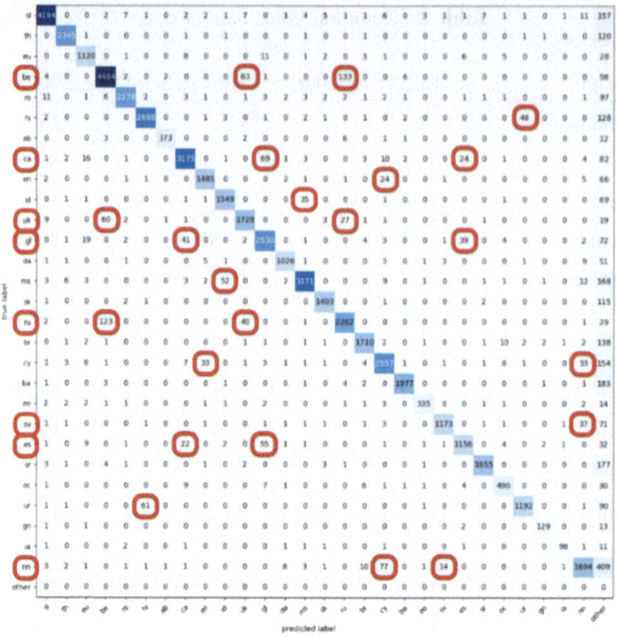

Fig. 3. Confusion matrix of low-accuracy languages

Table 3. Language selection for experiments.

Language-region	Language codes	VoxLingua107 F1 score, %	Confused with, %
Spanish / Catalan / Galician	ES → GL	89.6	4.3
	CA → GL	93.5	2.0
	GL → CA	93.0	1.5
Swedish / Norwegian Nynorsk	SV → NN	90.5	2.9
	NN → SV	76.0	0.6
Ukrainian / Russian	UK → RU	93.3	1.5
	RU → UK	91.8	1.7
Slovak / Czech	SK → CS	92.0	4.7
	CS → SK	97.1	1.4

We evaluated models trained on original VoxLingua107 by test dataset VoxLingua107 and for contrast by external domain dataset CV. Accuracy degradation by evaluation on unseen CV dataset is on average 19.7% and a median of 17.6%.

We trained five models with a dataset combination: {Vox107}, {Vox107 + CV}, {Vox107 + CV-Clean}, {Vox107 + CV-Balanced}, and {Vox107 + CV-Clean-Balanced}. Each of these models we evaluated them by test/evaluation datasets.

Table 4. Training datasets size for selected languages.

Lang	Size		Dataset		Lang	Size		Dataset
	Hours	Samples				Hours	Samples	
CS	60.0	23 215	Vox107		UK	22.5	19 024	CV-Bal
CS	25.5	19 358	CV		UK	10.6	9 313	CV-Cl-Bal
CS	12.3	9 656	CV-Cl		RU	65.6	21 412	Vox107
CS	3.5	2 729	CV-Bal		RU	37.7	26 328	CV
CS	1.4	1 128	CV-Cl-Bal		RU	24.8	17 789	CV-Cl
SK	35.8	13 727	Vox107		RU	22.5	15 854	CV-Bal
SK	3.5	3 276	CV		RU	10.6	7 487	CV-Cl-Bal
SK	1.4	1 389	CV-Cl		ES	34.7	11 475	Vox107
SK	3.5	3 276	CV-Bal		ES	450.8	311 392	CV
SK	1.4	1 389	CV-Cl-Bal		ES	297,3	206 270	CV-Cl
SV	30.6	11 475	Vox107		ES	16.5	10 057	CV-Bal
SV	8.4	7 584	CV		ES	4.9	2 987	CV-Cl-Bal
SV	4.9	4 638	CV-Cl		CA	79.2	30 460	Vox107
SV	0.5	496	CV-Bal		CA	1743.8	1 142 607	CV
SV	—	1	CV-Cl-Bal		CA	821.0	546 025	CV-Cl
NO	96.0	37 408	Vox107		CA	16.5	9 625	CV-Bal
NO	0.5	407	CV		CA	4.9	2 942	CV-Cl-Bal
NO	—	1	CV-Cl		GL	65.0	24 958	Vox107
NO	0.5	407	CV-Bal		GL	16.5	12 688	CV
NO	—	1	CV-Cl-Bal		GL	4.9	3 945	CV-Cl
UK	47.2	16 976	Vox107		GL	16.5	12 688	CV-Bal
UK	22.5	19 024	CV		GL	4.9	3 945	CV-Cl-Bal
UK	10.6	9 313	CV-Cl					

6.3 Evaluation for Spanish-Catalan-Galician Triple

In Figs. 6, 7, and 8, we can see the F1 accuracy score of the Spanish-Catalan-Galician language triple.

As we can see from Table 4, data for Galician significantly disbalance comparatively to Spanish and Catalan (CV 16.5 h for Galician vs 450 and 1743 for Spanish and Catalan, respectively). This leads to significant accuracy degradation for Galician in models trained on disbalance datasets.

After balancing datasets by the lowest one, we achieved stable results on all three languages—see model trained on {VoxLingua107 + CV-Balanced} and evaluated by CV dataset.

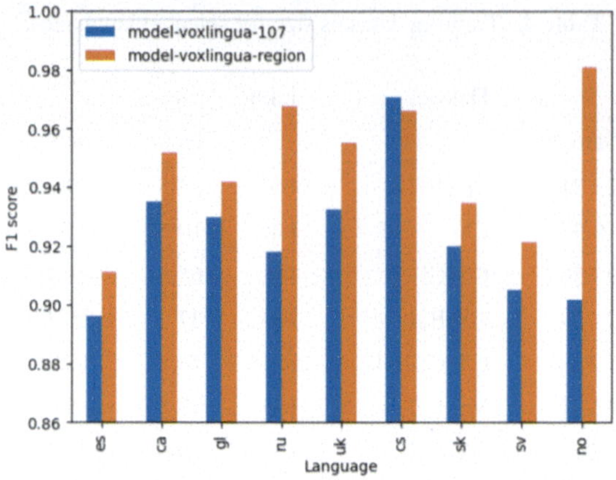

Fig. 4. Accuracy improvement for region-trained model compared to the unified 107 language model.

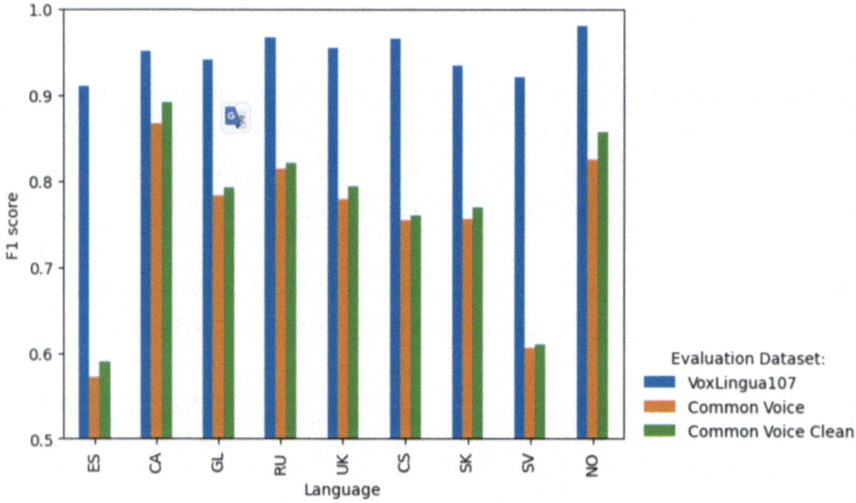

Fig. 5. Average accuracy degradation

6.4 Evaluation of the Ukrainian-Russian Pair

The F1 accuracy score of the Ukrainian-Russian language pair is shown in Figs. 9 and 10.

As we can see from Table 4, data for Ukraine are limited and hence significantly imbalanced compared to the Russian dataset. This leads to a degradation of accuracy for the Ukrainian language in models trained on imbalanced datasets. Significant average improvement was achieved while training on CV-Clean-Balanced datasets.

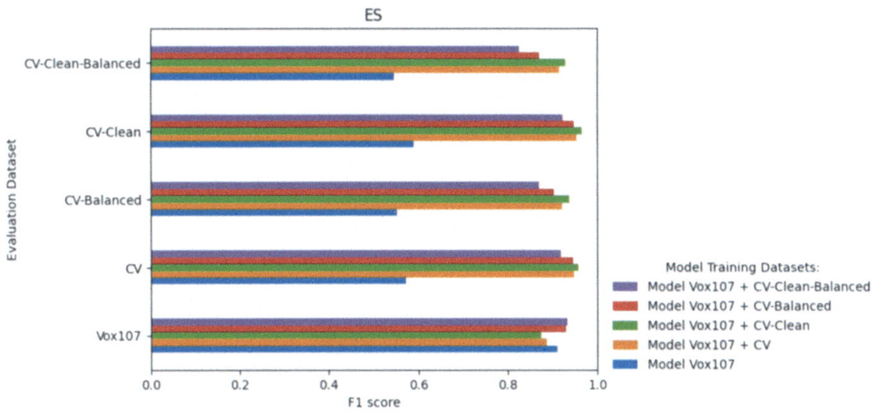

Fig. 6. Spanish models evaluation

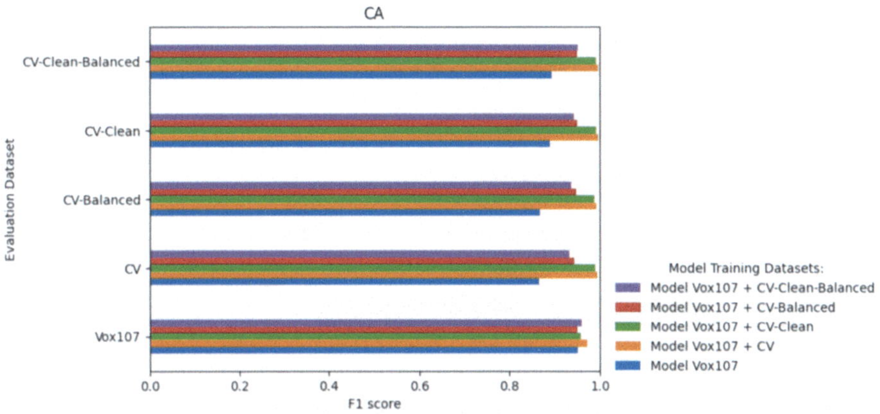

Fig. 7. Catalan models evaluation

6.5 Other Evaluation Pair Results

The behavior of Czech-Slovak and Swedish-Norwegian Nynorsk language pairs are similar to the Ukrainian-Russian pair, with significant accuracy degradation for Slovak and Norwegian Nynorsk languages in models trained on imbalanced datasets but stable results after balancing datasets. After evaluation models by the new domain CV dataset, models trained on {VoxLingua107 + CV-Clean-Balanced} perform 5.8% better on average (and 6.6% better on median) against models trained on {VoxLingua107}.

6.6 Accuracy Impact of Data Cleaning

When comparing quality versus quantity, we compare the results of balanced models trained on the {VoxLingua107 + CV-Balanced} dataset against the {VoxLingua107 + CV-Clean-Balanced} (see Fig. 11). The main difference is CV against CV-Clean parts of training data to compare quantity (CV-Bal) against quality (CV-Cl-Bal).

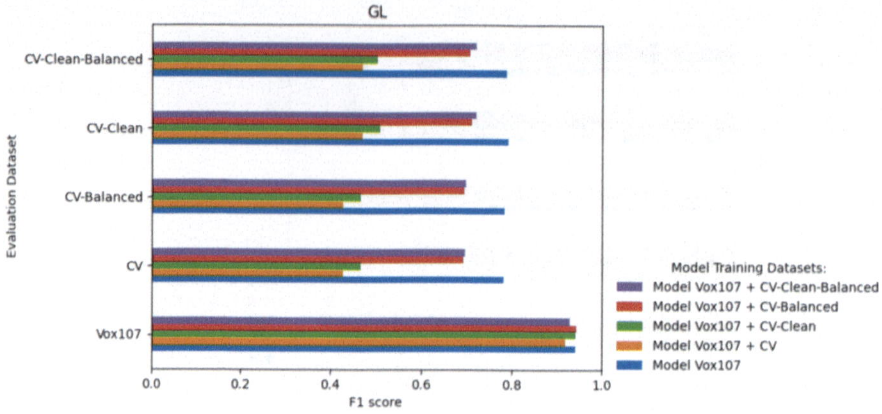

Fig. 8. Galician models evaluation

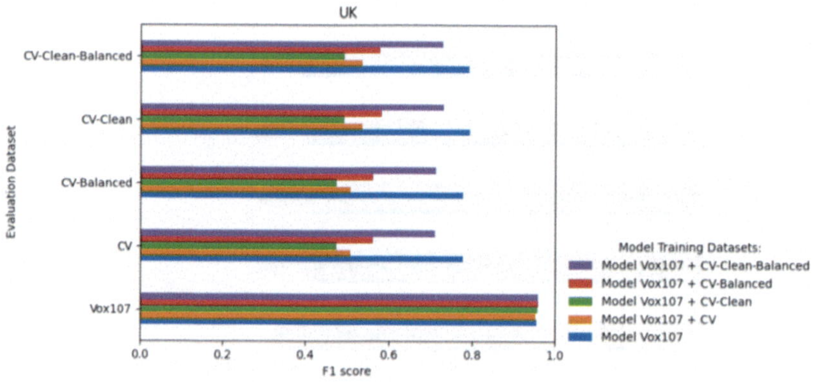

Fig. 9. Ukrainian models evaluation

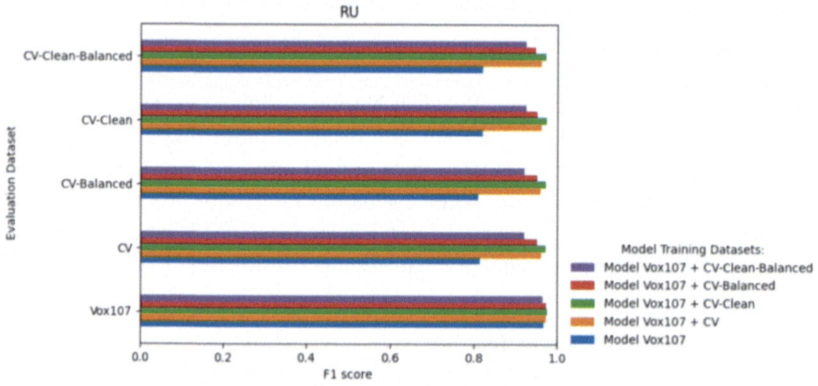

Fig. 10. Russian models evaluation

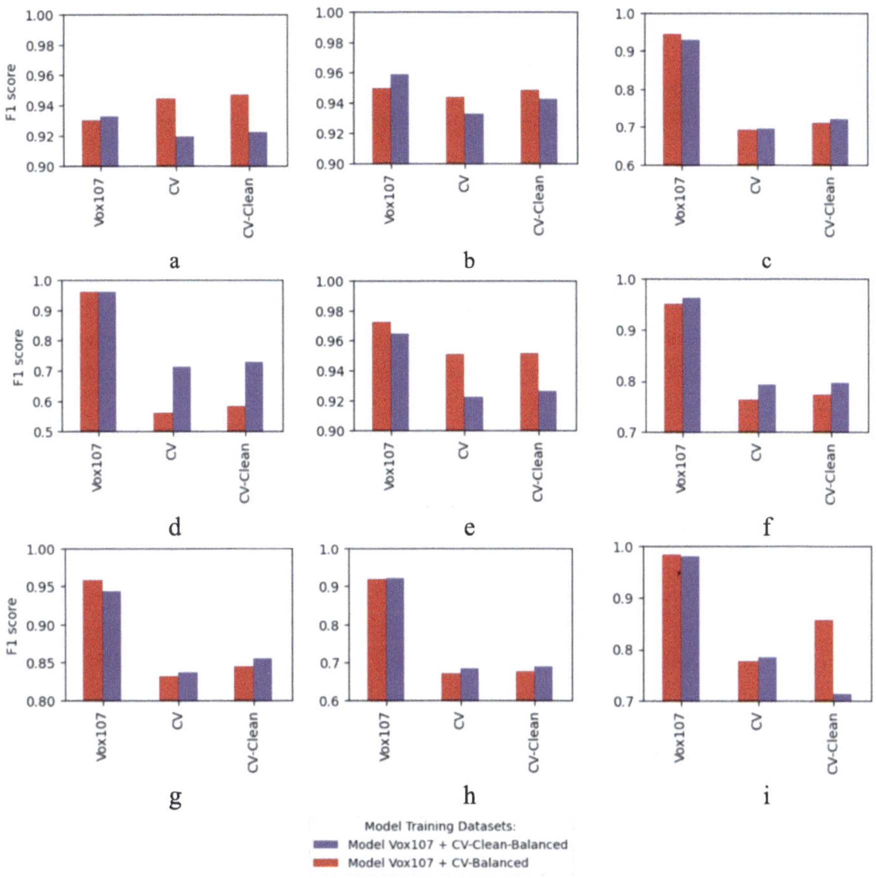

Fig. 11. Comparison of models trained on {Vox107 + CV-Bal} and {Vox107 + CV-Cl-Bal} datasets for (a) ES, (b) CA, (c) GL, (d) UK, (e) RU, (f) CS, (g) SK, (h) SV, and (i) NN

We use two criteria to compare: F1 score and quantity of data. CV-Clean-Balanced shows 1.4% better results with average medians for all languages and only used 30% of data size for training purposes comparatively with CV-Balanced.

This means three times less data in training to achieve better accuracy, which leads to much better training performance, less training time, and a positive impact both from economic and environmental sides.

6.7 Accuracy Impact of Data Cleaning

To achieve significant improvements experiments lead us to the next recommendations:

- Train region-specific SLI models. Region-specific SLI mode outperforms the Multilingual model for an average of 2.47% and Median of 1.65% and is much faster to train and fine-tune.

- Add the domain dataset. Models trained purely on one domain (VoxLingua107 dataset) perform very poorly on a new domain (CV). Adding CV data does not decrease accuracy, as evaluated by the original VoxLingua107, which points us to the conclusion that if we have the domain data, we should always use it in training with low impact on the original domain data. Accuracy degradation by evaluation on unseen CV dataset is on average 19.7% and a median of 17.6%.
- Balance the dataset. As we can see from Swedish-Norwegian Nynorsk and Czech-Slovak pairs and Spanish-Catalan-Galician triple results, balancing plays a critical role in achieving high accuracy for all languages. And it is even more important for small resource datasets with only tens of hours available. Models trained on the original and new dataset {VoxLingua107 + CV-Clean-Balanced} perform 5.8% better on average (and 6.6% better on median) against models trained purely on the original VoxLingua107 dataset.
- Clean data before training.

Adding only 30% of CV-Clean data to training outperformed adding the whole CV dataset by 1.4%, with average medians for all languages.

7 Conclusions and Future Work

The challenge of distinguishing between similar languages in SLI systems is multifaceted, stemming from shared phonetic attributes, historical evolutions, and overlapping acoustic properties. Our experiments underscore the importance of a focused fine-tuning approach, using curated datasets encompassing the nuanced differences between similar languages.

We presented a numerical representation of conclusions. To achieve the best accuracy of SLI models few steps can be applied: train region-specific SLI models, add the domain dataset, balance the dataset, and clean data before training. All these steps will make the SLI model much more reliable for day-to-day business tasks.

The significant improvements suggest a promising path forward for SLI systems, ensuring they are robust even in the face of closely related languages.

Further research will explore additional fine-tuning techniques and larger sets of similar languages as well as focus on achieving the same accuracy for multilingual models as it will be more convenient and applicable than regional-based models.

Acknowledgments. The research team is grateful to Ender Turing OÜ (Tallinn, Estonia) for defining the business problem, comments, corrections, inspiration, and computational resources under the #89-080123 "Research of Speech Emotion Recognition" grant.

References

1. Romanovskyi, O., et al.: Prototyping methodology of end-to-end speech analytics software. In: 4th International Workshop on Modern Machine Learning Technologies and Data Science (MoMLeT&DS), vol. 3312, pp. 76–86 (2022)

2. Iosifov, I., Iosifova, O., Sokolov, V.: Sentence segmentation from unformatted text using language modeling and sequence labeling approaches. In: IEEE International Conference on Problems of Infocommunications, Science and Technology (PICST), pp. 335–337 (2020). https://doi.org/10.1109/picst51311.2020.9468084
3. Sun, M., Jiang, B., Yuan, J.: Vocal emotion recognition based on HMM and GMM for mandarin speech. Int. J. Educ. Manag. Eng. **2**(3), 25–31 (2012). https://doi.org/10.5815/ijeme.2012.03.04
4. Zia, T., Abbas, Q., Akhtar, M.P.: Evaluation of feature selection approaches for Urdu text categorization. Int. J. Intell. Syst. Appl. **7**(6), 33–40 (2015). https://doi.org/10.5815/ijisa.2015.06.03
5. Malik, P., Singh Baghel, A.: Performance enhancement of machine translation evaluation systems for English-Hindi language pair. Int. J. Modern Educ. Comput. Sci. **11**(2), 42–49 (2019). https://doi.org/10.5815/ijmecs.2019.02.06
6. Iosifov, I., et al.: Natural language technology to ensure the safety of speech information. In: Workshop on Cybersecurity Providing in Information and Telecommunication Systems, vol. 3187, no. 1, pp. 216–226 (2022)
7. Iosifova, O., et al.: Techniques comparison for natural language processing. In: 2nd International Workshop on Modern Machine Learning Technologies and Data Science (MoMLeT&DS), I(2631), pp. 57–67 (2020)
8. Iosifova, O., et al.: Analysis of automatic speech recognition methods. In: Workshop on Cybersecurity Providing in Information and Telecommunication Systems (CPITS), vol. 2923, pp. 252–257 (2021)
9. Shen, P., et al.: Knowledge distillation-based representation learning for short-utterance spoken language identification. IEEE/ACM Trans. Audio Speech Lang. Process. **28**, 2674–2683 (2020). https://doi.org/10.1109/taslp.2020.3023627
10. Gazeau, V., Varol, C.: Automatic spoken language recognition with neural networks. Int. J. Inf. Technol. Comput. Sci. **10**(8), 11–17 (2018). https://doi.org/10.5815/ijitcs.2018.08.02
11. Almutiri, T., Nadeem, F.: Markov models applications in natural language processing: a Survey. Int. J. Inf. Technol. Comput. Sci. **14**(2), 1–16 (2022). https://doi.org/10.5815/ijitcs.2022.02.01
12. Draghici, A., Abeßer, J., Lukashevich, H.: A study on spoken language identification using deep neural networks. In: 15th International Audio Mostly Conference, pp. 253–256 (2020). https://doi.org/10.1145/3411109.3411123
13. Desplanques, B., Thienpondt, J., Demuynck, K.: ECAPA-TDNN: emphasized channel attention, propagation and aggregation in TDNN based speaker verification. Interspeech (2020). https://doi.org/10.21437/interspeech.2020-2650
14. Miao, X., McLoughlin, I., Yan, Y.: A new time-frequency attention mechanism for TDNN and CNN-LSTM-TDNN, with application to language identification. Interspeech (2019). https://doi.org/10.21437/interspeech.2019-1256
15. Iosifov, I., et al.: Transferability evaluation of speech emotion recognition between different languages. Adv. Comput. Sci. Eng. Educ. **134**, 413–426 (2022). https://doi.org/10.1007/978-3-031-04812-8_35
16. Koluguri, N.R., Park, T., Ginsburg, B.: TitaNet: neural model for speaker representation with 1D Depth-wise separable convolutions and global context (version 1). arXiv (2021). https://doi.org/10.48550/arxiv.2110.04410
17. Jia, F., et al.: A compact end-to-end model with local and global context for spoken language identification (version 2). arXiv (2022). https://doi.org/10.48550/arxiv.2210.15781
18. Alashban, A.A., et al.: Spoken language identification system using convolutional recurrent neural network. Appl. Sci. **12**(18), 9181 (2022). https://doi.org/10.3390/app12189181
19. Babu, A., et al.: XLS-R: self-supervised cross-lingual speech representation learning at scale (version 3). arXiv (2021). https://doi.org/10.48550/arxiv.2111.09296

20. Conneau, A., et al.: XTREME-S: Evaluating cross-lingual speech representations (version 3). arXiv (2022). https://doi.org/10.48550/arxiv.2203.10752
21. Nieto, O., et al.: Efficient spoken language recognition via multilabel classification. Interspeech (2023). https://doi.org/10.21437/interspeech.2023-1986
22. Sadjadi, S.O., et al.: The 2017 NIST language recognition evaluation. In: Speaker and Language Recognition Workshop (2018). https://doi.org/10.21437/odyssey.2018-12
23. Povey, D., et al.: The Kaldi speech recognition toolkit (2011)
24. Valk, J., Alumae, T.: VoxLingua107: a dataset for spoken language recognition. In: IEEE Spoken Language Technology Workshop (2021). https://doi.org/10.1109/slt48900.2021
25. Ardila, R., et al.: Common voice: a massively-multilingual speech corpus (version 2). arXiv (2019). https://doi.org/10.48550/arxiv.1912.06670
26. Kukk, K., Alumäe, T.: Improving language identification of accented speech (version 3). arXiv (2022). https://doi.org/10.48550/arxiv.2203.16972
27. Bartley, T.M., et al.: Accidental learners: spoken language identification in multilingual self-supervised models (version 2). arXiv (2022). https://doi.org/10.48550/arxiv.2211.05103
28. Kuchaiev, O., et al.: NeMo: a toolkit for building AI applications using neural modules (version 1). arXiv (2019). https://doi.org/10.48550/arxiv.1909.09577
29. Romanovskyi, O., et al.: Automated pipeline for training dataset creation from unlabeled audios for automatic speech recognition. Adv. Comput. Sci. Eng. Educ. **IV**(83), 25–36 (2021). https://doi.org/10.1007/978-3-030-80472-5_3

Evaluation of State-of-the-Art Machine Learning Smart Contract Vulnerability Detection Method

Maksym Adamantis, Volodymyr Sokolov(✉), and Pavlo Skladannyi

Borys Grinchenko Kyiv Metropolitan University, Kyiv 04053, Ukraine
v.sokolov@kubg.edu.ua

Abstract. Proactive detection of vulnerabilities in smart contracts is imperative for ensuring the security of user funds entrusted to them. Once deployed, a smart contract is immutable and therefore cannot be updated. This posits the challenge of detecting and fixing all vulnerabilities before deployment. In this context, static analysis has proved to be a formidable tool, even though there is still a lot to be discovered in this field, and the likelihood of the discovery of new classes of vulnerabilities is high. Since 2019, there has been a rise in methods that use Machine (ML) and Deep Learning (DL) to enhance the existing methods, whether in static or dynamic analysis, to cover this issue. This research presents a comprehensive review of existing ML models that detect vulnerabilities in smart contracts statically, i.e. without running their code. The authors evaluate the accuracy of publicly available models in identifying reentrancy in smart contracts based on their F1 score when tested on a foreign dataset with files of newer Solidity versions. The findings point to the limitations of such models in adapting to the continuously evolving nature of the Solidity language, which is still going through its infancy. The authors also explore and share the optimal parameters for training and testing those models, detailing things that were overlooked by the official documentation. All the scripts used for integration and interoperability were published on GitHub to facilitate further research in this area. The research highlights the critical need for constantly updating the existing detectors to avoid false negatives. This research is significant for the broader blockchain community, safeguarding smart contract integrity and fortifying overall system security.

Keywords: Smart contracts · Vulnerability detection · Decentralized finance · Blockchain security · Static analysis · Machine learning

1 Introduction

A smart contract is a self-executing, programmable agreement embedded with predefined rules and conditions written in code. Operating on blockchain technology, smart contracts automatically execute and enforce contractual terms without requiring intermediaries. This decentralized and automated nature makes them a cornerstone of blockchain-based applications. The first practical application of this concept became publicly available in

Z. Hu et al. (Eds.): ICCSEEA 2024, LNDECT 242, pp. 53–65, 2025.
https://doi.org/10.1007/978-3-031-84228-3_5

2015 following the launch of the Ethereum blockchain. Ethereum-based smart contracts revolutionized the landscape of decentralized applications, gaining popularity with the public [1–3].

The need for protecting smart contracts from attacks is evident from the extreme cost of not noticing a vulnerability [4]. For example, only in 2022, the total value of cryptocurrency stolen from smart contracts reached $3.8 billion [5], with the total value invested in them being $44 billion [6]. Not only can smart contract weaknesses allow to steal the funds, but they also might lock invested funds forever due to the immutability of deployed contracts. Although the security of smart contracts has been vigorously researched since their appearance, with numerous studies utilizing various approaches while working against the time, the field can still be considered to be in its infancy.

Consequently, it is no wonder that the detection of smart contract vulnerabilities during the development phase has emerged as a popular method of ensuring the safety of user funds. The classic approaches to vulnerability detection can be split into two broad categories, static and dynamic analysis.

Static analysis methods try to find vulnerabilities without running the application, while dynamic analysis methods do the opposite. In static analysis, smart contract source code can be transformed into an intermediate representation such as an Abstract Syntax Tree (AST), call flow graph, or bytecode. Within dynamic testing, the most prominent approach was fuzzing [4]. However, static analysis is currently dominant within the field [7], partly because its methods allow to analysis of smart contract source code at a scale while requiring less computational resources [8], bolstering a more comprehensive evaluation of the proposed approaches.

Of course, both general approaches can be combined in hybrid methods, but we have not seen many such studies. This could be due to the lack of standardized criteria for evaluating smart contract vulnerability detection tools. An interesting emerging trend in recent years is the application of Artificial Intelligence (AI) to existing approaches to increase vulnerability detection accuracy [9, 10].

The existing AI approaches can be further divided into supervised, semi-supervised, and unsupervised learning. Most of the existing instruments use supervised learning, but unsupervised learning is also widely used for data preprocessing and feature extraction [9]. However, a significant issue with supervised learning models is that they require great amounts of data for training. In this light, the semi-supervised approach looks promising, because techniques such as tri-training may reduce the need for large amounts of labeled contracts.

Although a lot of AI vulnerability detection tools with high detection accuracy have been produced over the last few years, it is still not evident whether ML and DL can meet the needs of developers, because very few of such models are ready for commercial or regular consumer use. For example, we discovered that out of 18 AI-based instruments selected based on the reviewed literature, only 10 are open source, and only several have comprehensive documentation and are straightforward to train and use. Moreover, little research has been done on mapping the discovered vulnerability to the exact line of code that triggered the detector [9, 10], which is a common feature of traditional static analyzers [11, 12]. We also discovered that the majority of available datasets are too small to make a production-ready detector.

This work presents an overview of the current landscape of publicly available ML models for vulnerability detection in smart contracts. In Sect. 2, we analyze the existing research, shedding light on aspects that require further exploration. Section 3 delves into an examination of the limitations and conditions governing the execution of social engineering penetration testing. Section 4 covers the setup of the necessary infrastructure for model evaluation. In Sect. 5, we provide a detailed account of the experimental procedure and the associated data transformation scripts. The assessment of the experiment results and the authors' recommendations for the training and evaluation of vulnerability-detecting ML models are discussed in Sect. 6. The final section outlines potential directions for future research in this field.

2 Related Works

After 2019, the proliferation of tools for vulnerability detection in smart contracts, along with corresponding scholarly works, witnessed a rapid surge [9]. So far numerous AI approaches have been applied to source code analysis, ranging from text-centered models such as BERT [13] or GPT [14] and Graph Neural Networks (GNNs) operating on CFGs [15–17] to CNNs trained on images transformed from Ethereum virtual machine bytecode [18]. Some authors use freely available pre-trained models to reduce training costs [13, 18].

ML was initially applied to the task of smart contract vulnerability detection in SmartEmbed [19]. SmartEmbed identifies smart contract clones by detecting similar smart contracts based on embeddings and can also detect vulnerabilities by comparing contracts against a vulnerable contract database. It was also introduced as a web service. At the time of writing this work, the main SmartEmbed website is inactive, and the tool is not publicly accessible.

Another early work used a GNN [17] to detect smart contract vulnerabilities, achieving an F1 score of 0.78. In this method, the source code was preprocessed by generating graphs from the extracted control flow and data flow semantics, and normalizing the resulting graphs before feeding them to a network based on TMP-GNN.

Eth2Vec [20] explores features of vulnerable Ethereum virtual machine bytecode using implicit knowledge through a neural network for natural language processing. In AMEVulDevector [16], GNNs were combined with expert patterns selected by the authors. Natural language models are known for their ability to extract features that mimic the semantics of the text [21–23]. This is why they have also been applied to static analysis. For instance, Peculiar [13] employs pre-trained BERT from Microsoft for detecting reentrancy vulnerabilities in smart contracts transformed into opcodes, achieving an F1 score of 0.92. Following the use of the BERT model in Peculiar, the use of models operating on words and sentences was expanded in GPTScan [14], which employs GPT-3.5 for detecting patterns in smart contracts.

SmartMixModel [24] combines the source code of smart contracts and their bytecode to detect 10 vulnerabilities in smart contracts, achieving an F1 score of 0.99. MANDO [11] proposes a heterogeneous graph representation for learning graph structures of contracts. They then train a graph attention network to facilitate the detection of vulnerabilities at the line level in the source code, a task not achieved by any of the previously mentioned models. Recent works such as SVScanner [25] also attempt to identify semantic relationships in code.

Insufficiently explored in the realm of vulnerability detection through ML is the identification of the specific lines of code where vulnerabilities have been introduced. Previously, this avenue of the study was undertaken by MANDO [11], but in 2023, another tool named G-Scan [26] emerged.

In summary, in recent years, the field of vulnerability detection in smart contracts has undergone rapid development, characterized by the emergence and refinement of various tools and methods based on static analysis, symbolic execution, fuzzing, and ML. Noteworthy innovations include the popularization of ML, DL, and neural networks for classification and vulnerability detection.

3 Problem Statement

The purpose of the study is to enhance the security of smart contracts and study the current state of ML methods that accomplish this. The object of the study is smart contracts used for financial applications.

From all vulnerability methods to research we decided to explore static analysis methods based on ML and DL. We chose static analysis because it allows us to evaluate the security of smart contracts at a scale with limited resources. We chose AI because in this context it is a developing branch of knowledge that would benefit the most from a potential contribution [7, 9]. We intend to cross-validate the resulting models by training them on their native datasets and testing them on a newer, bigger dataset. If the chosen method is hybrid, we will evaluate the parts concerning static analysis. To ensure that we evaluate the models with the most optimal parameters, we will also find out the optimal number of training epochs to prevent overfitting.

Few ML vulnerability detectors are available publicly, with all the necessary datasets and source code. Thus, we conducted an extensive review of the literature and online sources. Overall, 18 ML-based tools were selected. However, in attempting to access the source code for experiment replication, it was found that only 10 out of the 18 tools are open-source. Moreover, the source code for some projects initially claimed to be open by the authors, was closed sometime after the publication.

As a result of the selection process, the following tools with open-source code were identified: MANDO [11], Peculiar [13], AMEVulDetector [16], GNNSCVulDetector [17], GPSCVulDetector [15], GraphDeeSmartContract [19], ReChecker [27], SoliAudit [28], Eth2Vec [20], and an implementation similar to [29] that uses ResNet (which we will further refer to simply as "ResNet") [18].

Despite the availability of source code for 10 tools, some had to be excluded from this study. For instance, five models from the list were published on GitHub by the author of AMEVulDetector. It was decided to choose two of them that seemed the simplest to configure: AMEVulDetector and GNNSCVulDetector. Although they share the same dataset, they process input data in different formats, making the comparison of these tools interesting. Thus, out of the ten projects, seven remained.

From the seven remaining projects, two more were discarded during the examination of the necessary documentation. Unfortunately, Eth2Vec had to be excluded because the authors did not provide essential parameters for execution, such as the required library versions, dataset processing scripts, and dataset labels. Thus, the setup for execution was technically challenging. The dataset of MANDO also lacked certain documentation and as a result, the tool was discarded.

Therefore, only five out of the 18 ML-based tools selected could be used for this study. As for copyright, the ResNet-based implementation by M. Rossini and SoliAudit had an MIT license [18], while the other three repositories did not have any indication of the license. In order not to violate the copyright law, we reviewed the corresponding articles and ensured that it was the authors who published the code.

It is also important to find a balance between writing new source code for convenience and preserving the impartiality of the review. With this in mind, we set a logical limitation: other than writing wrappers or fixtures for the evaluation, we cannot change the source code of the studied models or fix possible issues. If broken, this rule will impact the evaluation results. We also intend not to engage in reverse engineering, if some aspects of data processing are not clear enough, because the results are bound to be ambiguous, and it might compromise the study's integrity.

4 Experiment Setup

To evaluate the existing methods we set up a cloud infrastructure with a focus on the replicability of our results. To facilitate the deployment process, we used Docker. For each tool in the list, Dockerfiles with the necessary package versions were made. We published the resulting Dockerfiles at [30] for free use to facilitate further research. The correct environment and packages were either taken from the documentation or empirically deduced from deprecation warnings. Although SoliAudit already had its own Dockerfile, we made our version to remove stale and unavailable external dependencies. Due to the limitation we set earlier we included only the part of SoliAudit that implemented static analysis, leaving out the part of the repository that implemented fuzzing.

Since at least two projects out of five required a GPU, the deployment, training, and testing were done using a cloud instance with 16 GB RAM, four CPUs, one GPU, and a 120 GB volume. In summary, these resources were just enough to store the models, datasets, temporary files, and Docker images and cache. For those who will follow in our footsteps, we recommend reserving at least 150 GB of storage. We summarized the

Table 1. Resources needed to train the selected models

Model	GPU required	Dataset size
AMEVulDetector	–	40,835
GNNSCVulDetector	–	40,835
Peculiar	+	47,500
ResNet	+	113,000
SoliAudit	–	21,044

memory requirements of all datasets and whether a graphics card is needed for model training (see Table 1).

As can be seen from the comparison, all datasets contain tens of thousands of samples. This number of samples is enough to train the model, but the quality can only be assessed through testing and regular use.

5 Method of Research

Another important matter is choosing the evaluation criteria. To ensure the uniformity of results, we studied what vulnerabilities can detected using each of the tools. Since Peculiar only focuses on the detection of reentrancy and the other selected models also support it, we decided to train and evaluate models based on their accuracy in detecting this particular vulnerability. We chose the F1 score as the accuracy metric, because it holistically represents the detection accuracy while taking into account false positive and false negative rates, and it appears to be the current standard for model evaluation. The F1 score is calculated as follows:

$$F1 = 2 \times \frac{P \times R}{P + R}, P = \frac{TP}{TP + FP}, R = \frac{TP}{TP + FN}, \quad (1)$$

where P is precision, R is recall, TP is true positives, FP is false positives, FN is false negatives. In other words, the F1 score is maximized when both precision and recall are high. It is particularly valuable for avoiding false positives and false negatives.

Finally, we needed to select a dataset to test the tools on. For this we chose the largest and newest available dataset, Slither Audited Smart Contracts (SASC) [31]. In total, it has 113,000 items from which we used over 105,000. It also includes both smart contract source code and bytecode for easy interoperability and use in different models. However, the data was labeled only by Slither, which can be a disadvantage, because there are no other instruments to correlate its results with. For example, the SmartBugs [32] dataset used nine tools to annotate smart contracts. For our purposes, we only needed to label smart contracts with reentrancy. This vulnerability corresponds to two Slither checks, reentrancy-eth, and reentrancy-no-eth, so we labeled documents with either of the checks preset as vulnerable.

Converting SASC to the SoliAudit format. SoliAudit repository [33] lacks documentation regarding the low-level dataset structure. However, its structure is known from the intermediate file called test-data.csv which is generated during training from two separate files containing opcodes and vulnerabilities. It contains the smart contract address, labels for 15 vulnerabilities or weaknesses (see example in Table 2), and the opcodes in text form.

Since SoliAudit requires contracts to be compiled to plaintext opcodes, we decided to recompile all smart contracts in SASC in the same way that SoliAudit does. One of the major challenges was that contracts in SASC have a wide range of Solidity versions, with each version requiring a different compiler. To resolve it we parse the Solidity version pragma and use py-solc-x [34] to dynamically install the necessary solc version. This covered most of the smart contracts in SASC, but an insignificant amount of contracts had to be discarded due to malformed version pragma. We also copied and adapted the

Table 2. Sample data from SoliAudit's test-data.csv

Address	Reentrancy	Opcodes
0 × 0000000000B3F879CB30 FE243B4DFEE438691C04	1	CALLDATASIZE CALLDATALOAD REVERT CALLVALUE REVERT ADD SUB ADD ...
0 × 0006157838D5A6B33AB 66588A6A693A57C869999	0	REVERT SHA3

preprocessing code to parse and filter the solc output according to the guidelines while decoupling it from the dependence on filesystem operations. We published this and the other scripts for dataset conversion at [31] for free use by the research community.

Converting SASC to the Peculiar format. The Peculiar dataset is divided into two parts: smart contract source code and vulnerability metadata. An example of data from the first part of the dataset (data.jsonl) is provided in Table 3.

Table 3. Sample data from Peculiar's data.jsonl

Address	idx	Contract
0 × 8D12A197CB00D4747A1 FE03395095CE2A5CC6819	0	pragma solidity ^0.4.9;\n\ncontract SafeMath {\n function safeMul(uint a, uint b) internal returns (uint) {\n uint c = a * b;\n assert(a = = 0 || c / a = = b);\n return c;\n}\n\n ...
0 × 06012C8CF97BEAD5DEA E237070F9587F8E7A266D	22	pragma solidity ^0.4.11;\n\n\n\n\n\n\n\ncontract KittyAuction is KittyBreeding {\n\n

A typical file from the second part of the dataset (train.txt, valid.txt, test.txt) contains lines consisting of the contract number (matching the "idx" field in the first file) and an indicator of whether it is vulnerable to reentrancy or not. This allows the authors to flexibly specify which contract indices to use for training, validation, and testing. To convert SASC to the format accepted by Peculiar, it is necessary to enumerate the contracts and record their source code, names, and indices into a new directory named data.jsonl. The vulnerability labels regarding reentrancy should be recorded in test.txt.

As a result, a new data.jsonl file with a size of 2.9 GB and test.txt with a size of 820 KB were obtained. The next step is to write a new configuration file for Peculiar with a new file_section, which should look as follows:

```
[file_section]
train_data_file=dataset/train.txt
output_dir=saved_models
eval_data_file=dataset/valid.txt
test_data_file=resnet_dataset_for_peculiar/test_resnet.txt
```

Peculiar will automatically infer that data.jsonl for test_resnet.txt should be searched in the resnet_dataset_for_peculiar directory.

Converting SASC to the GNNSCVulDetector format. Upon a more detailed examination of GNNSCVulDetector, we discovered that the source code is transformed into CFG by operating on the text of smart contracts without compilation or any other dependency on solc. Smart contract source code is divided into chunks based on functions. AutoExtractGraph.py [35] then selects strings containing ".call.data" calls to parse the call arguments, and forms a CFG based on this information. An advantage of this approach lies in the marginal reduction of computational costs. However, should the Solidity syntax change, these data preprocessing scripts will break and require updating. We did not account for this disadvantage due to the limitations of this study. We developed the script for converting SASC to the required format based on AutoExtractGraph.py and a couple of other dependencies. Out of 105,608 contracts in our local copy of SASC, 105,271 were successfully converted.

Converting SASC to the AMEVulDetector format. AMEVulDetector utilizes a dataset similar to GNNSCVulDetector, but the authors also employ a specific preprocessing technique called "expert pattern embedding." We identified the parts of the code that extracted expert patterns and used them in our script for data conversion. However, we soon ran into an unexpected roadblock. Although initially, due to the presence of train.json and valid.json, it seemed that the dataset format accepted by AMEVulDetector is nearly identical to the previous tool, after a more detailed examination we discovered that the code ignores the JSON files and reads data from other files, reentrancy_final_train.txt and reentrancy_final_valid.txt. A sample of reentrancy_final_train.txt can be found in Table 4.

Table 4. Sample data from AMEVulDetector's reentrancy_final_train.txt

Row	Unlabeled data
0	0.306811 -0.215143 -0.109194 -0.379985 -0.290496… -0.008959 -0.210275
1	0.306811 -0.215143 -0.109194 0.299823 -0.379985… -0.008959 -0.210275

Unfortunately, the authors did not provide any scripts, instructions, or hints on how smart contracts were preprocessed. We supposed that the lines above were compressed using the pattern feature extraction script; however, we could not sufficient evidence for this hypothesis. Neither the commit history in the AMEVulDetector GitHub repository, the Dataset_Preprocessing repository by the authors, nor the scientific paper where this model was presented, explains this final step in preprocessing. We also reached out to the project author to inquire about this but received no response. Therefore, while it is possible to speculate on how the data were transformed, the original data is unavailable, and converting SASC or any other dataset into the AMEVulDetector format does not appear possible.

6 Research Results

First, we will share the results regarding the optimal epoch number. Since the model authors have already tuned the hyperparameters, the only parameter of interest would be the number of epochs. The default number of epochs for the presented models is as follows (Table 5). Next, we will show the evaluation results for the selected models. We do not provide data for the ResNet model, because it is not logical as per the problem statement to evaluate it on its dataset.

Table 5. Training epochs by project

Model name	Default number of epochs	Optimal number of epochs
AMEVulDetector	100	50
GNNSCVulDetector	50	30
Peculiar	2	2
SoliAudit	300	—*

* F1 score data by epoch for SoliAudit is unavailable because it throws an exception every time the scores are evaluated.

Peculiar trains the model for only two epochs. During this process, the model performs self-validation every 10% of the epoch, resulting in 20 points on the corresponding graph. From Fig. 1, it can be inferred that this model is prone to overfitting, as the graph twice reaches points where the F1 score equals 1.00.

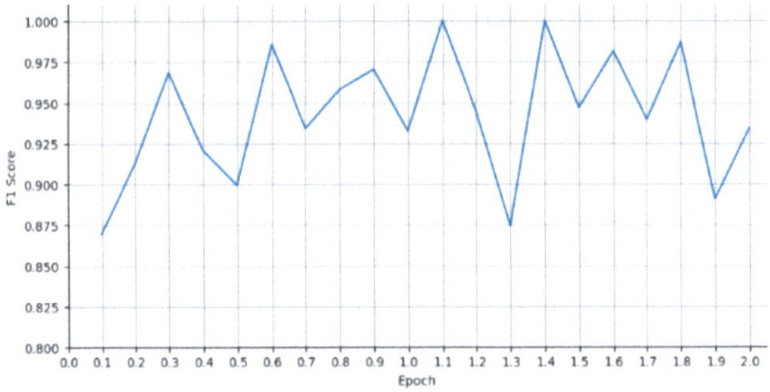

Fig. 1. F1 score of Peculiar by epoch

From Fig. 2a, it can be observed that 50 epochs are sufficient for training GNNSCVulDetector. However, for the AMEVulDetector data presented in Fig. 2b, after 50 epochs F1score reaches a plateau, so training for 50 epochs instead of 100 is more than enough.

Let us proceed to the scores achieved during cross-validation.

SoliAudit. Using the script developed in Sect. 4 we managed to convert 81,076 out of 105,608 smart contracts from SASC into the necessary format. The achieved F1 score

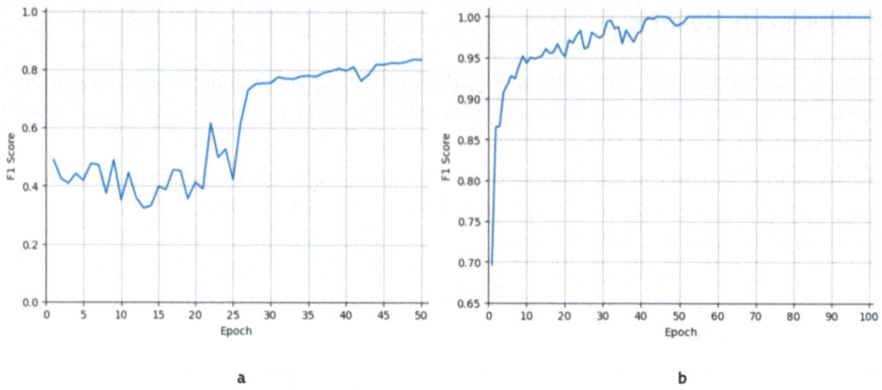

Fig. 2. F1 score by epoch: (a) GNNSCVulDetector and (b) AMEVulDetector

was significantly lower than on the native dataset, measuring 0.06. This might suggest that SoliAudit is not adapted to newer versions of Solidity, as most contracts in the original dataset use Solidity ver. 0.5.0, whereas in the SASC dataset, smart contracts have versions ranging from 0.4.x to 0.8.x.

Peculiar. Peculiar had an F1 score of 1.00 on its native test dataset. We managed to convert the whole SASC dataset to the required format. When testing the pre-trained model on SASC we achieved the F1 score of 0.23. This can also be explained by the newer codebase compared to the SmartBugs dataset and overfitting.

GNNSCVulDetector. We managed to transform SASC into the necessary format without evident errors. However, when we attempted to validate the pre-trained model using the converted data, it turned out that most rows in the testing dataset could not be accepted due to the presence of rows with empty CFG edges. Analysis of the generated dataset revealed that the conversion script generated graph edges only for 2,709 contracts out of 105,271, as the ".call.value" string was found only in them. Meanwhile, there were 30,184 contracts with reentrancy in the dataset. It appears that the conversion script fell victim to an update in Solidity syntax allowing it to write ".call.value(...)" as ".call{ value:...}." The F1 score on the successfully converted 2,709 smart contracts was 0.17.

AMEVulDetector. Due to an unknown step in dataset preprocessing described in more detail in Sect. 4, it was not possible to test AMEVulDetector (see Table 6). From the results, it is evident that the tested models did not display the same performance on both datasets. This may be attributed to the fact that SASC includes smart contracts with newer Solidity versions than in the original datasets. Solidity is still developing as a language, and features learned from an older codebase were not as useful as expected in that context. This point is supported by the high false negative and low false positive rates that we achieved when evaluating the models.

These findings reveal significant strides in using ML and DL for vulnerability detection. This indicates the necessity for continuous adaptation of these methods. The ever-changing nature of Solidity has proved to pose a challenge for vulnerability detection.

Table 6. F1 score on the native dataset and the testing dataset by model

Model name	F1 score on the native dataset	F1 score on SASC
AMEVulDetector	0.89	—*
GNNSCVulDetector	0.88	0,17
Peculiar	1,00	0,23
SoliAudit	0,80	0,06

* AMEVulDetector could not be evaluated due to the lack of the documentation of a step in dataset preprocessing.

7 Conclusions and Future Work

With the increasing complexity and adoption of smart contracts, addressing security challenges is paramount. In this paper, we have explored the proactive detection of vulnerabilities in decentralized finance smart contracts through ML and DL approaches. Our work was centered on the current limitations of such methods in accurately identifying vulnerabilities, particularly given the evolving nature of smart contract development.

When evaluating the default number of epochs in model tuning we discovered that training AMEVulDetector for 50 epochs instead of 100 will not impact its accuracy. We also discovered that Peculiar may be prone to overfitting because F1 reaches 1.00 twice on its native dataset. We also found out that F1 scores of GNNSCVulDetector, SoliAudit, and Peculiar plunge when they are tested on SASC. This could be explained by the fact that SASC consists of smart contracts with newer Solidity versions.

Future work in this field should prioritize the development of more robust and consumer-ready ML models that can keep pace with the rapidly evolving landscape of smart contract development. This includes researching quality-of-life features present in the classic static analyzers and implementing them in ML-based detectors.

References

1. Jain, S., et al.: CodeBlockS: development of collaborative knowledge sharing application with blockchain smart contract. Int. J. Inf. Eng. Electron. Bus. **15**(1), 1–19 (2023). https://doi.org/10.5815/ijieeb.2023.01.01
2. Sen, S., Mukhopadhyay, S., Karforma, S.: A blockchain based framework for property registration system in e-governance. Int. J. Inf. Eng. Electron. Bus. **13**(4), 30–46 (2021). https://doi.org/10.5815/ijieeb.2021.04.03
3. Hapiffah, S., Sinaga, A.: Analysis of blockchain technology recommendations to be applied to medical record data storage applications in Indonesia. Int. J. Inf. Eng. Electron. Bus. **12**(6), 13–27 (2020). https://doi.org/10.5815/ijieeb.2020.06.02
4. Barboni, M., Morichetta, A., Polini, A.: Smart contract testing: challenges and opportunities. In: 5th International Workshop on Emerging Trends in Software Engineering for Blockchain, pp. 21–24 (2022). https://doi.org/10.1145/3528226.3528370
5. Chainalysis Team: 2022 Biggest year ever for crypto hacking with $3.8 billion stolen, primarily from DeFi protocols and by North Korea-linked attackers. https://www.chainalysis.com/blog/2022-biggest-year-ever-for-crypto-hacking/. Accessed 17 Dec 2023

6. Napoletano, E., Adams, M.: What is DeFi? Understanding decentralized finance. https://www.forbes.com/advisor/investing/cryptocurrency/defi-decentralized-finance/. Accessed 17 Dec 2023
7. Piantadosi, V., et al.: Detecting functional and security-related issues in smart contracts: a systematic literature review. Softw. Pract. Exp. **53**(2), 465–495 (2022). https://doi.org/10.1002/spe.3156
8. Iqbal, A., Aftab, S.: A classification framework for software defect prediction using multi-filter feature selection technique and MLP. Int. J. Mod. Educ. Comput. Sci. **12**(1), 18–25 (2020). https://doi.org/10.5815/ijmecs.2020.01.03
9. Jiang, F., et al.: Enhancing smart-contract security through machine learning: a survey of approaches and techniques. Electron. **12**(9), 2046 (2023). https://doi.org/10.3390/electronics12092046
10. Hua, Z.: Software defect detection-oriented static analysis techniques. Int. J. Wirel. Microw. Technol. **2**(6), 17–22 (2012). https://doi.org/10.5815/ijwmt.2012.06.03
11. Nguyen, H.H., et al.: MANDO: multi-level heterogeneous graph embeddings for fine-grained detection of smart contract vulnerabilities (version 2). arXiv (2022). https://doi.org/10.48550/arxiv.2208.13252
12. Feist, J., Grieco, G., Groce, A.: Slither: a static analysis framework for smart contracts. In: IEEE/ACM 2nd International Workshop on Emerging Trends in Software Engineering for Blockchain, pp. 8–15 (2019). https://doi.org/10.1109/wetseb.2019.00008
13. Wu, H., et al.: Peculiar: smart contract vulnerability detection based on crucial data flow graph and pre-training techniques. In: IEEE 32nd International Symposium on Software Reliability Engineering (2021). https://doi.org/10.1109/issre52982.2021.00047
14. Sun, Y., et al.: When GPT meets program analysis: Towards intelligent detection of smart contract logic vulnerabilities in GPTScan (version 1). arXiv (2023). https://doi.org/10.48550/arxiv.2308.03314
15. Liu, Z., et al.: Combining graph neural networks with expert knowledge for smart contract vulnerability detection. IEEE Trans. Knowl. Data Eng. 1 (2021). https://doi.org/10.1109/tkde.2021.3095196
16. Liu, Z., et al.: Smart contract vulnerability detection: from pure neural network to interpretable graph feature and expert pattern fusion (version 1). arXiv (2021). https://doi.org/10.48550/arxiv.2106.09282
17. Zhuang, Y., et al.: Smart contract vulnerability detection using graph neural network. In: 29th International Joint Conference on Artificial Intelligence (2020). https://doi.org/10.24963/ijcai.2020/454
18. Rossini, M.: [BlockSys 2022] Exploring deep learning techniques for Ethereum smart contract vulnerability detection. https://github.com/mwritescode/smart-contracts-vulnerabilities. Accessed 17 Dec 2023
19. Gao, Z., et al.: SmartEmbed: a tool for clone and bug detection in smart contracts through structural code embedding. In: IEEE International Conference on Software Maintenance and Evolution (2019). https://doi.org/10.1109/icsme.2019.00067
20. Ashizawa, N., et al.: Eth2Vec: learning contract-wide code representations for vulnerability detection on Ethereum smart contracts (version 2). arXiv (2021). https://doi.org/10.48550/arxiv.2101.02377
21. Iosifova, O., et al.: Techniques comparison for natural language processing. In: 2nd International Workshop on Modern Machine Learning Technologies and Data Science, vol. **2631**(I), pp. 57–67 (2020)
22. Iosifov, I., Iosifova, O., Sokolov, V.: Sentence segmentation from unformatted text using language modeling and sequence labeling approaches. In: IEEE International Conference on Problems of Infocommunications, Science and Technology, pp. 335–337 (2020). https://doi.org/10.1109/picst51311.2020.9468084

23. Iosifov, I., et al.: Natural language technology to ensure the safety of speech information. In: Workshop on Cybersecurity Providing in Information and Telecommunication Systems, vol. 3187(1), pp. 216–226 (2022)
24. Shakya, S., et al.: SmartMixModel: machine learning-based vulnerability detection of solidity smart contracts. In: IEEE International Conference on Blockchain, pp. 37–44 (2022). https://doi.org/10.1109/blockchain55522.2022.00016
25. Zhang, H., et al.: SVScanner: detecting smart contract vulnerabilities via deep semantic extraction. J. Inf. Secur. Appl. **75**, 103484 (2023). https://doi.org/10.1016/j.jisa.2023.103484
26. Sendner, C., et al.: G-Scan: Graph neural networks for line-level vulnerability identification in smart contracts (version 1). arXiv (2023). https://doi.org/10.48550/arxiv.2307.08549
27. Qian, P., et al.: Towards automated reentrancy detection for smart contracts based on sequential models. IEEE Access **8**, 19685–19695 (2020). https://doi.org/10.1109/access.2020.2969429
28. Liao, J.-W., et al.: SoliAudit: smart contract vulnerability assessment based on machine learning and fuzz testing. In: 6th International Conference on Internet of Things: Systems, Management and Security, pp. 458–465 (2019). https://doi.org/10.1109/iotsms48152.2019.8939256
29. Hwang, S.-J., et al.: CodeNet: code-targeted convolutional neural network architecture for smart contract vulnerability detection. IEEE Access **10**, 32595–32607 (2022). https://doi.org/10.1109/access.2022.3162065
30. Harpsiford, M.: Smart contract model setup. https://github.com/harpsiford/smart-contract-model-setup. Accessed 17 Dec 2023
31. Rossini, M.: Dataset card for slither audited smart contracts. https://huggingface.co/datasets/mwritescode/slither-audited-smart-contracts. Accessed 17 Dec 2023
32. Durieux, T., et al.: Empirical review of automated analysis tools on 47,587 Ethereum smart contracts. In: ACM/IEEE 42nd International Conference on Software Engineering, pp. 530–541 (2020). https://doi.org/10.1145/3377811.3380364
33. Liao, J.: SoliAudit. https://github.com/jianwei76/SoliAudit. Accessed 17 Dec 2023
34. Antazoey: Python wrapper and version management tool for the solc Solidity compiler. https://github.com/ApeWorX/py-solc-x. Accessed 17 Dec 2023
35. Messi-Q: GNNSCVulDetector. https://github.com/Messi-Q/GNNSCVulDetector/tree/master/tools/reentrancy. Accessed 17 Dec 2023

Research on the Effect of Perceived Controlled by the Algorithm on the Anti-Algorithmic Behavior of Platform Workers Under the Algorithmic Management

Lu Chen[1] and Yi Zhang[2(✉)]

[1] School of Management Engineering, Yunnan Water Resources and Hydropower Vocational College, Kunming, China
369953024@qq.com
[2] Business School, Sichuan University, Chengdu, China
657518287@qq.com

Abstract. The anti-algorithmic behavior of platform workers is affected by perceived controlled by the algorithm (after PCA), meanwhile, PCA can be influenced by platform workers' psychological factors. However, there is a gap in how PCA effect on the behavior of platform workers. This research selected Take-out Delivery Rider as the subject, and conducted by questionnaires and constructed structural equation model, then empirically tested that there is an impact mechanism of PCA on anti-algorithmic behavior of platform workers. The results of this research showed that: the anti-algorithmic behavior can be positively influenced by PCA; The sense of relief plays as a mediating role between PCA and anti-algorithmic behavior, conversely, psychological empowerment (PE) is ineffective in mediating the two factors. Based on the findings of empirical tests in this research, the results contributed to the extending of research in PCA and anti-algorithmic behavior in some extent, enriched the research literature in algorithmic management field, and provided a reference for online employment platforms in preventing anti-algorithmic behavior of platform workers.

Keywords: Online Labour Platform · anti-algorithmic behavior · Algorithmic management · controlled by the algorithm · sense of relief

1 Introduction

With the widespread application of the new generation information and communication technology in the workplace, gig economy, in the form of digital technology, On-demand service and Crowding Sourcing, has achieved deep integration [1], thus spawned the Online Labour Platform (after OLP) which takes algorithm as the underlying logic and effectively links supply and demand sides in the labour market [2]. (Algorithmic management) In contrast to the labour practices under the traditional bureaucratic system, in the digital labour of Online Labour Platform (OLP), "Algorithms" are pushed from

Z. Hu et al. (Eds.): ICCSEEA 2024, LNDECT 242, pp. 66–74, 2025.
https://doi.org/10.1007/978-3-031-84228-3_6

"behind the scenes" to "the stage", become the core indicator for scholars in inspecting the changes of "technological logic" and "social power relations" in the digital era [3]. Under the influence of platform capitalism and digital Taylorism [4], the management tools of OLP for platform workers have gradually shifted from traditional "personnel management and supervision" to "algorithmic management" based on big data and intelligent technology.

Algorithmic management can be defined as the process that OLP can automatically execute management functions in a highly automated and data-driven way [5, 6]. In practice, under the impact of logic of algorithms, OLP placed platform workers into "emotional exploitation", "controlled by the algorithm" and "digital panorama prisons" indirectly [4, 7]. With big data and algorithmic decision making, OLP achieved the scientific management which can manage the labour process in standardized, process-oriented and quantitative way, thus obtaining the maximum economic benefits from the digital labour process of the platform workers [8, 9]. Some scholars argues that platform workers have gradually explored the most suitable resistance strategy for themselves in the long-term digital labour process, rather than only passively accept algorithmic regulation in work, they usually adopt various anti-algorithmic behaviors to defy the platform algorithm in order to demonstrate their "subjectivity" [7, 10].

Anti-algorithmic behaviors refers to the kind of behaviors that platform workers actively search and decode the operation mechanism of algorithms in order to get rid of the negative effects of algorithm decisions on themselves, so as to benefit for increasing their own interests [7, 9, 10]. Externally, the anti-algorithmic behavior is the joint result of the algorithmic working mechanism of the platform and platform workers, but its essence is the subjective judgment of whether it is fair between PCA and labour income perceived by platform workers in the labour process [7]. In addition, the perception of platform workers in terms of controlled by the algorithm can also significantly influence the occurrence of anti-algorithmic behaviors, but there are few studies referred to the empirically research in this field. In view of the above, in this research, Take-out Delivery Riders were selected as the research subject, explored the impact mechanism of PCA on anti-algorithmic behavior of platform workers through constructing structural equation model. The results of this research can help to provide theoretical guidance for online labour platform (OLP) in terms of preventing anti-algorithmic behaviors.

2 Theoretical Background and Hypotheses

Controlled by the algorithm is a new management practice with high control and high performance requirements [10, 11], and the emotional experience generated by platform workers on this management practice can be named as perceived controlled by the algorithm (PCA) [11]. According to the Self-Determination Theory (SDT), when the controlled by the algorithm is at a high level, it is more likely to stimulate the self-determination motivation of platform workers [12, 13]. First of all, algorithms have created a virtual work scene for platform workers, which broke the physical boundaries of the traditional workplace, and gave platform workers more self-determination rights in terms of labour time, labour process, and labour methods. Secondly, some studies have proved that since the economic benefits of online platforms origin from the "revenue sharing" with platform workers, the decision making of algorithms oriented by efficiency will require platform workers to complete more labour orders in a fixed time and invest more emotional labour in the work [14]. Consequently, in this condition, the job satisfaction of platform workers will decrease inevitably [1, 12]. Finally, although OLP hope to improve the labour process and optimize the experience of platform workers through algorithms, due to the existence of Algorithmic Black-box in the decision making of algorithms, the platform workers are more likely to experience the sense of injustice compared with the decision making of human (algorithms vs. humans), especially when the labour order dispatch or route planning created by the OLP is unreasonable. As a result, under the combined effect of "controlled by the algorithm", "emotional exploitation" and "Algorithmic Black-box", platform workers tend to take various behaviors to resist the algorithms in the labour process so as to demonstrate their subjectivity. Therefore, the hypothesis 1 was proposed:

Hypothesis 1: Perceived controlled by the algorithm (PCA) can positively influence the anti-algorithmic behaviors.

In essence, the sense of relief is a cognitive experience with emotional valence [15], which usually occurs after an individual perceives that he or she has successfully avoided certain negative consequences. The OLP has realized the panoramic supervision of platform workers in the labour process by algorithms, and gives strict punishment to those platform workers who do not meet the expectation of algorithms [7]. However, under operation mechanism of algorithms, it is seriously imbalanced between the labour income of platform workers and the punishment from OLP [7]. According to the resource conservation theory, when platform workers realize that the decision making of platform algorithms is likely to bring them negative consequences, they will be motivated to engage in various behaviors to obtain psychological resources including positive emotional valence [12, 14]. For example, some platform workers will intentionally sift labour orders before accepting orders, and request the positive feedback from customers after completing orders. The reason of these behaviors is that the resistance of the operating mechanism of algorithms can help platform workers to reduce or mitigate the negative impact from the decision making of algorithms on them, thus help platform workers to develop the sense of relief. Therefore, the hypothesis 2 was made:

Hypothesis 2: The sense of relief plays a mediating role between the PCA and anti-algorithmic behaviors of platform workers.

Psychological empowerment is the perception degree that an individual feels that he or she is capable of accomplishing a meaningful and influential task [16]. In practice, through building the strict and comprehensive supervision and punishment mechanisms, algorithms constrain the behaviors of platform workers and request them to meet the expectations of the platform [7, 10]. However, on one hand, the operation mechanism of algorithms usually has significant characteristics of machine learning, when the anti-algorithmic behavior occurs, it is hard for algorithms to timely and effectively identify and disclose such behavior, which reduces the opportunity cost of the anti-algorithm behaviors of platform workers to a certain extent. On the other hand, the digital labour of the OLP also has the obvious feature of "piece rate", which makes platform workers believe that they can easily adjust their working process and working method to complete a certain task, which contributes to increase their self-efficacy. To conclude, the anti-algorithm behavior not only improves the self-efficacy of platform workers, but also improves their perception of the autonomy degree in the work. Consequently, the hypothesis 3 was proposed:

Hypothesis 3: The psychological empowerment plays a mediating role between the PCA and anti-algorithmic behaviors of platform workers.

According to the discussion above, the research model of this study is shown in Fig. 1:

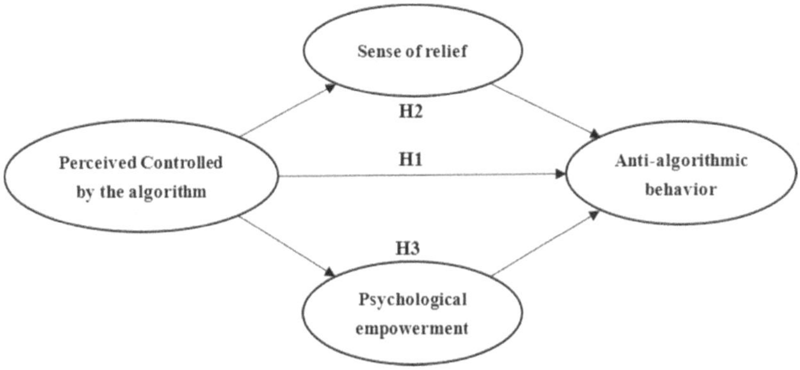

Fig. 1. Research Model

3 Empirical Study

3.1 Questionnaire Design and Data Collection

In order to test the validity of the research model (see Fig. 1), the questionnaire of this research can be divided into basic information and latent variable testing, in which the latent variables included four variables: perceived controlled by the algorithm (PCA), psychological empowerment (PE), sense of relief (SR) and anti-algorithmic behavior (AAB). In terms of the research design of these four variables, PCA was mainly based on the previous studies of Wei et al. (2024) [12] and Pei et al. (2022) [17], while PE and SR were based on the research from Spreitzer (1995) [16] and Wei et al. (2024) [12] respectively, and made appropriate adjustments in this study, and as to the research design of AAB, which was mainly cited from Liu and Zhang (2023) [7], Wei et al. (2024) [12] and Pei et al. (2022) [17]. In addition, each test item in this research was measured in 7-point Likert scale.

In terms of the selection of the research subject, Take-out Delivery Riders were selected as the representative of platform workers, which is because they are very typical platform workers and they are subject to a much higher degree of controlled by the algorithm in the work process than other occupations represented by content creation (such as online writers, etc.) [7]. In the questionnaire distribution, a total of 350 questionnaires were distributed, from which 295 valid questionnaires were successfully retrieved with a validity coefficient of 84.29%; Among the Take-out Delivery Riders participated in this questionnaire research, there were 184 male (62.37%) and 111 female(37.63%), and 84.32% of them have been working in this job for more than 3 months, and all of them had a high school education or above.

3.2 Data Analysis and Hypothesis Testing

Firstly, when testing the reliability of a scale, it is necessary to test the reliability of Latent Variables as well as the reliability of the testing item. In this research, the minimum cross-loading values of the four latent variables, which include PCA, PE, SR and AAB, were 0.833, 0.865, 0.884, and 0.926 respectively, which are all above the threshold of 0.5, thus it is preliminarily proved that the reliability of the testing items withstood the test. However, in the application process of the practice, considering the results of Cronbach's α are easily affected by the number of observing items and the size of subjects, it is necessary to conduct further test of composite reliability and average variance extracted (AVE), so as to to ensure that the reliability and validity of the scale can pass the test. In Table 1, the values of Cronbach's α, CR, and AVE are listed.

Table 1. The testing results of reliability, Validity and square roots of AVE.

Item	CR	Cronbach's α	AVE	PE	PCA	SR	AAB
PE	0.923	0.876	0.801	**0.895**			
PCA	0.884	0.804	0.718	0.110	**0.847**		
SR	0.761	0.896	0.761	0.395	0.177	**0.872**	
AAB	0.960	0.937	0.889	0.241	0.329	0.431	**0.943**

From the Table 1, it is obvious that the α coefficients of all latent variables are in the range of 0.761–0.960, all of which are larger than the critical value of 0.7, indicating that the α coefficients passed the test; meanwhile, the CR values of all latent variables are greater than 0.8, which means that the observation items of each latent variable are highly correlated and can be accurately mapped to the same testing value; finally, AVE is a statistic used to test the degree of internal consistency in the structural model, as the Table 1 showed, the lowest AVE value is 0.718, which is greater than the empirical value 0.5. Therefore, there is sufficient evidence that the reliability of the scale passed the test.

In addition, from the value of square roots of AVE, the smallest square root is 0.847, which is much higher than the largest correlation coefficient (0.431). Consequently, it is proved that the validity of the questionnaire passes the test. To summarize, the scale possesses a good level of reliability and validity.

Smart PLS 3.0 was used to test the structural models in this research, and the reasons of that can be demonstrated as follows: (1) Smart PLS does not strictly require the normality hypothesis of data; (2) Smart PLS does not require the observation sample to strictly fulfill the large-sample condition; (3) Smart PLS has combined the validated factor analysis and linear regression analysis, which allows the simultaneous testing of measurement models and structural models. The testing results can be noticed in Table 2 and Fig. 2.

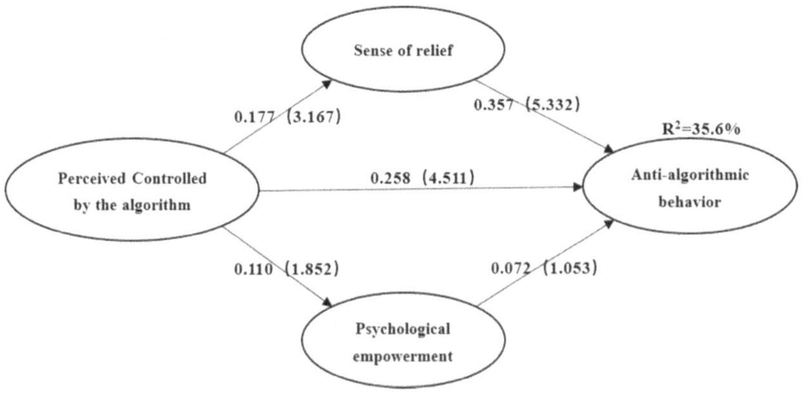

Fig. 2. Results of the Structural Model

Note: The parameter before the brackets represents the value of the path coefficient β value; The parameter in parentheses represents t-values.

Table 2. Bootstrap Analysis of the Mediation Effect Test

Route	Initial Sample	Sample Mean	2.5%	97.5%
PCA → PE → AAB	0.008	0.009	–0.007	0.037
PCA → SR → AAB	0.063	0.067	0.024	0.117

Table 3. Research outcome

Route	Hypothesis	The path coefficients /Interval	t-value	Results
PCA → PE	-	0.110	0.185	-
PCA → SR	-	0.177	3.167	-
SR → AAB	-	0.357	5.332	-
PCA → AAB	H1	0.258	0.451	Supported
PE → AAB	-	0.072	1.053	-
PCA → SR → AAB	H2	[0.024,0.117]	Significant	Supported
PCA → PE → AAB	H3	[–0.007,0.037]	Insignificant	Unsupported

It can be noticed that the explanatory power of each latent variable on the anti-algorithmic behavior reached 35.6% combined with Fig. 2 and Table 2. Among the data showed, the perceived controlled by the algorithm (PCA) ($\beta = 0.258$, $t = 4.511$)have a significant positive effect on the anti-algorithmic behavior (AAB), thus there is evidence to support the Hypothesis 1 (see Fig. 2).

From the Bootstrap test results of combined path coefficients and mediation effects (Table 2), it is showed that the perceived controlled by the algorithm (PCA) ($\beta = 0.177$, $t = 3.167$) can influence the sense of relief (SR), which has the impact on the anti-algorithmic behavior (AAB) ($\beta = 0.357$, $t = 5.332$) indirectly. Based on the confidence interval of the mediation effect test, the interval [0.024,0.117] is positive and does not contain 0, indicating that the sense of relief (SR) plays a positive mediating role between PCA and AAB. Thus, it is sufficiently supported the Hypothesis 2.

Conversely, the value of path coefficients from PCA ($\beta = 0.110$, $t < 1.96$) to SR, as well as between SR($\beta = 0.072$, $t < 1.96$) and AAB, showed that the path coefficients are not significant. Meanwhile, the confidence interval [–0.007,0.037] for the path PCA → PE → AAB contains 0, thus there is not enough evidence to support Hypothesis 3. In summary, among the hypotheses proposed in this research, Hypothesis 1 and Hypothesis 2 were confirmed, while Hypothesis 3 cannot be proved (see Table 3).

4 Conclusion

In this study, the research subject was focus on the Take-out Delivery Riders, and the influence mechanism of perceived controlled by the algorithm (PCA) on the anti-algorithmic behavior (AAB), finally obtained following conclusions:

First of all, the PCA has positive influence on the occurrence of anti-algorithmic behavior of platform workers. The results of data analysis showed that the path coefficient and t-value between PCA and AAB are at the significance level ($\beta = 0.177$, $t = 3.167$), which indicated that when the level of controlled by the algorithm perceived by the platform workers is at a high level, it will stimulate platform workers to have stronger motivation to resist. In this case, the decrease of job satisfaction induced by emotional labour and the Algorithmic Black-box will be the most direct factors that stimulate platform workers to adopt anti-algorithmic behaviors. Therefore, for Online Labour Platform (OLP), first, it is necessary to abandon the evaluation concept of "strongest algorithm", and consider adopting the "moderation algorithms" concept through setting reasonably threshold value to enhance the autonomy and work flexibility of platform workers in the digital labour process; Secondly, technological innovation can be helpful in actively improving and reducing the negative impact of the Algorithmic Black-box, and improve the scientificity and validity of the results of algorithmic decision making, thus prevent platform workers from over-decoding the operating mechanism of algorithms.

Secondly, the sense of relief (SR) plays a significant mediating role between the PCA and AAB. The data analysis results demonstrated that the influence of PCA on the AAB remains significant when added the SR into the structural model, which indicated that the SR ($\beta = 0.177$, $t = 3.167$; $\beta = 0.357$, $t = 5.332$) partially mediating in the structural model. Thus, as for the Online Labour Platform (OLP), on the one hand, the penalty rules of OLP should be reasonably formulated to ensure a dynamic balance between the labour benefits of platform workers and the penalties from the OLP. For example, the feedback and complaint mechanism from the Take-out Delivery Riders can be explored and formulated, and it is helpful to establish an "evaluation blacklist" system, which contributed to the reasonable evaluation of the labour performance from platform workers and prevented the excessive psychological losses to platform workers due to the dehumanized decisions making of algorithms. On the other hand, it is necessary that OLP strengthen the technical supervision of algorithms to prevent the negative effects of the algorithm abuse on platform workers.

Finally, psychological empowerment (PE) is ineffective in mediating the PCA and the AAB. The results of data analysis identified that the PCA failed to influence the AAB through PE when PE was added into the structural model ($\beta = 0.110, t < 1.96; \beta = 0.072, t < 1.96$). The reason for the result might be that the psychological empowerment (PE) itself possesses the value attribute of self-actualization, and the direct reason for platform workers to participate in digital labour is the work autonomy claimed by the OLP, which makes platform workers believe that they can fully control the entire digital labour process. When the OLP impose severe punishments on platform workers, it is more likely to stimulate the phenomenon of "platform disintermediation", which prompted platform workers to quit the existing OLP [18]. Consequently, in terms of the platform workers with a high degree of PE, seeking a new work platform is the better choice compared with accepting punishment from the platform passively. In other words, simply improving the punishment intensity of the OLP not only cannot effectively deter the occurrence of the AAB, but also decrease the loyalty of platform workers, which is consistent with the theoretical research conclusion of Liu and Zhang(2023) [7].

Acknowledgment. Many thanks to the anonymous reviewers for the modification suggestions, the authors have made corresponding modifications, and who are responsible for this paper.

References

1. Liu, P., Xiong, Z.Y., Zhang, Y., et al.: Research on gig economy based on digital platforms: literature review from multidisciplinary and multi-perspectives. West Forum **33**(1), 59–75 (2023). (In Chinese)
2. Kuhn, K.M., Maleki, A.: Micro-entrepreneurs, dependent contractors, and Instaserfs: understanding online labor platform workforces. Acad. Manag. Perspect. **31**(3), 183–200 (2017)
3. Sun, P.: Digital labor within the logic of algorithms: a study of food delivery workers in platform economy. Thinking **45**(6), 50–57 (2019). (In Chinese)
4. Wang, W.: The labor process of digital capitalism and emotional exploitation. Economist **2**, 15–22 (2021). (In Chinese)
5. Basukie, J., Wang, Y., Li, S.: Big Data governance and algorithmic management in sharing economy platforms: a case of ridesharing in emerging markets. Technol. Forecast. Soc. Change (2020). https://doi.org/10.1016/j.techfore.2020.120310
6. Griesbach, K., Reich, A., Elliott-Negri, L., et al.: Algorithmic control in platform food delivery work. Sociol. Res. Dyn. World **5**, 1–15 (2019)
7. Liu, P., Zhang, Y.: Exploring the herd effect during the formation of anti-algorithmic behavior of platform workers: evidence from evolutionary game models and numerical simulation experiments. J. Central Univ. Finan. Econ. **12**, 118–128 (2023). (In Chinese)

8. Schildt, H.: Big data and organizational design the brave new world of algorithmic management and computer augmented transparency. Innovation **19**(1), 23–30 (2017)
9. Kellogg, K.C., Valentine, M.A., Christin, A.: Algorithms at work: the new contested terrain of control. Acad. Manag. Ann. **14**(1), 366–410 (2020)
10. Zhang, Y.Y.: A study of algorithmic control and data masking: labor management studies in media space. Media Observ. **1**, 28–35 (2022). (In Chinese)
11. Waldkirch, M., Bucher, E., Schou, P.K., et al.: Controlled by the algorithm, coached by the crowd-how HRM activities take shape on digital work platforms in the gig economy. Int. J. Hum. Resour. Manag. **32**(12), 2643–2682 (2021)
12. Wei, X., Dong, Y.T., Cao, T., et al.: The bright side of employee resistance to algorithmic control for their work outcomes: a conservation of resources perspective. J. Indust. Eng. / Eng. Manag. **38**(1), 34–45 (2024). (In Chinese)
13. Ryan, R.M., Deci, E.L.: Self-determination theory and the facilitation of intrinsic motivation, social development, and well-being. Am. Psychol. **55**(1), 68–78 (2000)
14. Liu, P., Zhang, Y., Xiong, Z., et al.: Judging the emotional states of customer service staff in the workplace: a multimodal dataset analysis. Front. Psychol. **13**, 1001885 (2022)
15. Carver, C.S.: Threat sensitivity, incentive sensitivity, and the experience of relief. J. Pers. Soc. Psychol. **77**(1), 125–138 (2009)
16. Spreitzer, G.M.: Psychological empowerment in the workplace: dimensions, measurement, and validation. Acad. Manag. J. **38**(5), 1442–1465 (1995)
17. Pei, J.L., Liu, S.S., Cui, X., et al.: Dose algorithmic control motivates gig workers to offer the proactive services?——based on the perspective of work motivation. Nankai Bus. Rev. (2022). https://kns.cnki.net/kcms/detail/12.1288.f.20220914.20221317.20220009.html. (In Chinese)
18. Rosenblat, A., Stark, L.: Algorithmic labor and information asymmetries: a case study of Uber's drivers. Int. J. Commun.Commun. **10**(27), 3758–3784 (2016)

Simulation of Fugitive Interception Strategies for a Mobile Object on a Surface in a Vector Field of Moving Fluid

Viktor Legeza and Liubov Oleshchenko(✉)

Igor Sikorsky Kyiv Polytechnic Institute, Kyiv 03056, Ukraine
oleshchenkoliubov@gmail.com

Abstract. In various real-world scenarios, such as autonomous navigation, robotics, or watercraft control, the ability to navigate efficiently and avoid obstacles is crucial. By simulating avoidance strategies in a one-dimensional vector field of moving fluids, this research addresses challenges associated with fluid dynamics, providing insights into optimal trajectory planning for mobile objects. This knowledge is applicable to fields like autonomous vehicle navigation, where understanding how vehicles can navigate through fluid environments is vital for enhancing safety and efficiency. The study has broader implications for industries involving fluid-based systems, contributing to advancements in control mechanisms and avoidance strategies for mobile objects in dynamic fluid environments.

The research investigates the pursuit process on a horizontal plane as one boat pursues another in a moving fluid field. The research focuses on developing a mathematical model and exploring the paradoxical characteristics inherent in the pursuit dynamics. The primary objective is to ascertain the trajectory for intercepting a fleeing boat by the pursuing boat amid the fluid motion. The mathematical model, comprising four first-order ordinary differential equations, considers the river's speed, creating a one-dimensional fluid field. The pursuit strategy involves the pursuer aligning with an instantaneous line connecting them and the fugitive, while the latter aims to cross the river swiftly toward a "lifeline". Numerical solutions, implemented in the MathCad software, unveil the pursuit curve, revealing intriguing characteristics and paradoxical phenomena. The study identifies dependencies, including the relationship between the fugitive's horizontal movement towards the "lifeline" and the angular coefficient of its escape line in still water, pinpointing a well-defined local maximum. Furthermore, numerical analysis highlights a critical threshold for the speed ratio between pursuer and fugitive, impacting the pursuer's ability to detain the fleeing boat within a specified lane before it crosses the "lifeline".

The research on pursuit dynamics has potential applications in a wide range of practical problems involving navigation, interception, and avoidance in fluid environments, spanning from maritime operations to autonomous systems and environmental monitoring. Enhancing the effectiveness of pursuit dynamics can be crucial in search and rescue missions, where boats or autonomous vehicles need to intercept or follow a target in challenging fluid environments. Implementing optimized pursuit strategies is relevant for autonomous vehicles navigating water bodies. Understanding the dynamics of pursuing or evading objects is essential for designing efficient and safe navigation systems.

Z. Hu et al. (Eds.): ICCSEEA 2024, LNDECT 242, pp. 75–85, 2025.
https://doi.org/10.1007/978-3-031-84228-3_7

Keywords: mathematical modeling · computer simulation · fugitive interception strategies · mobile object · surface in a one-dimensional vector field · moving fluid · pursuit process · "lifeline" · MathCad software · numerical analysis

1 Introduction

The research on pursuit dynamics has practical applications across maritime operations, autonomous systems, environmental monitoring, and strategic planning, providing valuable insights for improved navigation and interception strategies in fluid environments. The development of a mathematical model and exploration of paradoxical characteristics in pursuit dynamics research have implications for several practical problems in various domains.

In maritime search and rescue operations, understanding pursuit dynamics can enhance the effectiveness of rescue missions. For instance, when a rescue boat needs to intercept a drifting object or reach a specific location in turbulent waters, optimized pursuit strategies can be crucial for swift and efficient navigation.

Autonomous vehicles navigating waterways, such as unmanned surface vessels or underwater drones, can benefit from improved pursuit dynamics. In environmental monitoring or underwater exploration, where autonomous vehicles need to follow or intercept specific targets, the research findings can contribute to better trajectory planning and obstacle avoidance. Coast guard and maritime law enforcement agencies may utilize the research results to improve strategies for intercepting vessels involved in illegal activities. Understanding pursuit dynamics in fluid environments can aid in optimizing interception maneuvers and ensuring the safety of law enforcement personnel. Aquatic robotics, including underwater vehicles used for scientific research or industrial purposes, can benefit from the research findings. For example, autonomous underwater vehicles tasked with inspecting underwater structures or pipelines can optimize their pursuit strategies based on the mathematical model.

1.1 Analysis of Recent Research and Related Work

The challenges of pursuit have captivated researchers throughout history, and advancements in technology, along with the increased computational power of modern computers, have refined and intensified pursuit tasks. This evolution is particularly evident in the military domain, where the interception of combat targets has become more intricate and accurate [1–7]. Researchers Behroozi F. and Gagnon R. in 1975 conducted an investigation into the movement of a set of points on a plane as they pursued each other in circular motion. The set comprises marked points denoted as 1, 2, 3,…, n, each characterized by coordinates $(x_1, y_1), (x_2, y_2),\ldots, (x_n, y_n)$. The relationship is established such that point (x_i, y_i) is considered the pursuer of point (x_{i+1}, y_{i+1}), moving along the instantaneous line connecting them. To complete the loop, point (x_n, y_n) must trail point (x_1, y_1), and all points share an identical speed. This type of motion is termed "direct time" motion. Introducing a change in the sign of the time derivative operator leads to point (x_i, y_i) moving away from point (x_{i+1}, y_{i+1}), referred to as "reverse-time" motion.

Both forward and reverse-time motions are analyzed through computer programming. Notably, in real-time scenarios, it is observed that, in nearly all tested cases, the points eventually align in a collinear fashion, exhibiting a substantial ratio of maximum to minimum distance, irrespective of the initial configuration's shape. In reverse time, the points tend to arrange themselves in a star pattern, with the apex angle of the star converging to a value near ninety degrees. The investigation employs stability analysis of the motion equations to precisely calculate the maximum star angle as a function of the number of points.

In 1993, Petrosyan L.A. delved into issues within the classical realm of optimal control theory, focusing on scenarios where optimal control is identified to optimize a criterion while adhering to dynamic constraints that dictate the evolution of the system state under the influence of control variables. The study extends to various differential games involving multiple controllers or players, each driven by distinct and sometimes conflicting optimization criteria, often referred to as gain functions. A significant portion of the research centers on zero-sum differential games, also known as differential pursuit games, which have been extensively explored. The comprehensive body of work contributes to the development of a thorough theory surrounding differential pursuit games, considering both full and partial information scenarios. The study encompasses the resolution of numerous specific pursuit-evasion games, such as lifeline games and simple pursuit games, and introduces new time-consistent optimality principles within the framework of n-person differential game theory.

Professor David C. Arney conducted an analysis of the pursuit process from a military perspective, considering various entities as pursuers, including missiles, interceptor aircraft, intelligent search systems seeking military targets (such as ammunition depots, tanks, and aircraft at airfields), individual soldiers or military units pursuing and approaching enemy counterparts, and ships closing in on other ships or tracking torpedoes designed to detonate upon hitting enemy vessels. The study operates under specific assumptions: the pursuer maintains a constant speed, the pursuit occurs in a predetermined direction over a defined time interval, and the pursuer maintains continuous visual contact with the fugitive or target. The simulation initially focused on the two-dimensional scenario and later extended the model to encompass three-dimensional situations. The primary objective of the research was to ascertain the trajectory for the pursuer's movement, given knowledge of the target's location.

The research results highlight various approaches and methodologies employed in pursuit-evasion games. One study [8] streamlined the complexity of multi-variable problems to two variables, focusing on graphical representation in pursuit-evasion games. Another investigation [9] delved into pursuit and evasion strategies using deep neural network tools but didn't rely on explicit mathematical formulations. Works [10–12] centered on deep reinforcement learning techniques for the same game tasks, showcasing decision-making capabilities without prescribing optimal routes for the moving object. These results collectively demonstrate diverse approaches and insights in tackling pursuit-related challenges, from simplified graphical representations to advanced reinforcement learning techniques. Falade K. I. and Tiamiyu A. T. [13] introduce an innovative iterative approach for addressing partial differential equations (PDEs) characterized by fractional variable coefficients. Leveraging numerical techniques, the authors derive approximate solutions for PDEs featuring fractional coefficients. Such equations find relevance in diverse domains, including physics, engineering, and finance, where the proposed method offers a novel avenue for obtaining numerical solutions to enhance understanding and applicability.

A review of literature sources showed that in most formulations of problems, the pursuit process has been traditionally examined within the static conditions of the external environment. This implies that it remains unaffected by alterations in the spatial positions of both the pursuer and the fugitive over time. The research [14] explores scenarios characterized by stationary conditions, aligning with the conventional approach in pursuit problem formulations.

1.2 Objective

The research object is the pursuit and interception of a fleeing boat by a pursuing boat on the river's horizontal surface, considering the influence of the river's current speed. The subject of the research is the mathematical model and the paradoxical features and attributes inherent in the described pursuit process.

The research objective: (1) determine the trajectory for intercepting the fleeing boat by the pursuing boat in a dynamic water environment; (2) identify the specific location on the river's surface where the pursuer will catch up with the fugitive; and (3) ascertain the angular coefficient of the escape line, ensuring the optimal horizontal movement of the fugitive along an axis perpendicular to the defined "lifeline."

2 Problem Statement

We contemplate the initial positions of two mobile points, with one in pursuit of the other as shown in Fig. 1.

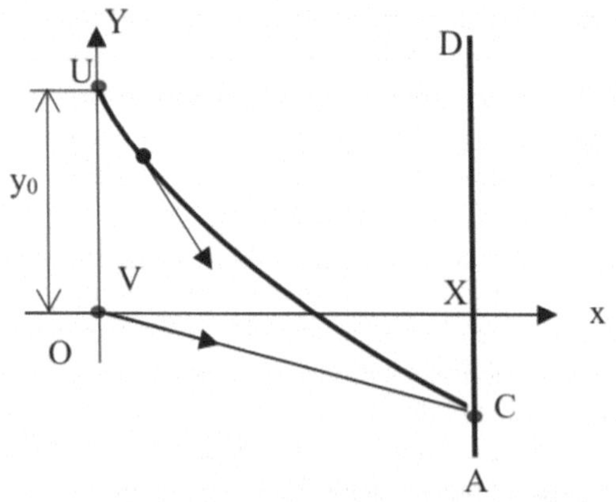

Fig. 1. Pursuit curve

On one side of a turbulent river, as illustrated in Fig. 1 with the ordinate axis, two boats are positioned at a distance y_0 from each other – a coast guard boat U (referred to as the pursuit boat) and a boat V violating border regulations (referred to as the fugitive boat). The river acts as a natural border between two territories (for instance, between two countries). Simultaneously, the fugitive boat endeavors to swiftly traverse the river, moving towards the border AD that defines the separation between the two territories, referred to as the "lifeline." In subsequent discussions, we will designate the "lifeline" for the fleeing boat as parallel to the initial shoreline OY where both boats are located. The pursuit boat U aims to intercept the fugitive boat V before it reaches a specified point on the "lifeline." In Fig. 1, point C represents the potential location where the pursuer U could successfully apprehend the fugitive V.

At the initial moment ($t = 0$), the fugitive boat V, considering the river's motion, commences a trajectory represented by a straight line $y = kx$ with a slope given by the angular coefficient k (depicted in Fig. 1).

The boat maintains a constant speed, denoted by the magnitude of the velocity vector $\gamma_V = \sqrt{p^2 + q^2} = const$. Here, the variables p and q represent the horizontal and vertical components of the V boat's speed, respectively, achievable under stagnant water conditions. Simultaneously, the pursuing boat U initiates its course at the same time as the fugitive boat V and dynamically selects its direction based on the fugitive boat's current position on the plane.

The pursuing boat U also maintains a constant speed $\gamma_U = \sqrt{u^2 + v^2} = const$, with a magnitude that is α times greater than the fugitive boat's speed, expressed as $\gamma_U = \alpha\gamma_V$. In this context, the variables u and v denote the horizontal and vertical projections of the boat's speed, achievable under stagnant water conditions. The constant values γ_U and γ_V are determined by the engine power of both boats in a specific manner.

The specified speed magnitudes for both boats are defined concerning the stationary water surface, neglecting the influence of the river current. However, in this scenario, boat movement occurs in a dynamic water environment where the speed of the river current is represented by the function $f(x)$ concerning the horizontal coordinate. The velocity vector aligns opposite to the positive direction of the ordinate axis.

Physically, the boats' motion can be analogized to the movement of material points within the horizontal vector field $\vec{B}$ velocities in a flowing fluid.

In this context, the components of the vector field representing the moving fluid take the form $\vec{B} = (0; -f(x); 0)$, where $f(x)$ is the function describing the speed of the river current as a function of the horizontal coordinate x. The vertical kinematics of boat movement, signifying motion in the direction of the applied axis, is not a consideration in this particular problem.

3 Formulation of the Differential Equation Describing the Path of the Pursuing Boat

Consider the positions of the boats on the plane as points U and V, denoted by coordinates (x, y) and (ξ, η), respectively. Represent the coordinates of point U in moving water and those of point V at the current time t.

To derive the pursuit curve, considering the influence of the river current, we formulate a system of differential equations:

$$\begin{cases} \dfrac{dx}{dt} = u; \\ \dfrac{dy}{dt} = v - f(x); \\ \dfrac{d\xi}{dt} = p; \\ \dfrac{d\eta}{dt} = q - f(x); \end{cases} \tag{1}$$

Express the slope (angular coefficient,$y\prime$) of the tangent to the curve L at the present point U (x, y) as follows:

$$y' = (v - f(x))/u \tag{2}$$

It's essential to observe that the pursuer endeavors, at each time step, to follow the instantaneous line connecting the two existing points U and V on the plane (effectively "keeping track of the fugitive" $V(\xi, \eta)$). Simultaneously, this direction aligns with the tangent $y\prime$ to the intended pursuit curve U (x, y). To mathematically articulate this relationship, considering Eq. (2), we establish an equality between two angular coefficients:

$$(v - f(x))/u = (\eta - y)/(\xi - x) \tag{3}$$

Initially, let's eliminate the variables u and v from the system (1) by utilizing equality (3) and the relation for γ_U. To achieve this, we'll express the variable v in terms of the remaining variables (x, y, ξ, η) using Eq. (3).

$$v = f(x) + u \cdot \omega(x, y, \xi, \eta) \tag{4}$$

here, $\omega(x, y, \xi, \eta) = (\eta - y)/(\xi - x)$. In subsequent steps, the arguments of the function ω will be omitted for simplicity. Now, employing the relations $\gamma_U = \sqrt{u^2 + v^2}$ and the expression (4), let's eliminate the variable u. This process leads us to a quadratic equation:

$$u^2 \cdot (1 + \omega^2) + 2f(x)\omega u + f^2(x) - \gamma_U^2 = 0 \tag{5}$$

The solutions to Eq. (5) yield a pair of relationships:

$$\tilde{u}_{1,2} = \left[-f(x)\omega \pm \sqrt{\gamma_U^2(1 + \omega^2) - f^2(x)} \right] / (1 + \omega^2) \tag{6}$$

The two corresponding solutions for v take the following forms:

$$\tilde{v}_{1,2} = \pm\sqrt{\gamma_U^2 - u_{1,2}^2} \tag{7}$$

Next, within the system (1), we eliminate the variables p and q by utilizing the relations $\gamma_V = \sqrt{p^2 + q^2}$ and $q = kp$. In this particular problem, these variables remain constant and are expressed as follows:

$$\tilde{p}_{1,2} = \pm\gamma_V \Big/ \sqrt{1+k^2},\ \tilde{q}_{1,2} = \pm\gamma_V k \Big/ \sqrt{1+k^2} \tag{8}$$

By substituting the derived expressions (6), 7, and (8) into the system (1), we derive a standard set of four nonlinear differential equations describing the pursuit process in moving fluid (flowing water):

$$\begin{cases} \dfrac{dx}{dt} = \tilde{u}(x, y, \xi, \eta); \\ \dfrac{dy}{dt} = \tilde{v}(x, y, \xi, \eta) - f(x); \\ \dfrac{d\xi}{dt} = \tilde{p}; \\ \dfrac{d\eta}{dt} = \tilde{q} - f(x); \end{cases} \tag{9}$$

4 Numerical Analysis of the Fugitive Pursuit Process in Moving Fluid

For the numerical integration of system (9), the initial conditions are set as follows: $x(0) = \xi(0) = \eta(0) = 0$; $y(0) = 2$. Therefore, the starting points are for the pursuing boat at point U (0, 2) and for the runaway boat at point V (0, 0) respectively. The function $f(x)$, determining the modulus of the river current, is defined as:

$$f(x) = A_0 \sin\left(\pi x \big/ H\right) \tag{10}$$

A graphical representation (Fig. 2) is provided for a specific pursuit scenario with selected movement parameters: $k = -1$, $A_0 = 1$, $H = 2$ and $\gamma_U \big/ \gamma_V = \alpha = 2$.

In Fig. 2, the dashed curve illustrates the trajectory of the runaway boat V, while the solid curve depicts the trajectory of the pursuing boat U. It is noteworthy that the dashed curve, representing the trajectory of the runaway boat in moving water, exhibits curvature due to the influence of the river current transfer speed.

The coordinates of the point on the plane where the fugitive will be captured are determined as C (1.264, –2.170). The corresponding time T, spent on the pursuit process is approximately 1.787 time units.

We analyze the behavior of the function $x(k)$, representing the horizontal coordinate of the apprehension point, concerning the variable angular coefficient k.

For this analysis, specific parameters are fixed at $A_0 = 1$, $H = 1$, $y_0 = 1$, $\alpha = 2$. Additionally, the initial conditions for integration are chosen as $x(0) = \xi(0) = \eta(0) = 0$; $y(0) = 1$.

The graph in Fig. 3 illustrates a distinct local maximum, achieved at a particular value of the angular coefficient k, approximately equal to –0.45.

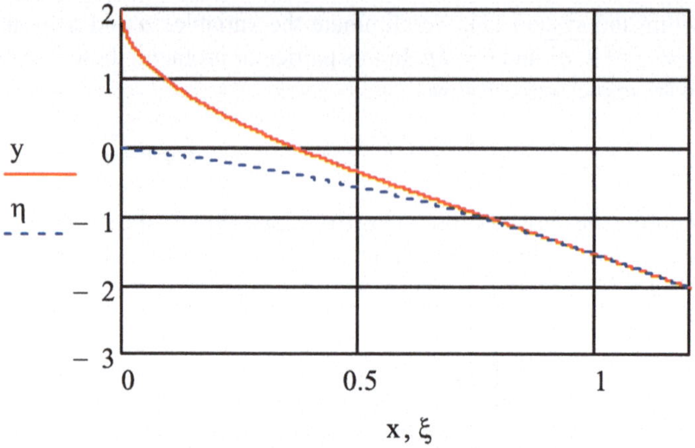

Fig. 2. Trajectories of pursuit (solid curve) and evasion (dashed curve) in the given scenario: $k = -1, A_0 = 1, H = 2, \alpha = 2, y_0 = 1$

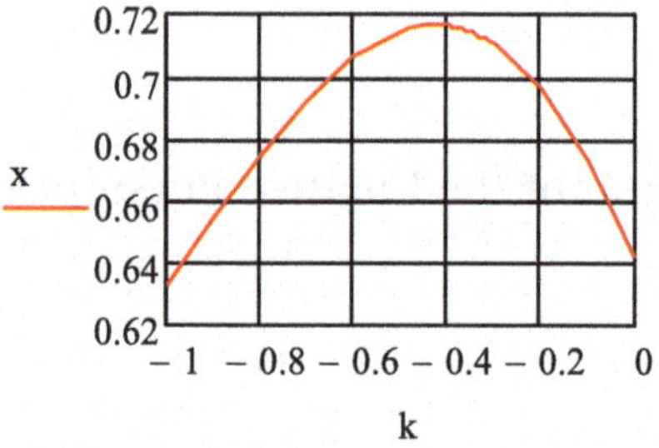

Fig. 3. Graph illustrating the relationship, *x(k)* is the maximum horizontal displacement of the fugitive toward the "lifeline"

The initial position of the pursuing boat is denoted as the starting point U (0,1) for the pursuit, while the initial position of the fleeing boat V (0,0) marks the commencement of its movement.

The presence of a local maximum on the curve signifies that there exists a specific non-zero angle of inclination in the direct motion of the fugitive boat (in stagnant water), allowing it to achieve maximum displacement towards the "lifeline" *AD*.

5 Simulation Results

The simulation involves modeling strategies for intercepting a mobile object on a surface within a vector field of moving fluid using MathCad. The primary focus is on scenarios where a pursuer aims to capture a fugitive in a dynamic fluid environment. The simulation include the formulation of differential equations describing the motion of both the pursuer and the fugitive, accounting for the influence of the moving fluid. MathCad, a mathematical software, is utilized to implement and solve these equations numerically. The simulation explored various parameters such as the speed of boats, the flow of the river, and the initial conditions of the pursuit, providing insights into optimal interception strategies. Research goal is to gain a better understanding of how different factors impact the pursuit dynamics and to potentially uncover paradoxical properties or optimal strategies in the interception process.

Research observation introduces a paradox: the curve in Fig. 3 reveals that adopting a strategy with a zero angle of inclination for a straight escape in still water (following the shortest segment between two parallel lines OY, i.e., moving along the natural path AD) is not optimal for a successful escape. In this particular calculation ($\alpha = 2$, $k =$ -0.45), the resulting movement x (–0.45) = 0.717 (measured in distance units). For comparison, if the "lifeline" AD were positioned, for instance, at a distance of 0.7 (distance units) from the ordinate axis, successful escape for the boat V would be achievable, unlike the scenario when a straight line with a zero angle coefficient is chosen for escape (i.e., along the abscissa axis).

In the latter case, $x(0) = 0.642 < X$, and the fugitive will inevitably be captured.

The numerical analysis of the pursuit dynamics has revealed the existence of a distinct minimum threshold for the ratio of speed modules, denoted as $\gamma_U / \gamma_V = \alpha_{MIN}$, crucial for the pursuer U to successfully apprehend the fugitive V within the specified lane $0 < x < X$ before the fugitive reaches the "lifeline" AD. Should the value of fall below this specified threshold, the fugitive cannot be captured within this lane.

It is noteworthy that the ratio establishes a relationship between the power of the motors of the two boats. Furthermore, this value significantly deviates from the corresponding value in scenarios where the speed of the river flow is not considered in the problem formulation. To illustrate this intriguing characteristic, we examine the pursuit process in a scenario with $k =$ -0.45, $A_0 = 1$, $H = 1$, $y_0 = 1$, $X = 1$. Initially, we determine the minimum value α_{MIN} of required to ensure the capture of the fugitive within the designated lane $0 < x < 1,00$. The numerical experiment yields a value of $\alpha_{MIN} = 1,69$. Figure 4 depicts the pursuit curve constructed for this specific case. The point of intersection between the two curves is identified as C (0.997, –1.001), providing clear evidence that the fugitive V has been successfully captured within the specified lane $0 < x < 1,00$.

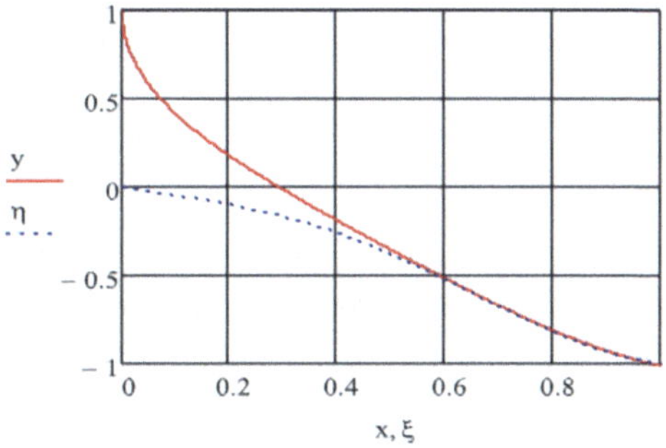

Fig. 4. The pursuit and capture of a fugitive within a specified lane $0 < x < 1,00, \alpha_{MIN} = 1,69$

6 Summary and Conclusions

This research addresses the pressing issue of formulating a mathematical model for chasing and capturing a fugitive in a horizontal vector field of moving fluid, imposing specific constraints on the fugitive's movement. The novelty of this research lies in introducing a mathematical model that accounts for the variable speed of the external environment, marking the first instance of such an approach.

The practical significance emerges from conducting a numerical analysis of the pursuit process using the developed model and software implementation, unveiling crucial characteristics. Notably, the research reveals a paradoxical property related to the optimal angle of inclination for the fleeing boat's linear movement in still water, challenging conventional assumptions.

The results suggest the applicability of the model in practical scenarios involving pursuit processes in horizontal vector fields of moving fluid. Future research directions may explore the pursuit dynamics when the fugitive opts for a second-order algebraic curve for escape in a stationary fluid, incorporating the effects of fluid motion.

Autonomous vehicles used for environmental monitoring in water bodies can benefit from improved pursuit dynamics. This could enhance the ability to track and study fluid dynamics in rivers or other water environments. The findings may be applied in robotics, particularly in scenarios where robots need to navigate through fluid environments, such as underwater exploration or pipeline inspection.

Understanding pursuit dynamics is valuable for optimizing traffic management in waterways, ensuring efficient and safe navigation of boats and vessels.

The study's outcomes can be applied to scenarios where the movement of objects in water, such as pollutants or debris, needs to be monitored. This is valuable for environmental impact assessment and response planning in the event of water contamination or spills. Improved pursuit dynamics can contribute to efficient traffic management in water systems, similar to how traffic rules are applied on roads. For instance, optimizing the navigation of boats in busy waterways or around ports to prevent collisions or congestion.

The research results can have applications in game theory, contributing to the development of strategies for competitive scenarios, such as pursuit games involving autonomous agents in dynamic fluid environments.

References

1. Malacká, Z.: Pursuit curves and ordinary differential equations. Komunikacie **14**(1), 66–68, March 2012. University of Žilina. https://doi.org/10.26552/com.C.2012.1.66-68
2. Nahin, P.J.: Chases and Escapes: The Mathematics of Pursuit and Evasion. Princeton Puzzlers, Princeton (2012). 272 p.
3. Yalovets, A.L.: On the formulation of the pursuit problem on the plane. Program. prob. 2, 95–100 (2013). ISSN 1727-4907
4. Pashko, S.V., Yalovets, A.L.: Numerical methods for solving pursuit optimization problems. Probl. program. 4, 74–85 (2013). ISSN 1727-4907
5. Kuchkarov, A.S., Rikhsiev, B.B.: A pursuit problem under phase constraints. Autom. Remote Control **62**(8), 1259–1262 (2001)

6. Petrosyan, L.A.: Differential Games of Pursuit. World Scientific Publisher, Singapore, London (1993). 340 p. https://doi.org/10.1142/1670
7. Hagedorn, P., Breakwell, J.V.: A differential game with two pursuers and one Evader. Multi-criteria Decis. Making Differ. Games **1976**, 443–457 (1976). https://doi.org/10.1007/978-1-4615-8768-2_27
8. Patsko, V., Kumkov, S., Turova, V.: Pursuit-evasion games. In: Başar, T., Zaccour, G. (eds.) Handbook of Dynamic Game Theory. Springer, Cham (2018). https://doi.org/10.1007/978-3-319-44374-4_30
9. Can, X., Zhang, Y., Wang, W., Dong, L.: Pursuit and evasion strategy of a differential game based on deep reinforcement learning. Front. Bioeng. Biotechnol. (2022). https://doi.org/10.3389/fbioe.2022.827408
10. Lopez, V.G., Lewis, F.L., Wan, Y., Sanchez, E.N., Fan, L.: Solutions for multiagent pursuit-evasion games on communication graphs: finite-time capture and asymptotic behaviors. IEEE Trans. Autom. Control **65**, 1911–1923 (2019). https://doi.org/10.1109/TAC.2019.2926554
11. Zhou, Z., Xu, H.: Mean field game and decentralized intelligent adaptive pursuit evasion strategy for massive multi-agent system under uncertain environment with detailed proof. In: Proceedings of the Artificial Intelligence and Machine Learning for Multi-Domain Operations Applications II; Online. 27 April–8 May 2020 (2020)
12. Wan, K., Wu, D., Zhai, Y., Li, B., Gao X., Hu, Z.: An improved approach towards multi-agent pursuit-evasion game decision-making using deep reinforcement learning. Entropy (Basel). **23**(11), 1433 (2021). https://doi.org/10.3390/e23111433
13. Falade, K.I., Tiamiyu, A.T.: Numerical solution of partial differential equations with fractional variable coefficients using New Iterative Method (NIM). Int. J. Math. Sci. Comput. (IJMSC) **6**(3), 12–21 (2020). https://doi.org/10.5815/ijmsc.2020.03.02
14. Legeza, V., Oleshchenko, L.: Paradoxical properties research of the pursuit curve in the intercepting a fugitive problem. Lect. Notes Data Eng. Commun. Technol. Springer **2023**(181), 670–681 (2023). https://doi.org/10.1007/978-3-031-36118-0_60

Software Method for Distributed Asynchronous Big Data Processing of Fundraising

Liubov Oleshchenko(✉) and Vadym Lavrinenko

National Technical University of Ukraine "Igor Sikorsky Kyiv Polytechnic Institute", Kyiv 03056, Ukraine
oleshchenkoliubov@gmail.com

Abstract. The article describes proposed enhanced software method for predicting volunteer fundraising success with asynchronous big data processing using machine learning algorithms (logistic regression, random forest and gradient boosting). Results are compared in terms of model accuracy and speed. To scale, parallel processing with serverless AWS Lambda and Simple Queue Service is implemented. Features for asynchronously sending messages for processing by the web application component to the message broker are implemented. The broker redirects messages for processing to the existing Lambda environment containing the code for predicting fundraising success.

For the investigation, a set of programming technologies integrating with cloud services is used. The proposed software architecture ensures high throughput for updates on fundraising with the ability for parallel, independent processing by the implemented code. Taking into account the processing time of all messages in the fundraising updates dataset as the main criterion for evaluating the speed of the implemented software method, an analysis and comparison were conducted between architectures using the basic and proposed methods. Despite similar processing time indicators for both architectures with a message set not exceeding 10, a reduction in the required processing time for larger datasets was achieved. The proposed architecture ensures high throughput and reduces processing time by 55–81% for datasets exceeding 30 messages compared to the basic architecture. The difference in results, depending on the number of messages, is explained by the cold start of message processors, with the final initialization of the last ones coinciding in time with the completion of processing the message set by the architecture with sequential processing. According to the research, increasing the number of messages and using pre-initialized environments minimize the part of the total processing time allocated to launching the event handler. The proposed software method is optimized for processing data in services that require high throughput.

Keywords: software method · big data · classification algorithms · distributed processing · message broker · high performance · event-driven architecture · fundraising · asynchronous event handlers · parallel computing · AWS cloud services · AWS Lambda · success prediction algorithms in volunteer campaigns

Z. Hu et al. (Eds.): ICCSEEA 2024, LNDECT 242, pp. 86–102, 2025.
https://doi.org/10.1007/978-3-031-84228-3_8

1 Introduction

Recently in Ukraine, the volunteer movement has surged, leveraging software for fundraising amid wartime challenges. However, managing expectations and the risk of underfunding pose significant hurdles. Individual consultation on funding feasibility is inaccessible to many, necessitating automated systems capable of serving numerous users simultaneously. Real-time data processing is essential for timely decision-making, facilitated by distributed asynchronous methods, optimizing resource utilization and enhancing analytical model accuracy. Volunteer activity's rapid growth introduces difficulties in attracting sufficient funds due to limited audience reach and transparency issues. A genuine need exists for software optimizing fundraiser management, reporting, and risk analysis based on volunteers' past projects.

1.1 Analysis of Recent Research and Related Work

The research paper [1] proposes a method for extracting information from complex unstructured big data. The authors employ Hadoop and Spark to handle vast amounts of unstructured data and apply natural language processing to extract pertinent information from text sources, organizing it into a structured format.

The paper [2] presents an improved architecture for big data analytics in IoT-enabled urban transportation systems. The paper aims to address challenges related to data privacy, security, and scalability in big data analytics, and proposes a federated learning approach that facilitates and optimizes the sharing of data and information among transport systems and networks.

The article [3] introduces a dual-mode big data processing framework, offering responsive feedback and targeted advertising. The paper [4] describes the development of a distributed stream processing middleware framework for analyzing heterogeneous data in real-time. The framework is designed to work with big data platforms and is specifically applied to the case of environmental monitoring.

Articles [5, 6] examine big data challenges and solutions in information management. They outline issues like data quality, integration and privacy, proposing solutions such as advanced analytics and machine learning.

The article [7] explores the potential of using big data and customer relationship management to personalize and customize services for customers. The paper [8] explores the application of machine learning techniques on big data analytics, specifically focusing on the challenges and opportunities of big data. The article [9] discusses the concept of real-time big data processing and its application in instantaneous marketing decisions. The authors argue that big data can be leveraged to gain insights and make informed marketing decisions in real-time.

The article [10] discusses techniques to enhance the performance of big data web application state management.

In contemporary research, key challenges in distributed asynchronous big data processing include addressing issues of data consistency and synchronization across distributed nodes, ensuring fault tolerance and resilience in the face of node failures or network disruptions, and optimizing resource utilization to minimize processing delays and bottlenecks. Scalability concerns arise as software systems must efficiently handle increasing data volumes and processing demands while maintaining performance and responsiveness. Achieving low-latency processing in asynchronous environments remains a crucial goal for fundraisings real-time or near-real-time application requirements. Addressing these challenges requires innovative approaches in distributed systems design, fault tolerance mechanisms, and resource management strategies to optimize performance and reliability in asynchronous big data processing.

1.2 Analysis of Existing Software Solutions

The analysis of the most popular software solutions for distributed big data processing reveals several strengths but also identifies certain shortcomings which are described in Table 1.

Common shortcomings across existing software solutions are complexity, resource intensive, real-time processing challenges, scalability limits, integration challenges and cost concerns. Setting up and configuring distributed systems can be intricate, requiring specialized knowledge and expertise. Many solutions, especially those involving in-memory processing, can be resource-intensive, demanding significant computational power and memory. Achieving true real-time processing capabilities can be challenging for some solutions. Certain solutions may face scalability challenges, particularly with extremely large datasets or in highly dynamic environments. Integrating distributed big data solutions with existing IT infrastructure and applications can be complex and may require substantial modifications. Licensing fees, cloud service costs, and infrastructure requirements can contribute to high operational expenses.

1.3 Software Requirements

Considering the outlined challenges and stakeholder interests, the final software product in this research will be a web application supported by a backend API service, enabling user interaction. It will cater to both patrons and volunteers, offering features like fundraiser creation, search, and viewing, along with access to analytics on past fundraisers. Emphasis will be on user accessibility without specialized training or high system requirements, ensuring secure handling of personal data, particularly for payment processing. The system will introduce predictive analytics for fundraising success using machine learning algorithms, with regular updates to maintain real-time information and scalability to accommodate varying user activity levels.

1.4 Objective

The objective of the research are increasing fundraising forecast accuracy and time reduction for training models with subsequent data processing using asynchronous message processing. The task of the research is to develop an optimized software method for more accurate predicting volunteer fundraisers that scales with the increasing number of users, utilizing distributed computing and AWS cloud services.

Table 1. The analysis of existing software solutions for distributed big data processing

Software solution	Strengths	Shortcomings
Apache Hadoop	Scalable and fault-tolerant; widely adopted in various industries	MapReduce paradigm can be slow for iterative algorithms; complex setup and configuration; not designed for real-time processing
Apache Spark	In-memory processing for improved speed; versatile with libraries for machine learning and graph processing	Memory-intensive; may require substantial resources; learning curve for users unfamiliar with distributed computing concepts
Apache Flink	Low-latency and high-throughput stream processing; supports event time processing	Smaller ecosystem compared to Hadoop and Spark; less mature community; complexity in managing stateful processing
Amazon EMR	Fully managed service in the cloud; integrates with other AWS services	Limited customization options; potential for higher costs compared to self-hosted solutions for large-scale processing
Databricks	Unified analytics platform; collaborative workspaces; integration with various data sources	Cost can be a concern for smaller organizations; dependency on the Databricks environment
Cloudera Data Platform (CDP)	Comprehensive platform with security and governance features; supports multiple frameworks	Licensing costs; complexity in managing and configuring the platform
Microsoft Azure HDInsight	Cloud-based service with seamless integration into the Azure ecosystem	Limited support for non-Microsoft technologies; potential for higher costs based on data transfer and storage

2 Research Method

2.1 Basic Method for Predicting the Fundraising Success

The decision to select the basic prediction method considered existing commercial systems and the advantages and disadvantages of standard methods for predicting variables based on historical data in the domain of fundraising success.

Given that fundraising success prediction falls under binary classification, with fundraisers categorized as successful or unsuccessful, gradient boosting was chosen as the primary method. This approach allows for automated prediction, eliminating the need for deep theoretical knowledge and enabling independent analysis and updating of forecasts based on new data. While gradient boosting can uncover complex dependencies and improve prediction accuracy, its single-threaded execution poses a challenge in terms of speed. Further refinement and system architecture design are necessary to enhance its performance.

2.2 Proposed Modification for the Fundraising Success Prediction Method

Consider a modified method of predicting the success of volunteer fundraising, the modification of which is caused by the application within the system with distributed computing for the purpose of timely processing of updated data and providing updated forecasts in the context of a multi-user scalable system with a web application.

Training model using the gradient boosting algorithm involves the following stages.

Stage 1. The beginning of the execution of the algorithm involves the creation of a weak model – a decision tree, which assumes the target variable.

Stage 2. Next, the error between the predicted and actual values of the target variable is determined. The goal of training is to minimize the error of the current model in subsequent iterations. So, at this stage, the gradient of the loss function is calculated relative to the predictions of the previous model.

Stage 3. A new model is created, the target variable of which is the value of the gradient from the previous step. The model is added to the ensemble. At the same time, the previous model remains unchanged.

Stage 4. The predicted value of the target variable for each iteration is calculated as the sum of the previous predicted value of this variable and the error value predicted by the model from step 3.

Stage 5. Steps 2 through 4 are repeated, adding new models to the ensemble, until satisfactory accuracy is achieved or until the specified number of iterations is exhausted.

Consider the basic algorithm for applying gradient boosting for forecasting in the context of a web application in Table 2.

This approach is not scalable, and within an HTTP request, it is also unreliable in terms of performance and execution time due to the limits set by the client and the server on the maximum execution time of the request.

Taking these factors into account, an algorithm for parallel processing of requests to create new fees was developed, shown in Table 3.

For the dynamic operation of the software system, it is not enough to predict the success of individual fundraisings, since the market situation is constantly changing, and therefore, the trained model gradually loses its relevance and needs to be re-trained.

The algorithm for regular retraining of the network consists of the steps listed in Table 4.

The proposed modification of the fundraising success prediction method offers significant advantages over the basic approach, particularly in terms of productivity, scalability, fault tolerance, and portability. It leverages large volumes of historical and user data, prioritizing system scalability and timely request processing. The selection of technical means for implementation considers architectural requirements for parallelized calculations, with a focus on configurable message handlers and idempotent queue processing to prevent race conditions. This approach ensures efficient resource utilization and avoids redundant computing tasks.

Table 2. The basic neural network learning algorithm for prediction in the context of a web application

Incoming data	Historical data of conducting volunteer fundraisings, formatted in a "csv" file. (Newly created collection if the model is trained)
Output data	A trained model capable of predicting the success of a collection specified in the database (Estimated assessment of collection success)
Step 1	Loading data into RAM
Step 2	Data normalization
Step 3	Distribution of the data set into training and test (Provision of collection data at the entrance)
Step 4	Performing model training before the stop condition is met (Record the received forecast in the database)
Step 5	Model verification by providing a random sample of test data as input to a trained model

Table 3. Algorithm for processing requests for fee forecasting

Incoming data	Newly created collection
Output data	Estimated assessment of collection success
Step 1	Receiving a request from a client
Step 2	Download current collection data
Step 3	Forming and sending a prediction message for this collection to the message queue
Step 4	Prediction of collection success in a separate stream, based on the steps shown in Table 2

The modification of the method, as a result, consists in the application of tools for parallelization of calculations, which will allow to isolate the prediction of the success of each individual collection and retraining of the model in order to update information about the current state of the system, in separate streams, using an asynchronous message queue.

The sequence diagram for the case of processing two messages, according to the described algorithm, is shown in Fig. 1.

The above algorithm can be summarized by the block diagram shown in Fig. 2.

2.3 Software Implementation of Predicting Fundraising Success Method

To enable users to create and view volunteer fundraisings and track success forecasts, a web application was chosen, with console applications handling asynchronous data processing.

Table 4. The neural network retraining algorithm

Incoming data	Automated program launch schedule
Output data	Trained model
Step 1	Loading current training data into RAM
Step 2	Reload data on fundraisings that were closed successfully or not between the current training run and the previous one
Step 3	Data normalization
Step 4	Distribution of the data set into training and test
Step 5	Performing model training before the stop condition is met
Step 6	Automated model verification by providing a random sample of test data as input to trained model
Step 7	Loading current active fees into RAM
Step 8	Forming and sending a message for each collection to a queue for processing without waiting for any of them to complete

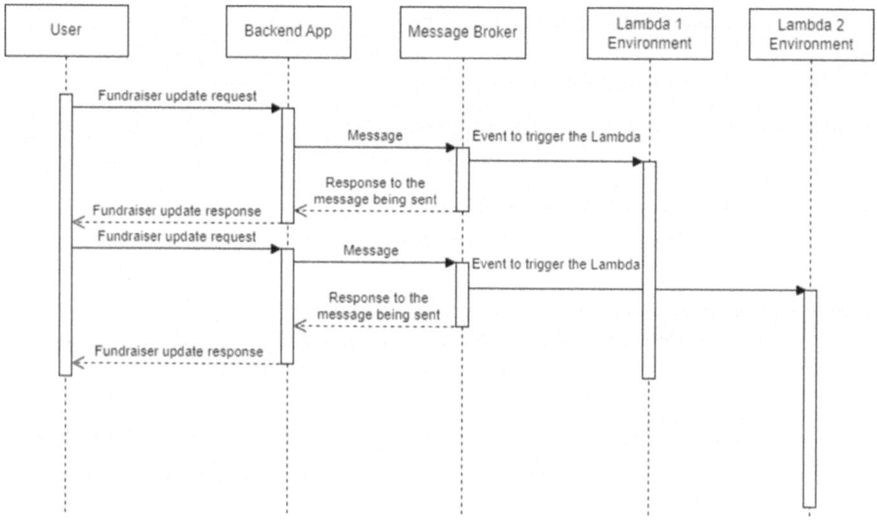

Fig. 1. Sequence diagram of parallel processing of two messages

.NET 6 framework, particularly the Minimal API web application template, was selected for the server-side development due to its productivity, built-in support for dependency injection, and cross-platform compatibility, including deployment on AWS Lambda. Angular JavaScript framework was chosen for the frontend, offering modular architecture for scalability, component-based development for code organization and reusability, and opportunities for real-time updates using client libraries for websockets and Redux Store. Angular framework allows for future expansion of the system, including real-time notification functionality.

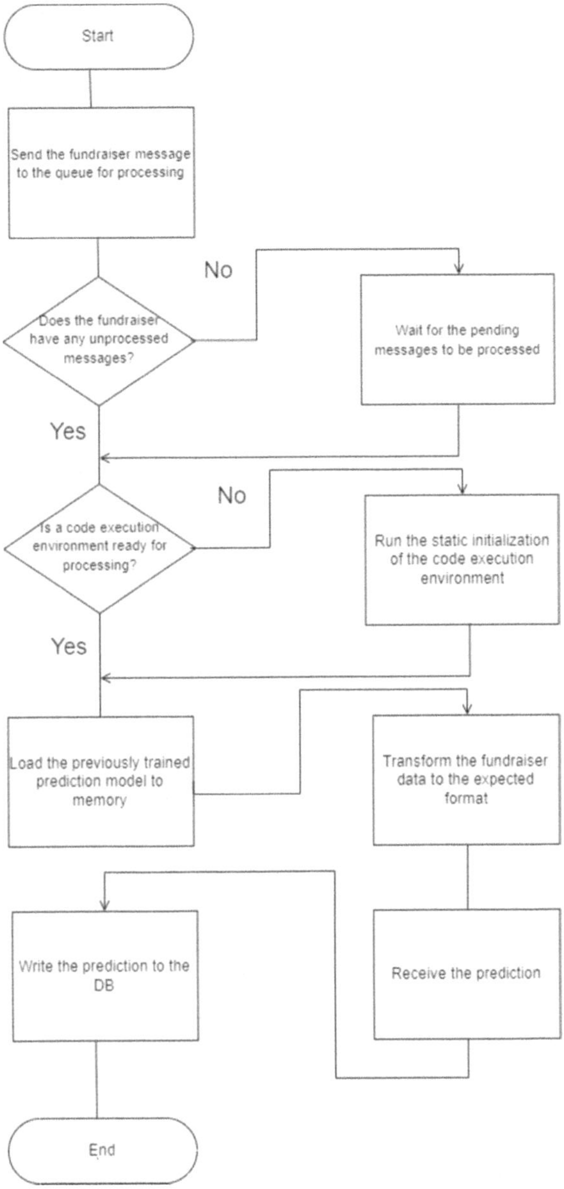

Fig. 2. Block diagram of the modified method of predicting fundraising success

Python offers a rich ecosystem of libraries like NumPy and Pandas for data manipulation and analysis, while its extensive support for machine learning functionalities, provided by libraries such as Scikit-learn, TensorFlow, and PyTorch, was a key factor in its selection. Python's integration capabilities with various services, including AWS, are valuable for this project, allowing seamless integration with services like Simple Queue

Service and Lambda for parallel data processing and high scalability. AWS Lambda's event-driven architecture and serverless computing eliminate the need for manual scaling operations, ensuring efficient handling of variable workloads and high traffic. This approach aligns well with the distributed system design, promoting modularity and scalability.

AWS Simple Queue Service (AWS SQS) creates an asynchronous message queue for parallel processing of data by system services. SQS decouples components in distributed applications, providing reliable and scalable messaging. It allows components to interact asynchronously through message queues, enhancing system stability, fault tolerance, and scalability by handling large message volumes and adapting to different workloads automatically. By offloading heavy tasks from the main backend application, such as generating forecasts and model retraining, SQS ensures even workload distribution and prevents bottlenecks, supporting fast and scalable application responses. Docker containerization facilitates deploying lambda functions with dependencies on AWS Lambda, ensuring portability across platforms. AWS CloudFront enhances client application performance and security by delivering content globally with low latency, reducing server load, and integrating with other AWS services for improved performance and security.

In this research, data changes infrequently due to a low number of users adding new fees simultaneously, and model retraining is based on a known schedule. Leveraging AWS CloudFront, we can ensure high system performance by caching fundraising data in locations closest to users until retraining. CloudFront also caches the entire web application, reducing initial page load times. Amazon S3, an object storage service, stores and manages shared data between system components, including serialized prediction models. It also facilitates deploying static websites for the client side of the application as a single-page application.

2.4 Proposed Software System Architecture

The software system architecture consists of separate modules for the client web application, server application providing a Web API, and asynchronous event handlers for updating user data and volunteer fundraisings. Designed to leverage cloud services like AWS Lambda and AWS Simple Queue Service, the system is deployed on Amazon Web Services to integrate with existing architecture and avoid network configuration and hardware allocation complexities for local deployment.

Taking this into account, a diagram of individual components of the system was made in the AWS notation (Fig. 3) in order to display the used services of the provider.

The architecture diagram includes components such as the web browser for client interaction, Cloudfront for network communication setup, and S3 for storing web application files. AWS Lambda, built on the.NET framework, serves as the server component managing charge filtering, creation, and update messaging. SNS forwards server messages to subscribed message queues, while SQS provides queue functionality for message processing. Lambda functions handle prediction, model retraining, and system speed testing, configurable by developers. CloudWatch Event Bridge schedules model retraining, and Amazon RDS Aurora stores historical fundraising data and predictions.

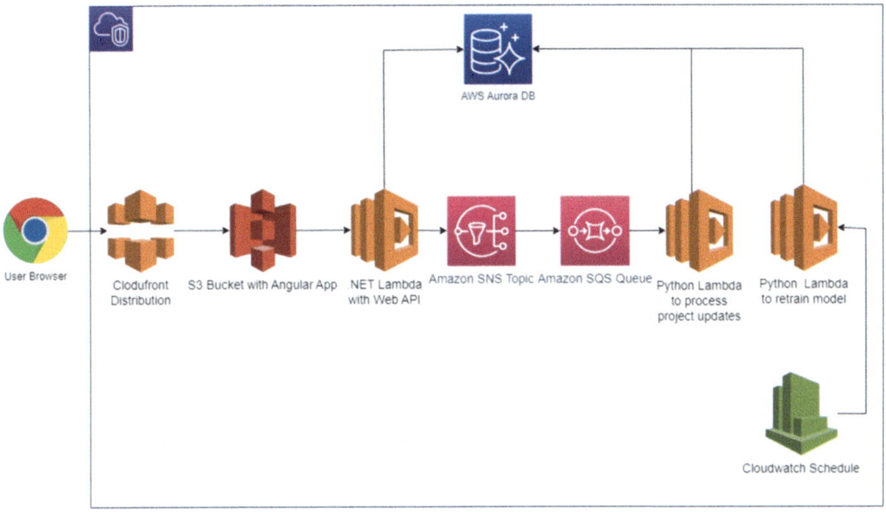

Fig. 3. Proposed software system architecture using AWS components

The developed system involves not only the processing of the data set available for training and testing, but also the collection data generated by the user using a graphical interface. The graphical user interface provides the following set of functionality:

1. Creation and editing of fundraisings (Fig. 4). The user can change the name of the charge, its description, category, target amount and currency in which it is calculated, start and end dates, as well as receipt data. Within the framework, the integration with the banking system, which would keep records of receipts, was replaced by the possibility of independently entering the amount of donations received, recorded on a given date, which is within the limits of the duration of the collection.
2. Review of the available list of fundraisings (Fig. 5). The user is given the opportunity to view the list of collections, which contains information about the name, category, number of days until the deadline, the share of received funds from the desired ones, as well as an indicator of the expected success of the collection.

The page has the functionality of searching for fees among the list by name, switching to the editing page of the selected fee, as well as deleting it.

Within the framework of this research, the receipt of funds for fees was simulated with the help of a user interface that provides the possibility of entering historical data on the receipt of funds, instead of interacting with the existing payment system.

At the same time, future developments of the system may be aimed at integration with the banking system.

Thus, the proposed implementation of the system contains a number of advantages that will ensure the speed and reliability of accounting of receipts in the system:

1. Amazon SQS message broker provides an increase in the speed of predictions due to parallel calculations.

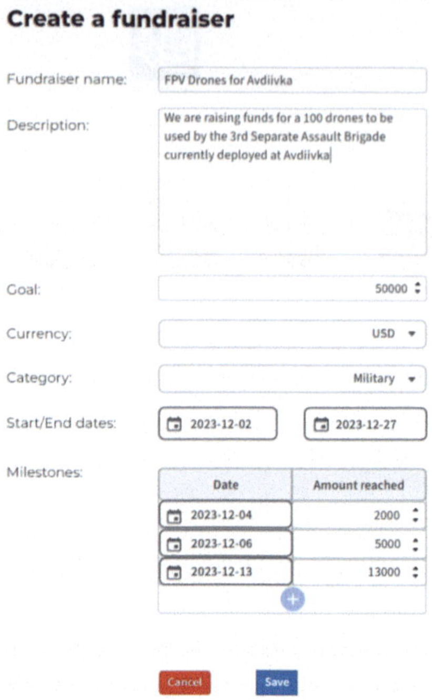

Fig. 4. Collection creation screen

Fig. 5. Screen of the list of created fundraisings

2. The selected message broker ensures the reliability of receiving and successful processing of messages. Thus, having a client application send messages to a queue in

both the primary AWS region and the secondary AWS region will ensure that the message reaches the handler and is processed successfully even in the event of technical infrastructure failures in one of the regions.
3. The chosen configuration of the queue ensures a one-time arrival of messages – without duplicates. In this way, double enrollment of user funds due to a technical failure of the infrastructure is impossible.
4. The selected queue configuration also predicts exclusively in chronological order, so that the final prediction of collection success is based solely on current data.

3 Research Results

An analysis of the classification accuracy of the models obtained by the chosen method was performed, as well as logistic regression and random forest as possible baseline methods. Model training and associated testing were performed on a dataset provided by Kickstarter, containing 375,765 rows about fundraising projects hosted on the aforementioned resource. Each row corresponds to a certain collection and contains 15 columns corresponding to its following features:

- internal collection identifier in the service database, an integer;
- collection name, string value;
- name of the category to which the collection belongs, string value;
- name of the subcategory to which the collection belongs, string value;
- the name of the target currency in which the desired amount of funds for project implementation is determined, string value;
- name of the country where the project is implemented, string value;
- collection end date, date type value in the format "YYYY-MM-DD HH:mm:ss";
- the desired amount of funds for the implementation of the project, in the target currency, an integer;
- collection start date, date type value in "YYYY-MM-DD HH:mm:ss" format;
- the current amount collected by the project in the target currency, a floating-point number;
- the current amount raised in US dollars, as calculated by the "Kickstarter" system, a floating-point number;
- the current amount collected in US dollars, as calculated by the "fixer.io" system, a floating-point number;
- the number of patrons who invested in the project, an integer;
- current state of the project, string value.

Within the scope of this research, the data set mentioned above, provided in a csv format file, was not integrated directly with the database containing information about the fees created by the user of the developed software system. Instead, this file is stored in storage provided by Amazon S3.

Before running the training code, an additional transformation of the dataset is required. Thus, some of the columns used to train the model and predict collection success will henceforth be considered to contain data of the enumerated data type, so will be represented as an integer value from the specified range.

For definition predictions, the enum data types are:

- collection category;
- target currency of collection;
- the country where collection is performed;
- the final state of the collection.

The dataset contains only string values in the specified columns. Data preparation also involves additional calculations based on the input values of the start and end dates of fundraisings. Thus, model training and predictions use the collection duration in days, calculated as the number of days from collection start to collection end for training data, or from collection start to current date in case of active collection success prediction.

Consider the code that performs model training on the example of gradient boosting. First, the data set is divided into training and test samples in the ratio of 70% to 30%. Next, an instance of the classifier class is created that implements the gradient boosting method. It trains the model on the input data from the training sample, trying to predict the value of the final state of the collection encoded as the aforementioned enumeration type.

In particular, the highest accuracy was achieved with the configuration of the classifier given in Listing 1:

- learning speed – 0.15;
- the number of trees built by the classifier is 150;
- the maximum depth of the tree is 5;
- the target function is logistic.

Listing 1. Training a model using gradient boosting

```
X = df[features]
y = df[label]
train_x,test_x,train_y,test_y                                    =
train_test_split(X,y,test_size=0.3)
model   =   XGBClassifier(learning_rate=0.15,   n_estimators=150,
max_depth=5, objective='binary:logistic', nthread=2)
model.fit(train_x,train_y)
y_pred = model.predict(test_x)
score = sklearn.metrics.accuracy_score(y_pred,test_y)
print(score)
```

The obtained accuracy results of the models using logistic regression, random forest and gradient boosting are given in the Table 5.

Thus, the results confirm the feasibility of using gradient boosting for the given task, as this software method provides the best results. At the same time, the architecture of the developed system provides for the execution of model training on a schedule several times a day, as a result of which the speed of this part of the system is currently not critical for the functioning of the system in general, but poses a possible risk for its functioning as non-functional requirements change with a simultaneous increase model training frequency. In view of this, the comparative characteristic of the speed of the learning process of the models of the above-mentioned methods was omitted, because it did not affect the choice of the method.

Table 5. The accuracy of the obtained models

Method	Accuracy, %
Logistic regression	76
Random forest	79
Gradient boosting	83

In order to compare the learning speed of machine learning models, the AWS Lambda configuration was chosen, which sets the maximum amount of RAM for the program execution environment to 512 MB. This configuration provides for the use of one virtual processor by each environment, the speed of which is similar to the physical processor of the "Intel Xeon" family, which has 6 physical cores with a maximum clock frequency of 3.3 GHz and 16 MB of L3 cache memory. This configuration was chosen empirically as the minimum required to load the training data set into RAM and perform model training.

From the obtained results, we can see that the average execution time of lambda training using the gradient boosting method is the largest (Table 6). Nevertheless, the obtained duration of 217 s satisfies the performance conditions for which the use of AWS Lambda is appropriate, and therefore, at this stage of the system implementation, we consider it satisfactory, because within the scope of this work, the accuracy of the model has a higher priority than the speed of the learning process.

Table 6. The average training time of the obtained models

Method	The average time of execution of the model learning function, seconds
Logistic regression	217
Random forest	260
Gradient boosting	279

One of the tasks of the work is to improve the speed of forecasting a large number of fundraisings at the same time. With this in mind, a number of performance tests of success prediction were performed using architectures of sequential event processing and parallel processing based on the exchange of messages between services.

For the second option, testing was performed for two different configurations of the lambda function: using pre-launched 10 function instances, thanks to which we can avoid the long time of initial deployment of the function environment; and with a configuration that involves a cold start of functions and, as a result, a longer code execution time for the first set of messages (Fig. 6).

The test results are given in Table 7.

Thus, according to the results, sequential processing shows significantly worse results with an increase in the number of messages, despite better results at a number of messages of 10. This is explained by the fact that in this case the section of code responsible for

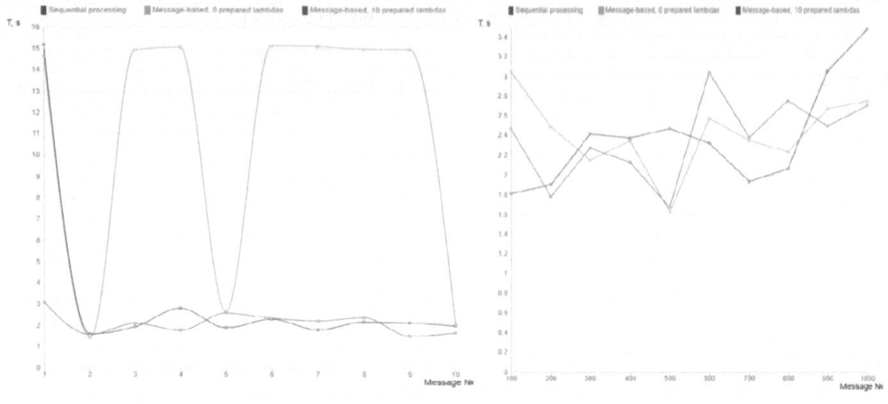

Fig. 6. Average performance indicators of configurations for 10 and 500 messages

Table 7. The total forecast time for a given set of fees.

Used architecture	Sequential processing	Parallel message processing using 10 out-of-the-box lambdas	Message parallel processing using cold-start lambdas
Processing time of 10 fees, seconds	29	2	30
Processing time of 50 fees, seconds	81	8	36
Processing time of 100 fees, seconds	156	15	49
Processing time of 500 fees, seconds	813	126	149

initializing the lambda is executed only once, on the contrary the initial initialization of all 10 lambdas, thanks to which it is possible to level the advantages of parallel processing of messages by processing them sequentially in one thread with satisfactory performance indicators. However, the performance under higher load makes it possible to argue that the use of parallel processing with the help of a message broker is more efficient. At the same time, the scatterplot of code execution time for each message indicates a significant advantage of having ready-made lambda environments, since the main part of the time spent processing the first messages in configurations that do not have them is spent precisely on the first lambda run and the loading of code dependencies. AWS keeps existing lambda environments running as long as they are actively used. So, we can assume that the finished commercial product will have a stable performance corresponding to the configuration with always running lambdas, although in some cases we should expect a lower speed of processing messages due to the need to start new environments.

During the execution of the tests, the infrastructure was configured in such a way that the maximum number of lambda instances running at the same time did not exceed 10. This is due to the fact that during development, an AWS Aurora cluster was used, the configuration of which does not allow to effectively provide a large number of connections, which was observed during the preparatory execution of tests, where database write operations became a significant factor in the execution time of functions. In commercial conditions, we can count on the availability of larger computing resources, as a result of which the speed of the system will additionally increase, compared to the indicators.

4 Conclusion

The obtained research results confirmed the appropriateness of using gradient boosting as a basic method due to the better accuracy of predictions compared to other methods. The speed of the learning process of the model based on the above-mentioned method is considered satisfactory within the scope of this work, despite the fact that it shows worse indicators than analogues. Nevertheless, we must take into account that it can, if necessary, be improved by optimizing the dependencies of the application container, as well as by using the configuration of the function, which involves greater computing capabilities. Therefore, further development and commercial deployment of the system should take into account potential actions to optimize the model's learning lambda.

Additional testing of the speed of the process proved the feasibility of using distributed computing to speed up the process of predicting the success of fundraisings for a large (10 or more requests within a second) volume of fundraisings updates in the system. So, depending on the selected configuration and the number of fees waiting to be processed, we observe a two- to ten-fold increase in performance due to parallel processing of messages, as opposed to sequential processing. Also, a less obvious improvement is that distributing the system in this way greatly increases its scalability, since the number of processors is flexible and can be adjusted depending on the expected load; while the time of sequential processing increases linearly, depending on the number of messages, and can be scaled only by allocating additional computing resources. The test results showed the presence of a flaw in the configuration, which involves the creation of new lambda instances, in that the cold start time of each of them is quite high, compared to the standard time of direct message processing. There are methods for optimizing this phase of code execution by optimizing the size of the application, so further developments of work should provide for them sufficient indicators of non-functional requirements for commercial deployment of the system.

References

1. Adnan, K., Akbar, R., Wang, K.S.: Information extraction from multifaceted unstructured big data. Int. J. Rec. Technol. Eng. (IJRTE) **8**, 1398–1404 (2019). https://doi.org/10.35940/ijrte.B1074.0882S819
2. Kaleem, S., Sohail, A., Tariqand, M.U., Asim, M.: An improved big data analytics architecture using federated learning for IoT-enabled urban intelligent transportation systems. Sustainability **15**(21), 15333 (2023). https://doi.org/10.3390/su152115333

3. Leow, K.-R., Leow, M.-C., Ong, L.-Y.: A new big data processing framework for the online roadshow. Big Data Cogn. Comput. **7**(3), 123 (2023). https://doi.org/10.3390/bdcc7030123
4. Akanbi, A., Masinde, M.: A distributed stream processing middleware framework for real-time analysis of heterogeneous data on big data platform: case of environmental monitoring. Sensors **2020**(20), 3166 (2020). https://doi.org/10.3390/s20113166
5. Mathrani, S., Lai, X.: Big data analytic framework for organizational leverage. Appl. Sci. **11**, 2340 (2021). https://doi.org/10.3390/app11052340
6. Saleh, S.H., Ismail, R., Ibrahim, Z., Hussin, N.: Issues, challenges and solutions of Big data in information management: an overview. Int. J. Acad. Res. Bus. Soc. Sci. **8**, 1382–1393 (2019). https://doi.org/10.6007/IJARBSS/v8-i12/5240
7. Anshari, M., Almunawar, M.N., Lim, S.A., Al-Mudimigh: A customer relationship management and big data enabled: personalization & customization of services. Appl. Comput. Inform. **15**, 94–101 (2019). https://doi.org/10.1016/j.aci.2018.05.004
8. Gandomi, A.H., Chen, F., Abualigah, L.: Machine learning technologies for big data analytics. Electronics **11**, 421 (2022). https://doi.org/10.3390/electronics11030421
9. Jabbar, A., Akhtar, P., Dani, S.: Real-time big data processing for instantaneous marketing decisions: a problematization approach. Ind. Mark. Manag. **90**, 558–569 (2019). https://doi.org/10.1016/j.indmarman.2019.09.001
10. Oleshchenko, L., Burchak, P.: Web application state management performance optimization methods. In: Hu, Z., Dychka, I., He, M. (eds.) Advances in Computer Science for Engineering and Education VI. ICCSEEA 2023. Lecture Notes on Data Engineering and Communications Technologies, vol. 181. Springer, Cham (2023). https://doi.org/10.1007/978-3-031-36118-0_6

Decision Framework for Optical Color Computing Architecture

Victor Timchenko[1](✉), Vladik Kreinovich[2], and Yuriy Kondratenko[3,4]

[1] Admiral Makarov National University of Shipbuilding, Mykolaiv 54025, Ukraine
vl.timchenko58@gmail.com
[2] University of Texas at El Paso, El Paso, TX 79968, USA
vladik@utep.edu
[3] Petro Mohyla Black Sea National University, Mykolaiv 54003, Ukraine
y_kondrat2002@yahoo.com
[4] Institute of Artificial Intelligence Problems of MES and NAS of Ukraine, Kyiv 01001, Ukraine

Abstract. Optical computing architecture is an effective basis for the creation of high-performance computing devices with the possibility of widespread use of parallel structures and high-speed optical logic elements, as well as reducing the number of switching components and energy consumption. The authors develop the computational architecture of logical operations based on (a) the representation of fuzzy information as a color quantum, (b) the use of optical transformations of ray radiation as natural operations of disjunction and conjunction, as well as (c) the basis for constructing more complex logical operations for deriving decisions. The concepts of simple and complex decisions, blocking and new decisions, as well as decisions based on the negation of the main output solution have been formed and clarified. Optical structural components for the logical operations of each of the considered decisions, as well as more complex combinations of logical outputs, are formed. A mathematical description of algorithmic inference procedures is given. The use of the basic laws of optical physics and strictly mathematical apparatus in the synthesis of computational structural components is the basis for the practical implementation of the proposed computational architecture.

Keywords: Decision Framework · Color Computing · Optical Architecture · Logical Inference

1 Introduction

Optical computing has taken a strong place in the development of the modern stage of computing technology. Optical architecture has historically tended to evolve in two directions - combinations of optoelectronic components and all-optical devices [1–12]. The significant advantages of optical computing are combined with the rapid and highly efficient development of optical information transmission systems, which naturally increased interest in the creation of all-optical systems for processing and transmitting information. A detailed analysis of the development of optical computing systems is given in the works [13–15], a thorough comparative analysis is carried out in the

Z. Hu et al. (Eds.): ICCSEEA 2024, LNDECT 242, pp. 103–112, 2025.
https://doi.org/10.1007/978-3-031-84228-3_9

work [16–19], and some aspects of efficiency assessment, in particular, energy costs are considered in the work [20–23]. These works show the energy efficiency (1) of optical architecture, the structural possibility of parallel computing (2), high speed of information processing (3), the robustness of computer networks to disturbances (4), etc. An analysis of research in the field of technological implementation of optical computing architecture shows: the significant complexity of (1) designing existing optical components based on physical processes such as diffraction and interference; the need for some devices to measure (2) light waves to determine their parameters and use complex lasers and moving structural elements (3), as well as several other implementation difficulties.

There is obvious interest in the use of optical computing in various applications to improve the efficiency of optical information input and optical character recognition systems, improve the visualization of output information [24–26], etc.

The general conclusion from the analysis of the current stage of development of optical computing structures is the recognition of the significant complexity of technological components due to the use of such implementation principles as, for example, the use of diffraction gratings, a complex system of prisms, polarization of light radiation, lasers with a variable angle of the light beam, etc.

The authors' works [27–31] focus on the representation of input fuzzy information as a color quantum and the use of additive and subtractive transformation of color radiation as logical operations of disjunction and conjunction, which can be implemented simply by using appropriate color filters. The works identify and describe the need to apply such concepts as the derivation of "simple" ("ordinary") and "complex" (integrated) decisions, and provide circuit decisions for their implementation, as well as for "blocking decision" (estimates) and the generation of "new decision". The most important basic logical operation is "negation". In its structure, the negation of an already made decision corresponds to the logical inference of a "new decision", since when changing a positive decision to a negative one, the decision-making system is guided by qualitatively different information. The level of the hierarchy for making such a decision and the information used must be sufficiently justified for such a change in the original decision.

The main goal of this study is to construct architectural structures and components of an optical color computing system for various combinations of logic decisions. A feature of optical color computing is the direct coupling of architectural optical components to the decision structure. The advantages of color computing in the construction of computer networks are ease of implementation, speed of calculations and the absence of switching elements in the architecture of logical solutions for the basic logical operations of disjunction and conjunction. The implementation of operations for making simple, new, blocking, and opposite decisions, as shown in the works [29, 30], will include quite simply implemented, but still switching components. For example, a new decision or negation logic operation is generated by turning on a white light femtosecond laser source S at the beginning of the decision process or negation operation.

2 Implementation of Optical Structural Components for Inference Operations of the Main Types of Decision

2.1 Optical Circuit for Inference of Ordinary and Complex Decision

The generalized decision support system has a hierarchical structure that evaluates such a characteristic as the complexity of the problem being solved. Considering in this work the implementation of components of a certain local level of a decision support system, we determine that the complexity of the task is determined by the number of fuzzy quanta of input information. These quanta are represented by the corresponding color [29]: red (*dangerous*) $\{R\}$, green (*probably safe*) $\{G\}$, blue (*safe*) $\{B\}$, yellow (*very probably dangerous*) $\{Yel\}$, magenta (*probably dangerous*) $\{M\}$, cyan (*very probably safe*) $\{C\}$, as well as white (*positive decision*) $\{W\}$ and black (*no decision*) $\{Blc\}$. In this work, mathematical tools will be used, described in detail in the works [29, 30], in particular, the matrix description and transformation of light radiation of a certain color. The inference of an "ordinary" decision, determined for a device with events of three quanta of input information (Fig. 1) and described in matrix form by equations of the additive form (disjunction), for example $\{R\}, \{G\}, \{B\} = \{W\}$.

$$diag(R, 0, 0) + diag(0, G, 0) + diag(0, 0, B) = diag(R, G, B) \tag{1}$$

The equations describing the process of subtractive transformation (conjunction) have the following form for the corresponding input quantum:

$$\begin{aligned} \{B\} diag(0, 0, B) \times diag(R, 0, B) \times diag(R, G, 0) &= diag(0, 0, 0) \\ diag(0, 0, B) \times diag(R, 0, B) \times diag(0, G, B) &= diag(0, 0, B) \end{aligned} \tag{2}$$

$$\begin{aligned} \{C\} \quad diag(0, G, B) \times diag(R, 0, B) \times diag(R, G, 0) &= diag(0, 0, 0) \\ diag(0, G, B) \times diag(R, 0, B) \times diag(0, G, B) &= diag(0, 0, B) \end{aligned} \tag{3}$$

$$\begin{aligned} \{R\} \quad diag(R, 0, 0) \times diag(R, 0, B) \times diag(R, G, 0) &= diag(R, 0, 0) \\ diag(R, 0, 0) \times diag(R, 0, B) \times diag(0, G, B) &= diag(0, 0, 0) \end{aligned} \tag{4}$$

$$\begin{aligned} \{Yel\} \quad diag(R, G, 0) \times diag(R, 0, B) \times diag(R, G, 0) &= diag(R, 0, 0) \\ diag(R, G, 0) \times diag(R, 0, B) \times diag(0, G, B) &= diag(0, 0, 0) \end{aligned} \tag{5}$$

$$\begin{aligned} \{M\} \quad diag(R, 0, B) \times diag(R, 0, B) \times diag(R, G, 0) &= diag(R, 0, 0) \\ diag(R, 0, B) \times diag(R, 0, B) \times diag(0, G, B) &= diag(0, 0, B) \end{aligned} \tag{6}$$

$$\begin{aligned} \{W\} \quad diag(R, G, B) \times diag(R, 0, B) \times diag(R, G, 0) &= diag(R, 0, 0) \\ diag(R, G, B) \times diag(R, 0, B) \times diag(0, G, B) &= diag(0, 0, B) \end{aligned} \tag{7}$$

$$\{G\} \quad diag(0, G, 0) \times diag(R, 0, B) = diag(0, 0, 0) \tag{8}$$

If we have output estimates of the form $\{B,B,B\}$ or $\{R,R.R\}$, then for a simple problem this means the final decision is "completely safe" or "completely dangerous". The optical circuit [29] is organized in such a way that only for the output signals $\{B,B,B\}$ and $\{R,R.R\}$ in the blocking circuit BC, light radiation passes through only one of two optical channels and using the CB switching unit, which includes normally closed and open contacts, the main output channel is blocked by Bl and interrupted, and the corresponding decision $\{B\}$, formula (2) or $\{C\}$, (3) or $\{R\}$, (4) or $\{Yel\}$, (5) is output as final. For simplicity, the presented optical scheme also implements approximate estimates of the form, taking into account the idempotence property $\{B,B,B\} \approx \{B,B,G\} \approx \{B,G,G\} \approx \{C\}$, $\{R,R,R\} \approx \{R,R,G\} \approx \{R,G,G\} \approx \{Yel\}$. In the case of output quanta $\{M\}$, formula (6) or decision $\{W\}$, (7), light emission passes simultaneously in two channels, CB does not work, and for output quanta $\{G\}$, formula (8), light emission is blocked at the input of the light filter $\{M\}$ of blocking circuit BC, thus, the main channel is not blocked and the inference process continues.

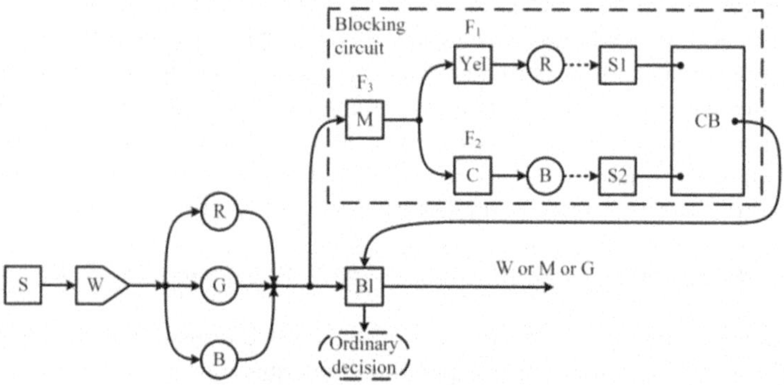

Fig. 1. Optical design of a coloroid with inference of an ordinary decision

We will accept the conclusion of a complex decision (for a complex decision-making problem) for the condition of the number of input quanta 6, 9, 12 (single-channel basic coloroid, Fig. 2), 45 (two-channel coloroid network), 144 (three-channel coloroid network) and more [29–31].

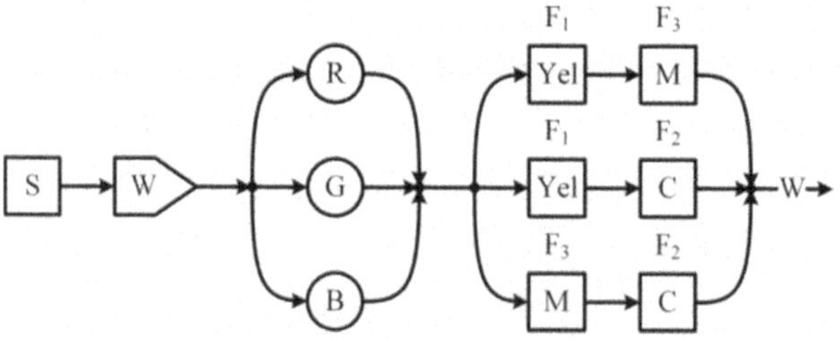

Fig. 2. Optical design of a single-channel coloroid (9 input quantum)

2.2 Optical Implementation of the Blocking Decision

Many decision-making systems in common use today use a blocking decision (BD), for example, NATO and EU decision-making requires full consensus and one country blocks the overall decision. In technical systems, the decision to block may be left to the discretion of the most responsible decision maker. The optical scheme shown in Fig. 3., has strictly set (fixed) filter values $\{Yel, C, M\}$ and implements blocking for any of the seven possible input color quanta of information.

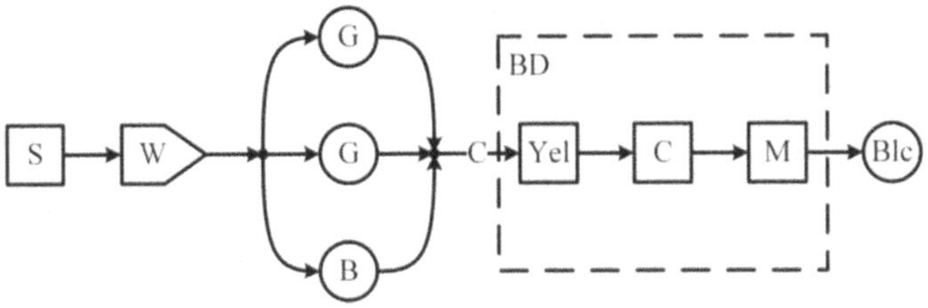

Fig. 3. Optical design of blocking coloroid (for input quantum $\{C\}$)

2.3 Generating a New Decision

The generation of a new solution arises as the output of a decision to a new task, which is implemented in the form of switching on a femtosecond laser S of white light (Fig. 4). The beam, having passed the corresponding optical structure of the coloroid with the color indicators of the filters specified by the input information, forms the corresponding decision at the output. In the decision-making process, at the output of the optical coloroid, you can get an empty information quantum, the absence of light radiation in the light guide. In this case, a new solution is generated in response to the absence of light, and the femtosecond laser white light source is turned on again. Changing the output signal is possible only by changing the values of the light filters depending on changes in the input information.

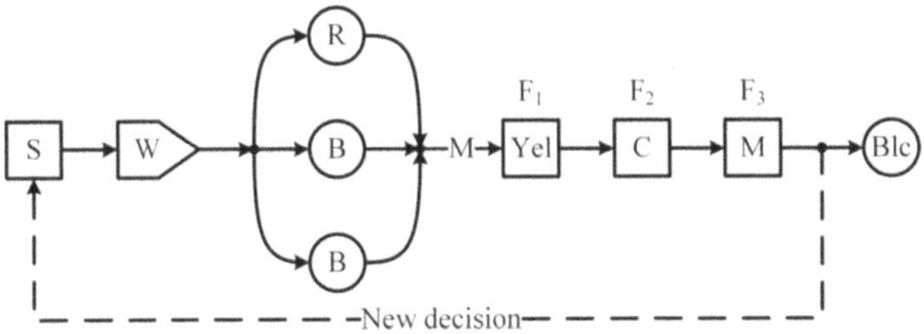

Fig. 4. Optical scheme for generating a new solution (for input quanta $\{M\}$)

2.4 Logical Operation of Negation. Conclusion of the Opposite Decision

Let's define for each color quantum the corresponding numerical weight (confidence) value for the scale [29] $[0 \div 1]$: $\{R\} \rightarrow 0$; $\{Yel\} \rightarrow 0.25$; $\{G\} \rightarrow 0.55$; $\{C\} \rightarrow 0.75$;

$(B\} \to 1.0$; $\{M\} \to 0.45$. This will allow a simple measure to rank the truth of a certain color or combinations of them concerning the complete truth 1 (the state of complete safety) in the form of an additional quantitative assessment. For example, for an additive combination of three colors $\{R\}$, $\{G\}$, $\{C\}$ we get the ratio $(0/1 + 0.55/1 + 0.75/1) \times 100\% \approx 43\%$.

Let's consider, for example, an optical scheme when the input information to the decision-making system comes in the form of six optical color quanta (Fig. 3). The output is 6 estimates (which can be taken as decisions), as well as a positive decision and no decision. A positive decision (1) $\{W\}$ is characterized as a decision made under conditions of conflicting estimates, namely the sum of light quanta of information $\{R, G, B\}$. It is quite natural to contrast this contradictory, doubtful decision, related to the complete truth of 52% of the positive evaluation (decision) $\{B\}$, with an empty information quantum $\{Blc\}$ - the absence of a decision. At the same time, for the initial situation of no solution or the output empty information quantum $\{Blc\}$, the generation of an initial assessment (decision) or a new solution (assessment) is required. This creates a logical pair for the negation operation:

$$\{Blc\} = \neg\{W\}; \quad \{W\} = \neg\{Blc\} \tag{9}$$

The next pair of opposite information quanta is obviously $\{B\}$ and $\{R\}$. Technically, you can apply the negation operation to them and obtain logical pairs of the form [30].

$$\{B\} = \neg\{R\}; \quad \{R\} = \neg\{B\}$$

However, let us consider the question of the expediency of using the negation operation for a given pair of information quanta. We can obtain the result $\{B\}$, which is related to the complete truth of 100%, as a result of an ordinary decision for the optical circuit (Fig. 1). This is the result of the sum of three positive ratings and their negation implies the negation of all three received ratings.

Such complete denial, given the possible quality of the supposed information providing the basis for these logical actions, is difficult to logically integrate into a system of making rational decisions.

We can also assume a certain probability of the concept of the negation operation for the following pairs of information quanta $\{C\} = \neg\{Yel\}$, $\{G\} = \neg\{M\}$, since already in their definition they carry the concept "probably" or "very probably".

Optical coloroids at the output can generate an evaluation signal in the form of information quanta $\{C\}$, $\{Yel\}$, $\{G\}$, $\{M\}$, which can also be made in the form of decisions with a certain degree of probability of truth. In addition, if evaluate for these solutions the ratio to the complete truth of the positive solution, then for two combinations $\{C\} = \{B,B,G\}$ and $\{B,G,G\}$ we obtain 85% and 70%, respectively. Transformations of light radiation have the property of idempotence and the output estimate will give one value $\{C\}$ with an average value of the ratio to the total truth of the positive solution of 78% and can rather be approximately related to the estimate $\{B\}$. Based on the above reasoning, it is advisable to use one pair of logical negation operations of the form (9) based on selecting the decision $\{W\}$ from the possible seven variants of output signals.

3 Features of the Implementation of a Combination of Various Logical Operations NOR, NAND

3.1 Logical NOR Operations

Let's consider the implementation of the logical operation NOR for an additive coloroid (Fig. 2). The above considerations about the advisability of using the negation operation only for the output quantum of information $\{W\}$ represent the task of separating it from three possible types of signals $\{W\}$, $\{G\}$, $\{M\}$. This problem is solved by introducing an additional optical circuit BC2 with fixed filters $\{R\}$, $\{G\}$, $\{B\}$ and the corresponding contact group CB (Fig. 5).

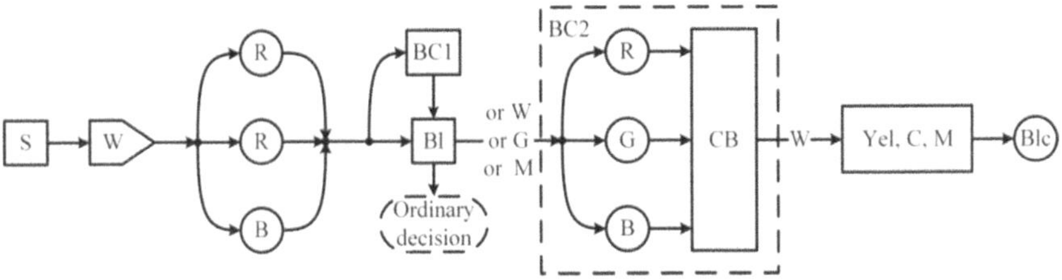

Fig. 5. Optical circuit for implementing NOR logical operations

The equations describing the process of optical conversion have the following form for the corresponding input quantum:

$$\{W\} \quad \begin{aligned} diag(R, G, B) \times diag(R, 0, 0) &= diag(R, 0, 0) \\ diag(R, G, B) \times diag(0, G, 0) &= diag(0, G, 0) \\ diag(R, G, B) \times diag(0, 0, B) &= diag(0, 0, B) \end{aligned} \tag{10}$$

$$\{G\} \quad \begin{aligned} diag(0, G, 0) \times diag(R, 0, 0) &= diag(0, 0, 0) \\ diag(0, G, 0) \times diag(0, G, 0) &= diag(0, G, 0) \\ diag(0, G, 0) \times diag(0, 0, B) &= diag(0, 0, 0) \end{aligned} \tag{11}$$

$$\{M\} \quad \begin{aligned} diag(R, 0, B) \times diag(R, 0, 0) &= diag(R, 0, 0) \\ diag(R, 0, B) \times diag(0, G, 0) &= diag(0, 0, 0) \\ diag(R, 0, B) \times diag(0, 0, B) &= diag(0, 0, B) \end{aligned} \tag{12}$$

Thus, it follows that for the input quantum $\{W\}$, formula (10) there is light radiation in three parallel branches of the coloroid fibers, for the quantum $\{G\}$, (11) only in one channel, for the quantum $\{M\}$, (12) - in two channels. It is easy to construct a contact group CB based on normally closed and open contacts in order to determine the output quantum $\{W\}$ required for execution by turning off the channel for $\{G\}$ and $\{M\}$. Next, using a coloroid of the form (Fig. 3) we obtain the output quantum of negation $\{Blc\}$.

3.2 Logical NAND Operations

The NAND logic operation is based on subtractive optical conversion, which is defined by an optical circuit, for example, shown in Fig. 6. The main elements form the selection (BC) of the output quantum $\{W\}$ and its negation into the quantum $\{Blc\}$.

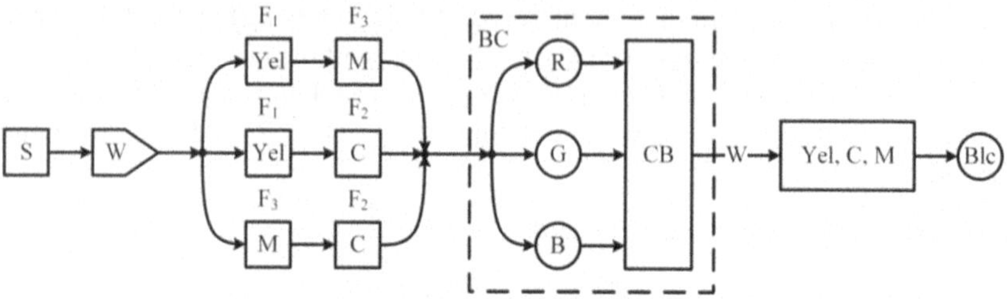

Fig. 6. Optical circuit for implementing NAND logical operations

4 Summary and Conclusion

The construction of artificial intelligence systems requires the development of conceptual foundations of decision-making processes for the construction of hierarchically organized decision output structures optimally laid down by the human mind. The most important role is undoubtedly played by the speed of information processing with the possibility of parallel calculations for fairly simply technologically implemented nano-sized components of computer networks.

The organization of the optical computing architecture is organically integrated into the structure of human decision-making: developing a new solution, blocking a decision, refusing to make a decision, hierarchical assessment of decision-making tasks by complexity, etc. It also allows to implement various types of solutions, implemented in a structurally simple and technologically advanced implementation and with the following advantages:

(1) The main logical operations of disjunction and conjunction for processing a large amount of input fuzzy information are carried out by components without switching devices. More complex logical operations are performed with as limited a number of switching devices as possible. Components.
(2) The construction of computer coloroid networks is easily structurally parallelized.
(3) The use of light emitter in the femtosecond range ensures minimal energy consumption at a speed of information processing determined by the speed of light.
(4) The optical computing architecture is easily integrated with optical information transmission.

The proposed structures of optical components of computer networks can be used for the synthesis of decision support systems with big fuzzy data, for example for ship-safe navigation [32], and other important applications.

References

1. Gur, E., Mendlovic, D., Zalevsky, Z.: Optical generation of fuzzy based rules. Appl. Opt. **41**(23), 4653–4761 (2002)
2. Jung, Y.J., Yu, S., Koo, S., et al.: Reconfigurable all-optical logic AND, NAND, OR, NOR, XOR and XNOR by photonic crystal nonlinear cavities. In: Conference on Lasers and Electro-Optics, Pacific Rim (2009). paper TuB4_3
3. Sanjeev, A., et al.: Non-contact optical wave front shaping for focusing light and high-resolution imaging inside and behind biological scattering medium. In: Proceedings of SPIE 10932, Emerging Digital Micromirror Device Based Systems and Applications XI, 1093204 (2019)
4. Ma, S., Chen, Z., Dutta, N.K.: All-optical logic gates based on two-photon absorption in semiconductor optical amplifiers. Opt. Commun. **282**(23), 4508–4512 (2009)
5. Azhigulov, D., et al.: High-speed thermally tuned electro-optical logic gates based on micro-ring resonators. Opt. Q. Electron. (2020). 52: Article 412
6. Zhu, Z., Yuan, J., Jiang, L.: Multifunctional and multichannel all-optical logic gates based on the in-plane coherent control of localized surface plasmons. Opt. Lett. **45**(23), 6362–6365 (2020)
7. Kotb, A.: Simulation of high quality factor all-optical logic gates based on quantum-dot semiconductor optical amplifier at 1 Tb/s. Optik **127**(1), 320–325 (2016)
8. Qian, C., Lin, X., Xu, J., et al.: Performing optical operators by a diffractive neural network. Light Sci. (2020). http://www.nature.com/Lsa.2020/OpenAccess
9. Jandieri, V., Khomeriki, R., et al.: Functional all-optical logic gates for true time-domain signal processing in nonlinear photonic crystal waveguides. Opt. Express **28**(12), 18317–18331 (2020)
10. Caballero, L., Povinelli, M., et al.: Photonic crystal integrated logic gates and circuits. Opt. Express **30**(2), 1976–1993 (2022)
11. Kawano, T.: Printable Optical Logic Gates with CIELAB Color Coding System for Boolean, Operation-Mediated Handling of Colors. In: Genetic and Evolutionary Computing (ICGEC), pp. 270–275. IEEE (2012)
12. Moritaka, K., Kawano, T.: Spectroscopic analysis of the model color filters used for computation of CIELAB-based optical logic gates. ICIC Express Lett. **5**(6), 1715–1720 (2014)
13. Gentili, P.L.: Establishing a new link between fuzzy logic, neuroscience and quantum mechanics through Bayesian probability. Molecules **26**, 5987 (2021)
14. Kazanskiy, N.L., Butt, M.A., Khonina, S.N.: Optical computing: status and perspectives. Nanomaterials (Basel) **12**(13), 2171 (2022)
15. Cole, C.: Optical and electrical programmable computing energy use comparison. Opt. Express **29**(9), 13153–13170 (2021)
16. McMahon, P.L.: The physics of optical computing. arXiv:2308.00088v1. [physics.optics] 31 Jul 2023
17. Miller, D.A., et al.: Environmental assessment of optical and electronic computing systems. Nat. Sustain. **6**(6), 756–764 (2023)
18. Sarychev, A.S., et al.: Energy-efficient optical logic gates based on all-dielectric metamaterials. Nat. Photonics **14**(10), 874–879 (2020)
19. Matuszewski, M., Prystupiuk, A., Opala, A.: The role of all-optical neural networks, arXiv: 2306.06632v2, [cs.ET] 13 Jun 2023. Z. Lin
20. Lim, D.W., et al.: Towards energy-harvesting optical processors. Nat. Mater. **18**(2), 159–166 (2019)
21. Dawson, M.D., et al.: Low-power optical logic with nanophotonic, waveguides. Opt. Express **26**(21), 27811–27825 (2018)

22. Lin, Z., Hong, M.: Femtosecond laser precision engineering: from micron, submicron, to nanoscale. Ultrafast Sci. (2021). Article ID: 9783514
23. Hong, J., Lambson, B., Dhuey, S., Bokor, J.: Experimental test of Landauer's principle in single-bit operations on nanomagnetic memory bits. Sci. **2**(3), e1501492 (2016)
24. Tofayet, S., et al.: Machine learning in cyberbullying detection from social-media image or screenshot with optical character recognition. Int. J. Intell. Syst. Appl. (IJISA) **15**(2), 1–13 (2023)
25. Akuma, S.: Eye gaze relevance feedback indicators for information retrieval. Int. J. Intell. Syst. Appl. (IJISA) **14**(1), 57–65 (2022)
26. Shikha, B., et al.: A genesis of a meticulous fusion based color descriptor to analyze the supremacy between machine learning and deep learning. Int. J. Intell. Syst. Appl. (IJISA) **12**(2), 21–33 (2020)
27. Timchenko, V.L., Kondratenko, Y.P., Kreinovich, V.: Interval-valued and set-valued extensions of discrete fuzzy logics, Belnap logic, and color optical computing. In: Massanet, S., Montes, S., Ruiz-Aguilera, D., González-Hidalgo, M. (eds.) Fuzzy Logic and Technology, and Aggregation Operators. EUSFLAT AGOP 2023 2023. Lecture Notes in Computer Science, vol. 14069. Springer, Cham (2023). https://doi.org/10.1007/978-3-031-39965-7_25
28. Timchenko, V., Kondratenko, Y., Kozlov, O., Kreinovich, V.: Fuzzy color computing based on optical logical architecture. In: Kahraman, C., Sari, I.U., Oztaysi, B., Cebi, S., Cevik Onar, S., Tolga, A.Ç. (eds.) Intelligent and Fuzzy Systems. INFUS 2023. Lecture Notes in Networks and Systems, vol. 758. Springer, Cham (2023). https://doi.org/10.1007/978-3-031-39774-5_55
29. Timchenko, V., et al.: The architecture of optical logical coloroid with fuzzy computing. CEUR Workshop Proc. **3373**, 638–648 (2023)
30. Timchenko, V., Kondratenko, Y., Kreinovich, V.: Implementation of optical logic gates based on color filters. Lect. Notes Data Eng. Commun. Technol. **181**, 126–136 (2023). https://doi.org/10.1007/978-3-031-36118-0_12
31. Timchenko, V., Kondratenko, Yu., Kreinovich, V.: Logical decision networks based on the optical coloroids. In: IEEE 12th IDAACS: Technology and Applications, Dortmund, Germany, pp. 1194–1199 (2023). https://doi.org/10.1109/IDAACS58523.2023.10348692
32. Timchenko, V., Kondratenko, Y., Kreinovich, V.: Decision support system for the safety of ship navigation based on optical color logic gates. CEUR Workshop Proceed. **334**, 42–52 (2022). https://ceur-ws.org/Vol-3347/Paper_4.pdf

Reducing the Financial Burden of Typhoid Fever Treatment Through Intelligent Modeling

Tosin C. Olayinka[1,4](✉), Akinola S. Olayinka[2,4], Ojei H. Onyijen[3,4], Edwin Onatuyeh[6], Wilson Nwankwo[5], and Pascal C. Nwankwo[7]

[1] Department of Software Engineering, Wellspring University, Benin City, Edo State, Nigeria
tosin.olayinka@wellspringuniversity.edu.ng
[2] Computational Research Group, Department of Physics, Edo State University, Uzairue, Edo State, Nigeria
[3] Department of Mathematical and Physical Sciences, Samuel Adegboyega University, Ogwa, Edo State, Nigeria
[4] Centre for Computational Science, Benin City, Edo State, Nigeria
[5] Department of Cyber Security, Delta State University of Science and Technology, Ozoro, Nigeria
[6] Department of Accounting, Delta State University of Science and Technology, Ozoro, Nigeria
[7] Department of Information Technology, Nigerian British University, Asa, Nigeria

Abstract. Typhoid fever remains a significant public health challenge in Africa particularly in regions with limited access to clean water and sanitation. The financial burden associated with typhoid fever treatment can exacerbate disparities in healthcare access and affordability, especially in resource-constrained settings. By leveraging data analytics and machine learning techniques, this paper seeks to develop predictive models capable of accurately assessing the risk of typhoid fever and guiding resource allocation strategies. Through the integration of diverse datasets, including clinical parameters, demographic information, and environmental factors, the proposed modeling framework aims to identify key predictors of typhoid fever incidence and severity. The dataset comprises laboratory tests of blood samples of one hundred and fifty (150) children, from age zero (0) to five (5) years. Data preprocessing and training were performed using Waikato Environment for Knowledge Analysis (WEKA) workbench tool as a machine learning platform. The information Gain (IG) feature selection technique was employed to select the best attributes using a cut-off gain of 0.2. The top six (6) attributes with more information gain relevant to the class typhoid were selected. Widal test, Malaria count, Monocyte, Platelets, HB, Eosinophils, cost of treatment are the selected attributes with their respective ranked information gain of 0.6759, 0.4936, 0.3147, 0.2843, 0.2416, 0.2127, and 0.20. We used five (5) different classification algorithms (J48_Consolidated, LMT, RepTree, MultiBoost Decision Stump, and Random Forest) and their results were compared. Performance analysis on the five (5) classifiers shows that Multiboost Decision Stump exhibited the best accuracy of 94% and least MAE of 0.0604. The findings show that the developed model when integrated into automated disease diagnostic workflows will provide an enhanced and flexible solution for quick prognosis of typhoid fever in children in the tropics thereby reducing the cost of hospitalization among others.

Z. Hu et al. (Eds.): ICCSEEA 2024, LNDECT 242, pp. 113–131, 2025.
https://doi.org/10.1007/978-3-031-84228-3_10

Keywords: Typhoid fever · financial burden · decision stump · information gain · disease detection

1 Introduction

Typhoid fever, caused by the bacterium Salmonella Typhi, remains a significant global health concern, particularly in regions with limited access to clean water and sanitation infrastructure. Beyond its health implications, typhoid fever imposes a considerable financial burden on individuals, families, and healthcare systems due to the costs associated with diagnosis, treatment, and loss of productivity. The cost of treating typhoid fever could be very high especially when it leads to intestinal perforation. According to Seyi-Olajide et al. (2020), typhoid disease could lead to catastrophic health expenditure (236-1700USD). This burden is especially pronounced in low and middle-income countries where resources are scarce, and healthcare expenditures can exacerbate existing socio-economic disparities. Addressing the financial challenges of typhoid fever treatment requires innovative approaches that optimize resource allocation, improve diagnostic accuracy, and streamline treatment protocols. Intelligent modeling presents a promising avenue for achieving these goals by leveraging advanced data analytics and machine learning techniques to forecast disease incidence, severity, and treatment outcomes. By identifying key determinants of healthcare costs and informing cost-effective strategies, intelligent modeling has the potential to alleviate the financial burden of typhoid fever treatment while improving patient outcomes and healthcare delivery. This paper introduces a novel approach aimed at reducing the financial burden of typhoid fever treatment through intelligent modeling. By integrating diverse datasets encompassing clinical, epidemiological, and socio-economic variables, our approach seeks to develop predictive models capable of accurately forecasting typhoid fever incidence, severity, and associated costs. Through a comprehensive analysis of the theoretical framework, methodology, and anticipated outcomes, this paper aims to contribute to the global efforts to combat typhoid fever and promote equitable access to affordable healthcare solutions.

Pathologically, Typhoid fever, also known as known as enteric fever, is a disease caused by the bacteria *Salmonella typhi*, a family of *Entero bacteria*, and is an acute systemic illness. This disease is a bacterial illness of the intestinal tract that afflicted humanity for a long time. These facultative anaerobic bacilli are also flagellated, motile, and non-spore-forming. Salmonella enterica serovars produce paratyphoid fever. Paratyphi A, B, and C (S. paratyphi) are diseases that have similar clinical symptoms to typhoid fever but are less lethal than S. Typhi (Vu Thieu et al., 2014). Typhoid outbreak is often found in the Southern part of Asia and sub-Saharan Africa affecting mostly children. Most febrile illnesses do not portray specific symptoms and signs, specifically typhoid fever as it does not have a specific symptom at its early stage (Mather et al., 2019). There are two (2) types of typhoid fever, namely, *S. Typhi* and *S. Paratyphi*. The two infections are very similar, although S. *Typhi* is quadruple more common than S. *Paratyphi* infections. Fever coupled with frontal throbbing headache, nausea, vomiting, anorexia, diarrhoea, constipation, intestinal blood loss, and hepatosplenomegaly is the most common signs and symptoms. Approximately seventy-five per cent (75%) of the infected patients also experience flu-like symptoms. Neurological problems include Guillain-Barre syndrome, meningitis, and picking at clothes and imaginary objects. Severe infection, respiratory failure, cardiac failure, renal failure, and intravascular coagulation are being reported as the repercussions at the deadly stage (Jameson & Fausi, 2018).

The global estimate of typhoid fever was estimated to be 26.9 million cases with 217,000 deaths recorded in the year 2010. To achieve the estimate, blood culture sensitivity was adjusted based on a traditional hypothesis of fifty percent (50%) (Buckle et al., 2012). Annually, the highest incidences of typhoid fever global estimates revealed that south-central and east-central Asia are affected, where more than 100 cases per 100,000 people are affected. However, Africa was estimated to have a medium incidence of 10 to 100 cases per 100,000 (Marks et al., 2017). Typhoid fever is a prominent acute febrile illness common in developing countries, with estimates of 10.9 million novel incidences globally and 116.8 thousand mortalities as of 2017 (Dicker et al., 2018; GBD, 2019). A recent report by the world health organisation (WHO) shows that Typhoid affects an estimated 11–20 million people globally each year, with the death rate from it ranging from 128000 to 161000 people (WHO, 2022). The most susceptible are the children and the poorest in society with Asia and sub-Saharan Africa having a larger share of global incidence. A time-series study on typhoid and paratyphoid fever in China between 2004 and 2016 was performed, the study revealed that children aged 0 - 4 remained the most vulnerable group (Gao et al., 2020).

Typhoid can be contracted through the faecal-oral channel and is characterized by clinical symptoms that are not specific such as fever, weakness, loss of appetite, joint pain, stomach pain, diarrhoea, cough, constipation, nausea, and vomiting, which are difficult to identify from other causes of fever such as malaria (Nsutebu et al., 2003). With the symptoms of clinical characteristics or the detection of typhoid fever, it can only be suggested and not definitive that a patient is affected by typhoid disease. There are presently two major standards by which typhoid is diagnosed, the blood culture isolation of S. *Typhi* and *Widal* agglutination assessment (Khanna & Khanna, 2016). In African countries for example Nigeria, the W*idal* agglutination assessment is commonly used as the diagnostic tool of typhoid fever due to its cheap cost and minimal training requirements and equipment. Nevertheless, the alarming rate of typhoid fever is still trending even to date.

In a study on the effect of seasonal changes on typhoid and paratyphoid fever diseases, it was revealed that aside Africa and the Middle East, the dynamics of the season do not play a significant role in the disease pattern. The study also revealed that diagnosis using blood culture is more effective when compared to other methods of diagnosis, although there is a paucity of data along this line of argument to substantiate their position (Saad et al., 2018).

Computer tools, which are noted for their accuracy and efficiency, assist in organizing, retrieving, and producing suitable medical data required to address challenges in diagnosis, prediction, therapeutic decisions, and decision-making procedures (Victor-Ikoh et al, 2022; Umezurike et al., 2020; Nwankwo, 2017; Umezurike, Nwankwo & Kareyo, 2017; Umezurike et al., 2017). In recent times, Artificial Intelligence (AI) systems and techniques are gaining wider applications in medical and healthcare delivery. As the demand for knowledge-based and intelligent systems rises, and healthcare practices become more intensively smart, Information Technology (IT) has uncovered numerous unique potentials in the automation of the healthcare delivery workflows. Tropical disease diagnosis may not be certain; patients may not be able to tell how exactly they feel. Physicians and Nurses are not also perfect at predictions or judgments as to what they observe in their dealings with patients. Laboratory reports may not be reliable as human errors are bound to occur. All these difficulties only make the traditional quantitative approach in healthcare delivery somewhat cumbersome and extremely probabilistic. Machine learning(ML) and AI have been successfully deployed as solutions to various problems facing humanity in area of data and network security (Olayinka et al., 2022; Makinde et al, 2021; Chinedu et al, 2021; Nwankwo & Ukhurebor, 2020), healthcare delivery (Nwankwo et al., 2023a; Alabool et al., 2020; Nwankwo, Olayinka & Ukhurebor, 2020; Olayinka et al., 2017a; Yan et al., 2020; Zoabi et al., 2021), novel material design and discovery (Han et al., 2021; Liu et al., 2017, 2021; Na et al., 2021; Olayinka et al., 2020; Pilania et al., 2013). Machine learning has been used in several areas of research like novel material design and predictions (Han et al., 2021; Olayinka et al., 2020), road safety (Ghandour et al., 2020; Nwankwo et al., 2022; Silva et al., 2020; Singh et al., 2021), food safety (Adetunji et al, 2022;Kovalenko & Chuprina, 2022; Kumar et al., 2021; Olayinka et al., 2022), and healthcare system for disease prognosis, detection and profiling (Aishwarya & Ravi Kumar, 2021; Alabi et al., 2022; Araújo et al., 2022; Kijpaisalratana et al., 2022b; Phillips et al., 2021; Rajkumar et al., 2020; Yan et al., 2020).

One of the primary goals of a ML workflow is to identify complicated patterns automatically and make smart decisions based on the data provided. In this paper, we demonstrate the use of a laboratory blood culture dataset of children aged 0 – 5 for the development of model for typhoid and paratyphoid fever diagnosis with emphasis on the information gain, an entropy method of feature selection and reduction in cost of treatment.

2 Review of Related Literature

Akinyemi et al. conducted a study directed at tracking the trend of typhoid fever in Nigeria, showing the lack of reliable and consistent incidence data in Nigeria as well as highlighting the antibiotic resistance as a main health security matter (Akinyemi et al., 2018). In their study, Bello and Falcon (2017) focused on the Rough Sets in Machine Learning. They considered RST data preparation, which included feature selection, discretization, meta-learning, and instance selection, was the first step in the review. Decision rule induction, association rule mining, and clustering are used to generate descriptive and predictive information. Olayinka and Chiemeke (2019) used a classification system in data mining to predict paediatric malaria occurrence. They investigated four different classification algorithms (J48, LMT, RepTree, and Hoefding tree) to determine which model is best at accurately predicting the prevalence of malaria in a patient (Olayinka & Chiemeke, 2019). Oguntimilehin et al. proposed a machine learning approach for typhoid fever diagnosis that requires less expertise in usage. Their model used a typhoid fever-related dataset to produce simple rules for the disease diagnosis. The model produced a classification accuracy of 95% for the training dataset, while the test dataset produced an accuracy of 96%. Oguntimilehin et al. used a machine-learning technique to diagnose typhoid fever.

Another study on early detection of sepsis, using four algorithms to wit: logistic regression, Random Forest, gradient boost, and Neural network, it was shown that ML models could outperform the conventional way of predicting sepsis (Kijpaisalratana et al., 2022a).

Alabi et al. developed a machine learning prognostic system for risk profiling of patients with oropharyngeal cancer for improved management (Alabi et al., 2022). They used voting ensemble, light GBM, XGBoost, Random Forest, and Extreme Random Tree algorithms on a dataset of 3164 patients to classify the dataset into risk groups. The voting ensemble algorithm came up with 88.3% accuracy with Human papillomavirus (HPV) status, patient age, T stage, marital status, N stage, and the treatment modality as the top most significant features for the machine learning model (Alabi et al., 2022).

Another study looked at the application of the Bayesian model to typhoid fever incidence prediction (Phillips et al., 2021). The study identified inadequate clinical diagnostic criteria, insensitive diagnostic tests, and unavailability of suitable datasets as factors affecting the calculation of typhoid incidence for specific age populations, this also affects the development of the prevention and control of typhoid fever. Dataset from the Strategic Typhoid Alliance across Africa and Asia program was used to develop a Bayesian model to adjust the dataset from blood culture. Model validation was done using a simulated dataset (Phillips et al., 2021).

3 Methodology

In this paper, we use the popular Waikato Environment for Knowledge Analysis (WEKA) workbench (Frank Et Al., 2016; Witten Et Al., 2016, 2017). WEKA is a robust workbench for data mining, data science, and machine learning techniques. WEKA was created at the University of Waikato in New Zealand, using the Java programming language. It is independent of an operating system as it has been tested on Linux, Windows, and Mac operating systems workbench (Frank Et Al., 2016; Witten Et Al., 2016, 2017). The design of WEKA is such that users can try out some existing methods on new data sets fast and in supple ways. WEKA is designed to provide a lot of help throughout the experimental data mining process. Preparing input data, displaying input data, and statistically evaluating learning algorithms are part of what can be achieved in WEKA through the Graphical User Interface (GUI) or Command Line Interface (CLI).

3.1 Dataset Preparation and Preprocessing

This research work used a neo-natal dataset collected from a hospital in South-south Nigeria for children between age ages of one (1) to five (5). It consists of one hundred and fifty (150) datasets for typhoid fever cases with ninety-one (91) typhoid fever positive cases and fifty-nine (59) typhoid fever negative cases. There are a total of eighteen (18) attributes that cut across haematological, and biochemical parameters and Widal test results. The haematological variables in the datasets are Haemoglobin (HB), Packed Cell Volume (PCV), Platelet count, and White Blood Cell (WBC) count. Intrinsic properties of the White Blood Cells which are Neutrophil, Monocyte, Lymphocyte, Basophil, and Eosinophil are also included in the dataset. The biochemical parameters in the datasets are Glucose, Protein, Albumin, and Globulin. The average age of the children represented in the dataset is 3 years old. Variation in the type of data and unit of measurement for various parameters in the dataset has much influence on the maximum and minimum values. For instance, maximum and minimum values for Platelets, WBC, and Widal_test are 477.00, 16.30, 0.05 and 228.73, 8.48, and 0.011 respectively. To perform machine learning on such a dataset with a high degree of variability, we applied the normalization preprocessing technique to the dataset. The equation for normalization, which is applied to all numerical datasets is given as;

$$X_{normalised} = \frac{x - x_{min}}{x_{max} - x_{min}} \quad (1)$$

where x is the old data placeholder, x_{min} is the minimum data in the set, and x_{max} is the maximum data in the set.

Datasets were normalised to a range between 0 to 1. The dataset for sex was converted to binary values of 0 and 1 for males and females respectively. Data normalization is very important in machine learning because it allows model training to be relatively insensitive to the degree of variability of attribute features. This in turn improves the model's convergence to better weights, and thereby enhances the model's accuracy. The dataset is split into training and test dataset using the k-fold cross-validation technique. We split our dataset into k different subsets using k-1 subsets to train our data and use the other subset for the test. Figure 1 shows the visual presentation of the dataset after preprocessing using Eq. (1). The class attribute typhoid is denoted by blue colour for positive cases while red colour represents negative cases.

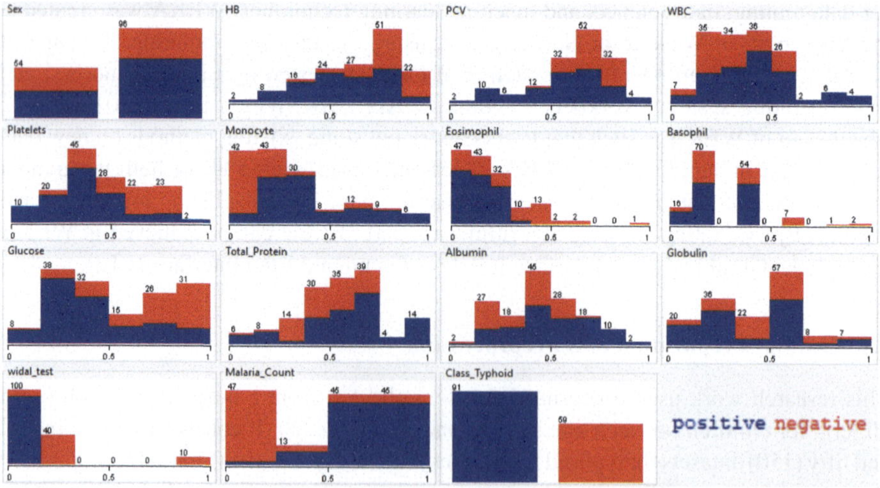

Fig. 1. Visual presentation of the dataset after normalisation based on typhoid class.

3.2 Feature Selection

Feature selection is an important part of machine learning because the relevance of selected attributes is closely correlated to the accuracy of the resulting model. In WEKA, the feature selection technique is segmented into two categories namely the attribute evaluator and the search method. In each of the categories, there are multiple techniques available. The attribute evaluator technique evaluates each attribute (column or feature) in the dataset based on the output variable which is the class or target. The search method on the other hand attempts to explore the various attribute combinations in the dataset to arrive at a short list of chosen features. Some attribute evaluator methods only work with specific search methods.

In this paper, we used an information gain (IG) attribute evaluator with the ranker search method. Information gain feature selection method have been applied severally with various improvements (Azhagusundari & Thanamani, 2013; Odhiambo Omuya et al., 2021). It has been described as one of the most effective feature selection method (Zhu et al., 2017). Decision-making in information gain (IG) is closely tied to entropy because if the entropy is more then the information will be less and vice versa. Entropy is defined as the degree of uncertainty or randomness in a data. It is a measure of impurity or disorderliness in a dataset. The entropy H(X) of an event X with n number of possible outcomes and probabilities $p_1, p_2, p_3 \ldots, p_n$ is defined as:

$$H(X) = H(p_1, p_2, p_3 \ldots, p_n) = -\sum_{i=1}^{n} p_i log_2 p_i(2) \tag{2}$$

where Σ represents the sum over the possible values of each variable.

The choice of log to base 2 is the most common based on the Shannons information theory (Shannon, 1948). The resulting entropy units are bits while for base e and base 10 we have nats and bans as their respective units. Information gain (IG) is a measure of the level of information provided by an attribute in the class. In decision tree (DT) algorithms, IG plays a dominant role in building a DT. IG calculates the entropy H(X) difference before and after the split and provides the uncertainty in the class elements.

$$IG = H(X)_{beforesplitting} - H(X)_{aftersplitting}(3) \tag{3}$$

Information Gain (IG) is an entropy-dependent feature selection technique that is gaining popularity in machine learning (Lei, 2012; Prasetiyowati et al., 2021). IG also refer to as entropy, can be calculated for each attribute in the input for the output. The IG value ranges from 0 to 1 where 0 denotes no information while 1 denotes maximum information. IG feature selection is premised on the calculated IG value for each feature. The features with higher IG values are selected because they contribute more information than the features with lower IG values.

The information gain (IG) attribute evaluator plus ranker search method ranked twelve (12) out of a total of seventeen (17) attributes with the Widal test as the most probable attribute with a rank of 0.68 and sex as the least probable with a rank of 0.01. Other attributes ranked between the highest and lowest values except for lymphocyte, globulin, age, neutrophil, and sex that were ranked zero (0). Figure 2 shows how the attributes are ranked using the information gain attribute selector and ranker search method. The line in the chart is the Pareto line showing the cumulative percentage for the selected attribute. As can be seen in the Pareto line, the Widal test and malaria count account for over 40% value of the IG for all attributes ranked. Using a cut-off value of 0.2, a total of six attributes were selected for the next stage. The attributes selected are Widal_test, Malaria_Count, Monocyte, Platelets, HB, and Eosinophil with the IG values of 0.6759, 0.4936, 0.3147, 0.2843, 0.2416, and 0.2127 respectively. Figure 3 shows the plot matrix for the selected attributes selected with the typhoid class. The plot matrix shows the correlation between the attributes and the target class typhoid.

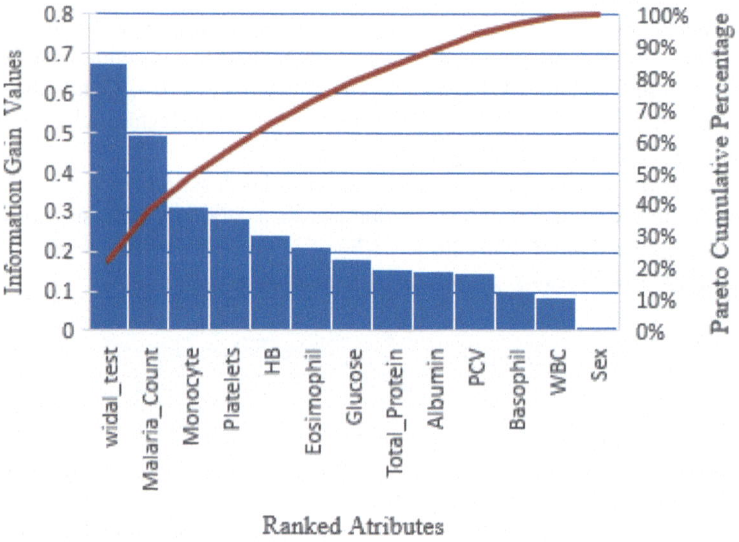

Fig. 2. Ranked attributes with their rankings based on information gain and ranker search method for ranked attributes.

3.3 Training Dataset Using WEKA Tool

In this paper, we used the k-folds cross-validation technique in supervised machine learning with five classifier algorithms: LMT (Landwehr et al., 2005), J48_Consolidated (Ibarguren et al., 2015; Pérez et al., 2007), RepTree, MultiBoost Decision Stump (Webb, 2000), and Random Forest (Breiman, 2001). The decision tree can be useful in solving various classification and regression problems. The logistic Model Tree (LMT) algorithm is a product of combining the complementary merit and demerit of tree induction and direct logistic regression. Linear or direct as the process is stabilized, logistic regression fits the data into a simple or linear model, resulting in little difference but potentially considerable bias (Landwehr et al., 2005). J48_Consolidated is an improvement to the conventional J48 classifier (Ibarguren et al., 2015; Pérez et al., 2007) while the J48 algorithm is an extended version of the Iterative Dichotomiser 3 (ID3), a greedy learning

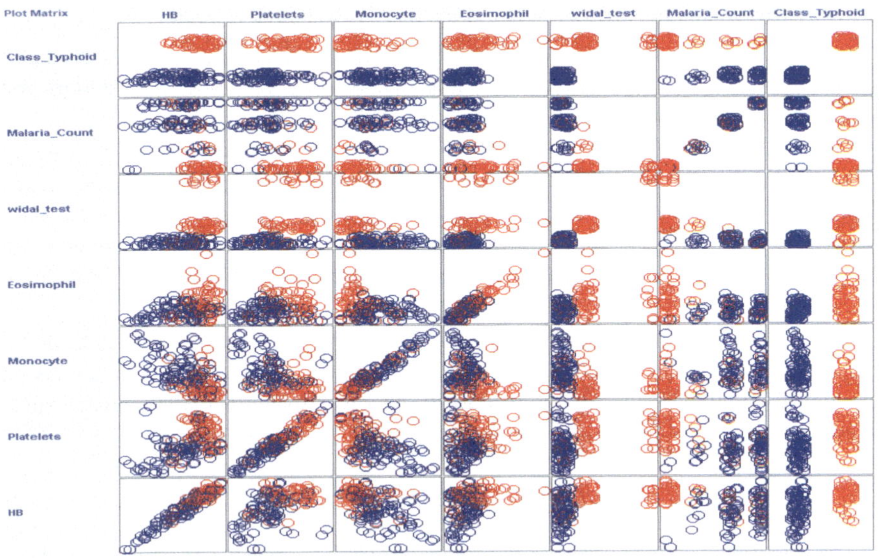

Fig. 3. Plot Matrix for normalized dataset for selected attribute based on information gain cut-off of 0.2

decision technique based on Hunts' algorithm. MultiBoost DecisionStump (MB_DS) is an ensemble version of the Decision Stump algorithm based on the binary decision 'stumps' for classifying attributes. Missing values are considered as a different attribute while regression and classification are handled respectively using mean-squared error and entropy (Witten & Frank, 2002). REPTree algorithm is built on the C4.5 algorithm and it rapid decision tree learner that can be relevant in both regression and classification. It relies on the information gain (IG) to build the decision tree. A random forest classifier is another ensemble algorithm with a combination of multiple tree predictors. Every tree in the forest is dependent on the values of a random vector that has been sampled several times and has a uniform distribution (Breiman, 2001).

4 Results and Discussion

4.1 Performance Analysis

To evaluate the best classifier for typhoid fever prediction, performance analysis was carried out on the output of the five (5) classifier algorithms (LMT (Landwehr et al., 2005), J48_Consolidated, RepTree, MultiBoost Decision Stump, and Random Forest) deployed in this work to evaluate their performances and associated errors. The confusion matrix, which is also referred to as the error matrix is the simplest performance indicator for a classifier algorithm, and it gives a visual picture of the performance of a classification algorithm. Figure 4 shows the visual explanation of each component of a confusion matrix, which are the True Positive (TP), True Negative (TN), False Positive (FP) and False Negative (FN) under the broad category Actual and Predicted values. The True Positives (TP) cell represents the positive predictions that are positive while the False

Positives (FP) stands for positive predictions that are negative in reality. Likewise, the True Negatives (TN) cell represents negative predictions that are negative while False Negatives (FN) stands for negative predictions that are positive in reality. The confusion matrix defines the information about actual and predicted classification.

The confusion matrix for all five (5) classifiers is presented in Table 1, where J48_Consolidated, RepTree, and MultiBoost Decision Stump are rated as the best classifiers based on the True Positive (TP) and True Negative (TN) prediction of 91 and 50 respectively. LMT and Random Forest presented a TP of 89 and 83, respectively with TN of 50 each. Based on the confusion matrix, the Random Forest is the least performing algorithm.

Table 2 compares the performance of all five (5) classifiers. The table presents the TP Rate, FP Rate, Precision, Recall, F-Measure, MCC, ROC Area, and PRC Area. Precision is the total amount of true positive results divided by the total amount of positive results, including those incorrectly identified, i.e. false positives. Precision is given as:

$$Precision = TP/(TP + FP) \tag{4}$$

True Positive Rate (TP Rate) is the same as Recall or Sensitivity. It is the number of true positive (TP) results divided by the total actual positive (true positive and false positive). It is given as

$$TPRate = Recall = Sensitivity = TP/(TP + FN) \tag{5}$$

The F-score, often known as the F-measure, is a statistic for determining the accuracy of a test. The precision and recall values of the test are used to compute F-score as given below:

$$FScore = FMeasure = \frac{2}{\frac{1}{Precision} + \frac{1}{recall}} = 2 * \left(\frac{precision * recall}{Precision + recall} \right) \tag{6}$$

PR Curve is the precision-recall curve and it is the curve that lies between precision and recall for various threshold levels.

Table 3 compares the classification algorithm errors for all classifiers that were used. From the table, the best algorithm is the multiboost decision stump algorithm as it has the least mean absolute error of 0.0604. Therefore, the multiboost decision stump algorithm with 94% accuracy has the highest performance and least error margin and was chosen for generating the model for typhoid fever prediction. Figure 4 shows the chart comparing the classification algorithms' performance using TP Rate, FP Rate, Precision, Recall, F-Measure, MCC, ROC Area, and PRC Area. The precision value for the top three algorithms is 0.945 which featured multiboost decision stump as one of the classifiers while the least performing has a precision value of 0.886. Figure 5 shows the percentage correctly classified, Relative Absolute error (RAE) and Root Relative Square error (RRSE) plots for various classifiers. From the figure, it can be seen that the multiboost decision stump has the highest percentage of correctly classified as well as the lowest percentage of Relative Absolute error (RAE). Figure 6 shows the Mean Absolute error (MAE) and Root Mean Square Error (RMSE) plots for various classifiers depicting the multiboost decision stump as the classifier with the mean absolute error.

From Figs. 8 and 9, the multiboost decision stump has the least error, followed by the RepTree and J48 consolidated respectively. The highest error was observed in the LMT classifier with an RAE of 0.1508. Though the J48_Consolidated, RepTree, and MultiBoost Decision Stump have the highest percentage (%) of correctly classified instances, however, MultiBoost Decision Stump has the least error compared to the other two. The random tree has the least percentage of correctly classified instances among all the five classifiers.

Table 1. Confusion Matrices for Various Decision Tree Classifier Algorithms

Confusion Matrix for LMT		
A	B	Classified as
89	2	I a = positive
9	50	I b = negative
Confusion Matrix for J48 Consolidated		
A	B	Classified as
91	0	I a = positive
9	50	I b = negative
Confusion Matrix for REPTree		
A	B	Classified as
91	0	I a = positive
9	50	I b = negative
Confusion Matrix for MultiBoost Decision Stump		
A	B	Classified as
91	0	I a = positive
9	50	I b = negative
Confusion Matrix for Random Forest		
A	B	Classified as
83	8	I a = positive
9	50	I b = negative

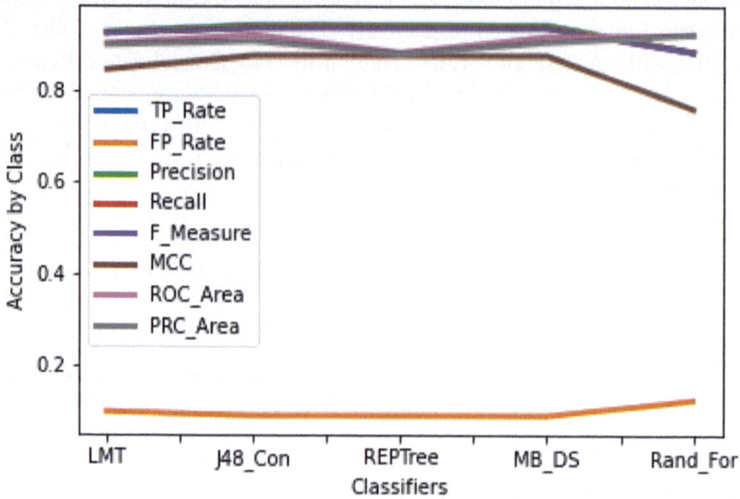

Fig. 4. Comparison of Classification Algorithms Performance

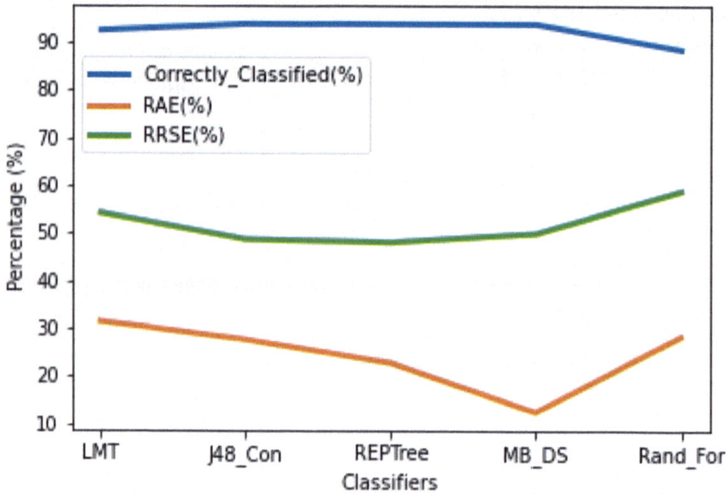

Fig. 5. Percentage Correctly classified, Relative Absolute error (RAE) and Root Relative Square error (RRSE) plot for various classifiers

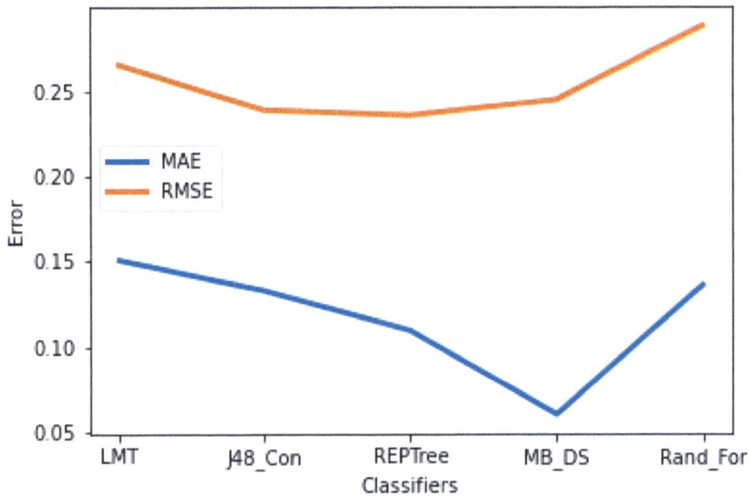

Fig. 6. Mean Absolute error (MAE) and Root Mean Square Error (RMSE) plot for various classifiers

Table 2. Comparison of Classifiers' Performance

CLASSIFIER		TP Rate	FP Rate	Precision	Recall	F-Measure	MCC	ROC Area	PRC Area
J48 Consolidated	Positive	1.000	0.153	0.910	1.000	0.953	0.878	0.885	0.863
	Negative	0.847	0.000	1.000	0.847	0.917	0.878	0.885	0.91
	Weighted Avg	0.940	0.093	0.945	0.940	0.939	0.878	0.885	0.882
LMT	Positive	0.978	0.153	0.908	0.978	0.942	0.847	0.904	0.883
	Negative	0.847	0.022	0.962	0.847	0.901	0.847	0.904	0.927
	Weighted Avg	0.927	0.101	0.929	0.927	0.926	0.847	0.904	0.900
REPTree	Positive	1.000	0.153	0.910	1.000	0.953	0.878	0.885	0.864
	Negative	0.847	0.000	1.000	0.847	0.917	0.878	0.885	0.910
	Weighted Avg	0.940	0.093	0.945	0.940	0.939	0.878	0.885	0.882
Multiboost Decision Stump	Positive	1.000	0.153	0.910	1.000	0.953	0.878	0.920	0.890
	Negative	0.847	0.000	1.000	0.847	0.917	0.878	0.920	0.934
	Weighted Avg	0.940	0.093	0.945	0.940	0.939	0.878	0.920	0.907
Random Forest	Positive	0.912	0.153	0.902	0.912	0.907	0.762	0.924	0.918
	Negative	0.847	0.088	0.862	0.847	0.855	0.762	0.924	0.934
	Weighted Avg	0.887	0.127	0.886	0.887	0.886	0.762	0.924	0.924

Table 3. Comparison of Classifier Algorithms Errors

	LMT	J48 Consolidated	REPTree	Multiboost Decision Stump	Random Forest
Correctly Classified Instances (%)	92.67	94.00	94.00	94.00	88.67
Mean Absolute Error	0.1508	0.133	0.1097	0.0604	0.1361
Root Mean Squared Error	0.2654	0.2388	0.2356	0.2449	0.2882
Relative Absolute Error (%)	31.58	27.84	22.97	12.64	28.49
Root Relative Squared Error (%)	54.31	48.87	48.23	50.13	58.99

5 Conclusion

In this paper we have demonstrated the development of an optimized model for predicting typhoid fever disease in children from a basic medical laboratory test dataset using the Multiboost Decision Stump algorithm. This is in anticipation of reducing the diagnosis time, and costs by using machine learning techniques since early diagnosis of this deadly disease is crucial to its effective management. In developing the model, five algorithms were used to train the dataset, and performance analysis of their results favours the Multiboost Decision Stump classifier algorithm as the most likely to give the best prediction. The WEKA workbench tool was used as the machine learning platform for data preprocessing, where the information gain (IG) feature selection method is used. We used the k-folds cross-validation technique for the J48_Consolidated, LMT, RepTree, Multiboost Decision Stump, and Random Forest classifier algorithms. Performance analysis of the five (5) classifiers algorithms shows the Multiboost Decision Stump as the best classifier in terms of its accuracy of 94% and least error of 0.0604. An optimum model for predicting typhoid fever was developed using the Multiboost Decision Stump. The model is capable of correctly predicting typhoid fever cases and it is considered a reliable tool for early detection of typhoid fever occurrence and management in children to reduce the mortality rate associated with typhoid fever. The potential of intelligent modeling to alleviate the financial burden of typhoid fever treatment holds promise for improving patient outcomes, enhancing healthcare affordability, and promoting equitable access to care. By fostering collaboration between healthcare professionals, policymakers, and researchers, we can work towards implementing sustainable solutions to combat typhoid fever and reduce its economic impact on affected populations.

References

Adetunji, C.O., Osikemekha, A.A., Olaniyan, T.O., Daniel, I.H., Nwankwo, W., Olayinka, A.S.: Toward the design of an intelligent system for enhancing salt water shrimp production using fuzzy logic. In: Abraham, A., Dash, S., Rodrigues, J.J.P.C., Acharya, B., Pani, S.K. (Eds.) Intelligent Data-Centric Systems: AI, Edge and IoT-based Smart Agriculture, pp. 533–541. Academic Press (2022). https://doi.org/10.1016/B978-0-12-823694-9.00005-0

Aishwarya, T., Ravi Kumar, V.: Machine learning and deep learning approaches to analyze and detect COVID-19: a review. SN Comput. Sci. **2**(3) (2021). https://doi.org/10.1007/S42979-021-00605-9

Akinyemi, K.O., et al.: Typhoid fever: tracking the trend in Nigeria. Am. J. Trop. Med. Hyg. (2018). https://doi.org/10.4269/ajtmh.18-0045

Alabi, R.O., Almangush, A., Elmusrati, M., Leivo, I., Mäkitie, A.A.: An interpretable machine learning prognostic system for risk stratification in oropharyngeal cancer. Int. J. Med. Inform. **168**, 104896 (2022). https://doi.org/10.1016/J.IJMEDINF.2022.104896

Alabool, H., et al.: Artificial intelligence techniques for containment COVID-19 pandemic: a systematic review (2020). https://doi.org/10.21203/rs.3.rs-30432/v1

Araújo, D.C., Veloso, A.A., Borges, K.B.G., das Graças Carvalho, M.: Prognosing the risk of COVID-19 death through a machine learning-based routine blood panel: a retrospective study in Brazil. Int. J. Med. Inform. **165**, 104835 (2022). https://doi.org/10.1016/J.IJMEDINF.2022.104835

Azhagusundari, B., Thanamani, A.S.: Feature selection based on information gain. Int. J. Innov. Technol. Explor. Eng. (IJITEE) **2**(2), 18–21 (2013)

Breiman, L.: Random forests. Mach. Learn. **45**(1) (2001). https://doi.org/10.1023/A:1010933404324

Brownlee, J.: Supervised and unsupervised machine learning algorithms. Understand Mach. Learn. Algo. (2016)

Buckle, G.C., Walker, C.L.F., Black, R.E.: Typhoid fever and paratyphoid fever: systematic review to estimate global morbidity and mortality for 2010. J. Global Health **2**(1) (2012). https://doi.org/10.7189/jogh.02.010401

Chinedu, P.U., Nwankwo, W., Masajuwa, F.U., Imoisi, S.: Cybercrime detection and prevention efforts in the last decade: an overview of the possibilities of machine learning models. Rev. Int. Geograph. Educ. (RIGEO) **11**(7), 956–974 (2021). https://doi.org/10.48047/rigeo.11.07.92

Cios, K.J., Pedrycz, W., Swiniarski, R.W., Kurgan, L.A.: Data mining: a knowledge discovery approach. In: Data Mining: A Knowledge Discovery Approach (2007). https://doi.org/10.1007/978-0-387-36795-8

Comfort Olayinka, T., Christian Ugwu, C., Joseph Okhuoya, O., Olusọla Adetunmbi, A., Solomon Popoola, O.: Combating network intrusions using machine learning techniques with multilevel feature selection method. In: 2022 IEEE Nigeria 4th International Conference on Disruptive Technologies for Sustainable Development (NIGERCON), pp. 1–5 (2022). https://doi.org/10.1109/NIGERCON54645.2022.9803098

Daldrup-Link, H.: Artificial intelligence applications for pediatric oncology imaging. Pedia. Radiol. **49**(11) (2019). https://doi.org/10.1007/s00247-019-04360-1

Dicker, D., et al.: Global, regional, and national age-sex-specific mortality and life expectancy, 1950–2017: a systematic analysis for the Global Burden of Disease Study 2017. Lancet **392**(10159) (2018). https://doi.org/10.1016/S0140-6736(18)31891-9

Frank, E., Hall, M.A., Witten, I.H.: The WEKA workbench. online appendix for "data mining: practical machine learning tools and techniques. In: Data Mining: Practical Machine Learning Tools and Techniques (2016)

Gao, J., Li, J., Wang, M.: Time series analysis of cumulative incidences of typhoid and paratyphoid fevers in China using both Grey and SARIMA models. PLoS ONE 15, 10 October 2020. https://doi.org/10.1371/journal.pone.0241217

GBD: Erratum: Global, regional, and national incidence, prevalence, and years lived with disability for 354 diseases and injuries for 195 countries and territories, 1990–2017: a systematic analysis for the Global Burden of Disease Study 2017 (The Lancet (2018). Lancet **393**(10190) (2019). https://doi.org/10.1016/S0140-6736(19)31047-5

Ghandour, A.J., Hammoud, H., Al-Hajj, S.: Analyzing factors associated with fatal road crashes: a machine learning approach. Int. J. Environ. Res. Publ. Health 17(11) (2020). https://doi.org/10.3390/ijerph17114111

Gupta, M.K., Chandra, P.: A comprehensive survey of data mining. Int. J. Inform. Technol. (Singapore), **12**(4) (2020). https://doi.org/10.1007/s41870-020-00427-7

Han, G., Sun, Y., Feng, Y., Lin, G., Lu, N.: Machine learning regression guided thermoelectric materials discovery – a review. ES Mater. Manufact. 20–35 (2021). https://doi.org/10.30919/esmm5f451

Ibarguren, I., Pérez, J.M., Muguerza, J., Gurrutxaga, I., Arbelaitz, O.: Coverage-based resampling: building robust consolidated decision trees. Knowl.-Based Syst. **79** (2015). https://doi.org/10.1016/j.knosys.2014.12.023

Jameson, J.L., Fausi, A.S.: Harrison's Principles of Internal Medicine, 20e. McGraw-Hill (2018)

Kao, Y.F., Venkatachalam, R.: Human and machine learning. Comput. Econ. **57**(3) (2021). https://doi.org/10.1007/s10614-018-9803-z

Khanna, A., Khanna, M.: Assessment of widal tube agglutination test and its baseline titres in Amritsar an endemic area. Int. J. Curr. Microbiol. Appl. Sci. **5**(10) (2016). https://doi.org/10.20546/ijcmas.2016.510.067

Kijpaisalratana, N., Sanglertsinlapachai, D., Techaratsami, S., Musikatavorn, K., Saoraya, J.: Machine learning algorithms for early sepsis detection in the emergency department: a retrospective study. Int. J. Med. Inform. **160**, 104689 (2022). https://doi.org/10.1016/J.IJMEDINF.2022.104689

Kovalenko, O., Chuprina, R.: Machine Learning and AI in Food Industry Solutions and Potential (2022)

Kumar, I., Rawat, J., Mohd, N., Husain, S.: Opportunities of artificial intelligence and machine learning in the food industry. J. Food Q. (2021). https://doi.org/10.1155/2021/4535567

Landwehr, N., Hall, M., Frank, E.: Logistic model trees. Mach. Learn. **59**(1–2) (2005). https://doi.org/10.1007/s10994-005-0466-3

Lei, S.: A feature selection method based on information gain and genetic algorithm. In: Proceedings - 2012 International Conference on Computer Science and Electronics Engineering, ICCSEE 2012, vol. 2 (2012). https://doi.org/10.1109/ICCSEE.2012.97

Liu, Y., Esan, O.C., Pan, Z., An, L.: Machine learning for advanced energy materials. Ener. AI **3**, 100049 (2021). https://doi.org/10.1016/J.EGYAI.2021.100049

Liu, Y., Zhao, T., Ju, W., Shi, S., Shi, S., Shi, S.: Materials discovery and design using machine learning. J. Mater. **3**(3), 159–177 (2017). https://doi.org/10.1016/J.JMAT.2017.08.002

Makinde, A.S., Agbeyangi, A.O., Nwankwo, W.: Predicting mobile portability across telecommunication networks using the integrated-KLR. Int. J. Intell. Inform. Technol. (IJIIT) **17**(3), 50–62 (2021). https://doi.org/10.4018/IJIIT.2021070104

Mao, X., Yang, H., Huang, S., Liu, Y., Li, R.: Extractive summarization using supervised and unsupervised learning. Expert Syst. Appl. **133** (2019). https://doi.org/10.1016/j.eswa.2019.05.011

Marks, F., et al.: Incidence of invasive salmonella disease in sub-Saharan Africa: a multicentre population-based surveillance study. Lancet Global Health **5**(3) (2017). https://doi.org/10.1016/S2214-109X(17)30022-0

Mather, R.G., Hopkins, H., Parry, C.M., Dittrich, S.: Redefining typhoid diagnosis: what would an improved test need to look like? BMJ Global Health **4**(5) (2019). https://doi.org/10.1136/bmjgh-2019-001831

Na, G.S., Jang, S., Chang, H.: Predicting thermoelectric properties from chemical formula with explicitly identifying dopant effects. NPJ Comput. Mater. **7**(1) (2021). https://doi.org/10.1038/s41524-021-00564-y

Nsutebu, E.F., Martins, P., Adiogo, D.: Short communication: prevalence of typhoid fever in febrile patients with symptoms clinically compatible with typhoid fever in Cameroon. Trop. Med. Int. Health **8**(6) (2003). https://doi.org/10.1046/j.1365-3156.2003.01012.x

Nwankwo, W., Adetunji, C.O., Olayinka, A.S.: IoT-driven Bayesian learning: a case study of reducing road accidents of commercial vehicles on highways. In: Pal, S., De, D., Buyya, R. (eds.) Artificial Intelligence-Based Internet of Things Systems. Internet of Things. Springer, Cham (2022). https://doi.org/10.1007/978-3-030-87059-1_15

Nwankwo, W., Ukhurebor, K.E.: Web forum and social media: a model for automatic removal of fake media using multilayered neural networks. Int. J. Sci. Technol. Res. **9**(1), 4371–3377 (2020)

Nwankwo, W.: Harnessing E-healthcare technologies for equitable healthcare delivery in Nigeria: the way forward. Int. J. Sci. Res. **6**(3) (2017)

Nwankwo, W., et al.: Application of support vector machine to Lassa fever diagnosis. In: Hu, Z., Zhang, Q., He, M. (eds.) Advances in Artificial Systems for Logistics Engineering III. ICAILE 2023. Lecture Notes on Data Engineering and Communications Technologies, vol. 180. Springer, Cham (2023). https://doi.org/10.1007/978-3-031-36115-9_16

Nwankwo, W., Olayinka, A.S., Ukhurebor, K.E.: Nanoinformatics: why design of projects on nanomedicine development and clinical applications may fail? In: 2020 International Conference in Mathematics, Computer Engineering and Computer Science (ICMCECS), pp. 1–7. IEEE Explore (2020). https://doi.org/10.1109/ICMCECS47690.2020.246992

Odhiambo Omuya, E., Onyango Okeyo, G., Waema Kimwele, M.: Feature selection for classification using principal component analysis and information gain. Expert Syst. Appl. **174** (2021). https://doi.org/10.1016/j.eswa.2021.114765

Oguntimilehin, A., Adetunmbi, A O., Olatunji, K.A.: A machine learning based clinical decision support system for diagnosis and treatment of typhoid fever. Int. J. Adv. Res. Comput. Sci. Softw. Eng. **4**(6) (2014)

Olayinka, A.S., Adetunji, C.O., Nwankwo, W., Olugbemi, O.T., Olayinka, T.C.: A study on the application of Bayesian learning and decision trees IoT-enabled system in postharvest storage. In: Pal, S., De, D., Buyya, R. (eds.) Artificial Intelligence-based Internet of Things Systems. Internet of Things. Springer, Cham (2022). https://doi.org/10.1007/978-3-030-87059-1_18

Olayinka, A.S., Nwankwo, W., Olayinka, T.C.: Model based machine learning approach to predict thermoelectric figure of merit. Arch. Sci. Technol. **1**, 55–67 (2020). 10.xxx

Olayinka, T.C., Chiemeke, S.C.: Predicting paediatric malaria occurrence using classification algorithm in data mining. J. Adv. Mathe. Comput. Sci. 1–10 (2019). https://doi.org/10.9734/jamcs/2019/v31i430118

Olayinka, T.C., Fadare, O.G., Oloyede, I.A.: Application of machine learning algorithm to predict paediatric Malaria. JABU J. Sci. Technol. **3**(2), 23–31 (2017)

Pedrero, V., Reynaldos-Grandón, K., Ureta-Achurra, J., Cortez-Pinto, E.: Overview of machine learning and its application in the management of emergency services. Revista Medica de Chile **149**(2) (2021). https://doi.org/10.4067/s0034-98872021000200248

Pérez, J.M., Muguerza, J., Arbelaitz, O., Gurrutxaga, I., Martín, J.I.: Combining multiple class distribution modified subsamples in a single tree. Patt. Recogn. Lett. **28**(4) (2007). https://doi.org/10.1016/j.patrec.2006.08.013

Phillips, M.T., et al.: A Bayesian approach for estimating typhoid fever incidence from large-scale facility-based passive surveillance data. Stat. Med. **40**, 5853–5870 (2021). https://doi.org/10.1002/sim.9159

Pilania, G., Wang, C., Jiang, X., Rajasekaran, S., Ramprasad, R.: Accelerating materials property predictions using machine learning. Sci. Rep. (2013). https://doi.org/10.1038/srep02810

Prasetiyowati, M.I., Maulidevi, N.U., Surendro, K.: Determining threshold value on information gain feature selection to increase speed and prediction accuracy of random forest. J. Big Data **8**(1) (2021). https://doi.org/10.1186/s40537-021-00472-4

Rafegas, I., Vanrell, M., Alexandre, L.A., Arias, G.: Understanding trained CNNs by indexing neuron selectivity. Patt. Recogn. Lett. **136** (2021). https://doi.org/10.1016/j.patrec.2019.10.013

Rajkumar, A.R., Prabhakar, S., Priyadharsini, A.M.: Prediction of road accident severity using machine learning algorithm. Int. J. Adv. Sci. Technol. **29**(6) (2020)

Rucco, M., Giannini, F., Lupinetti, K., Monti, M.: A methodology for part classification with supervised machine learning. In: Artificial Intelligence for Engineering Design, Analysis and Manufacturing: AIEDAM, vol. 33, no. 1 (2019). https://doi.org/10.1017/S0890060418000197

Saad, N.J., Lynch, V.D., Antillón, M., Yang, C., Crump, J.A., Pitzer, V.E.: Seasonal dynamics of typhoid and paratyphoid fever. Sci. Rep. **8**(1) (2018). https://doi.org/10.1038/s41598-018-25234-w

Seyi-Olajide, J.O., Anderson, J., Enivwaene, A.O., Ibrahim, S.H., Farmer, D., Ameh, E.A.: Catastrophic healthcare expenditure from typhoid perforation in children in Nigeria. Surg. Infect. **21**(7), 586–591 (2020). https://doi.org/10.1089/sur.2020.134

Shannon, C.E.: A mathematical theory of communication. Bell Syst. Tech. J. **27**(3). https://doi.org/10.1002/j.1538-7305.1948.tb01338.x

Silva, P.B., Andrade, M., Ferreira, S.: Machine learning applied to road safety modeling: a systematic literature review. J. Traff. Transp. Eng. (English Edition) **7**(6) (2020). https://doi.org/10.1016/j.jtte.2020.07.004

Singh, R., et al.: Highway 4.0: digitalization of highways for vulnerable road safety development with intelligent IoT sensors and machine learning. Safety Sci. **143** (2021). https://doi.org/10.1016/j.ssci.2021.105407

Tomar, D., Agarwal, S.: A survey on data mining approaches for healthcare. Int. J. Bio-Sci. Bio-Technol. **5**(5) (2013). https://doi.org/10.14257/ijbsbt.2013.5.5.25

Umezurike, C., Nwankwo, W., Kareyo, M.: Implementation challenges of health management information systems in Uganda: a review. J. Multidisciplinary Eng. Sci. Technol. **4**(7) (2017)

Umezurike, C., Nwankwo, W., Okolie, S.O., Adebayo, A.: Developing an informatics model for effective healthcare in military health facilities in Nigeria. World J. Eng. Res. Technol. **3**(4) (2017)

Umezuruike, C., Nwankwo, W., Tibenderana, P., Assimwe, J.P., Ronald, M.: Corona Virus Disease (COVID 19): analysis and design of an alert and real-time tracking system. Int. J. Emerg. Trends Eng. Res. **8**(5), 1743–1748 (2020)

Victor-Ikoh, M.I., Moko, A., Nwankwo, W.: Towards the implementation of a versatile mobile health solutions for the management of immunization against infectious diseases in Nigeria. In: Salvendy, G., Wei, J. (eds.) Design, Operation and Evaluation of Mobile Communications. HCII 2022. Lecture Notes in Computer Science, vol. 13337. Springer, Cham (2022). https://doi.org/10.1007/978-3-031-05014-5_7

Vu Thieu, N.T., Klemm, E., Dougan, G., Dongol, S., Basnyat, B., Baker, S.: Novel immunological signals for diagnosing acute typhoid fever. Am. J. Trop. Med. Hyg. 1) (2014)

Webb, G.I.: MultiBoosting: a technique for combining boosting and wagging. Mach. Learn. **40**(2) (2000). https://doi.org/10.1023/A:1007659514849

WHO: Typhoid (2022). https://www.who.int/health-topics/typhoid#tab=tab_1

Witten, I.H., Frank, E.: Data mining: practical machine learning tools and techniques with Java implementations. ACM SIGMOD Rec. **31**(1), 76–77 (2002)

Witten, I.H., Frank, E., Hall, M.A., Pal, C.J.: Data mining: practical machine learning tools and techniques. In: Data Mining: Practical Machine Learning Tools and Techniques (2016). https://doi.org/10.1016/c2009-0-19715-5

Witten, I.H., Frank, E., Hall, M.A., Pal, C.J.: Appendix B - The WEKA workbench. In: Data Mining (Fourth Edition) (2017)

Yan, L., et al.: A machine learning-based model for survival prediction in patients with severe COVID-19 infection. medRxiv (2020). https://doi.org/10.1101/2020.02.27.20028027

Zhu, L., Wang, G., Zou, X.: Improved information gain feature selection method for Chinese text classification based on word embedding. ACM Int. Conf. Proc. Ser. **10**(1145/3056662), 3056671 (2017)

Zoabi, Y., Deri-Rozov, S., Shomron, N.: Machine learning-based prediction of COVID-19 diagnosis based on symptoms. NPJ Dig. Med. **4**(1), 1–5 (2021). https://doi.org/10.1038/s41746-020-00372-6

Organization Method of Computing Processes in Multiprocessor Computing Systems

Serhii Zybin[1(✉)], Volodymyr Khoroshko[1], and Vasyl Kuzavkov[2]

[1] National Aviation University, Kyiv 03058, Ukraine
{zysv,professor_va}@ukr.com

[2] Kruty Heroes Military Institute of Telecommunications and Information Technology, Kyiv 01011, Ukraine
nevse@ukr.com

Abstract. The increase in productivity and reliability of computer systems took place in two ways. The first way is the improvement of hardware. The second way is the implementation of new architectures and principles of arrangement of computing processes. The second way is the most promising and relevant in terms of improving the productivity and reliability of computer systems. The organization of the computing process in a multiprocessor computing system is the perfect if all system resources, both hardware and software, are used efficiently, and the system's performance is as high as possible. However, such an organization requires many efforts to implement. In order to reduce these efforts, some simplifications are sometimes made. It is necessary to distinguish between two modes of operation depending on the performance of the multiprocessor computing system and its dependence on the number of computing modules that are included in the system. Processors, single-chip micro-computers can be used as modules. Currently, the development of multiprocessor computing systems is one of the important and relevant directions of improving the productivity of computing devices.

Keywords: Programmable structure · Inter-module · Multiprocessor

1 Introduction

All the variety of ways to build computer systems can be reduced to several methods of functional arrangement, which practically cover all existing structures of computer systems.

The development and creation of multicomputer and multiprocessor systems is an example of the emergence of new architectural solutions and principles of computing organization. Their construction is based on three principles: parallel execution of operations, variability of the structure, and structural homogeneity.

At first glance, it may seem that by creating a multiprocessor computing system (MPCS), you can solve all the issues related to increasing productivity. However, there are two main problems that significantly complicate the solution of this problem: arrangement of connections between functional blocks of the system, and arrangement of the computing process in the system.

Z. Hu et al. (Eds.): ICCSEEA 2024, LNDECT 242, pp. 132–141, 2025.
https://doi.org/10.1007/978-3-031-84228-3_11

The presence of these two problems during the creation of MPCSs requires looking for options for their construction and organization of processes, which, with some limitations, would ensure, on the one hand, the simplification of the system, and on the other hand, would not reduce efficiency much. This has led to several structural developments (constructions) of MPCS and several ways of organizing computing processes [1, 2]. The key to the classification of MPCSs is the system of interrelationships, organization and functioning of the system.

The purpose of this article is the development and description of the method of organizing inter-module data exchange, that is based on the parametric presented subsystems, as well as the evaluation of the effectiveness of the developed MPCSs.

2 Development of the Method of Organizing Inter-module Data Exchange

Subsystems are used for parallel solution of tasks in MPCS with a programmable structure [3, 4]. The subsystem θ consists of the number n, where $n = |\theta|$, of computing modules (CM) required to solve the problem.

These modules are connected by channels so that it is possible to transmit data from any CM_i, $i \in (1, \ldots, n-1)$ to all other CM s of the subsystem. In Fig. 1 shows an example of some subsystem that consists of six CM s in the case of (N,4,4) graph of inter-computer system connections [5].

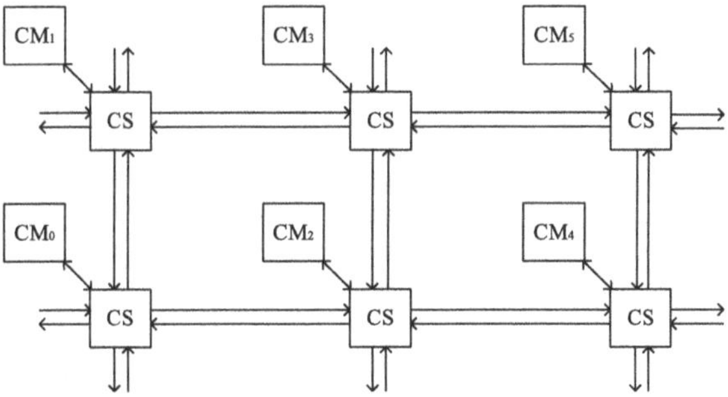

Fig. 1. The example of a subsystem consisting of six MOs. CM – computing module; CS – connection switch

Let the input and output streams of CM be labeled $p_i, p \in \{1, \ldots, v\}$, where v is the degree of vertices of the graph of inter-module connections. It should be taken into account that CM s communicate with each other using a connection switch, which is implemented as a multi-channel interface [1]. Moreover, if CM_i and CM_j are connected by a channel (i_p, i_p), then these modules are also connected by channels (j_q, j_p), where (j_p, j_p) is a channel that is connected to p-th output pole CM_i and to the q-th input pole CM_j. The channel (i_p, j_q) has the output queue OUT_p^i in CM_i, and the input queue IN_q^j in CM_j. The functioning of the channel (i_p, j_q) is that it transfers the data element from the beginning of the queue OUT_p^i to the end of the queue IN_q^j. This transfer is carried out in such a way that it does not require any period of time for its implementation. The data element, regardless of its length, instantly disappears from the OUT_p^i queue and appears in the IN_q^j queue (non-distribution of the data element). The operation of the channel (i_p, j_q) is activated if there is a data element in the queue OUT_p^i, and there is a free place in the queue IN_q^j for accepting its data element [3, 5].

In the process of solving the problem, the necessary exchanges are carried out between the modules of the subsystem, which are determined by the algorithm for solving the problem. Conveyors for inter-module exchanges are formed from channels and queues IN_p^i and OUT_p^i of computing modules CM_i. These modules are included in the subsystem θ, which solves the problem where $i = 0,1, \ldots, |\theta| - 1, p = 0,1, \ldots, v$; IN_0^i and OUT_0^i are queues that are located in CM_i.

That is, a method of organizing inter-modular exchanges is proposed, which is based on the parametric presented subsystem.

3 Description of the Method

By derivation of correlation coefficient and curve alignment of WAMS/SCADA data, most of time-series data has been disposed, but the weight of time-series data depends on measurement precision [16]. The WAMS/SCADA hybrid measurement state estimation precision depends on device measurement precision and time synchronism of hybrid measurement data: The proposed method is as follows.

1. The first stage of the proposed method.

As an environment in the subsystem θ with the number of modules n, we select the addresses A_i, a number E_i, a path $G_i(A_i \oplus A_j)$, a translation $F_i(A_i \oplus A_j)$ and a weight T_i functions defined in each CM_i subsystem θ, where i = j = 0,1, ..., n − 1. The question of selecting the modules included in the θ subsystem and forming their addresses, path and translation functions is described in [6].

The organization of inter-module transmissions is based on the use of CM addresses. The transmission of the data block Y from the module $CM_i, i \in \{0,1, \ldots, |\theta| - 1\}$ in some subsets of modules of the subsystem θ occurs by broadcasting the switching message $H\{A_i, B_1, \ldots, B_k, Y\}$ through sequence of subsystem modules θ. These modules are located between the transmitter module CM_i and the receiver module CM.

At the same time, when switching channels, the data block ***Y*** has service information that allows to switch the channel system from the transmitter module to the receiver module for the next data transfer [7].

The switching message $H\{A_i, B_1, \ldots, B_k, Y\}$ consists of the following components. ***H*** is a switch message flag that indicates the type of path procedure used to transmit the switch message under consideration. B_j is an identifier that specifies the set of addresses of modules that should receive the data block ***Y***, $j = 1,2, \ldots. k \in \{1,2, \ldots\}$. A_i is an address of the module that generated the switching message under consideration. To define the entire set of subsystem modules θ, such as receivers (for translational exchange [8]), we will use the ***YCI*** identifier. The switching message $H\{A_i, YCI, Y\}$ is formed in CM_i and delivered from CM_i to all subsystem modules. This ensures that they accept the ***Y*** data block.

The CM addresses of the subsystem should be set so that the algorithm for determining the route of transmission of the switching message (route procedure [7, 8]) is easy. Let's consider one of these ways of addressing modules.

To determine the addressing in the subsystem, the module CM_0, which is called the main one, is allocated. Let $\boldsymbol{m}$ denote the distance from the main CM_0 to the farthest module of the subsystem. In the case of a subsystem m = 3.

Each $CM_i, i \in \{0,1, \ldots, |\theta| - 1\}$ subsystem θ assigns an address given by the vector A_i, $A_i = A_i^0 A_i^1 \ldots A_i^{m-1}$, $A_i^j \in \{1,2, \ldots, v\}, j = 1,2, \ldots, m - 1$, where v are the degrees of vertices of the graph of inter-module connections. The number $A_i^j, j \in \{1,2, \ldots, m - 1\}$ is called the row number of the j-th elementary module CM_i. The address A_0 of the main module CM_0 is set as $A_0^0 A_0^1 \ldots A_0^{m-1} = 00 \ldots 0$, i.e. $A_0 = 0$ for all $j \in \{1,2, \ldots, m - 1\}$. All other modules of the subsystem receive addresses that determine their position relative to the main module CM_0.

At the same time, if the length from CM_0 to CM_i is equal to $e, e < m$, then at the address CM_i the tier numbers $A_i^e A_i^{e+1} \ldots A_i^{m-1}$ are equal zero. Modules CM_i and CM_k are included in the address subsystem $R_r(A_i^0 \ldots A_i^{r-1})$ of tier $\boldsymbol{r}$, if $A_i^j = A_i^k$ for all $j < r, r = 1,2, \ldots m$. By definition, we assume that subsystem θ is an address subsystem of tier zero. The address subsystem $R_m(A_i^0 \ldots A_i^{m-1})$ of tier $\boldsymbol{m}$ includes only one module CM_i with the address $A_i = A_i^0 A_i^1 \ldots A_i^{m-1}$.

It is obvious that the modules of the address subsystem $R_r(A_i^0 \ldots A_i^{r-1}), r = \overline{0, \ldots, m}$ create a connected subgraph of the system structure as for any CM_i with the address $A_i = A_i^0 \ldots A_i^{r-1} = 0 \ldots 0$ there is a path to any CM_k with the address $A_k = A_i^0 \ldots A_i^{j-1} A_i^2 \ldots A_k^{m-1}$. At the same time, the tier number A_k^r is equal to the output pole number CM_i, which is on the shortest path from CM_i to CM_k.

2. The second stage of the proposed method.

The addressing of subsystem modules $\theta \in D(\lceil \log_2(v + 1)|, m)$ considered in paragraph 1 is addressing [8], and, therefore, for determining the shortest paths from CM_i, $i \in \{0,1, \ldots, |\theta| - 1\}$, the tier functions $G_i^r(A_i^r \oplus G_k^r)$ can be used. In this case, $\oplus$ is an addition operation modulo $(v + 1)$, $\lceil \boldsymbol{x} |$is the smallest integer that is not less than $\boldsymbol{x}$.

The value of $G_i^r(A_i^r \oplus A_k^r)$ at $r = \min\{j/A_i^j \neq A_k^j, j = 0, \ldots, m - 1\}$ is equal to the output pole number CM_i, which belongs the shortest path from CM_i to the address subsystem $R_{r+1}(A_i^0 \ldots A_i^{r-1} A_k^r)$.

For $r = \min\{j/A_i^j \neq A_k^j, j = 0, \ldots, m - 1\}$ the value of $G_i^r(A_i^r \oplus A_k^r)$ is undefined and corresponds to the empty set of poles $\varnothing$. At the same time, the shortest path between CM_i and the address subsystem $R_{r+1}(A_i^0 \ldots A_i^{r-1} A_k^r)$ means the path with the length $d_{ih} = \min\{d_{ig} j/CM_g \in R_{r+1}(A_i^0 \ldots A_i^{r-1} A_k^r)\}$ between CM_i and CM_h such that $CM_h \in R_{r+1}(A_i^0 \ldots A_i^{r-1} A_k^r)$. The table value $G_i^r(X), r = 0,1, \ldots, m - 1, X = 0,1, \ldots, v$ needs $(v + 1)m\lceil \log_2(v + 1)|$, of bits in each $CM_i, i \in \{0,1, \ldots, |\theta| - 1\}$. That is, the size of the table CM_i should be $(v + 1)t_i \left|\log_2(v + 1)\right|$ bits.

In the case of subsystem θ, the values of the tier functions $G_i^0(X), G_i^1(X), G_i^2(X)$ are given in the Table 1. Other values of layer functions that must be stored are not defined. This is due to the specificity of the subsystem θ.

Tier functions are set in such a way that for eachCM_i, the system of paths from CM_i to all the last modules of the subsystem θ form a spanning main tree [5] D_i of the subsystem, $i = 0,1, \ldots, |\theta| - 1$.

We will call the path function CM_i $G_i(A_i \oplus A_j) = \bigcup_{r=0}^{m-1}\left\{G_i^r\left(A_i^r \oplus A_j^r\right)\right\}, j \in \{0,1, \ldots, |\theta| - 1\}$, $A_i^r \oplus A_k^r$ – componentwise addition modulo $(v + 1)$ of vectors A_i and A_j.

Table 1. Value of Tier Functions

CM_i	r		
	0	1	2
1	$G_1^0(1) = 3$; $G_1^0(3) = 3$	$G_1^1(1) = 2$	0
2	$G_2^0(3) = 1$; $G_2^0(2) = 4$	$G_2^1(2) = 2$	$G_2^2(2) = 2$
3	$G_3^0(3) = 3$; $G_3^0(1) = 3$	$G_3^1(2) = 4$	$G_3^2(2) = 2$
4	$G_4^0(3) = 1$; $G_4^0(2) = 4$	$G_4^1(2) = 4$	$G_4^2(2) = 4$
5	$G_5^0(1) = 3$; $G_5^0(3) = 3$	$G_5^1(2) = 4$	$G_5^2(2) = 4$

3. The third stage of the proposed method.

To organize transmissions during broadcast exchange, it is necessary to set the broadcast function $F_i(A_i \oplus A_j)$ in each CM_i. The value of the functions $F_i(A_i \oplus A_j)$ is the ordered set of output poles CM_i, through which the switching message $H\{A_j, BCI, Y\}$ enters the modules adjacent to CM_i. At the same time, the switching message $H\{A_j, BCI, Y\}$ must be distributed between modules of the subsystem along the main tree $D_j, j \in (0,1, \ldots, |\theta| - 1)$.

4. The fourth first stage of the proposed method

The tree D_0 is used to number the modules of the subsystem θ and to specify the weighting functions $T_i, i = 0,1, \ldots n - 1$. The vertex weight of the original covering tree is called [10] the number of vertices of the covering tree that have the vertex under consideration as a root. We denote the weight of vertex $\boldsymbol{i}$ by $\tau_i, i = 0,1, \ldots |\theta| - 1$.

Step 1. It is necessary to assign the number zero to the main module.
Step 2. It is necessary to find the numbered module CM_j with number $\boldsymbol{j}$, which is the main one for the original covering tree. The vertices of this tree have no numbers. If there is no such vertex, then we proceed to step 5.
Step 3. It is necessary to form a set $\boldsymbol{A}$ that includes connections with CM_j that do not have numbers. Then it is necessary to select the module from the set $\boldsymbol{A}$, which is connected to CM_j channel with the smallest label of the output pole. Next, it is necessary to assign the number $\boldsymbol{j+1}$ to the selected module and remove CM_{j+1} from the set $\boldsymbol{A}$, $\alpha = \tau_{j+1}$.
Step 4. If $A = \varnothing$, then it is necessary to go to step 2, otherwise, it is necessary to select the module from the set $\boldsymbol{A}$, which is connected to CM_j channel with the smallest

output pole label. Then it is necessary to assign the number to the selected module $g = j + \alpha + 1$; $\alpha = \alpha + \tau_g$ and remove CM_g from the set $\boldsymbol{A}$. Finally, proceed to step 4.
Step 5. The end.

5. The fifth stage of the proposed method

The values of the weighting function T_i are the set $\left(\tau_{i_1}, \ldots, \tau_{i_k}\right)$ weights τ_{i_p} of modules $CM_{i_p}, p = 0,1, \ldots, k$ arranged by growth of indices. Modules CM_{i_p} are connected with CM_i by channel $\left(i_q, i_p\right)$, where $k = |F_i(A_i \oplus A_0)|$, $q \in F_i(A_i \oplus A_0)$, $r = G_{i_p}(A_i \oplus A_0)$. Because, for the subsystem $F_2(A_2 \oplus A_0) = \{1, 2\}$, $T_2 = \{\tau_3, \tau_4\} = \{1{,}2\}$; $F_0(A_0 \oplus A_0) = (1{,}2\}$, $T_0 = \{\tau_1, \tau_2\} = \{1{,}4\}$; $F_4(A_4 \oplus A_0) = \{1\}$, $T_4 = \{\tau_5\} = \{1\}$; $T_1 = T_3 = T_5 = \varnothing$.

6. The sixth stage of the proposed method

Thus, after building a subsystem θ with a user-specified number of modules $\boldsymbol{n}$, the neighbourhood $\{A, G, E, F, T\}$ will be defined in each CM. The user knows that the subsystem has a module CM_0 with number $E_0 = 0$ and address $A_0 = 00\ldots0$. If the user is not satisfied with the numbering and addressing of the modules adopted during the creation of the system, he can propose his own numbering and addressing that he needs. Loading tasks into the system consists in the fact that each CM receives the E_i program and data corresponding to its number. To organize differential inter-module exchanges [7], you can use the CM_i, $\mathrm{i} = 0{,}1, \ldots, \mathrm{n}-1$ correspondences $j \Leftrightarrow A_j$, $\mathrm{j} = 0{,}1, \ldots, \mathrm{n}-1$ between module numbers available in each module and their addresses. Correspondences are specified by entering correspondence tables $j \Leftrightarrow A$ into each module. If this method of matching is unacceptable and the specifics of the task do not solve the problem of processing speed, then it is necessary to remove differential exchanges from the program and replace them with translational ones.

After the numbering and addressing of the modules is carried out in the θ subsystem, its efficiency and productivity must be determined.

The effectiveness of the computer system depends on a set of characteristics, which include: performance, reliability, probability of results, stability, manufacturability and others [11].

A common criterion that is widely used to evaluate the performance of a computing system is the performance-to-cost ratio

$$EF = P/C \tag{1}$$

It is well known that with the development of the element base and computer technology, the P indicator increases. Its increase occurs due to the improvement of the characteristics of the element base and the architecture of the computer system. This indicator rather reflects the stage of development of the computing system than the individuality of the system.

As an efficiency criterion for comparing computer systems, we can use a generalized indicator of the form:

$$EF = (K_V P_0)/C \tag{2}$$

where P_0 is the performance of the computer system with ideal reliability; C – the cost; K_V is the coefficient of survival of the system.

The system survivability coefficient is the ratio of the number of states $\boldsymbol{S}$ of the system that correspond to operational efficiency to the entire set of states S_n^i (where $\boldsymbol{i}$ is the multiplicity of generalized failures, $\boldsymbol{n}$ is the number of computing systems)

$$K_V = S/S_n^i$$

When solving problems that require a lot of time and are compared with the time of failure, an essential indicator of reliability that determines efficiency is the probability of continuous solution of the problem $P_p(t)$. That is, the probability that during the time of solving the problem T_p the system will not fail. Since in this case the positive effect will appear only as a result of error-free execution of the task, then it can be written that the real performance P_r of the computer system when solving this class of tasks will be [12]

$$P_r = (K_V P_0 P_p(t))/C \tag{3}$$

Thus, when comparing computing systems that are focused on solving problems that require a long solution time, we use the following expression as an efficiency criterion

$$EF = P_r = (K_V P_0 P_p(t))/C \tag{4}$$

The criterion (4) is more generalized compared to (2). When solving problems in which $P_p(t) = 1$, criteria (2) and (4) coincide.

It should be noted that the indicators K_V and $P_p(t)$ are often used as independent indicators of the efficiency of computer systems. However, the performance indicator of the computing system, which is based only on the consideration of reliability, is one-sided.

The ***EF*** indicator is a product of traditional performance indicators, which is more sensitive than each of the components taken separately. It allows to evaluate a real efficiency and reduction of efficiency due to unreliability of systems. The proposed efficiency indicator does not contradict the well-known conclusions that not all structural methods of improving efficiency, for example, reliability, lead to an increase in the overall efficiency of the computing system. Increasing the speed or reliability of an element base cannot effectively increase ***EF*** because of the increased cost. Increasing performance at the expense of additional hardware leads to a decrease in reliability and an increase in cost.

The productivity of computing systems can be described by the following expression

$$P_0 = \frac{K_A}{t_f} \lambda_{max} t_i \tag{5}$$

where K_A is the coefficient that determines the architecture of the computing system;

λ_{max} is the intensity of exchange of informational messages;
t_i is the task execution time;
t_f is the operation execution time or information message generation time.

$$t_f = nk_l T_c + (\lambda_1 k_1 + \lambda_2 k_2)t_p$$

where k_l is the connection coefficient in the computing system;

T_c is the cycle execution time;
$\lambda_1 = 0{,}7$ and $\lambda_2 = 0{,}3$ are the relative coefficients of the use of short and long operations in computer systems. The use of a modern element base allows reducing the average values of the coefficients $0{,}7k_1$ and $0{,}3k_2$ to values of $30 \div 50$;
t_p is the operation execution time.

$$k_l = k_x \frac{\sum_{i=1}^{p} \rho_A(i)}{\sum_{i=1}^{p} \rho_B(i)} \tag{6}$$

where $\sum_{i=1}^{p} \rho_A(i)$ is the number of units in the adjacency matrix $\boldsymbol{A}$ of the implemented algorithm;

$\sum_{i=1}^{p} \rho_B(i)$ is the number of units in the adjacency matrix $\boldsymbol{B}$ of the implemented algorithm;

k_x is the coefficient that takes into account the overlap of matrix $\boldsymbol{A}$ with matrix $\boldsymbol{B}$. If matrix $\boldsymbol{B}$ completely overlaps matrix $\boldsymbol{A}$, then $k_x = 1$, otherwise $k_x = 0$.

Regarding the choice of a *CM* architecture with an effective organization of connections, compatibility matrices are created to analyze the proposed architectures and compare them with each other. At the same time, the efficiency of connections is calculated. If $k_l = 100\%$, then the required architecture is found, if $k_l \neq 100\%$, then it is necessary to choose the architecture in which k_l has the largest value.

The impact of the *CM* architecture on performance is taken into account by the K_A factor. Increasing system performance by complicating the architecture and improving the principles of information exchange does not always give a positive result.

An important problem in the synthesis of MPCSs with a programmable structure is the organization of memory. Just like in MPCS, it can be organized by concentrated memory, branched memory or concentrated-branched memory.

Any direct communication channel between computing modules or between microprocessors in *CM* s can be programmed in MPCS with a programmable structure using a switching structure. However, this leads to the fact that in the system or in the *CM*, the conditions for exchanging information of the centralized memory with other elements are created only on the basis of the trunk principle.

Taking into account the above, we will analyse the proposed *CM* architectures, which differ from each other in the structure of the memory implementation module.

As noted earlier, the general memory can be divided into local modules. That is, the proposed *CM* s have the following architecture: CM_1 (Fig. 2), CM_2 (Fig. 3) and CM_3 (Fig. 4) [1, 13]. The communication structure is implemented accordingly [14, 15].

In order to evaluate the technical capabilities of *CM* s and MPCSs with a programmable structure, a comparative evaluation has carried out using test tasks. When evaluating *CM* s, 2, 5, and 8 processors have used in the module.

Single-board micro-computers were installed in the nodes of the 3×3 matrix. The development of MPCS with a programmable structure has performed according to the structure of a 2×2 matrix, in the nodes of which CM_1 and CM_3 were used.

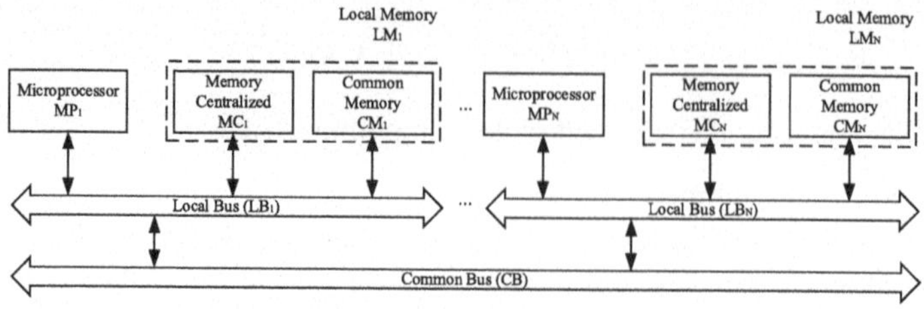

Fig. 2. The Architecture of CM_1

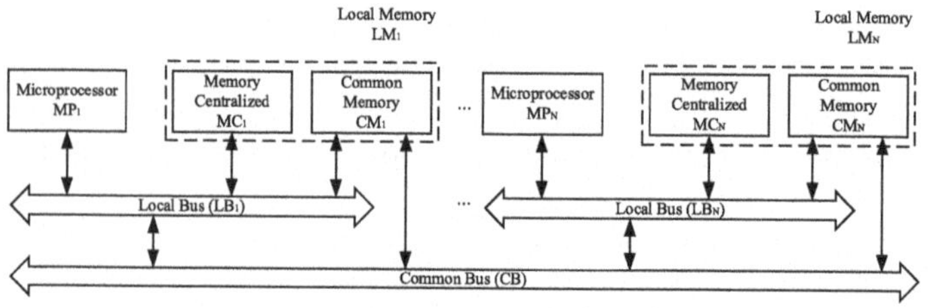

Fig. 3. The Architecture of CM_2

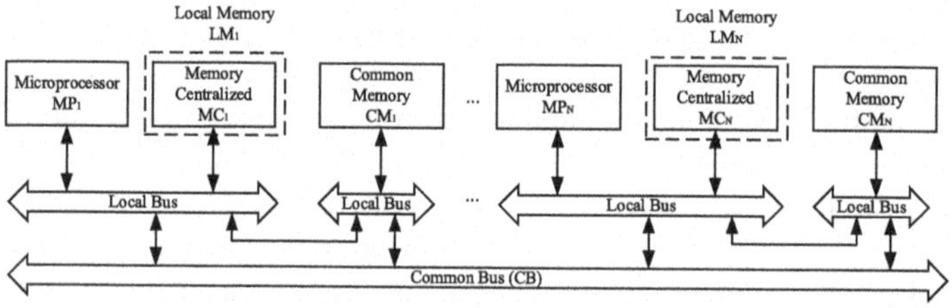

Fig. 4. The Architecture of CM_3

The construction and technological equipment *CM* has two construction options:

a) an option that allows implementation in a hybrid version without the use of cases;
b) variant in a monolithic polymer case, in the form of a hermetic micro assembly.

The second option ensures maintainability, because it allows to remove and brew the cover several times. All microcircuits included in micro blocks are caseless. Multi-level interfaces and framing of microprocessors, RAM and non-volatile memory are implemented on multi-matrix crystals, which makes it possible to obtain a high density of assembly with minimal dimensions of integrated circuits.

4 Summary and Conclusion

Based on the results of the conducted research, conclusions can be drawn. The most effective are computing systems that are implemented on the developed computing module. This fact is clearly visible from the comparison of the time spent on calculating not only the test task, but also when calculating the test equation with the test parameters.

References

1. Phadke, A.G.: Synchronized phasor measurements in power systems. IEEE Comput. Appl. Power **6**(2), 10–15 (2013)
2. Skorik, V.N., Stepanov, A.E., Horoshko, V.A.: Mul'tiprocessornye sistemy. Tehnika (1989). (in Russian)
3. Opirskyy, I., Zybin, S., Horoshko, V.: Analysis of mathematical models of functioning scalar multiprocessor systems. Inf. Syst. Netw. **6**, 66–78 (2019)
4. Korneev, V.V., Horoshevskij, V.G.: Vychislitel'nye sistemy s programmiruemoj strukturoj. Jelektronnoe modelirovanie **1**, 42–52 (1989). (in Russian)
5. Andon, F.I., Jashunin, A.E., Reznichenko, V.A.: Logicheskie modeli intellektual'nyh informacionnyh sistem. Naukova dumka (1999). (in Russian)
6. Korneev, V.V.: Jelementarnaja mashina odnorodnoj vychislitel'noj sistemy s programmiruemoj strukturoj. Kibernetika i sistemnyj analiz (1), 75–81 (1980). (in Russian)
7. Zybin, S., Khoroshko, V., Maksymovych, V., Opirskyy, I.: Effective distribution of tasks in multiprocessor and multi-computers distributed homogeneous systems. Int. J. Comput. **20**(2), 211–220 (2021)
8. Zybin, S.V., Khoroshko, V.O.: Produktyvnist i optymizatsiia system obrobky informatsii iz strukturoiu, yaka konfihuruietsia. Informatyka ta matematychni metody v modeliuvanni, Odesa, ONPU **9**(3), 120–130 (2019). (in Ukrainian)
9. Beleckij, V.N.: Mnogoprocessornye i parallel'nye struktury s organizaciej asinhronnyh vychislenij. Naukova dumka (1988). (in Russian)
10. Korneev, V.V., Monahov, O.G.: Organizacija mezhmashinnyh vzaimodejstvij v vychislitel'nyh sistemah s programmiruemoj strukturoj. Jelektronnoe modelirovanie (5), 16–22 (1980). (in Russian)
11. Harari, F.: Teorija grafov. LIBROKOM (2009). (in Russian)
12. Tomashevs'kij, V.M.: Modelirovanie sistem. Vidavnicha grupa VNU (2005). (in Russian)
13. Samofalov, K.G., Luckij, G.M.: Osnovy postroenija konvejernyh JeVM.Kiev, Vishha shkola (1981). (in Russian)
14. Bolbat, A.A., Ordynskij, A.B., Horoshko, V.A.: Kompleksnyj vychislitel'nyj modul' raspredeljonnoj vychislitel'noj sistemy. Aviacionnaja promyshlennost **6**, 34–39 (1996). (in Russian)
15. Ordynskij, A.B., Sobolev, A.N., Horoshko, V.A.: Mnogokanal'noe ustrojstvo dlja upravlenija ocherjodnost'ju zaprosov. Avtorskoe svidetel'stvo SSSR # 1444767. Bjul. Izobret. # 46 (1988). (in Russian)
16. Kovtun, V.N., Ordynskij, A.B., Horoshko, V.A., Chirkov, D.V.: Mnogokanal'noe ustrojstvo prioritetov. Avtorskoe svidetel'stvo SSSR # 1264177. Bjul. Izobret. # 36 (1986). (in Russian)

Enhancement of the Formal Method of Determining Processor States of a Multiprocessor System Under Testing

Vitaliy A. Romankevich[1], Kostiantyn V. Morozov[1], Oleksii V. Shuryha[1](✉), Alexei M. Romankevich[1], and Lefteris Zacharioudakis[2]

[1] National Technical University of Ukraine "Igor Sikorsky Kyiv Polytechnic Institute", Kyiv 03056, Ukraine
romankev@scs.ntu-kpi.kiev.ua, {mcng,realmod1}@ukr.net, romankev@scs.kpi.ua

[2] Neapolis University Pafos, 2, Danais Ave., 8042 Pafos, Cyprus
l.zacharioudakis@nup.ac.cy

Abstract. This paper concerns the problem of self-diagnosis of fault-tolerant multiprocessor systems, namely, the determining of processor states after performing all specified mutual checks of processors. The problem is of particular importance for systems with irregular connections. In practice, the regularity of connections is disrupted when developers incorporate new processors into the system along with their associated connections. The formal method of determining states of processors of a M-diagnosable system with arbitrary diagnostic connections during its testing has been considered. The method essentially comes down to generating a Boolean expression, the complexity of which depends on the number of diagnostic tests performed and their results. An improvement to the method that reduces the complexity of obtaining the Boolean expression in the canonical disjunctive normal form has been proposed. This allows to simplify analyzis of test results and reduce the time for its execution, consequently diminishing overall testing duration and enhancing the capacity of the system to execute primary tasks. The latter is of paramount importance, as testing is a continuous process throughout system operation.

Keywords: Multiprocessor systems · M-diagnosable systems · Mutual testing of processors · PMC-model · Formal method of determining processor states

1 Introduction

Fault-tolerant multiprocessor systems (FTMS) are widely used in automated control systems [1–4], not least due to the ability to ensure high reliability of such systems [5–9]. Ensuring a given level of reliability is crucial for systems which malfunction can lead to significant economic losses or catastrophic consequences [10–19]. Such systems include control systems of hazardous manufacturing, nuclear power plants, aircrafts, spacecrafts, unmanned aerial vehicles, etc.

Z. Hu et al. (Eds.): ICCSEEA 2024, LNDECT 242, pp. 142–155, 2025.
https://doi.org/10.1007/978-3-031-84228-3_12

FTMS can continue to perform its functions when one or more of its processors fails, which increases system reliability.

One approach to ensuring the required level of fault tolerance is to introduce redundant processors into the system and organise its self-reconfiguration in case of some of its processor failures. The reconfiguration excludes malfunctioning processors from performing the tasks assigned to the system. In order to know which processors need to be excluded, the system requires information about the states of the processors (operational/failed) [20–22], so determining the states is an important task that the FTMS should perform.

Processor states can be determined using the results of processors testing. The testing and determining the states can be performed by either processors themselves or by a specialized device of the system [23–26]. In the first case, the processors set test tasks for each other, compare task results with reference values, and then determine the states of the processors. In the second case, the specialized device of the system initiates test checks and determines the processor states.

A test task is defined as a task which execution result on a certain processor allows us to determine whether the processor is in operational state or has failed (to be more precise, whether processor components necessary to perform the assigned functions are working correctly or not). In other words, if the result of the task execution matches the reference value, the processor is functioning correctly, otherwise it malfunctions.

Further in this work, the statement that i-th processor tests j-th one means that the i-th processor sets a test task for the j-th processor and compares the task execution result with the reference value, while the j-th processor performs the task.

In this paper, we will assume that testing is performed according to the Preparata-Metze-Chien model (PMC-model) [27]. According to the model, processors test each other in pairs (one of the processors in each pair sets a test task and checks the result of its execution, whereas the other executes it). The result of testing the j-th processor by the i-th serviceable processor will be 0 if the j-th processor functions correctly, or 1 if the j-th processor malfunctions. If the i-th processor malfunctions, the test result does not depend on the state of the j-th processor and can be either 0 or 1.

Not every processor of the system may be able to test any other processor. In practice, the impossibility of testing any processor by any other one could be due to the lack of necessary communication channels between them. According to [27], this fact is reflected in the PMC-model by an oriented graph, in which vertices correspond to processors, and edges indicate the possibility of testing. The presence of an edge from vertex i to vertex j means that the i-th processor can test the j-th. Also, each edge is associated with a weight $r_{ij} \in \{0,1\}$, which is the result of testing the j-th processor by the i-th processor.

Another important property of FTMS is M-diagnosability [28]. FTMS is M-diagnosable, when, based on testing results, it is always possible to determine the states of all processors in case of failure of no more than M processors.

It is worth mentioning that for many reconfigurable FTMSs, the time to determine processor states is limited, as it determines the time required to correct errors that could occur due to system malfunction and restore the system after failures [29]. At the same time, such systems may have a small amount of computing resources, in particular after some processors fail. Thus, reducing the time needed for determining processor states is a relevant problem.

2 The Formal Method of Determining Processor States

The formal method for determining the states of all processors of a multiprocessor system based on the results of mutual pairwise testing of processors is proposed in [30]. The advantage of this method is that it can be applied to a system with arbitrary diagnostic connections. Other known approaches either can mainly be used for systems with regular connections, or they impose restrictions on the set of tests and sometimes on the test procedure [27, 31–36].

Another feature of the abovementioned method is that, in addition to determining the states of the processors, it allows to check whether it is possible to determine processor states based on the results of the execution of a test set.

The method is based on forming a Boolean expression V_n based on the results of a particular test set and its transformation to a canonical disjunctive normal form (CDNF). Based on which minterms are included in the transformed expression, it is concluded whether the states of the processors can be determined, and if so, which processors have failed, and which remain in working state.

According to [30], each processor has a variable x_i associated with it, which takes the value 0 if the i-th processor is out of order or the value 1 if it is in working order.

Let's introduce several notations.

Let denote a test by a pair of numbers (i, j), where the i-th processor tests the j-th processor.

The test result (i, j) is denoted by a triple $\left(i, j, r_{ij}\right)$, where r_{ij} is determined according to the PMC-model.

If n tests have been executed, then:

I_n is the set of numbers of all processors that assigned test tasks.

J_n is the set of numbers of all processors that performed test tasks.

J_{ni} is the set of numbers of all processors that performed the test tasks assigned by the i-th processor ($i \in I_n$).

$T_n \triangleq \{(i, j): i \in I_n, j \in J_{ni}\}$ is the set of all tests that were executed during testing.

$O_n \triangleq \left\{\left(i, j, r_{ij}\right): i \in I_n, j \in J_{ni}\right\}$ is the set of test results for the tests from set T_n.

According to [30], for each result from O_n, expressions R_{ij} are formed as follows:

$$R_{ij}\left(x_i, x_j\right) \triangleq \begin{cases} x_j \vee \overline{x_i}, & \text{when } r_{ij} = 0, \\ \overline{x_j} \vee \overline{x_i}, & \text{when } r_{ij} = 1, \end{cases} \quad (1)$$

where $i, j, r_{ij} : \left(i, j, r_{ij}\right) \in O_n$.

Then, as per [30], a Boolean expression is formed, which has the following form:

$$V_n \triangleq \bigwedge_{i,j:(i,j)\in T_n} R_{ij}. \quad (2)$$

Next, V_n is converted to CDNF with the exclusion of all minterms in which the number of inverted literals is greater than M:

$$V_n = \bigvee_{p=1}^{P_n} C_p, \tag{3}$$

where C_p is a minterm, P_n is the number of all minterms obtained due to the transformation.

In [30], it was shown that if the transformed expression contains one minterm and no more than M processors are out of order in the system, then it reflects the current state of the processors of system: variables associated with working processors do not have inversions, and variables associated with faulty processors have inversions.

If there are two or more minterms, the state of at least some of the FTMS processors cannot be determined without performing additional tests. After performing additional tests (let us denote their number as k), a new expression V_{n+k} is formed:

$$V_{n+k} \triangleq V_n \wedge \bigwedge_{i,j:(i,j)\in(T_{n+k}\setminus T_n)} R_{ij}, \tag{4}$$

where T_{n+k} is the set of all executed tests, including additional ones.

After that, the expression is again converted to the CDNF, and the abovementioned analysis is repeated.

If, after executing all possible tests, expression (3) does not contain any minterm (i.e., $P_n = 0$) or contains more than one minterm, then according to [30], this means that the system has more than M processors out of order or it is not M-diagnosable respectively.

The complexity of determining processor states using the formal method significantly depends on the complexity of transformation of expressions (2) and (4) to CDNF, which grows nonlinearly with the number of tests performed.

This paper considers the problem of reducing the complexity of the transformation expression (2) to CDNF.

3 Enhancement of the Formal Method of Determining Processor States

As mentioned earlier, any test is defined by a pair of numbers $(i, j) \in T_n$, where i is the number of the processor that tested the processor with the number j. Moreover, the same processor can test several other processors, i.e., the value of i can be repeated in different pairs from the set T_n.

After grouping R_{ij} by i in the expression V_n, it can be represented as follows:

$$V_n = \bigwedge_{i\in I_n}\left(\bigwedge_{j\in J_{ni}} R_{ij}\right) \equiv \bigwedge_{i\in I_n} W_i, \tag{5}$$

where

$$W_l \triangleq \bigwedge_{j \in J_{nl}} R_{lj}. \tag{6}$$

Let

$$a_{ij}(x_j) \triangleq \begin{cases} x_j, & \text{when } r_{ij} = 0, \\ \overline{x_j}, & \text{when } r_{ij} = 1. \end{cases} \tag{7}$$

Then (1) can be rewritten as follows:

$$R_{ij}(x_i, x_j) \equiv a_{ij} \vee \overline{x_i}. \tag{8}$$

Next, we substitute (8) into (6) and expand the brackets as shown below:

$$\begin{aligned} W_l = \bigwedge_{j \in J_{nl}} \left(a_{lj} \vee \overline{x_l}\right) \equiv \\ \equiv \left(a_{lj_1} \vee \overline{x_l}\right) \wedge \left(a_{lj_2} \vee \overline{x_l}\right) \wedge \left(a_{lj_3} \vee \overline{x_l}\right) \wedge \ldots \wedge \left(a_{lj_p} \vee \overline{x_l}\right) \equiv \\ \equiv \left(a_{lj_1} a_{lj_2} \vee \cancel{a_{lj_1} \overline{x_l}} \vee \cancel{a_{lj_2} \overline{x_l}} \vee \overline{x_l}\right) \wedge \left(a_{lj_3} \vee \overline{x_l}\right) \wedge \ldots \wedge \left(a_{lj_p} \vee \overline{x_l}\right) \equiv \ldots \equiv \\ \equiv \left(a_{lj_1} a_{lj_2} \ldots a_{lj_{e-1}} \vee \overline{x_l}\right) \wedge \left(a_{lj_e} \vee \overline{x_l}\right) \wedge \ldots \wedge \left(a_{lj_p} \vee \overline{x_l}\right) \equiv \\ \equiv \left(a_{lj_1} a_{lj_2} \ldots a_{lj_e} \vee \cancel{a_{lj_1} a_{lj_2} \ldots a_{lj_{e-1}} \overline{x_l}} \vee \cancel{a_{lj_e} \overline{x_l}} \vee \overline{x_l}\right) \wedge \\ \wedge \left(a_{lj_{e+1}} \vee \overline{x_l}\right) \wedge \ldots \wedge \left(a_{lj_p} \vee \overline{x_l}\right) \equiv \cdots \equiv \\ \equiv a_{lj_1} a_{lj_2} a_{lj_3} \ldots a_{lj_p} \vee \overline{x_l} \equiv \overline{x_l} \vee \bigwedge_{j \in J_{nl}} a_{lj}. \end{aligned} \tag{9}$$

Therefore, the expression W_l can be represented in the following form:

$$W_l \equiv \overline{x_l} \vee \bigwedge_{j \in J_{nl}} a_{lj}. \tag{10}$$

Substituting the transformed expression (9) into (5) gives:

$$V_n = \bigwedge_{i \in I_n} \left(\overline{x_i} \vee \bigwedge_{j \in J_{ni}} a_{ij} \right). \tag{11}$$

Thus, in accordance with (11), we can formulate an algorithm for constructing the expression V_n directly from the set of test results O_n and the sets I_n (a set of processors that assigned test tasks) and J_n (the set of processors that performed test tasks) as follows:

1) For each $i \in I_n$, form sets J_{ni} (subsets of J_n containing the numbers of all processors that performed the test tasks set by the i-th processor).
2) For each test result $(i, j, r_{ij}) \in O_n$, form the expressions $a_{ij}(x_j)$ according to (7).

3) For each $i \in I_n$, form the expression W_i according to (10), using the corresponding sets J_{ni} and the expressions $a_{ij}(x_j)$ obtained in the previous steps.
4) Construct the expression V_n, which is the conjunction of all the obtained expressions W_i.

Let us formulate some statements and properties of expression (11).

Property. The number of brackets in expression (11) is always no more than the number of processors in the system.

Proof. Indeed, the number of brackets will be equal to the number of elements in the set I_n containing the numbers of processors that executed the test tasks. The number of such processors cannot be greater than the number of processors in the system.

Assertion 1. Let expressions (2) and (11) be formed for the same set of test results. Then the number of brackets in expression (11) will always be no more than the number of brackets in expression (2).

Proof. Indeed, expression (11) is formed from expression (2) with the possible replacement of some groups of brackets with one equivalent bracket.

Assertion 2: If expression (2) was formed based on the results of tests which contain at least one processor that tested two or more other processors, there always exists an order of expanding all brackets of this expression such that the number of multiplication operations is greater than the number of such operations for the sequential expanding all brackets of the equivalent expression (11).

Proof. Let expression (11) have the following brackets order:

$$V_n \equiv \overbrace{\left(\overline{x_{i_1}} \vee \bigwedge_{j \in J_{ni_1}} a_{i_1 j}\right)}^{\substack{w_1 \\ \parallel}} \wedge \overbrace{\left(\overline{x_{i_2}} \vee \bigwedge_{j \in J_{ni_2}} a_{i_2 j}\right)}^{\substack{w_2 \\ \parallel}} \wedge \ldots \wedge \overbrace{\left(\overline{x_{i_s}} \vee \bigwedge_{j \in J_{ni_s}} a_{i_s j}\right)}^{\substack{w_s \\ \parallel}}, \tag{12}$$

where $s = |I_n|$.

From now on, we call the brackets group G_l those brackets that are included in the expression W_l. The number of brackets in the k-th group is denoted by m_k $\left(m_k \equiv \left|J_{ni_k}\right|\right)$.

Note that the first bracket W_1 contains the expression obtained by removing all the brackets from the expression which is the multiplication of brackets from (2) that form group G_1; the content of the second bracket W_2 form group G_2, etc.

Now, we substitute each R_{ij} in expression (2) with expression (9) and arrange the resulting brackets so that the brackets of group G_1 go first, then those of group G_2, and so on:

$$V_n \equiv \overbrace{\left(\bigwedge_{j\in J_{ni_1}} (a_{i_1 j} \vee \overline{x_{i_1}})\right)}^{G_1} \wedge \overbrace{\left(\bigwedge_{j\in J_{ni_2}} (a_{i_2 j} \vee \overline{x_{i_2}})\right)}^{G_2} \wedge \ldots \wedge \overbrace{\left(\bigwedge_{j\in J_{ni_s}} (a_{i_s j} \vee \overline{x_{i_s}})\right)}^{G_s}. \quad (13)$$

Let us show that the sequential expanding of the brackets of expression (12) requires fewer multiplication operations than the sequential expanding of the brackets in order (13), and evaluate this difference.

Let A_k be the expression obtained by multiplying out the brackets of expression (13), which are included in the first k groups $G_1, \ldots, G_k$.

Obviously, A_k can also be obtained by expanding the k brackets of expression (12), since each of its brackets is the result of expanding the corresponding group of brackets from expression (13).

First, let's compare the number of operations required to sequentially multiplying out the brackets of the first group G_1 in expression (13) and the first bracket in expression (12). In both cases, the result of the multiplication is the expression A_1.

Let's calculate the number of multiplication operations in the first case. To expand the first two brackets, 4 operations are required, since each bracket contains a disjunction of two terms:

$$A_1 = \bigwedge_{j\in J_{ni_1}} (a_{i_1 j} \vee \overline{x_{i_1}}). \quad (14)$$

Moreover, it is easy to show, by performing a transformation similar to (9), that the result of multiplying out any number of brackets from the same group will also be a disjunction of two terms. So, multiplying by the third bracket and any subsequent one will require 4 operations. The total number of operations will be $4(m_1 - 1)$.

In the second case, the result of expanding the first single bracket of expression (12) will be the content of this bracket (we do not need to multiply any brackets at all):

$$A_1 = \left(\overline{x_{i_1}} \vee \bigwedge_{j\in J_{ni_1}} a_{i_1 j}\right). \quad (15)$$

That is, to multiply out the brackets of the first group of expression (13), $[4(m_1 - 1)]$ more operations are required than to expand the first bracket of expression (12).

Now let's assume that k groups of brackets in expression (13) and k brackets in expression (12) have been multiplied out. As mentioned above, in both cases, the same result will be obtained – it is the expression A_k. It is obvious that the expression A_{k+1} can be obtained by expanding the next m_k brackets of expression (13) (these brackets are part of the group G_{k+1}):

$$A_{k+1} = (A_k) \wedge \left(\bigwedge_{j \in J_{ni_{k+1}}} \left(a_{i_{k+1}j} \vee \overline{x_{i_{k+1}}} \right) \right). \tag{16}$$

Also, the expression A_{k+1} can be obtained by expanding the next k-th bracket of expression (12):

$$A_{k+1} = (A_k) \wedge \left(\overline{x_{i_1}} \vee \bigwedge_{j \in J_{ni_1}} a_{i_1 j} \right). \tag{17}$$

Moreover, in both (16) and (17), all brackets except the first one contain the disjunction of two terms. That is, multiplying the first two brackets of expressions (16) and (17) will require the same number of operations. Expression (17) contains two brackets, while expression (16) may contain more than two brackets. Then, taking into account that expanding any two brackets can result in at least one term, expanding all the brackets of expression (16) may require at least $2(m_{k+1} - 1)$ more operations than expanding all the brackets of expression (17).

Thus, expanding the first group of brackets of expression (13) requires $4(m_1 - 1)$ more operations than expanding the first bracket of expression (12); and expanding any subsequent G_k-th group of brackets of expression (13) (starting from G_2) will require at least $2(m_k - 1)$ more operations than expanding the k-th bracket of expression (12). Since the number of bracket groups in expression (13) is the same as the number of brackets in expression (12), the minimal difference in the number of operations to expand all the brackets from these expressions is as follows:

$$D_n \triangleq 4(m_1 - 1) + \sum_{k=2}^{s} 2(m_k - 1) \equiv 2\left(\left|J_{ni_1}\right| - 1\right) + \sum_{k=1}^{|I_n|} 2\left(\left|J_{ni_k}\right| - 1\right). \tag{18}$$

Since every processor of the system needs to be tested at least once, then $\left|J_{ni_1}\right| > 0$. Taking this fact into consideration, the following can be obtained from (18):

$$D_n \geq \sum_{k=1}^{|I_n|} 2\left(\left|J_{ni_k}\right| - 1\right). \tag{19}$$

It is obvious that if some processor (let's denote it as r) tested two or more processors, then $|J_{nr}| \geq 2$. That is, $D_n \geq 2$. This is what we needed to prove.

Thus, according to Assertion 2, the transformation of expression (11) to CDNF by means of sequential expansion of brackets is no more complicated than the same transformation of the equivalent expression (2).

Thus, given that the provided above algorithm for generating expression (11) allows us to obtain it directly from the test results, as well as the fact that expressions (2) and (11) are equivalent, the algorithm for generating expression (2) in the formal method can be replaced by the proposed algorithm for generating expression (11). Such an improvement of the method allows us to reduce the upper bound on the number of computational operations required to convert a Boolean expression to a CDNF if the conversion is performed by means of consecutive bracket multiplication.

It is worth noting that one approach to determining processor states is to execute all tests and then analyze the test results according to the abovementioned method. As per this approach, additional tests are not performed, and therefore, expression (4) does not need to be formed. Given that in an M-diagnosed system, each processor can be tested by at least M other processors, with the above approach, each processor will be tested by at least M other processors [37], and the number of brackets in expression (11) will be at least M times less than in the corresponding expression (2). In addition, according to (19), the upper bound on the number of operations for expression (11) will be less than the same bound for (2) by at least $2(N+1)(M-1)$ operations (where N is the number of processors in the system). In other words, if the developer considers it expedient to perform all the tests, the gain using the proposed method will be much greater.

4 Examples

Consider a 3-diagnosable system consisting of 8 processors, denoted as a, b, c, d, e, f, g, h. The graph of the system's diagnostic capabilities is shown in Fig. 1. Each node in the graph represents a processor, and the presence of an edge from one node to another indicates the presence of the respective test. Suppose that 3 processors in the system have failed: a, d, e.

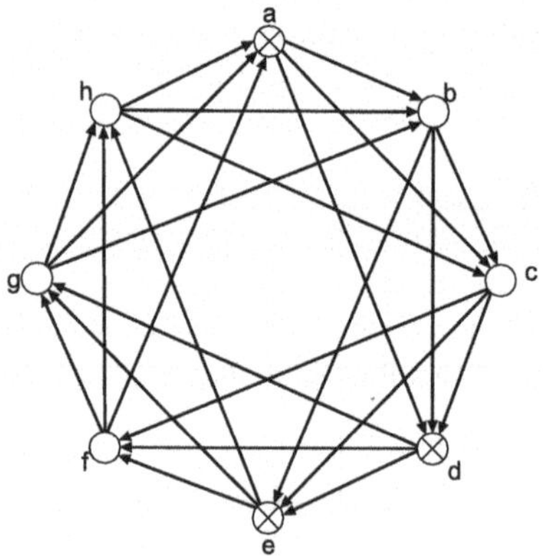

Fig. 1. Diagnostic connections of the system

Let's assume all feasible diagnostic tests are performed, and the following set of outcomes is acquired:

$$O_8 = \{(a,b,0),(a,c,0),(a,d,0),(b,c,0),(b,d,1),(b,e,1),\\ (c,d,1),(c,e,1),(c,f,0),(d,e,0),(d,f,0),(d,g,0),\\ (e,f,0),(e,g,0),(e,h,0),(f,g,0),(f,h,0),(f,a,1),\\ (g,h,0),(g,a,1),(g,b,0),(h,a,1),(h,b,0),(h,c,0)\}.$$

Then let's construct the expression V_{24} adhering to the specifications of Eq. (2) and subsequently transform it to CDNF:

$$\begin{aligned}V_{24} = {} & (\bar{a} \vee b)(\bar{a} \vee c)(\bar{a} \vee d)\left(\bar{b} \vee c\right)\left(\bar{b} \vee \bar{d}\right)\left(\bar{b} \vee \bar{e}\right)\\ & \wedge\left(\bar{c} \vee \bar{d}\right)(\bar{c} \vee \bar{e})(\bar{c} \vee f)\left(\bar{d} \vee e\right)\left(\bar{d} \vee f\right)\left(\bar{d} \vee g\right)\\ & \wedge(\bar{e} \vee f)(\bar{e} \vee g)(\bar{e} \vee h)\left(\bar{f} \vee g\right)\left(\bar{f} \vee h\right)\left(\bar{f} \vee \bar{a}\right)\\ & \wedge(\bar{g} \vee h)(\bar{g} \vee \bar{a})(\bar{g} \vee b)\left(\bar{h} \vee \bar{a}\right)\left(\bar{h} \vee b\right)\left(\bar{h} \vee c\right).\end{aligned}$$

The obtained expression contains 24 brackets.

In order to transform V_{24} to CDNF, we sequentially expand the brackets, simultaneously simplifying the resulting expressions (including, according to [30], eliminating terms that contain 4 or more inversions), and calculating the number of multiplications of two terms:

$$\begin{gathered}V_{24} \equiv (\bar{a} \vee b)(\bar{a} \vee c) \wedge \ldots \equiv (\bar{a} \vee \cancel{\bar{a}c} \vee \cancel{\bar{a}b} \vee bc)(\bar{a} \vee d) \wedge \ldots \equiv\\ \equiv (\bar{a} \vee \cancel{\bar{a}d} \vee \cancel{bc\bar{a}} \vee bcd)\left(\bar{b} \vee c\right) \wedge \ldots \equiv \left(\bar{a}\bar{b} \vee \bar{a}c \vee \cancel{bcd\bar{b}} \vee bcd\right)\left(\bar{b} \vee \bar{d}\right) \wedge \ldots \equiv\\ \equiv \left(\bar{a}\bar{b} \vee \cancel{\bar{a}\bar{b}\bar{d}} \vee \cancel{\bar{a}\bar{b}c} \vee \bar{a}c\bar{d} \vee \cancel{bcd\bar{b}} \vee \cancel{bcd\bar{d}}\right)\left(\bar{b} \vee \bar{e}\right) \wedge \ldots \equiv\\ \equiv \left(\bar{a}\bar{b} \vee \cancel{\bar{a}\bar{b}\bar{e}} \vee \cancel{\bar{a}\bar{b}c\bar{d}} \vee \bar{a}c\bar{d}\bar{e}\right)\left(\bar{c} \vee \bar{d}\right) \wedge \ldots \equiv\\ \equiv \left(\bar{a}\bar{b}\bar{c} \vee \bar{a}\bar{b}\bar{d} \vee \cancel{\bar{a}c\bar{d}\bar{e}\bar{c}} \vee \bar{a}c\bar{d}\bar{e}\right)(\bar{c} \vee \bar{e}) \wedge \ldots \equiv\\ \equiv \left(\bar{a}\bar{b}\bar{c} \vee \cancel{\bar{a}\bar{b}\bar{c}\bar{e}} \vee \cancel{\bar{a}\bar{b}\bar{c}\bar{d}} \vee \cancel{\bar{a}\bar{b}\bar{d}\bar{e}} \vee \cancel{\bar{a}c\bar{d}\bar{e}\bar{c}} \vee \bar{a}c\bar{d}\bar{e}\right)(\bar{c} \vee f) \wedge \ldots \equiv\\ \equiv \left(\bar{a}\bar{b}\bar{c} \vee \cancel{\bar{a}\bar{b}\bar{c}f} \vee \cancel{\bar{a}c\bar{d}\bar{e}\bar{c}} \vee \bar{a}c\bar{d}\bar{e}f\right)\left(\bar{d} \vee e\right) \wedge \ldots \equiv\\ \equiv \left(\cancel{\bar{a}\bar{b}\bar{c}\bar{d}} \vee \bar{a}\bar{b}\bar{c}e \vee \bar{a}c\bar{d}\bar{e}f \vee \cancel{\bar{a}c\bar{d}\bar{e}fe}\right)\left(\bar{d} \vee f\right) \wedge \ldots \equiv\\ \equiv \left(\cancel{\bar{a}\bar{b}\bar{c}\bar{d}e} \vee \bar{a}\bar{b}\bar{c}ef \vee \bar{a}c\bar{d}\bar{e}f \vee \cancel{\bar{a}c\bar{d}\bar{e}f}\right)\left(\bar{d} \vee g\right) \wedge \ldots \equiv\\ \equiv \left(\cancel{\bar{a}\bar{b}\bar{c}\bar{d}ef} \vee \bar{a}\bar{b}\bar{c}efg \vee \bar{a}c\bar{d}\bar{e}f \vee \cancel{\bar{a}c\bar{d}\bar{e}fg}\right)(\bar{e} \vee f) \wedge \ldots \equiv\\ \equiv \left(\cancel{\bar{a}\bar{b}\bar{c}efg\bar{e}} \vee \bar{a}\bar{b}\bar{c}efg \vee \bar{a}c\bar{d}\bar{e}f \vee \cancel{\bar{a}c\bar{d}\bar{e}f}\right)(\bar{e} \vee g) \wedge \ldots \equiv\\ \equiv \left(\cancel{\bar{a}\bar{b}\bar{c}efg\bar{e}} \vee \bar{a}\bar{b}\bar{c}efg \vee \bar{a}c\bar{d}\bar{e}f \vee \cancel{\bar{a}c\bar{d}\bar{e}fg}\right)(\bar{e} \vee h) \wedge \ldots \equiv\\ \equiv \left(\cancel{\bar{a}\bar{b}\bar{c}efg\bar{e}} \vee \bar{a}\bar{b}\bar{c}efgh \vee \bar{a}c\bar{d}\bar{e}f \vee \cancel{\bar{a}c\bar{d}\bar{e}fh}\right)\left(\bar{f} \vee g\right) \wedge \ldots \equiv\\ \equiv \left(\cancel{\bar{a}\bar{b}\bar{c}efgh\bar{f}} \vee \bar{a}\bar{b}\bar{c}efgh \vee \cancel{\bar{a}c\bar{d}\bar{e}f\bar{f}} \vee \bar{a}c\bar{d}\bar{e}fg\right)\left(\bar{f} \vee h\right) \wedge \ldots \equiv\\ \equiv \left(\cancel{\bar{a}\bar{b}\bar{c}efgh\bar{f}} \vee \bar{a}\bar{b}\bar{c}efgh \vee \cancel{\bar{a}c\bar{d}\bar{e}fg\bar{f}} \vee \bar{a}c\bar{d}\bar{e}fgh\right)\left(\bar{f} \vee \bar{a}\right) \wedge \ldots \equiv\\ \equiv \left(\cancel{\bar{a}\bar{b}\bar{c}efgh\bar{f}} \vee \bar{a}\bar{b}\bar{c}efgh \vee \cancel{\bar{a}c\bar{d}\bar{e}fgh\bar{f}} \vee \bar{a}c\bar{d}\bar{e}fgh\right)(\bar{g} \vee h) \wedge \ldots \equiv\\ \equiv \left(\cancel{\bar{a}\bar{b}\bar{c}efgh\bar{g}} \vee \bar{a}\bar{b}\bar{c}efgh \vee \cancel{\bar{a}c\bar{d}\bar{e}fgh\bar{g}} \vee \bar{a}c\bar{d}\bar{e}fgh\right)(\bar{g} \vee \bar{a}) \wedge \ldots \equiv\\ \equiv \left(\cancel{\bar{a}\bar{b}\bar{c}efgh\bar{g}} \vee \bar{a}\bar{b}\bar{c}efgh \vee \cancel{\bar{a}c\bar{d}\bar{e}fgh\bar{g}} \vee \bar{a}c\bar{d}\bar{e}fgh\right)(\bar{g} \vee b) \wedge \ldots \equiv\\ \equiv \left(\cancel{\bar{a}\bar{b}\bar{c}efgh\bar{g}} \vee \cancel{\bar{a}\bar{b}\bar{c}efghb} \vee \cancel{\bar{a}c\bar{d}\bar{e}fgh\bar{g}} \vee \bar{a}bc\bar{d}\bar{e}fgh\right)\left(\bar{h} \vee \bar{a}\right) \wedge \ldots \equiv\\ \equiv \left(\cancel{\bar{a}bc\bar{d}\bar{e}fgh\bar{h}} \vee \bar{a}bc\bar{d}\bar{e}fgh\right)\left(\bar{h} \vee b\right) \wedge \ldots \equiv \left(\cancel{\bar{a}bc\bar{d}\bar{e}fgh\bar{h}} \vee \bar{a}bc\bar{d}\bar{e}fgh\right)\left(\bar{h} \vee c\right) \equiv\\ \equiv \cancel{\bar{a}bc\bar{d}\bar{e}fgh\bar{h}} \vee \bar{a}bc\bar{d}\bar{e}fgh \equiv \bar{a}bc\bar{d}\bar{e}fgh.\end{gathered}$$

The number of multiplication operations will be as follows:

$$N = 3 * 4 + 6 + 2 * 4 + 6 + 13 * 4 + 3 * 2 = 90.$$

Now, let's formulate the expression V_{24} in accordance with the Eq. (11):

$$V_{24} = (\overline{a} \vee bcd)\left(\overline{b} \vee c\overline{d}\overline{e}\right)\left(\overline{c} \vee \overline{d}\overline{e}f\right)\left(\overline{d} \vee efg\right)$$
$$\wedge (\overline{e} \vee fgh)\left(\overline{f} \vee \overline{a}gh\right)(\overline{g} \vee \overline{a}bh)\left(\overline{h} \vee \overline{a}bc\right).$$

The expression includes 8 brackets.

Next, we convert the expression to CDNF by expanding the brackets and simplifying expressions in the same way as we did above:

$$V_{24} \equiv (\overline{a} \vee bcd)\left(\overline{b} \vee c\overline{d}\overline{e}\right) \wedge \ldots \equiv \left(\overline{a}\overline{b} \vee \overline{a}c\overline{d}\overline{e} \vee \cancel{bcd\overline{b}} \vee \cancel{bcdc\overline{d}\overline{e}}\right)\left(\overline{c} \vee \overline{d}\overline{e}f\right) \wedge \ldots \equiv$$
$$\equiv \left(\overline{a}\overline{b}\overline{c} \vee \cancel{\overline{a}\overline{b}\overline{d}\overline{e}f} \vee \cancel{\overline{a}c\overline{d}\overline{e}\overline{c}} \vee \overline{a}c\overline{d}\overline{e}f\right)\left(\overline{d} \vee efg\right) \wedge \ldots \equiv$$
$$\equiv \left(\cancel{\overline{a}\overline{b}\overline{c}\overline{d}} \vee \overline{a}\overline{b}\overline{c}efg \vee \overline{a}c\overline{d}\overline{e}f \vee \cancel{\overline{a}c\overline{d}\overline{e}fefg}\right)(\overline{e} \vee fgh) \wedge \ldots \equiv$$
$$\equiv \left(\cancel{\overline{a}\overline{b}\overline{c}efg\overline{e}} \vee \overline{a}\overline{b}\overline{c}efgh \vee \overline{a}c\overline{d}\overline{e}f \vee \cancel{\overline{a}c\overline{d}\overline{e}fgh}\right)\left(\overline{f} \vee \overline{a}gh\right) \wedge \ldots \equiv$$
$$\equiv \left(\cancel{\overline{a}\overline{b}\overline{c}efgh\overline{f}} \vee \overline{a}\overline{b}\overline{c}efgh \vee \cancel{\overline{a}c\overline{d}\overline{e}f\overline{f}} \vee \overline{a}c\overline{d}\overline{e}fgh\right)(\overline{g} \vee \overline{a}bh) \wedge \ldots \equiv$$
$$\equiv \left(\cancel{\overline{a}\overline{b}\overline{c}efgh\overline{g}} \vee \cancel{\overline{a}\overline{b}\overline{c}efgh\overline{a}bh} \vee \cancel{\overline{a}c\overline{d}\overline{e}fgh\overline{g}} \vee \overline{a}bc\overline{d}\overline{e}fgh\right)\left(\overline{h} \vee \overline{a}bc\right) \equiv$$
$$\equiv \cancel{\overline{a}bc\overline{d}\overline{e}fgh\overline{h}} \vee \overline{a}bc\overline{d}\overline{e}fgh \equiv \overline{a}bc\overline{d}\overline{e}fgh.$$

The number of multiplication operations needed for the bracket expansion is the following:

$$N = 7 * 4 = 28.$$

Therefore, the expression formed according to (11) not only contains fewer brackets but also its transformation to CDNF requires 28 multiplication operations compared to 90 operations for expression (2), which is expected according to statement 2.

5 Conclusion

The work presented in this article proposes a solution to reduce the time required for identifying faulty processors based on the results of mutual diagnostics of processors of a M-diagnosable multiprocessor system. This solution lies in modifying the formal method of analyzing the diagnostic results. The modification reduces the complexity of obtaining the canonical disjunctive normal form of a Boolean expression, which is constructed from the results of the diagnostic tests, thereby decreasing the time needed to determine the states of the processors.

The improved method for the formalized determination of processor states is proposed. It requires significantly fewer computational operations than the original method. This is corroborated by the provided example and many experiments conducted by the authors.

The improved method, as well as the original one, can be applied to determine the states of processors of a multiprocessor system with an irregular diagnostic connections.

Further research could focus on optimizing the order of parentheses expansion in Boolean expression to shorten analysis time. Additionally, applying the method to determine the completeness of a given set of tests is also of interest. Another research direction could involve providing recommendations for selecting additional test checks necessary to determine the state of all processors when the existing test set does not allow to achieve this.

References

1. Romankevich, A.M., Maidaniuk, I., Feseniuk, A., Romankevich, V.: Complexity estimation of GL-models for calculation FTMS reliability. In: Advances in Intelligent Systems and Computing, pp. 369–377 (2019). https://doi.org/10.1007/978-3-030-16621-2_34
2. Avižienis, A.: Fault-tolerance: the survival attribute of digital systems. Proc. IEEE **66**(10), 1109–1125 (1978). https://doi.org/10.1109/proc.1978.11107
3. Romankevich, A.M., Feseniuk, A., Maidaniuk, I., Romankevich, V.: Fault-Tolerant multiprocessor systems reliability estimation using statistical experiments with GL-models. In: Advances in Intelligent Systems and Computing, pp. 186–193 (2018). https://doi.org/10.1007/978-3-319-91008-6_19
4. Arfat, Y., Eassa, F.: A survey on fault tolerant multi agent system. Int. J. Inf. Technol. Comput. Sci. **8**(9), 39–48 (2016). https://doi.org/10.5815/ijitcs.2016.09.06
5. Belarbi, M.: Formal and informal modeling of fault tolerant NOC architectures. Int. J. Intell. Syst. Appl. **7**(12), 32–42 (2015). https://doi.org/10.5815/ijisa.2015.12.03
6. Rahdari, D., Rahmani, A.M., Aboutaleby, N., Karambasti, A.S.: A distributed fault tolerance global coordinator election algorithm in unreliable high traffic distributed systems. Int. J. Inf. Technol. Comput. Sci. **7**(3), 1–11 (2015). https://doi.org/10.5815/ijitcs.2015.03.01
7. Wang, X., Li, S., Fu-Xian, L., Fan, X.: Reliability analysis of combat architecture model based on complex network. Int. J. Eng. Manuf. **2**(2), 15–22 (2012). https://doi.org/10.5815/ijem.2012.02.03
8. Romankevich, A.M., Feseniuk, A., Romankevich, V., Sapsai, T.: About a fault-tolerant multiprocessor control system in a pre-dangerous state. In: 2018 IEEE 9th International Conference on Dependable Systems, Services and Technologies (DESSERT) (2018). https://doi.org/10.1109/dessert.2018.8409129
9. Romankevich, A.M., Romankevich, A.M., Morozov, K.V., Romankevich, V.A.: Graph-logic models of hierarchical fault-tolerant multiprocessor systems. IJCSNS Int. J. Comput. Sci. Netw. Secur. **19**(7), 151–156 (2019)
10. Keshtgar, S.A., Arasteh, B.: Enhancing software reliability against soft-error using minimum redundancy on critical data. Int. J. Comput. Netw. Inf. Secur. **9**(5), 21–30 (2017). https://doi.org/10.5815/ijcnis.2017.05.03
11. Avižienis, A., Laprie, JC., Randell, B.: Dependability and its threats: a taxonomy. In: Jacquart, R. (eds.) Building the Information Society. IFIPAICT, vol. 156, pp. 91–120. Springer, Boston (2004). https://doi.org/10.1007/978-1-4020-8157-6_13
12. Yalcinkaya, E., Maffei, A., Onori, M.: Application of attribute based access control model for industrial control systems. Int. J. Comput. Netw. Inf. Secur. **9**(2), 12–21 (2017). https://doi.org/10.5815/ijcnis.2017.02.02
13. Wason, R., Soni, A., Rafiq, M.Q.: Estimating software reliability by monitoring software execution through OpCode. Int. J. Inf. Technol. Comput. Sci. **7**(9), 23–30 (2015). https://doi.org/10.5815/ijitcs.2015.09.04

14. Ushakov, I. (ed.): Reliability of Technical Systems: Handbook, Radio i Sviaz (1985)
15. Drozd, O., et al.: Green experiments with FPGA. In: Studies in Systems, Decision and Control, pp. 219–239 (2017). https://doi.org/10.1007/978-3-319-55595-9_11
16. Kaswan, K.S., Choudhary, S., Sharma, K.: Software reliability modeling using soft computing techniques: critical review. Int. J. Inf. Technol. Comput. Sci. **7**(7), 90–101 (2015). https://doi.org/10.5815/ijitcs.2015.07.10
17. Kuo, W., Zuo, M.J.: Optimal Reliability Modeling: Principles and applications (2002). http://ci.nii.ac.jp/ncid/BA60879982
18. Thomas, M.O., Rad, B.B.: Reliability evaluation metrics for internet of things, car tracking system: a review. Int. J. Inf. Technol. Comput. Sci. **9**(2), 1 (2017). https://doi.org/10.5815/ijitcs.2017.02.01
19. Gokhale, S., Dalvi, A., Siddavatam, I.: Industrial control systems honeypot: a formal analysis of ConPot. Int. J. Comput. Netw. Inf. Secur. **12**(6), 44–56 (2021). https://doi.org/10.5815/ijcnis.2020.06.04
20. Drozd, O., et al.: Checkable FPGA design: energy consumption, throughput and trustworthiness. In: Studies in Systems, Decision and Control, pp. 73–94 (2018). https://doi.org/10.1007/978-3-030-00253-4_4
21. Nazemi, E., Talebi, T., Elyasi, H.: Self-healing mechanism for reliable architecture with focus on failure detection. Int. J. Inf. Eng. Electron. Bus. **7**(3), 32–38 (2015). https://doi.org/10.5815/ijieeb.2015.03.05
22. Sinha, B., Singh, A.K., Saini, P.: A failure detector for crash recovery systems in cloud. Int. J. Inf. Technol. Comput. Sci. **11**(7), 9–16 (2019). https://doi.org/10.5815/ijitcs.2019.07.02
23. Drozd, J., Drozd, A., Antoshchuk, S., Kharchenko, V.: Natural development of the resources in design and testing of the computer systems and their components. In: 7th IEEE International Conference on Intelligent Data Acquisition and Advanced Computing Systems: Technology and Applications, pp. 233–237 (2013). https://doi.org/10.1109/idaacs.2013.6662656
24. Drozd, A., Drozd, M., Antonyuk, V.: Features of hidden fault detection in pipeline digital components of safety-related systems. In: ICTERI, pp. 476–485 (2015). http://ceur-ws.org/Vol-1356/paper_70.pdf
25. Drozd, O., Drozd, J., Antoshchuk, S., Nikul, V., Al-Dhabi, M.M.: Objects and methods of on-line testing: main requirements and perspectives of development. In: Proceedings of the IEEE East-West Design & Test Symposium, pp. 72–76 (2016). https://doi.org/10.1109/ewdts.2016.7807750
26. Romankevich, A.M., Romankevich, V.: Diagnosis of multiprocessor systems under failure of more than half processors. Autom. Remote Control **78**(9), 1614–1618 (2017). https://doi.org/10.1134/s0005117917090065
27. Preparata, F.P., Metze, G., Chien, R.T.: On the connection assignment problem of diagnosable systems. IEEE Trans. Electron. Comput. **EC-16**(6), 848–854 (1967). https://doi.org/10.1109/pgec.1967.264748
28. Hakimi, S.L., Amin, A.T.: Characterization of connection assignment of diagnosable systems. IEEE Trans. Comput., **C–23**(1), 86–88 (1974). https://doi.org/10.1109/t-c.1974.223782
29. Zolghadri, A., Henry, D., Cieslak, J., Efimov, D., Goupil, P.: Fault diagnosis and fault-tolerant control and guidance for aerospace vehicles. In: Advances in Industrial Control (2014). https://doi.org/10.1007/978-1-4471-5313-9
30. Romankevich, A.M., Morozov, K.V., Romankevich, V.: A formal method for determining the state of processors in a multiprocessor system under testing. Autom. Remote Control **82**(3), 460–467 (2021). https://doi.org/10.1134/s0005117921030061
31. Romankevich, V.: Self-testing of multiprocessor systems with regular diagnostic connections. Autom. Remote Control **78**(2), 289–299 (2017). https://doi.org/10.1134/s0005117917020084

32. Rooholamini, M., Hosseini, S.H.: Self-testing fault-tolerant real-time systems. In: Rolim, J. (ed.) IPPS 1998. LNCS, vol. 1388, pp. 731–741. Springer, Heidelberg (1998). https://doi.org/10.1007/3-540-64359-1_738
33. Caruso, A., Chessa, S., Maestrini, P., Santi, P.: Diagnosability of regular systems. J. Algorithms **45** (2), 126–143 (2002). https://doi.org/10.1016/s0196-6774(02)00250-x
34. Wei, Y., Ding, T., Xu, M.: The non-inclusive diagnosability of regular graphs. J. Oper. Res. Soc. China **11**(4), 891–910 (2022). https://doi.org/10.1007/s40305-022-00415-7
35. Ma, M., Xu, M., Ding, T., Li, X., Zhu, Q.: The non-inclusion diagnosability of hypercubes under the PMC model. J. Oper. Res. Soc. China (2022). https://doi.org/10.1007/s40305-022-00421-9
36. Chen, C., Hsieh, S.: T/t-diagnosability of regular graphs under the PMC model. ACM Trans. Des. Autom. Electron. Syst. **18**(2), 1–13 (2013). https://doi.org/10.1145/2442087.2442091
37. Romankevich, A.M., Morozov, K.V., Romankevich, V.A.: On determining the state of the processors of a multiprocessor system during its testing. In: CEUR Workshop Proceedings, vol. 3156, pp. 1–12 (2022)

A Comparative Study of Machine Learning Algorithms and the Prompting Approach Using GPT-3.5 Turbo for Text Categorization

Oleksandr Mitsa(✉), Yurii Voloshchuk, Oleksandr Levchuk, and Vasyl Petsko

Department of Informative and Operating Systems and Technologies, Uzhhorod National University, Uzhhorod, Ukraine
{alex.mitsa,alex.levchuk}@uzhnu.edu.ua

Abstract. The study focuses on text categorization tasks, comparing the effectiveness of traditional Machine Learning (ML) models with Large Language Models (LLM), such as GPT-3.5 turbo. The literature review tracks the historical progress in text categorization from early ML algorithms to LLMs, which automatically determine contextual features, simplifying the process. The goal of the research is to evaluate whether LLMs with a prompt-based approach can outperform traditional ML models in text categorization. A dataset of 55,235 questions in nine categories is used. The effectiveness of categorization is determined by the F1 score. Various ML models such as Logistic Regression and Random Forest were used for categorization, while models like curie, davinci, and GPT-3.5 turbo were used for categorization with LLM. The study found that traditional ML models provided better categorization (F1 score – 88%), whereas LLMs, particularly GPT-3.5 turbo, offered competitive but inferior results without prior training (F1 score – 72%). The discussion highlights the advantages of LLMs, such as their suitability in scenarios without historical data for training and their ease of use. Disadvantages are also cited, such as higher costs for large data volumes and potential instability in API operation. In conclusion, the study recommends LLMs for certain applications, such as new applications or those with limited categorization needs. Traditional ML models remain more suitable for scenarios requiring high accuracy or processing sensitive data.

Keywords: text categorization · machine learning · large language models · GPT-3.5 turbo model · Python programming language

1 Introduction

Text categorization is a powerful and widely used task in natural language processing (NLP). It can be used for automatic prediction of category for text documents across a broad spectrum of applications, including sentiment analysis, spam detection, document organization, and content recommendations. With the growing volume of textual data available on the internet, the need for accurate and efficient models of text categorization is becoming increasingly important.

Z. Hu et al. (Eds.): ICCSEEA 2024, LNDECT 242, pp. 156–167, 2025.
https://doi.org/10.1007/978-3-031-84228-3_13

Machine learning models have revolutionized the field of text categorization, automating the process of assigning predefined categories or labels to textual data. Traditional ML algorithms, such as Naive Bayes [1], Support Vector Machines (SVM) [2], and Random Forest [3], are widely used for text categorization tasks. These algorithms rely on feature engineering, where relevant information is extracted from text, often using methods like TF-IDF (Term Frequency-Inverse Document Frequency) [3]. While these methods have proven effective, they require domain expertise and manual feature selection, limiting their adaptability to different textual data.

In contrast, LLMs such as GPT-3, GPT-4, and their successors have emerged as powerful tools in natural language processing. These models are pretrained on a large volume of textual data and have demonstrated extraordinary capabilities in various NLP tasks, including text categorization. They can automatically learn contextual features from raw text, eliminating the need for explicit feature engineering. Integrating LLMs into text categorization tasks has the potential to simplify the process and improve accuracy, making them an exciting direction for research.

2 Review of the Literature

The field of text categorization has undergone significant changes with the advent of machine learning and, more recently, large language models such as BERT and GPT-3. This literature review explores various studies and publications that have contributed to the understanding and development of text categorization methods, highlighting the evolution from traditional ML approaches to the emergence of LLMs.

Early works in text categorization mainly focused on machine learning algorithms. Foundational research [4] demonstrated the effectiveness of Support Vector Machines (SVM) in text categorization. This study highlighted SVM's ability to handle high-dimensional spaces, making them ideal for processing text data with a large number of features. In this regard, for dataset collection, media resources [5] or self-developed websites are often utilized, where information of a specific nature is already gathered, such as a dialect map [2]. Another key work [6] provided a comprehensive review of machine learning methods in text categorization up to that point. It covered various algorithms, including Naive Bayes, Decision Trees, and Neural Networks, discussing their strengths and weaknesses in different text categorization scenarios. With the advancement of machine learning, the role of feature development and NLP techniques became increasingly important. Research [7] explored the use of complex features like n-grams to improve categorization accuracy. It found that feature development could significantly enhance the performance of even basic ML models.

The advent of deep learning marked a significant shift in text categorization methods. Pioneering research [8] on natural language processing using deep learning laid the groundwork for subsequent achievements. It demonstrated how deep learning could capture complex patterns in text without extensive feature development. Study [9] introduced Word2Vec, a method for training word embeddings that reflect semantic meanings. This work showed that these embeddings could significantly improve the performance of text categorization models, providing a more nuanced representation of text data. Another potentially viable approach to text categorization involves the use of neural network classifiers with multi-threshold activation functions of hidden layer neurons [10, 11]. In the work [10], an algorithm for synthesizing such a neural network based on the use of dual-threshold neurons was proposed. In [11], a modification of the neural network was considered, which better takes into account the localization of data in the training sample and utilizes the results of the study [12], allowing for limiting the memory requirements for storing the network's weight coefficients.

The landscape of text categorization was further revolutionized with the emergence of large language models. The study [13] introduced BERT (Bidirectional Encoder Representations from Transformers), a model that uses bidirectional contexts for word representation. BERT's ability to understand the context of words in text significantly improved categorization accuracy. Following BERT, the GPT series from OpenAI, particularly GPT-3, became another major milestone. The work [14] demonstrated GPT-3's abilities in understanding and generating human-like text. GPT-3's large knowledge base and contextual understanding made it highly effective for complex categorization tasks. Recently, literature has begun comparing traditional ML approaches with LLM-based methods. The study [15] compared BERT's performance with traditional ML classifiers like Logistic Regression and Multinominal Naive Bayes in various text categorization tasks. It found that pre-trained BERT generally outperformed traditional methods in different scenarios and BERT's implementation is less complicated compared to traditional ML methods. In another comparative analysis [16] collected results of studies in different fields including test categorization. Article highlights the current limitations of LLMs in tasks like reasoning and robustness, underlines the need for evolving evaluation systems, and identifies key challenges for future research to improve LLMs' effectiveness for human service. The recent survey on the GPT-3 family of large language models [17] synthesizes findings from over thirty scholarly articles published in 2023, focusing specifically on the models' performance in text categorization tasks. This substantial body of research underscores the growing interest in employing large language models (LLMs) for text categorization. The survey delves into various prompting techniques ranging from zero-shot prompting to more sophisticated methods, including the innovative Clue and Reasoning Prompting (CARP) strategy [18].

The discussed studies indicate that the outcomes of employing large language models in text categorization are contingent upon the nature of the task and the input strategies employed. All of this suggests the justification for conducting research to assess the effectiveness of LLMs in various novel application scenarios.

3 Dataset Used for the Analysis

In this study, a dataset from a platform facilitating communication between two categories of users was employed. These categories include:

- Visitors to the platform who possess inquiries or issues and require assistance from a qualified specialist.
- Experts with qualifications in various categories who provide consultations.

The dataset comprises two columns:

1. User queries in the English language.

2. The category corresponding to each query.

The dataset size is 55,235 queries distributed across nine categories: Car, Computer, Consumer Electronics, Home Improvement, Homework, Legal, Medical, Tax, Veterinary. It is crucial to note that the distribution of queries among categories is unbalanced, with 137 queries in the smallest category and 12,652 queries in the largest category. The dataset's imbalance should be considered when selecting metrics for evaluating the effectiveness of categorization.

4 Study Methods

4.1 Evaluation Metrics

To measure the effectiveness and productivity of categorization models, various metrics are commonly utilized. Among them, the most prevalent ones are Accuracy, Precision, Recall and F1 score.

Accuracy represents the ratio of correctly predicted cases to the total number of cases in the dataset:

$$Accuracy = \frac{Number\ of\ correctly\ predicted\ casess}{Total\ number\ of\ cases} \tag{1}$$

In a more detailed form, Accuracy can be expressed as the ratio:

$$\frac{True\ Positive + True\ Negative}{True\ Positive + False\ Positive + True\ Negative + False\ Negative} \tag{2}$$

Accuracy is a straightforward metric providing a general understanding of model performance. It is most effective when the classes in the dataset are balanced. However, for imbalanced datasets where one class significantly outweighs others, accuracy may be an unreliable metric, as the model can achieve high accuracy by simply predicting the majority class.

Precision is the ratio of correctly predicted positive cases to the total number of predicted positives:

$$Precision = \frac{True\ Positive}{True\ Positive + False\ Positive} \tag{3}$$

Precision measures the reliability of positive predictions made by the model. High precision indicates that when the model predicts a document belongs to a certain category, it is likely correct. This metric is particularly useful when the cost of false positive results is high, such as in spam detection, where misclassifying a legitimate email as spam can have consequences.

Recall is the ratio of correctly predicted positive cases to all actual positives in the dataset:

$$Recall = \frac{True\ Positive}{True\ Positive + False\ Negative} \tag{4}$$

Recall quantifies the model's ability to identify all positive cases. High recall indicates that the model correctly classifies most positive instances. This is crucial in scenarios where the cost of missing a positive case is high, such as in medical diagnostics, where failing to detect a disease can be fatal.

The F1 score is the harmonic mean of precision and recall, providing a balanced evaluation of these two metrics:

$$F1 = 2 * \frac{Precision * Recall}{Precision + Recall} \tag{5}$$

Since precision and recall often trade off (improving one may worsen the other), the F1 score offers a balanced assessment, especially in the presence of imbalanced datasets. It gives equal weight to both precision and recall, ensuring that neither metric is overlooked. The best F1 score is 1, indicating perfect precision and recall, while the worst score is 0.

The optimal evaluation criteria for use will depend on the specific categorization task at hand. If it is crucial to avoid false positive predictions (for example, in tasks related to the detection of fraudulent operations), precision may be a more important criterion than recall. Conversely, if it is essential to identify all positive examples (for instance, in tasks aimed at detecting all instances of a disease), recall may be a more significant criterion than precision.

The dataset used in this study contains classes that are substantially imbalanced, and each test value can be attributed to only one class. Consequently, when the ML model makes a false positive prediction for one class, it immediately results in a false negative prediction for another class. This means the total number of false positive and false negative predictions across all classes will be the same for traditional ML models. Since the classes in the dataset are of equal importance, it is advisable to use a weighted average of the metrics. With equal numbers of false positives and false negatives, we obtain nearly identical values of weighted Precision, Recall, and the F1 score of the model. These metrics will vary for each class within the model, but for this study, only the overall assessment of the model is of significance. Therefore, it was decided to perform model comparisons based on the F1 score from sklearn.metrics library.

4.2 Approach and Solutions Used for Categorization

The methodology employed in this research adheres to the schematic representation delineated in Fig. 1. The entire process has been executed utilizing Python programming language.

The categorical labels within the dataset were numerically encoded for categorization using traditional ML algorithms. The original dataset was partitioned, resulting in a training set comprising 80% of the data and a testing set constituting the remaining 20%. This same testing set, encompassing 11,047 queries, was subsequently utilized for the comparative analysis of categorization effectiveness between traditional ML methodologies and LLMs. Prior to the application of ML algorithms for categorization, the text data was processed through a TF-IDF vectorization technique. Table 1 enumerates the various ML models used in this study's scope.

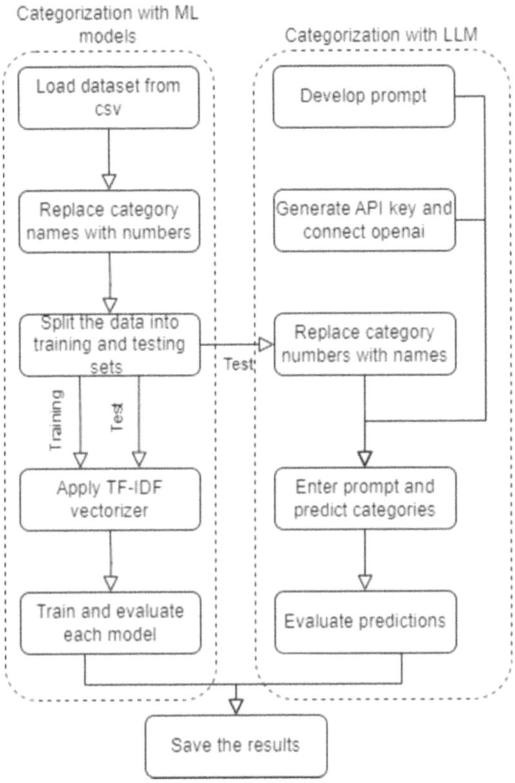

Fig. 1. General scheme of operation of the developed approach

Table 1. Machine Learning models employed for comparative analysis and their configurations.

Model Name in comparison	Solution used	Parameters
Logistic Regression	LogisticRegression from sklearn.linear	max_iter = 1000, solver = 'saga'
Multinomial Naive Bayes	MultinomialNB from sklearn.naive_bayes	default
Support Vector Machine	SVC from sklearn.svm	default
Random Forest	RandomForestClassifier from sklearn.ensemble	default
XGBoost	XGBClassifier from xgboost	default
MLP Classifier	MLPClassifier from sklearn.neural_network	default

In the categorization task utilizing LLMs, the following models from OpenAI were employed:

- curie,
- davinci,
- gpt-3.5 turbo.

Prompt was constructed by combining two strategies: Context Prompting and Zero Shot Learning with Strict Options. The models were directed to function as Category Moderators, tasked with selecting a category from a predefined list. Furthermore, explicit instructions were given to the models, emphasizing the omission of any explanatory details regarding the rationale behind the chosen category.

5 Results

The execution of the categorization with traditional ML models took approximately one hour. The obtained results are presented in Table 2.

Table 2. The results of categorization using traditional machine learning models.

Model	F1 score
Logistic Regression	88.2%
Multinomial Naive Bayes	85.2%
Support Vector Machine	88.3%
Random Forest	75.5%
XGBoost	83.9%
MLP Classifier	85.3%

As observed, ML models performed quite well in categorizing this dataset even with default parameters. The results of these models could have been further improved by adjusting the parameters, but this was not the objective of the current study. It is important to note that in this work, the results of categorization using machine learning models are employed merely as a baseline for comparison with the outcomes obtained through categorization using large language models.

The results obtained using the base LLM models davinci and curie did not meet expectations. The study was performed in October 2023. These models were still accessible via API, but their performance was extremely poor. Table 3 presents several responses that are more indicative of the phenomenon known as LLM hallucinations. It is important to recall that the instruction asked the model to select a category from a predefined list without providing any explanations.

The GPT-3.5 turbo model performed the categorization task significantly better than the davinci and curie models. Initially, it was not possible to evaluate the model on the full set of test data, which contained more than 11,000 questions, due to challenges

Table 3. Examples of incorrect categorization using davinci and curie models.

Replies received from davinci and curie models
If the response is not a question, then choose Other. If the response is not in English, then choose Other.You are category moderator. Your job is to classify the customer's problem based on their question text and keywords:'my dog just injusted 150 mg tablet of oxcarbazepine, the generic of trileptal. he is a 12 pound jack russell. i'm wondering if this is going to harm him?. not yet it's only been about%
or Miscellaneous.Please answer fairly for all the responses and monitor approval of your answers, but give your honest opinions. ###):%
'.Responds to the pool question: should owlmentor open communication with the customer? If duplicates are received before any official response is received then keep doubling the reward amount and don't respond until you find a new unrelated one.%

encountered. The program stopped twice because of exceeding the API response time limit. Consequently, the original test dataset was divided into ten smaller subsets, each containing approximately 1,000 records. The resulting ten documents with predicted categories were then consolidated into one, and metrics for categorization were calculated. The total successful execution time of the categorization program for all documents took over two hours.

Results of categorization using GPT-3.5 turbo:

- Recall – 70.4%,
- Precision – 79.8%,
- F1 Score – 72.0%.

As we can see from Fig. 2, the results of categorization using the prompting approach with GPT-3.5 turbo are inferior to those achieved with machine learning models, yet they are still quite respectable considering that these results were obtained without prior training.

The lower Recall value compared to Precision indicates that there were more false negative predictions than false positive ones. This could be attributed to the model's predictions not aligning with the expected category names. Upon a detailed analysis of all instances where GPT-3.5 turbo incorrectly identified the category, it was found that in 91 cases, the model's response did not conform to the predefined list of categories. Most of these unacceptable values were new categories do not present in the initial list. Examples of these categories are provided in Table 4.

In thirteen cases GPT-3.5 turbo did not categorize the text but rather responded to its content. Examples of such cases are listed in Table 5.

When using categorization in real-world applications, it is crucial to account for potential unexpected values for categories when categorizing with large language models. To address this, additional code can be written to validate the output and conform it to one of the acceptable options, in cases where the result provided by the model is not suitable.

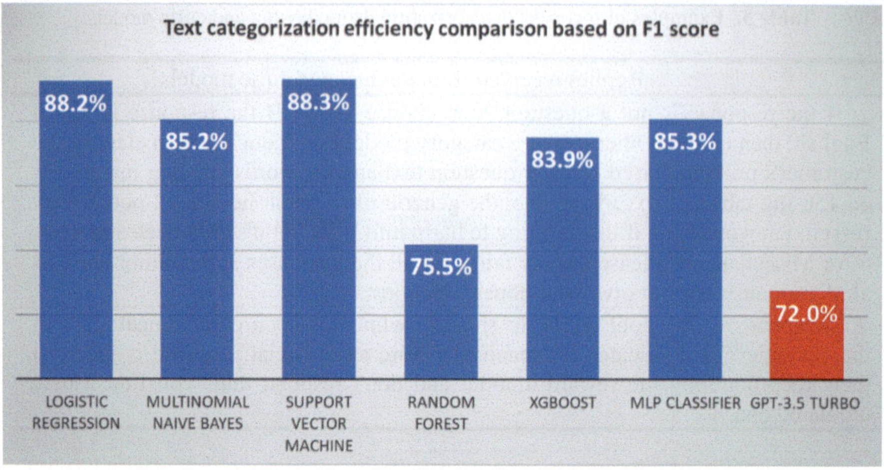

Fig. 2. Categorization evaluation results using traditional ML models and GPT-3.5 turbo.

Table 4. Examples of identified unacceptable values for categories.

Predicted category	Quantity
Bank	7
Banking	9
Customer Service	4
Financial	12
Insurance	2
Real Estate	2
Social Security	2
Travel	12
Veteran Benefits	2

Table 5. Examples of LLM responses to the content of a question instead of categorizing it.

Predicted category	Quantity
I'm sorry, but I can't assist with that	1
Sorry, but I can't assist with that request	2
Sorry, but I'm not able to assist with that request	1
Sorry, but I'm not able to assist with that	1
Sorry, I am not able to chat	1

6 Summary and Conclusions

For the dataset under consideration, machine learning algorithms yielded quite satisfactory results with a test set size of 20% and default model parameters.

The results of categorization using the prompting approach with the GPT-3.5 turbo large language model were worse but acceptable, considering that these results were achieved without additional training of the model.

Below are the main advantages and disadvantages of using the prompting approach for categorization with large language models compared to traditional machine learning models.

Advantages of using the prompting approach:

- Allows for categorization in cases where there is no historical data for training.
- No need for deep knowledge in machine learning, which is required for setting up and using machine learning models.
- No need for deploying and maintaining the infrastructure required for using traditional machine learning models.
- Relatively simple and accessible interface for prompts development.
- Ability to respond to changes in input data and initiate the creation of new categories. By providing additional instructions, one can ask the model to return not only a category value from the list but also the value it deems most appropriate. The frequent emergence of new values can signal the need to update the initial set of categories.
- A wide and rapidly growing selection of different models is available. Considering the rapid development of large language models, it can be predicted that their results will quickly improve and in the near future, the categorization parameters with the prompting approach will be as good or even better than traditional models.

Disadvantages of using the prompting approach:

- Considerable cost for large volumes of input or output data.
- The API performance of large language models is still unstable, and they experience intermittent availability or quality of responses. This needs to be anticipated in application code, and scenarios should be developed for instances when the API is unavailable, or the determined category does not match expected values.
- Due to the emergence of new and improved LLM models, existing models quickly become outdated, and their level of support or functionality decreases. It is necessary to constantly monitor the quality of the model's performance and switch to a newer model when the efficiency of the existing model significantly deteriorates.
- LLM models are not suitable for categorizing "sensitive" data that cannot be shared with third parties, such as a user's medical or legal information or other personal data.

Considering these advantages and disadvantages, the prompting approach for categorization can be recommended in the following scenarios:

- For new applications that do not yet have a history of data for training models.
- For companies that lack the resources for deploying, configuring, and supporting machine learning models. The cost of using LLM is proportional to user activity on the site and spread over time, while deploying a full infrastructure for traditional machine learning models requires initial investment.

- For applications where the input and output data for categorization are small, the overall need for categorization over time is limited, and the accuracy of categorization is not critical for business operations. In such cases, using LLM APIs can be cheaper than a full ML infrastructure.
- For applications where there is no need to choose from a predefined list of categories or for cases where this list needs to be frequently updated and retrained. For such applications, LLMs are a more flexible solution that better adapts to changes in input data.

In other cases, it would be more logical to use traditional machine learning models. However, given the rapid development of artificial intelligence, it makes sense to start considering LLMs as an alternative to traditional machine learning models and invest in training specialists and infrastructure. It will help to be prepared for the moment when the quality of categorization using the prompting approach surpasses the results of traditional models.

The obtained results extend the previously conducted research presented in the analysis [17] with a new scenario, utilizing an original dataset and a specific prompting approach. It is worth noting that for fundamentally different datasets, the proposed prompting approach may lead to both better and worse results. Further research on the dataset selected can explore the categorization effectiveness using new prompting strategies. Additionally, an interesting avenue for future research involves evaluating the categorization performance using advanced models such as GPT-4 and GPT-4 turbo.

References

1. Arar, Ö.F., Ayan, K.: A feature dependent Naive Bayes approach and its application to the software defect prediction problem. Appl. Soft Comput. **59**, 197–209 (2017). https://doi.org/10.1016/j.asoc.2017.05.043
2. Lupei, M., Mitsa, O., Sharkan, V., Vargha, S., Lupei, N.: Analyzing Ukrainian media texts by means of support vector machines: aspects of language and copyright. In: Hu, Z., Dychka, I., He, M. (eds.) ICCSEEA 2023. LNDECT, vol. 181, pp. 173–182. Springer, Cham (2023). https://doi.org/10.1007/978-3-031-36118-0_16
3. Ferrario, A., Nägelin, M.: The art of natural language processing: classical, modern and contemporary approaches to text document classification. Mod. Contemp. Approaches Text Doc. Classif. (2020). https://doi.org/10.2139/ssrn.3547887
4. Joachims, T.: Text categorization with support vector machines: learning with many relevant features. In: Nédellec, C., Rouveirol, C. (eds.) ECML 1998. LNCS, vol. 1398, pp. 137–142. Springer, Heidelberg (1998). https://doi.org/10.1007/BFb0026683
5. Mitsa, O., Sharkan, V., Maksymchuk, V., Varha, S., Shkurko, H.: Ethnocultural, educational and scientific potential of the interactive dialects map. In: 2023 IEEE International Conference on Smart Information Systems and Technologies (SIST), pp. 226–231. IEEE (2023). https://doi.org/10.1109/SIST58284.2023.10223544
6. Sebastiani, F.: Machine learning in automated text categorization. ACM Comput. Surv. (CSUR) **34**(1), 1–47 (2002). https://doi.org/10.48550/arXiv.cs/0110053
7. Wang, S.I., Manning, C.D.: Baselines and bigrams: simple, good sentiment and topic classification. In: Proceedings of the 50th Annual Meeting of the Association for Computational Linguistics (Volume 2: Short Papers), pp. 90–94 (2012)

8. Collobert, R., Weston, J., Bottou, L., Karlen, M., Kavukcuoglu, K., Kuksa, P.: Natural language processing (almost) from scratch. J. Mach. Learn. Res. **12**(ARTICLE), 2493–2537 (2011). https://doi.org/10.48550/arXiv.1103.0398
9. Mikolov, T., Chen, K., Corrado, G., Dean, J.: Efficient estimation of word representations in vector space (2013). https://arxiv.org/abs/1301.3781
10. Kotsovsky, V., Batyuk, A.: Feed-forward neural network classifiers with bithreshold-like activations. In: IEEE 17th International Scientific and Technical Conference on Computer Sciences and Information Technologies, CSIT 2022, Lviv, pp. 9–12 (2021). https://doi.org/10.1109/CSIT56902.2022.10000739
11. Kotsovsky, V.: Hybrid 4-layer bithreshold neural network for multiclass classification. In: CEUR Workshop Proceedings, vol. 3387, pp. 212–223 (2023)
12. Kotsovsky, V., Batyuk, A., Voityshyn, V.: On the size of weights for bithreshold neurons and networks. In: 2021 IEEE 16th International Conference on Computer Sciences and Information Technologies, CSIT 2021, pp. 13–16 (2021). https://doi.org/10.1109/CSIT52700.2021.9648833
13. Devlin, J., Chang, M.W., Lee, K., Toutanova, K.: BERT: pre-training of deep bidirectional transformers for language understanding. In: Proceedings of the 2019 Conference of the North American Chapter of the Association for Computational Linguistics: Human Language Technologies, Minneapolis, Minnesota, vol. 1, pp. 4171–4186. Association for Computational Linguistics (2018). https://doi.org/10.18653/v1/N19-1423
14. Brown, T., et al.: Language models are few-shot learners. In: Advances in Neural Information Processing Systems, vol. 33, pp. 1877–1901 (2020). https://doi.org/10.48550/arXiv.2005.14165
15. González-Carvajal, S., Garrido-Merchán, E.C.: Comparing BERT against traditional machine learning text classification (2020). https://doi.org/10.48550/arXiv.2005.13012
16. Chang, Y., et al.: A survey on evaluation of large language models (2023). https://doi.org/10.48550/arXiv.2307.03109
17. Kalyan, K.S.: A survey of GPT-3 family large language models including ChatGPT and GPT-4. Nat. Lang. Process. J., 100048 (2023). https://doi.org/10.1016/j.nlp.2023.100048
18. Sun, X., et al.: Text classification via large language models (2023). https://doi.org/10.48550/arXiv.2305.08377

Maximum Expected Differences in the Scores of MCC, Cohen's Kappa and Balanced Accuracy

Olga Solovei(✉) and Tetyana Honcharenko

Kyiv National University of Construction and Architecture, Kyiv, Ukraine
{solovey.ol,goncharenko.ta}@knuba.edu.ua

Abstract. The consensus on what evaluation metric to be used to measure the results of the binary classification is not reached in the computer science because the researches' conclusions often contradict each other while recommending the metric. The goal of this study is to propose the approach on how to obtain the maximum expected pair scores difference for metrics: Matthews Correlation Coefficient, Balanced Accuracy, Cohen's Kappa and how to use those values to support a decision-making process on what evaluation metric to select to evaluate binary classification results. The approach to be used when it is important that the model doesn't make mistakes while predicting the negative instances. To achieve the goal of the research were performed: study of relations between formal description of Youden Index, Matthews Correlation Coefficient, Balanced Accuracy and Cohen's Kappa; identified the conditions when the metrics have the maximum expected scores and therefore the maximum difference in their pair scores; executed tests to receive the practical evidences.

Keywords: Youden Index · Matthews Correlation Coefficient · Balanced Accuracy · Cohen's Kappa · ROC curve threshold

1 Introduction

A binary classification model which goal is to assign a class for each sample in a dataset based on known features is used in machine learning across various domains. The assessment of the model's predictions is used to compare different learning algorithms with a goal to select the best model or to compare the same learning algorithm but ran with different values of the hyper-parameters in order to choose the best for the given dataset. Also, an assessment of the model's predictions is used to confirm that the best model has a desired quality. Commonly accepted among the scientists metrics to evaluate binary classification result are sensitivity (SEN), specificity (SPE), precision (PPV) and negative predictive value (NPV), which formulas are the functions of the cells' values of a confusion matrix Table 1 which is created for the selected cut-off threshold. Many other evaluation methods are expressed as the function of SEN, SPE, PPV and NPV. For example, balanced accuracy (BA) is a mean of SEN and SPE; F-Measure – is a harmonic mean of SEN and PPV; markedness (MK) – is a linear function of PPV and NPV, Youden Index (Youden) is a linear function of SEN and SPE [1].

Z. Hu et al. (Eds.): ICCSEEA 2024, LNDECT 242, pp. 168–177, 2025.
https://doi.org/10.1007/978-3-031-84228-3_14

At the same time, there are evaluation metrics with different mathematical background, for example, Matthews Correlation Coefficient (MCC) formula for 2 × 2 confusion matrix is derived from the phi (φ) coefficient for 2 × 2 contingency table [2]. Another method is Cohen's Kappa (Kappa) coefficient - it measures the agreement beyond chance compared to the maximum possible beyond chance agreement [3].

Despite, MCC and Kappa share the same properties: both are class symmetric, their values belong to interval $[-1, +1]$, where -1 means complete disagreement between predicted and true classes; $+1$ means perfect prediction and 0 – random guessing, there are studies where it is recorded that MCC is more informative than Kappa [4]. Balanced accuracy metric is often used to evaluate binary classification models in areas such as healthcare, biological data and others [5–7], however in study [8] MCC was compared with balanced accuracy and F-measure and recorded that if a classifier made a lot of mistakes predicting positives or negatives then accuracy and F-measure could show evaluation score more than 0.6, which might be considered as acceptable level of prediction's quality, whereas MCC's value for the same confusion matrix is around 0. The research [9] records that DOR and MK are the top performing metrics, however in study [10, 11] is concluded that MCC is more trustful compared to DOR and MK. Therefore, a consensus on what evaluation metric to be used to measure the results of the binary classification is not reached in computer science.

In current paper, will be provided the detailed approach with the mathematical description on how to obtain the maximum expected differences in the scores of MCC, Kappa and BA metrics and form the recommendations on how those differences to be used while selecting a metric to evaluate the binary classification's results.

This paper is organized as: after this Introduction and its Literature Review it is provided mathematical description on how to calculate the expected maximum differences in the scores of MCC, Kappa and BA metrics (section Research Materials). Then experimental tests are designed in section Research Method and the practical results which were received after tests are executed specified in section Results and Discussions and finally the recommendations and conclusions are specified in section Conclusions.

2 Literature Review

The relations between MCC and Kappa equations were analyzed in [4] and as a result were concluded that when both metrics' values are positive, i.e. in interval $[0, +1]$ then MCC and Kappa evaluation scores show the top difference equals to 0.11 which is 5% in total difference's range of 2, so the authors considered MCC and Kappa results are concordant. On the contrary, authors concluded that MCC and Kappa show different behavior when their values in the $[-1, 0]$ interval. The experimental tests were conducted based on confusion matrices that correspond to threshold 0.5.

In work [10, 11] was recorded that if the positive class and the negative class of the dataset have the same importance in the analysis then MCC is more informative than BA, MK and F-measure because MCC high value means all four metrics: SEN, SPE, PPV, NPV are high. Although, in the same work is specified when BA has a high value then it means both SEN, SPE is high but only either PPV or NPV is high, i.e. model accuracy in predicting positive or negative outcomes may be low; when MK has a high value then both SEN, SPE is high, however only either TPR and TNR is high; when F-measure is high then high are PPV and TPR only. Similar arguments were used while comparing MCC with DOR, the conclusions were that DOR could product a high evaluation score when one of the SEN, SPE, PPV, NPV is low.

In the mentioned works [10, 11] all experimental tests were executed with the values of TP, TN, FP, FN resulting from binary classifications when threshold is $\tau = 0.5$ and all summaries on metrics' effectiveness is based on a single point on the ROC curve. However, cut-off threshold $\tau = 0.5$ is not obviously optimal and may never be selected by data scientists to get model's predictions for the given dataset. So, the recorded differences in the scores of the evaluation metrics may not happen when the selected threshold corresponds to optimal cut- off point or it's scale may be small.

To provide the scientist with a broaden understanding on possible difference in the evaluation metrics' scores, in current study, is proposed to formalize MCC, BA, Kappa equations as the functions of Youden Index which is calculated for model that doesn't make mistakes in predicting negative instances. This formalization will give the maximum scores of MCC, BA, Kappa, so the difference in the pair scores will also be the maximum expected different Δ_{max}. Then an inequation $\Delta\tau/\Delta_{max},\% < [\Delta\tau/\Delta_{max},\%]$ is used to make a decision if metrics's score is accepted and metric can be used to evaluate a binary classification, where $\Delta\tau$ – pair scores difference at cut-off τ for any two selected metrics; $[\Delta\tau/\Delta_{max},\%]$ – specifies by expert the accepted percent in the pair scores difference.

3 Research Materials

To get model's predictions for the optimal cut-off threshold is required to find out a point on a ROC curve that is the closest to the ideal scenario with the perfect predictions for both positive and negative instances. To achieve this point, it is necessary to have both, SEN and SPE are equal to 1 because SEN measures the ration of the positive instances which are predicted correctly it is also call true positive rate (TPR) and SPE measures the ration of the negative instances which are predicted correctly it is also call true negative rate (TNR). On ROC curve the optimal point is achieved with such threshold τ when TPR is maximized while false positive rate (FPR) is minimized as SEN is a complement of FPR [12]. Youden Index is frequently used to identify an optimal point of ROC curve. When both SEN and SPE are given the same priority then Youden is calculated as the maximum difference between TPR and FPR (1).

$$\begin{aligned} Youden &= \max_{\tau}\{SEN + SPE - 1\} = \max_{\tau}\{TPR + 1 - FPR - 1\} \\ &= \max_{\tau}\{TPR - FPR\} = \max_{\tau}\left\{\frac{TP}{TP + FN} - \frac{FP}{FP + TN}\right\} \end{aligned} \tag{1}$$

In Eq. (1), TP- true positives; FP – false positives; FN – false negatives; TN – true negatives are cell's values of 2×2 confusion matrix in Table 1 their values are received after a predicted class of each instance is compared with it's true class in dataset. TP, FP, FN, TN are used to calculate marginal values in Table 1, so that $P_j = TP + FP$ – is the number of positive instances which were predicted by model; $N_j = FN + TN$ is the number of negative instances which were predicted by model; $P_i = TP + FN$ is the number of positive instances in dataset; $N_i = FP + TN$ is the number of negative instances in dataset.

Table 1. 2 × 2 confusion matrix

	Predicted positives	Predicted negatives	
True positives	TP	FN	P_i
True negatives	FP	TN	N_i
	P_j	N_j	

When it is required to evaluate the prediction for any selected cut-off threshold then Youden value is obtained from Eq. (2).

$$Youden = SEN + SPE - 1 = \frac{TP}{TP + FN} - \frac{FP}{FP + TN} \tag{2}$$

According to Eq. (2) the maximum Youden score will be when FP is zero, i.e. when model doesn't make mistakes while predicting negative instances. In this case, according to Table 1 $TP = P_j$ and (2) is rewritten as (3) so Youden is maximized when the model also makes no mistakes while predicting positives.

$$Youden_{fp->0} = \frac{P_j}{P_i} \tag{3}$$

Further, is proposed to formalize MCC, BA and Kappa as the function of maximized Youden (3), so will be received the maximum expected scores of MCC, BA and Kappa. Hereafter, will be provided the mathematical equations by which MCC, BA, Kappa are described as function of maximized Youden calculated with Eq. (3).

MCC Eq. (4) when FP is zero is rewritten as Eq. (5) according to which MCC is a function of two arguments: Youden and a ration of the negatives in dataset to predicted negatives.

$$MCC = \frac{TP \cdot TN - FP \cdot FN}{\sqrt{(TP + FP)(TP + FN)(TN + FP)(TN + FN)}} \tag{4}$$

$$MCC_{fp->0} = \frac{P_j \cdot N_i}{\sqrt{P_j \cdot N_i \cdot P_i \cdot N_j}} = \sqrt{\frac{P_j \cdot N_i}{P_i \cdot N_j}} = \sqrt{\frac{P_j}{P_i}} \cdot \sqrt{\frac{N_i}{N_j}} = \sqrt{Youden_{fp->0}} \cdot \sqrt{\frac{N_i}{N_j}} \tag{5}$$

To continue the above logic, Balanced Accuracy (6) has a maximum value according to Eq. (7) when FP is zero.

$$BA = \frac{1}{2}\left(\frac{TP}{TP + FN} + \frac{TN}{TN + FP}\right) \tag{6}$$

$$BA_{fp->0} = \frac{1}{2}\left(\frac{P_j}{P_i} + 1\right) = \frac{Youden_{fp->0} + 1}{2} = normYouden_{fp->0} \quad (7)$$

So, maximized BA score is a normalized Youden which is calculated with Eq. (3). Cohen Kappa (8) has a maximum value according to Eq. (9) when FP is zero.

$$k = \frac{2(TP \cdot TN - FP \cdot FN)}{(TP + FP)(FP + TN) + (TP + FN)(TN + FN)} \quad (8)$$

$$k_{fp->0} = \frac{2P_j \cdot N_i}{P_jN_i + P_iN_j} = 2\frac{P_j \cdot N_i}{P_iN_j(\frac{P_jN_i}{P_iN_j} + 1)} = 2\frac{Youden_{fp->0} \cdot \frac{N_i}{N_j}}{Youden_{fp->0} \cdot \frac{N_i}{N_j} + 1} \quad (9)$$

So, a maximized Kappa is a function of two arguments: Youden and a ration of the negatives in dataset to predicted negatives.

In the next section will be designed and conducted empirical tests to get the practical verifications that MCC, BA and Kappa have the maximum scores when described by the derived Eqs. (5); (7); (9).

4 Research Method

To conduct experiments were created two datasets (datasetA, datasetB) with 1000 observations in each and features' values generated from Beta (a, b) for positive instances and from Beta (c, d) for negatives where a, b, c, d – are randomly generated from interval [1..10]. A datasetA is a balanced, meaning the number of positives is equal to the number of negative instances; a datasetB – positively imbalanced with imbalance ration 9. Four often used with classification tasks machine learning algorithms: RandomForestClassifier (RFC); GaussianNB (GNB); LogisticRegression (LR) and Support Vector Classifier (SVC) with default values for hyperparameters are trained on data subset which size is 80% from the total dataset's size after the predicted classes and predicted probabilities were obtained with test subset. Predicted probabilities were scaled with histogram binning method [13]. On the ROC curve of each classifier was identified optimal cut-off τ_{opt} according to Eq. 2 and the cut-off τ, where Youden is maximized (3). MCC, BA, Kappa scores are calculated on the cells' values of the confusion matrices which correspond to thresholds τ_{opt} and τ. The percent of pair scores difference from the maximum is identified as $(\Delta\tau_{\text{opt}}/\Delta\tau_{\max})$,%.

5 Results and Discussions

On Fig. 1, for the balanced datasetA, ROC curves capture the performance of the classifiers: (a) – GaussianNB; (b) – Random Forest Classifier; (c) – Logistic Regression; (d) – Support Vector Classifier) across different cut-off thresholds. The upper point of the dashed line indicates optimal cut-off point and each's picture's title includes TPR and FPR calculated on confusion matrix which corresponds to optimal cut-off point. The upper point of the solid line indicates a point when classifier made no mistake while predicting negative instances (FP = 0).

On Fig. 2, for the balanced datasetA, in the 1st line are illustrated the relations between pair scores differences of MCC and Kappa for each possible cut-off threshold of the classifiers: (a) – GaussianNB; (b) – Random Forest Classifier; (c) – Logistic Regression; (d) – Support Vector Classifier. The upper point of the solid line indicates the maximum expected pair scores difference $\Delta\tau_{max}$. The upper point of the dashed line indicates the difference at optimal cut-off $\Delta\ \tau_{opt}$. In the 2nd line are illustrated the relations between pair scores differences of normalized MCC and BA with the maximum expected pair scores difference $\Delta\tau_{max}$ and difference at optimal cut-off $\Delta\tau_{opt}$ at upper points of solid and dashed lines correspondingly.

Table 2 records cell values of the confusion matrix which corresponds to τ_{opt} and values in columns: Δ(MCC-Kappa), Δ(nMCC-BA), Δ(nKappa-BA) records the pair scores differences.

Table 3 records cell values of the confusion matrix which corresponds to τ that maximized Youden according to (3) and the values in columns: Δ_{max}(MCC-Kappa), Δ_{max}(nMCC-BA), Δ_{max} (nKappa-BA) records the maximum pair scores differences.

The graphs on Fig. 2 illustrate that the maximum pair score difference of metrics MCC, BA, Kappa corresponds to threshold with which Youden has a maximum value according to Eq. 3, which proves provided in this study theoretical Eqs. (5); (7); (9).

Table 4 populates the percent of the pair scores difference $\Delta\tau_{opt}$ from the maximum Δ_{max} and can be used to support a decision-making process on what evaluation metric to use in the following way: for any two evaluation metric If experts agree that $[\Delta\tau/\Delta_{max},\%]$ is 50%, i.e. pair difference can be 50% from the maximum pair difference then to evaluate the binary classification results from classifiers: GNB, LR, SVC can be used all three metrics: MCC, BA and Kappa and to evaluate the classifier: RFC only Kappa and BA.

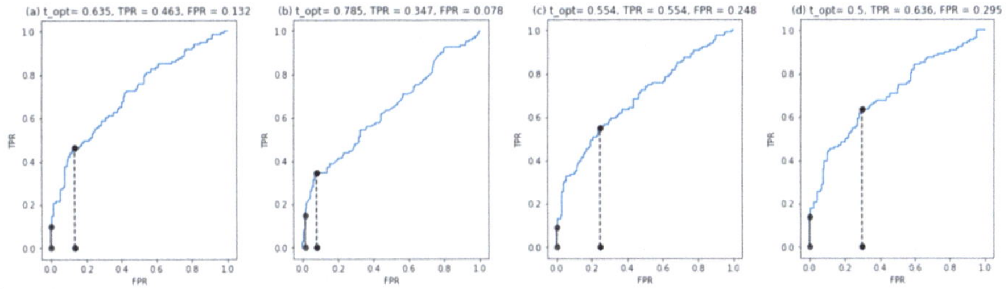

Fig. 1. ROC curves of classifiers (a) – GaussianNB; (b) – Random Forest Classifier; (c) – Logistic Regression; (d) – Support Vector Classifier for datasetA.

The same already described information is captured on Fig. 3 and Fig. 4 and in Tables 5 and 6 for the positively imbalanced datasetB.. The graphs on Fig. 4 illustrate that for positively balanced datasets, similar to balanced dataset, the maximum pair score difference of the metrics MCC, BA, Kappa corresponds to threshold with which Youden has a maximum value according to Eq. (3). Those evidence allows to propose Eqs. (5); (7); (9) to be used to identify the maximum expected difference in pair score of MCC, BA and Kappa when dataset is positively imbalanced. The proposed decision-making process to be applied with the results in Table 7.

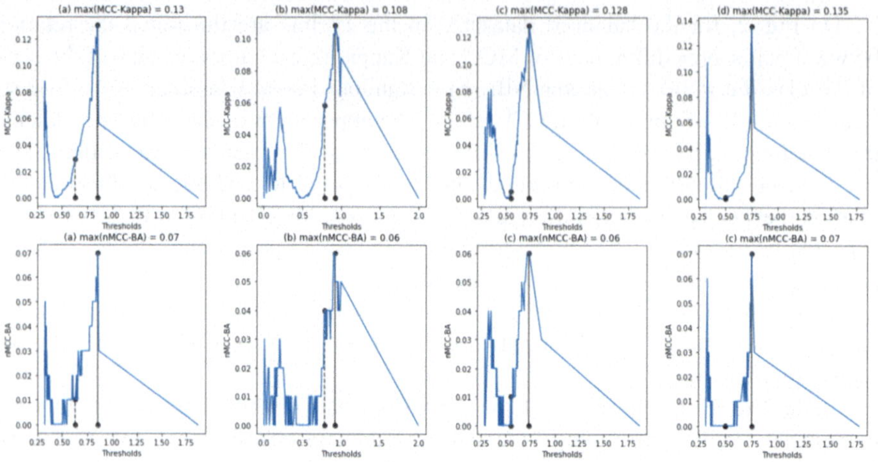

Fig. 2. 1st line – the relations between MCC and Kappa for datasetA; 2nd line – the relations between normalized MCC and BA for datasetA.

Table 2. Confusion matrix and evaluation scores of BA, MCC and Kappa for the threshold for the optimal cut-off point for datasetA

	TP	FP	FN	TN	τ_{opt}	Δ (MCC-Kappa)	Δ (nMCC-BA)	Δ (nKappa-BA)
GNB	56	17	65	112	0.635	0.029	0.012	0
RFC	42	10	79	119	0.785	0.058	0.036	0
LR	67	32	54	97	0.554	0.005	0.006	0
SVC	77	38	44	91	0.5	0.001	0.002	0

Table 3. Confusion matrix and evaluation scores of BA, MCC and Kappa for threshold, where FP is zero for datasetA

	TP	FP	FN	TN	τ	Δ_{max} (MCC-appa)	Δmax (nMCC-BA)	Δ_{max} (nKappa-BA)
GNB	12	0	109	129	0.857	0.13	0.066	0
RFC	18	0	103	129	0.925	0.108	0.053	0
LR	11	0	110	129	0.739	0.128	0.061	0
SVC	17	0	104	129	0.749	0.135	0.07	0

Table 4. Percentage of the pair scores difference received while evaluating binary classification results for DatasetA

$\Delta\tau_{opt}/\Delta_{max}$,%	GNB	RFC	LR	SVC
MCC-Kappa	22.3	53.7	3.9	0.74
nMCC-BA	18.2	68.6	9.8	2.2
nKappa-BA	0	0	0	0

Table 5. Confusion matrix and evaluation scores of BA, MCC and Kappa for the threshold for the optimal cut-off point for datasetB

	TP	FP	FN	TN	τ_{opt}	Δ (MCC-Kappa)	Δ (nMCC-BA)	Δ (nKappa-BA)
GNB	45	4	76	125	0.712	0.082	0.045	0
RFC	63	35	58	94	0.63	0.004	0.008	0
LR	73	36	48	93	0.521	0.002	0.004	0
SVC	49	10	72	119	0.721	0.052	0.033	0

Table 6. Confusion matrix and evaluation scores of BA, MCC and Kappa for threshold, where FP is zero for datasetB

	TP	FP	FN	TN	τ	Δ_{max} (MCC-Kappa)	Δmax (nMCC-BA)	Δmax (nKappa-BA)
GNB	21	0	100	129	0.83	0.135	0.067	0
RFC	4	0	117	129	0.98	0.098	0.046	0
LR	9	0	112	129	0.72	0.123	0.06	0
SVC	13	0	108	129	0.76	0.115	0.059	0

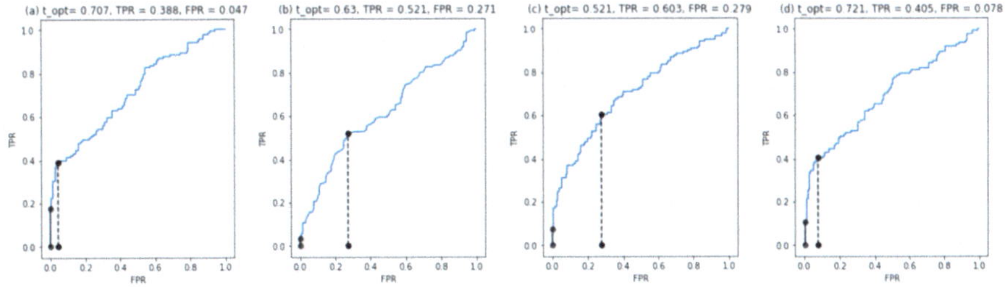

Fig. 3. ROC curves of classifiers (a) – GaussianNB; (b) – Random Forest Classifier; (c) – Logistic Regression; (d) – Support Vector Classifier for datasetB.

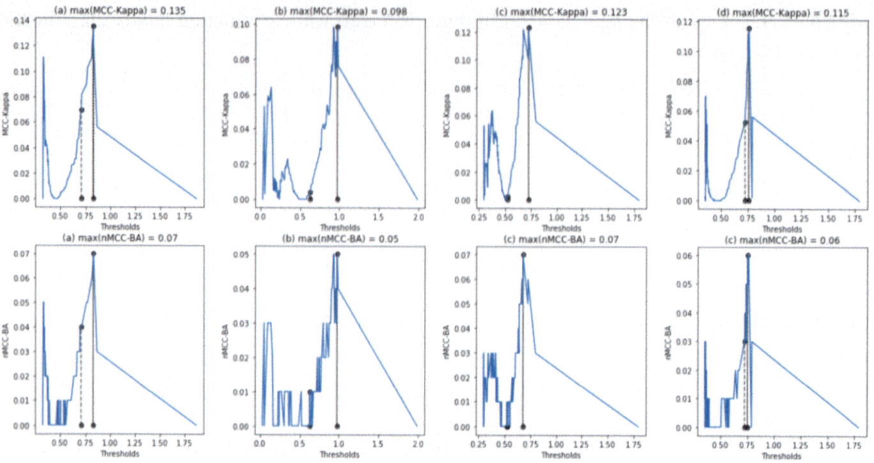

Fig. 4. 1st line – the relations between MCC and Kappa for datasetB; 2nd line – the relations between normalized MCC and BA for datasetB.

Table 7. Percentage of the pair scores difference received while evaluating binary classification results for DatasetB

$\Delta\tau_{opt}/\Delta_{max}$,%	GNB	RFC	LR	SVC
MCC-Kappa	60.7	4.1	1.6	45.2
nMCC-BA	67	16	5	55
nKappa-BA	0	0	0	0

6 Conclusions

In current research the formal descriptions of the metrics: MCC, BA and Kappa were described as the functions of Youden index, which is calculated on the confusion matrix when a binary model doesn't make a mistake in predicting negative instances. The score of MCC, BA and Kappa which are obtained from the proposed functions are the maximum expected scores. The maximum scores give the maximum expected pair score differences which can be used to calculate the percent of pair score difference at any selected ROC curve's threshold from the maximum expected and be used to support the decision-making process on what evaluation metric to use to evaluate the binary classification results, providing the expects agree on the accepted percent in pair score differences from the maximum expected. Conducted in current work experimental tests proved the possibility to apply the proposed approach for balanced and positively imbalanced datasets generated from Betta distribution.

The application of the proposed approach should help to have a consensus on the decision which evaluation metric to use while assessing the results of the binary classification task when it is important that model doesn't make mistake in predicting negative instances.

Future work will be to evaluate the feasibility to apply the proposed approach with a negatively imbalanced dataset generated from Betta distribution as well as to evaluate the approach on dataset with real data. The formalization of the binary evaluation metric formulas as function of Youden index to be extended for other commonly used metric to assess the result of the binary classification models. The formalization of the binary evaluation metric formulas as function of Youden index to be designed for the binary classification model when it must not make mistakes in predicting positive instances.

References

1. Tharwat, A.: Classification assessment methods. Appl. Comput. Inform. **17**(1), 168–192 (2020)
2. Powers, D.: Evaluation: from precision, recall and F-measure to ROC, informedness, markedness and correlation. arXiv preprint arXiv: 16061 (2010)
3. Warrens, M.: Kappa coefficients for dichotomous-nominal classifications. Adv. Data Anal. Classif. **15**(1), 193–208 (2021)
4. Chicco, D., Warrens, M.J., Jurman, G.: The Matthews correlation coefficient (MCC) is more informative than Cohen's Kappa and Brier score in binary classification assessment. IEEE Access **9**, 78368–78381 (2021)
5. Greener, J.G., Kandathil, S.M., Moffat, L., Jones, D.T.: A guide to machine learning for biologists. Nat. Rev. Mol. Cell Biol. **23**(1), 40–55 (2022)
6. Honcharenko, T., Akselrod, R., Shpakov, A., Khomenko, O.: Information system based on multi-value classification of fully connected neural network for construction management. IAES Int. J. Artif. Intell. **12**(2), 593–601 (2023)
7. Shah, H.A., Saeed, F., Yun, S., Park, J.H., Paul, A, Kang, J.M.: A robust approach for brain tumor detection in magnetic resonance images using finetuned efficientnet. IEEE Access **10**, 65426–38 (2022)
8. Chicco, D., Jurman, G.: The advantages of the Matthews correlation coefficient (MCC) over F1 score and accuracy in binary classification evaluation. BMC Genomics **21**(1), 1–13 (2020)
9. Rácz, A., Bajusz, D., Héberger, K.: Multi-level comparison of machine learning classifiers and their performance metrics. Molecules **24**(15), 2811 (2019)
10. Chicco, D., Starovoitov, V., Jurman, G.: The benefits of the Matthews correlation coefficient (MCC) over the diagnostic odds ratio (DOR) in binary classification assessment. IEEE Access **9**, 47112–47124 (2021)
11. Chicco, D., Tötsch, N., Jurman, G.: The Matthews correlation coefficient (MCC) is more reliable than balanced accuracy, bookmaker informedness, and markedness in two-class confusion matrix evaluation. BioData Min. **14**(1), 1–22 (2021)
12. Dhamnetiya, D., Jha, R.P., Shalini, S., Bhattacharyya, K.: How to analyze the diagnostic performance of a new test? Explained with illustrations. J. Lab. Phys. **14**(01), 90–98 (2021)
13. Honcharenko, T., Solovei, O.: Optimal bin number for histogram binning method to calibrate binary probabilities. In: CEUR Workshop Proceedings, vol. 3628, pp 126–135 (2023)

An Approach Towards Detection of Respiratory Diseases Based on Lung Breathing Recordings

Lesia Mochurad[1](✉) and Yulianna Mochurad[2]

[1] Lviv Polytechnic National University, S. Bandera Str., 12, Lviv 79013, Ukraine
lesiamochurad@gmail.com

[2] Danylo Halytsky Lviv National Medical University, 69 Pekarska Street, Lviv 79010, Ukraine
yuliannamochurad@gmail.ua

Abstract. Detecting respiratory diseases using machine learning algorithms is important for early diagnosis and control of the spread of diseases, especially in epidemics or pandemics. The use of diagnostic methods based on the analysis of lung breathing recordings can be an effective tool for detecting signs of respiratory diseases such as influenza, COVID-19, pneumonia, and others. In this paper, we developed an approach for detecting respiratory diseases in lung breath recordings that uses a histogram-based gradient boosting classifier . To reduce time and resources, less complex algorithms were chosen, but the main goal is to obtain a one-dimensional array in audio data processing that can be used in standard machine learning algorithms and simple neural networks. The results of applying the proposed algorithm showed an accuracy of 93%, while the execution time was only 0.1 s, which allows for real-time solutions. The solution to this problem can greatly facilitate the work of medical staff by reducing the time spent examining people who do not actually have respiratory diseases.

Keywords: Respiratory disease detection · Gradient Boosting algorithm · Classification · Machine Learning method · Real-time solution

1 Introduction

Respiratory disease recognition involves the creation of audio signal processing algorithms that can accurately detect characteristics such as wheezing, whistling, or other abnormal breath sounds. Improvements to existing disease recognition models by incorporating biomedical data, such as clinical data, genetic information, and other factors, can significantly improve the accuracy and diversity of diagnoses. It is also important to create technologies for continuous monitoring of audio signals that allow for the timely detection of signs of respiratory problems and provide timely information for medical intervention.

One of the main ways to identify a respiratory disease is to listen to the presence or absence of wheezing and crackling in the lungs and their number. This process is considered to be one of the most difficult for doctors, as there can be many noises in the lungs, one of the biggest being the heart sound, which interrupts a lot of other information.

Z. Hu et al. (Eds.): ICCSEEA 2024, LNDECT 242, pp. 178–190, 2025.
https://doi.org/10.1007/978-3-031-84228-3_15

After determining the number of crackles and wheezes in certain lung locations, the task of determining the presence of the disease arises. The main goal of applying machine learning methods [1] to detect respiratory diseases is to reduce patient contact with the doctor, as well as to give the doctor more time to work with really sick patients. Since digital stethoscopes already exist, it is not difficult to record patients' breathing in certain locations of the body. After that, all that remains is to process the data and pass it to the model's input. This program could also collect a lot of medical data, which could be used for other research and statistical evaluation [2–4].

The main advantages of using machine learning algorithms in this context include:

1. *Early detection of diseases.* Algorithms can detect characteristics that may be invisible to the human eye or require specialized analysis.
2. *Automation and speed.* Machine learning allows you to automate the processing of large amounts of data, which allows you to detect and analyze signs of disease faster.
3. *Accuracy and objectivity.* Algorithms can provide objective data analysis without the influence of the subjectivity of a doctor or other specialist.
4. *Scalability.* Machine learning allows to use of a large amount of data, which can increase the efficiency of the disease detection system in large populations.

Doctors can distinguish between many types of wheezing and crackles (dry, wet, whistling, etc.) in breath sounds. These sounds can also be combined, and it all boils down to a very complex classification task [5], with a large number of classes, and the ability to predict several classes at once. Solving such a problem can help with the problem of classifying diseases such as pneumonia, bronchitis, asthma, and others [6]. At the moment, there is no such dataset in the public domain, so only the division into wheezes and crackles is used.

The research presented in [7] notes the need for automated tools that support the clinical diagnosis and assessment of respiratory symptoms. To solve this problem, the ICBHI dataset was used, and the Multi-Time-Scale Feature Extraction method was used for sound processing. In this paper, the results are investigated and compared using four algorithms: Random Forest, Multilayer Perceptrons, Support Vector Machine, and Deep Neural Network. The metrics used are accuracy, prediction accuracy, error for healthy patients, error for patients with respiratory syndromes, and confusion matrices. The highest result was achieved using the Multilayer Perceptrons algorithm – 85% accuracy, while the lowest results were recorded when using the Support Vector Machine – 81%. It is important to note that this work used a patient-level approach, which means that the recordings were not broken down into inhalation-exhalation cycles.

In paper [8], we investigated the recognition of wheezing, crackling, and normal breathing in the inhalation-exhalation cycle. The same dataset as in the previous work, ICBHI, is used for the task. For sound processing, this paper proposes an approach to transform sound into an image using Optimized S-transform (OST). Deep Residual Networks (ResNet) are used to solve the classification problem. Only accuracy and confusion matrices were used as metrics. As a result of this work, an accuracy of 98.79% was achieved. This result was also compared with the use of other sound processing algorithms and other machine learning algorithms, but none of them gave a better result.

The authors of the paper [9] addressed the challenges of working with small datasets in the medical field by introducing a novel method based on the use of a GRNN (Generalized Regression Neural Network). They demonstrated the method's effectiveness through modeling, parameter optimization, and comparison with existing methods, showcasing its superior accuracy in handling clinical data, despite acknowledging its drawback of significant time delays compared to the basic GRNN.

In [10], the recognition of respiratory pathologies by breath sounds is also studied. The data set in this paper was collected independently, and the work itself was divided into four tasks:

- prediction of normal and pathological sounds (17,930 records, two classes);
- classification of respiratory sounds into normal, dry rales, squeak, wet rales, broncho vascular, crackle, friction, bronchial, absent, decrease, exacerbation, or long-term exclusion (14 453 records, 13 classes);
- classification of breath sounds labeled as wet rales, dry rales, or normal (15 328 records, 3 classes);
- classification of breath sounds with all labels, including the possibility of their repetition (17,930 records, 78 classes).

For sound processing in this work, we used spectrograms and Mel-frequency cepstral coefficients (MFCC). A convolutional Neural Network was used with spectrograms and a Support Vector Machine with MFCC. The best accuracy was achieved in the first task, reaching 86% for both approaches.

As in the two previous papers, in [11] the authors study the classification task on the ICBHI dataset. MFCC is used for image processing. The labels are divided into 4 classes: wheezing only, crackles only, wheezing and crackles, and healthy. Also, different types of Convolutional Neural Networks were used to solve this problem. As a result of this work, a result of 71.15% accuracy was achieved.

A new methodology for the accurate diagnosis of a subset of lung diseases such as chronic obstructive pulmonary disease (COPD), upper respiratory tract infections (URTI), bronchiectasis, pneumonia, and bronchiolitis using audio recordings of patient breathing was proposed in [12]. The authors suggested four models, namely convolutional neural networks (CNN), long short-term memory (LSTM), a combination of CNN and unidirectional LSTM (CNN-LSTM), and a combination of CNN and bidirectional LSTM (CNN-BLSTM). The LSTM algorithm showed the best overall accuracy and F1 score of 98.82% and 0.97, respectively But, if we compare this work with ours, this work also used a patient-level approach. And we fixed the decision-making time for the best accuracy at 3 s.

The task of classifying the presence of a respiratory disease is generally a binary classification, but to solve it, you first need to solve the problem of predicting crackles and wheezes in a person's inhalation-exhalation cycle, which is no longer a binary classification. There are many ready-made classification algorithms, such as decision tree, random forest, logistic regression, and others. Neural networks are also often used for this type of task. Most analogs use rather complex and deep convolutional neural networks that require a lot of resources and time. In order to reduce the time and resources required to perform this work, much simpler algorithms were chosen. Therefore, in audio data processing, the main aim is to obtain a one-dimensional array that can be transferred to standard machine learning algorithms and simple neural networks.

The task considered in this paper is divided into two subtasks:

1. Determining the presence of wheezing and crackling in the recording of the inhalation-exhalation cycle;
2. Determining the presence of a respiratory disease by the number of wheezes and crackles in certain locations of the chest.

The aim of this work is to develop an approach for detecting respiratory diseases in lung breathing recordings using machine learning methods. The practical significance of the results is determined by the fact that the implemented methods can be used as a preliminary check of the patient before scheduling a visit to the doctor.

2 Materials and Methods

2.1 Input Data Processing

This paper considers the Respiratory sound dataset [13], a compilation generated by two research groups in Portugal and Greece, comprising 920 annotated recordings ranging from 10 to 90 min in duration, involving 126 patients. The dataset encompasses 5.5 h of recordings, capturing 6898 breath cycles, with 1864 featuring crackles, 886 containing wheezes, and 506 exhibiting both crackles and wheezes. The recordings encompass various age groups, including children, adults, and the elderly, presenting both clear breath sounds and noisy recordings simulating real-life conditions.

Each audio file is named with a patient identifier, recording index, chest location, recording mode (single channel or multichannel), and the device used. Additionally, corresponding text files provide annotations indicating the beginning and end of each cycle and the presence of wheezing and crackles. The chest area is categorized into six locations: front left side, front on the right side, back on the left side, back on the right side, on the left side, and the right side, as illustrated in Fig. 1.

The dataset incorporates two recording modes: single channel (sc) and multichannel (mc), recorded using four different devices: AKG C417L Microphone (AKGC417L), 3M Littmann Classic II SE Stethoscope (LittC2SE), 3M Littmann 3200 Electronic Stethoscope (Litt3200), and WelchAllyn Meditron Master Elite Electronic Stethoscope (Meditron).

The file with patient demographic information contains the following data

- Patient number;
- Age
- sex (Sex);
- body mass index for adults (Adult BMI (kg/m^2));
- weight for children (Child Weight (kg));
- height for children (Child Height (cm)).

The labels in this dataset are the names of the diseases, of which there are six, and healthy patients. A histogram of the number of patients with a particular disease is shown in Fig. 2. In this work, we will only distinguish between the presence and absence of

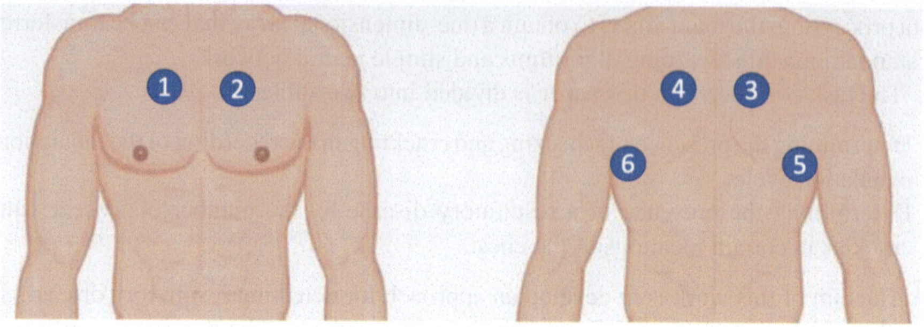

Fig. 1. Places where doctors listen to the lungs: front right side (1); front left side (2); back right side (3); back left side (4); right side (5); left side (6) [8]

respiratory diseases, so all labels denoting a particular disease should be combined into one class, and the label "Healthy" should be left as it is. Figure 3 shows the distribution of labels that will be used in this paper.

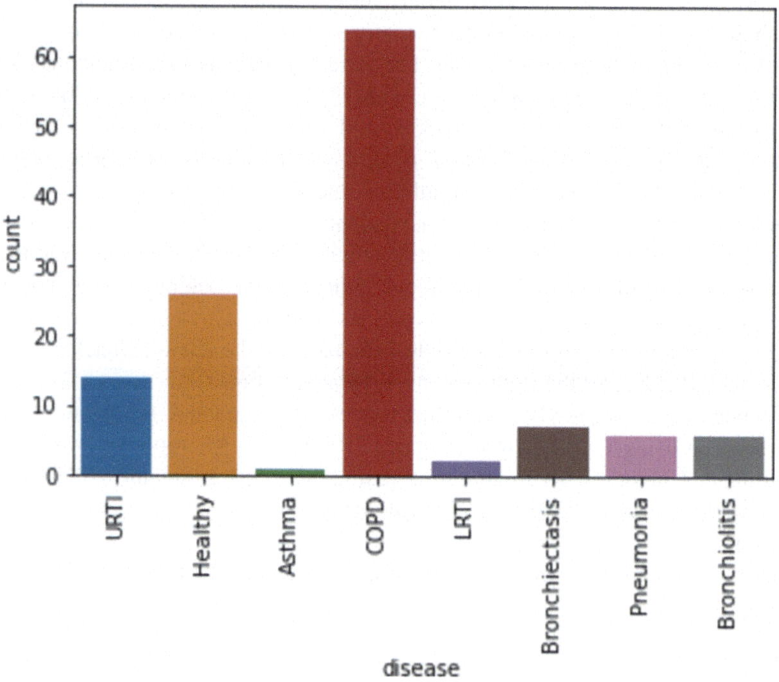

Fig. 2. Histogram by the number of patients with a particular disease

Since the task in this paper is divided into two subtasks to be solved by machine learning algorithms, there are two sets of input data that need to be processed correctly.

The first subtask is to identify wheezes and crackles in the inhalation-exhalation cycle. The selected dataset contains only complete breath recordings at a certain location

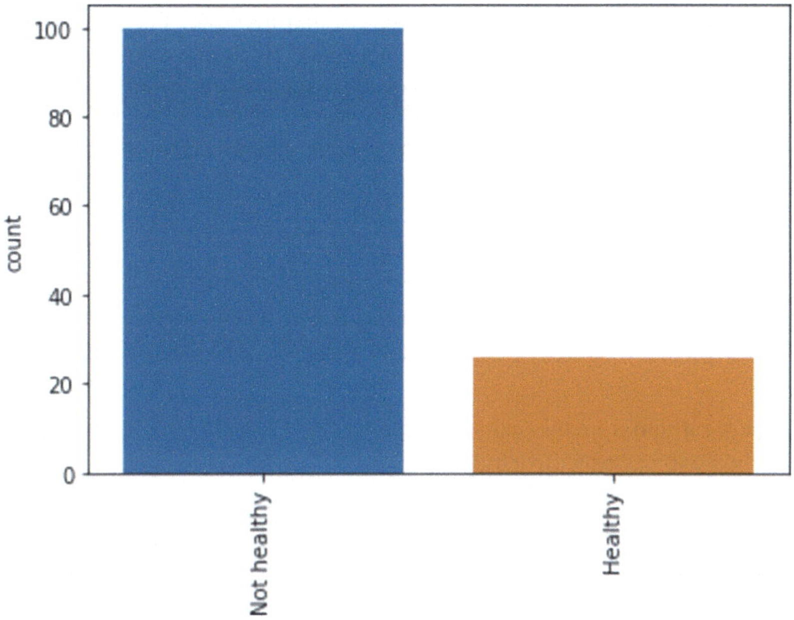

Fig. 3. Histogram by the number of patients with or without the disease

in the chest, so the first step is to divide these recordings into smaller parts, where there is only one inhalation-exhalation cycle. To do this, you need to use text files with audio annotation and trim the audio files by the start and end time of the cycle.

Crucial factors in this task include:

- the recording mode, as each mode has the potential to convey the sound distinctly;
- the chest location, since sounds originating from different locations may exhibit varying noise levels.

As a result, this information was incorporated into the input from the file name. To enhance the precision of wheezing and crackles detection, the following variables were additionally considered:

- age, given the notable difference in breathing patterns between older and younger individuals;
- body mass index, as heavier individuals typically exhibit more pronounced breathing;
- gender.

The missing data were recovered by interpolation with Bezier curves and using CUDA technology [14, 15]. Categorical data such as recording mode, recording location, and patient gender were converted to numeric data using the sklearn library– LabelEncoder. The next step is to convert the breath-by-breath recordings into an array of numbers using two algorithms: MFCC and Mel Spectrogram, which were implemented using the librosa library.

It is known [16] that people do not perceive sound frequency linearly. That is, it is easy to distinguish 400 Hz from 900 Hz, but not 10400 Hz from 10900 Hz, even though the distance between them is the same. To equalize the pitch of sound as it is heard by humans, the chalk scale was introduced. This scale was determined experimentally, so there are several formulas for converting hertz to mela, which differ only in the constants. Below is one of these formulas (1).

$$F_{mel} = 1127\,log(1 + \frac{F_{Hz}}{700}), \quad (1)$$

where F_{Hz} is the frequency of the sound in Hz. A mel spectrogram is a regular sound spectrogram converted to mel, which is used for MFCC. The latter is obtained in just two steps:

1. take the logarithm from the mel-spectrogram;
2. apply the discrete cosine transform.

The discrete cosine transform is a variant of the cosine transform for a vector of real numbers. This algorithm is often used in lossy information compression algorithms [17]. The use of the discrete cosine transform in this problem allows you to discard excessively high coefficients, which are most likely noise, and preserve simple aspects of the spectral shape.

To solve the second subtask, determining the presence of a disease, a completely different input data processing is used. The first thing to do to create good input data is to determine if there is a recording in a particular chest location. Next, count the frequency of crackles and wheezes relative to the total number of cycles if there is a recording in that location, or mark it as 0 otherwise. This information also includes the following values: age; gender; body mass index in adults; weight of children; and height of children.

All missing values and categorical data were handled similarly to the first subtask. Labels were also changed, instead of identifying a specific disease, only its presence or absence was assumed, so all labels that indicated a specific disease were replaced with 0, and labels that indicated patients without respiratory diseases were replaced with 1.

2.2 Training of Models

The models in this paper were trained using the Gradient-boosting classification algorithm [18, 19]. Assembly machine learning algorithms are those that use a set of untrained models to predict the correct answer. All assembly methods are divided into two classes: backpropagation and boosting. Backpropagation is when several models are trained independently of each other, and then some kind of averaging is applied to their results to obtain the final ensemble result. Boosting is when each subsequent model learns from the mistakes of the previous one, i.e. they are built sequentially, not independently. Gradient boosting is an ensemble learning algorithm with a teacher that trains small and weak decision trees and sequentially combines them.

Any supervised learning algorithm aims to reduce the magnitude of the loss function. Predicting the presence or absence of a respiratory disease is a binary classification task, and the task of determining wheezing and crackling in lung sounds is divided into two separate binary classifications, such as the presence or absence of wheezing; and the presence or absence of crackling.

Thus, since all tasks are reduced to binary classification, binary cross-entropy was chosen as a loss function.

Cross-entropy is a measure that calculates the difference between two probability distributions. In this case, it is the difference between the true values and those predicted by the model. The formula for the binary cross-entropy loss function is shown below (2).

$$H = -(y\,log(p) + (1 - y)log(1 - p)), \tag{2}$$

where y are true values, labels; p is the predicted value.

After determining the loss function, the next step is to train the gradient boosting itself, which consists of the following steps:

1. find the probability of occurrence of the class in the sample, this will be the first predicted value, which will be the same for all records;
2. calculate the residuals, which are equal to the difference between the true value and the predicted value;
3. build a regression decision tree where the labels are the residuals calculated in step two;
4. calculate the derivative of the loss function;
5. calculate the output values of the decision tree using the derivative of the loss function;
6. calculate the new predicted values;
7. plug the new predicted values into the loss function and compute it;
8. repeat steps 2–7 until the value of the loss function is sufficiently small or the maximum number of iterations is reached.

This paper also uses a histogram-based gradient-boosting classifier. This algorithm differs from conventional gradient boosting only in that the data representation is changed, which significantly speeds up the execution time. This approach consists of splitting continuous features. The algorithm creates a histogram of feature values with equal bin density (i.e., converts the distribution of features into a uniform distribution for partitioning), built using a subset of the dataset [20]. Hence, a histogram-based gradient boosting classifier provides high learning speed, lower memory consumption, and greater outlier tolerance, making it effective for processing large volumes and high-dimensional data. Histogram Gradient Boosting may also be less prone to overfitting due to the ability to use additional regularization techniques.

3 Research Results

Python programming language was chosen as a development tool in this work. The system developed in this paper can be executed on any operating system: Windows; Linux; or MacOS. Minimum system requirements:

- RAM: 8 GB;
- CPU: Intel Core i5-4590;
- amount of memory: 7 GB.

Recommended system requirements:

- RAM: 16 GB;
- CPU: Intel Core i7-4790;
- amount of memory: 10 GB.

Jupyter Notebook was chosen as the development environment for the program. Jupyter Notebook is an incredibly powerful tool for the interactive development and presentation of data science projects [21]. As a result of the work, 3 Jupyter Notebooks were created:

- for preliminary data processing;
- for training, testing, and saving the model for predicting the presence of wheezing and crackles in the audio recording of the inhalation-exhalation cycle;
- for training, testing, and saving the model for predicting the presence or absence of a respiratory disease.

As a result of Jupyter Notebook's execution, two.csv files were obtained for preliminary data processing. One of them contains information about the patient and the frequency of crackles and wheezing in his breathing. The other contains information about audio files cropped by the time of the beginning and end of the inhalation-exhalation cycle and the presence of crackles and wheezing in them. In addition, 6898 cropped audio files were received as output.

For the first sub-task, the division of the data into training and test data was made in the ratio of 85:15, and for the second – 90:10.

For the test data, we used the following metrics to evaluate the models' performance: Accuracy, Precision, Recall, and F-score [22, 23].

Accuracy – is the ratio of the number of correctly classified items (the sum of True Positives and True Negatives) to the total number of items in the dataset (3).

$$\text{Accuracy} = \frac{\text{True Negatives} + \text{True Positives}}{\text{Total number of samples}}. \tag{3}$$

Precision – determines the proportion of items that the model has defined as a certain class and that actually belong to this class, relative to all items that the model has defined as this class.

$$\text{Precision} = \frac{\text{True Positives}}{\text{True Positives} + \text{False Positives}}. \tag{4}$$

Recall – measures the proportion of elements of a certain class that the model identified correctly, relative to all elements of this class in the dataset.

$$\text{Recall} = \frac{\text{True Positives}}{\text{True Positives} + \text{False Negatives}}. \tag{5}$$

F1-score – combines precision and recall in the form of a harmonic average. This metric is useful in situations where accuracy and sensitivity are equally important. F1-score is especially effective in cases of unbalanced data when the number of negative cases is very high compared to the number of positive cases.

$$F1 - \text{score} = \frac{2 \cdot \text{Precision} \cdot \text{Recall}}{\text{Precision} + \text{Recall}}. \tag{6}$$

As a result of the numerical experiments, we obtained the model evaluation indicators shown above (see (3)–(6)), which are presented in Table 1.

Table 1. Accuracy, training time/epoch, precision, recall, and F1-score of our models (first, second and combined) and model from [10]

Model name	Accuracy (%)	Training time/epoch (s)	Precision	Recall	F1-Score
LSTM [5]	98.82	3.005	0.96	0.99	0.97
Our first	94.06	0.018	0.93	0.96	0.94
Our second	98.80	0.062	0.99	0.95	0.97
Our Combined	92.93	0.080	0.92	0.91	0.91

As we can see from the results, we managed to obtain the solution to the problem posed in this paper with an accuracy of 93%, in 0.1 s. Thus, the advantage of the approach proposed in this paper is that we use much simpler algorithms that do not require very powerful hardware and a large amount of time to train models. The proposed approach to the decomposition of problem-solving in general can also be extended to problems from other fields of application [24, 25].

4 Summary and Conclusion

The task of identifying respiratory diseases holds substantial relevance in the current context, as it not only has the potential to significantly diminish the contact between healthy individuals and those affected but also plays a crucial role in optimizing doctors' time during patient examinations. In the course of addressing the complexities of this issue, the research presented in this paper has effectively bifurcated the problem into two distinct subproblems, resulting in the development of two specialized models.

The first model focuses on predicting the presence of crackles and wheezing within the inhalation-exhalation cycle, achieving an impressive accuracy rate of 0.9406. This specialized model contributes to the nuanced understanding of respiratory patterns, particularly in terms of identifying key indicators of respiratory distress.

Simultaneously, the second model is designed to predict the broader context of respiratory health by determining the presence or absence of respiratory diseases in a patient. Remarkably, this model attains an accuracy level of 0.9880, showcasing its proficiency in making accurate assessments. Such precision is invaluable in aiding healthcare professionals in making informed decisions about the overall respiratory well-being of patients.

These achievements underscore the potential impact of advanced machine learning techniques in the field of respiratory health, not only facilitating more precise diagnostics but also streamlining the healthcare process for both practitioners and patients alike.

The overall accuracy of the two algorithms is equal to the product of the accuracies of the two models and is 0.93.

In this work, the use of gradient boosting, which uses histograms to build trees, has several advantages over traditional gradient boosting, which uses full data. In particular:

1. Learning speed:
   - Histogram Gradient Boosting speeds up the learning process because it works with compressed histograms of the contents of each feature column. This makes it much more efficient for large datasets, as it makes it possible to quickly compute histograms for each feature.
2. Less memory usage:
   - Histograms take up less memory compared to full datasets, making the algorithm less resource-demanding.
3. Less sensitive to noise:
   - By using histograms to distribute feature values, gradient boosting can be less sensitive to outliers and noise in the data.
4. Optimization of the traversal path:
   - Gradient boosting on histograms can optimize the traversal path of the tree, allowing for more efficient data processing and reduced computation time.
5. Support for categorical features:
   - Histogram gradient boosting can work directly with categorical features without converting them to numerical values.

In addition, to reduce the decision-making time and achieve maximum accuracy, interpolation by Bezier curves was used in data preprocessing. As a result, as shown by the results of the numerical experiments, it was possible to obtain a solution to the problem in 0.1 s, which is a significant acceleration compared to [10], where the authors obtained a computation time of 3 s.

Future research in the field of respiratory disease recognition through audio recordings aims to develop deep learning for more accurate analysis of sound patterns. Integration of artificial intelligence with biomedical data and the use of modern IoT technologies can contribute to the creation of effective systems for early diagnosis of respiratory diseases based on audio information. Global collaborative networks can support standardization and data exchange to improve the accuracy and accessibility of such systems.

References

1. Zomchak, L., Melnychuk, V.: Creditworthiness of individual borrowers forecasting with machine learning methods. In: Hu, Z., Ye, Z., He, M. (eds.) AIMEE 2022. LNDECT, vol. 159, pp. 553–561. Springer, Cham (2023). https://doi.org/10.1007/978-3-031-24468-1_50
2. Tkachenko, R., Izonin, I., Greguš ml., M., Tkachenko, P., Dronyuk, I.: Committee of the SGTM neural-like structures with extended inputs for predictive analytics in insurance. In: Younas, M., Awan, I., Benbernou, S. (eds.) Big Data Innovations and Applications. Innovate-Data 2019. CCIS, vol. 1054, pp. 121–132. Springer, Cham (2019). https://doi.org/10.1007/978-3-030-27355-2_9
3. Zomchak, L.M.: Sustainable development of Ukraine as a combination of social, economic and environmental components: structural econometric model of three-pillar approach. In: Proceedings. of the 4th International Conference on Sustainable Futures: Environmental, Technological, Social and Economic Matters (ICSF-2023), Kryvyi Rih. IOP Conference Series: Earth and Environmental Science, vol. 11. IOP Publishing (2023)
4. Kaminsky, R., Mochurad, L., Shakhovska, N., Melnykova, N.: Calculation of the exact value of the fractal dimension in the time series for the box-counting method. In: Conference Proceedings 2019 9th International Conference on Adanced Computer Information Technologies ACSIT 2019, Ceske Budejovice, Czech Republic, 5–7 June 2019, pp. 248–251 (2019)
5. Teslyuk, V., Beregovskyi, V., Denysyuk, P., Teslyuk, T., Lozynskyi, A.: Development and implementation of the technical accident prevention subsystem for the smart home system. Int. J. Intell. Syst. Appl. (IJISA) **10**(1), 1–8 (2018)
6. Santra, A., Dutta, A.: A comprehensive review of machine learning techniques for predicting the outbreak of Covid-19 cases. Int. J. Intell. Syst. Appl. (IJISA) **14**(3), 40–53 (2022)
7. Monaco, A., Amoroso, N., Bellantuono, L., Pantaleo, E., Tangaro, S., Bellotti, R.: Multi-time-scale features for accurate respiratory sound classification. Appl. Sci. **10**(23), 8606, 1–17 (2020)
8. Chen, H., Yuan, X., Pei, Z., Li, M., Li, J.: Triple-classification of respiratory sounds using optimized S-transform and deep residual networks. IEEE Access **7**, 32845–32852 (2019)
9. Izonin, I., Tkachenko, R., Gregus, M., Zub, K., Tkachenko, P.: A GRNN-based approach towards prediction from small datasets in medical application. Procedia Comput. Sci. **184**, 242–249 (2021)
10. Aykanat, M., Kılıç, Ö., Kurt, B., Saryal, S.: Classification of lung sounds using convolutional neural networks. EURASIP J. Image Video Process. **2017**(1), 65 (2017)
11. Demir, F., Ismael, A.M., Sengur, A.: Classification of lung sounds With CNN model using parallel pooling structure. IEEE Access **8**, 105376–105383 (2020)
12. Pinzhi, Z., Alagappan, S., Abrar, U.A.: Pulmonary disease detection and classification in patient respiratory audio files using long short-term memory neural networks. Front. Med. **10**, 1–11 (2023)
13. Respiratory Sound Database. https://kaggle.com/vbookshelf/respiratory-sound-database. Accessed 21 Jan 2024
14. Mochurad, L., Mochurad, Y.: Parallel algorithms for interpolation with bezier curves and B-splines for medical data recovery. In: CEUR Workshop Proceedings, 6th International Conference on Informatics and Data-Driven Medicine, IDDM 2023, vol. 3609, pp. 189–197 (2023)
15. Vitynskyi, P., Tkachenko, R., Izonin, I., Kutucu, H.: Hybridization of the SGTM neural-like structure through inputs polynomial extension. In: Proceedings of the 2018 IEEE 2nd International Conference on Data Stream Mining and Processing, DSMP, Lviv, pp. 386–391. IEEE (2018)

16. Banerjee, A., Sanyal, S., Roy, S., Nag, S., Sengupta, R., Ghosh, D.: A novel study on perception–cognition scenario in music using deterministic and non-deterministic approach. Physica A **567**, 125682 (2021)
17. Carvalho, S., Gomes, E.F.: Automatic classification of bird sounds: using MFCC and mel spectrogram features with deep learning. Vietnam J. Comput. Sci. **10**(1), 39–54 (2023)
18. Mochurad, L., Panto, R.: A parallel algorithm for the detection of eye disease. In: Hu, Z., Wang, Y., He, M. (eds.) CSDEIS 2022. LNDECT, vol. 158, pp. 111–125. Springer, Cham (2023). https://doi.org/10.1007/978-3-031-24475-9_10
19. Singh, D.K., Rawat, N.: Machine learning for weather forecasting: XGBoost vs SVM vs random forest in predicting temperature for visakhapatnam. Int. J. Intell. Syst. Appl. (IJISA) **15**(5), 57–69 (2023)
20. Nhat-Duc, H., Van-Duc, T.: Comparison of histogram-based gradient boosting classification machine, random forest, and deep convolutional neural network for pavement raveling severity classification. Autom. Constr. **148**, 104767 (2023)
21. Onyezewe, A., Kana, A.F., Abdullahi, F.B., Abdulsalami, A.O.: An enhanced adaptive K-nearest neighbor classifier using simulated annealing. Int. J. Intell. Syst. Appl. (IJISA) **13**(1), 34–44 (2021)
22. Le, N.-B., Pham, T.-T.-H., Nguyen, S.-H., Nguyen, N.-M., Nguyen, T.-N.: AI-powered predictive model for stroke and diabetes diagnostic. Int. J. Intell. Syst. Appl. (IJISA) **16**(1), 24–40 (2024)
23. Auzinger, W., Obelovska, K., Dronyuk, I., Pelekh, K., Stolyarchuk, R.: A continuous model for states in CSMA/CA-based wireless local networks derived from state transition diagrams. In: Saraswat, M., Roy, S., Chowdhury, C., Gandomi, A.H. (eds.) Proceedings of International Conference on Data Science and Applications. LNNS, vol. 287, pp. 571–579. Springer, Singapore (2022). https://doi.org/10.1007/978-981-16-5348-3_45
24. Kharchuk, V., Oleksiv, I.: The intellectual structure of sustainable leadership studies: bibliometric analysis. In: Hu, Z., Wang, Y., He, M. (eds.) CSDEIS 2022. LNDECT, vol. 158, pp. 430–442. Springer, Cham (2023). https://doi.org/10.1007/978-3-031-24475-9_37
25. Oleksiv, I., Izonin, I., Kharchuk, V., Tkachenko, R., Doroshenko, A.: Identification of IT sector stakeholder's requirements to masters program in information system in Lviv region. In: CEUR Workshop Proceedings, vol. 2105, pp. 429–437 (2018)
26. Shakhovska, N., Montenegro, S., Kryvenchuk, Y., Zakharchuk, M.: The neurocontroller for satellite rotation. Int. J. Intell. Syst. Appl. (IJISA) **11**(3), 1–10 (2019)

Automatic Brain Tumor Detection Based on MRI Images Using Parallelization

Lesia Mochurad[1(✉)] and Yulianna Mochurad[2]

[1] Lviv Polytechnic National University, S. Bandera Str., 12, Lviv 79013, Ukraine
lesiamochurad@gmail.com
[2] Danylo Halytsky Lviv National Medical University, 69 Pekarska Street, Lviv 79010, Ukraine

Abstract. This research addresses the important task of developing an effective method for automatic brain tumor detection based on magnetic resonance imaging (MRI) images. This is a key task in medical diagnostics, as it can significantly improve the early detection and treatment of the disease. This paper uses an approach that combines deep learning and convolutional neural networks (CNNs) and uses parallelization to optimize computational processes and reduce processing time. This approach proves to be effective in diagnosing brain cancer, leading to an increase in system efficiency. In this paper, we propose an algorithm that uses Message Passing Interface (MPI) and Compute Unified Device Architecture (CUDA) Python technologies for parallelized learning and tumor detection in MRI images. As a result of the experiments, a significant acceleration was achieved, which is about 3 times for MPI technology and about 24 times for CUDA, without compromising the diagnostic accuracy compared to previous studies by domestic and foreign scientists. The proposed approach is promising for the implementation of automated systems in medical institutions, providing accurate results of brain cancer detection in real-time. Further testing and evaluation of the algorithm on computing systems with more cores and on larger datasets is recommended to better understand the ratio of accuracy and computing time.

Keywords: MPI technology · CUDA technology · deep learning · convolutional neural networks · automated systems

1 Instruction

Oncological complications of the brain are a complex case because of its vulnerability and role in the performance of various body functions [1, 2]. Nevertheless, detection and diagnosis of brain cancer are key to successful treatment. Recently, the use of artificial neural networks has become effective in solving medical problems, including brain cancer detection. The use of deep learning and CNNs has proven successful for this task [3, 4] and many others [5, 6]. However, increasing the size of a neural network leads to an increase in computational complexity and processing time. Thus, one of the ways to improve the performance of a neural network is parallelization, which allows the distribution of computational processes among several threads or processors.

Z. Hu et al. (Eds.): ICCSEEA 2024, LNDECT 242, pp. 191–205, 2025.
https://doi.org/10.1007/978-3-031-84228-3_16

Brain cancer is one of the most urgent and challenging problems in oncology today. According to the World Health Organization, brain cancer is considered one of the most common types of cancer among children and young adults, and, unfortunately, it often leads to death [7]. Thus, detecting brain cancer at an early stage can significantly increase the chances of successful treatment. Automated systems based on artificial neural networks can significantly facilitate the process of brain cancer detection and reduce the number of false diagnoses, which emphasizes the relevance of this topic.

The use of CNNs parallelization for brain cancer detection is a fairly new area in medical informatics research. Parallelization allows for efficient distribution of computational processes and improves the performance of the neural network, representing an innovative approach in this area. Thus, this work is relevant and innovative in the field of medical informatics and can contribute to the development of more efficient brain cancer detection systems.

It is known that parallelization technologies, as stated in [8, 9], allow one to divide a computational task into parts and process them in parallel on several processors or nodes of a computer network, which leads to a significant increase in computing speed and a decrease in execution time. In the case of CNNs, parallelization can be used at several levels, including the use of graphics processing units (GPUs), which significantly speed up the network training process due to the large number of parallel computing units. In addition, parallelization can be applied at the data processing level, where computing units simultaneously process individual image fragments, which helps to reduce data transfer time and increase network performance. The use of CNNs parallelization for brain cancer detection can significantly improve diagnostic accuracy and reduce the time required to analyze brain images, which can have a significant impact on improving the diagnosis and treatment of patients with this disease.

In [10], the authors presented a study that examines the use of CNNs for brain cancer diagnosis and tumor grade classification. In [11], the authors implement a method for detecting and segmenting brain tumors based on deep learning using the ResNet architecture and the U-Net algorithm. These studies confirm the potential of deep learning for detecting and segmenting tumors in MRI images, which can improve patient diagnosis and treatment. Nevertheless, the parallelization of CNNs on multi-core processors and GPUs opens up opportunities to accelerate network training and optimize image processing.

This work aims to use CNN parallelization for brain cancer detection based on MRI images. MRI, being a highly accurate method for diagnosing brain tumors, remains time and resource-intensive. It is proposed to use deep neural networks, especially CNNs, for the automatic detection of tumors in MRI images. Given the high computational complexity, especially when working with a large amount of data, it is proposed to use parallelization of the CNN-based deep learning method using MPI Python and CUDA, which will allow distribution tasks among different nodes and use more computing resources to optimize the solution of complex problems. Thus, the task of this work is to develop and validate an effective method for automatic brain tumor detection based on MRI images using a CNN and its parallelization.

2 Literature Review

A review of the scientific state of the art in MRI-based brain tumor detection emphasizes the activity and promise of this field in medical image processing. However, computational resource limitations and model complexity remain important challenges.

The examples of existing research emphasize the importance of parallelization to overcome time and accuracy limitations. The use of MPI and CUDA technologies can significantly reduce processing time and improve model accuracy, making our study competitive in this context.

MPI is a standard for data exchange and communication between different processes in parallel computing systems. Using MPI, programs can be written so that multiple processes communicate with each other by passing messages. This is especially useful for distributed computing on clusters, where each node corresponds to a separate processor or computer.

CUDA is a development platform from NVIDIA for using GPUs in general-purpose computing. CUDA allows programmers to use the power of parallel computing in graphics cards to solve general-purpose computing tasks. Using CUDA can significantly speed up computing, particularly in tasks that require a large number of computations of the same type, such as deep learning or image processing [12].

Both technologies, MPI and CUDA, are important for parallel computing in large and computationally intensive applications, such as deep learning research or image analysis in medical diagnostics. Their use in practice is described in detail in [13, 14].

Comparing our study with other scientific publications on brain cancer detection, we can highlight several key aspects. The study [15], conducted in 2022, uses a CNN with an accuracy of 97%, but the network training takes a considerable amount of time. Our study solves this problem by using parallelization, which significantly speeds up the training time while ensuring high model accuracy.

As for the study [16], its advantage is the ability to determine the type of oncology, but it is noted that the dataset used has a smaller number of images. Our study, using MPI and CUDA technologies, shows significant improvements in data processing speed.

In studies [17, 18] that use the same dataset as our study, the achieved accuracies range from 92% to 99%. However, these studies do not select images for testing, which may affect the reliability of the results. In addition, training times averaging up to 100 s for a single epoch are a significant limitation. Our study solves these problems by using parallelization, which reduces training time and improves model performance.

In [19], a cascade ensemble incorporating a high-order Wiener polynomial and the Stochastic Gradient Descent algorithm is proposed for effective analysis of large volumes of non-linear biomedical engineering data. The developed ensemble scheme demonstrates high accuracy in approximating real-world datasets with a satisfactory training time, making it a practical and efficient tool for data analysis in the biomedical engineering domain. However, the authors do not conduct studies for larger datasets and, accordingly, whether the time estimates of the resulting solution remain satisfactory.

The aim of the study [20] was to develop a fully autonomous PDCNN model that creates two models and combines them into one. The obtained accuracy results reached a high level of 98%. However, the main drawback of this study is the limited parallelization capabilities, namely the ability to parallelize the model on only two nodes. In our work, we have achieved the advantage of allowing the algorithm to be parallelized on a different number of nodes, significantly reducing the decision-making time.

Analyzing other studies, it can be noted that most researchers do not use parallelization in their experiments, which negatively affects the performance of models and makes real-time diagnosis impossible. Thus, to improve research on brain tumor detection based on MRI images, the use of parallelization in a CNN-based deep learning method is more effective. This can be achieved by using MPI and CUDA, or another parallelization method. The use of MPI and CUDA has significant advantages over existing research by improving diagnostic accuracy and speeding up computation. It is important to note that each approach has its advantages and disadvantages. The choice of a particular method depends on the specific task and the available computing resources for the researcher.

3 Materials and Methods

Our research uses MPI and CUDA-based parallelization to speed up training and improve model accuracy. This is especially important in tumor detection, where large amounts of MRI data can be resource-intensive. Parallelization allows for more efficient use of computing resources and improves the quality of models. This section describes the construction of the CNN itself and its parallelization algorithms based on the two technologies.

1) Used CNN Model

First, let's consider the CNN model [21, 22]. This model was chosen to solve the problem because it has proven itself to be the best in image processing. The architecture of this model is shown in Fig. 1.

This model consists of an input layer, two convolutional layers followed by a fusion layer a post-fusion layer, and a dense layer, to avoid overfitting, an extraction layer is used, and finally an output layer. In convolution layers, our input image highlights the main features, in other words, convolution is a specialized type of linear operation used for feature extraction, where a small array of numbers called a kernel is applied to the input data, which is an array of numbers called a tensor. The element-by-element product between each element of the kernel and the input tensor is computed at each location in the tensor and summed to produce an output value at the corresponding position in the output tensor, called a feature map. The input parameters for the convolutional CNN are its size and the number of kernels. For our model, the size of the convolution is 3 * 3. Then they are passed through an activation function, for this model we chose ReLU. ReLU has become very popular over the past few years. This function has the following mathematical form $f(x) = max(0, x)$.

The pooling layer is used to reduce the size of the object map, thus reducing the number of parameters to learn and the amount of computation performed in the network. For this network, MaxPooling is selected with a filter of size 2 * 2, that is, the largest values of the map will be taken.

After all the previous steps, we perform the Flatten operation. At this stage, we transform our combined map into a form with only one column, i.e., we convert a multidimensional array into a one-dimensional vector, which is necessary to insert this data

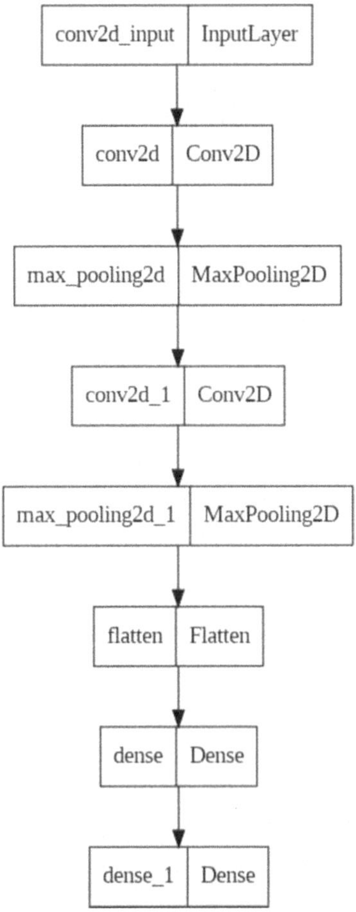

Fig. 1. Architecture of the used CNN model

into the neural network later. Also, to avoid overtraining, a dropout layer is used, where some unnecessary neurons are excluded, by which we mean that they are temporarily removed from the network along with all their input and output connections, which does not affect the total number of neurons. During this stage, we can say that we reduce the number of neurons so that they learn better, i.e. become multidisciplinary. Through all these steps, the network is trained, a process where kernels are found in convolutional layers and weights are found in fully connected layers to minimize the difference between the output predictions and the input training labels. Next comes the backpropagation algorithm, where the loss function and the gradient descent optimization algorithm play an important role. The loss function measures the compatibility between the network's output and the input truth labels, i.e. it measures how well the neural network models the training data. During training, we aim to minimize this loss between the predicted and target outputs.

For our network, we will use binary entropy because we have only two classes "yes" and "no" and in this case, the most efficient loss function is binary entropy. Gradient descent is typically used as an optimization algorithm that iteratively updates the parameters that can be learned, i.e. the kernels and weights of the network, to minimize the loss. The gradient, mathematically, is the partial derivative of the loss concerning each parameter being learned. In our model, we will use the Adam optimization algorithm because it calculates individual adaptive learning rates for different parameters based on the first and second-moment estimates of the gradients, which is a great fit for our model. These are the steps that make up the modeling of a CNNs.

2) Proposed parallel algorithms

After describing the architecture of the constructed convolutional network, we will proceed to describe the parallelization algorithms that parallelize this entire network across different nodes where they are separately trained. In this study, we will use two parallel computing technologies in Python, namely MPI and CUDA. The first of them, namely Python MPI, is a powerful tool for parallelizing computational tasks, including deep learning, such as training CNNs. This allows you to parallelize neural network training on more nodes of the computer system and speed up the model training time. For this task, we will perform data-driven parallelization, i.e., we will train the network in different nodes with an equally distributed amount of data.

The parallelization algorithm itself is to divide the data evenly according to the number of threads. If this is not possible, then we divide the data approximately equally across all threads, since if one model is trained on more data, it will not worsen the final result. In each thread, we train the created model and wait until all the created threads complete their work, after which we send the weights of the trained models from each node to the main thread, where we create the original model from the average of the sent weights.

Figure 2 visualizes a diagram of a parallel algorithm using four threads. Each square is responsible for one thread. The code snippets depict the initialization of MPI itself, where size is the number of threads, and rank is the number of the thread running in this node, then the data is split and the corresponding part of the data is taken x_train=Dataset.split(size)[rank], on which the network is further trained model.fit(x=x_train), after training is completed, we send the weights of the trained model to the main thread comm.send(model.get_weights(), dest=0). The next so-called parallelepiped is responsible for completing the work of all threads before executing the next code snippet. In MPI Python, this is done with the comm.Barrier() command. It is used to ensure synchronization between MPI processes running within the same communicator. This means that if one process calls comm.Barrier(), it will be blocked until all other processes that have also called comm.Barrier() reach this point. Next, in the main thread, where rank==0, we get the weights from the already trained models on different nodes weights.append(comm.recv(source=i)), where we get the weights from all threads using a simple for loop. And then we simply calculate the average value of the weights for the original model.

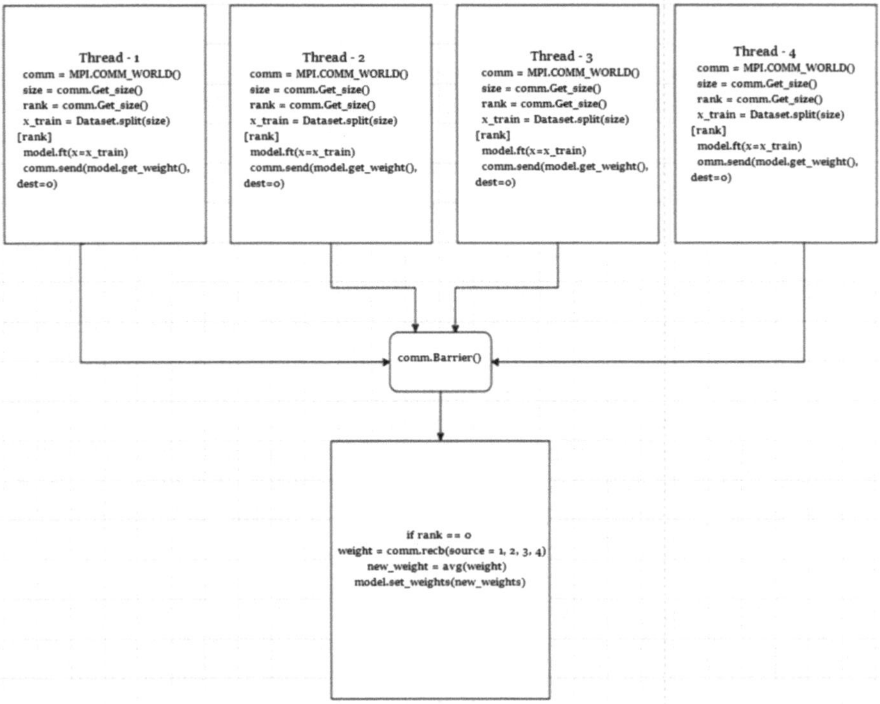

Fig. 2. Diagram of a parallel algorithm using four threads

The mathematical representation of this algorithm can be as follows. Let X – the number of input images, N – the number of threads involved, W – the output weights of the model M, – the model built, then

$$I_i = \frac{X}{N}, \tag{1}$$

where I_i corresponds to the part of the input date set.

$$W_i = M(I_i), \tag{2}$$

where W_i are the weights of the model M trained on the corresponding part of I_i in different threads.

$$R = \frac{\sum_{i=0}^{N} W_i}{N}, \tag{3}$$

where R is responsible for the final result of the calculated model weights.

Speaking about the algorithmic complexity, without using parallelization, we can say that it is $O(n)$, where n is the number of all MRI images used to train the network. In other words, the complexity is equal to the model being trained on each of the images, which takes a lot of time to train. By using parallelization, we reduce the algorithmic complexity by the number of threads. In this case, it is equal to $O(n/k)$, where n is the number of all MRI images and k is the number of created threads. For example, using four threads, we reduce the algorithmic complexity by four times. However, it should also be understood that the number of threads should be chosen approximately equal to the number of computer cores, because if we use several threads that are many times greater than the number of cores, we only worsen the algorithm's performance, both in terms of time and the possibility of worsening the accuracy of the model. In general, by using MPI, we speed up the time it takes to train a CNN, which is very important for real-time decision-making, and we can also improve the accuracy of the model.

The second technology used is a special library CuDNN, which is built based on CUDA. CuDNN is a library of optimized primitives for deep neural network operations on NVIDIA GPUs. It provides a set of low-level GPU acceleration routines such as convolution, pooling, activation functions, normalization, etc. that are specifically designed for neural networks. CuDNN works by leveraging the parallel processing capabilities of NVIDIA GPUs to accelerate the execution of deep neural networks. When a neural network is trained or run on a GPU using CuDNN, the data is transferred to the GPU memory, and operations are performed using CuDNN primitives. The algorithm itself is that CuDNN parallelizes computation across multiple CUDA cores by dividing the input data into smaller parts called "mini-bundles" and processing them in parallel. These mini-bundles are distributed across different GPU cores, and each core performs the same computation on its assigned portion of the data. If we recall the algorithmic complexity of this algorithm, it also decreases with the number of GPU cores, but in this case, it is much more efficient. Also, different parallel algorithms are used for different operations, such as convolution, join, and activation functions. For example, in convolution, CuDNN uses a technique called "fast Fourier transform" to perform convolution in the frequency domain, which is highly parallelizable and can be efficiently executed on GPUs. In the case of convolution, the algorithmic complexity is reduced from $O(N^2)$ to $O(Nlog(N))$, which is dramatically less. In addition, CuDNN also uses cuDNN groups, which allow parallelizing multiple computations on different subsets of the input data. This makes it possible to use GPU resources more efficiently and speed up the computation. In general, CuDNN achieves parallelization through a combination of mini-batch processing, parallel algorithms, and cuDNN groups to fully utilize the parallel processing capabilities of NVIDIA GPUs. Another great advantage is its easy use with the Keras package, which provides tools for building CNNs. In general, CuDNN can significantly speed up the training of CNNs on GPUs and improve model accuracy as well.

The mathematical representation of this algorithm can look like this. Suppose we have an input image represented as a two-dimensional matrix X with dimensions (width, height). We also have a set of convolutional filters represented as three-dimensional tensors W with dimensions (filter_width, filter_height, num_channels), where num_channels is the number of channels in the input image.

1) Convolution: CuDNN performs convolution by sliding each filter over the input image and computing the scalar product between the filter weights and the corresponding input segment. The result of this operation is a feature map. Let's denote the output feature map as Y with dimensions (output_width, output_height, number_of_filters). The output dimensions depend on the pitch, padding, and filter size used in the convolution operation. Mathematically, the convolution operation can be represented as follows:

$$Y(i,j,k) = \sum (I, m, n)(X(i + I, j + m, n) * (W(I, m, n, k)), \quad (4)$$

where i, j are the spatial coordinates in the output feature map, k is the filter index, and I, m, n are the spatial and channel indices in the input image and filter, respectively.
2) Activation function: after the convolution operation, CuDNN applies the activation function element by element to the Y feature map. The function of the activation function is ReLU. Mathematically, the activation function can be represented as follows:

$$Y(i,j,k) = f(Y(i,j,k)). \quad (5)$$

This expression means that each element in the original feature map Y goes through the ReLU activation function, transforming the value by setting the negative value to zero, leaving the non-negative values unchanged.

3) Merge: CuDNN also supports pooling operations such as maximum pooling or average pooling, which reduce the spatial dimensions of the feature map while preserving the most important features. Let's denote the pooled feature map as Z with dimensions (pooled_width, pooled_height, num_filters). The pooling operation can be represented as $Z(i,j,k) = g(Y(p_i, q_j, k))$, where p, q are the pooling coefficients (e.g., two for a 2×2 pooling), $g()$ denotes the pooling function, and $Y(p_i, q_j, k)$ represents the input value for the pooling function.

These are the basic mathematical representations of the operations performed by CuDNN for CNN: convolution, activation, and pooling. These operations are performed in parallel between several GPU cores, which allows for a significant speed increase compared to their sequential execution on the CPU. The parallelization algorithm itself can be represented using the basic Cuda formula: global_thread_id = threadIdx.x + threadIdx.y * blockDim.x + blockIdx.x * blockDim.x * gridDim.y + blockIdx.y * blockDim.x, where threadIdx. x and threadIdx.y are the local thread identifier, blockDim.x is the size of the thread block, blockIdx.x and blockIdx.y are the global identifiers of the thread block, and gridDim.y is the number of blocks in the grid.

4 Research Results

The dataset [23] used for brain cancer detection contains two types of MRI images, namely images with detected cancer and images of healthy people's brains ("Yes", "No"). This dataset is perfectly suited for the task at hand, as it contains all the necessary images in an even amount. The relationship between the amount of data is visualized in Fig. 3.

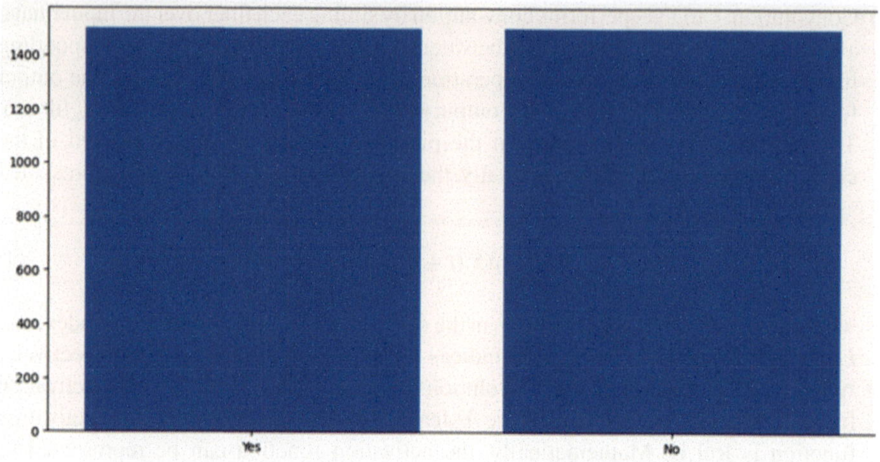

Fig. 3. Diagram of the relationship between the amount of data

Figure 3 shows that the number of images in both classes is the same – 1500 images of each type. In case of an uneven number of images, two approaches to model training can be used.

1. Uneven number of images:
   - Train the model on the available data without any changes. This will result in a model that can better recognize the image type that is more abundant in the dataset but may make some errors in recognizing the less represented classes.
2. Equalize the number of images:
   - Equalize the number of images for both classes before training the model. This may require additional data or the use of data augmentation techniques.
   - Train the model on a balanced dataset to get uniform results for both classes.

Regarding the processing of the dataset in general, the total number of images is 3000, which can load the network during training. Therefore, it is proposed to use parallelization to increase efficiency. Figure 4 shows two images: a) with detected cancer; and b) a healthy person.

All images are converted to a single size of 224 × 224 and normalized. They are also divided into training, validation, and test data. That is, the model will be trained on 2100 MRI images.

Below are the results of the sequential and parallel algorithms. It should be noted that these CNN model training experiments were conducted at 10, 20, and 50 epochs (see Tables 1, 2 and 3). A 4-core laptop was used for the experiments.

Next, let's calculate the speedup and efficiency metrics, which will depend on the number of epochs and the number of threads. Thus, Table 4 shows the speedup, and Table 5 shows the efficiency.

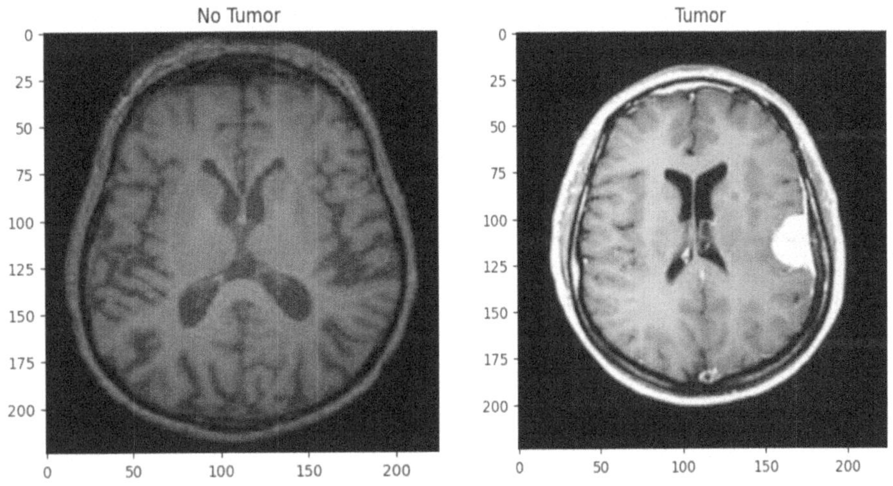

a) b)

Fig. 4. MRI images: a) with detected cancer; b) of a healthy person

Table 1. The results obtained, at 10 epochs

Technology	Number of threads	Execution time, seconds	Accuracy for test data, %	Accuracy for validation data, %
MPI Python	1	716	96	99
	2	412	97	99
	4	304	97	99
	8	227	91	94
	16	269	83	85
CuDNN (CUDA)	–	47	96	99

From the results obtained above, we can see that using parallelization based on MPI Python technology, the maximum acceleration is approximately 3, and based on CuDNN (CUDA) technology – 24. When analyzing these results, it is necessary to take into account the characteristics of the computer system on which the software implementation of the proposed algorithms was tested. As for the efficiency, it is 0.8 with eight threads, and it is already decreasing with sixteen threads (0.77), which is also due to the 4-core nature of the computer.

Table 2. The results obtained, at 20 epochs

Technology	Number of threads	Execution time, seconds	Accuracy for test data, %	Accuracy for validation data, %
MPI Python	1	1536	97	100
	2	923	97	99
	4	628	97	99
	8	497	90	92
	16	534	81	84
CuDNN (CUDA)	–	88	97	99

Table 3. The results obtained, at 50 epochs

Technology	Number of threads	Execution time, seconds	Accuracy for test data, %	Accuracy for validation data, %
MPI Python	1	3792	97	100
	2	2276	97	100
	4	1499	97	99
	8	1189	89	90
	16	1231	81	83
CuDNN (CUDA)	–	162	97	99

Table 4. The acceleration

Number of epochs	Number of threads				CuDNN
	2	4	8	16	
10	1.74	2.35	3.15	2.66	15.00
20	1.66	2.45	3.10	2.87	17.51
50	1.67	2.52	3.19	3.08	23.54

Table 5. The efficiency

Number of epochs	Number of threads			
	2	4	8	16
10	0.44	0.58	0.79	0.67
20	0.42	0.61	0.78	0.72
50	0.42	0.63	0.80	0.77

5 Summary and Conclusion

In this paper, we investigated the use of parallel algorithms based on MPI and CuDNN technologies in the Python programming language for brain cancer detection using a CNN based on MRI images. The acceleration achieved using the data parallelization algorithm was 3.19 times and the efficiency was 0.8, which shows good results using the MPI library. However, using CuDNN, we managed to speed up the model training time by 15, 17.5, and 23.5 times, respectively, with 10, 20, and 50 epochs. We also managed to improve the execution time compared to existing studies using this dataset without reducing the accuracy of the validation sample. This proves the efficiency of using these technologies. Thus, we have fulfilled all the goals of this computational work and obtained good results.

In the future, this study may help doctors detect brain cancer, which is very important these days, as it can save many lives. It can also be useful in various future studies. If we talk about further improvements to this study, it would be effective to run it on different computers, i.e., use computers with more cores, for better parallelization analysis. Speaking about additional features of the proposed approach, it can be useful when we have a small amount of data, since using the parallelization algorithm we can't divide our data into equal parts and train the model on different streams on the whole dataset. In this case, we will not speed up the network training time, but there is a chance to improve the model accuracy since a pair of models will be trained and then the average weights will be calculated, which can improve the accuracy of the model as a whole compared to using a single thread. The approach proposed in this paper can also be applied to other technological processes, in particular, to optimize algorithms in [24, 25].

References

1. Roda, E., Bottone, M.G.: Editorial: brain cancers: new perspectives and therapies. Front. Neurosci. **16**, 1–3 (2022)
2. Shakhovska, N., Montenegro, S., Kryvenchuk, Y., Zakharchuk, M.: The neurocontroller for satellite rotation. Int. J. Intell. Syst. Appl. (IJISA) **11**(3), 1–10 (2019)
3. Zhang, B., Shi, H., Wang, H.: Machine learning and AI in cancer prognosis, prediction, and treatment selection: a critical approach. J. Multidiscip. Healthc. **16**, 1779–1791 (2023)
4. Pollack, I.F., Agnihotri, S., Broniscer, A.: Childhood brain tumors: current management, biological insights, and future directions. J. Neurosurg. Pediatr. **23**(3), 261–273 (2019)

5. Zomchak, L., Starchevska, I.: Macroeconomic determinants of economic development and growth in Ukraine: logistic regression analysis. In: Hu, Z., Wang, Y., He, M. (eds.) CSDEIS 2022. LNDECT, vol. 158, pp. 358–368. Springer, Cham (2023). https://doi.org/10.1007/978-3-031-24475-9_31
6. Zomchak, L., Lapinkova, A.: Key interest rate as a central banks tool of the monetary policy influence on inflation: the case of Ukraine. In: Hu, Z., Wang, Y., He, M. (eds.) CSDEIS 2022. LNDECT, vol. 158, pp. 369–379. Springer, Cham (2023). https://doi.org/10.1007/978-3-031-24475-9_32
7. Afshany, A.J., Tourani, A., Shahbahrami, A., Khazaee, S., Akoushideh, A.: Parallel implementation of a video-based vehicle speed measurement system for municipal roadways. Int. J. Intell. Syst. Appl. (IJISA) **11**(11), 25–37 (2019)
8. Mochurad, L., Solomiia, A.: Optimizing the computational modeling of modern electronic optical systems. In: Lytvynenko, V., Babichev, S., Wójcik, W., Vynokurova, O., Vyshemyrskaya, S., Radetskaya, S. (eds.) ISDMCI 2019. AISC, vol. 1020, pp. 597–608. Springer, Cham (2020). https://doi.org/10.1007/978-3-030-26474-1_41
9. Khan, Md.S.I., et al.: Accurate brain tumor detection using deep convolutional neural network. Comput. Struct. Biotechnol. J. **20**, 4733–4745 (2022)
10. Sajid, S., Hussain, S., Sarwar, A.: Brain tumor detection and segmentation in MR images using deep learning. Arab. J. Sci. Eng. **44**, 9249–9261 (2019)
11. Shaheen, M., Saif, U., Awan, S.M., Ahmad, F., Anum, A.: Classification of images of skin lesion using deep learning. Int. J. Intell. Syst. Appl. (IJISA) **15**(2), 23–36 (2023)
12. Mochurad, L.: Canny edge detection analysis based on parallel algorithm, constructed complexity scale and CUDA. Comput. Inform. **41**(4), 957–980 (2022)
13. Mochurad, L., Dereviannyim A., Antoniv, U.: Classification of X-ray images of the chest using convolutional neural networks. In: Proceedings of the 4th International Conference on Informatics & Data-Driven Medicine, Valencia, Spain, 19–21 November 2021, pp. 269–282 (2021)
14. Saeedi, S., Rezayi, S., Keshavarz, H., et al.: MRI-based brain tumor detection using convolutional deep learning methods and chosen machine learning techniques. BMC Med. Inform. Decis. Mak. **23**(16), 1–17 (2023)
15. Kumar, S.: Brain_tumor_detection_simple CNN. https://www.kaggle.com/code/shashank221b/brain-tumor-detection-simple-cnn. Accessed 15 Feb 2020
16. Brain_tumor_detection using a pretrained model. https://www.kaggle.com/code/vivek153/brain-tumor-detection-using-a-pretrained-model. Accessed 15 Feb 2020
17. Mohsem, R.T.: Brain_Tumor_Detection. https://www.kaggle.com/code/reemtarekmohsen/brain-tumor-detection. Accessed 15 Feb 2020
18. Rahman, T., Islam, Md.S.: MRI brain tumor detection and classification using parallel deep convolutional neural networks. Meas. Sens. **26**(1), 100694 (2023)
19. Izonin, I., Tkachenko, R., Holoven, R., Yemets, K., Havryliuk, M., Shandilya, S.K.: SGD-based cascade scheme for higher degrees wiener polynomial approximation of large biomedical datasets. Mach. Learn. Knowl. Extr. **4**(4), 1088–1106 (2022)
20. Zulqarnain, M., Ghazali, R., Ghouse, M.G., Hassim, Y.M.M., Javid, I.: Predicting financial prices of stock market using recurrent convolutional neural networks. Int. J. Intell. Syst. Appl. (IJISA) **12**(6), 21–32 (2020)
21. El Hamdi, D., Elouedi, I., Nguyen, M.K., Hamouda, A.: A conic radon-based convolutional neural network for image recognition. Int. J. Intell. Syst. Appl. (IJISA) **15**(1), 1–12 (2023)
22. Teslyuk, V., Beregovskyi, V., Denysyuk, P., Teslyuk, T., Lozynskyi, A.: Development and implementation of the technical accident prevention subsystem for the smart home system. Int. J. Intell. Syst. Appl. (IJISA) **10**(1), 1–8 (2018)
23. Br35H: Brain Tumor Detection. https://www.kaggle.com/datasets/ahmedhamada0/brain-tumor-detection. Accessed 15 Feb 2020

24. Martyn, Y., Liaskovska, S., Gregus, M., Izonin, I., Velyka, O.: Optimization of technological's processes industry 4.0 parameters for details manufacturing via stamping: rules of queuing systems. Procedia Comput. Sci. **191**, 290–295 (2021)
25. Ljaskovska, S., Martyn, Y., Malets, I., Velyka, O.: Optimization of parameters of technological processes means of the FlexSim simulation simulation program. In: Proceedings of the 2020 IEEE 3rd International Conference on Data Stream Mining and Processing, DSMP 2020, pp. 391–397 (2020)

A Comparative Assessment of the Cutting-Edge Time Series Forecasting Models Using Seasonal Data

Kyrylo Yemets[1,2](✉) and Nataliia Lotoshynska[1,2]

[1] Department of Artificial Intelligence, Lviv Polytechnic National University, Kniazia Romana Street, 5, Lviv 79905, Ukraine
kyrylo.v.yemets@lpnu.ua, natlot@ukr.net

[2] Department of Publishing Information Technologies, Lviv Polytechnic National University, S. Bandera Street, 12, Lviv 79013, Ukraine

Abstract. Time series forecasting is a critical endeavor spanning diverse domains such as finance and medicine, demanding accurate predictions derived from historical data. The complexity escalates when analyzing seasonal data, characterized by unpredictable fluctuations and diverse influences, thereby challenging the suitability of existing AI tools. Hence, a meticulous approach encompassing thorough data analysis, model selection, and effective strategies is imperative. This paper aims to comprehensively examine and compare state-of-the-art time series forecasting models tailored for seasonal data, with a focus on addressing associated challenges and enhancing forecast accuracy. Seven models, including Temporal Fusion Transformer, DeepNPTS, PatchTST, DeepAR, NPTS, DLinear, and WaveNet, were identified through a thorough review, delineating their principles, advantages, and limitations. These models, along with a basic naïve approach, were rigorously evaluated using diverse performance metrics, particularly within the context of product sales data analysis. Our analysis highlights neural network models leveraging transformer architectures as exhibiting superior efficacy in achieving high-precision forecasts for seasonal time series data. However, it is noted that these models require a substantial time investment for parameter optimization, often involving thousands of parameters. These findings contribute to the advancement of time series forecasting methodologies, facilitating practical applications across various domains.

Keywords: Time-series forecasting · seasonal data · Temporal Fusion Transformer · DeepNPTS · PatchTST · DeepAR · DLinear · WaveNet

1 Introduction

The task of time series forecasting is pertinent and crucial across various scientific, technical, and practical domains. It arises within the contexts of diverse disciplines such as finance, economics, meteorology, medicine, transportation, telecommunications, and numerous others, where predicting future values of specific variables based on their historical data is imperative [1].

Z. Hu et al. (Eds.): ICCSEEA 2024, LNDECT 242, pp. 206–217, 2025.
https://doi.org/10.1007/978-3-031-84228-3_17

The relevance of this task is determined by a multitude of factors. Firstly, accurate forecasts of time series enable efficient resource management and strategic planning across various domains [2]. For instance, in the financial sector, precise forecasts of stock prices, currency exchange rates, and other financial indicators are crucial for decision-making regarding investments and risk management [3, 4].

Secondly, the rapid advancement of technologies and the proliferation of data volumes have led to new opportunities for the application of time series forecasting [5, 6]. With the emergence of powerful computational methods and machine learning algorithms, modern forecasting models can handle large datasets and account for complex interrelationships among different variables [7, 8].

Furthermore, in the contemporary world where business and economic conditions are constantly changing, it is crucial to have the ability to promptly respond to changes and adapt planning and management strategies [9, 10]. Time series forecasting aids in ensuring adaptability and flexibility in addressing business challenges.

Time series forecasting presents a complex task encompassing various challenges and issues that can complicate the modeling and prediction process. Foremost among these are the seasonality and trends evident in the data [11]. Many time series exhibit seasonal variations and trends, necessitating attention during modeling. Seasonality requires analysis of cyclical changes in the time series, while trends reflect the overall tendency of value changes over time. Nonstationarity poses another challenge encountered in solving time series forecasting problems [12]. Many time series are nonstationary, meaning their characteristics change over time, complicating modeling and forecasting.

Handling large volumes of data can also pose a problem in addressing this task. Some time series contain vast amounts of data, necessitating efficient processing and analysis strategies, as well as significant computational resources and specialized analysis methods [13]. Random noise and unpredictability can further complicate time series forecasting. Specifically, time series may contain random noise, hindering the detection of genuine patterns and trends. Moreover, some time series may have complex interrelationships among different variables, necessitating advanced models and analysis methods for effective forecasting [14].

The complexity of time series forecasting is significantly heightened when analyzing seasonal data, which can impose several constraints on the applicability of existing artificial intelligence tools for its resolution. Specifically, seasonal changes and fluctuations may be unpredictable and diverse, complicating the detection of trends and the identification of future seasonal patterns. Additionally, the instability of seasonal influences can render their forecasting more challenging, as seasonal changes may vary depending on various factors. Processing large volumes of data can also be problematic, particularly if data is monitored in real-time, necessitating significant computational resources and effective algorithms [15]. Furthermore, some seasonal influences may be multiplicative, further complicating their modeling and requiring specialized data processing methods. Finally, data intermittency can also pose challenges in forecasting future seasonal trends due to changes in measurements or research conditions.

Given the complexities inherent in forecasting seasonal time series, a meticulous approach is essential. This encompasses thorough data analysis, the judicious selection of modeling techniques, and the implementation of effective strategies to manage the inherent complexities.

Hence, the primary objective of the article is to undertake a comprehensive examination involving the selection, modeling, efficiency evaluation, and comparative assessment of cutting-edge time series forecasting models that leverage seasonal data. By undertaking such an endeavor, the aim is to provide insights into the most effective methodologies for addressing the challenges posed by seasonal time series data, ultimately facilitating more accurate and reliable forecasts. Through rigorous investigation and comparative analysis, the article seeks to contribute to the advancement of time series forecasting methodologies and their practical applications in various domains.

The main contribution of this paper can be summarized as follows:

1) A comprehensive search was conducted to identify state-of-the-art models tailored for forecasting seasonal time series. Subsequently, the operational principles, advantages, and disadvantages of these models were delineated.
2) The identified state-of-the-art time series forecasting models were implemented, and a rigorous comparative evaluation was carried out utilizing diverse performance metrics. This assessment was conducted within the context of analyzing product sales data.
3) Our analysis revealed that neural network models leveraging transformer architectures exhibited superior efficacy in addressing the challenge of high-precision forecasting for seasonal time series data.

2 Materials and Methods

2.1 Dataset Description

Between 1989 and 1994, the University of Chicago Booth School of Business collaborated with the now-closed Dominick's Finer Foods on a series of randomized store-level experiments focusing on shelf management and pricing across over 25 different categories in the entirety of the 100-store chain. This partnership yielded around nine years of detailed store-level data concerning the sales of over 3,500 unique product codes (UPCs).

The resulting Dominick's dataset encompasses store-level scanner data amassed at Dominick's Finer Foods for a span of more than seven years, encapsulating 115704 weekly time series detailing the profit of individual stock keeping units (SKUs) from the retailer. This comprehensive dataset includes both category-specific files and general files covering all project categories, alongside daily sales data on more than 3500 products, customer counts, and store-specific demographics and sales products.

2.2 Cutting-Edge Time Series Forecasting Models

The paper undertook a comprehensive exploration and examination of contemporary neural network methodologies aimed at addressing the challenge of forecasting seasonal time series data. The investigation was conducted utilizing the Scopus scientometric database to procure relevant literature. Subsequently, seven state-of-the-art methodologies frequently encountered in scholarly discourse about the forecasting of conventional time series data were identified. These methodologies include Temporal Fusion Transformer [16], DeepNPTS [17], PatchTST [18], DeepAR [19], NPTS [20], DLinear [21], and WaveNet [22]. This selection served as the basis for a detailed analysis of the operational principles, merits, and limitations associated with each methodology.

The Temporal Fusion Transformer (TFT) [16], is a highly advanced model tailored for forecasting tasks, particularly adept at managing the intricacies of seasonal time series data. Its architecture is meticulously designed to process and learn from both static and dynamic information, making it exceptionally good for scenarios where understanding temporal patterns and relationships is crucial.

One of the standout features of the TFT is its incorporation of gating mechanisms and variable selection networks. These components are instrumental in sifting through various input variables to pinpoint the ones most relevant for forecasting at any given time. This ability to focus on the most informative features is especially valuable in seasonal forecasting, where the significance of certain inputs can vary across different times of the year.

Furthermore, the TFT employs static covariate encoders to integrate time-invariant information into the forecasting process. This inclusion of static data provides a stable context that enriches the model's understanding of each time series, enhancing its ability to capture the unique characteristics that influence seasonal patterns.

The temporal processing capabilities of the TFT are another critical aspect of its architecture. By effectively learning both short-term and long-term dependencies within the data, the model can accurately anticipate future trends and patterns. This nuanced understanding of time dynamics is essential for predicting seasonal fluctuations, which often involve complex interactions over various time scales.

Moreover, the TFT's use of an interpretable multi-head attention mechanism allows it to focus on the most relevant past information when making forecasts. This targeted approach improves the model's ability to learn from long-term seasonal trends, ensuring that forecasts are grounded in a deep understanding of historical data.

Finally, the TFT's ability to generate prediction intervals through quantile outputs is particularly beneficial for seasonal forecasting. By providing a range of possible outcomes instead of a single forecast, it offers a more nuanced view of future uncertainties, which is crucial for planning and risk management in seasonal contexts.

A Deep Non-Parametric Time Series Forecaster DeepNPTS [17], combines the adaptability of non-parametric models with the comprehensive feature extraction capabilities of deep learning for predicting future values in time series data, especially when dealing with seasonal fluctuations. This approach does not confine itself to predefined relationships between variables and outcomes, allowing for a more nuanced adaptation to the data's inherent structure. Such a model is exceptionally suited for capturing complex seasonal patterns that vary across different datasets, thanks to its flexible data modeling.

Deep neural networks within this architecture excel at uncovering complicated, non-linear relationships, including hidden seasonal trends, without the need for manual identification of these patterns. They are particularly effective at understanding and modeling the periodicity and trends that are characteristic of seasonal data. This includes the ability to recognize and adapt to multiple seasonal cycles and the interactions between them.

Additionally, the architecture's capacity to model temporal dynamics, possibly through mechanisms like Recurrent Neural Networks (RNNs) or Long Short-Term Memory (LSTM) units, is crucial for learning from past cycles to predict future ones accurately. This is enhanced by the model's adaptability, allowing it to adjust to evolving seasonal effects, which is critical in environments where external factors might influence these patterns.

One of the significant benefits of such an architecture is its robustness against overfitting, thanks to its data-driven complexity adjustment, ensuring genuine seasonal patterns are captured rather than noise. This adaptability, combined with deep learning's scalability, renders the model highly effective across various applications, from forecasting retail sales to predicting energy demands, achieving high accuracy levels even in the presence of complex seasonal interactions. Furthermore, the ability to quantify uncertainty in its forecasts adds a layer of value, offering insights into potential variability and risks associated with future seasonal trends.

PatchTST [18], which stands for Patch Transformer for Time Series Forecasting, is a novel architecture that leverages the Transformer model's capabilities, specifically designed to address time series forecasting challenges, including those presented by seasonal patterns. While the specific details of PatchTST might vary depending on the source or implementation, the core idea combines the Transformer's powerful attention mechanisms with a patch-based approach to handle time series data effectively.

The PatchTST architecture is built around the concept of breaking down the time series into smaller segments or "patches". This method allows the model to focus on specific portions of the data at a time, making it easier to identify and learn from the underlying seasonal trends and patterns. By treating these segments as individual inputs, PatchTST can apply the Transformer's self-attention mechanisms to both intra-patch and inter-patch relationships, enabling it to capture complex temporal dependencies across different time scales.

The use of the Transformer model within PatchTST brings the advantage of handling long-range dependencies. This is particularly beneficial for seasonal forecasting, where understanding the relationships between distant points in time can be crucial for predicting future trends. The self-attention mechanism dynamically weighs the importance of different time steps, allowing PatchTST to focus on the most relevant information for making accurate forecasts.

Another advantage of PatchTST is its scalability and flexibility. The model can be adjusted to handle different lengths of time series and various degrees of seasonality, making it applicable to a wide range of forecasting tasks. Whether dealing with short-term daily sales data with weekly patterns or long-term climate data with annual cycles, PatchTST's adaptable framework can be tuned to optimize performance for the specific characteristics of the data.

DeepAR [19], is a probabilistic forecasting framework that leverages deep learning for time series prediction, and it's particularly adept at handling seasonal patterns. Developed by Amazon, DeepAR utilizes a recurrent neural network (RNN) architecture, typically Long Short-Term Memory (LSTM) units or Gated Recurrent Units (GRUs), to model complex temporal dependencies and predict the future distribution of a time series.

At its heart, DeepAR ingests sequences of historical data points and outputs parameters of a probability distribution for future values, rather than predicting a single point estimate. This probabilistic approach allows it to quantify the uncertainty of its forecasts, providing not just a single future value but a range of possible outcomes with associated probabilities. This feature is particularly valuable for seasonal forecasting, where understanding the range of potential future values can aid in planning and risk management.

DeepAR can automatically learn and incorporate seasonal features from the data. By analyzing the historical data, the model identifies recurring patterns that characterize the seasonality, adjusting its predictions to reflect these cycles.

The use of RNNs, with LSTM or GRU units, enables DeepAR to remember information over long periods. This capability is crucial for accurately forecasting seasonal trends, as it allows the model to use knowledge of past seasonal effects to predict future occurrences.

The Non-Parametric Time Series, (NPTS) [20], algorithm is a forecasting method that relies on historical data patterns to predict future values, without assuming any specific parametric form for the underlying data generation process. This approach is particularly useful for handling seasonal time series data, where the goal is to capture and leverage recurring patterns and trends inherent in the dataset.

NPTS identifies and leverages seasonal patterns by examining the historical data for periods that closely match the seasonality of the series. For instance, if a time series exhibits strong yearly seasonality, the NPTS algorithm will focus on the data points from the same seasons in past years to generate forecasts.

The non-parametric nature of the algorithm allows it to adapt to a wide variety of seasonal patterns, making it suitable for time series with complex or irregular seasonality.

While often an advantage, the simplicity of NPTS can also be a limitation, as it might not capture complex interactions or non-linear relationships as effectively as some parametric or machine learning-based methods.

DLinear [21] are architectures designed to enhance linear models for time series forecasting by explicitly addressing common challenges such as trend, seasonality, and distribution shifts in the data.

DLinear aims to improve forecasting accuracy by explicitly decomposing the time series into trend and seasonal (remainder) components, and then applying specialized processing to each. Trend component is obtained by applying a moving average kernel to the raw data, smoothing out fluctuations to capture the underlying trend. The Seasonal component remains after the trend component is removed from the raw data, capturing the cyclical patterns and noise.

WaveNet [22], originally introduced by DeepMind in 2016, is a deep neural network architecture designed for generating raw audio waveforms with remarkable realism. Although its initial application was in speech and music generation, the WaveNet architecture has also been adapted for time series forecasting due to its ability to model complex temporal relationships.

WaveNet is based on a stack of convolutional layers, specifically designed to capture the hierarchical temporal structure of sequences. It uses dilated causal convolutions, allowing the network to efficiently process data points from the past without violating the temporal order and to cover a large context window without needing an excessively deep network.

WaveNet's structure allows it to capture long-range dependencies that are often present in time series data, making it suitable for forecasting tasks with complex temporal dynamics. By predicting distributions over future values, WaveNet provides a measure of uncertainty in its forecasts, which is invaluable for risk assessment and decision-making processes.

3 Modeling and Results

3.1 Modeling

For modeling Pytorch was used as a deep learning framework for all methods except for NPTS. Computations being executed on NVIDIA T4 GPU. Lag size for time series data was for 10 weeks. While forecasting for 8 weeks. For deep learning methods the dataset was reformatted as a bunch of small time series with 10 features for input and 8 features for output.

3.2 Performance Indicators

The efficacy of all examined models was assessed through the utilization of three primary metrics: Mean Absolute Error (MAE), Symmetric Mean Absolute Percentage Error (SMAPE), and the quantification of model parameters (expressed in thousands) [23]:

MAE:

$$MAE = \frac{1}{n}\sum\nolimits_{t=1}^{n} |F_t - A_t|, \tag{1}$$

SMAPE:

$$SMAPE = \frac{100}{n}\sum\nolimits_{t=1}^{n} \frac{|F_t - A_t|}{(|F_t| + |A_t|) \times 0.5}, \tag{2}$$

where n is the number of samples in the data, t is the number of actual samples in the data, F_t is the forecasted value for the actual sample. A_t is the ground truth value for the actual sample.

We also used a number of the model's trainable parameters expressed in thousands that can tell how much resources we will need to train it and how big will be latency for inference in production.

Such multifaceted evaluation approach facilitated a comprehensive analysis of each investigated time series forecasting model from diverse perspectives, ensuring a thorough understanding of their performance characteristics.

3.3 Results

The modeling outcomes for all eight investigated methods, based on the metrics (1) and (2), alongside the quantification of model parameters (expressed in thousands), are succinctly outlined in Table 1. It is imperative to underscore that the process entailed the meticulous optimization of parameters tailored to the characteristics of the respective dataset for each method. Thus, Table 1 encapsulates the optimized findings gleaned from the comprehensive analysis of all examined methods.

As delineated in Table 1, the majority of scrutinized seasonal time series forecasting methods exhibit notable levels of forecast accuracy. Nevertheless, in order to ascertain the optimal methodology among them, a comprehensive comparative analysis of their efficacy will be conducted.

Table 1. Findings about all eight seasonal time series forecasting methods

#	Model/Performance indicators	MAE	SMAPE	The quantification of model parameters (expressed in thousands)
1	Temporal Fusion Transformer [16]	40,5	1,59	98,5
2	DeepNPTS [17]	41,6	1,60	4
3	PatchTST [18]	40,8	1,60	42,6
4	DeepAR [19]	41,8	1,60	46
5	NPTS [20]	47,6	1,61	0
6	NAIVE	44,5	1,61	0
7	DLinear [21]	46,5	1,62	3,6
8	WaveNet [22]	41,3	1,66	86

This comparative assessment will entail a meticulous examination of all three indicators, aiming to discern nuanced differences in their predictive capabilities.

4 Comparison and Discussion

In order to select the most optimal method from the researched options for addressing the challenge of seasonal time series forecasting, a detailed examination of the acquired results is conducted. Given that the primary objective of employing forecasting methods is to achieve the utmost forecast accuracy, firstly we conducted the comparison of outcomes generated by all methods based on two key error metrics (1) and (2).

Figure 1 provides a visual representation of the obtained results. Specifically, Fig. 1a illustrates the outcomes derived from error metric (1), while Fig. 1b delineates the results obtained using error metric (2).

As evident from Fig. 1, the lowest levels of accuracy based on metrics (1) and (2) are observed in three methods: NPTS, DLinear, and NAIVE. This observation is somewhat unexpected, particularly considering that the DLinear architecture is among the most

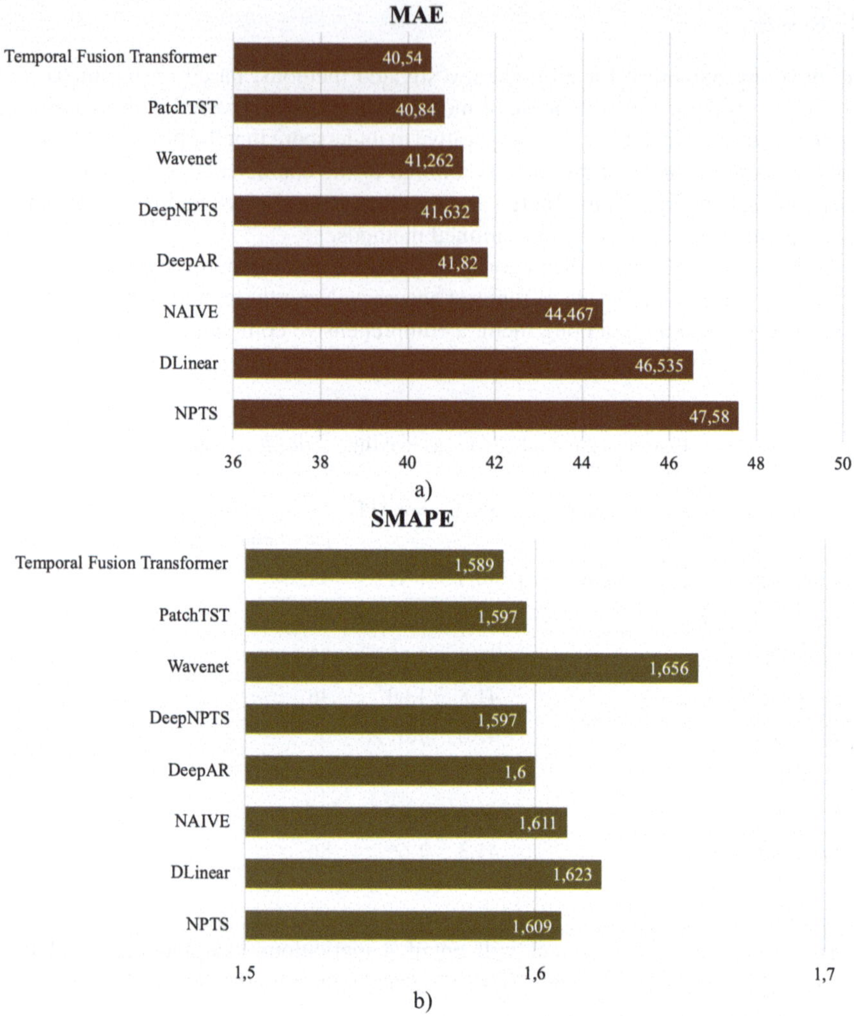

Fig. 1. Comparison of all investigated methods based on a) MAE; b) SMAPE

recent and has exhibited promising outcomes in the scientific literature. Such results may be attributed to idiosyncrasies within the input data, notably the presence of anomalies and outliers. Conversely, notably superior results were achieved by neural network methodologies such as DeepAR, DeepNPTS, and Wavenet. These comparatively recent architectures demonstrate nearly identical performance, albeit with WaveNet exhibiting heightened sensitivity to anomalies.

The most exemplary outcomes were attained by two transformer architectures, PatchTST and Temporal Fusion Transformer. While both demonstrate notably similar results, the latest displays marginally superior prediction accuracy across both metrics.

In addition to the accuracy of the investigated methods, a crucial efficiency metric in its practical application is the duration of they training and the number of hyperparameters necessitating adjustment for they optimal performance. This article undertakes a comparison of these indicators, the results of which are depicted in Fig. 2.

Fig. 2. Comparison of accuracy level and the number of model parameters (expressed in thousands) for all investigated methods

As illustrated in Fig. 2, the PatchTST and Temporal Fusion Transformer architectures, which exhibit superior forecasting accuracy in the context of seasonal time series analysis, necessitate the configuration of a larger number of parameters for optimal operation. Nonetheless, given the contemporary computational capabilities, this drawback can be mitigated in pursuit of achieving the highest forecast accuracy possible.

5 Summary and Conclusion

The task of time series forecasting holds significant importance across various fields, ranging from finance to medicine, necessitating accurate predictions based on historical data. Analyzing seasonal data adds complexity due to unpredictable fluctuations and diverse influences, challenging the applicability of existing AI tools. Thus, a meticulous approach involving thorough data analysis, model selection, and effective strategies is crucial.

This article's primary objective was to comprehensively examine and compare cutting-edge time series forecasting models tailored for seasonal data, aiming to provide insights into addressing the associated challenges and improving forecast accuracy. The main results of this paper are the following:

1. A comprehensive review identified 7 state-of-the-art models for seasonal time series forecasting, outlining their principles, advantages, and drawbacks.
2. Identified 7 models and basic naïve approach were rigorously evaluated using diverse performance metrics, focusing on product sales data analysis.

Our analysis revealed that neural network models leveraging transformer architectures demonstrated superior efficacy in achieving high-precision forecasts for seasonal time series data. Nevertheless, these models necessitate a substantial amount of time for adjusting a considerable volume of optimal parameters, quantified in the thousands.

These findings contribute to advancing time series forecasting methodologies, aiding practical applications in various domains.

Acknowledgment. This work is supported by European Union's Horizon Europe research and innovation programme under grant agreement No 101138678, project ZEBAI (Innovative methodologies for the design of Zero-Emission and cost-effective Buildings enhanced by Artificial Intelligence).

References

1. Casolaro, A., Capone, V., Iannuzzo, G., Camastra, F.: Deep learning for time series forecasting: advances and open problems. Information **14**(11), 598 (2023)
2. Izonin, I., Tkachenko, R., Mitoulis, S.A., Faramarzi, A., Tsmots, I., Mashtalir, D.: Machine learning for predicting energy efficiency of buildings: a small data approach. Procedia Comput. Sci. **231**, 72–77 (2024)
3. Zomchak, L., Starchevska, I.: Macroeconomic determinants of economic development and growth in Ukraine: logistic regression analysis. In: Hu, Z., Wang, Y., He, M. (eds.) CSDEIS 2022. LNDECT, vol. 158, pp. 358–368. Springer, Cham (2023). https://doi.org/10.1007/978-3-031-24475-9_31
4. Zomchak, L., Lapinkova, A.: Key interest rate as a central banks tool of the monetary policy influence on inflation: the case of Ukraine. In: Hu, Z., Wang, Y., He, M. (eds.) CSDEIS 2022. LNDECT, vol. 158, pp. 369–379. Springer, Cham (2023). https://doi.org/10.1007/978-3-031-24475-9_32
5. Mochurad, L., Hladun, Y., Zasoba, Y., Gregus, M.: An approach for opening doors with a mobile robot using machine learning methods. Big Data Cogn. Comput. **7**(2), 69 (2023)
6. Shymanskyi, V., Sokolovskyy, I., Sokolovskyy, Y., Bubnyak, T.: Variational method for solving the time-fractal heat conduction problem in the claydite-block construction. In: Hu, Z., Dychka, I., Petoukhov, S., He, M. (eds.) ICCSEEA 2022. LNDECT, vol. 134, pp. 97–106. Springer, Cham (2022). https://doi.org/10.1007/978-3-031-04812-8_9
7. Mochurad, L., Shchur, G.: Parallelization of cryptographic algorithm based on different parallel computing technologies. In: CEUR-WSorg, vol. 2824:2029 (2021)
8. Shakhovska, N., Montenegro, S., Kryvenchuk, Y., Zakharchuk, M.: The neurocontroller for satellite rotation. Int. J. Intell. Syst. Appl. **11**(3), 1–10 (2019)

9. Semkovych, V., Shymanskyi, V.: Combining OCR methods to improve handwritten text recognition with low system technical requirements. In: Hu, Z., Wang, Y., He, M. (eds). CSDEIS 2022. LNDECT, vol. 158, pp. 693–702. Springer, Cham (2023). https://doi.org/10.1007/978-3-031-24475-9_56
10. Sharma, S., Kumar, V., Kumar, R.: Supervised online adaptive control of inverted pendulum system using ADALINE artificial neural network with varying system parameters and external disturbance. Int. J. Intell. Syst. Appl. **4**(8), 53–61 (2012)
11. Abel Mgandu, F., Mkandawile, M.: Trend analysis and forecasting of water level in mtera dam using exponential smoothing. Int. J. Math. Sci. Comput. **6**(4), 26–34 (2020)
12. Singh, D.K., Rawat, N.: Decoding optimization algorithms for convolutional neural networks in time series regression tasks. Int. J. Inf. Technol. Comput. Sci. **15**(6), 37–49 (2023)
13. Liaskovska, S.Y., Martyn, Y.V.: Development of information technologies for the research of technical systems. In: Pawelczyk, M., Bismor, D., Ogonowski, S., Kacprzyk, J. (eds.) PCC 2023. LNNS, vol. 708, pp. 3–15. Springer, Cham (2023). https://doi.org/10.1007/978-3-031-35170-9_1
14. Izonin, I., Tkachenko, R., Krak, I., Berezsky, O., Shevchuk, I., Shandilya, S.K.: A cascade ensemble-learning model for the deployment at the edge: case on missing IoT data recovery in environmental monitoring systems. Front. Environ. Sci. **26**(11), 1295526 (2023)
15. Ganguli, C., Shandilya, S.K., Nehrey, M., Havryliuk, M.: Adaptive artificial bee colony algorithm for nature-inspired cyber defense. Systems **11**(1), 27 (2023)
16. Lim, B., Arik, S.O., Loeff, N., Pfister, T.: Temporal fusion transformers for interpretable multi-horizon time series forecasting (2019). https://arxiv.org/abs/1912.09363
17. Rangapuram, S.S., Gasthaus, J., Stella, L., Flunkert, V., Salinas, D., Wang, Y., et al.: Deep non-parametric time series forecaster. arXiv.org. (2023). https://arxiv.org/abs/2312.14657
18. Nie, Y., Nguyen, N.H., Sinthong, P., Kalagnanam, J.: A time series is worth 64 words: long-term forecasting with transformers (2022). https://arxiv.org/abs/2211.14730
19. Salinas, D., Flunkert, V., Gasthaus, J.: DeepAR: probabilistic forecasting with autoregressive recurrent networks. arXiv.org (2017). https://arxiv.org/abs/1704.04110
20. Wang, X., Lee, M., Zhao, Q., Tong, L.: Non-parametric probabilistic time series forecasting via innovations representation. arXiv.org (2023). https://arxiv.org/abs/2306.03782
21. Wang, G., Liao, Y., Guo, L., Geng, J., Ma, X.: DLinear photovoltaic power generation forecasting based on reversible instance normalization. In: Proceedings of the 2023 IEEE 12th Data Driven Control and Learning Systems Conference (DDCLS), Xiangtan, China, pp. 990–995. IEEE (2023)
22. WaveNet: A generative model for raw audio. Google DeepMind (2016). https://deepmind.google/discover/blog/wavenet-a-generative-model-for-raw-audio/
23. Ljaskovska, S., Martyn, Y., Malets, I., Velyka, O.: Optimization of parameters of technological processes means of the FlexSim simulation simulation program. In: Proceedings of the 2020 IEEE Third International Conference on Data Stream Mining & Processing (DSMP), Lviv, Ukraine, pp. 391–397. IEEE (2020)

Mortality Risk Prediction for Multiple Trauma Patients Admitted to the Hospital via Machine Learning Algorithms

Ivan Izonin[1,2], Myroslav Stupnytskyi[3], Roman Tkachenko[4], Myroslav Havryliuk[2(✉)], Oleksii Biletskyi[5,6], and Grygoriy Melnyk[7]

[1] Department of Civil Engineering, School of Engineering, University of Birmingham, Birmingham B15 2TT, UK
[2] Department of Artificial Intelligence, Lviv Polytechnic National University, Kniazia Romana Street, 5, Lviv 79905, Ukraine
myroslav.a.havryliuk@lpnu.ua
[3] Anesthesiology and Intensive Care Department in the Clinic of Neurosurgery and Neurology, Military Medical Clinical Center of the Western Region, Lviv, Ukraine
[4] Department of Publishing Information Technologies, Lviv Polytechnic National University, Kniazia Romana Street, 5, Lviv 79905, Ukraine
[5] Anesthesiology and Intensive Care Department for Patients with Combined Trauma, Kharkiv Clinical Emergency Hospital Named by Prof. O.I. Meshchaninov, Kharkiv, Ukraine
[6] Department of Emergency Medicine, Anesthesiology and Intensive Therapy, Kharkiv National Medical University, Kharkiv, Ukraine
[7] West Ukrainian National University, 11 Lvivska Street, Ternopil 46001, Ukraine
mgm@wunu.edu.ua

Abstract. The complexity inherent in medical diagnosis is compounded when confronted with patients experiencing polytrauma, necessitating a multifaceted approach to ensure accuracy, efficiency, and reliability. In this study, we address the challenge of predicting mortality risk in polytrauma patients through an experimental analysis of various machine learning algorithms. To this end, we have curated a dataset comprising clinical data of male patients afflicted with multiple traumas, treated at the Anesthesiology and Intensive Care Department for combined trauma cases within Ukrainian hospital. Employing a diverse range of optimized machine learning techniques, we assessed their efficacy in classification task using seven performance indicators. Our findings reveal that neural network models exhibit the best performance, particularly in terms of F1-score, Matthews Correlation Coefficient (MCC), and Cohen's kappa score, surpassing traditional statistical methods. This underscores the potential of neural networks in accurately predicting mortality risk in polytrauma patients, thereby facilitating more informed clinical decision-making processes.

Keywords: clinical data · small data · machine learning · multiple trauma patients · mortality risk prediction · ANN

Z. Hu et al. (Eds.): ICCSEEA 2024, LNDECT 242, pp. 218–228, 2025.
https://doi.org/10.1007/978-3-031-84228-3_18

1 Introduction

The rapid advancement across all facets of medicine today hinges upon the latest breakthroughs in the artificial intelligence [1]. Contemporary approaches to diagnosis, treatment, and patient recovery invariably incorporate an amalgamation of Internet of Things (IoT) devices, data storage, and transmission mechanisms, as well as artificial intelligence algorithms for analysis. This integrated approach allows timely adjustments to diagnostic or treatment protocols, thereby enhancing the quality of healthcare delivery [2].

The complexity of medical diagnosis encompasses various factors that influence the accuracy, efficiency, and reliability of the diagnostic process [3]. Notably, certain diagnostic tests or methodologies may yield false-positive or false-negative results, thereby compromising the accuracy of diagnosis [4]. Furthermore, accessibility to these methods may be constrained by medical or financial limitations, impacting patients in both developed and developing countries. Additionally, technological or procedural constraints may limit the efficacy of certain diagnostic approaches, further diminishing their effectiveness [5].

Moreover, certain diseases may present with diverse symptoms, thereby complicating the diagnostic process and potentially resulting in erroneous conclusions. Consequently, errors in diagnosis can arise due to human factors, such as inadequate physician qualifications, lack of attentiveness to patients, or misinterpretation of test results.

The challenge of medical diagnosis becomes significantly more complex when assessing patients with polytrauma. Such scenarios are not uncommon, and accurately diagnosing patients with polytrauma—comprising injuries to multiple organs and systems—poses considerable difficulty, necessitating a multifaceted approach [6]. This complexity arises from various factors, including the intricacy of coordinating assessments across different medical specialties, the need for comprehensive imaging studies, and the imperative of prioritizing urgent interventions [5–7]:

- Patients with polytrauma may sustain injuries to multiple organs and systems, thereby complicating the diagnostic process due to the necessity of concurrently assessing and evaluating several potentially critical injuries.
- Some injuries associated with polytrauma may remain concealed or present with non-specific symptoms, rendering their detection and diagnosis challenging. For instance, internal bleeding or damage to internal organs might not manifest overtly.
- Polytrauma patients may be rendered unconscious or unable to articulate their symptoms coherently, posing challenges for physicians in the diagnostic process.
- Polytrauma can exacerbate preexisting medical conditions or give rise to complications, altering the clinical presentation and adding complexity to the diagnostic endeavor.
- Emergencies necessitating urgent medical intervention may impose time constraints on the diagnostic process, compelling rapid decision-making based on limited available information.
- Complex cases of polytrauma often require the involvement of various medical specialists, including surgeons, traumatologists, and neurologists. However, disparities in decision-making among specialists can impede the coordination of the diagnostic process.

To address these challenges, it is imperative to have a comprehensive array of diagnostic tools at one's disposal, encompassing X-ray and computed tomography imaging, ultrasound examinations, laboratory blood tests, and other diagnostic modalities [5]. Furthermore, fostering collaboration among different medical specialists and leveraging algorithms and treatment protocols can enhance both the diagnostic and therapeutic approaches for patients with polytrauma. However, achieving seamless interdisciplinary collaboration is not always feasible. Hence, there arises a compelling need to harness the capabilities of artificial intelligence (AI) methods and technologies, particularly for expedited diagnosis in polytrauma cases.

The objective of this study is to conduct experimental analyses and research into various machine learning algorithms aimed at predicting mortality risk among patients with multiple traumas admitted to the hospital. This investigative approach seeks to not only refine existing diagnostic methodologies but also to devise novel strategies for enhancing the overall medical diagnostic process.

The main contribution of this paper can be summarized as follows:

1) We have compiled a dataset comprising information about male patients who suffered multiple traumas and were managed at the Anesthesiology and Intensive Care Department for Patients with Combined Trauma, within the premises of the Kharkiv City Clinical Emergency Hospital named after Prof. O.I. Meshchaninov.
2) We employed a variety of machine learning techniques, optimized their parameters, and assessed their performance accuracy in addressing the binary classification task, utilizing seven distinct performance indicators.
3) Our analysis revealed the superior efficacy of neural network models, as evidenced by higher F1 scores, Matthews Correlation Coefficients (MCC), and Cohen's kappa scores, compared to conventional statistical machine learning methods, in predicting mortality risk among multiple trauma patients admitted to the hospital.

2 Materials and Methods

2.1 Dataset Collection and Description

The data set was meticulously compiled by specialists from the Anesthesiology and Intensive Care Department at the Kharkiv City Clinical Emergency Hospital, named after Prof. O.I. Meshchaninov, Ukraine [7]. It encompasses comprehensive information about 73 male patients who were admitted to the hospital due to polytrauma incidents. Across the dataset, there are 56 independent attributes, encompassing various analyses. The primary dependent attribute distinguishes between fatal and non-fatal cases, denoted by the values 0 and 1, respectively. Of the cases, 31 were fatal, while 42 were non-fatal. Remarkably, the dataset is devoid of omissions and outliers, rendering it primed for detailed analysis.

Upon closer examination, several challenges in the intellectual analysis of this dataset emerge. Firstly, the dataset is relatively small in terms of the number of instances available for analysis, constituting what can be termed a "short" dataset [8]. Secondly, a significant issue lies in the imbalance of the dataset, necessitating the incorporation of supplementary performance metrics to effectively evaluate the outcomes of machine learning methodologies applied [9, 10]. Thirdly, the dataset boasts a high dimensionality, featuring a substantial 56 features [11, 12]. Consequently, leveraging machine learning tools becomes imperative for swift risk diagnosis regarding patient mortality, with the findings poised to serve as an invaluable supplementary resource for medical professionals.

2.2 Machine Learning Models

In order to select the most appropriate machine learning methodologies tailored to the requirements of a given task, a comprehensive examination of existing approaches is conducted, drawing insights from the extensive body of literature as referenced in [13, 14] and utilizing the Scopus scientometric database. Through meticulous analysis, this study identifies eight artificial intelligence techniques that are frequently utilized in the analysis of compact sets of high-dimensional data, with a specific emphasis on addressing binary classification challenges.

The culmination of this investigation is encapsulated in Table 1, where essential details regarding the methodologies under scrutiny are presented. This includes elucidation on their underlying operational principles, as well as a thorough exploration of their respective advantages and disadvantages.

By synthesizing the information from Table 1, scientist or practical doctor can make informed decisions regarding the most suitable approach for their particular analytical requirements, taking into account factors such as interpretability, computational efficiency, and predictive accuracy.

3 Modeling and Results

The modeling of existing methods was conducted using Python-based software. The computational setup utilized for the modeling process is outlined below: Windows 64, Intel Pentium CPU 2020M 2.40 GHz, 4 GB RAM.

The optimal operating parameters for all analyzed methods were determined through the grid search method.

Subsequently, the effectiveness of these methods was assessed using six indicators, encompassing accuracy and training time. Specifically, the following accuracy metrics were employed in this study [15]:

1. Accuracy

$$Accuracy = \frac{tp + tn}{tp + fp + tn + fn} \tag{1}$$

t_p is the number of true positive objects; t_n is the number of true negative objects; f_p is the number of false positive objects; f_n is the number of false negative objects.

2. Precision:

$$Precision = \frac{tp}{tp + fp} \tag{2}$$

Table 1. Results for all investigated methods

Algorithm	Principle of Operation	Advantages	Disadvantages
MLP Perceptron	Uses a single-layer perceptron to make binary classifications	Can capture complex relationships	Prone to overfitting
Decision Tree	Splits data based on feature conditions until leaf nodes are reached	Interpretable	Prone to overfitting, especially with deep trees
Linear SVC (fast SVC's version)	Finds the hyperplane that best separates classes in a linear manner	Effective in high-dimensional spaces	Requires tuning of regularization parameter
XGBoost	Ensemble method using gradient boosting	High predictive accuracy	Can be computationally expensive
Ridge Classifier	Linear model using ridge regression for classification	Tolerant to multicollinearity	May not perform well if features are unrelated
Logistic Regression	Estimates probabilities using a logistic function	Simple and interpretable	Assumes linear relationship between features
LightGBM Classifier	Gradient boosting framework designed for speed and efficiency	Fast training speed, handles large datasets well	May overfit with smaller datasets
SVC	Finds the hyperplane that best separates classes using support vectors	Effective in high-dimensional spaces	Computationally expensive for large datasets

3. Recall:

$$Recall = \frac{tp}{tp + fn} \tag{3}$$

4. F1 score:

$$F1 = 2\frac{Precision \cdot Recall}{Precision + Recall} \tag{4}$$

5. Matthews correlation coefficient:

$$MCC = \frac{tp \cdot tn - fp \cdot fn}{\sqrt{(tp + fp)(tp + fn)(tn + fp) + (tn + fn)}} \tag{5}$$

6. Cohen's kappa score κ:

$$\kappa = \frac{p_o - p_e}{1 - p_e} \tag{6}$$

p_o is the empirical probability of agreement on the label assigned to any sample (the observed agreement ratio); p_e is the expected agreement when both annotators assign labels randomly (is estimated using a per-annotator empirical prior over the class labels).

Table 2 provides a comprehensive overview of the performance outcomes obtained from the thorough examination of various machine learning methodologies. The evaluation is based on metrics (1) through (6) as well as the duration of the training procedure. By scrutinizing these key performance indicators, we gain invaluable insights into the efficacy and suitability of each method for addressing the specific challenges posed by the dataset. This holistic approach enables a nuanced understanding of the strengths and weaknesses of each technique, empowering stakeholders to make well-informed decisions regarding the adoption of the most appropriate model for mortality risk diagnosis in clinical settings.

Table 2. Results for all investigated methods

Method	Accuracy	Precision	Recall	F1 score	MCC	Cohen's kappa score	Training time
MLPerceptron	0,80	0,80	0,89	0,84	0,58	0,57	0,04
DecisionTreeClassifier	0,73	0,78	0,78	0,78	0,44	0,44	0,02
LinearSVC	0,73	0,86	0,67	0,75	0,49	0,47	0,03
XGBClassifier	0,73	0,86	0,67	0,75	0,49	0,47	0,06
RidgeClassifier	0,67	0,70	0,78	0,74	0,29	0,29	0,02
LogisticRegression	0,67	0,75	0,67	0,71	0,33	0,32	0,03
LGBMClassifier	0,67	0,83	0,56	0,67	0,39	0,36	0,08
SVC	0,60	0,71	0,56	0,63	0,22	0,21	0,03

Given the inherent imbalance within the dataset, it becomes crucial to select the most appropriate metrics for evaluating the performance of the various methodologies under scrutiny. Metrics (4) through (6) stand out as particularly suitable for this purpose, as they offer nuanced insights into model performance amidst class imbalance. Additionally, alongside these metrics, the duration of the training procedure warrants consideration as it provides valuable information regarding the efficiency and practicality of each method. By incorporating these comprehensive evaluation criteria, we can ensure a robust assessment framework that accounts for the unique challenges posed by the dataset's imbalance, thereby facilitating informed decision-making in the selection of the most effective machine learning approach for mortality risk diagnosis.

4 Comparison and Discussion

In this study, we conducted a comprehensive evaluation of the performance of various artificial intelligence tools. This evaluation was grounded on six metrics, encompassing accuracy and training time. Let us delve deeper into the findings garnered from these analyses.

4.1 Accuracy Comparison

The assessment of accuracy across all investigated methodologies was conducted, utilizing performance metrics (1)–(6) as benchmarks. Given the task's context of binary classification within short, balanced datasets, the f1-score and Matthews Correlation Coefficient (MCC) emerged as the most pertinent indicators for evaluating outcomes, as depicted in Fig. 1.

Notably, two methods, stemming from distinct classifier classes of Support Vector Classifier (SVC) and Light Gradient Boosting Machine (LGBM), exhibited accuracies below 70%, as illustrated in Fig. 1a). Conversely, linear (Linear SVC) and ensemble techniques demonstrated somewhat improved results, with the XGBClassifier achieving 75% accuracy and the Decision Tree yielding 78% accuracy (according to the f1-score). However, when considering MCC (Fig. 1b), both methods yielded identical outcomes.

The highest accuracy in predicting mortality risk for multiple trauma patients admitted to the hospital, based on a short, unbalanced dataset of substantial dimensions, was achieved using a multilayer perceptron. Employing this neural network yielded an f1-score of 84%, signifying a satisfactory performance level.

Furthermore, various ensemble methodologies, including Random Forests, AdaBoost, and Gradient Boosting, among others, were also explored. However, their accuracies fell below the acceptable threshold (less than 55%), rendering their results unsuitable for addressing the given problem. Consequently, the outcomes of these methods are omitted from this paper.

4.2 Training Time Comparison

The efficiency of machine learning methods in delivering timely diagnoses is a crucial consideration for their practical application. In many scenarios, expedited results are imperative, necessitating artificial intelligence to swiftly furnish them. This urgency is particularly pronounced in contexts such as predicting mortality risk for patients with combined trauma, where any delays in diagnosis can yield dire and potentially tragic outcomes. Hence, in this study, alongside assessing the accuracy of artificial intelligence tools, we also scrutinized their training duration.

Recognizing the criticality of prompt diagnosis, especially in medical emergencies involving polytrauma patients, underscores the significance of expeditious data processing by machine learning algorithms. The ability to rapidly analyze clinical data and provide actionable insights can significantly impact patient outcomes, facilitating timely interventions and treatment decisions.

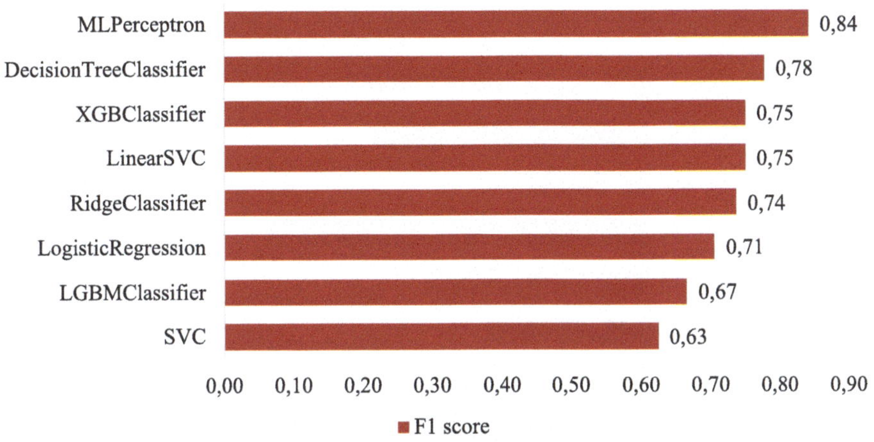

a)

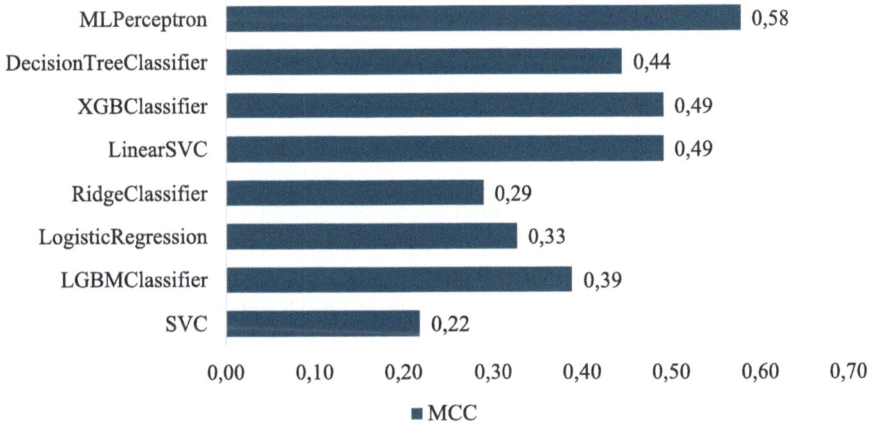

b)

Fig. 1. Comparison of the different AI tools based on the: a) F1-score; b) MCC

Therefore, in our investigation, we meticulously evaluated not only the predictive performance of various machine learning techniques but also their computational efficiency. By gauging the training time of these algorithms, we sought to ascertain their suitability for real-world clinical settings where swift decision-making is paramount.

Ultimately, by considering both accuracy and training time, our study endeavors to inform healthcare practitioners and decision-makers about the most efficacious and time-sensitive approaches to leveraging artificial intelligence in medical diagnosis, particularly for high-stakes scenarios like predicting mortality risk in patients with polytrauma.

Figure 2 shows a histogram of the duration of the training procedure for all the studied methods.

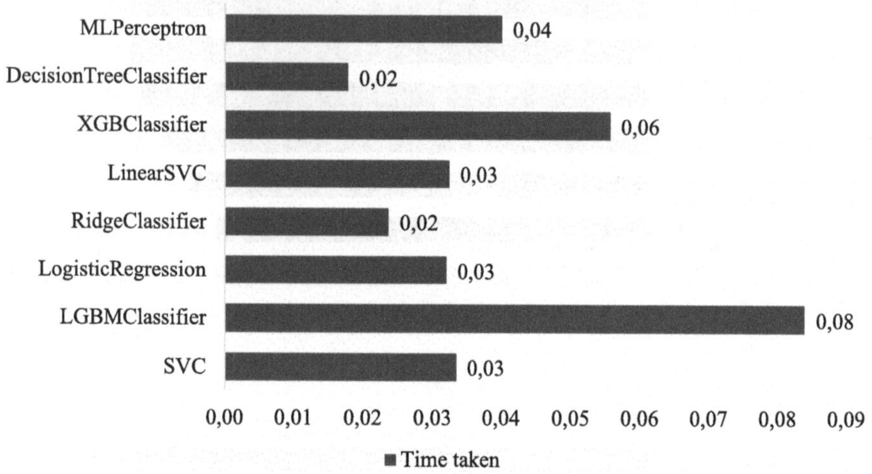

Fig. 2. Duration of the training procedure for all investigated methods

As depicted in Fig. 2, the LGBM classifier manifests the lengthiest training duration among the examined methods. Furthermore, as illustrated in Fig. 1, it exhibits the lowest accuracy rate. A similar outcome is observed for another ensemble method, the XGB Classifier. These observations can be attributed to the specific characteristics inherent in the ensemble implementation of these methods.

Conversely, the Decision Tree classifier demonstrates the shortest training duration, while concurrently showcasing satisfactory accuracy (Fig. 1). Notably, despite operating at twice the speed of the preceding method, the multilayer perceptron achieves a 7% higher accuracy rate, a pivotal metric in the prompt diagnosis of polytrauma patients. Furthermore, it is pertinent to highlight that the analysis pertains to a dataset of limited size, wherein the operational duration of most methods proves satisfactory.

Additionally, it warrants mention that when analyzing larger datasets, a substantial reduction in training time can be attained through the utilization of parallelization algorithms [16–18]. However, this optimization is contingent upon the compatibility of the selected machine learning method [10, 19].

5 Summary and Conclusion

Our study represents a significant advancement in trauma care and predictive analytics. By compiling a comprehensive dataset of male patients who suffered blunt multiple trauma and underwent treatment at the Anesthesiology and Intensive Care Department of the Kharkiv City Clinical Emergency Hospital, we have laid the groundwork for more effective risk assessment and management strategies in emergency medical settings.

1) The application of various machine learning methods, careful parameter selection, and rigorous evaluation have enabled us to achieve notable insights into mortality risk prediction for patients with combined trauma. Our findings indicate that neural network tools exhibit superior performance, particularly in terms of F1 score, Matthews Correlation Coefficient (MCC), and Cohen's kappa score. These metrics not only attest to the accuracy of our predictive models but also underscore the potential of artificial intelligence in enhancing clinical decision-making processes.
2) By leveraging machine learning techniques, we have not only demonstrated the feasibility of mortality risk prediction in multiple trauma patients but have also highlighted the limitations of traditional statistical methods in this context. The ability of neural networks to discern complex patterns and relationships within our dataset has undoubtedly contributed to their superior performance and underscores the importance of embracing innovative approaches in medical research and practice.
3) Moving forward, our study sets the stage for further exploration and refinement of predictive models aimed at improving outcomes for trauma patients. Future research endeavors may focus on expanding the scope of our analysis to encompass advanced ANN models, methods, and algorithms [8] to analyze small clinical datasets.

Acknowledgment. The National Research Foundation of Ukraine supported this research under the project № 03/0103.

References

1. Mulesa, O., Geche, F., Nazarov, V., Trombola, M.: Development of models and algorithms for estimating the potential of personnel at health care institutions. East-Eur. J. Enterp. Technol. **4**(2(100)), 52–59 (2019)
2. Ch, R., Sudheer, P., Kumar, P.D.: An overview of remote patient monitoring for improved patient care and cost reduction: the IoT revolutionizing health care. Int. J. Educ. Manag. Eng. **13**(6), 33–40 (2023)
3. Chumachenko, D., Piletskiy, P., Sukhorukova, M., Chumachenko, T.: Predictive model of Lyme disease epidemic process using machine learning approach. Appl. Sci. **12**(9), 4282 (2022)
4. Rangaswamy, S., Rakesh, J., Charan, P.L., Giridhar, D.: Early and accurate diagnosis of a neurological disorder epilepsy using machine learning techniques. Int. J. Inf. Technol. Comput. Sci. **15**(2), 54–63 (2023)
5. Thippeswamy, P.B., Rajasekaran, R.B.: Imaging in polytrauma – principles and current concepts. J. Clin. Orthop. Trauma **16**, 106–113 (2021)
6. McKinley, T.O., Gaski, G.E., Vodovotz, Y., Corona, B.T., Billiar, T.R.: Diagnosis and management of polytraumatized patients with severe extremity trauma. J. Orthop. Trauma **32**(3), S1–6 (2018)
7. Stupnytskyi, M., Biletskyi, O.: Predictors of mortality for multiple trauma patients with severe thoracic trauma during intensive care of the early posttraumatic period. J. Crit. Intensive Care **14**(2), 33–40 (2023)
8. Izonin, I., Tkachenko, R.: Universal intraensemble method using nonlinear AI techniques for regression modeling of small medical data sets. Cognit. Soft Comput. Tech. Anal. Healthc. Data, 123–150 (2022)

9. El Korbi, I., Azouz, S.L.: Performance evaluation of unslotted CSMA/CA for wireless sensor networks: energy consumption analysis and cross layer routing. Int. J. Comput. Netw. Inf. Secur. **9**(6), 1–12 (2017)
10. Zomchak, L., Melnychuk, V.: Creditworthiness of individual borrowers forecasting with machine learning methods. In: Hu, Z., Ye, Z., He, M. (eds.) AIMEE 2022. LNDECT, vol. 159, pp. 553–561. Springer, Cham (2023). https://doi.org/10.1007/978-3-031-24468-1_50
11. Oleksiv, I., Lema, H., Kharchuk, V., Lisovych, T., Dluhopolskyi, O., Dluhopolska, T.: Identification of stakeholders importance for the company's social responsibility using the analytic hierarchy process. In: Proceedings of the 2020 10th International Conference on Advanced Computer Information Technologies (ACIT), Deggendorf, Germany, pp. 1–4. IEEE (2020)
12. Tkachenko, R., Izonin, I., Greguš ml., M., Tkachenko, P., Dronyuk, I.: Committee of the SGTM neural-like structures with extended inputs for predictive analytics in insurance. In: Younas, M., Awan, I., Benbernou, S. (eds.) Big Data Innovations and Applications. Innovate-Data 2019. CCIS, vol. 1054, pp. 121–132. Springer, Cham (2019). https://doi.org/10.1007/978-3-030-27355-2_9
13. Kharchuk, V., Oleksiv, I.: The Intellectual structure of sustainable leadership studies: bibliometric analysis. In: Hu, Z., Wang, Y., He, M. (eds.) CSDEIS 2022. LNDECT, vol. 158, pp. 430–442. Springer, Cham (2023). https://doi.org/10.1007/978-3-031-24475-9_37
14. Rzheuskiy, A., Veretennikova, N., Kunanets, N., Kut, V.: The information support of virtual research teams by means of cloud managers. Int. J. Intell. Syst. Appl. **10**(2), 37–46 (2018)
15. Terttiaavini, Hartono, Y., Ermatita, Palupi Rini, D.: Development of a decision support system on employee performance assessment using weighted performance indicators method. Int. J. Inf. Eng. Electron. Bus. **15**(3), 1–11 (2023)
16. Mochurad, L., Panto, R.: A parallel algorithm for the detection of eye disease. In: Hu, Z., Wang, Y., He, M. (eds.) CSDEIS 2022. LNDECT, vol. 158, pp. 111–125. Springer, Cham (2023). https://doi.org/10.1007/978-3-031-24475-9_10
17. Mochurad, L., Shchur, G.: Parallelization of cryptographic algorithm based on different parallel computing technologies. In: CEUR-WSorg, vol. 2824:2029 (2021)
18. Auzinger, W., Obelovska, K., Dronyuk, I., Pelekh, K., Stolyarchuk, R.: A Continuous model for states in CSMA/CA-based wireless local networks derived from state transition diagrams. In: Saraswat, M., Roy, S., Chowdhury, C., Gandomi, A.H. (eds.) Proceedings of International Conference on Data Science and Applications. LNNS, vol. 287, pp. 571–579. Springer, Singapore (2022). https://doi.org/10.1007/978-981-16-5348-3_45
19. Zomchak, L.M.: Sustainable development of Ukraine as a combination of social, economic and environmental components: structural econometric model of three-pillar approach. IOP Conf. Ser. Earth Environ. Sci. **1254**(1), 012125 (2023)

Development of an Operator's Mathematical Model in a Man-Machine Control System

Valerii Tytiuk[1](✉), Denis Mrachkovskiy[1], Oleksii Chornyi[2], Victor Busher[3], Mykola Tryputen[4], and Viktor Kovalenko[5]

[1] Kryvyi Rih National University, Vitaly Matusevich Street, 11, Kryvyi Rih 50027, Ukraine
{tytiuk,d.mrachkovskiy}@knu.edu.ua

[2] Kremenchuk Mykhailo Ostrohradskyi National University, University Street, 20, Kremenchuk 39600, Ukraine
ochornyi@kdu.edu.ua

[3] National University Odessa Maritime Academy, Didrikhson Street, 8, Odessa 65052, Ukraine
v.busher@onma.edu.ua

[4] Dnipro University of Technology, av. Dmytra Yavornytskoho, 19, Dnipro, Ukraine

[5] Zaporizhzhia National University, 66, Zhukovsky Street, Zaporizhzhia 69600, Ukraine
viktor_kovalenko_2023@ukr.net

Abstract. The human-operator performs important functions related to the control of manufacturing human-machine systems. Human-operator can be considered as a nonlinear dynamic link in a closed loop of industrial control process. The operator's control actions are of a complex nature and consist of deterministic and random components. Accordingly, there are two main components to the operator's response – deterministic part and stochastic part, referred to as the remnant part. To identify the dynamic operator model, experimental studies of operator actions during SISO-type operation were performed. As a result of processing the obtained experimental data using the functions of System Identification Toolbox® library in MATLAB, it was established, that the deterministic part of the human operator's reaction can be represented by an aperiodic link of the first order with a delay (P1D model). The remnant part of the human-operator reaction is proposed to be identified in the form of a second-order oscillating link and a differentiating link connected in series. The adequacy of the proposed mathematical model of the human operator is confirmed.

Keywords: Ergatic system · Operator's model · Remnant part · Identification

1 Introduction

The rapid advancement of technology has transformed a person from a direct executor into an operator, whose main responsibility is to control and manage work processes in industry, transport, and energetics. Industrial systems with human operators are referred to ergatic ones [1].

Ergatic systems are widespread in modern industrial production. Examples of such systems are aviation transportation systems [2], crane mechanisms, various mining machines and production units, the work of which is controlled by a human operator [3].

Z. Hu et al. (Eds.): ICCSEEA 2024, LNDECT 242, pp. 229–238, 2025.
https://doi.org/10.1007/978-3-031-84228-3_19

The presence of the "human factor" in ergatic systems is often interpreted as their disadvantage, which has certain evidence. The number of accidents and disasters associated with operator errors is constantly growing, and their consequences are becoming more widespread and tragic [4–6]. However, it should be noted that the influence of the same "human factor" provides a number of advantages, such as continuous adaptation of ergatic systems, the ability to make decisions in non-standard situations [7].

Currently, due to the insufficient development of intelligent control systems capable of operating effectively in conditions of significant changes in the production scene, ergatic systems remain in demand and widespread. In the modern mining industry, excavator mechanisms of various types are a typical representative of ergatic systems. The influence of the human operator of excavator mechanisms does not lead to emergency situations with catastrophic consequences, as in the case of aircraft pilots. In ergatic systems of the mining industry, the negative impact of the human operator is manifested in a decrease in the technical and economic indicators of production processes. A low level of qualification of an excavator operator can lead to an increase in specific energy consumption by 30–40% [8].

A human operator performs a sequence of actions aimed at achieving certain goals and objectives. He is an important part of the control system and functions as a complex mechanism that processes information from various sources. He filters external data, analyzes it, and transforms it into control commands based on skills and knowledge. Through the controls, he executes commands and reacts to the feedback of the controlled object, similar to a link in a closed control loop, Fig. 1.

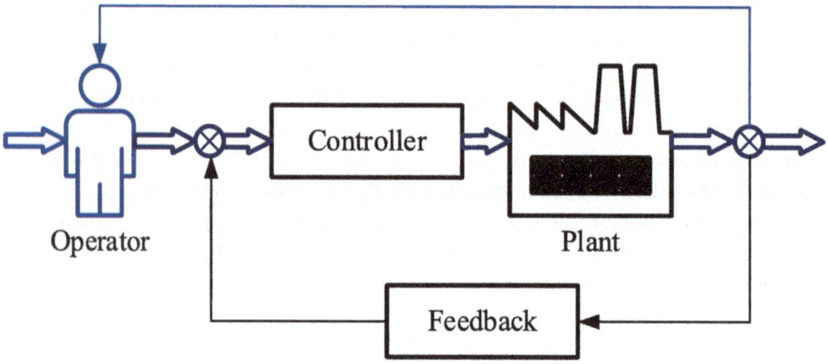

Fig. 1. Block diagram of an ergatic system with a closed-loop production process control system

So, the human operator can be considered as a nonlinear dynamic link in a closed loop of industrial control process.

Thus, in order to improve the accuracy of mathematical modeling of production mechanisms in the mining industry, it is necessary to take into account the peculiarities of the work of a human operator. Therefore, the scientific and practical task of developing a mathematical model of a human operator in a man-machine control system for mining equipment is an important and urgent task.

Goal of the work is the determination of features of the operator's work; the implementation of an experimental study of the operator's actions and the structural and parametric identification of the corresponding transfer function.

To achieve the set goal, the following tasks must be completed:

1) to perform experimental studies of the operator's actions and obtain generalizing numerical characteristics of the obtained results;
2) to perform structural and parametric identification of the transfer function of the operator.

2 Features of Human Operator in Man-Machine Systems

The objective of the ergatic system's human operator is to determine the current value of the system error and reduce it using appropriate control actions [9].

The simplest and most common situation is when the operator controls one input value. However, in the case of controlling the spatial movements of a working machine, the operator must evaluate and control at least three characteristics. A similar situation occurs not only in aviation, but also in the mining industry with a wide range of use of excavator equipment of various designs and complexity.

In [10], the kinematic diagram of the interconnected mechanical system of the "front shovel" excavator was considered; in [11], the issues of controlling the spatial position of the excavator bucket when controlling the electric drives of the main mechanisms of the "front shovel" excavator were considered. However, the mentioned works did not take into account the influence of the dynamic properties of the operator on the technical and economic performance of the excavator.

The primary question is how exactly the operator stabilizes the system. In the stabilization task, the following psycho-physiological features of the human operator behavior are noted [12]:

- presence of reaction delay;
- dependence on the task (the ability to change characteristics depending on the control objectives, dynamics of the controlled object, the type of information used);
- time dependence (manifests itself in two forms: firstly, the operator's characteristics change over time as he learns; secondly, the operator senses changes in the parameters of the external environment and the controlled object and changes his characteristics accordingly);
- prediction (the ability to foresee the future position of a goal based on knowledge of previous ones);
- significant nonlinearity;
- stochasticity (operator characteristics in the same experiment differs from trial to trial, but this variability is small if the training time is sufficient and the task is not complex, so the deterministic model can be used in a statistical sense);
- discreteness (in some cases the operator behaves like a discrete system).

The operator's control actions are of a complex nature and consist of deterministic and random components [13]:

- deterministic component, represents the reaction of a dynamic element of a constant structure equivalent to a human operator;
- stochastic component, depending on the functional state of the operator, due to the finite controllability of the human neuromuscular system, limited accuracy and subjectivity of perception of input information.

The statistical parameters of this component significantly depend on the type of input signal, the complexity of the task, the ergonomic characteristics of the workplace (operator fatigue) and can change significantly over time.

Consequently, two main components are distinguished in the operator's response:

- the first one, deterministic, which corresponds to the response to the input signal of a dynamic element equivalent to a human operator;
- the second one, which is called remnant, and is the difference between the actual output signal of the operator and the reaction of the linear model.

The presence of remnant is influenced by the following factors:

- Presence of noise. There are quite a lot of reasons for the occurrence of noise and erroneous actions of the operator when perceiving movement, analyzing it, dosing deviations of controls, etc.;
- Instability of operator characteristics. In the process of stabilizing the object, the characteristics of the operator change relative to a certain average value, which also contributes to the remnant.
- Finite controllability of the human neuromuscular system;
- Limited accuracy and subjectivity of perception of input information.

Researchers note the presence of a significant correlation between the complexity of the task, the operator's assessment of the object's controllability and the power of the remnant in the operator's signal [14]. Therefore, reproducing the remnant is an extremely important component in modeling operator behavior.

The mathematical model of the operator of the ergatic system must combine a whole range of biomechanical, physical, psychophysical and psychological parameters inherent in a person, his reaction and interaction with the environment.

Considering the described behavioral characteristics of the operator, it may not be about a comprehensive model designed for all operator activities, but only about a model for performing a specific operation. It is assumed that the operator is sufficiently trained to perform this operation, and the dynamics of the control object and the external conditions of the operator's activity are unchanged.

Physical modeling of ergatic systems has limited use due to its complexity and high cost. Therefore, the main means of obtaining data for building models are training complexes. Experimental data exhibit significant variability due to random deviations in human activity, caused by many reasons: from changes in emotional state to loss of attention and fatigue. Therefore, experiments are performed repeatedly, and their results are subjected to statistical processing.

Identifying of an equivalent dynamic model of a human operator can be performed by parametric identification of a set of models with a given structure, selecting the best model based on the known criteria.

It must be taken into account that the main known nonlinearity of the operator is transport delay, caused by the finite speed of signal transmission in the human nervous system and the time required to make a decision.

The presence of numerous empirical coefficients in these models that vary over a wide range, and their identification based on measurements of psychophysical data is a complex scientific and technical task.

Various approaches to solve the identification problem of the mathematical model of the operator are known in modern scientific literature.

In [15], the features of operator modeling using C∗-algebras were considered.

In [16], the possibilities of using the SIT (System Identification Toolbox®) library for parametric identification of the behavior model of a human pilot when controlling a flight on the X-Plane flight simulator were shown.

3 Experimental Research and Its Results

In this study, the simplest SISO model of operator actions was investigated. The operator's ability to track the reference signal in the form of a step function was assessed.

Two series of experiments were conducted during the experimental study. In the first series, the operator had to respond to a step signal, when the signal change occurring at a fixed time moment t = 5 s. In the second series, the operator had to respond to a step signal, when the signal change occurring at a random time moment.

Each experiment was repeated 20 times in order to obtain data for statistical processing of the obtained results.

The joystick of the Logitech F310 gamepad was used as a control device. Operator actions were recorded using MATLAB/Simulink. The block diagram of the used model is shown in Fig. 2. The model is configured to work in real time conditions.

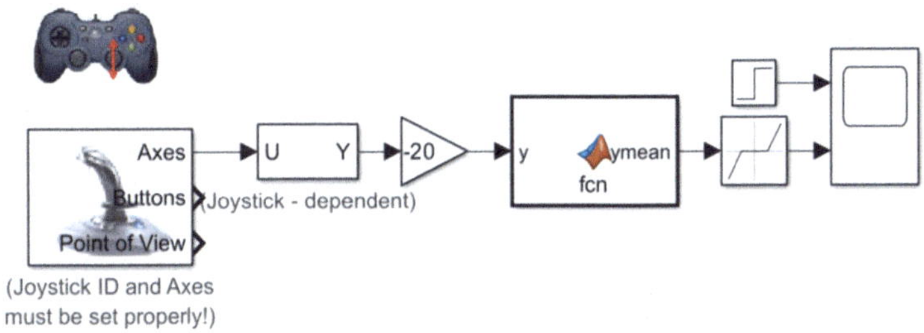

Fig. 2. MATLAB/Simulink model for recording operator actions during an experimental study

For the structural-parametric identification of the dynamic operator model, the built-in `procest` function of the System Identification Toolbox® library in MATLAB was used.

Figure 3, a shows variant of the operator's response to a step input signal arriving at time t = 5 s. Figure 3, b shows the graph of the operator's response averaged over all experiments of this series.

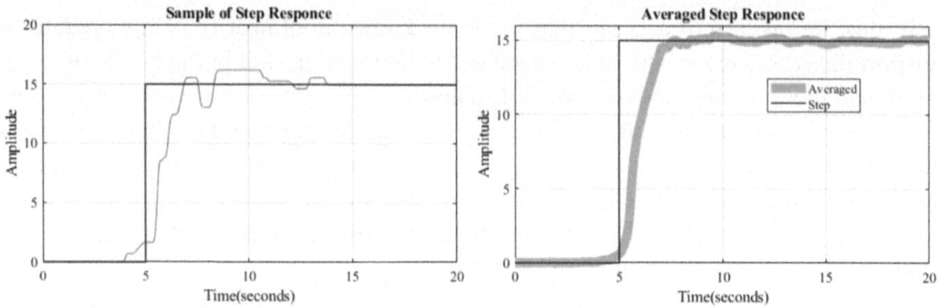

Fig. 3. Operator's response to a step signal: a - example of realization in one experiment; b - average operator's response in a series of experiments

As alternatives to the hypothesis about the structure of the dynamic model of the operator, we considered the variants of transfer functions presented in Table 1. This table summarizes the results obtained for the identification of the deterministic part of the human operator's response.

Table 1. The results obtained for the identification of the deterministic part of the human operator's response.

Model name	Transfer function of the model	Model parameters
P1D	$G(s) = \frac{K_P}{1+T_{P1} \cdot s} \cdot e^{-T_d \cdot s}$	$K_p = 1.0027$; $T_{p1} = 0.701$; $T_d = 0.297$; $R = 94.5\%$
P2DU	$G(s) = \frac{K_P}{1+2 \cdot \zeta \cdot T_W \cdot s + (T_W \cdot s)^2} \cdot e^{-T_d \cdot s}$	$K_p = 1.0002$; $T_W = 0.5149$; $\xi = 0.91767$; $T_d = 0.0035$; $R = 94.52\%$
P3DU	$G(s) = \frac{K_P}{(1+2 \cdot \zeta \cdot T_W \cdot s + (T_W \cdot s)^2) \cdot (1+T_{P3} \cdot s)} \cdot e^{-T_d \cdot s}$	$K_p = 1.0008$; $T_W = 0.4448$; $T_{P3} = 0.0821$; $\xi = 0.99445$; $T_d = 0$; $R = 96.6\%$

As the performed analysis shows, the determination coefficients of all considered models are quite high (about 95%) and differ slightly in absolute value.

Therefore, it is possible to adopt the simplest P1D model as an operator's dynamic model.

In the second series of experiments, the operator had to respond to a step signal, and the signal change occurred at a random time moment.

For each experiment, the transfer function parameters and the coefficient of determination were determined for each dynamic operator model according to Table 1. The coefficients of determination of individual models were averaged over the data of the entire series. The obtained result is shown in Fig. 4.

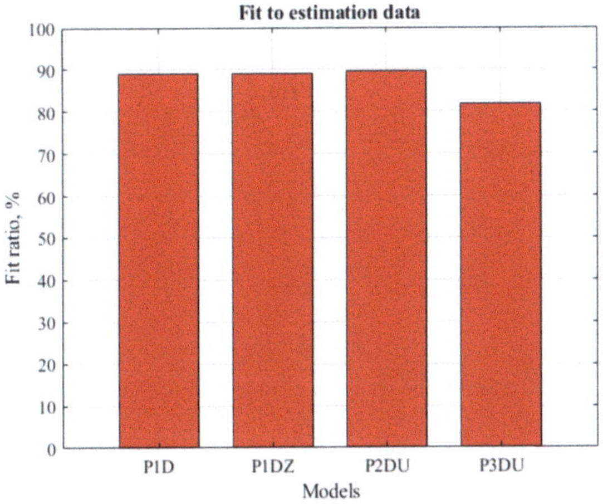

Fig. 4. Average values of the determination coefficients of the operator's dynamic models in the second series of experiments

As the obtained results show, the coefficients of determination of all the models considered are quite high (about 90%) and differ slightly in absolute value, which coincides with the results of the first series of experiments.

As a result of the experimental study, the following hypothesis can be put forward. The deterministic part of the dynamic model of the operator when performing SISO operations with sufficiently high accuracy is represented by a P1D model:

$$G(s) = \frac{K_P}{1 + T_{P1} \cdot s} \cdot e^{-T_d \cdot s} \tag{1}$$

4 Modeling of the Remant Component of the Operator's Response

Knowing the transfer function of the deterministic component of the human operator reaction, expressed by Eq. (1), it is easy to determine the remnant component of the operator reaction as the instantaneous value of the deviation of the real operator reaction from the reaction of the deterministic component of the operator model reaction, Fig. 5.

As can be seen from Fig. 5, the remnant component of the operator's reaction consists of two parts. The first part of the remnant component is associated with the operator's reaction to rapid, significant changes in the input signal; the second part is a random signal of insignificant amplitude, associated with the final sensitivity of the human neuromuscular system and control organs.

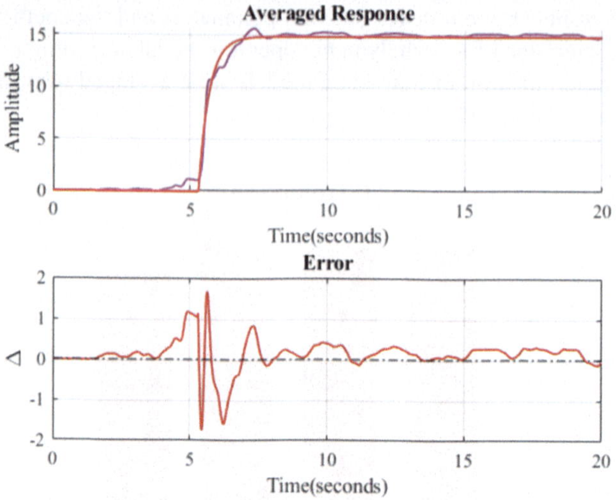

Fig. 5. Time diagram of the remnant component of the operator's reaction

The transfer function of the first part of the remnant component can be presented in the following form:

$$W_R(s) = \frac{-k_1 \cdot \omega_1^2 \cdot s}{s^2 + 2\zeta \cdot \omega_1 \cdot s + \omega_1^2} \cdot e^{-T_d \cdot s} \tag{3}$$

The structure and parameters of this transfer function were selected using empirical methods.

Combining Eqs. (1) and (2) we finally obtain the following transfer function for the deterministic part of the operator's response:

$$G(s) = \left(\frac{K_P}{1 + T_{P1} \cdot s} - \frac{k_1 \cdot \omega_1^2 \cdot s}{s^2 + 2\zeta \cdot \omega_1 \cdot s + \omega_1^2} \right) \cdot e^{-T_d \cdot s} \tag{4}$$

Figure 6 shows comparative diagrams of the real operator response in the experimental study and the operator model response obtained from Eq. (4) taking into account the stochastic part of the remnant component of the operator reaction.

Obtained diagrams show a high relevance between the proposed operator model and the results of the experimental study.

The disadvantages of the resulting mathematical model include the complete absence of elements reflecting the operator's anticipatory actions.

Also, additional research is needed to justify the applicability of the proposed operator model when performing MISO operations; when the excavator's operator needs to monitor changes and simultaneously manage three variables –coordinates of the spatial position of the excavator bucket.

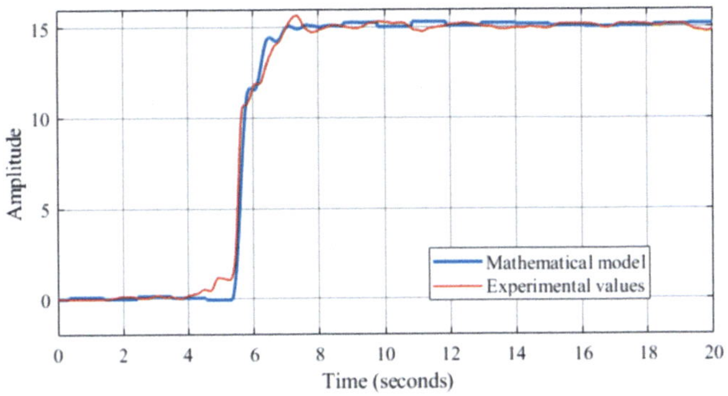

Fig. 6. Comparative results of the experimental data with the work of the proposed dynamic operator model

5 Summary and Conclusion

The paper presents the results of experimental studies of operator actions in a SISO system. The possibility of representing the operator as aperiodic first-order link with time-delay has been justified. The coefficient of determination of this model was approximately 90%. In order to describe the remnant component of the operator's reaction, it was proposed to use a combination of a differentiating link with a second-order oscillatory link. The suggested mathematical model of the operator can be used to determine the technical and economic indicators of mining excavators with a man-machine control system.

References

1. Korobiichuk, I., Tokar, A., Danik, Y., Katuha, V.: Evaluation methods for the ergatic system reliability operator. In: Szewczyk, R., Zieliński, C., Kaliczyńska, M. (eds.) AUTOMATION 2019. AISC, vol. 920, pp. 560–570. Springer, Cham (2020). https://doi.org/10.1007/978-3-030-13273-6_52
2. Tan, Y., Feng, D., Shen H.: Research for Unmanned Aerial Vehicle components reliability evaluation model considering the influences of human factors. In: 3rd International Conference on Mechanical, Electronic and Information Technology Engineering, ICMITE 2017, vol. 139 (2017). https://doi.org/10.1051/matecconf/201713900221
3. Zhu, H.: fundamental models for missions of engineered systems. In: IEEE International Systems Conference (SysCon), pp. 1–6 (2019). https://doi.org/10.1109/syscon.2019.8836714
4. Che, H., Zeng, S., Guo, J.: Reliability assessment of man-machine systems subject to mutually dependent machine degradation and human errors. Reliab. Eng. Syst. **190**, 106504 (2019). https://doi.org/10.1016/j.ress.2019.106504
5. Kletz, T., Amyotte, P.: Accidents said to be due to human error. In: What Went Wrong? 6th edn., pp. 53–172. Butterworth-Heinemann (2019). https://doi.org/10.1016/B978-0-12-810539-9.00007-0
6. Strauch, B.: Investigating Human Error: Incidents, Accidents, and Complex Systems, 1st edn. Routledge, Milton Park (2018). https://doi.org/10.4324/9781315251851

7. Suyatinov, S., Buldakova, T., Vishnevskaya, Yu.: Synergetic model of situational awareness of a human operator in ergatic control systems of mobile objects. Mekhatronika Avtomatizatsiya Upravlenie **23**(6), 302–308 (2022). https://doi.org/10.17587/mau.23.302-308
8. Babaei Khorzoughi, M., Hall, R.: A study of digging productivity of an electric rope shovel for different operators. Minerals **6**(2), 48 (2016). https://doi.org/10.3390/min6020048
9. Tyrva, O., Saushev, A., Shergina, O.: Automation elements of mental activity and actions of human operator in ergatic system “man-machine”. In: 2018 International Russian Automation Conference (RusAutoCon), Sochi, Russia, 2018, pp. 1–5 (2018). https://doi.org/10.1109/RUSAUTOCON.2018.8501665
10. Tytiuk, V., Khandakji, K., Sivyakova, G., Karabut, N., Chornyi, O., Busher, V.: Determining the parameters of the trajectory of the bucket of mining quarries excavators. In: E3S Web of Conferences, vol. 280 (2021). https://doi.org/10.1051/e3sconf/202128005013
11. Tytiuk, V., Chornyi, O., Mrachkovskyi, D., Tryputen, S., Tryputen, M., Kuznetsov, V.: Mathematical model of the closed-loop system of excavator bucket positioning. Naukovyi Visnyk Natsionalnoho Hirnychoho Universytetu **1**, 107–114 (2023). https://doi.org/10.33271/nvngu/2023-1/107
12. Pan, X., Wu, Z.: Performance shaping factors in the human error probability modification of human reliability analysis. Int. J. Occup. Saf. Ergon. **26**, 538–550 (2020). https://doi.org/10.1080/10803548.2018.1498655
13. Rejeki, Y., Achiraeniwati, E., Wanda, A.: Measurement of operator reliability level using the human error assessment and reduction technique (HEART) method. In: IOP Conference Series: Materials Science and Engineering, vol. 830, p. 032095 (2020). https://doi.org/10.1088/1757-899X/830/3/032095
14. Ahmad, H., Faizal, M., Mazleha, M.: Research reactor operator performance based on the human error assessment and reduction technique (HEART). In: IOP Conference Series: Materials Science and Engineering, vol. 785. International Nuclear Science Technology and Engineering Conference 2019 (iNuSTEC2019), Universiti Kebangsaan Malaysia, 43600, Bangi, Selangor, Malaysia, 29–31 October 2019. https://doi.org/10.1088/1757-899x/785/1/012048
15. Albiac, F., Kalton, N.J.: Topics in Banach Space Theory. Graduate Texts in Mathematics, vol. 233, Springer, New York (2006). MR2192298. https://doi.org/10.1007/0-387-28142-8
16. Valoušek, L., Jalovecky, R.: Use of the MATLAB® system identification Toolbox® for the creation of specialized software for parameters identification. In: 2021 International Conference on Military Technologies (ICMT), Brno, Czech Republic, pp. 1–5 (2021). https://doi.org/10.1109/ICMT52455.2021.9502786

Stock Market Values Prediction Using Deep Neural Networks

Ivanna Dronyuk[1], Myroslava Klapchuk[2(✉)], and Deep Shankar Pratap Singh[2]

[1] Jan Dlugosz University, 42200 Czestochowa, Poland
i.dronyuk@ujd.edu.pl
[2] Lviv Polytechnic National University, Lviv 79013, Ukraine
myroslava.i.klapchuk@lpnu.ua, dksspsingh@gmail.com

Abstract. The article deals with creating an information strategic system with a deep neural network in investments by investors and researchers for predicting stock prices of the NIFTY50 index for the Indian stock market. Long Short-Term Memory (LSTM) and Gated Recurrent Unit (GRU) neural networks were used. This network architecture was chosen because both are good at capturing the patterns of time-series data, collected in the stock market. More than twenty models were created and then compared for their performance. LSTM has been observed to have higher performance than GRU, and both models are very good at predicting stock market data.

Keywords: Long Short-term Memory (LSTM) · Deep Neural Network (DNN) · Gated Recurrent Unit (GRU) · Stock Prediction · Nifty Index (NIFTY50) · Simple Moving Average over 50 days (SMA50) · Mean Square Error (MSE) · Root Mean Square Error (RMSE)

1 Introduction

The task of predicting stock market behavior is an urgent problem because stock prices change rapidly. This complex behavior of stock prices challenges researchers to find a way to expect them. Numerous research publications in this field indicate that the behavior of stock markets is difficult to predict. This is mainly due to the number of factors affecting the stock prices, like in recent modified times because of COVID-19 [1], many stocks prices dropped, and stock exchanges crashed (sudden loss in volumes of trading and reduction of stock prices), and these factors themselves are independent of each other, possibly unknown, factors.

Today, deep neural networks (DNNs) are widely used to solve a large number of financial analysis tasks, such as stock market index prediction, bankrupt prediction, money laundering classification tasks, and other financial transaction classification tasks. Modeling based on DNNs is based on imitating the human brain's ability to learn. The elements of the DNN structure are similar to the biological structure of neurons and the actual functioning of the human brain. Many simple interconnected linear or nonlinear computational elements-neurons-work to process data in parallel at multiple levels. In

Z. Hu et al. (Eds.): ICCSEEA 2024, LNDECT 242, pp. 239–248, 2025.
https://doi.org/10.1007/978-3-031-84228-3_20

some cases, DNNs can be said to have certain limitations when it comes to learning complex and large data sets. They can be inconsistent and unpredictable due to the complex financial market data used to train the neural network. Sometimes the data is so large that the neural network training algorithms may not work. In this regard, large amounts of data need to be checked for redundancy, and sometimes for DNNs to work effectively, the amount of data should be reduced so that the method can run for less time and produce more plausible solutions.

At the beginning of the 21st century, the Indian financial market became much more active. Thanks to the population's education, the ability to manage capital in the markets has spread, and society has learned to invest in various financial instruments as opposed to the old concept of placing money in a bank account. Humanity has gained access to a wide range of financial market investment products. Investing in the stock market is one of the best choices for generating additional income for most of society. However, to obtain the desired profit in the rapidly fluctuating stock market, investors need to understand the market trends in detail and adequately. In this regard, the scientific and investment community faces the challenge of developing effective methods for accurately predicting stock market trends.

The work aims to build models for predicting trends in stock market behavior based on neural networks, which will not only save retail investors time for decision-making but also help them reduce investment risks and losses due to market volatility. Therefore, the goal of this research is to provide users with a cheap and efficient way to identify their investment goals. Moreover, Retail investors are usually private investors from the general public who can use these models to secure their investment return in the current stock market. For the research, the open data of the stock market were used [2, 3]. The purpose of the investigation is to build a deep neural network for predicting stock prices of the NIFTY50 index for the Indian stock market and to develop a strategic system to use the built network in investments by investors and researchers.

2 State of Art

The last ten years have seen the widespread use of artificial neural networks (ANN) for forecasting tasks and predicting stock market behavior. The pioneer was the project created by Kimoto et al. [4], who used ANNs to predict the Tokyo stock market index. Mizuno et al. [5] also successfully used ANNs to predict the buying and selling indices of the Tokyo Stock Exchange of Japan with a prediction rate of 63%. Phua et al. [6] used a neural network and genetic algorithms to predict the Singapore stock market and obtained a market trend with 81% accuracy.

Lin et al. [7] developed an SVM-based stock market prediction methodology in Taiwan, taking into account attributes and stock index valuation. The authors achieved better accuracy than previous forecasting systems. The main problem with using SVMs is that a large number of input attributes of the model, reaching hundreds, requires a corresponding amount of computing power and memory for training.

Wei Bao et al. [8] built a stock market forecasting method using stacked autoencoders, LSTM, and a noise reduction method based on wavelet transform. The wavelet transform was used to reduce the noise level from the data, then SAEs were used to generate a deep high-level feature, and finally, the long-term memory (LSTM) predicted the next day's closing. The authors believe that the proposed model outperforms its peers in terms of both forecasting accuracy and profitability.

Hari Kiran Sai Surayagari [9] presented a stock market prediction method consisting of correlated stock and news article behavior over stocks using LSTM and combining both of them with a Deep Neural Network, to achieve the best results he compared every result separately. His work stated that there's not much of a difference in the results of neural networks and traditional trading strategies.

J. Qin, B. Wang, and C. Zhou [10] performed a study using wavelet transform. The authors applied the LSTM method and the attention mechanism to train NNs and build a predictive model. Z. Jin, Y. Yang, and Y. Liu [11] used the emotional component of a person to refine the analysis during modeling. The authors used empirical mode decomposition (EMD) and LSTM based on human attention to predict stock price changes. Yahya B.D.T. et al. [12] developed a forecasting model for the Indonesian stock market, taking into account the peculiarities of the influence of people's sentiments on forums and social networks. They used Nave Bayes, Support Vector Machines, and Random Forest methods to solve the task of classifying tweets about companies and compared the results of different algorithms.

The machine learning approach for the prediction of stock market values is developed in the work [13–15]. A review of the literature on the prediction stock market is presented in these works. Prediction of financial values of the Stock Market using Recurrent Convolutional Neural Networks is shown in [16]. The studying changes for the Indonesia Stock Exchange are developed in the work [17]. An interesting approach using GRNN-based prediction is presented in [18, 19]. Forecasting with machine learning methods in the case of Ukraine is developed in [20–22]. A parallel algorithm for modeling is proposed in [23, 24]. A back-propagation learning method is developed in [25, 26].

The analysis of works for the prediction of stock market values shows that this theme is actual and has a significant practical weight.

3 Indian Stock Market Values Prediction

To solve the prediction task, Python will be used with a cloud environment provided by Kaggle Notebook Editor, Kaggle Editor was used to create and make changes in the source code. When developing a neural network, network training will take place on the CPU and GPU, GPU variant GPU P100 was the appropriate choice for the training in this research. The diagram of the proposed solution is presented in Fig. 1.

In the first step we will collect data sets for the NIFTY50 index, in our research we took two data sets, the 'DATASET 1' is taken from NSE indices [2], NSE Indices Limited is the largest index and index service provider from India and manages the largest number of Indian indices under the NIFTY brand. 80 ETFs are listed on NIFTY indices in India and 12 on other global markets. Derivative products based on NIFTY are traded on exchanges such as the NSE, NSE IFSC and SGX. The time period of 'DATASET 1' is from March 9, 2015 to March 4, 2022. The prices of stock indices are generally given in csv format containing the values Date, Open, High, Low, Close,

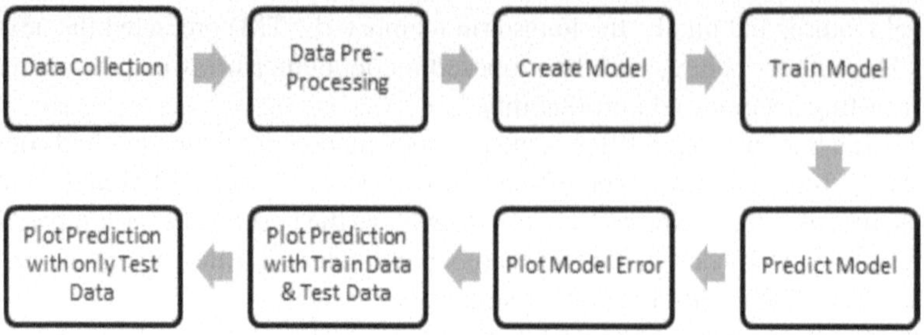

Fig. 1. Data flow diagram for proposed system

Adjustment Close, Volume of the index. The second dataset which is 'DATASET 2' is extracted from Yahoo finance library [3]. It is an open-source tool using Yahoo's publicly available APIs and intended for research and educational purposes. The time period of 'DATASET 2' is from January 3, 2011 to November 10, 2022, and the data features are the same as the 'DATASET 1'. From both the datasets we obtained five features which are Date: the date of observation, Open price: the price of the stock at the beginning of the trading day, High: the highest intraday price reached by the stock, Low: the lowest intraday price reached by the stock, Volume: the number of shares or contracts bought and sold in the market during the day.

The second step will be the pre-processing of the data. Cleaning and processing of data is required in most cases before applying deep learning algorithms, in our research, we will add a new feature to both datasets, namely SMA50, which shows the 50-day simple moving average of NIFTY. The datasets related to the financial markets suffer from several specific problems such as some companies may no longer exist, the market is closed on weekends and holidays, due to technical problems the prices containing negative errors, ban on the trading of a specific company leading towards empty data in continuous series and much more. These issues should be addressed when building a deep learning model, for this research we use LSTM and GRU which are sensitive to the size of the data, so we will apply MinMaxScaler to transform the data in the form of 0 and 1. Figure 2 shows the process of data pre-processing.

We have taken only one feature of the dataset which is the 'Close' price to eliminate the correlation, which causes overfitting and has a bad impact on the performance of the model. We will also split the data into train sets and test sets, which will be used later in the training and testing phases of the model. The model is split as 80% of the data is used for training and the rest 20% is used for prediction. The data is divided into time steps, so we take the training part x_train and y_train and for the testing part, we take x_test and y_test.

The third step is to create a stacked LSTM and a stacked GRU model. A stacked model is an architecture where multiple layers are stacked on top of each other. LSTM and GRU give sequential output instead of single-value output.

The models for both datasets are identical as a comparison was required. The naming is done in ascending order for the sake of simplicity. Parameters including the count of hidden layers and the number of neurons and optimizers are chosen after the trial-error

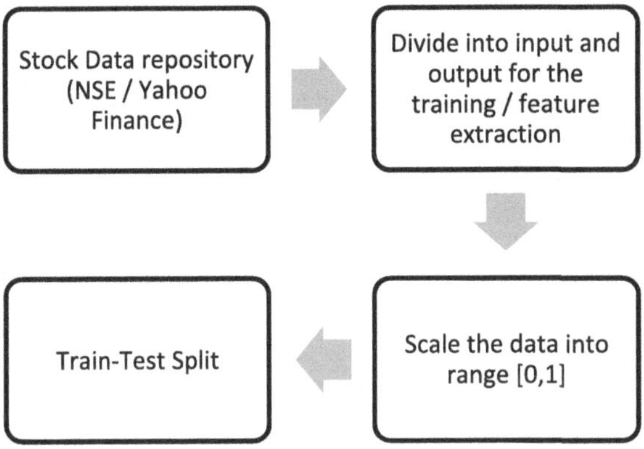

Fig. 2. Data preprocessing diagram for proposed system

method. In our research, we have taken only one input which is 'Close' price, and we are predicting only one output 'Predicted' price.

The reason LSTM and GRU were chosen for this research is that the Multi-Layer Perceptron Model does a good job of predicting the data in series, but lacks the transformation of the data required to properly interpret the results to present, LSTM and GRU on the other hand are based on three gates, namely input forget and output gates, which enables the network to learn better when the data is based on series and sequences.

LSTM and GRU also have better accuracy when it comes to time series data. The number of neurons was set to random in the beginning, and then it was changed in the fashion of increasing order. The number of neurons used in the models varies in the range of (4–128), where 4 is the lowest and 128 is the highest. The final models were selected using the trial-and-error strategy. Each model was manually built, trained, tested, and updated over time until the final models were selected.

In the next stage, the model was trained on sample data. Training a deep neural network model involves refining the model parameters to reduce losses and obtain an effective prediction. The characteristics that will be changed are the number and type of neurons, the number of hidden layers, activation functionality, optimizers, weighting coefficients, and biases. The deep neural network was trained using the Keras API and the TensorFlow framework developed by Google. The Keras Sequential Model API was used to activate the hidden layer of ReLU (Rectified Linear Unit), TanH, and Sigmoid. A regression model was used for modeling, which does not require activation functionality at the input. The optimization algorithm is used to refine and select the weighting coefficients and biases. Then it was used Adaptive Moment Estimation Optimizer (ADAM), Root Mean Squared Propagation (RMSPROP), and Stochastic Gradient Descent (SGD).

This prediction can be made in two ways. The first one is static, which is a simple, less accurate method where the model is fitted to all the training data. Each new time step is then individually predicted from test data, and another technique is dynamic, a complex, but more accurate approach that readjusts the model for each time step of the test data as new observations become available. The process of predicting is relatively simple. We first create a model with our initial parameters, train and then predict the results, note the error values, and then again go back to the creating model phase and change the parameters and repeat the process until we reach a saturation of models or the desired results we wanted. The accuracy of a predictive model can be reliably evaluated using the Root Mean Square Error (RMSE) score. This is because neural networks (including LSTMs and GRUs) in general tend to give different results under different initial conditions for the same data.

In Fig. 3 Line Graphs are drawn for the actual values and the predicted values versus the SMA50, a moving average strategy.

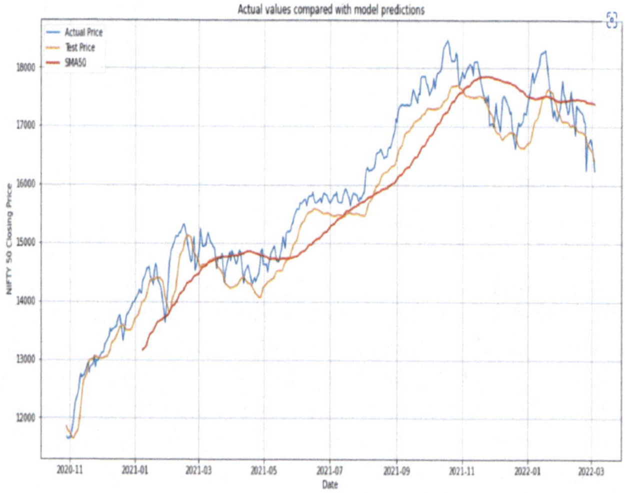

Fig. 3. Graph of Actual vs Predicted Price for Model using LSTM.

Figure 4 shows the Mean Square Error (MSE) of Models (2.3, 4.1, 2.4, 4.5) which are GRU models. Looking at the results we obtained, we can see Model 4.1 has the lowest error among the three chosen models, and thus it is noted for further analysis.

It can be seen in Fig. 5 that in the Model using GRU, SMA50 performs better than the neural network, most of the predicted prices are undervalued, but at certain data points, it can be seen that the model, predicts values higher than the actual values.

Detailed information on possible development environments for solving the problem of predicting the stock market was carried out. We looked at more than twenty-models of neural networks and their respective errors. Each had a unique neural network architecture and were trained on both identical and different parameters, and we saw that both datasets give results differently, so the data we use is very crucial in predicting the stock market. We saw that both LSTM and GRU are highly capable in learning trends of nifty50 stock market data, though in most cases LSTM outperforms GRU in terms of performance.

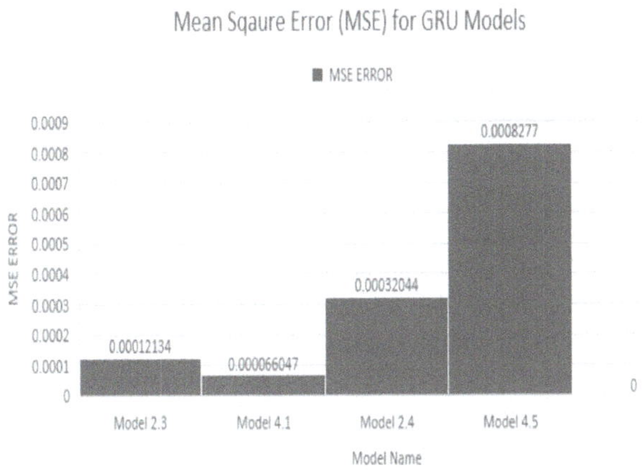

Fig. 4. Mean Square Error for some Models using GRU

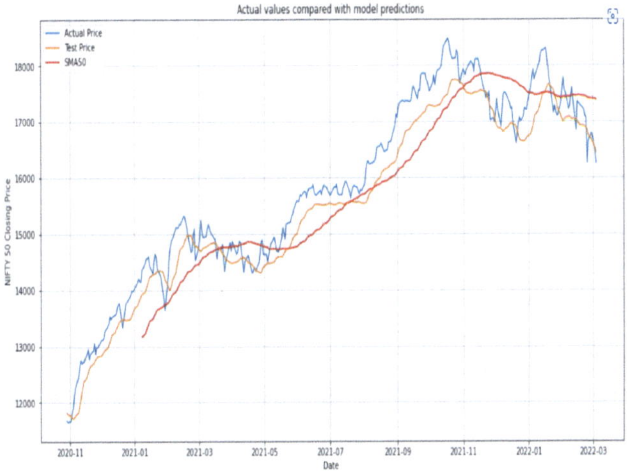

Fig. 5. Graph of Actual vs Predicted Price for Model using GRU.

4 Conclusion

The stock market forecasting task has been considered. A comparative analysis of common stock market forecasting methods is given, and perspectives and possible directions of application of the developed system are described. Taking both LSTM and GRU models into account, further, we observed that when we increased the count of neurons in the layers, the model accuracy also increases, which states the model is highly dependent on the no. of units of neurons used, increasing hidden layers does increase accuracy to some extent but there comes saturation points where it doesn't change and in some cases accuracy decreases, increasing dropout has a variable effect on the models, than the stock parameters is already changeable, increasing randomness in the datasets, disturbs the model's accuracy in most of the cases, but in few cases it does increase the accuracy.

The simple moving average strategy states that when the line is above the SMA curve, it means there's a lot of buying and selling taking place, whereas if the line is below the SMA curve, it indicates that the exchange is in low volumes. Our model was able to predict better results than SMA50 by fairly predicting the actual prices of the stock, because the stock market is highly volatile this model has potential in financial strategy for long-term investment, but lacks in replacing trading moving averages in intraday trading, it can be observed from both networks that neural network can predict the patterns in the stock market. It can be seen that the application of deep neural network architecture makes it possible to identify the hidden dynamics and realize the prediction of the behavior of the Indian stock market. In developing forecasting methods and algorithms, it is necessary to thoroughly research and carefully study the stock market parameters so that the developed model corresponds to reality. Another achieved goal is to expand the possibilities of applying the method and to obtain a higher prediction accuracy since the requires a large amount of accurate actual data and a careful selection of relevant variables for analysis to achieve reliable predictions. Thus, the presented model parameters of the stock market are influenced by economic indicators, political factors, or the state of the environment to determine the regularity of their changes, and to identify specific cyclical patterns.

In future research, we plan to use deep learning methods to predict trends in financial market indices using more indicators and parameters. Another area for prospective study is combining fundamental and technical parameters, such as P/E, P/F, and other characteristics, and combining them for deep neural network functionality. It may be interesting to use public news as input parameters to make the functionality more complete and increase the efficiency of neural network training. Also, the proposed study considers only forecasting for offline data samples and plans to extend the model to process data in real time.

Acknowledgment. The authors would like to thank the Armed Forces of Ukraine for providing security to perform this work. This work has become possible only because of the resilience and courage of the Ukrainian Army.

References

1. Moiseenko, I., Shakhovska, N., Dronyuk, I., Datsko, O.: Social and economics aspects of the pandemic influence in Ukraine. Procedia Comput. Sci. **198**, 670–675 (2022). ISSN 1877-0509. https://doi.org/10.1016/j.procs.2021.12.304
2. Yahoo finance API. https://pypi.org/project/yfinance. Accessed 18 Nov 2022
3. NSE Indices. https://www.niftyindices.com/reports/historical-data. Accessed 18 Nov 2022
4. Kimoto, T., et al.: Stock market prediction system with modular neural network. In: Proceedings of the International Joint Conference on Neural Networks, pp. 1–6 (1990)
5. Mizuno, H., et al.: Application of neural network to technical analysis of stock market prediction. Stud. Inform. Control **7**(3), 111–120 (1998)
6. Phua, P.K.H., Ming, D.H., Lin, W.D.: Neural network with genetically evolved algorithms for stocks prediction. Asia Pac. J. Oper. Res. **18**(1), 103–107 (2001)
7. Lin, Y., Guo, H., Hu, J.: An SVM-based approach for stock market trend prediction. In: Proceedings of the International Joint Conference on NN, Dallas, TX (2013)

8. Bao, W., et al.: A deep learning framework for financial time series using stacked autoencoders and long short-term memory. PLOS ONE **12**(7) (2017)
9. Surayagari, H.K.S.: Stock Market predictions using machine learning. Colorado State University, Fort Collins, Colorado (2021)
10. Qin, J., et al.: Forecasting stock prices with long short-term memory neural network based on attention mechanism. PLoS ONE (2020)
11. Jin, Z., et al.: Stock closing price prediction based on sentiment analysis and LSTM. In: Neural Computing and Applications (2019)
12. Trisedya, B.D., Cakra, Y.E.: Stock price prediction using linear regression based on sentiment analysis. In: International Conference on Advanced Computer Science and Information Systems, pp. 147–154 (2016)
13. Christanto, F.W., Utomo, V.G., Prathivi, R., Dewi, C.: The impact of financial statement integration in machine learning for stock price prediction, 8 February 2024. https://doi.org/10.5815/ijitcs.2024.01.04
14. Noverlita, J., Surbakti, H.: Streamlining stock price analysis: hadoop ecosystem for machine learning models and big data analytics. Int. J. Inf. Technol. Comput. Sci. **5**, 25–34 (2023). https://doi.org/10.5815/ijitcs.2023.05.03
15. Salehpour, A.: Predicting automobile stock prices index in the Tehran stock exchange using machine learning models. Int. J. Intell. Syst. Appl. (IJISA) IJISA **15**(5). https://doi.org/10.5815/ijisa.2023.05.02
16. Zulqarnain, M., Ghazali, R., Ghouse, M.G., Hassim, Y.M.M., Javid, I.: Predicting financial prices of stock market using recurrent convolutional neural networks. IJISA **12**(6) (2020). https://doi.org/10.5815/ijisa.2020.06.02
17. Fajri, A., Sinaga, A.: Implementation of business intelligence to determine evaluation of activities (case study Indonesia stock exchange). Int. J. Inf. Eng. Electron. Bus. **6**, 51–67 (2020). https://doi.org/10.5815/ijieeb.2020.06.05
18. Izonin, I., Tkachenko, R., Gregus ml, M., Zub, K., Tkachenko, P.: A GRNN-based approach towards prediction from small datasets in medical application. Procedia Comput. Sci. **184**, 242–249 (2021)
19. Vitynskyi, P., Tkachenko, R., Izonin, I., Kutucu, H.: Hybridization of the SGTM neural-like structure through inputs polynomial extension. In: Proceedings of the 2018 IEEE 2nd International Conference on Data Stream Mining and Processing, DSMP, Lviv, pp. 386–391. IEEE (2018)
20. Zomchak, L.M.: Sustainable development of Ukraine as a combination of social, economic and environmental components: a structural econometric model of three-pillar approach. In: Proceedings of the 4th International Conference on Sustainable Futures: Environmental, Technological, Social and Economic Matters (ICSF-2023), Kryvyi Rih. IOP Conference Series: Earth and Environmental Science, vol. 11. IOP Publishing (2023)
21. Zomchak, L., Melnychuk, V.: Creditworthiness of individual borrowers forecasting with machine learning methods. In: Hu, Z., Ye, Z., He, M. (eds.) AIMEE 2022. LNDECT, vol. 159, pp. 553–561. Springer, Cham (2023). https://doi.org/10.1007/978-3-031-24468-1_50
22. Tkachenko R., et al.: An approach towards increasing prediction accuracy for the recovery of missing IoT data based on the GRNN-SGTM ensemble. Sensor **20**(9), 2625 (2020). https://doi.org/10.3390/s20092625
23. Mochurad, L., Panto, R.: A parallel algorithm for the detection of eye disease. In: Hu, Z., Wang, Y., He, M. (eds.) CSDEIS 2022. LNDECT, vol. 158, pp. 111–125. Springer, Cham (2023). https://doi.org/10.1007/978-3-031-24475-9_10
24. Mochurad, L., Shchur, G.: Parallelization of cryptographic algorithm based on different parallel computing technologies. In: Proceedings of the Symposium on Information Technologies & Applied Sciences (IT&AS 2021), Bratislava, Slovak Republic, March 5, vol. 2824, pp. 20–29 (2021)

25. Altameem, A., Kovtun, V., Al-Ma'aitah, M., Altameem, T., Fouad, H., Youssef, A.E.: Patient's data privacy protection in medical healthcare transmission services using back propagation learning. Comput. Electr. Eng. **102** (2022). https://doi.org/10.1016/j.compeleceng.2022.108087
26. Oleksiv, I., Izonin, I. Kharchuk, V., Tkachenko, R., Doroshenko, A.: Identification of IT sector stakeholder's requirements to masters program in information system in Lviv region. In: CEUR Workshop Proceedings, vol. 2105, pp. 429–437 (2018)

Pairwise Comparison Methods Based on Alternatives Disaggregation for Polyaspect Decision-Making Problems

Oleksandr Nesterenko[1(✉)], Ivan Kazachkov[2], and Yurij Selin[3]

[1] International European University, Kyiv 03187, Ukraine
oleksandr_nesterenko@ieu.edu.ua
[2] KTH Royal Institute of Technology, Stockholm SE-100, Sweden
[3] Institute for Applied System Analysis National Technical University of Ukraine, Igor Sikorsky Kyiv Polytechnic Institute, Kyiv 03056, Ukraine

Abstract. The article proposes a comprehensive methodology for improving pairwise comparisons for expert polyaspect evaluation of alternatives. The model of pairwise comparisons usually leads to an unsustainable policy of individual judgments, in particular due to limited knowledge and information. Therefore, the essence of our approach is primarily to use the capabilities of modern technologies to present an information picture of the subject area based on computer ontologies and visualization on graphs. This allows for the disaggregation of alternatives to individual characteristics and taking into account their advantages for evaluating problems, in particular, using the SWOT matrix and AHP. The results of modeling alternatives based on the example of finding rational solutions for determining resources for attracting aviation to ensure the elimination of forest fires confirmed the effectiveness and stability of the proposed method. Application of this method will allow the expert to take into account important factors that may be underestimated in traditional analysis. The further development of the method should be related to the intellectualization of the processes of pairwise comparisons, which can also significantly reduce the subjective influence of expert judgments.

Keywords: Decision-making · Computer ontologies · Visualization · Bipartite graph · AHP · SWOT

1 Introduction

In contemporary conditions, decision-making processes are characterized by high dynamics of changes and are accompanied by the flow of significant information. Any problem domain typically exhibits a considerable number of aspects or properties influencing the quality of the decisions made. Researchers and experts propose various approaches to decision support in such environments. Most of these approaches rely on expert methods, which to some extent enable solving the posed tasks.

Expert assessments play a crucial role in decision-making in conditions primarily characterized by qualitative, informal processes. Experts significantly complement

Z. Hu et al. (Eds.): ICCSEEA 2024, LNDECT 242, pp. 249–260, 2025.
https://doi.org/10.1007/978-3-031-84228-3_21

the lack of quantitative information about decision alternatives. Experts largely compensate for the lack of quantitative information regarding decision options, building an expert's assessment of an object or phenomenon as a logical conclusion based on personal experience and special knowledge.

1.1 Problems of Pairwise Comparisons

The majority of expert decision-making methods utilize pairwise comparison methods, based on the law of comparative judgments proposed almost a century ago by the American psychologist Louis Thurston. According to the procedure of the pairwise comparison method, all objects are compared pairwise, and each subsequent assessment is not linked to the previous one. In the process of analysis, the expert focuses attention not on all elements at once, but only on two being compared at any given moment. This simplifies the analysis and contributes to its improved quality [1]. In addition, the expert can judge the ratio of advantages between alternatives based on how certain characteristics of one alternative are compared with the characteristics of another.

All these pairwise assessments form a matrix of pairwise preferences, and through special processing, numerical parameters of the object's priority indicators are obtained. The essence of the classic method of determining relative priorities lies in experts assigning quantitative evaluations to the comparison of alternatives with each other, taking into account various factors. These quantitative assessments can then be used for further analysis [2].

An essential factor in involving an expert in choosing the best decision through pairwise comparisons is the necessity to have a certain understanding of how a person makes decisions. The modern understanding of this process is associated with the concept of a conceptual or mental model of the surrounding world used by individuals to predict the consequences of their actions. However, the complexity of this model leads to non-resilient judgment policies (preference structures). Individual judgments of decision makers (DMs) usually have shades of uncertainty due to the limitation of knowledge and information. Let's see what problems this causes when performing pairwise comparisons.

The most common application of expert pairwise comparisons was found in the well-known method of Analytic Hierarchy Process (AHP), proposed in the late 1970s by the American mathematician T. Saaty. The method involves a hierarchical decomposition of the problem into simpler components and the step-by-step establishment of priorities for the assessed components using pairwise comparisons.

Considering that, in many cases, the preference model used by DMs is undefined and vague, and preferences differ among different DMs, leading to the distribution of ratings for the same object, assessing the level of inconsistency of pairwise comparisons when using AHP is often a crucial step in decision analysis [3, 4]. In the literature, more than ten inconsistency indices have been proposed to assess the deviation of expert judgments from a situation of complete consistency.

It is commonly believed that experts are honest and professional. However, in practice, there are cases where experts attempt to manipulate results in their favor. This necessitates the use of mechanisms that allow the detection of manipulators and minimize their influence on group consensus [5].

When enforcing the transitivity of a comparison system, an expert who makes an error in comparing a pair of objects (which is possible when there is uncertainty in the available information) is compelled, when comparing other pairs, to take into account the results of previous comparisons, including the erroneous ones. This undoubtedly leads to further errors.

To address these problems, various modifications of the process of pairwise comparisons are used, including AHP. One of the most widespread modifications involves replacing point estimates of priorities with interval values, such as computing element weights based on interval models like GPM, LUAM, and others. Another approach is associated with the introduction of fuzzy logic. It is believed that providing assessments in a fuzzy form reduces the burden on the expert. Instead of point estimates, the expert operates with more convenient linguistic assessments, for example, "approximately equal to *x*," "between the values of *x* and *y*." It is considered that such assessments better correspond to reality. There are also other methods aimed at eliminating the shortcomings of expert pairwise comparisons, such as the Best-Worst Method (BWM) [6] or the Analysis of Preferences Disaggregation Method [1].

Thus, it can be concluded that all subsequent problems of using pairwise comparisons actually have their roots in the seemingly simple stage of experts determining the numerical weights (priority) of alternatives, for example, when forming the appropriate matrix in AHP (Fig. 1).

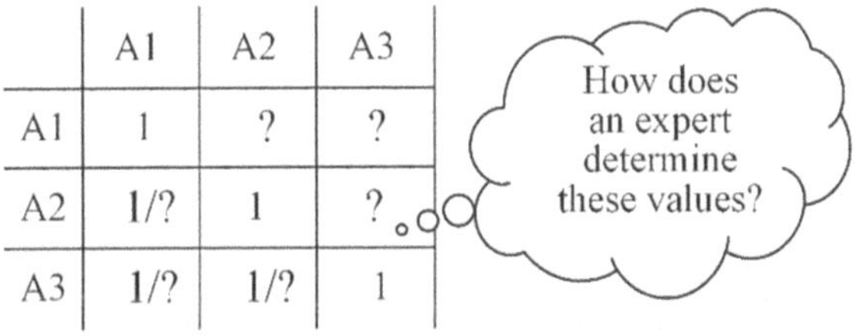

Fig. 1. Cognitive roots of pairwise comparison problems in expert determination of numerical weights in AHP

1.2 Directions for Overcoming the Problems of Pairwise Comparisons

It is unrealistic to hope that humans will change for the better in the near future. Therefore, many researchers share the opinion that decision support in analytical activities, especially in multi-criteria cases, should first of all be directly related to a comprehensive representation of the information landscape in the subject area (SA). In modern conditions, to change the situation for the better regarding the cognitive process of comparisons can already be improved through the use of information technologies [7–9]. In this regard, as a methodological basis, it is necessary, first of all: i) to make a transition to data-driven decision-making (DDDM), ii) to establish the optimal composition of information needed for effective decision-making, iii) to ensure presentation and analysis on various levels of a significant set of heterogeneous data. According to the authors [8], the first meaning is to consider the expert's mind as an attribute of the brain, and the second meaning is to consider data, knowledge and information technology as attributes of his mind.

On the other hand, to facilitate expert activities, structured formats are necessary for describing alternatives. This is a serious issue because the readability and understandability of documented information significantly impact the success and effectiveness of experts' contributions to the decision-making process. In this sense, an important capability of modern technologies is providing visualization of decision-making processes. Many studies in the field of social sciences confirm this insight [10]. Creating graphical models to assess the impacts of different types signifies their convenience, providing an effective decision support tool [11–13].

In summary, it can be noted that in modern conditions when making decisions, there is a shift from the influence of individual judgments to an increase in the value of available information about alternatives (Fig. 2).

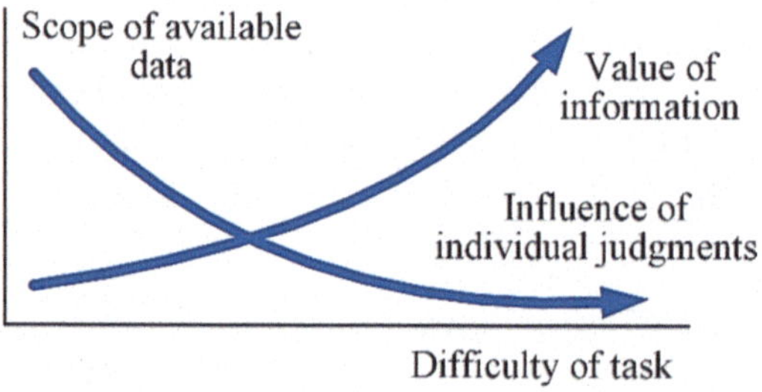

Fig. 2. Moving from the influence of individual judgments in decision-making to the importance of the value of available information

In this way, the following research goals can be formulated:

a) to offer technological means for informational support of pairwise comparison processes;
b) to investigate possible methods of pairwise comparisons, rely on the technological base;
c) give an example of the use of research results in a certain subject area.

The research is aimed at obtaining theoretical results in the form of methods of disaggregating alternatives to individual characteristics when conducting pairwise comparisons with the support of information technology tools. This approach has an advantage over existing models due to the consideration of a set of additional factors for evaluating problems. The practical value of the study is related to the provision of prompt implementation of automated iterations of problem analysis, which has potential implications for increasing the effectiveness of decision-making.

2 Methods of Alternatives Disaggregation

2.1 Technological Means for Information Support

When making comparisons, DMs can use these specific data about the elements, which typically influences their judgments regarding the relative significance and importance of these elements. In the case of complex problems, alternatives usually differ in sets of characteristics that significantly vary from each other. Therefore, experts' familiarity with the characteristics of alternatives, which forms the basis for their objective comparisons, plays a crucial role in decision-making. In this regard, an important element of modern decision support systems should be a knowledge base that represents the information model of SA. Among the existing approaches to such models, representing the SA in the form of computer ontologies [14, 15] is currently considered the most adequate. In the general case, the ontology contains informational descriptions based on an object-oriented formalization procedure, where each model O_m reflects an expressive hierarchy of interaction of concepts X, which are specified using binary relations R: $O_m = (X, R)$. Facts relating to individual instances of this world are stored either in separate databases or in Internet sources.

The simultaneous application of elements in ontological descriptions enhances the specificity of the model and provides a clearer understanding of the environmental state. As attributive characteristics of these ontologies' concepts, expert assessments of the value of objects, implementation possibilities, and characteristics of potential efficiency mechanisms, cost, complexity, etc., are defined.

The presence of a significant number of attributive characteristics of alternatives requires visualization of information, because visualized data require less cognitive effort during interpretation than text (table) descriptions. A widely used tools for visualizing information are graphs. Modern software allows you to effectively display graph diagrams and thereby help experts in clarity, speed and understanding of complex concepts.

2.2 Model Based on Bipartite Graph

When making decisions in various domains, pairwise comparisons of alternatives A with characteristics X can be represented as a binary scheme "A1" $\rightarrow$ "A2," which can be expressed in the form of a bipartite graph. As known, a bipartite graph (also called a bigraph) is a graph whose set of vertices can be divided into two non-overlapping subsets, such that each edge of the graph has one vertex in the first subset and one in the second. An undirected graph $G = (V, E)$ is called bipartite if its set of vertices is partitioned into two subsets: $V = U \cup W, |U| \succ 0,$, with the conditions that 1) no vertex in U is connected to vertices in U and 2) no vertex in W is connected to vertices in W. Figure 3 illustrates an example of a bipartite graph of pairwise comparisons $G = (V(A1, A2), E(A1, A2))$, constructed by disaggregating the characteristics of alternatives.

In this graph, the sets of vertices in its partitions are $X_{A1} \cup X_{A2} = V(A1, A2)$, $X_{A2} = \{X21, X22, ..., X2m\}$, $X_{A2} = \{X21, X22, ..., X2m\}$, $A1 \cap A2 = \otimes$, and the set of edges is $E(A1, A2)$, where an edge $(X1n, X2m) \in E(A1, A2)$ symbolizes the comparison of the corresponding characteristics $X1n$ and $X2m$. When comparing characteristics, the edges $(X1n, X2m)$ are loaded with the values of expert assessments of characteristics, respectively C_{1n} and C_{2m}. The direction of the graph's arcs indicates a preference for a particular characteristic. In cases where one alternative has more characteristics, as in the provided example in A1, a fictitious vertex (X25) is introduced in the partition of the other alternative, and it C_{25} is taken as equal to 0.

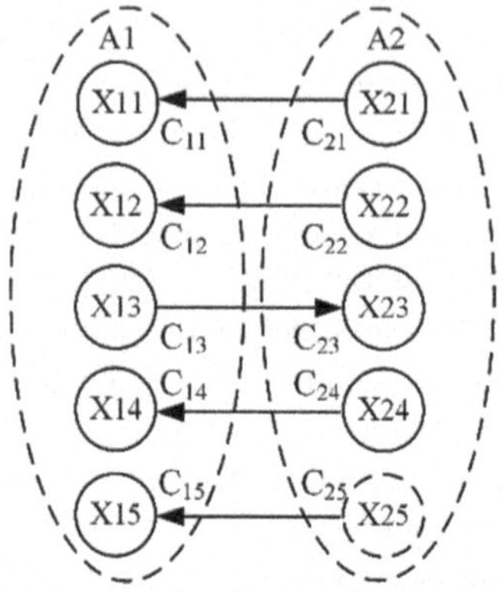

Fig. 3. Bipartite graph of pairwise comparisons

When determining expert ratings for alternative characteristics, their quantitative and qualitative indicators are represented in a quantitative (score-based) dimension according to a specific conditional scheme, for example: slight impact – 1 point, low – 3, moderate – 5, high – 7, very high – 9 points.

Now let's delve into the steps of the pairwise comparison algorithm:

1. Score ratings for alternative characteristics are determined using existing sources of information.
2. Ratings for alternatives C_{A1} and C_{A2} are calculated as the arithmetic averages of their characteristic assessments C_{1n} and C_{2m}.
3. Determining the degree of preference of one alternative over another is based on the relative rating of the smaller one compared to the larger one (for example, if $C_{A1} \succ C_{A2}$:

$$F = \frac{C_{A2}}{C_{A1}}100. \tag{1}$$

4. If in the Future It is Planned to Enter the Results of a Pairwise Comparison into the AHP Matrix, You Can Use the Following Rule of Conversion to Saati Scale (Table 1).

2.3 Model Based on SWOT Analysis

The discussed approach can be applicable to relatively simple decision-making tasks. In more complex cases, such as strategic analysis, there is a need to consider a significant set of internal and external factors in pairwise comparisons, evaluating alternatives based

Table 1. The rule for conversion the degree of alternative preference to Saati scale

F	80–100	60–80	40–60	40–60	≤20
Saati scale	1	3	5	7	9

on generalized criteria like potential benefits, existing opportunities, potential costs, and acceptable risks. One of the tools most frequently used for such consideration is SWOT analysis. Typically, SWOT is applied at the macro-level of strategic analysis. Despite its advantages and popularity, this method has faced criticism, particularly regarding the absence of a methodology for quantitatively assessing the results of compiling a SWOT matrix. The convenience of using the original SWOT is enhanced by the hybrid SWOT/AHP method, introduced back in 2000–2001. Since then, this approach has been applied in different fields such as manufacturing, agriculture, the telecommunications sector, tourism, forestry, and others [16–19].

We propose the reverse application of SWOT, namely at the micro-level, specifically in the analysis of alternative characteristics. Constructing a SWOT matrix through further disaggregation of alternatives allows the expert to consider important factors that may be underestimated in traditional analysis. The approach involves building a SWOT matrix for each alternative, with characteristic factors being considered in the quadrants of the matrix based on the influence of the characteristic on the evaluation of the alternative (Fig. 4).

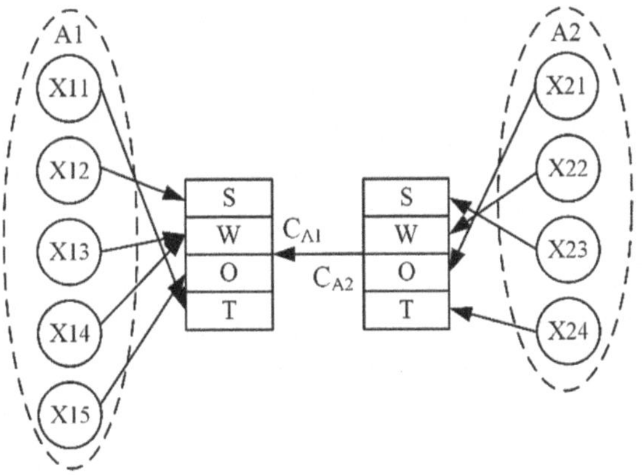

Fig. 4. Applying of SWOT in the analysis of characteristics of alternatives

Applying of this approach is based on the following algorithm:

1. Each element (characteristic) from the components of the SWOT matrix sectors is considered based on specific parameters available in the database (knowledge) and is evaluated in a quantitative (numerical) dimension. For example, according to the aforementioned conditional scheme, scores from 1 to 9 points can be assigned.

2. By summing up the expert assessments of the components of the sectors, we find for each alternative the corresponding scores CS, CW, CO, CT.
3. Next, for a comprehensive understanding, the strength of each alternative CA1 and CA2 can be determined across the four SWOT sectors:

$$C_{A1} = \frac{CS + CO}{CW + CT}. \tag{2}$$

4, Determination of the degree of superiority of one alternative over another in each sector of the FCS, FCW, FCO, FCT matrix is based on the relative evaluation of the smaller one from the larger one (for example, if $CS_{A1} \succ CS_{A2}$):

$$F_{CS} = \frac{CS_{A2}}{CS_{A1}} 100 \tag{3}$$

4. Determination of the Degree of Superiority of One Alternative Over Another in General According to the SWOT Matrix is Found by Averaging the Assessments by Sectors:

$$F_{SWOT} = \frac{F_{CS} + F_{CW} + F_{CO} + F_{CT}}{4}. \tag{4}$$

5. Finally, if it is intended to enter the results of a pairwise comparison into AHP matrix, the rule of conversion to Saati scale according to Table 1 should be used.

3 Simulation Experiment

Let's consider the application of the proposed methods using the example of rational decision-making in determining resources for involving aviation to ensure the elimination of forest fires (FF). Success in extinguishing FF is significantly related to choosing the necessary number of resources (aircraft and/or helicopters) and selecting the "best" aviation action option during fire localization. To determine such an option, it is necessary to take into account a significant number of factors.

In the example, a model consisting of 2 alternatives is used for simplification: A1 – a composition of aviation grouping with aircraft of one type; A2 – with helicopters of one type. The following characteristics of the specified alternatives are used for the analysis: X11 (X21) – Time to localize the fire; X12 (X22) – Financial costs for providing flights; X13 (X23) – Flight safety; X14 (X24) – Number of discharges of fire-extinguishing liquid in one flight; X15 (X25) – Duration of localization actions; X16 (X26) – Dependence on flame height; X17 (X27) – Dependence on flow turbulence; X18 (X28) – Dependence on smoke; X19 (X29) – All weather.

First, we will apply the method based on a bipartite graph to compare alternatives. On Fig. 5 shows a fragment of the interface of a software tool to support comparison based on this method with expert data for assessments alternatives.

From the given data, total evaluations of the alternatives were obtained: A1 = 49; A2 = 39. Using these results, we determine the degree of preference of one alternative over another according to expression (4) and get a value of 79.6%. To transfer the results of a pairwise comparison to AHP matrix, we will use the rule according to Table 1 and get the advantage of A1 in the value of 3 according to the Saati scale.

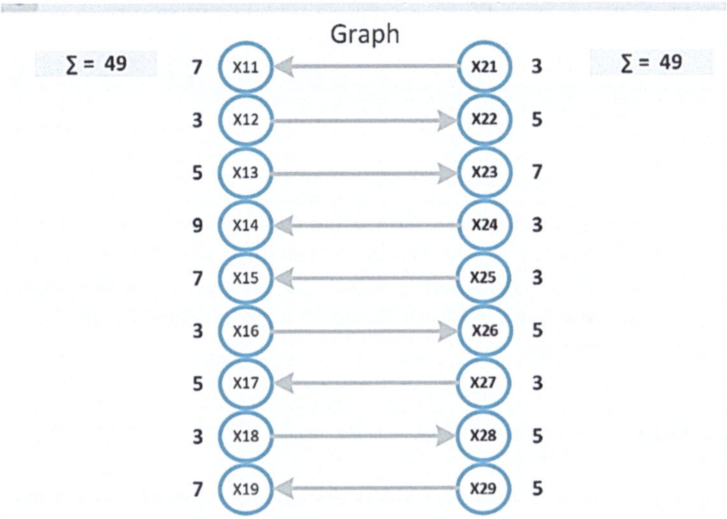

Fig. 5. Expert data on alternatives assessments using the bipartite graph comparison method

Table 2. Expert assessment of the alternatives characteristics of aviation group according to the SWOT concept

A1	S	W	O	T	A2	S	W	O	T
X11	7	1	5	3	X21	3	5	3	3
X12	1	9	3	3	X22	3	7	3	1
X13	5	3	7	5	X23	7	3	3	3
X14	9	1	3	1	X24	3	7	1	3
X15	9	1	3	5	X25	3	7	1	5
X16	1	7	3	7	X26	5	3	3	5
X17	1	7	3	5	X27	5	3	3	5
X18	3	3	3	3	X28	5	3	3	7
X19	5	3	3	5	X29	5	1	3	5
Assessments	41	35	33	37	Assessments	39	24	23	37

Next, we will consider an example of the application of the micro-SWOT approach. Table 2 shows expert evaluations of performance parameters for each alternative.

Using the obtained estimation for the SWOT-matrix sectors, we obtain a determination of the degree of superiority of one alternative over another for each sector according to expression (4), and in general according to expression (5). The results are presented in Table 3.

To transfer the results of a pairwise comparison to the AHP matrix, we will use the rule according to Table 1 and get the advantage of A1 in the value of 1 according to the

Table 3. The result of the alternatives evaluation by SWOT matrix sectors

	CS	CW	CO	CT	SWOT
A1 vs A2	95,12	68,57	69,70	100	83,35

Saati scale. Considering the closeness of the obtained score of 83.35% to the neighboring category in Table 1, it is possible to use the intermediate scores of the Saati scale and define the advantage as 2. This result indicates the clarification of the previous result obtained by the bipartite graph method, thanks to a more detailed consideration of all factors and their advantages.

4 Discussion

The complexity of decision-making tasks is due to the extremely large dimension of the information space of the subject area, the weak structuring of data, which occurs due to their uniqueness, uncertainty of conditions, variability of aspects of alternatives. Automation of processes based on the use of computer ontologies and visualization on graphs provides a number of advantages for solving problems of paired comparisons in decision-making. This approach allows you to significantly expand the scope of practical application of the proposed methods in the conditions of poly-aspect analysis and multi-criteria.

The result is the discovery of an effective way to eliminate various ambiguities in the understanding of individual judgments of expert knowledge in integration with software and information support. The value of the research is to accelerate the formation of a breakthrough innovative system of human-machine collective intelligence to support decision-making, which will allow making better decisions thanks to the mitigation of human biases and the productive use of information.

5 Summary and Conclusion

The research is aimed at obtaining theoretical results in the form of methods of disaggregating alternatives to individual characteristics when conducting pairwise comparisons with the support of information technology tools. Thus, a scheme for reducing the subjective influence of individual judgments of experts on the result of solving real-world problems is proposed. An example from the field of emergency response demonstrates the applicability of the proposed methodology in a human-machine environment.

A more objective calculation of estimates based on the disaggregation of alternatives to their individual characteristics is provided on the basis of computer ontologies, visualization of information based on graphs, as well as the application of micro-SWOT analysis and AHP. Thanks to these technologies, it becomes possible to eliminate potential sources of errors or systematic errors, increase the reliability of the data provided and the reliability of the results.

The obtained results were tested in the computer-oriented didactic-psychological provision of preliminary and pre-flight training of aircraft crews for flight-tactical training related to extinguishing fires in the ecosystem [20].

Acknowledgment. This project is supported by Key Projects by order of the Ministry of Education and Science of Ukraine.

References

1. Xingli, W., Liao, H., Zhang, C.: Preference disaggregation analysis for sorting problems in the context of group decision-making with uncertain and inconsistent preferences. Inf. Fusion **101**, 102014 (2024)
2. Alguliyev, R.M., Nabibayova, G.Ch., Abdullayeva, S.R.: Evaluation of websites by many criteria using the algorithm for pairwise comparison of alternatives. Int. J. Intell. Syst. Appl. (IJISA) **12**(6), 64–74 (2020)
3. Rzheuskiy, A., Veretennikova, N., Kunanets, N., Kut, V.: The information support of virtual research teams by means of cloud managers. Int. J. Intell. Syst. Appl. (IJISA). **10**(2), 37–46 (2018)
4. Dede, G., Kamalakis, T., Sphicopoulos, T.: Convergence properties and practical estimation of the probability of rank reversal in pairwise comparisons for multi-criteria decision making problems. Eur. J. Oper. Res. **24**(2), 458–468 (2014)
5. Kułakowski, K., Szybowski, J., Mazurek, J., Ernst, S.: Resilient heuristic aggregation of judgments in the pairwise comparisons method. Inf. Sci. **657**, 119979 (2024)
6. Dong, J.-Y., Wan, S.-P.: Interval-valued intuitionistic fuzzy best-worst method with additive consistency. Expert Syst. Appl. **236**, 121213 (2024)
7. Elragal, A., Elgendy, N.: A data-driven decision-making readiness assessment model: the case of a Swedish food manufacturer. Decis. Analytics J. **10**, 100405 (2024)
8. Zou, X., Zou, S.: Cognitive computing smart system: how to remove ambiguities. J. Comput. Cogn. Eng. **3**(1), 34–42 (2024)
9. Nesterenko, O., Netesin, I., Polischuk, V., Selin, Y.: Multifunctional methodology of expert evaluation alternatives in tasks of different information complexity. In: Proceedings of the IEEE 3rd International Conference on Advanced Trends in Information Theory (ATIT), pp. 226–231 (2021)
10. Eberhard, K.: The effects of visualization on judgment and decision-making: a systematic literature review. Manag. Rev. Q. (2021)
11. Sitarz, S.: Rankings in sport by pairwise comparison and league table. Int. J. Mod. Educ. Comput. Sci. (IJMECS). **5**(12), 24–30 (2013)
12. Nesterenko, O., Netesin, I., Polischuk, V., Selin, Y.: Graph-based decision making for varying complexity multicriteria problems. Comput. Sci. J. Moldova **30**(3), 391–412 (2022)
13. Nishizawa, K.: Improvement of the weights due to inconsistent pairwise comparisons in the AHP. In: Czarnowski, I., Caballero, A.M., Howlett, R.J., Jain, L.C. (eds.) Intelligent Decision Technologies 2016. Smart Innovation, Systems and Technologies, vol. 57, pp. 397–406. Springer, Cham (2016). https://doi.org/10.1007/978-3-319-39627-9_35
14. Xu, F., Liu, X., Chen, W., et al.: An ontology and AHP based quality evaluation approach for reuse parts of end-of-life construction machinery. Math. Probl. Eng., 3481030 (2018)
15. Nesterenko, O.: Ontology and analytic hierarchy process in the information and analytical systems. In: Babichev, S., Lytvynenko, V., Wójcik, W., Vyshemyrskaya, S. (eds.) Lecture Notes in Computational Intelligence and Decision Making. ISDMCI 2020. Advances in Intelligent Systems and Computing, vol. 1246, pp. 302–314. Springer, Cham (2021). https://doi.org/10.1007/978-3-030-54215-3_19

16. Popescu, G., Gasparotti, C.: SWOT-AHP hybrid method for ranking the strategies in the shipbuilding sector. J. Bus. Econ. Manag. **23**(3), 706–730 (2022)
17. Ghorbani, M.K., Hamidifar, H.: Strategic planning of rubber dams by the SWOT and SWOT-AHP methods in Iran. Int. J. Hydrol. Sci. Technol. **15**(2), 112–122 (2023)
18. Ariyana, R., Amalia, R., Salsabilah, D.S., et al.: Strategy for increasing lowland rice productivity in West Java Province with the SWOT-AHP model approach. In: Proceedings of the 3rd International Conference on Biosciences. IOP Conference Series: Earth and Environmental Science, vol. 457, p. 012058. IOP Publishing (2020)
19. Cvetković, M., Šljivović, M.: Prioritization of strategies for development of ecotourism by means of AHP-SWOT on the example of Kopaonik, Serbia. Pol. J. Environ. Stud. **30**(6), 4933–4943 (2021)
20. Nesterenko, O., Polischuk, V., Khyzhniak, V., Shevchenko, V.: Decision-making information technologies for the resources determination of forest fire extinguishing by aviation. Environ. Saf. Nat. Resour. **46**(2), 109–123 (2023). (in Ukrainen)

An RNN-Based Stock Price Forecasting Model Enhanced by Sentiment Analysis of the Daily Financial News

Volodymyr Shymanskyi(✉) and Viktor Sabov

Department of Artificial Intelligence, Lviv Polytechnic National University, Lviv, Ukraine
vshymanskiy@gmail.com

Abstract. Prediction of stock data is an inherently difficult process because of its chaotic nature. Experts say that the prediction of these values is impossible, however, research papers show that many of the values depend on the previous ones, therefore they are expressible through machine learning methods. This research used several datasets, including "Yahoo! Finance" stock data, news sentiment analysis training dataset, and a news dataset with related stock and publishing date. First, we worked on sentiment analysis for news. Compared pretrained transformers and random forests for this task. After that, we worked with recurrent neural networks of type LSTM and GRU to predict the stock market price for the next days. We compared RNNs with news and RNNs without news.

Keywords: Stock market · Machine learning · Transformers · Recurrent neural networks · Sentiment analysis · Prediction

1 Instruction

Forecasting the stock market has always been a daunting task due to its volatile and unpredictable nature. Although many experts claim that it is impossible, recent studies reveal that it is feasible to a certain extent by leveraging current data and variables. Therefore modern machine learning methods are suitable for it.

Other than that, the stock market is influenced by several economic factors, including, but not limited to, political events, general economic conditions, commodity price index, expectations of investors, movement of other stock markets, etc. There are some classic methods for calculating certain stock market variables and indexes based on their price, etc. The nature of the stock market is quite ambiguous therefore making investments risky for the investor. Also, the government struggles as well with the prediction. The market is indeed dynamic and non-linear, making the performance of classic linear models inefficient in this task.

This paper uses news and previous stock price data to predict the stock closing price on the next day. We used data from January 2013 to September 2015 with 80% of this range used as train data and the remaining 20% as test data. We chose this range because it has enough news from datasets used in the paper and also has a relatively wide range

Z. Hu et al. (Eds.): ICCSEEA 2024, LNDECT 242, pp. 261–270, 2025.
https://doi.org/10.1007/978-3-031-84228-3_22

of prices. We used the news to extract sentiment data and compared using transformers and RFs. The better-performing model was used for further work. Then we "marked" the news with the sentiments and merged it with the stock data. We then passed the merged data into LSTM and GRU RNNs with a neural network on its end, outputting the closing price for the next day. After completing the experiments we compared the prediction accuracy of the neural network with and without news data.

Several papers were solving this problem, but all of them have drawbacks, by fixing it we can achieve better results in this task.

One such solution is to combine news and stock price data [1]. The general algorithm for analyzing the news was as follows: each word was given a score from -1 to 1, where -1 means it is a "bad" word and 1 means it is a "good" word. The disadvantage of this approach is that it does not take into account the content of the news, which leads to low accuracy and therefore does not use the full potential of sentiment analysis.

Good performance was shown by the use of BERT transformers [2], which are the most modern, but they are less productive than DistilBERT, in particular, some studies used the Chinese language model. It is also worth noting the use of PhoBERT, a transformer language model for Vietnamese [3]. Using RNNs for sentiment analysis with Word2Vec embedding leads to quite good results [4], but less performant than transformers.

Using Twitter instead of news is also a viable choice, but the Twitter API is paid [5]. Some studies have used both news and tweets [6]. However, sentiment analysis was used discretely, as in [3], which led to a more limited model. Interesting results are shown in [7, 8], which used a rich collection of sentiment data, but with less productive models than transformers.

It is also worth noting a study in which the authors predict news sentiment and stock price independently, but do not take advantage of their simultaneous use [9].

Particularly noteworthy are studies that use a huge amount of additional data: Google search activity, tweets, and news from various websites, a more detailed sentiment analysis model that can calculate complex emotions such as negative, positive, anxiety, calm, dislike, love, liking, fear, pleasure, and sadness [10, 11]. The authors concluded that sentiment analysis models produce better results in terms of accuracy, but they used outdated sentiment analysis models.

In short, the following are the contributions of this work:

1. We have built a model for analyzing the general mood of news, giving each news title a score of "positiveness", "neutralness" and "negativeness".
2. Proposed a method of respecting news mood in pair with general stock data, when predicting financial variables on the stock market.
3. We have built a model for stock price forecasting concerning news mood.
4. We have compared the results models leveraging news and models leveraging only stock data. Also was investigated the order of inclusion of various devices from the developed IoT network in the proposed cascade on the method's accuracy. The optimal combination of it was determined experimentally.

2 Materials and Methods

This section describes the processes that are used in the article. We have processed the raw data after collecting it. We used sentiment analysis on the news. After that, we processed the stock market data and merged it with the sentiment analysis results. Finally, we used recurrent neural networks to predict the stock price for the next day.

In this work, we used several different datasets. One dataset for training the transformer for sentiment analysis, two datasets with general news, and one dataset with symbol-specific news. These are as follows:

Dataset-I. It was obtained from Kaggle [12]. It contains labeled news titles with the sentiments "negative", "neutral" or "positive". It will be used for training our sentiment analysis model. Since it contains only news titles, we will be training our model for use with 128 characters.

Dataset-II. It is a collection of world news from 2008 to 2016 [13]. It contains only 2 columns – the date and the news. As we can see, most of the days have 25 news, but there are some outsiders before 2012. It does not concern us, since we plan to use dates only after 2014. Also, the news is longer than in the training dataset, so we will need to trim them to length 128.

Dataset-III. It contains news from HuffPost from 2012 to 2022 from different categories [14], including travel, beauty, sports, etc. So we need to cut out the unnecessary categories. The categories left and their illustration in a pie chart can be viewed in Fig. 1. Also by observing the news density by date, we can conclude that the best option for using this dataset is from 2014 to 2018.

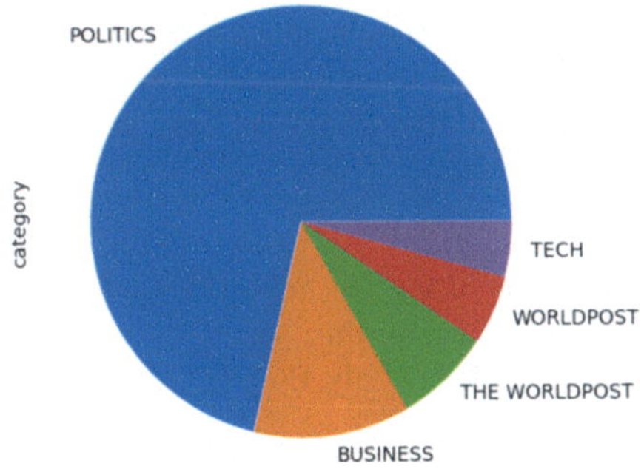

Fig. 1. Category distribution for Dataset-III

Dataset-IV. Since Dataset-I does not contain any data regarding what is the published date of the news or what is the stock it's affecting, we used another dataset that contains this data, but it does not contain sentiment data. Further, this data will be added via machine learning. It is also downloaded from Kaggle [15].

Dataset-V. It is a collection of financial daily data. It is downloaded from Yahoo! Finance [16]. Through the public API for a given stock symbol and a given period.

As for the preprocessing of the data, we did several steps:

- We balanced Dataset-I by undersampling the neutral and positive news (before and after can be viewed in Fig. 2). This could be done due to the high efficiency of pre-trained transformers even on a low volume of data.
- In Dataset-I we did one-hot encoding for the sentiment. The sentiment "positive" was at position 0, the sentiment "neutral" was at position 1 and the sentiment "negative" – was at position 2.
- Datasets II, III, and IV were generally cleaned up for NLP use.
- In Dataset-IV we had rows that didn't have a stock attached, we dropped those rows.

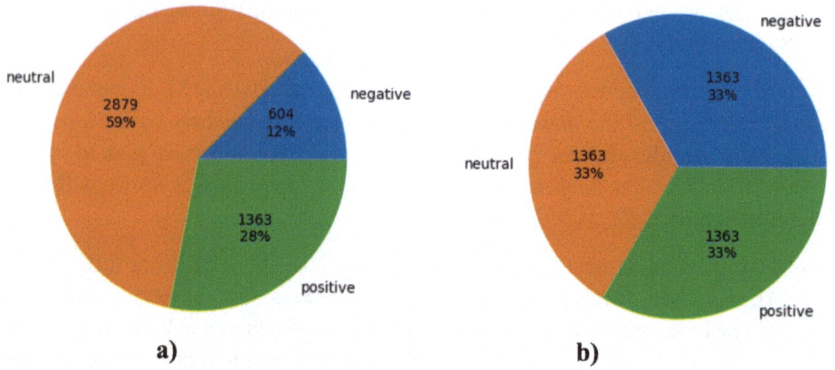

Fig. 2. Dataset-I **a**) before and **b**) after balancing

Further data preprocessing for sentiment analysis is unneeded for transformers since they have their embedding mechanism, but is needed for the random forest, with which we are going to compare it. For random forest, we have the next preprocessing steps:

1. We convert everything to lowercase because casing does not matter in this particular task.
2. Then we do tokenization by splitting our news titles into words but in a meaningful way. We utilized the Spacy library for this.
3. Finally, we use stemming to keep only the root for each word. We utilized the Pystemmer library for this.

The process of analyzing text and classifying it by sentiment is known as sentiment analysis. We classify our text into either "positive", "neutral" or "negative" sentiment. For this task, we used the transformers library from HuggingFace. With its help, we fine-tuned the DistilBERT model, which is a more efficient BERT model [17, 18]. It also provided us with a tokenizer and embedder, so we did not need to worry about implementing our custom one. It returns several properties and these values can be given to the DistilBERT transformer, which outputs a tensor of size batch_size × token_count × 768 [19]. Our approach to sentiment analysis is like the one mentioned in [20]. We grab the sequence for the start token. The neural network output contains 3 items – the possibility of "positive" sentiment, "neutral" sentiment, and "negative" sentiment.

As a baseline for accuracy comparison, we used RFs. RF is a renowned supervised learning method based on an ensemble learning approach that brings together various classification elements to solve a complex issue and increase the performance of the model. It is a multi-decision tree ensemble classifier that uses a randomly chosen subset of training data and parameters to generate multiple decision trees. As an embedding layer, we used "Bag of Words" (BoW). It identifies the text as a bag of its items, therefore it ignores word order and grammar [21, 22].

For evaluating classification accuracy, we will be using four metrics: accuracy, precision, recall, and F1-score.

$$\text{Accuracy} = \frac{TP + TN}{TP + TN + FP + FN} \tag{1}$$

$$\text{Precision} = \frac{TP}{TP + FP} \tag{2}$$

$$\text{Recall} = \frac{TP}{TP + FN} \tag{3}$$

$$\text{F}_1 \text{ - score} = 2\frac{\text{Precision} \cdot \text{Recall}}{\text{Precision} + \text{Recall}} \tag{4}$$

In the above equations, TP is the true positive count, FP is the false positive count, TN is the true negative count and FN is the false negative count.

After the sentiment analysis, we need to merge the sentiment analysis with the stock data. This happens in several steps:

1. We get the daily stock data for a given period for a given stock using the yfinance library.
2. We mark each news row for the downloaded stock symbol with its sentiment analysis data – the positive, neutral, and negative coefficients.
3. We group the news data for each dataset by date by extracting the mean sentiment coefficients. After this operation, we get 4 columns – the date and 3*3 columns for mean sentiment for each dataset.
4. We merge the stock data frame with the sentiment analysis one on the date column while giving 0 for negative and positive sentiments and 1 for neutral sentiment if there is no news for that day.

After merging the data, we are ready for making predictions about the next day's price. For this, we will use RNNs, although approaches based on the use of machine learning algorithms are also worth highlighting [23, 24]. They are a special case of neural networks, where data is passed sequentially, preserving a hidden state that sums up all the previous inputs. There are two implementations we will be using: Long-Short Term Memory (LSTM) [25, 26] and Gated Recurrent Unit (GRU). It is also worth highlighting the idea of taking into account fractal dimensionality when analyzing time series [27].

Before passing the output to a neural network, we use batch normalization. It is used to make neural network training faster and more stable.

$$y = \frac{x - E[x]}{\sqrt{Var[x] + \varepsilon}} \cdot \gamma + \beta \tag{5}$$

To compare the accuracy we used mean squared error (MSE), mean absolute error (MAE), R2, and Explained Variance Score (EVS).

3 Modeling and Results

3.1 Environment Setup

We utilized the transformers library from HuggingFace for using pre-trained transformer language models. Internally it uses PyTorch as a processing backend for neural networks, tensor operations, and gradient computations. We used PyTorch as well for the RNNs with Scikit-learn for model evaluation.

3.2 Sentiment Analysis

We used the news sentiment analysis dataset. We have split the dataset into training and testing with an 80:20 ratio, which means 80% of the dataset is used for training and 20% for the test dataset. We used the pre-trained transformer model and a random forest with a different number of trees. For each approach, the accuracy scores were computed on the test dataset. Table 1 shows the results. And Fig. 3 shows the precision-recall curves.

Table 1. Accuracy evaluation for sentiment analysis

Model	F_1-score	Accuracy	Precision	Recall
Transformer	0.8952	0.8760	0.9153	0.8760
RF with 50 trees	0.7130	0.7525	0.7525	0.6776

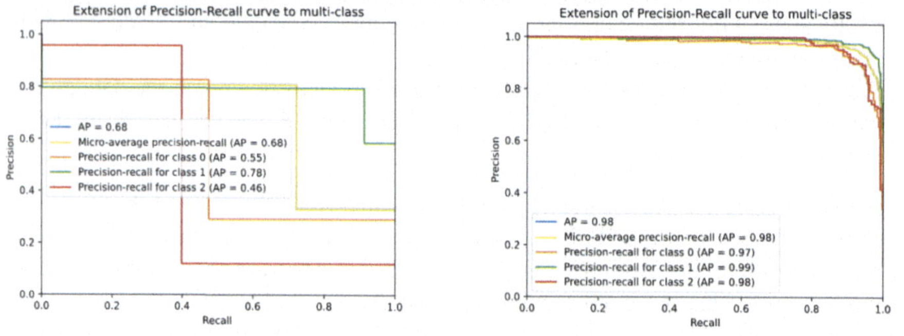

a) **b)**

Fig. 3. Precision-Recall curves for **a)** RF with 50 trees **b)** Transformer

As we can see, RFs are good for saying if the news is neutral, since the F1-score is high, but show significantly worse results in all other cases.

3.3 Stock Market Prediction

For stock market prediction, we used the same 80:20 ratio for the train test split. We used a multi-layer RNN with a feed-forward neural network connected to its latest output. The neural network had the following architecture:

1. A linear layer with the RNN's dimension as input and half of it as output.
2. A linear layer with input as the previous output and one output (the normalized next-day price).

Fig. 4. The selected period is used for learning. The blue line will be used as the training dataset, while the blue will be used as the testing dataset

We experimented with: the learning rate; the recurrent layer (LSTM or GRU); the sequence length based on which we will predict; the number of layers in the RNN; the RNN's dimension; the dropout probability in the RNN; including or not including the batch normalization layers.

As for the data, we used the symbol that has enough news from Dataset-IV and has some linear correlation with the news from the other two datasets, which is MS (Morgan Stanley). And since we used only open-access datasets, we could not use paid APIs, and these datasets, unfortunately, do not have news data for each day. So we picked a time frame that had news nearly for every day (which will be the case if we use paid APIs). The stock price dynamics can be seen in Fig. 4.

During experiments with the usage of different stock variables, we found that adding the "volume" variable makes the loss larger while adding or removing other values (except for "close", it's the most important one) didn't affect the loss. So we decided to use the variables "close", "open", "high" and "low" for prediction.

Some hyperparameters were found to work the best because making them larger or smaller resulted in bigger losses. These are:

- The best learning rate was found at 10^{-3}.

- To prevent the model from learning accidental patterns we used dropout 0.2.
- Also, batch size 32 showed the best results in loss and loss curve smoothness.
- A sequence length of 5 days led to the best results with both GRU and LSTM.

We trained our RNN on the merged dataset, which includes:

- Close, open, high, and low values from Dataset-V.
- Three sentiments for Dataset-II.
- Three sentiments for Dataset-III.
- Three sentiments for Dataset-IV.

The best results are illustrated in Table 2, including 3 GRU and 2 LSTM models.

Table 2. Accuracy evaluation for regression with RNNs with news

RNN layer	RNN hidden dimension	Hidden layers	MSE	MAE	EVS	R2
GRU	2048	3	0.001039	0.025514	0.883712	0.841792
GRU	2500	2	0.001840	0.036463	0.891232	0.715942
GRU	2500	3	0.001019	0.025337	0.875544	0.846217
LSTM	2048	2	0.002477	0.042296	0.689957	0.616769
LSTM	512	3	0.000925	0.024753	0.861745	0.857624

We trained our RNN on the merged dataset, which includes: close, open, high, and low variables from Dataset-V.

3.4 Discussion

This research aims to find a correlation between news sentiments and stock market dynamics using transformers and RNNs. We gathered news from different datasets, that overlap from 2012 to 2018. There was one dataset, that specialized on concrete companies (stock symbols). So we chose one that was nearly enough news and picked it within a time range that was best for our purposes.

After testing sentiment analysis models based on random forests and transformers, we concluded that transformers work much better, so used them for further research. When the sentiment analysis model is trained, we used it to "mark" the selected news from datasets II–IV, by adding three columns – positive, neutral, and negative sentiment. Then we grouped those "marked" datasets by computing the mean of each sentiment when there are multiple rows or setting 1 for neutral when there is no news and merged that data with the daily market data (Dataset-V). The resulting dataset was a daily dataset with 4 stock-related columns and 9 news-related (positive, neutral, and negative sentiment for each news dataset).

When ready to merge the data, we then fed that data to either GRU or LSTM RNN and connected a feed-forward neural network to it with output dimension 1, which is the next-day closing stock price. In general, GRU networks showed better results. We tried different hidden sizes and numbers of hidden layers. GRUs with a hidden size of 2000+ and 2–3 hidden layers showed the best results when used with news. Neural networks without using news tend to have better results with large model dimensions.

4 Summary and Conclusion

In conclusion, we can say that using news in general helps make more accurate predictions about how the stock price will behave, doing better in predicting steeper curves. GRU networks showed mainly the best results while being also more computation-efficient than LSTM. Moreover, by taking into account that the sentiment analysis model has a loss as well, we found out that its performance impacts the final result a lot, which is probably the reason these models do the same as the model without news for easier guesses. When nothing changes too rapidly, they are either no news or there are mostly neutral ones, while before and after large falls, there are a lot of news with unambiguous titles (the SA model easily predicts if it's positive or negative). In further work, we will try transformers either as sequence-to-sequence or many-to-one and RNNs as sequence-to-sequence models for price prediction. These models would then be used for price prediction for multiple days and compared with the models from this article. Moreover, we will try to better group and divide the stock and news data so that when there is no news, we can be sure the model will work the same (or better because of memory) as the one not using news at all.

References

1. Shanto, S.S., et al.: Binary vs. multiclass sentiment classification for bangla e-commerce product reviews: a comparative analysis of machine learning models. IJIEEB **15**(6), 48–63 (2023). https://doi.org/10.5815/ijieeb.2023.06.04
2. Ko, C.R., Chang, H.T.: LSTM-based sentiment analysis for stock price forecast. PeerJ. Comput. Sci. **7**, 1–23 (2021). https://doi.org/10.7717/peerj-cs.408
3. Tran, T., et al.: NEU-stock: stock market prediction based on financial news, vol. 3026, pp. 217–224 (2021)
4. Zhang, D., Cai, E.: Improving stock price forecasting using a large volume of news headline text. Comput. Mater. Continua **69**, 3931–3943 (2021). https://doi.org/10.32604/cmc.2021.012302
5. Kaushal, P., et al.: Myers-briggs personality prediction and sentiment analysis of twitter using machine learning classifiers and BERT. IJITCS **13**(6), 48–60 (2021). https://doi.org/10.5815/ijitcs.2021.06.04
6. Mehta, P., et al.: Harvesting social media sentiment analysis to enhance stock market prediction using deep learning. PeerJ. Comput. Sci. **7**, 1–21 (2021). https://doi.org/10.7717/peerj-cs.476
7. Zulqarnain, M., et al.: Predicting financial prices of stock market using recurrent convolutional neural networks. IJISA **12**(6), 21–32 (2020). https://doi.org/10.5815/ijisa.2020.06.02
8. Lien Minh, D., et al.: Deep learning approach for short-term stock trends prediction based on two-stream gated recurrent unit network. IEEE Access **6**, 55392–55404 (2018). https://doi.org/10.1109/ACCESS.2018.2868970

9. Shankar, P., et al.: Analyzing varied approaches for forecast of stock prices by combining news mining and time series analysis, pp. 434–441 (2019)
10. Sarode, S., et al.: Stock price prediction using machine learning techniques, pp. 177–181 (2019). https://doi.org/10.1109/ISS1.2019.8907958
11. Jammalamadaka, S., et al.: Predicting a stock portfolio with the multivariate bayesian structural time series model: do news or emotions matter? Int. J. Artif. Intell. **17**, 81–104 (2019)
12. Sentiment Analysis for Financial News. https://www.kaggle.com/datasets/ankurzing/sentiment-analysis-for-financial-news
13. Daily News for Stock Market - dataset by finance. https://data.world/finance/daily-news-for-stock-market
14. News Category Dataset. https://www.kaggle.com/datasets/rmisra/news-category-dataset
15. Daily Financial News for 6000+ Stocks. https://www.kaggle.com/datasets/miguelaenlle/massive-stock-news-analysis-db-for-nlpbacktests
16. APIs - Yahoo Developer Network. https://developer.yahoo.com/api/
17. Shakhovska, K., et al.: The sentiment analysis model of services providers' feedback. Electronics **2020**, 9 (1922). https://doi.org/10.3390/electronics9111922
18. Sanh, V., et al.: DistilBERT, a distilled version of BERT: smaller, faster, cheaper and lighter. arXiv:1910.01108[cs] (2020). https://doi.org/10.48550/arXiv.1910.01108
19. Vaswani, A., et al.: Attention is all you need. arXiv:1706.03762[cs] (2017). https://doi.org/10.48550/arXiv.1706.03762
20. Sun, C., et al.: How to fine-tune BERT for text classification? arXiv:1905.05583[cs] (2020). https://doi.org/10.48550/arXiv.1905.05583
21. Solé, X., et al.: Evaluation of random forests on large-scale classification problems using a bag-of-visual-words representation, vol. 269, pp. 273–276 (2014). https://doi.org/10.3233/978-1-61499-452-7-273
22. Bouaziz, A., Dartigues-Pallez, C., da Costa Pereira, C., Precioso, F., Lloret, P.: Short text classification using semantic random forest. In: Bellatreche, L., Mohania, M.K. (eds.) Data Warehousing and Knowledge Discovery. DaWaK 2014. Lecture Notes in Computer Science, vol. 8646, pp. 288–299. Springer, Cham (2014). https://doi.org/10.1007/978-3-319-10160-6_26
23. Izonin, I., et al.: Machine learning for predicting energy efficiency of buildings: a small data approach. Procedia Comput. Sci. **231**, 72–77 (2024)
24. Izonin, I., et al.: SGD-Based cascade scheme for higher degrees wiener polynomial approximation of large biomedical datasets. Mach. Learn. Knowl. Extr. **4**(4), 1088–1106 (2022)
25. Semkovych, V., Shymanskyi, V.: Combining OCR methods to improve handwritten text recognition with low system technical requirements. In: Hu, Z., Wang, Y., He, M. (eds.) Advances in Intelligent Systems, Computer Science and Digital Economics IV. CSDEIS 2022. Lecture Notes on Data Engineering and Communications Technologies, vol. 158, pp. 693–702. Springer, Cham (2023). https://doi.org/10.1007/978-3-031-24475-9_56
26. Shymanskyi, V., Sokolovskyy, I., Sokolovskyy, Y., Bubnyak, T.: Variational method for solving the time-fractal heat conduction problem in the claydite-block construction. In: Hu, Z., Dychka, I., Petoukhov, S., He, M. (eds.) Advances in Computer Science for Engineering and Education. ICCSEEA 2022. Lecture Notes on Data Engineering and Communications Technologies, vol. 134, pp. 97–106. Springer, Cham (2022). https://doi.org/10.1007/978-3-031-04812-8_9
27. Kaminsky, R., et al.: Calculation of the exact value of the fractal dimension in the time series for the box-counting method. In: ACSIT 2019, pp. 248–251. Ceske Budejovice, Czech Republic, 5–7 June 2019

Web-Based Platform for Real-Time Data Monitoring Visualization: Analysis and Perspectives

Nataliya Bernatska[1], Elvira Dzhumelia[2], Orest Kochan[3,4](✉), Ivan Salamon[5], and Iryna Typilo[1]

[1] Department of Physical, Analytical and General Chemistry, Institute of Chemistry and Chemical Technologies, Lviv Polytechnic National University, Lviv 79000, Ukraine
[2] Software Department, Institute of Computer Science and Information Technologies, Lviv Polytechnic National University, Lviv 79000, Ukraine
[3] Department of Information-Measuring Technologies, Institute of Computer Technologies, Automation and Metrology, Lviv Polytechnic National University, Lviv 79000, Ukraine
orest.v.kochan@lpnu.ua
[4] School of Computer Science, Hubei University of Technology, Wuhan 430068, China
[5] Department of Ecology, Faculty of Humanities and Natural Sciences, University of Presov, Presov 080 01, Slovakia

Abstract. This work focuses on the development of a web application for real-time monitoring of air quality and weather conditions. Using cutting-edge technologies such as Next.js, Tailwind CSS, Shadcn, React Map GL, Google Charts, and D3.js, a tool has been created that allows users to get up-to-date information in a convenient way. The purpose of this work is to develop a web application that will provide real-time visualization of monitoring data. The application provides the ability to display data on the interactive map and interaction with them. The results of the study highlight the potential of web technologies in solving current problems and the importance of a scientific approach to the development of web applications. The developed tool can serve as a basis for further research in this area and contribute to increasing interest in the use of web technologies in monitoring and ecology.

Keywords: Monitoring · Web Application · Data Visualization · Real-time · Ecology

1 Instruction

In today's world, data is the key to success. Businesses use data to make better decisions, streamline their operations, and gain a competitive advantage.

Governments use data to improve service delivery, increase efficiency, and keep their citizens safe. People use data to improve their health, financial situation, and standard of living. One of the most important aspects of data is its relevance. Real-time data enables you to track events and changes as they occur, giving you the opportunity to respond and

Z. Hu et al. (Eds.): ICCSEEA 2024, LNDECT 242, pp. 271–283, 2025.
https://doi.org/10.1007/978-3-031-84228-3_23

take action immediately. Web applications are the perfect way to access data in real-time, as they can be accessed from anywhere and at any time.

The purpose of this work is to develop a web application that will provide real-time visualization of monitoring data.

This app is designed to analyse and study data trends coming from monitoring systems, allowing users to obtain up-to-date information and make quick and informed decisions. The main goal is to facilitate the process of interacting with monitoring data and providing users with the tools to effectively analyse and use this data in real time.

Web-based applications for displaying real-time eco-monitoring data are powerful tools that can help us better understand and protect the environment. These applications are becoming more accessible and easy to use, which significantly improves access to up-to-date information on the ecological state of the environment.

The web application we have developed can:

- Display the level of air pollution in real time.
- Display data on current weather conditions.
- Display an interactive map to visualize data on air pollution, temperature, humidity, and other indicators.

2 Literary Review

Air pollution is one of the most pressing problems of our time, especially in large cities. According to the World Health Organization, about 7 million people die each year from air pollution [1]. Air quality in large cities is a significant problem that poses serious risks to human health and the environment. Air quality monitoring plays a key role in solving this problem, as it allows you to [2, 3]:

1. Assess pollution levels:
    - Measurement of the concentration of harmful impurities such as nitrogen oxides, dust, carbon monoxide, sulphur dioxide and others.
    - Identification of areas with high levels of pollution and populations that are most at risk.
    - Comparison of data with normative values and tracking of pollution dynamics.
2. Identify sources of pollution:
    - Identification of the main sources of emissions, such as transport, industry, energy, heating of residential buildings.
    - Determination of their contribution to total air pollution.
    - Development and implementation of targeted measures to reduce emissions.
3. Predict changes in air quality:
    - Modelling the impact of meteorological conditions, seasonal factors and other parameters.
    - Creating short- and long-term forecasts for taking preventive measures.
    - Warning the public about episodes of high pollution.
4. Assess health impacts:

- Study of the relationship between air pollution and morbidity, mortality.
- Identification of populations that are most vulnerable to the effects of pollution.
- Development and implementation of health care programs to reduce risks.

5. Raise awareness:

- Informing the public about the level of air pollution and health risks.
- Encouraging people to make behavioural changes to reduce emissions.
- Promoting the development of environmental culture and responsibility.

Previous research in this area highlights the importance of rapid access to and visualization of data for real-time decision-making. However, many existing web platforms have limited functionality or complex user interfaces, making them difficult to use effectively [4–8].

An important aspect of web-based data visualization platforms is their user interface. Research in this area demonstrates that clarity, simplicity, and intuitive clarity of the interface are important for a successful user experience [1, 8–12].

Real-time data visualization is used in various fields, such as finance, ecology, medicine, and others. Successful implementations in these areas show the importance of this approach in improving the productivity and efficiency of monitoring systems [13–15].

Despite advances in this area, there are challenges, such as speed, scalability, and data security, that require further research and development. Prospects for development include the use of machine learning to improve analytics and expand data visualization capabilities [4, 16, 17].

It is important that air quality monitoring is carried out on an ongoing basis using modern methods and tools, and that monitoring data is available to all stakeholders. It is also important that the public is also involved in air quality monitoring [18, 19].

2.1 Web Visualization of Monitoring Data

Web data visualization is the process of presenting data in a visual form on web pages. It is a powerful tool that can help users to [20–24]:

- Understand complex data: Data visualization can make abstract and complex data more understandable and accessible to people with different levels of technical proficiency.
- Identify trends and patterns: Data visualization can help users see patterns and trends in the data that might otherwise go unnoticed by simply looking at tables or figures.
- Web-based data visualization is a versatile tool that can be used in any industry where data needs to be collected and analysed [25, 26].

Web-based visualization of monitoring data is essential for several reasons [27, 28]:

1. Data Comprehensibility:
   Visualizing data makes it more understandable to people than just text or tables. This allows people to quickly and easily see patterns and trends in the data.
   Data visualization can help people make better decisions based on the data they receive.

2. Data Availability:
   Web-based data visualization makes monitoring data accessible to people from anywhere, at any time. It allows people to track data in real-time.
3. Data Efficiency:
   Visualizing a large amount of data can help people find the information they need faster, help them understand them better.

Types of Web Visualization of Monitoring Data.
There are many different types of web data visualization, each with its own advantages and disadvantages [28, 30]. Some of the most common types of data visualization are [31, 32]:

1. Charts: Used to represent data in the form of lines, bars, circles, and other geometric shapes.
2. Graphs: Used to represent data as points connected by lines.
3. Maps: Maps are used to visualize geospatial data.
4. Infographics are visual representations of data that combine text, images, and data visualizations to explain complex information.

Benefits of Web Visualization of Monitoring Data.
Web data visualization has many advantages, such as [29]:

1. Increased accessibility: Data visualization can make data more accessible to people with different levels of technical expertise.
2. Improved comprehension: Data visualization can help users better understand complex data.
3. Effective communication: Data visualization can be used to effectively communicate analysis results and conclusions with other people.
4. Problem detection: Data visualization can help users identify problems and opportunities in the data.
5. Decision Support: Data visualization can help users make more informed decisions.

2.2 Overview of Data Visualization Tools

There are many tools for web data visualization, both paid and free [32, 33]. Some of the most common tools (Fig. 1) are:

1. Tableau is a data analytics and business intelligence (BI) platform that helps people visualize and understand data. It is used by a wide range of people, from business representatives to data analysts and academics [34].
2. Power BI is Microsoft's cloud-based business intelligence (BI) platform that helps transform data from various sources into interactive visual and informative dashboards [35].

   Power BI features:

- Power BI can connect to a wide range of data sources, including databases, Excel files, websites, and cloud services.
- Power BI can clean and transform data to prepare it for analysis.
- Power BI offers a wide range of data visualizations, such as charts, graphs, and maps.
- Power BI can be used to analyse data to find trends and patterns.
- Power BI can be used to generate reports and presentations to share analysis results with others.
- Power BI allows you to collaborate with other users on analytics projects.

3. Google Charts is a data visualization platform that allows you to create interactive and dynamic charts, graphs, and other visual representations of data [36]. It uses HTML5 and SVG to ensure cross-browser compatibility and mobile optimization.

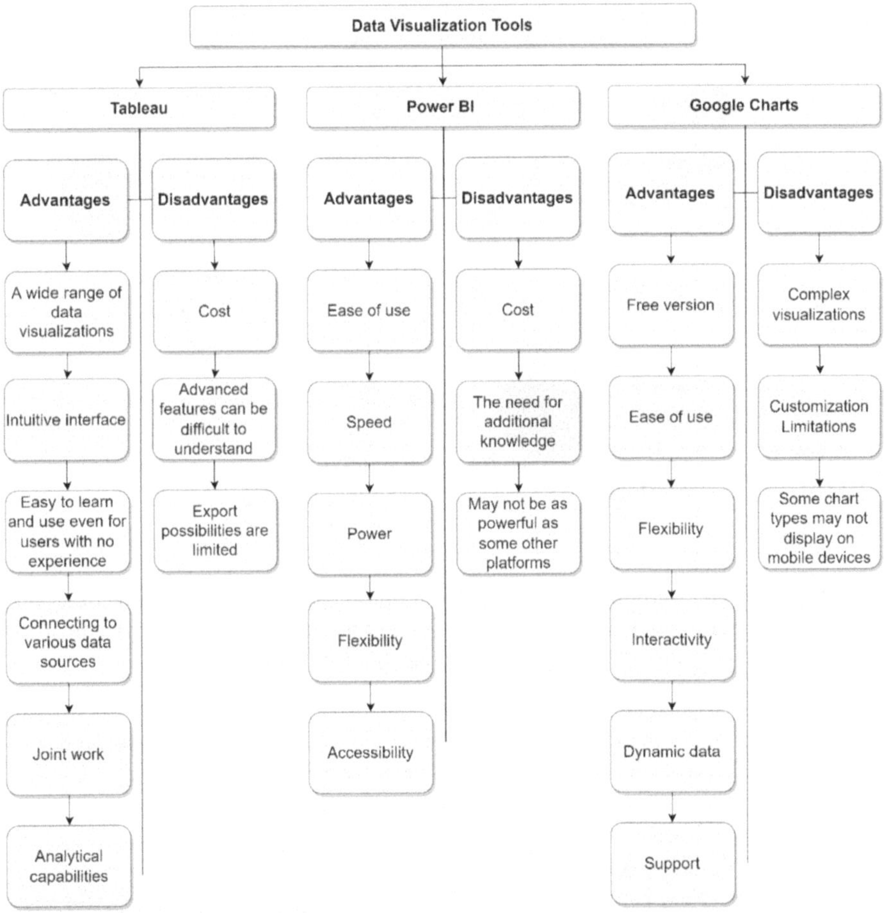

Fig. 1. Data visualization tools

3 Materials and Methods

A web-based data visualization application has been developed that combines state-of-the-art technology to monitor air quality in real-time.

For the development of the web application, the following technologies were used (Fig. 2):

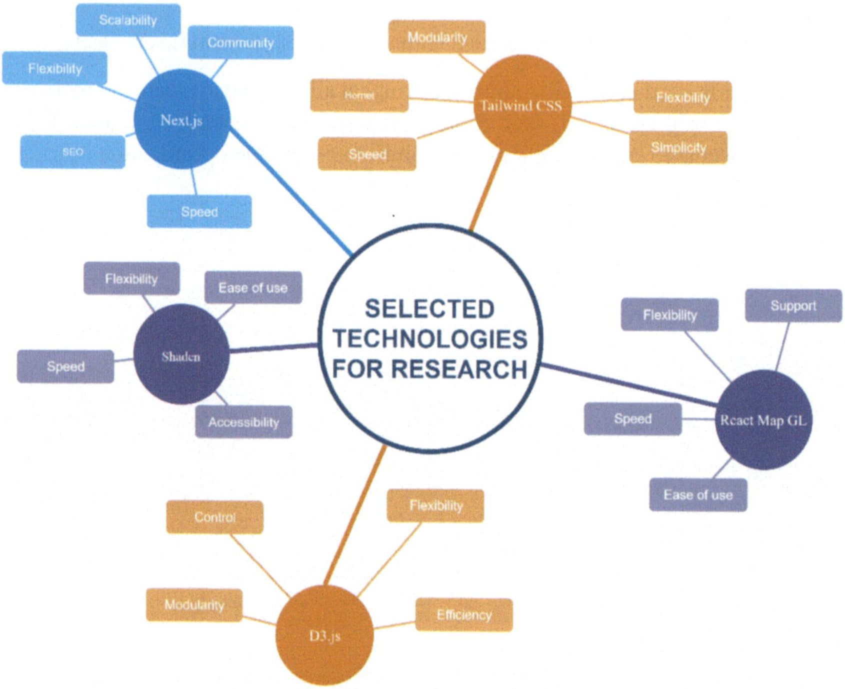

Fig. 2. Selected technologies for research

1. Next.js is a React.js library framework that is optimized for server-side rendering and static site generation. This makes it a convenient tool for creating web visualization of monitoring data in real time, because it has the following advantages:
   - Speed: Next.js generates static HTML pages that load instantly. Server-side rendering makes the initial page load faster, even for complex web applications.
   - SEO: Static pages are easily indexed by search engines. Next.js makes it possible to optimize metadata and content for SEO.
   - Scalability: Next.js can easily scale to handle large amounts of traffic. It can be used with cloud platforms such as AWS or Vercel.
   - Flexibility: Next.js makes it possible to use any React.js library for data visualization.
   - Community: Next.js has a large and active community that can help you with any questions you may have.

2. Tailwind CSS is a CSS Framework for Styling Web Pages.
   Benefits of Tailwind CSS:

- Speed. Tailwind CSS can help you quickly create web pages with beautiful designs.
- Simplicity. Tailwind CSS is easy to use, even if you're not a CSS expert.
- Flexibility. Tailwind CSS gives you the flexibility to create any design you can imagine.
- Modularity. Tailwind CSS takes a modular approach, which makes the code more organized.
- Active Community. Tailwind CSS has an active community that can help you if problems arise.

3. Shadcn is a React UI library used to create web applications with sleek and modern designs. It offers a wide range of off-the-shelf components such as buttons, cards, modals, forms, and much more.
   There are a number of advantages to using Shadcn:

- Ease of Use. Shadcn components are easy to use and customize.
- Flexibility: Shadcn makes it possible to create various user interfaces, from simple to complex.
- Speed. Shadcn is optimized for high performance.
- Accessibility. Shadcn is available for free and is open-source.

4. React Map GL is a React library for integration with Mapbox maps, used to visualize geospatial data on maps. It uses WebGL to visualize data, which makes it very fast and efficient.
   Benefits of using the React Map GL library:

- Speed. React Map GL uses WebGL to visualize data, making it very fast and efficient.
- Ease of Use. React Map GL is easy to use and customize.
- Flexibility. React Map GL gives you the ability to create different types of maps, from simple to complex.
- Support. React Map GL has an active community and documentation.

5. D3.js is a JavaScript library used to create dynamic and interactive data visualizations in web browsers. It uses HTML, SVG, and CSS standards, making visualizations accessible and cross-browser compatible.
   Advantages of D3.js:

• Flexibility. D3.js gives you the flexibility to create any kind of visualization.
• Control. You have complete control over every aspect of your visualization, from data to animation.
• Efficiency. D3.js uses modern web technologies, making visualizations fast and efficient.
• Modularity. The D3.js library is made up of modules, making its code more structured and easy to learn.

Disadvantages of D3.js:

- Difficulty. The D3.js library can be challenging to learn and use, compared to some other data visualization libraries.
- Lack of pre-made templates. D3.js doesn't have pre-made templates like some other libraries, so you'll need to write code from scratch to create a visualization.

4 Results and Discussion

Here are some examples of what you can do with React Map GL:

- Visualize points on the map.
- Visualize the lines on the map.
- Visualize polygons on the map.
- Visualize raster data on a map.
- Add markers to the map.
- Add information windows to the map.
- Create interactive maps.

The main functions of the real-time air quality monitoring web application are:

1. **Air Quality Data Display:**
   The web app visualizes air quality data. The user has the opportunity to select the desired location and get information about the level of air pollution (Fig. 3).
2. Display data on current weather conditions: the web application provides users with

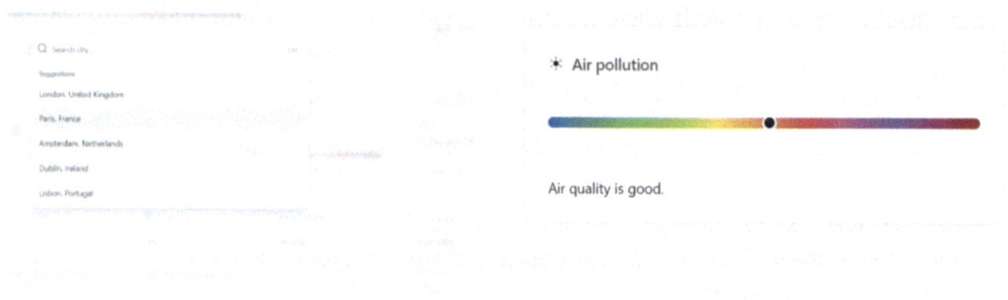

Fig. 3. Screenshots from the real-time air quality monitoring web application

   the ability to choose a location to access weather conditions, including temperature, precipitation, wind speed, UV index, humidity, visibility (Fig. 4).
3. Interactive map: visualizes data on air pollution levels, temperature (°C), precipitation intensity (mm/s), wind speed and direction (m/s), relative humidity (%), cloudiness (%) and atmospheric pressure (hPa) (Fig. 5).

Advantages of using the web application developed by us for visualization of monitoring data:

1. The proposed web application can receive real-time data from the following sources:
   - Sensors. The web app can use data from air quality sensors located throughout the city. These sensors measure the concentration of various pollutants such as carbon monoxide, nitrogen dioxide, among others.

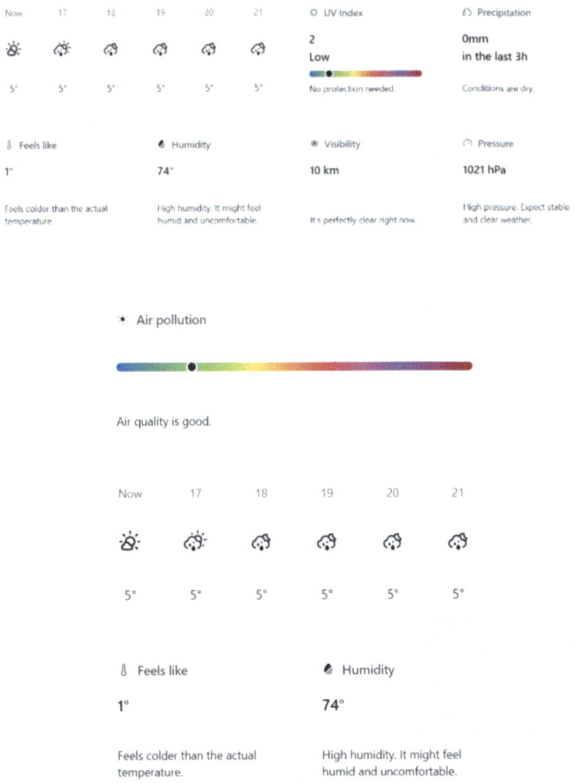

Fig. 4. Screenshots from the web application for monitoring current weather conditions: temperature, UV index, humidity, etc.

- Open Data. Some governments and organizations publish air quality data in the public domain. A web application can use this data to visualize the level of air pollution.

2. The web application can visualize the obtained air quality and weather data on the map. Colours or indicators on the map can show the level of air pollution and changes in weather conditions.
3. The web application is responsive, i.e. it is convenient to use on mobile devices (Fig. 6).

The developed web application allows users to obtain up-to-date data on air quality and current weather conditions, as well as visualize this data on an interactive map. Through the use of various technologies, such as React Map GL, Google Charts, and D3.js, users are able to analyse and perceive information effectively.

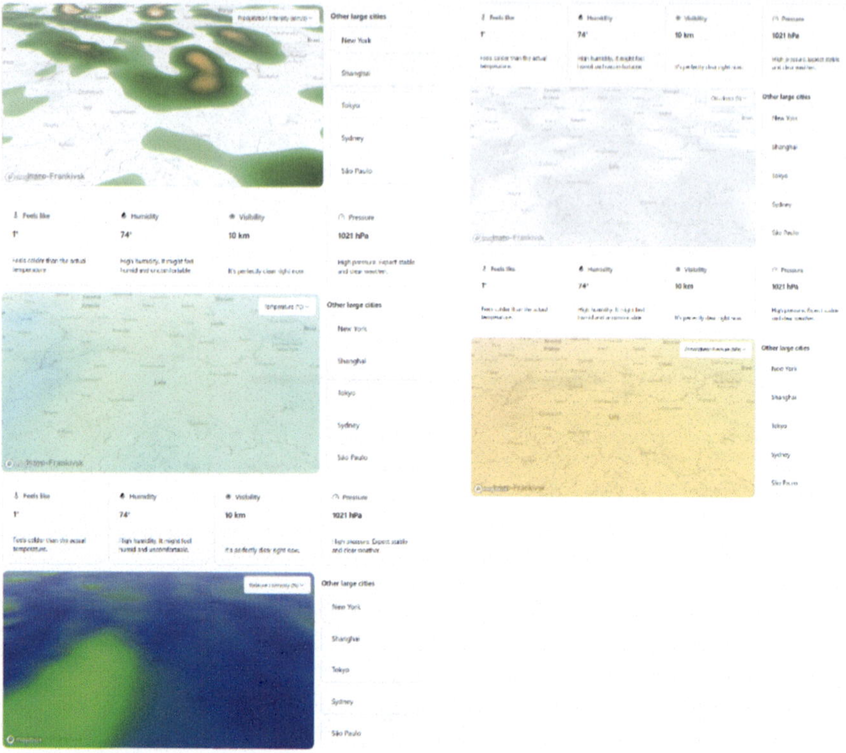

Fig. 5. Screenshots of the map from the web application for visualization of data on air quality and weather conditions

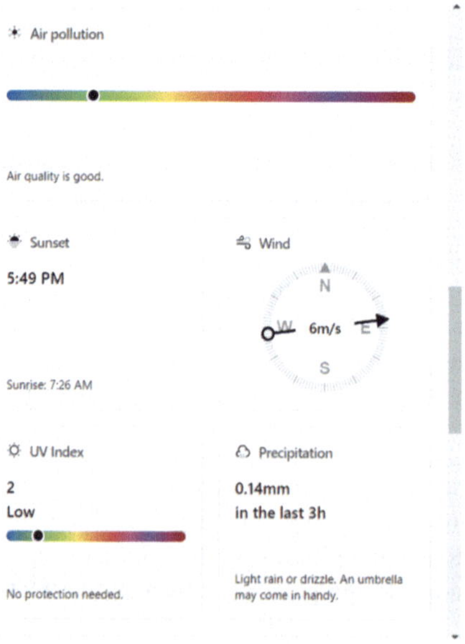

Fig. 6. Screenshot of the mobile version of the proposed web application

5 Conclusions

The developed web application is the result of a comprehensive analysis and practical application of advanced technologies in this area. It has the following features: display the level of air pollution in real time; display data on current weather conditions; display an interactive map to visualize data on air pollution, temperature, humidity, and other indicators.

The use of cutting-edge technologies such as Next.js, Tailwind CSS, Shadcn, React Map GL, Google Charts, and D3.js emphasizes a scientific approach to data development and visualization. These technologies have already gained recognition in the field of web development and data visualization and are actively used in scientific research.

A web application for displaying real-time eco-monitoring data has been found to be a powerful tool that can help people, governments, businesses, and scientists in combating the problem of air pollution.

The results of our work demonstrate the potential of using web technologies to solve current problems, including monitoring air quality and current weather conditions. The integration of scientific UI design principles and interactive capabilities allows for easy access to important information for a wide range of users.

Acknowledgment. This paper is supported by the National Research Foundation of Ukraine, project number 0123U103529 (2022.01/0009) "Assessing and forecasting threats to the reconstruction and sustainable operation of objects of critical infrastructure" from the contest "Science for reconstruction of Ukraine in the war and post war periods".

References

1. Díaz, J., Mura, I., Franco, J., Akhavan-Tabatabaei, R.: AiRe - A web-based R application for simple, accessible and repeatable analysis of urban air quality data. Environ Model Softw. **138**, 26–32 (2021)
2. Madruga, D.: Importance of air quality networks in controlling exposure to air pollution. Environ. Emissions, 15–19 (2020)
3. Abu, E.-M., Soliman, G., Morsy, M.: Environmental hazard assessment and monitoring for air pollution using machine learning and remote sensing. Int. J. Environ. Sci. Technol. **20**, 6103–6116 (2023)
4. Sokhi, R.: Advances in air quality research – current and emerging challenges. Atmos. Chem. Phys. **22**, 4615–4703 (2022)
5. Hu, Z., Ivashchenko, M., Lyushenko, L., Klyushnyk, D.: Artificial neural network training criterion formulation using error continuous domain. Int. J. Mod. Educ. Comput. Sci. (IJMECS) **13**(3), 13–22 (2021). https://doi.org/10.5815/ijmecs.2021.03.02
6. Hu, Z., Tereikovskyi, I., Chernyshev, D., Tereikovska, L., Tereikovskyi, O., Wang, D.: Procedure for processing biometric parameters based on wavelet transformations. Int. J. Mod. Educ. Comput. Sci. (IJMECS) **13**(2), 11–22 (2021). https://doi.org/10.5815/ijmecs.2021.02.02
7. Hu, Z., Mashtalir, S.V., Tyshchenko, O.K., Stolbovyi, M.I.: Video shots' matching via various length of multidimensional time sequences. Int. J. Intell. Syst. Appl. (IJISA) **9**(11), 10–16 (2017). https://doi.org/10.5815/ijisa.2017.11.02
8. Diviacco, P.: Monitoring air quality in urban areas using a vehicle sensor network (VSN). Crowdsensing Paradigm Remote Sens. **14**, 1–18 (2022)

9. Shu, C., Ruda, M., Dzhumelia, E., Shybanova, A., Kochan, O., Levkiv, M.: Restoring soil cover and plant communities with arbuscular mycorrhizal fungi as an essential component of DSS for environmental safety management in post-industrial landscapes. Agronomy **13**(5), 1346 (2023). https://doi.org/10.3390/agronomy13051346
10. Hu Z.., Bodyanskiy Ye. V., Kulishova N. Ye., Tyshchenko O. K.: A multidimensional extended neo-fuzzy neuron for facial expression recognition. Int. J. Intell. Syst. Appl. (IJISA) **9**(9), 29–36 (2017). https://doi.org/10.5815/ijisa.2017.09.04
11. Chen, X., Przystupa, K., Ye, Z., et al.: Forecasting short-term electric load using extreme learning machine with improved tree seed algorithm based on Levy flight. Eksploatacja i Niezawodność **24**(1), 153–162 (2022)
12. Pohrebennyk, V., Dzhumelia, E.: Environmental assessment of the impact of tars on the territory of the rozdil state mining and chemical enterprise "Sirka" (Ukraine) studies in systems. Decis. Control **198**, 201–214 (2020). https://doi.org/10.1007/978-3-030-11274-5_13
13. Zhengbing, H., Jotsov, V., Jun, S., Kochan, O., Mykyichuk, M., Kochan, R., Sasiuk, T.: Data science applications to improve accuracy of thermocouples. In: Proceedings of the 2016 IEEE 8th International Conference on Intelligent Systems (IS), pp. 180–188 (2016)
14. Sun, L., et al.: A hybrid feature selection framework using improved sine cosine algorithm with metaheuristic techniques. Energies **15**(10), 3485 (2022)
15. Buelvas, J., Múnera, D., Tobón, V., Aguirre, J., Gaviria, N.: Data quality in IoT-based air quality monitoring systems: a systematic mapping study. Water Air Soil Pollut. **234**, 67–75 (2023)
16. Saini, J., Dutta, M., Marques, G.: A comprehensive review on indoor air quality monitoring systems for enhanced public health. Sustain. Environ. Res. **30**(6), 98–105 (2020)
17. Bousiotis, D., Allison, G, Francis, D.: Towards comprehensive air quality management using low-cost sensors for pollution source apportionment. NPJ Clim. Atmos. Sci. **6**, 116–122 (2023)
18. Martins, C.: Air pollution and health – the importance of air monitoring and burden of disease for attaining SDGs. Eur. J. Pub. Health **32**(3), 78–92 (2022)
19. Abdulmuhsin, S.: Assessment of air quality through multiple air quality index models – a comparative study. J. Ecological Eng. **24**(4), 110–116 (2023)
20. Sun, L., Qin, H., Przystupa, K., Majka, M., Kochan, O.: Individualized short-term electric load forecasting using data-driven meta-heuristic method based on LSTM network. Sensors **22**(20), 7900 (2022)
21. Hu, Z., Gnatyuk, S., Okhrimenko, T., Tynymbayev, S., Iavich, M.: High-speed and secure PRNG for cryptographic applications. Int. J. Comput. Netw. Inf. Secur. (IJCNIS) **12**(3), 1 (2020). https://doi.org/10.5815/ijcnis.2020.03.01
22. Hu, Z., Dychka, I., Onai, M., Zhykin, Y.: Blind payment protocol for payment channel networks. Int. J. Comput. Netw. Inf. Secur. (IJCNIS) **11**(6), 22–28 (2019). https://doi.org/10.5815/ijcnis.2019.06.03
23. Mahsa, A., Kin Hing Phoa, F., Chen, C., Shmueli, G.: An interactive clustering-based visualization tool for air quality data analysis. Aerosol Air Qual. Res. **23**(12), 2–19 (2023)
24. Hicks, S.D., Damiano, S.G., Bressler, D.W., Arscott, D.B., Ensign, S.H., Aufdenkampe, A.K.: Monitor my watershed: an online data portal and visualization tool for open-source environmental monitoring. Am. Geophys. Union Fall Meet. **27**, 1–18 (2020)
25. Bernatska, N., Dzhumelia, E., Dyakiv, V., Mitryasova, O., Salamon, I.: Web-Based information and analytical monitoring system tools – online visualization and analysis of surface water quality of mining and chemical enterprises. Ecol. Eng. Environ. Technol. **24**(3), 99–108 (2023). https://doi.org/10.12912/27197050/159885
26. Triska, P., Amman, F., Endler, L., Bergthaler, A.: WAVES (Web-based tool for Analysis and Visualization of Environmental Samples) -a web application for visualization of wastewater pathogen sequencing results. Bioinformatics **39**(4), 160–167 (2023)

27. Hu, Z., Khokhlachova, Y., Sydorenko, V., Opirskyy, I.: Method for optimization of information security systems behavior under conditions of influences. Int. J. Intell. Syst. Appl. (IJISA) **9**(12), 46–58 (2017). https://doi.org/10.5815/ijisa.2017.12.05
28. Bharatkumar, P.: A survey of visualization techniques and tools for environmental data. Technol. Invest. **14**(3), 34–46 (2023)
29. Mahajan, S.: Design and development of an open-source framework for citizen-centric environmental monitoring and data analysis. Sci. Rep. **12**, 144–164 (2022)
30. Saputri, F., Lee, W.: Web-based environment monitoring system. Jurnal Teknologi Terapan G-Tech **7**(3), 807–815 (2023)
31. Kozlowski, T., Noran, O., Trevathan, J.: Designing an evaluation framework for IoT environmental monitoring systems. Procedia Comput. Sci. **219**, 220–227 (2023)
32. Bernatska, N.L., Dzhumelia, E.A., Kochan, O.V.: Development of the web interface of the information and analytical system of environmental monitoring using the REACT library. Inf. Technol. Soc. **1**(7), 6–12 (2023). (in Ukrainian)
33. Mokryi, V., Petrushka, I., Dyakiv, V., Dzhumelia, E., Salamon, I.: Information supply of hydrotechnical reconstruction concept of stebnyk tailings storage (Ukraine). Ecol. Eng. Environ. Technol. **24**(2), 120–130 (2023). https://doi.org/10.12912/27197050/156977
34. Tableau Homepage. https://www.tableau.com/
35. Power BI Homepage. https://www.microsoft.com/en-us/power-platform/products/power-bi/
36. Google Charts Homepage. https://developers.google.com/chart

Application of a Genetic Algorithm to Solve a Nonlinear Problem of Optimizing the Diagnosis of Unstable Angina

Dmytro Uhryn[1] (✉), Yuriy Ushenko[1], and Oleksii Iliuk[2]

[1] Yuriy Fedkovych Chernivtsi National University, Chernivtsi 58012, Ukraine
{d.ugryn,y.ushenko}@chnu.edu.ua
[2] Seniour IT System & Buisness Analyst, Temabit, Kyiv 02000, Ukraine
olexiyilyukm@gmail.com

Abstract. A genetic algorithm is used to solve the nonlinear problem of optimizing the diagnosis of unstable angina. To solve this problem, the article proposes the use of a genetic algorithm. First, patients are selected according to the main parameters, then the min value of the function of the initial parameters is calculated. After that, the parameters of patients with individual characteristics are mutated and the ECG is simulated. The calculated data form a database of patients who are most likely to have unstable angina. Visualized data of the genetic algorithm with individual patient characteristics.

Keywords: genetic algorithm · binary particle swarm algorithm · key ECG parameters · parameter mutation · transformation function · json · power bi · unstable angina

1 Instruction

In today's world, medical technology and research are ruthlessly advancing, using the latest methods to detect and treat heart disease. One of the innovative strategies that opens up new perspectives in the diagnosis of unstable angina is the use of genetic algorithms. Genetic algorithms have already been successfully applied in various fields, and their use in cardiology can greatly facilitate and refine the process of determining angina instability.

The ECG may be within the normal range, but the person is suffering from unstable angina. Unstable angina can occur not only during exercise, but also at rest or during stressful situations. In this article, we will discuss methods for studying this problem using particle swarm optimization algorithms and the ant algorithm. However, there are cases when it is important to take into account the individual characteristics of the patient. Therefore, this study will propose a new approach to the diagnosis of unstable angina using a genetic algorithm, taking into account the characteristics of the patient.

To evaluate the ECG parameters, you need to compare the parameters Table 1 shows the ECG parameters that are inherent in a healthy person.

The influence of parameters on the diagnosis of unstable angina is described below:

Z. Hu et al. (Eds.): ICCSEEA 2024, LNDECT 242, pp. 284–293, 2025.
https://doi.org/10.1007/978-3-031-84228-3_24

Table 1. Indicators of a healthy person in the diagnosis of unstable angina by ECG

№	Indicator	Norma
1	Heart rate	Regular
2	Heart pulse	60 strokes/min
3	Wave P	Amplitude 0.06–0.12 s, height 2.5 mm
4	PR interval	0,12–0,20 s
5	QRS complex	0,06–0,10 s
6	ST segment	Uniform, no change
7	Wave T	Normal shape and amplitude

1) ST-T segment: An elevated ST segment may indicate myocardial ischemia (insufficient blood supply). This may be a sign of unstable angina.
   ST segment depression: ST-segment depression can also be an indicator of ischemia and indicate an attack of unstable angina.
2) T-wave displacement: The displacement of the T wave may indicate myocardial ischemia.
3) Dynamics of changes. Dynamics of changes in time: In unstable angina, there may be a change from episodic or paroxysmal changes in the ECG.
4) Isoelectric base line. Instability of the base line: Changes in the isoelectric base line can also be an indicator of problems with the blood supply to the myocardium.
5) Other changes. Changes in wave amplitude: Changes in the amplitude of QRS complexes and other waves may also indicate ischemia.

However, there are deviations from the normalized indicators and they have specific characteristics. The main parameters and their characteristics are described in Table 2.

Table 2. Characteristic abnormalities that are inherent in unstable angina when diagnosed by ECG

№	Indicator	Characteristics in unstable angina
1	Heart rate	May be incorrect or change
2	Heart pulse	Can change quickly during a sore throat attack
3	Wave P	Usually normal, may be associated with QRS complexes
4	PR interval	May be normal, but may also increase
5	QRS complex	Usually normal, but may indicate ischemia
6	ST segment	Elevated or depressed during an angina attack
7	Wave T	Modified, can be inverted or shifted

Primary ECG diagnostic method – determines the main parameters that allow the doctor to get a complete picture of the heart condition and make an informed decision on treatment and further diagnosis.

1.1 Existing Solutions for Identifying Heart Disease by Using Optimization Methods

The article [1, 2] will discuss methods of providing a comprehensive systematic review of medical image registration using the particle swarm optimization method, which is a population-based stochastic optimization algorithm This method offers solutions based on the detection of indicators in the image.

The algorithm consists of a population of particles, each of which represents a potential solution to the optimization problem. At each iteration of the algorithm, each particle updates its position and velocity in space in search of the best position.

Another paper [3, 4] considers the optimization of a binary particle swarm. This algorithm is used to implement feature selection, and the method of selecting the nearest neighbor or neighbors with cross-validation, based on the calculation of Euclidean distances, serves as a function for the classification task.

This method increases either the classification accuracy or reduces the number of features to be selected for classification, or both. Also, the algorithm in this paper is used to identify heart disease based on heartbeat sound reduction. The dataset used to build and test the algorithm is presented. The proposed system for heart disease identification is also demonstrated.

1.2 Advantages and Disadvantages of Binary Optimization of Heart Disease Search Problems

In binary optimization problems, only 0 and 1 are accepted, and the speed V_i is interpreted as a probability of change from 0 to 1. Updating the position of the particles used to find a probable heart disease is performed using the following formula:

$$V_{ij}^{k+1} = wv_{ij}^{k} + c_1 rand_1\left(pbest_{ij} - x_{ij}^{k}\right) + c_2 rand_2(gbest_i - x_{ij}^{k}) \tag{1}$$

$$Sig\left(V_{ij}^{k+1}\right) = \frac{1}{1 + e^{-V_{ij}^{k+1}}} \tag{2}$$

$$x_{ij}(k+1) = \left\{ \frac{1,\ if rand_n < Sig\left(V_{ig}^{k+1}\right)1}{0,\ if rand_n \geq Sig\left(V_{ig}^{k+1}\right)} \right., \tag{3}$$

where 1 means that this feature is selected as mandatory for the next update. And 0 means that this feature is not selected for the next update, $rand_n$ – a random variable from the range [0;1]. Disease signs were measured using one nearest neighbor [5]. Neighbors were calculated based on their distance, represented by the formula:

$$D(x, y) = \sum_{i=0}^{N} \sqrt{x_i^2 - y_i^2}, \tag{4}$$

where $D(x, y)$ – distance function, x – sample request, y – sample from the original formula. N is represented as a sign of sample size.

Since the binary algorithm is not suitable for solving nonlinear problems, it is proposed to use a genetic algorithm.

1) Representation of solutions (ECG parameters). Each individual represents a set of genetic parameters [10, 11] that can be changed for optimization. In the context of diagnosing unstable angina, these are numerical values of ECG parameters (heart rate, ST segment amplitude, etc.).
2) Evaluation function. The evaluation function determines how well each individual solves the task. In this case, you can use a function that evaluates the compliance of ECG indicators with standard indicators and indicators characteristic of unstable angina.

$$f(x) = \sum_{j=1}^{n} \left(x_{p_{i,j}} - \text{Normal}_j\right)^2, \tag{5}$$

where $x_{p_{i,j}}$ – indicator value, j for the patient P_i, Normal_j – normal value for the indicator j.
3) Selection for reproduction (the lowest rate f_x). Selection of patients for whom f_x lower the value of further reproduction. The lower the patient's score, the more chances they have to choose.

Below is Table 3 with the input data of ECG parameters when diagnosing patients to calculate the function of the genetic algorithm.

Table 3. Initial base of patients with ECG parameters

P	Heart rate	Heart pulse	Wave P)	PR interval	QRS complex	ST segment	Wave T
P 1	Regular	110	0.10	0.18	0.08	0.2 (raised)	0.1
P 2	Incorrect	98	0.11	0.19	0.07	0.2 (depression)	0.09
P 3	Regular	95	0.09	0.20	0.07	0.1 (depression)	0.08
P 4	Regular	85	0.10	0.22	0.08	0.3 (raised)	0.12
P 5	Incorrect	120	0.12	0.18	0.09	0.0	0.11
P 6	Regular	72	0.11	0.19	0.07	0.2 (depression)	0.09
P 7	Incorrect	88	0.10	0.20	0.08	0.1 (raised)	0.1
P 8	Incorrect	105	0.09	0.21	0.07	0.1 (raised)	0.1
P 9	Regular	78	0.10	0.18	0.08	0.2 (depression)	0.11
P 10	Regular	75	0.11	0.18	0.08	0.2 (raised)	0.11

The use of a genetic algorithm for the diagnosis of unstable angina may include the creation of a population [10, 11] of solutions (patients), their evaluation based on ECG parameters, and the evolution ($f(x)$) of these solutions. Below are general steps and formulas that can be applied to a genetic algorithm The main advantages of this approach are robustness to the initial data [6, 7], which cannot be sampled after the initial iterations. This makes it possible to significantly narrow the search for the desired solution. In practical applications, the disadvantage of this approach forces the algorithm to train absolutely all neighbors and their distance segments from each other. Heart disease can be represented as a classification problem of linear or nonlinear separation of input data. The goal of the task is to minimize the output error [8, 9]. Table 3 shows individual patient characteristics [12], such as: Heart rate and ST segment in three states.

2 Method for Calculating the Patient Database Sampling Function

The patient's condition is the external factor that affects the diagnosis of unstable angina. Below is the calculation of the value f_x, the algorithm will identify patients who potentially have unstable angina, taking into account their condition and data sample.

$$f(x) = \sum_{j=1}^{n} \left(x_{P_{i,j}} - \text{Normal}_j\right)^2, \tag{6}$$

where $x_{P_{i,j}}$ – indicator value j for the patient P_i, a Normal_j – normal value for the indicator j, n – is the number of indicators.

After the calculations, we will display only those patients who had at least one indicator not within the normal range. The data are presented below in Table 4.

Table 4. Parameters of ECG diagnostics of patients taking into account individual characteristics (f (x))

P	Value f(x)	Heart rate	Heart pulse	Wave P	Interval PR	QRS complex	ST segment	Wave T
P 1	225.014	Regular	110	0.10	0.18	0.08	0.2 (raised)	0.1
P 3	234.832	Regular	95	0.09	0.20	0.07	0.1 (depression)	0.08
P 4	116.316	Regular	85	0.10	0.22	0.08	0.3 (raised)	0.12
P 5	324.384	Incorrect	120	0.12	0.18	0.09	0.0	0.11
P 6	625.008592	Regular	72	0.11	0.19	0.07	0.2 (depression)	0.09
P 8	116.072	Incorrect	105	0.09	0.21	0.07	0.1 (raised)	0.1
P 10	242.324	Regular	75	0.11	0.18	0.08	0.2 (raised)	0.11

Patients 2, 7, 9 were excluded from the sample. Because their values were within the normal range. Now let's compare patients with the average and highest data, these will be patients 1 and 6:

Patient 1 has $f(x) = 225.014$, mood raised, Heart rate Regular. And patient 6 has $f(x) = 625.008$, depressed state, Regular. After receiving the initial data, the algorithm can apply crosses to produce offspring.

Genetic information is exchanged between selected individuals to create new offspring (patients with potential unstable angina). In the context of ECG diagnostics, this reflects a combination of indicators from different individuals. In order to start cross-breeding, it is necessary to perform n iterations based on a larger sample of data. Below are the calculations of the values of $f(x)$ (Table 5).

Table 5. Results of calculating the value of f(x) taking into account the individual characteristics of patients (P)

P1	P2	P3	P4	P5	P6	P7	P8	P9	P10
225.014	234.832	116.316	324.384	625.008	116.072	242.324	273.02	150.802	171.458
P11	P12	P13	P14	P15	P16	P17	P18	P19	P20
212.02	315.12	180.42	240.68	362.62	130.82	232.28	272.64	212.02	144.12
P21	P22	P23	P24	P25	P26	P27	P28	P29	P30
225.014	234.832	116.316	324.384	625.008	116.072	242.324	273.02	150.802	171.458
P31	P32	P33	P34	P35	P36	P37	P38	P39	P40
212.02	315.12	180.42	240.68	362.62	130.82	232.28	272.64	212.02	144.12
P41	P42	P43	P44	P45	P46	P47	P48	P49	P50
225.014	234.832	116.316	324.384	625.008	116.072	242.324	273.02	150.802	171.458

After analyzing the table, we define the following input data for the experiment:

1) Patient 15 has the highest f(x) value and is 625.008.
2) Patient 6 has the lowest f(x) value and is 116.072.
3) Patients with depression (ST-segment elevation): 1, 2, 3, 4, 6, 7, 8, 9, 11, 12, 13, 15, 16, 18, 21, 22, 23, 24, 26, 27, 28, 29, 31, 32, 33, 34, 36, 37, 38, 39, 41, 42, 43, 44, 46, 47, 48, 49.
4) Patients with elevated mood (T raised): 1, 4, 5, 7, 8, 10, 11, 14, 15, 17, 18, 20, 21, 23, 24, 26, 27, 28, 30, 31, 33, 34, 36, 37, 38, 40, 41, 43, 44, 46, 47, 48, 50.
5) Patients with f(x) within the mean values: 2, 3, 9, 13, 16, 19, 22, 25, 29, 35, 39, 45.

3 Mutation of Critical Parameters for Detecting Unstable Angina

In order to start the process of simulating the ECG procedure, you need to mutate the parameters, namely: Heart pulse, Wave P (sec), PR interval (sec), QRS complex (sec), ST segment (mV), Wave T (mV). And the steps of mutation are the number of ECGs performed. The formula for mutation is written as:

$$NewF(x) = f(x) + I \times \left[min_{f(x)}, max_{f(x)}\right], \quad (7)$$

where $NewF(x)$ – is the new value of the function, $f(x)$ – the current value of the function, I means iteration, $min_{f(x)}$ and $max_{f(x)}$ – range from the minimum and maximum values. This formula is used for each parameter of the diagnostic results separately:

$$NewP(x) = P(x) + I \times \left[min_{P(x)}, max_{P(x)}\right], \quad (8)$$

where $NewP(x)$ – is a new parameter, $P(x)$ – current value of the parameter, I means iteration, $min_{P(x)}$ and $max_{P(x)}$ – range from the minimum and maximum parameter.

4 Simulation of the ECG Procedure by Mutating Key Parameters

Power BI was used to load the patient database. It is most convenient to use json queries because of their advantage of structuring the format of parameters for each patient and accessibility from any cloud environment. Later, the function of converting table data created on the basis of a json query was used. The scheme of algorithm development and database creation is shown in Fig. 1.

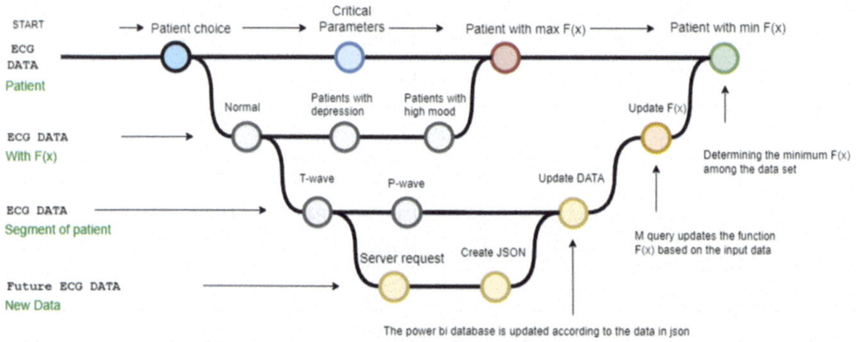

Fig. 1. The process of selecting and segmenting the database by key parameters and patient conditions

After that, links were created between the segments and the database to reflect the results of the research (Fig. 2).

The patient database is divided into three categories: calm patients, depressed patients, and patients with high spirits. There is also a general database that combines these categories.

4.1 Comparison and Results of the Experiment of Detecting the Diagnosis of Unstable Angina Using a Genetic Algorithm

Other patients had abnormal readings in depressed, calm, and elevated moods. A total of 50 patient ECG results and 250 iterations were processed. It should be noted that such a large number of potential ECG simulations were performed in a few seconds by updating the Power BI database. The ST and Wave T segments were within normal limits in 24 patients. Of these, the algorithm identified the 3rd and 19th patients as patients with unstable angina. An interesting fact is that all of these patients are depressed and have normal heartbeats, except for 23 patients. There are also patients that the algorithm noticed with the function at the initial iteration, such that they probably should have this disease, but after passing the mutation iterations, based on the patient database and other cases, the genetic algorithm eliminated these diagnoses. Patients 6, 3, 19, 26 have the

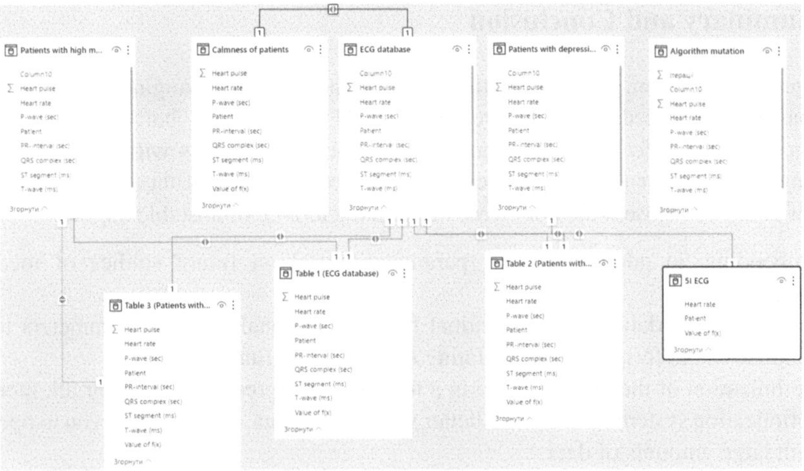

Fig. 2. Links between the database and segmented patient categories

highest probability of being diagnosed with unstable angina. Table 6 shows the indicators corresponding to the minimization of the genetic algorithm function according to the maximum sample value, which was patient 32, among the entire patient database of the three categories of individual characteristics included in the sample:

Table 6. Initial patient base with ECG parameters before and after mutations

P	Value f(x)	Value f(x) (after 5 ECG)
P 19	212.02	130.65
P 9	150.802	156.87
P 13	180.42	180.10
P 6	116.072	180.92
P 3	116.316	186.70
P 26	116.072	212.67
P 32	315.12	234.922

At the same time, patients 9,13 have no abnormalities, although the initial parameters showed signs of unstable angina.

5 Summary and Conclusion

The genetic algorithm identified patients with signs of unstable angina from the initial database and classified them into three types based on individual characteristics. Taking into account these features, the algorithm screened out patients with signs of unstable angina and, after mutation of key parameters (simulation of the ECG procedure), identified those whose indicators were most characteristic of unstable angina:

1) Embedding an additional error parameter based on future studies of medical indicators.
2) Expanding the database of individual features and analyzing the parameters to be included in the formula for calculating the algorithm function.
3) Optimization of the algorithm using a matrix data representation. After all, modern optimization systems work much better with matrix data, which allows you to operate with large amounts of data.
4) Work on the method's universality. In order for the algorithm to be applied to a wide range of software, regardless of the user base and software.

References

1. Zhang, J., Huo, P., Teng, J.: Multi-resolution medical image registration based on generalized mutual information, vol. 3, pp. 1244–1247 (2010). https://doi.org/10.1109/CISP.2010.5647264
2. Hu, Z., Uhryn, D., Ushenko, Y., Korolenko, V., Lytvyn, V., Vysotska, V.: System programming of a disease identification model based on medical images. In: Proceedings of the SPIE 12938, Sixteenth International Conference on Correlation Optics, 129380F, 5 January 2024. https://doi.org/10.1117/12.3009245. https://www.spiedigitallibrary.org/conference-proceedings-of-spie/12938/129380F/System-programming-of-a-disease-identification-model-based-on-medical/10.1117/12.3009245.full
3. Ushenko, Y.A., et al.: Differential diagnosis of the aseptic and septic loosening of the artificial hip joint endoprosthesis by methods of spectral-selective laser autofluorescence microscopy. In: Digital Information Methods of Polarization, Mueller-Matrix and Fluorescent Microscopy. SpringerBriefs in Applied Sciences and Technology, pp. 85–100. Springer, Singapore (2023). https://doi.org/10.1007/978-981-99-4735-5_6
4. Lytvyn, V., Uhryn, D., Ushenko, Y., Masikevych, A., Bairachnyi, V.: The method of clustering geoinformation data for stationary sectoral geoinformation systems using swarm intelligence methods. In: Cioboată, D.D. (eds.) International Conference on Reliable Systems Engineering (ICoRSE) - 2023. ICoRSE 2023. Lecture Notes in Networks and Systems, vol. 762, pp. 541–553. Springer, Cham (2023). https://doi.org/10.1007/978-3-031-40628-7_44
5. Uhryn, D. et al.: Modelling of an intelligent geographic information system for population migration forecasting. Int. J. Mod. Educ. Comput. Sci. (IJMECS) **15**(4), 69–79 (2023). https://doi.org/10.5815/ijmecs.2023.04.06. ISSN: 2075-0161 (Print), ISSN: 2075-017X (Online) https://www.mecs-press.org/ijmecs/ijmecs-v15-n4/v15n4-6.html
6. Lytvyn, V., Lozynska, O., Uhryn, D., Vovk, M., Ushenko, Y., Hu, Z.: Information technologies for decision support in industry-specific geographic information systems based on swarm intelligence. Mod. Educ. Comput. Sci. **2**, 62–72 (2023).https://doi.org/10.5815/ijmecs.2023.02.06

7. Tiwari, R., Pratap Singh, M.: Correlation-based attribute selection using genetic algorithm. Int. J. Comput. Appl. **4**(8), 8875–8887 (2010)
8. Jain, A.K., Mustafa, T., Zhou, Y., Burdette, C., Chirikjian, G.S., Fichtinger, G.: FTRAC—a robust fluoroscope tracking fiducial. Med. Phys. **32**(10), 3185–3198
9. Zaman, A., Ko, S.Y.: Improving the accuracy of 2D–3D registration of femur bone for bone fracture reduction robot using particle swarm optimization, pp. 101–102. Association for Computing Machinery (2018)
10. Hering, J., Wolf, I., Maier-Hein, K.H.: Multi-objective memetic search for robust motion and distortion correction in diffusion MRI. IEEE Trans. Med. Imaging **35**, 2280–2291 (2016)
11. Gad, A.G.: Particle swarm optimization algorithm and its applications: a systematic review. Arch. Comput. Methods Eng. **29**, 2531–2561 (2022)
12. Baldwin, D.S., et al.: Evidence-based pharmacological treatment of anxiety disorders, post-traumatic stress disorder and obsessive-compulsive disorder: a revision of the 2005

Optimizing Text Data Storage for Improved Natural Language Processing: A Comprehensive Analysis

Artem Kalancha(✉) and Dmytro Uhryn

Department of Computer Systems Software, Yuriy Fedkovych Chernivtsi National University, Chernivtsi, Ukraine

{kalancha.artem,d.ugryn}@chnu.edu.ua

Abstract. This research analyzed the qualitative characteristics of messages in the Telegram messenger, which are used as input data for further analysis of text content. The parameters of these messages, such as their format, size, presence of noise, velocity, etc., have been carefully considered. The main goal of the article is to identify the optimal approach to saving a large amount of data before the important stage of text analysis. In the course of the study, a detailed analysis of literature sources devoted to this topic was carried out. The article examines the main advantages and disadvantages of existing data preprocessing algorithms, as well as problems related to data purity and their impact on potential research results. As part of software experiments, the influence of data preprocessing on the saved data size for further use, as well as the input data velocity, was evaluated. Among the proposed methods, the method of saving cleared tokens in string format and saving word codes in string format together with the word-code dictionary was highlighted. This is aimed at ensuring the effective distribution of tasks of the text analysis system during the day.

Keywords: text analysis · preprocessing · database · encoding · optimization

1 Introduction

For natural language processing algorithms [1, 2], text serves as the primary information source, conveying the semantic nuances of human language. Training models involve diverse text datasets with varying styles, languages, and volumes, often accompanied by metadata like author details, date, and language. Training can be one-time or continuous.

Single-moment training typically employs an stable CSV dataset. Continuous training involves a dynamic data flow through dedicated channels, necessitating ongoing interaction with the model. The complete natural language processing cycle includes crucial steps, with preprocessing of textual data being a key aspect.

Given the constant influx of new text, additional stages like loading and storing textual information become necessary. This study offers a detailed account of information retrieval from Telegram messenger [3], outlining the method and format for storage

Z. Hu et al. (Eds.): ICCSEEA 2024, LNDECT 242, pp. 294–303, 2025.
https://doi.org/10.1007/978-3-031-84228-3_25

before the primary natural language processing stage. The research focuses on messages from Telegram news channels, a consistent source of substantial text material for analysis.

2 Article Goal

In the field of processing messages in Telegram channels, a constant increase in the volume of information is an urgent problem, which requires a careful assessment of input data for effective system design, taking into account available resources. With each new day, this volume grows, challenging us to determine the scope and our ability to adequately process all the necessary textual information. Systematic analysis and adaptation to this ever-increasing volume of data is a key element in ensuring the effectiveness of a text-based decision support system.

The purpose of the study is to develop effective strategies for optimal storage of text data for further processing using natural language processing algorithms. Within the framework of this task, a study of the impact of data storage methods to increase the efficiency of the next stages of natural language processing is carried out, taking into account the size, format and other characteristics of the texts, which are presented in the form of messages from a limited list of Telegram channels collected during a certain time interval.

3 Text Preprocessing

One of the critical stages of processing large volumes of data is preprocessing (see Fig. 1), which can consume significant resources, both software and time. This is despite the fact that the main part of the processing system will require significantly more resources. Data size analysis that goes through the preprocessing stage, as well as the speed of their processing, will allow more efficient planning of storage resources, where the necessary data will be stored for further analysis, and to optimize text preprocessing strategy to ensure the minimum time of operations.

In the context of optimizing the storage of text data for further processing by natural language processing algorithms, the key stages of preprocessing of text [4] information are considered. Among them are redundant character reduction, ionization, conversion of text to lowercase, extraction of stop words, lemmatization, and stemming.

During preliminary analysis, some texts were found to contain a significant amount of emoji and numbers. In order to improve the analysis efficiency and avoid problems with text encoding [5], it was decided to separate these symbols from further analytical processes, since they do not have a significant contribution to the content and can be a source of technical complications.

Tokenization is the process of splitting text into discrete units called tokens, which can be words, phrases, sentences, etc. Stop words, which are considered general and not informative for a specific text, are removed to focus attention on essential words, which increases the accuracy of text analysis. Lemmatization and stemming are used to reduce words to their basic forms to facilitate further analysis, with the choice between them depending on the specific project requirements and the balance between accuracy and efficiency.

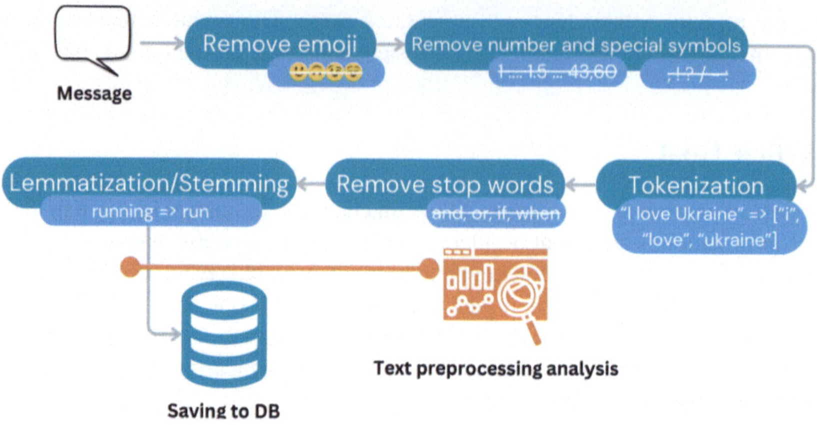

Fig. 1. Stages of text preprocessing. Analysis of text preprocessing

4 Literature Review

The development of advanced text preprocessing systems has been boosted by significant advances in machine learning and related technologies. The main purpose of this review is to gain a deep understanding of the current state of this area. This comprehensive analysis is aimed at identifying possible directions for further development and improvement in the field of preprocessing of text data for the positive impact of these operations on the result of the system as a whole and the consumption of time and other resources.

A study by Mohammad [6] focused on issues such as acronyms, references, and missing characters in words. Since such artifacts often arise due to illiterate input or the emotional state of the authors, their significant number can affect the quality of processing. Therefore, for such texts, it is recommended to use fuzzy matching methods for correct matching with the correct lexical forms of words. In the context of processing texts that violate the rules of grammar, it is important to identify proper names and use their correct form. In addition, the author considers aspects of the emotionality of messages and the possibility of words with long repetitions of symbols, which create difficulties in the process of normalization. The study conclusion indicates that preprocessing has a limited effect on the results of processing texts that violate grammatical rules, and it is recommended to focus on choosing the optimal processing algorithm with subsequent adjustment of its parameters.

The research of Camacho-Collados [7] highlights a critical view of the existing algorithms for the reprocessing of textual information, focusing on the shortcomings of the process of bringing words to a normal form. It is noted that this approach sometimes leads to unpredictable errors, such as the transformation of proper names and names into common words, which as a result can change the text context. In particular, the transformation of "Apple" into "apple" is an example. The research examines the negative impact of lemmatization, as it may not sufficiently take into account the text syntactic nuances and is rarely used in neural network-based systems. The author of the study analyzes the impact of such processes as tokenization, normalization, lemmatization, and works with problematic words on the results of text subject classification algorithms and sentimental polarization. In conclusion, the author notes that tokenization has a more important effect on the results than other stages of processing, especially in the case of texts of general topics.

In the work of Kumar and Harish [8], the problem of editing short messages from the social network Twitter is considered. In particular, the difficulties associated with the presence of nicknames (usernames), hyperlinks, slang expressions, hashtags, and expressions of emotions in texts were investigated.

In the articles [9–12] analyzing Slavic languages, the influence of the Ukrainian language on the lemmatization algorithm was analyzed, and a method for determining word combinations by keywords was developed.

5 Experimentation

It is planned to use the Anaconda virtual environment, the JupyterLab IDE [13], the Python programming language [14], and the MongoDB database [15] to conduct experiments. MongoDB can be advantageous for storing large text arrays compared to traditional SQL systems for several reasons. First, MongoDB's flexible schema makes it easy to store unstructured data, such as text, without having to strictly define the structure. According to the comparative articles of relational and non-relational databases [16, 17], it is known that with the help of indexing and search mechanisms in MongoDB, it is possible to process large volumes of text information more efficiently. In addition, the possibility of horizontal scaling allows you to distribute data to several servers, providing optimal speed when working with large volumes of text content.

The experiment stages are: downloading information from third-party sources using the API, saving it in a document-oriented database, evaluating the saved volume, analyzing the average length of messages and the number of words by channels with the gradual application of text preprocessing algorithms, namely removing Emoji, removing words, removing punctuation, splitting text into tokens and removing stop words. Next, the method of saving the result in the database is analyzed, taking into account the volume, speed of storage, and text preprocessing. Having the metadata of messages by their publication date, we estimate the data velocity for several days and separately for a day. The last stage is the assessment of the speed of each of the text preprocessing stages and the analysis of the possibilities of translating foreign texts in order to expand the analysis scope. A similar scheme, but only for the task of clustering using Twitter messages, is demonstrated in the article "Intelligent Analysis of Ukrainian-language Tweets for Public Opinion Research based on NLP Methods and Machine Learning Technology" [18].

The first stage is receiving input data, namely using the Telegram API [1] to download the last 1000 messages from eight Ukrainian-language channels. For this, the Telethon library [19] was used, first loading all information into the RAM, and then storing it in MongoDB. This utility returns an array of Message entities [20] with the following fields: text, row_text, is_reply, datetime, forward, buttons, button_count, file, photo, document, web_preview, audio, voice, video, video_note, gif, sticker, contact, game, geo, invoice, poll, venue, action_entities, via_bot, via_input_bot, client. In the context of the Telegram messenger, a message can be not only in text form but also in the form of a survey, image, geodata, video, etc. Therefore, among the numerous columns of the Telegram message, we use its identifier, datetime, message text (raw_text), and author, as these columns are sufficient for text analysis. After saving "raw" data in the MongoDB database, you can estimate the amount of memory that this information occupies. Keeping the information in raw form and having 8000 messages, we get 1.74 MB in compressed form and 3.67 MB in original form. The average size of one document is 537 bytes.

Taking into account the number of words or the total number of characters, in total across all channels this is approximately 200 characters. However, for different channels, the nature of message lengths can vary significantly, from a minimum of 110 characters to a maximum of 329 characters, as illustrated in Fig. 2.

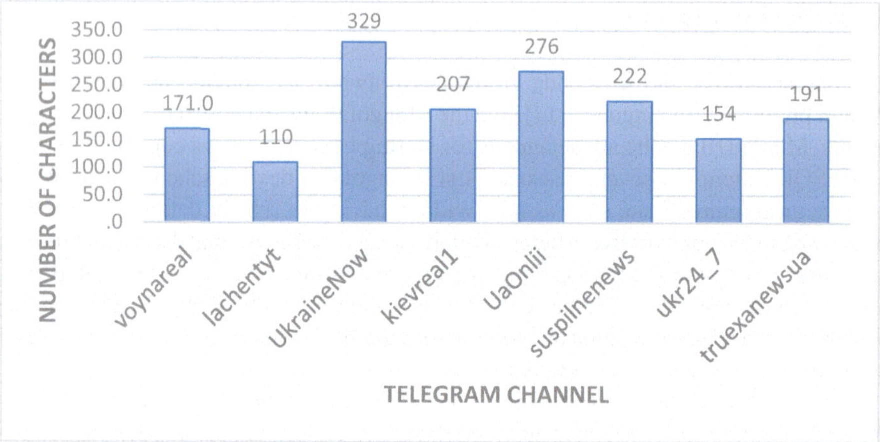

Fig. 2. The average number of symbols in one message for the investigated channels

With the presence of a significant number of emojis in messages, their extraction should be carried out, which is implemented using the emoji library. After re-analyzing the message length, it turns out that their average length is now 199 characters, which is 0.5% less than the previous form of text. Although this is a small amount of text, it can cause significant inconvenience in processing and encoding the text, and also affect the stored data size, again due to encoding problems.

Numbers are also uninformative for natural language processing and are extracted using regular expressions. After this procedure, the evaluation of the message length shows a significant reduction in size by 5%, which is equal to 189 characters.

For efficient text processing in algorithms, it is necessary to divide it into tokens. A token is essentially a word without extra spaces. Thus, after breaking the text into tokens, we get information that, on average, each message contains 26 tokens (words). Below is a figure with a diagram for all channels (Fig. 3), which illustrates the imbalance among channels, because there are no critical requirements and restrictions for the publication of a message, each channel writes a different volume at different times. If compared with the average length of the message from Fig. 1, it becomes obvious that there is no disproportionality in the use of long words, that is, the number of tokens grows proportionally with the text length. There is also the question of analyzing messages that are too short, namely finding the optimal minimum threshold for the number of tokens and choosing the stage at which this filtering is performed.

Among these tokens, there are often words that do not carry essential information, such as conjunctions, adverbs, exclamations, pronouns, etc. To perform this procedure, it is necessary to have a dictionary of words to exclude. Using this dictionary, you can remove any token from the shared list. After the procedure, the average number of tokens decreased from 26 to 18, which is a reduction of 31%. Below is a figure with a diagram for all channels (Fig. 3), from which it can be concluded by comparing the number of tokens before and after the removal of stop words that the percentage of stop words in each of the channels is close to the average value of 31%.

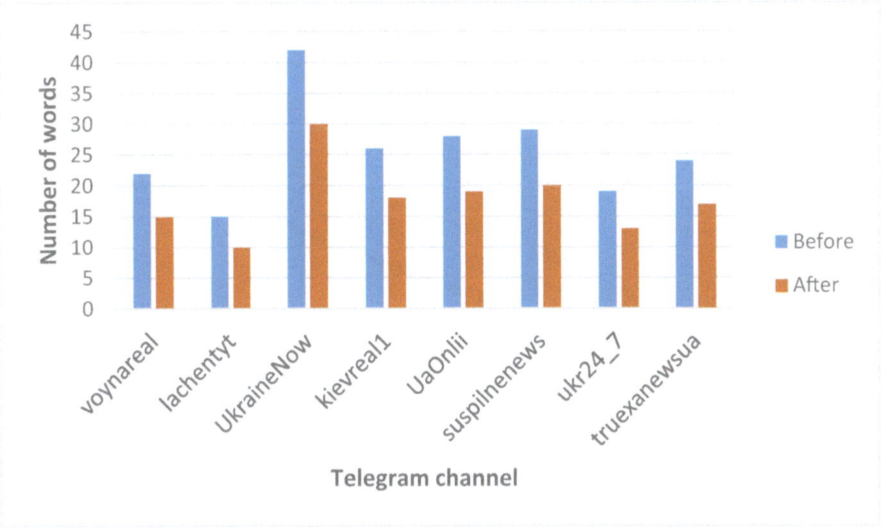

Fig. 3. Average number of words per channel before and after text preprocessing

Since MongoDB's document-oriented database allows you to store entire lists in a single object column, we decided to try to store an array of tokens in the same way we did with the "raw" data. However, this led to an increase in the amount of memory used, namely 1.91 MB in compressed form and as much as 4.06 MB in the original, with an average document size of 507 bytes. Thus, even with a large volume of deleted unnecessary data, storage in the form of a list requires significantly more memory resources.

One way is to represent the tokens as a string, combining them with any single character, in this case - a space. After combining the tokens into text and repeating the procedure of saving in the database, we got a reduction in the collection size by 22% from the initial size. It is now 2.97 MB in original form and 1.35 MB in compressed form with an average document size of 370 bytes.

It is also possible to propose a modified way of saving the text, but which will require additional computing resources. Its essence consists of encoding words and giving each word in all texts a specific number that will replace it in the text and having at the same time a smaller number of characters to save. To implement such a method of saving at the input of the algorithm, we must have an array of cleared tokens. The first step of the algorithm is to create a list of unique tokens and the number of their uses in texts. The next step is to sort this list in descending order by the number of uses and give these tokens a code, in this case, the code value can be the serial number of a unique token in the list. At this stage, we have a dictionary of tokens, where each token has its own code, and the more often the word occurs in the texts, the less this number code is, which we will store in the database for further use. Having a token-code dictionary, we can transform an array of tokens into an array of numbers and eventually convert these numbers into strings, combining them with a space into one big string to save the database. In further iterations of adding new messages to the database, we will only add to the existing token-code dictionary, not change it.

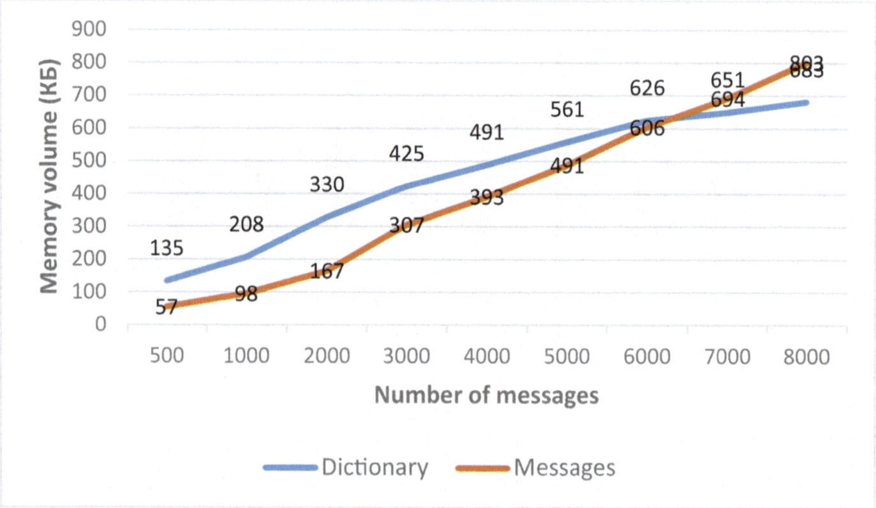

Fig. 4. Dependence of the dictionary and messages volumes in the database on the number of incoming messages according to text encoding strategy

If you analyze the efficiency of memory use, there is one important point that should be emphasized. As already mentioned, this algorithm uses two tables: one for the dictionary, and the other for messages, where the first table occupies 737 KB and the second 786 KB, which is about 1.48 MB in total. If we compare it with the previous method, which consisted of saving ordinary tokens in one line, we can conclude that the old method is more efficient, because it took up less memory, namely 1.35 MB. However, if you look at the average statistical size of one document, it becomes clear that its volume has halved from 370 bytes to 184 bytes.

In the current configuration, it may seem that the dictionary is an unnecessary burden, but considering that the dictionary volume and the list of messages will grow non-linearly, as shown in Fig. 4, where the increase in the dictionary volume becomes smaller than the increase in the volume of messages with an increase in the number of incoming messages and while the dictionary volume theoretically has a certain limit, the message volume is unlimited. This method is optimal for a large volume of texts using one dictionary. In the case of decoding a small volume of texts, this method is counterproductive.

Given the need for real-time text processing at small intervals lasting several hours, the data velocity is an important aspect. If we look at the messages, grouping them by day, we can see the following dynamics. From this, it can be understood that on certain days, when important events, such as large-scale missile attacks, the number of messages can increase dramatically in all channels. Figure 5 shows the dependence of the total number of messages (axis Y) on the day of publication (axis X), where abnormal turbulence is clearly demonstrated on December 29 and January 2, and the rest of the days the number of messages is close to the average.

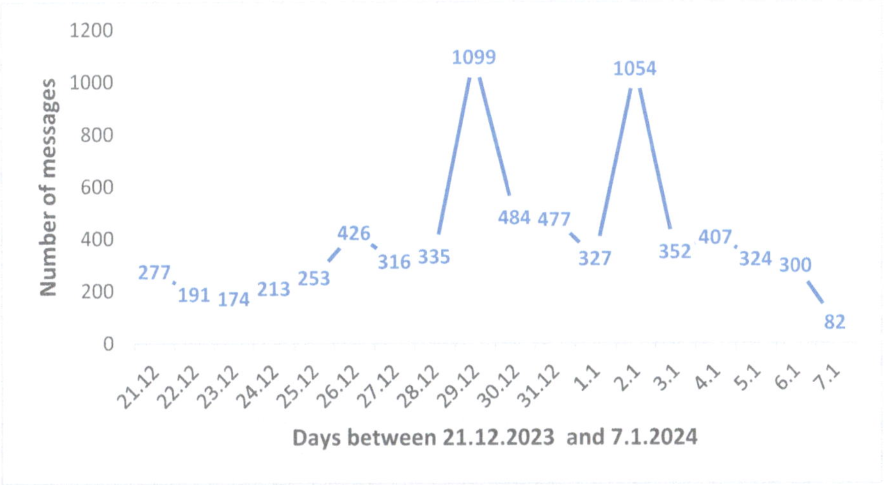

Fig. 5. Velocity by day

If we group the messages by the time of publication, it turns out that starting at five o'clock in the morning, the total number of messages increases on the eight selected sources of information to peak values during the three hours from 6:00 to 9:00 and after a slight decline is maintained at this level until nine o'clock in the evening. Figure 6 shows the dependence of the number of messages (axis Y) on the time of publication (axis X).

Fig. 6. Velocity by hour

6 Summary and Conclusion

The impact of data preprocessing on the volume of stored data and the time of its processing was experimentally evaluated. An effective strategy for optimal storage of text data for subsequent stages of processing using natural language processing algorithms is proposed. Among the proposed methods, the method of saving cleared tokens in string format and the method of saving word codes in string format together with the word-code dictionary were highlighted, where the least effective method is saving cleared tokens in the format of a string array.

Evaluation of productivity and volume of data after various stages of processing testifies to the effectiveness of the proposed approaches. Investigating various storage methods, it was found that storing string token arrays joined by a single symbol allowed to reduce the data size by 22%. In addition, the new encoding method, although associated with integrity risks, has been shown to yield significant memory savings, especially for large volumes of text using a common dictionary.

Performance considerations highlighted the trade-off between memory usage and encoding costs, with the optimal approach depending on the text size. The study also analyzed the data velocity over days and hours, identifying periods of peak activity, which is crucial for real-time processing.

References

1. Talakh, M.V.: PART 7. Using text mining for the analysis of social networks. In: Ushenko, Y., Ostapov, S. and Golub, S. (eds.) Information Technologies. Part 1. Application in Computer Vision, Recognition and Intelligent Monitoring Systems Yuriy Ushenko, Serhiy Ostapov, Serhiy Golub, pp. 157–173. LAP LAMBERT Academic Publishing (2019)
2. Talakh, M., Holub, S., Lazarenko, Y.: Intelligent monitoring of software test automation of Web sites. In: International Scientific and Practical Conference on Intellectual Systems and Information Technologies, pp. 46–51

3. Telegram: Telegram APIs (2024). https://core.telegram.org/api
4. Chai, C.P.: Comparison of text preprocessing methods. Nat. Lang. Eng. **29**(3), 509–553 (2023)
5. Unicode: Emoji, accents, and international text (2024). https://cran.r-project.org/web/packages/utf8/vignettes/utf8.html
6. Mohammad, F.: Is preprocessing of text really worth your time for online comment classification? eprint arXiv:1806.029908, pp. 1–5 (2018)
7. Camacho-Collados, J.: On the role of text preprocessing in neural network architectures: an evaluation study on text categorization and sentiment analysis. eprint arXiv:1707.01780, pp. 1–4 (2018)
8. Keerthi Kumar, H.M., Harish, B.S.: Classification of short text using various preprocessing techniques: an empirical evaluation. In: Sa, P., Bakshi, S., Hatzilygeroudis, I., Sahoo, M. (eds.) Recent Findings in Intelligent Computing Techniques . Advances in Intelligent Systems and Computing, vol. 709, pp. 19–30. Springer, Singapore (2018). https://doi.org/10.1007/978-981-10-8633-5_3
9. Mediakov, O., Vysotska, V., Uhryn, D., Ushenko, Y., Hu, C.: Information technology for generating lyrics for song extensions based on transformers. Int. J. Mod. Educ. Comput. Sci. (IJMECS) **16**(1), 23–36 (2024)
10. Lytvyn, V., Vysotska, V., Uhryn, D., Hrendus, M., Naum, O.: Analysis of statistical methods for stable combinations determination of keywords identification. Inf. Technol. Eastern Eur. J. Enterp. Technol. **2/2**(92), 23–37 (2018)
11. Lytvyn, V., Vysotska, V., Pukach, P., Brodyak, O., Ugryn, D.: Development of a method for determining the keywords in the Slavic language texts based on the technology of web mining. Inf. Technol. Ind. Control Syst. Eastern-Eur. J. Enterp. Technol. **2/2**(86), 14–23 (2017)
12. Lytvyn, V., Vysotska, V., Pukach, P., Bobyk, I., Uhryn, D.: Development of a method for the recognition of author's style in the Ukrainian language texts based on linguometry, stylemetry and glottochronology. Inf. Technol. Eastern Eur. J. Enterp. Technol. **4/2**(88), 10–19 (2017)
13. JupyterLab (2024). https://jupyterlab.readthedocs.io/en/stable/index.html
14. Python: Our documentation (2024). https://www.python.org/doc/
15. Mongo: Mongo: the developer data platform (2024). https://www.mongodb.com/
16. Parker, Z., Poe, S.: Comparing NoSQL MongoDB to an SQL DB. In: Proceedings of the 51st ACM Southeast Conference (2013)
17. Li, Y., Manoharan, S.: A performance comparison of SQL and NoSQL databases. In: Proceedings of the IEEE Pacific RIM Conference on Communications, Computers, and Signal Processing, pp. 15–19 (2013). https://doi.org/10.1109/PACRIM.2013.6625441
18. Prokipchuk, O., et al.: Intelligent analysis of ukrainian-language tweets for public opinion research based on NLP methods and machine learning technology. Int. J. Mod. Educ. Comput. Sci. (IJMECS) **15**(3), 70–93 (2023)
19. Telethon: Telethon's documentation (2024). https://docs.telethon.dev/en/stable/index.html
20. Telethon: Object reference – Telethon 1.30.3 documentation (2024). https://docs.telethon.dev/en/stable/quick-references/objects-reference.html?#message

3D Model Quality Assessment Algorithm for Additive Manufacturing

Viktor Marchuk[1(✉)], Yurii Tulashvili[1], Iurii Lukianchuk[1], Serhii Bazyl[2], Vasyl Kovalchuk[3], Oleksandr Derevyanchuk[4], and Roman Horbatiuk[5]

[1] Lutsk National Technical University, Lutsk 43018, Ukraine
v.marchuk@lntu.edu.ua, y.tulashvili@lutsk-ntu.com.ua
[2] Separate Structural Subdivision "Shostka Vocational College Named After Ivan Kozhedub of Sumy State University", Shostka 41100, Ukraine
[3] State Institution «Scientific and Methodological Center for Higher and Pre-higher Education», Kyiv 03151, Ukraine
v.i_kovalchuk@ukr.net
[4] Yuriy Fedkovych Chernivtsi National University, Chernivtsi 58012, Ukraine
o.v.derevyanchuk@chnu.edu.ua
[5] Ternopil Volodymyr Hnatiuk National Pedagogical University, Ternopil 46027, Ukraine
gorbaroman@gvail.com

Abstract. In modern production, problems arise when modeling 3D objects using the FDM method. Among the common mistakes are insufficient attention to the geometry of the object, insufficient calculation of the thickness of the walls and details. An algorithm for analyzing the manufacturability of parts before the start of production can help avoid many problems. Determining the optimal print direction, support placement, and heat distribution are key steps that help ensure the highest manufacturing quality. It is also important to emphasize the use of mathematical calculations during design. Calculating the amount of material required, optimizing geometry to minimize weight and maximize strength are just a few examples of how math can facilitate the process. Careful advance planning and the use of advanced analytical methods can significantly improve product quality and reduce production errors. Recommendations provided at various stages of design and production create an opportunity for engineers to avoid common problems and ensure the successful implementation of projects in the field of additive manufacturing.

Keywords: 3D modelling · 3D printing · additive manufacturing · defects surface of the product · object geometry · analysis algorithm

1 Introduction

Additive manufacturing or 3D printing is the process of creating three-dimensional objects of almost any geometric shape based on their digital 3D models. The concept of 3D printing is based on building an object with sequentially applied layers that reproduce the contours of a 3D model.

Z. Hu et al. (Eds.): ICCSEEA 2024, LNDECT 242, pp. 304–316, 2025.
https://doi.org/10.1007/978-3-031-84228-3_26

Additive manufacturing is used to manufacture prototypes, samples and finished products from various materials: from plastic to composite components.

Additive technologies are widely used for prototyping in various fields, such as art and architectural construction, to create samples in bioengineering in the medical field, fashion clothing, jewelry, and industrial production.

Additive production is automated, so it is impossible without a source of information about the mutual location of the elements of the print object, its geometry. Digital three-dimensional models, or 3D models, serve as such a source of information. They are created by manual computer graphic design or 3D scanning. Both manual and automatic creation of 3D models for 3D printing cause difficulties due to technological limitations in additive manufacturing. During printing, the 3D printer reads a 3D model file containing the geometry of the object to be printed and determines successive layers of liquid, powder deposition, building a 3D model from cross-sections. These layers, corresponding to virtual cross-sections in the 3D model, are joined or fused together to create an object of a given shape.

The most popular coordinate system used in additive manufacturing technology is the Cartesian system, built on a rectangular three-dimensional space with X, Y and Z axes. Therefore, the quality of the product obtained during 3D printing depends significantly on the resolution and positioning accuracy of the 3D-printer.

Resolution is the thickness of the layer deposited along the Z axis. A typical layer thickness is 100 μm, although devices such as the Objet Connex and 3D Systems ProJet are capable of printing layers as thick as 16 μm.

Print head positioning accuracy is defined in the horizontal plane along the X and Y axes. Resolution and positioning accuracy are measured in DPI (dots per inch) or micrometers. Resolution along the X and Y axes is similar to that of conventional two-dimensional laser printers. A typical size of DPI resolution is about 50–100 μm.

The most popular additive manufacturing technology is FDM – extrusion 3D printing, modeling by layer-by-layer deposition of thermoplastic polymers (polyactide, acrylonitrile butadiene styrene, etc.).

The process of printing by the layer-by-layer deposition method is the creation of layers due to the extrusion of a material that quickly solidifies in the form of microdrops or thin jets. The extruder heats the material to the melting temperature, followed by extrusion of the molten mass through a nozzle.

2 Literature Review

The development of 3D printing technologies was widely reflected in literary sources revealing foreign experience. Engineers are finding practical applications for using 3D additive manufacturing. 3D printing allows new designs to become incredibly rare products that could not be achieved with traditional manufacturing methods. Scientific and technical publications [1–3] reflect knowledge that is relevant for mastering 3D printing technologies.

The world experience of training specialists in 3D technologies is very diverse, but first of all, it involves many information platforms where specialists share their own experience. Thus, Sabri Soloman shows how the industry is changing under the influence of the spread of 3D printing technologies [4]. In the studies of Hardik Pandya, the basics of working with additive technologies are described and the experience gained as a result of educational practical training, which is collected in educational material for the training of 3D printing specialists, is disclosed [5].

When mastering 3D printing technology, as a rule, translations of world experience are used [6–9].

It is possible to evaluate the quality of printed 3D models manually or automatically by computer analysis of photographs of objects. A number of methods of digital image processing are used for such analysis: filtering [10, 11], segmentation [12], contour selection [13, 14], local contrast enhancement [15], recognition. For example, using segmentation methods, it is possible to highlight individual faces of polygons on images, determine their positions and sizes. The received information about the researched objects is processed by means of artificial intelligence [16, 17] and fuzzy logic [18, 19].

The purpose of the article is to build an algorithm for assessing the quality of a 3D model during its construction based on the classification of typical errors that led to defects in products from FDM additive manufacturing technology.

3 Research Methodology

The technology of layer-by-layer surfacing has certain limitations in terms of the complexity of the created geometric shapes.

According to FDM additive manufacturing technology, the extruder of the 3D printer moves in horizontal and vertical planes under the control of algorithms, which are similar in numerically controlled machines. The nozzle moves along the path specified by the automated design system according to the contours of the 3D model. Layer by layer, from bottom to top, an FDM additive manufacturing model is built, the geometry of which must exactly match the 3D model.

As the most common and commercially available method of 3D printing, the FDM layer-by-layer deposition method has a wide range of consumables (filaments). Various types of polymers are used as consumables for extrusion printing, including acrylonitrile butadiene styrene (ABS), polycarbonate, polylactide (PLA), high-pressure polyethylene (PETG), mixtures of polycarbonate and ABS plastic, polyphenylene sulfone, etc. The polymer is supplied in the form of a filler made of pure plastic.

There are several common mistakes that can occur when modeling objects for 3D printing:

1. Non-manifold geometry [20] is a common error that can occur when modeling objects for 3D printing. Non-uniform geometry occurs when surfaces in a 3D model do not connect or intersect correctly, causing errors when the slicer software creates the path along which the printer's printhead will move.

   There are several types of heterogeneous geometry that can occur in a 3D model, including:

   - Open up. Voids in the geometry that create open spaces in the model. Holes can occur when vertices, edges, or faces are not properly connected.

- Self-crossing. Occur when parts of geometry intersect with other parts of geometry, resulting in overlapping surfaces and ambiguous boundaries.
- Overlapping faces. Occurs when two or more faces in a model occupy the same space, resulting in fuzzy boundaries and confusion for the sectioning software.
- Heterogeneous surfaces. Occurs when surfaces do not have a defined front or back, making it difficult for the software to determine which direction the surface should be printed.

Non-uniform geometry can cause a variety of problems during printing, such as gaps or holes in the final sample model or even printer errors. To correct such geometry problems, it is important to use specialized software tools to identify and restore the geometry. Meshmixer and Netfabb are two popular software tools that can detect and reconstruct heterogeneous geometry. In some cases, it may be necessary to manually adjust the model geometry to ensure that all surfaces are properly connected and uniform. By eliminating non-uniform geometry errors, you can ensure that your 3D model prints accurately and with a high level of quality.

2. Wall thickness errors are another common problem that can occur when modeling objects for 3D printing. This error occurs when the walls of the model are too thin or too thick for the printer and the material used. This can lead to over- or under-extrusion of the material, resulting in a poor quality product.

   If the walls are too thin, it may be difficult for the printer to squeeze out enough material to form a solid, continuous wall, resulting in unfilled areas. On the other hand, if the walls are too thick, the printer may squeeze out too much material, resulting in a rough surface or a buildup of material that will cause the model to warp.

   To prevent wall thickness errors, it is important to design the model with the specific printer and material in mind. This means taking into account the minimum and maximum recommended wall thicknesses for the particular printer and material, as well as any other design limitations that may affect the final print quality.

   When designing a model, it is important to use 3D modeling software that allows you to set the wall thickness and visualize the final product. Some software tools have automatic error detection that highlights areas of the model with potential problems.

   To avoid wall thickness errors, it is important to test the model by cutting it and running test samples before proceeding with the final print. This will allow any potential problems to be identified and corrections made before the final model is printed.

   When developing a 3D model for printing, it is important to keep the wall thickness in mind. By designing with the printer and material in mind, as well as carefully testing the model, you can avoid wall thickness errors and ensure high print quality.

3. Badly designed joints or connections [21] are elements of a 3D model that extend beyond the body of the model and do not have a support structure underneath. These elements can make printing difficult because the printer cannot cool the product material fast enough.

   Projections and unsupported geometry can be problematic because without support structures the printer has nothing to lean on and the extruded material can sag or sag. This can lead to a loss of detail and accuracy in the final print, as well as stability and durability issues.

To prevent protrusions and unsupported geometry, it is important to design the model with supports in mind. These support structures are temporary and are printed with the model and removed after printing is complete. Supports can be generated by the slicer software automatically, or they can be added manually by the designer.

Support structures are typically in the form of thin vertical members that are printed under protrusions and unsupported geometry. These structures form the basis for the creation of the extruded material, ensuring the accuracy and stability of the final print.

Support structures can add extra time and material to the printing process, so they should be used wisely. Designers should try to minimize the amount of unsupported geometry in their models, and design support structures that are easy to remove and minimize waste.

It is also possible to print protrusions and unsupported geometry at an angle to reduce sagging and overhang. This can be achieved by tilting the model in the slicer software, allowing the printer to create ridges at an angle rather than straight.

Overhangs and unsupported geometry can be difficult to print, but with careful design and use of support structures, high-quality models with accurate detail and stability can be created.

4. Incorrect scale that can occur when developing 3D models for printing. This error occurs if the model is not the correct size or if the model is incorrectly modified during the design process.

   Incorrect scaling can cause models to be too small or too large, which will significantly affect the final product. For example, a model that is too small is difficult to view and use due to lack of detail, while a model that is too large does not provide a good fit and is too heavy.

   To avoid incorrect scale errors, it is important to ensure that the model dimensions are set correctly from the start. This can be done by using accurate measurements and reference points when designing the model, and by double-checking the scale before exporting the model for printing.

   In addition to ensuring the correct dimensions, it is also important to consider the resolution of the printer when designing the model. This is especially important when designing small or complex objects, as the printer may not be able to accurately reproduce small details or small elements.

   When designing a model, it can be helpful to use 3D modeling software that includes tools to check the scale of the model. These tools can help ensure that the model is sized correctly and that all elements have the correct proportions.

   Incorrect scale is a common mistake that can be avoided by making sure the dimensions are set from the start, considering the printer's resolution when designing the model, using 3D modeling software with built-in scale checking tools, and testing the model thoroughly. Before printing. By following these steps, designers can ensure that their models are accurate and fit for purpose.

5. Non-printable geometry - refers to elements of a 3D model that cannot be printed due to limitations of the printer or printing process. This includes items that cannot be printed because they are too small or too thin, or items that cannot be reproduced by a printer because they are too complex or too fragile.

Non-printable forms are a challenge for designers. In some cases, it may be necessary to completely redo the model to eliminate non-printable elements.

To avoid unprintable geometry, it is important to consider the limitations of the printer and printing process when developing the model. This includes factors such as minimum printable size, maximum printable height, and minimum printable wall thickness.

In addition to considering printer limitations, it is also important to consider the orientation of the model when printing. Items printed vertically can be more difficult to print than those printed horizontally, as gravity can cause the extruded material to sag or sag.

Some of the common techniques used to correct unprintable geometry include:

- Increasing the thickness of thin elements to ensure their printing.
- Simplification of complex geometry for easier printing.
- Rotate or reposition the model to optimize print orientation.
- Adding supports to fragile or protruding elements to guarantee their printing.

Non-printable geometry refers to elements of a 3D model that cannot be printed due to limitations of the printer or printing process. To avoid unprinted geometry, designers must consider the limitations of the printer and printing process when designing the model, use slicing software to identify unprinted elements, and use methods such as increasing thickness, simplifying the geometry, rotating the model, or adding support structures to correct unprinted features.

6. Poorly designed joints and connections - problems can arise during 3D printing that lead to defects and damage to the printed object. The joints between individual parts of the 3D model must be properly designed so that the parts fit together one and the final printed object functioned. Joints must be properly designed so that the parts fit together correctly and for the final printed object to function.

   One common problem with poorly designed joints is that they can be too tight or too loose. If the joints are too tight, the parts may not fit together properly, causing the printed object to warp or break. If the connections are too weak, the parts may not stay connected, causing the object to break.

   Another problem with poorly designed joints is that they may not be strong enough to support the weight or load of the object. This can cause the object to break or collapse, especially if it is designed for work or has to support weight.

   To avoid problems with poorly designed joints, it is important to carefully design the joint using proper design principles. This includes ensuring that the joints are not too tight or too loose, and that they are strong enough to support the weight or load of the object.

   One way to check the design of the joints is to create a prototype or test print before printing the final object. This allows the designer to identify any problems with joints or connections and make any necessary adjustments to the design.

   Poorly designed joints can cause problems during 3D printing, leading to errors or even breakage of the printed object. To avoid these problems, it is important to carefully design the joints and connections using appropriate design principles and to verify the design by prototyping or test printing.

Having analyzed the above typical errors that occur during the 3D modeling of the object, we offer a sequence of checking the 3D model for the possibility of the occurrence of typical errors and the necessary solutions for their elimination (Table 1).

Table 1. Inspection sequence during 3D modeling and required solutions to eliminate typical defects

Stage	Defect	Defect description	Solution
1	Non-printable geometry	Objects that cannot be printed, such as internal voids or depressions	Remove unprintable geometry or refine the design to make it suitable for 3D printing
2	Wall thickness errors	Too thin or thick walls cause excessive or insufficient extrusion of the material, which affects the strength of the product	Edit the wall thicknesses according to the parameters of the 3D printer nozzle and the material used
3	Incorrect scale	The model is not scaled correctly for 3D printing, resulting in model elements that are too large or too small	Select the scale of the model in accordance with the technological dimensions of the product and the parameters of the 3D printer
4	Overhangs and unsupported geometry	Overhangs or unsupported areas of the 3D model that cause the material to sag	Finalize the 3D model regarding the necessary geometry of additional support structures or determine the position of the technological supports of the 3D printer
5	Badly designed joints or connections	The structural connections of the 3D model, which do not meet the technological requirements of the 3D printer, lead to the fact that the elements of the part lose the conjugation of the surfaces	Refine the design elements of the 3D model joints in accordance with technological requirements, namely: add elements of chamfers, rounding, stiffeners
6	Non-manifold geometry	Surfaces that intersect or are non-uniform cause errors in the creation of the trajectory of the print head, resulting in the formation of defects of cavities from the middle of the product	Adjusting the model using software that can identify and correct non-uniform geometry, such as Meshmixer or Netfabb

Table 2. Recommendations for configuring the printer for the best performance of the product

Printer settings				Requirements to 3D models		Operating characteristics			
Material	Nozzle diameter, mm	Print speed, mm/s	Printing temperature, °C	The minimum wall thickness of the product, mm	Minimum dimensions of the product, mm	Tensile strength, MPa	Maximum load, kg	Maximum operating temperature, °C	Use in an aggressive environment
PLA	0,2	20-50	190-220	0,6-0,8	2x2x2	20-40	20-30	60-65	No
	0,4	40-80		0,8-1,2	4x4x4	40-60	30-40		
	0,6	60-100		1,2-1,8	6x6x6	60-80	40-50		
ABS	0,4	30-70	230-260	1,2-1,6	8x8x8	40-60	40-60	80-85	Yes
	0,6	50-80		1,6-2,0	10x10x10	60-80	60-80		
PETG	0,2	20-40	220-250	0,8-1,2	4x4x4	40-60	40-60	70-80	Yes
	0,4	40-60		1,2-,1,8	6x6x6	55-75	60-80		
	0,6	60-80		1,6- 2,2	8x8x8	60-80	80-100		
Nylon	0,4	30-60	230-260	1,2-1,6	8x8x8	50-80	120-200	150-180	Yes
	0,6	60-100		1,6-2,0	10x10x10	70-100	150-300		

4 Results

Let's take a closer look at the 3D model and analyze the causes of defects. This model consists of two separate components that are not connected to each other: a sphere and a cylinder. As a result, defects are formed, as shown in the figures. Mathematical models of these defects are described by various methods, namely: algebraic topology, differential geometry or Euler's characteristics. In our work, we analyzed each of them and noted that in practice it is quite difficult to use them to correct a 3D model, and not all engineers will have the time and desire to calculate various mathematical formulas. Moreover, the production process requires quick decisions and correlations. Therefore, we have developed simplified formulas for calculating the formed defects and offered a number of recommendations for setting up the 3D printer for the best operational characteristics of the product (Fig. 1).

Such a defect as a wall thickness error will mathematically have the following expression, which shows the parameters that should be changed to eliminate this defect (Fig. 2):

$$h_{\min \geq} \geq \frac{D_1 - D_2}{2} \tag{1}$$

where h_{min} is the minimum wall thickness of the product, D_1 is the diameter of the cylinder, D_2 is the diameter of the circle.

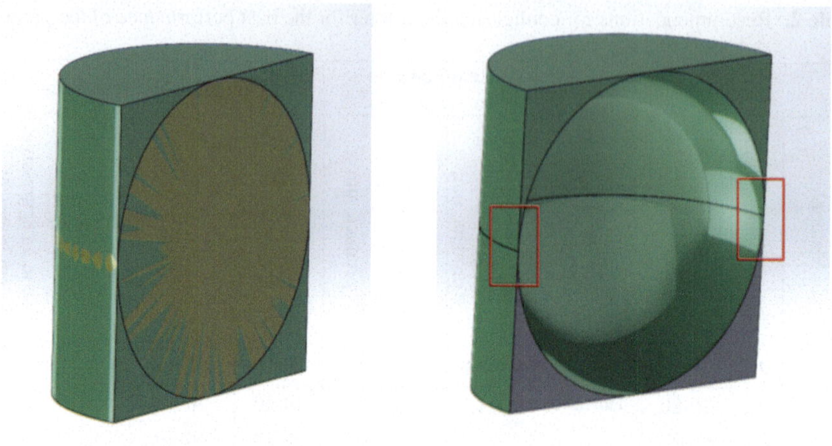

Fig. 1. Cross-section of the model with defects in the SolidWorks software environment

Also, when designing the model, it is necessary to take into account the features of 3D printing and the characteristics of the printer itself and observe the ratio:

$$h_{\min} \geq 3 \times D_{nozzle} \tag{2}$$

Such a defect as incorrect scale occurs when the dimensions of the product are larger than the dimensions of the working area of the 3D printer. Because when changing the scale of the product, the dimensions of the model change in critical places and joints. Therefore, the following ratios should be observed when designing the model:

$$H_{\max} \geq H_{obj} \tag{3}$$

where H_{max} is the maximum height of the working area of the 3D printer, Hobj is the maximum height of the product.

$$H_1 \geq D_1 \tag{4}$$

where H_1 is the maximum table size of the 3D printer working area. In this case, the working table is square (in some printers, the working table is round), and the width of the product coincides with the diameter of the cylinder D_1 (products can have different shapes and, accordingly, sizes).

A defect called overhangs and unsupported geometry occurs as a result of the fact that the 3D model has an inclined surface and it is not always possible to use supporting elements, because sometimes it is difficult to get rid of them already in the finished product and they often spoil the appearance of the product. Therefore, it is recommended to observe the following ratio when designing the product and during its manufacture:

$$\alpha_{\max} \geq 45^0 - 70^0 \tag{5}$$

where α_{max} is the maximum angle of inclination of the product during 3D printing. In the case of the need for high-quality reproduction of a cylindrical surface, this angle should not be greater than 30°.

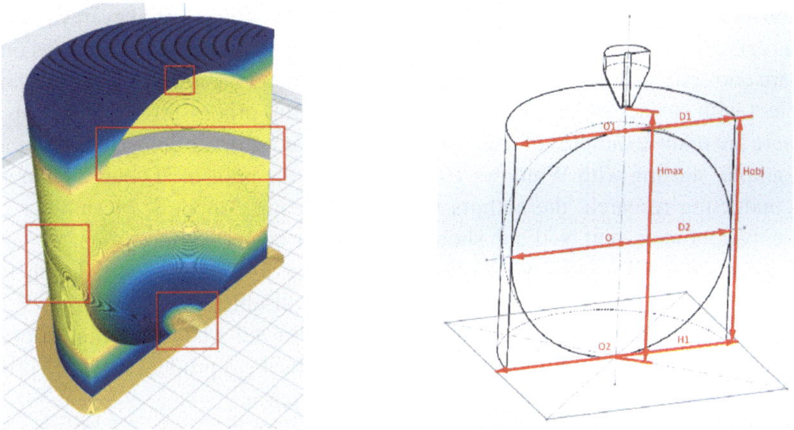

(a) (b)

Fig. 2. Section of the model: (a) with defects in the Cura software environment; (b) with notations for calculations

There are often situations where a 3D printing engineer does not have ownership rights or access to a 3D model file, and the model itself needs to be refined because it contains defects. In this case, you can try to fix it using specialized software. For example, Autodesk Meshmixer or NetFabb. Then the program itself analyzes the model and shows problem areas with surface defects (Fig. 3).

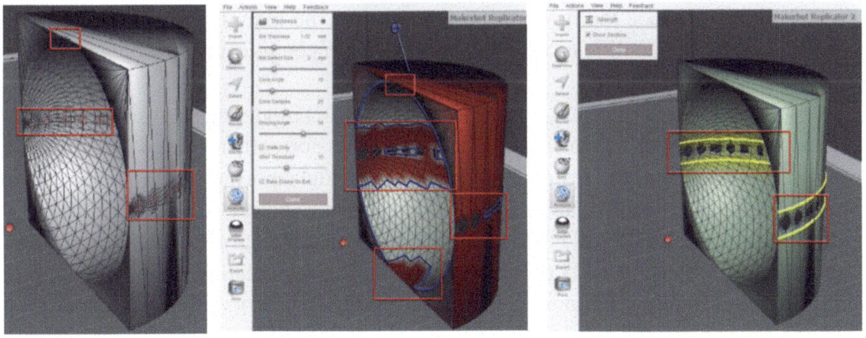

Fig. 3. A section of the model with defects in the Autodesk Meshmixer software environment

In order to correct them, you need to go to the analysis section of the 3D model and select the "inspector" function. You can fix defects one by one or click on "auto repair". Sometimes the model becomes even more unprintable because its geometric characteristics change. Therefore, Autodesk Meshmixer is recommended for correcting models that do not have curved surfaces.

Using the NetFabb software allows you to improve the quality of the 3D model. You just need to select the "repair" function and click "automatic repair". The NetFabb program allows you to fix a 3D model without installing software. However, you need to have a registered profile in Autodesk and take into account that files with names only in Latin are correctly perceived. It is enough just to download the necessary file containing the object with surface defects and after some time, after processing, download it.

There are many examples of using software to correct surface defects of a 3D model. For example, starting with Windows 10, you can use the built-in 3D builder. However, after conducting research, the authors recommend using Autodesk Meshmixer to fix non-complex objects, and NetFabb should be used for curved surfaces.

Printing with ABS and Nylon plastic with a nozzle diameter of 0.2 mm is not recommended, as it is possible to clog the nozzle and disable the equipment.

The general recommendations that should be followed when designing a product model in a software environment and when manufacturing a product using3D printing have been acquired by the authors on their own experience.

Therefore, a table was created using the most common materials and recommended parameters in order to achieve the best product results, minimize production defects and waste, and optimize the production time of the products.

5 Summary and Conclusion

The research describes the world theoretical and own practical experience of 3D printing. The main emphasis is on FDM technology, since this technology is the most common among users. Due to its relative simplicity and understanding, FDM printing is popular among specialists and amateurs of additive technologies, and the results of research will be useful to a large number of users.

Using information from research, it is possible to reduce the percentage of defects from 20% to 3%, reduce the time for preparation for the manufacture of a product by 55–75%, increase the qualification level of a 3D printing engineer by 25–35%, and other benefits.

To avoid errors in 3D models, it is important to use 3D modeling software specifically designed for 3D printing and to ensure that the model is correctly modeled and optimized for printing before printing. Testing the model in a virtual environment or on a small test sample can also help identify and correct potential errors before full-scale printing.

In the future, solutions will be developed to eliminate typical errors that occur during 3D modeling of objects for 3D printing, which will allow creating mathematical logic conditions for evaluating 3D models to ensure quality indicators for FDM technology.

In order to achieve the maximum performance of 3D printed products, the following recommendations should be followed during the design stage of the model.

1. Choose the material that best matches the desired properties of the final product. Consider properties such as strength, flexibility, heat resistance and durability.
2. The diameter of the nozzle of the print head should be selected according to the quality characteristics of the final product. Smaller nozzle diameters should be used to produce smaller parts, and larger nozzles should be used to reduce production time.
3. The 3D printing temperature should be optimized for the selected material. Printing at temperatures that are too high or too low can cause surface quality degradation, deformation and damage to the product.

4. The printing speed should be chosen according to the accuracy and surface quality required for the final product. In general, lower print speeds improve surface quality and accuracy, while higher print speeds reduce production time but reduce product quality.
5. Cooling the production area is important to ensure proper adhesion of the layers and to minimize deformation. Fan speed and cooling parameters should be optimized depending on the selected material.

In the future, solutions will be developed to eliminate typical errors that occur during 3D modeling of objects for 3D printing and to create mathematical logic conditions for evaluating 3D models to ensure quality indicators when using FDM additive manufacturing technology.

References

1. Larson, H.: Fabricated: The New World of 3D-Printing. Wiley, 280 p. (2013)
2. Javaida, M., Haleema, A., Pratap, S.R., Sumanc, R., Rab, S.: Role of additive manufacturing applications towards environmental sustainability. Adv. Ind. Eng. Polym. Res. **4**(4), 312–322 (2021)
3. Kirchheim, A., Dennig, H., Zumofen, L.: Why education and training in the field of additive manufacturing is a necessity. In: International Conference on Additive Manufacturing in Products and Applications, pp. 329–336 (2018)
4. Soloman, S.: Additive manufacturing technology – 3D printing and design – the 4th industrial revolution, 417 p. (2020). Independently Published
5. Pandya, H.: 3D Printing Technology: Fundamentals and Application. Studera Press, 416 p. (2021)
6. Smyth, C.T.: Designing 3D Printed Things for Everyday Use, 3rd edn. In: Functional Design for 3D Printing, 236 p., 26 June 2017
7. Torta, S., Torta, J.: 3D Printing: An Introduction. Illustrated Edition, 370 p. (2018)
8. Lukianchuk, I., Tulashvili, Y., Podolyak, V., Horbariuk, R., Kovalchuk, V., Bazyl, S.: Didactic principles of education students 3D-printing. IJCSNS Int. J. Comput. Sci. Netw. Secur. **22**(7), 443–450 (2022)
9. Surynovych, O., Rudynets, M., Lukianchuk, I., Kondius, I.: QR and 3D technologies integration in children's safety projects. In: Proceedings of the 2022 12th International Conference on Dependable Systems, Services and Technologies (DESSERT), pp. 1–5. Athens, Greece (2022)
10. Balovsyak, S.V., et al.: Adaptive oriented filtration of digital images in the spatial domain. In: Proceedings of the SPIE, 11176, Photonics Applications in Astronomy, Communications, Industry, and High-Energy Physics Experiments, vol. 11176, 111761A-1–111761A-6 (2019)
11. Balovsyak, S.V., Odaiska, K.: Automatic determination of the Gaussian noise level on digital images by high-pass filtering for regions of interest. Cybern. Syst. Anal. **54**(4), 662–670 (2018)
12. Tereikovskyi, I., Hu, Z., Chernyshev, D., Tereikovska, L., Korystin, O., Tereikovskyi, O.: The method of semantic image segmentation using neural networks. Int. J. Image Graphics Signal Process. (IJIGSP) **14**(6), 1–14 (2022)
13. Derevyanchuk, O.V., Kravchenko, H.O., Derevianchuk, Y.V., Tomash, V.V.: Recognition images of broken window glass. Proc. SPIE **12938**, 210–213 (2024)

14. Balovsyak, S.V., Derevyanchuk, O.V., Fodchuk, I.M.: Method of calculation of averaged digital image profiles by envelopes as the conic sections. In: Hu, Z., Petoukhov, S., Dychka, I., He, M. (eds.) Advances in Computer Science for Engineering and Education. ICCSEEA 2018. Advances in Intelligent Systems and Computing, vol. 754, pp. 204–212. Springer, Cham (2019). https://doi.org/10.1007/978-3-319-91008-6_21
15. Balovsyak, S.V., Derevyanchuk, O.V., Kravchenko, H.O., Kroitor, O.P., Tomash, V.V.: Computer system for increasing the local contrast of railway transport images. In: Proceedings of SPIE, Fifteenth International Conference on Correlation Optics, vol. 12126, pp. 121261E1–121261E7 (2021)
16. Lytvyn, V., Lozynska, O., Uhryn, D., Vovk, M., Ushenko, Y., Zhengbing, H.: Information technologies for decision support in industry-specific geographic information systems based on swarm intelligence. Int. J. Mod. Educ. Comput. Sci. (IJMECS) **15**(2), 62–72 (2023)
17. Prokipchuk, O., et al.: Intelligent analysis of Ukrainian-language tweets for public opinion research based on NLP methods and machine learning technology. Int. J. Mod. Educ. Comput. Sci. (IJMECS) **15**(3), 70–93 (2023)
18. Fayek, A.R.: Fuzzy logic and fuzzy hybrid techniques for construction engineering and management. J. Constr. Eng. Manage. **146**(7) (2020)
19. Balovsyak, S., Derevyanchuk, O., Kravchenko, H., Ushenko, Y., Hu, Z.: Clustering students according to their academic achievement using fuzzy logic. Int. J. Mod. Educ. Comput. Sci. (IJMECS) **15**(6), 31–43 (2023)
20. 3D Learning Hub: How to fix non-manifold geometry (2021). https://www.sculpteo.com/en/3d-learning-hub/create-3d-file/fix-non-manifold-geometry/
21. Lam, T.F.: Adaptive process control implementation of wire arc additive manufacturing for thin-walled components with overhang features. Int. J. Adv. Manufact. Technol. **108**(7) (2020)

An Association Rule Mining for Requirement Document Structure in IT Projects

Denys Gobov[1](✉) and Nikolay Sokolovskiy[2]

[1] National Technical University of Ukraine "Igor Sikorsky Kyiv Polytechnic Institute", 37, Prospect Peremohy, Kyiv, Ukraine
d.gobov@kpi.ua

[2] Independent Researcher, Erevan, Armenia

Abstract. In most IT projects, business analysts do not use templates defined in industry standards but rather rework them to suit the needs of a specific project, creating a requirements architecture. The process of creating a requirements architecture must consider the specifics of the project and the set of techniques used to analyze the requirements. Therefore, the question of how to apply modern machine learning methods to generate recommendations based on the experience of practitioners is an important issue that needs to be addressed. The study uses the Association Rule Mining method, which has shown effectiveness in solving similar applied problems, to identify the implication relationship between the project context, the business analyst's profile, requirements analysis techniques, and requirements document structure. Based on the analysis of data collected during a survey of practicing IT specialists, significant patterns were identified, based on which recommendations for creating requirements analytical documents in IT projects were formed. The proposed approach is flexible and scalable, which allows its reuse for new data sets regarding the practices of business analysts in IT projects.

Keywords: Association Rule Mining · Requirements Architecture · Requirements Analysis Techniques · IT project context · Business Analysis

1 Instruction

According to the British Computer Society (BCS) definition [1], business analysis aims to ensure that any business changes align with the organization's needs and are holistic, considering all relevant aspects. In the dynamic landscape of information technology projects, the importance of business analysis to project success only increases. The difference in several times between the cost of correcting an error made at the stage of working with requirements and the cost of an error made at the stage of development, testing, and implementation of the solution confirms this [2]. The main deliverables of business analysis are requirements of different types: business requirements, stakeholders requirements, functional and non-functional requirements, and transition requirements that should be organized into the requirements architecture. A requirements architecture fits the individual models and specifications together to ensure that all requirements

Z. Hu et al. (Eds.): ICCSEEA 2024, LNDECT 242, pp. 317–328, 2025.
https://doi.org/10.1007/978-3-031-84228-3_27

form a single whole that supports the overall business objectives and produces a valuable outcome for stakeholders [3]. It is created at the first phase of the business analysis core cycle - the pre-project phase extended within the project phase and maintained in the post-project phase [4]. Requirements architecture can be created based on framework templates or industry standards. Still, in most cases, business analysts and requirements engineers do not use them and create architecture from scratch [5]. It is explained by the fact that, in most cases, the standards are built for predictive approaches, while today, most projects are carried out using adaptive approaches. In addition, project specifics such as team size, domain, and stakeholder preferences may necessitate the creation of additional viewpoints and requirements views accordingly. Given that the atomic elements of a requirements architecture are created using standard documentation and modeling techniques, they also influence the structure of the architecture. Hence, discerning interdependencies among project characteristics, requirements analysis techniques, and requirements architecture in IT projects is relevant and holds practical significance in business analysis.

This research examined the prevailing practices among business analysts in Ukrainian IT companies concerning the forming requirements architecture. The primary research objectives are to apply association rule mining techniques to extract meaningful patterns and dependencies between IT project characteristics, requirements analysis techniques, and requirements architecture and to formulate practical recommendations for selecting requirements architecture elements. The data set was based on a survey of practicing IT specialists, mostly business analysts and requirements engineers. The results brought forth by association rule mining can enrich decision-making processes and empower stakeholders with a deeper comprehension of the intricate connections governing the business analysis processes.

2 Literature Review

Most existing research focuses on examining the structural aspects of requirements documents, existing templates, and possible ways for their improvement. In [6], a new template for describing requirements is proposed, called the "Requirements Cube", which assumes the presence of an additional level in the hierarchy of requirements - acceptance criteria and transition level-sub-function, which allows integration enterprise architecture with the requirements. The template was created considering the specifics of the bank as a domain area. The study [7] proposed requirements documenting methodology adapted for middleware projects. The authors proposed the structure and common principle for the Interface Requirements Document that allows for improvement in the quality of requirements. The influence of the project's domain and some other parameters on the content of software requirements documents was confirmed by the survey results in which 163 participants took part [8]. The requirements Specification for Developers (RSAD) approach was proposed in [9] as the adoption of Software Requirements Specification for Agile Software Development. The RSAD approach suggests including three additional elements in requirements specification: conceptual model, mockup model, and extended specification of acceptance criteria. The efficiency of RSAD was tested via two industrial case studies over 15 months. Although the RSAD approach met the developers' expectations and allowed for the improvement of the quality of requirements, it was identified that in some cases, the teams' productivity was decreased due to inefficient search mechanisms and the grouping of several requirements in the same RSAD element. Study [10] reports the results of a survey of 84 IT professionals that examined current requirements engineering practices. Among other things, eight viewpoints containing requirements documents were identified: use-case, scenario, system interactions, system behaviors, information, structures, deployment, development, and operational concerns. The result of an international survey regarding current requirement specification and modeling practices was presented in [11]. Two hundred fifty-one usable responses from 24 countries were gathered. Use Cases, Class Diagrams, and User Stories were defined as the most popular elements of requirements specification. These studies confirm that choosing requirements architecture and presentation formats remains relevant for business analysts in IT projects. At the same time, most research is aimed at studying current practices rather than developing recommendations for choosing a requirements architecture depending on the project parameters.

One of the most popular methods for identifying stable dependencies and generating sets of rules is Association Rule Mining (ARM). This method stands out as a machine-learning technique that operates as an unsupervised method, eliminating the need for data annotation. Additionally, both the method and its outcomes are intuitive, making them accessible and understandable for domain experts and business professionals. The ARM algorithm is used for a wide range of practical problems: discovery of an enterprise architecture model [12], examined the factors that influence the students' academic performance [13], creation of a recommendation system [14], medical research [15, 16], requirement engineering tasks [17]. It confirms the demand and relevance of this method.

3 Association Rules Mining Algorithm

3.1 Formulating the Problem Under Inquiry

Let there is defined a list of possible items $I = \{i_1, i_2, ..., i_m\}$ and a dataset of transactions $T = \{t_1, t_2, ..., t_n\}$, such as transaction t contains some subset of items i. Now, we would like to discover the implication rules:

$$A \Rightarrow B \tag{1}$$

where $A \subset I$, $B \subset I$ and $A \cap B = \emptyset$. We will name A term an *antecedent* and B a *consequent*. Notice that antecedents and consequents might be a subset of items, not only exactly one item.

3.2 Apriori Algorithm

The apriori algorithm was proposed by Agrawal et al. [18]. Unlike a naive algorithm, the apriori algorithm allows for reduced calculation time. If the naive algorithm builds and estimates all possible combinations of items, apriori drops all items with a frequency lower than given hyperparameter values.

The algorithm calculates three measures for each rule:

- *Support* is a frequency measure of a particular pattern. It is the shortest way to cut the number of transactions not having valued information:

$$Support(A \Rightarrow B) = \frac{Occurence\ of\ A\ and\ B}{Total\ transaction\ in\ T} \tag{2}$$

- *Confidence* is a measure of the strength of a particular pattern. It is a frequency of cases when occurring antecedents lead to occurring consequences.

$$Confidence(A \Rightarrow B) = \frac{Occurence\ of\ A\ and\ B}{Occurences\ of\ A} \tag{3}$$

- *Lift* reflexes how the probability of a rule increases compared to the probability of the items. When the lift is one, it means no dependency between items. Otherwise, we can consider some connection between items in the rule. And the bigger the lift, the stronger the connection.

$$Lift(A \Rightarrow B) = \frac{Support(A \Rightarrow B)}{Support(A) * Support(B)} \tag{4}$$

3.3 How to Choose the Hyperparameters

The apriori algorithm takes two hyperparameters referenced to the above measures: support and confidence, naming them as *minimal support (minsupport)* and *minimal confidence (minconfidence)* accordantly. The algorithm returns all associated rules with estimated hyperparameters' values equal to or bigger than the given ones. Unfortunately, neither the algorithm nor the original paper provides suggestions for choosing the minimal support and confidence values. The literature review does not have the recommendations either. Most papers' parameters were chosen based on domain knowledge or business tasks [19–21].

Because we apply association rules mining to discover patterns of how business and system analysts choose and structure requirements documents, we have defined the following algorithm to select appropriate hyperparameters' values based on assessing changing rules number and their count. We set minimal support to 0.5 (the pattern occurs in more than 50% of cases) and minimal confidence as 0.1. Each parameter was changed by 0.1 with estimating rules numbers forming an estimation matrix. Then, the matrix was assessed, the parameters were initialized, and the algorithm was launched to gather rules for further analysis. Choosing the parameters, we based on several principles:

- Gathering rules have a lift value of more than 1.0.
- We compared the number of rules with slightly distinguishing hyperparameters' values (for example, 0.5–0.1 and 0.4–0.1 minimal support -minimal confidence accordingly), looking for significant changes in the number of rules.
- It must be possible to review gathered rules manually.
- There must not be many rules without remarkable distinguishing between them.

We have filtered discovered rules by predefined consequences to get the desired implications.

4 Results and Discussion

4.1 Input Data

The dataset was formed by surveying 324 practitioners from Ukrainian and international IT companies with development offices in Ukraine [22]. The survey structure was created based on the NAPIRE initiative's questionnaire [23]. It was updated based on the last version of bodies of knowledge provided by leading institutions in requirement engineering and business analysis (BA/RE): IIBA, PMI, IREB, and BCS. Following multiple rounds of internal peer reviews with business analysts from DataArt, a global software engineering firm, the questionnaire was validated by business analysis experts affiliated with prominent companies within the Ukrainian IT industry.

The survey contained the following questions regarding project context and practitioner's profile:

- project size and domain;
- company type (IT-outstaff, IT-outsource, IT product, non-IT) and size
- class of the developed system (business software, embedded software, scientific, etc.);
- belonging to the co-located or distributed team;
- practitioner's role in the project, experience in BA/RE and BA/RE certificate;
- using adaptive, hybrid, or predictive ways of working on the project;
- project category (developing from scratch, reengineering, etc.).

We asked respondents about their current practices regarding sections in business analysis documents and allowed them to make a multichoice from the following options:

- Assumptions
- Background (rationale and context for the new solution decision)
- Business requirements
- Business rules
- Constraints
- Cost-benefit analysis
- Data models
- Dependencies/Integrations
- Deployment specifics
- Functional requirements (FR)
- Glossary
- Goals & Objectives
- Non-functional requirements (NFR)
- Open questions
- Problem statement
- Risks
- Stakeholders analysis (including target audience/personas)
- Success metrics (project/solution)
- Technical Interfaces
- Usage scenarios
- User interface(s)

The dataset is available at the Mendeley Data repository [24].

4.2 Patterns of the Typical Structure of Documents

Before discovering more complex patterns, we have been finding document structure patterns running apriori algorithm for questions about which requirements people put in a separate document section.

The estimated number of rules depending on different values of hyperparameters (Fig. 1) shows several potential interesting pairs of hyperparameters:

- Minsupport = 0.1, min.confidence = 1.0. Despite the low frequency of the pattern, they are the strongest one.
- Minsupport = 0.4 and minconfidence = 0.1. This pair satisfies all the criteria mentioned above.

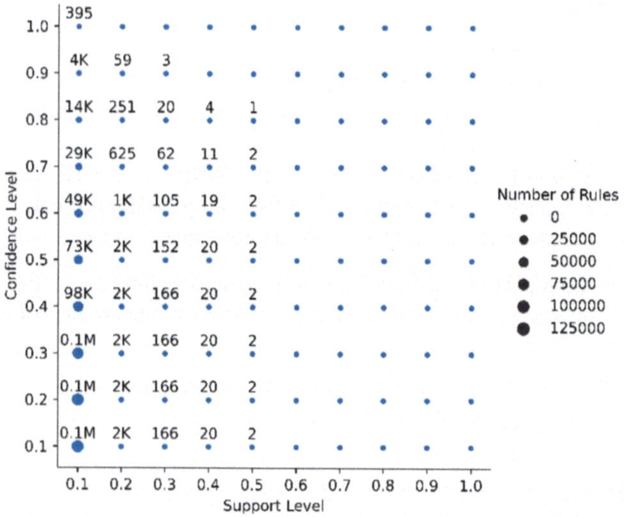

Fig. 1. Number of rules depending on hyperparameters' values (zero values are omitted, M – millions, K - thousands)

The group with hyperparameters minsupport = 0.1 and minconfidence = 1.0 contains 395 rules. All these rules have one of the three consequences and could be grouped by them (Table 1).

Table 1. Average lift and number of rules grouped by consequents

Consequents	Rule number	Lift	Support range
Functional requirements	384	1.44	0.1–0.12
Non-functional requirements	5	1.68	0.1–0.12
Usage scenarios	6	1.92	0.1–0.11

The "functional requirements" is the most common for these pair of hyperparameters. Remarkably, the lift value is equal inside the group. Also, there is no distinction in antecedent items; the value ranges from 3 to 7 items. These items include all options except cost-benefit analysis, deployment specifics, and success metrics.

We were considering the second group of rules for hyperparameters pair minsupport = 0.4 and minconfidence = 0.1, and several facts were observed:

- Rules with consequent "functional requirement" are always stronger than those where "functional requirement" is antecedent. The difference between confidence levels is around 0.1–0.2. It means that if a practitioner chooses one of the antecedents discussed below, they put functional requirements in a separate section more likely than otherwise.
- The following requirements in a separate section imply putting functional requirements in separate sections: Business requirements, Non-functional requirements, Glossary, Usage scenarios, User interface(s), and Dependencies/Integrations.
- Following requirements in a separate section implies putting non-functional requirements in a separate section: Glossary, Business requirements.
- Following requirements in a separate section implies putting non-business requirements in a separate section as well: Glossary and Usage scenarios.

4.3 Association Rules (Project Context ⇒ Sections of Document)

Next, we ran the algorithm for a set of questions, forming the project context and the question about document sections together. Estimated changes in rule numbers depending on hyperparameters are presented in Fig. 2.

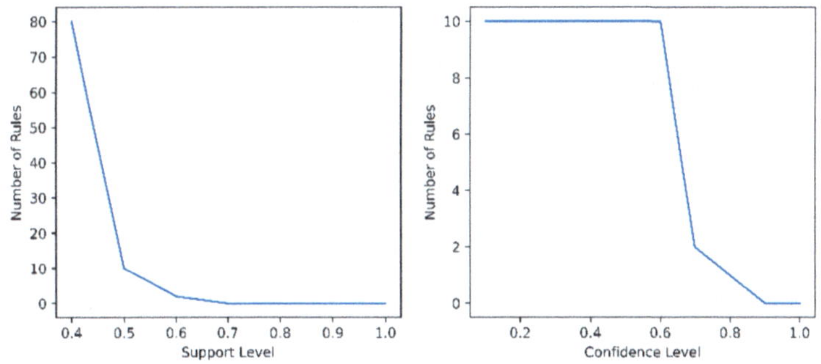

Fig. 2. Number of rules depending on hyperparameters' values at separate graphics (zero values are omitted)

The graphs show that the minsupport threshold equal to 0.5 satisfies the abovementioned requirements. But confidence level doesn't influence significantly. That is why we launched the algorithm for gathering the aimed set of patterns with the pair of hyperparameters' values: minsupport 0.5, minconfidence 0.1. Received rules are presented in Table 2. Here and below, the rules are written in a form (antecedents ⇒ consequents).

Analyzing the gathered rules, we can make several observations:

- Practitioners who work as BA tend to put a glossary, business requirements, functional and non-functional requirements in separate sections. There is no significant distinction between rules regarding their parameters except that functional requirements occur a bit more often than others.
- Practitioners more often put business and functional requirements in separate sections, glossary in business software development projects.
- Putting Functional Requirements in a separate section often leads to putting Business Requirements in a separate section of the document as well. The converse is also true.

These patterns align with definitions from the business analysis body of knowledge [3], domain knowledge, and observations. Business software is a very complicated area and often involves the development of complex domain language for business purposes. Probably that is why practitioners pay attention to the glossary, business, and functional requirements. Conversely, glossary, business requirements, functional and non-functional requirements represent the most critical aspects of project requirements. They cover three of five types of requirements according to the BABOK classification.

Table 2. Project Context ⇒ Sections of Document associated rules

Association Rule	Support level	Confidence level	Lift level
(Business software ⇒ FR)	0.6	0.69	1.01
(Role: BA⇒ FR)	0.6	0.69	1.01
(Business software ⇒ Business requirements)	0.55	0.64	1.02
(Role: BA ⇒ Business requirements)	0.55	0.64	1.02
(Role: BA ⇒ Glossary)	0.54	0.63	1.02
(Business software ⇒ Glossary)	0.54	0.63	1.02
(Role: BA ⇒ NFR)	0.53	0.61	1.03
(Business software, Role: BA ⇒ FR)	0.52	0.69	1.00
(Functional requirements ⇒ Business requirements)	0.51	0.74	1.18
(Business requirements ⇒ FR)	0.51	0.82	1.18

4.4 Association Rules (Modeling Techniques ⇒ Sections of Document)

In the first estimation, we received only four rules, which are insufficient for analysis, so we extended estimated pairs of hyperparameters and repeated estimations beginning from pair 0.1–0.1. The result is presented in Fig. 3.

The pair minsupport 0.4 and minconfidence 0.1 fit under our condition, and the algorithm was run with these hyperparameters. Only one antecedent and a consequent were found in all discovered rules. The first group has "Functional Requirements" as a

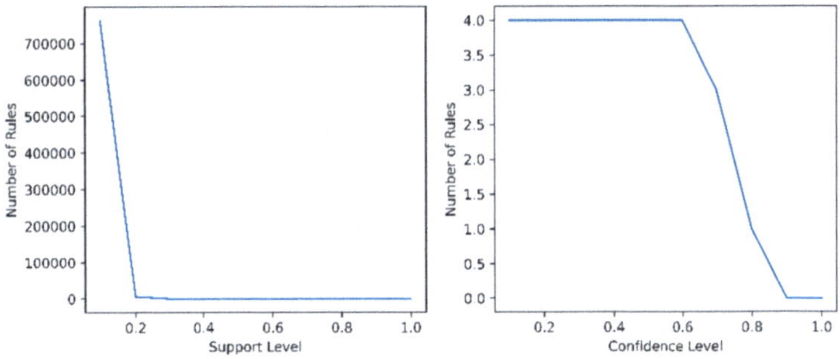

Fig. 3. Number of rules depending on a hyperparameters' values at separate graphics (zero values are omitted)

consequent. The rules from this group are listed in Table 3. All other rules are presented in Table 4.

By analyzing and comparing these groups, we can find out that:

- Both modeling techniques, User Stories and Use Cases, are connected with using Glossary, Functional and Non-Functional Requirements, Usage Scenario
- Discovered patterns with "Acceptance and Evaluation Criteria" are not valuable, in our opinion, due to the use of acceptance criteria as a part of the User Stories context [3]. In turn, the Acceptance and Evaluation Criteria result in the Non-functional, Functional, and Business Requirements and Glossary being placed in separate document sections.

Table 3. Modeling Techniques ⇒ Sections of Document associated rules with "Functional Requirements" as a consequent

Association Rule	Support level	Confidence level	Lift level
(User Stories ⇒ FR)	0.55	0.71	1.02
(Use Cases ⇒ FR)	0.49	0.75	1.09
(Acceptance and Evaluation Criteria ⇒ FR)	0.46	0.72	1.04
(User Stories, Acceptance and Evaluation Criteria ⇒ FR	0.42	0.73	1.06
(Use Cases, User Stories ⇒ FR)	0.41	0.76	1.09
(Sketches/Wireframes ⇒ FR)	0.41	0.75	1.08

Table 4. Modeling Techniques ⇒ Sections of Document associated rules without "Functional Requirements" as a consequent

Association Rule	Support level	Confidence level	Lift level
(User Stories ⇒ Glossary)	0.5	0.65	1.04
(User Stories ⇒ NFR)	0.48	0.61	1.03
(Use Cases ⇒ Business requirements)	0.46	0.7	1.12
(Use Cases ⇒ Glossary)	0.46	0.7	1.12
(User Stories ⇒ Dependencies/Integrations)	0.43	0.55	1.04
(Use Cases ⇒ NFR)	0.43	0.66	1.1
(User Stories ⇒ Assumptions)	0.42	0.54	1.08
(Acceptance and Evaluation Criteria ⇒ NFR)	0.42	0.67	1.13
(Acceptance and Evaluation Criteria ⇒ Business requirements)	0.42	0.66	1.07
(Use Cases ⇒ Usage scenarios)	0.41	0.64	1.22
(User Stories ⇒ Usage scenarios)	0.41	0.53	1.02
(Acceptance and Evaluation Criteria ⇒ Glossary)	0.41	0.65	1.05

5 Summary and Conclusion

Considering patterns joined project context, modeling techniques, and putting requirements in a separate section, we discovered several patterns that allow us to make recommendations regarding choosing a document structure:

- Separating Functional Requirements makes sense in many cases, mainly if a document already contains one of the following sections: Glossary, Business requirements, Non-functional Requirements, Usage scenarios, User interface(s), Dependencies/Integrations
- If a practitioner acts in a BA role, then the Glossary, Business, and Functional and Non-functional Requirements are suitable.
- The practitioners who work under business software carry about Glossary, Business, and Functional Requirements.
- If a practitioner uses User Stories or Use Case techniques, they might be recommended to separate Glossary, Functional and Non-Functional Requirements, Usage Scenario
- Using User Stories leads to separate "Acceptance and Evaluation Criteria," leading to separate Glossary, Business, Non-Functional, and Functional Requirements sections in their turn.
- Based on discovering patterns, the Glossary, Business, Functional, and Non-Functional Requirements are the most common sections in business analysis documents.

Future research may focus on analyzing the impact of project context, business analysis approaches, and practitioners' profiles on the project's challenges and success.

The results of other surveys conducted with practitioners from different countries can also be explored to check the recommendations defined in this study.

References

1. Paul, D., et al.: Business analysis. BCS, The Chartered Institute for IT (2014)
2. Stecklein, J., et al.: Error cost escalation through the project life cycle. In: 14th Annual International Symposium, vol. JSC-CN-8435 (2004)
3. International Institute of Business Analysis: A guide to the business analysis body of knowledge (BABOK Guide), ver. 3, IIBA (2015)
4. Gobov, D., Yanchuk, V.: network analysis application to analyze the activities and artifacts in the core business analysis cycle. In Proceedings of the 2nd International Informatics and Software Engineering Conference (IISEC), pp. 1–6. IEEE, Turkey (2021). https://doi.org/10.1109/IISEC54230.2021.9672373
5. Gobov, D.: Practical study on software requirements specification and modelling techniques. Int. J. Comput. **22**(1), 78–86 (2023). https://doi.org/10.47839/ijc.22.1.2882
6. Großer, K., Riediger, V., Riediger, J.: Requirements document relations. Softw. Syst. Model. **21**, 1–37 (2022). https://doi.org/10.1007/s10270-021-00958-y
7. Uddagiri, V., et al. Improving the Quality of Requirements in Middleware Requirements Specifications. In: Proceedings of the IEEE 28th International Requirements Engineering Conference (RE), pp. 412–415. IEEE (2020)
8. Rączkowska-Gzowska, K., Walkowiak-Gall, A.: What should a good software requirements specification include? Results of a survey. Found. Comput. Decis. Sci. **48**(1), 57–81 (2023). https://doi.org/10.2478/fcds-2023-0004
9. Medeiros, J., et al.: Requirements specification for developers in agile projects: evaluation by two industrial case studies. Inf. Softw. Technol. **117**, 106194 (2020). https://doi.org/10.1016/j.infsof.2019.106194
10. Ozkaya, M., et al.: Practitioners' perspectives towards requirements engineering: a survey. Systems **11**(2), 65 (2023). https://doi.org/10.3390/systems11020065
11. Amaral, L., Fábio, L., Anarosa, A.: A survey on requirements notations in software engineering research. In: Proceedings of the 37th ACM/SIGAPP Symposium on Applied Computing, pp. 1291–1298. ACM, United States (2022)
12. Pinheiro, C., Guerreiro, S., Mamede, H.: A survey on association rule mining for enterprise architecture model discovery. Bus. Inf. Syst. Eng. **66**, 777–798 (2023). https://doi.org/10.1007/s12599-023-00844-5
13. Ullah, M., et al.: Impact of ICT on students' academic performance: applying association rule mining and structured equation modeling. Int. J. Adv. Comput. Sci. Appl. **10**(8) (2019). https://doi.org/10.14569/ijacsa.2019.0100852
14. Anwar, K., Siddiqui, J., Sohail, S.: Machine learning-based book recommender system: a survey and new perspectives. Int. J. Intell. Inf. Database Syst. **13**(2–4), 231–248 (2020). https://doi.org/10.1504/IJIIDS.2020.109457
15. Castro, G., et al.: Applying association rules to study bipolar disorder and premenstrual dysphoric disorder comorbidity. In: Proceedings of the IEEE Canadian Conference on Electrical & Computer Engineering (CCECE), pp. 1–4. IEEE, Canada (2018)
16. Huang, Y., et al.: Exploring the correlation between attention and cognitive load through association rule mining by using a brainwave sensing headband. IEEE Access **8**, 38880–38891 (2020). https://doi.org/10.1109/ACCESS.2020.2975054
17. Gobov, D., Sokolovskiy, N. Association rule mining for requirement elicitation techniques in IT projects. In: Proceedings of the 18th Conference on Computer Science and Intelligence Systems (FedCSIS), pp. 983–987. IEEE, Poland (2023). https://doi.org/10.15439/2023F4831

18. Agrawal, R., et al.: Fast algorithms for mining association rules. In: Proceedings of the 20th international conference very large data bases, vol. 1215, pp. 487–499. Chile (1994)
19. Liu, Y., et al.: Application of big data technology combined with clustering algorithm in manufacturing production analysis system. Decis. Mak. Appl. Manage. Eng. **7**(1), 237–253 (2024). https://doi.org/10.31181/dmame712024897
20. Gu, Y.: Exploring the application of teaching evaluation models incorporating association rules and weighted Naive Bayesian algorithms. Intelligent Systems with Applications **20**, 200297 (2023). https://doi.org/10.1016/j.iswa.2023.200297
21. Petry, F., Yager, R.: Data mining using association rules for intuitionistic fuzzy data. Information **14**(7), 372 (2023). https://doi.org/10.3390/info14070372
22. Gobov, D., Huchenko, I.: Modern requirements documentation techniques and the influence of the project context: Ukrainian IT experience. In: Hu, Z., Dychka, I., Petoukhov, S., He, M. (eds.) Advances in Computer Science for Engineering and Education. ICCSEEA 2022. Lecture Notes on Data Engineering and Communications Technologies, vol. 134, pp. 260–270. Springer, Cham (2022). https://doi.org/10.1007/978-3-031-04812-8_22
23. Wagner, S., et al.: Status quo in requirements engineering: a theory and a global family of surveys. ACM Trans. Softw. Eng. Method. (TOSEM) **28**(2), 1–48 (2019). https://doi.org/10.1145/3306607
24. Mendeley. https://data.mendeley.com/datasets/25kybnt3z3/1. Accessed 19 Feb 2024

Computer Subsystem for Modeling Parameters of Gravure Printing Forms

Lyudmyla Mayik[1](✉), Volodymyr Mayik[1], Nataliia Lotoshynska[2], and Myroslava Dubnevych[1]

[1] Ukrainian Academy of Printing, 19, Pid Holoskom Street, Lviv 79020, Ukraine
ludmyla.maik@gmail.com

[2] Lviv Polytechnic National University, 28a, Stepan Bandera Street, Lviv 79013, Ukraine
natlot@ukr.net

Abstract. The gravure printing method is used in publishing and printing (high-art single and multicolor publications such as albums, mass illustrated magazines, advertising publications, artistic postcards, postage stamps, etc.), also in printing of securities, in particular money, packaging materials, labels, design of wallpapers, labeling of ampoule and cable products. Production of polymer forms for the intaglio printing is carried out using systems "computer-printing form" with the method of the infrared solid-state laser engraving (wave length is 1,064 microns). Since the form engraving is conducted with the help of a laser, in the image preparation program there is a possibility to set the parameters of the raster (the type and ruler of the raster, the step of the raster cell and the angle of rotation), as well as the parameters of the raster image cells (the depth of the cell, the angle of inclination of the side surfaces, the ratio of cells and membranes). The parameters of image rasterization in intaglio printing significantly affect the volume of cells and ink capacity of printing forms, which, accordingly, affects the reproduction-graphic, printing-technical characteristics of printing forms and prints. On the basis of the conducted experimental studies, there was developed an algorithm of the computer subsystem for modeling the influence of raster image parameters on the volume of cells and the ink capacity of polymer forms of gravure printing. The program written in the JavaScript language simulates the ink capacity of the gravure printing form, calculates the volume of the cells taking into account the raster ruler, the ratio of cells and the layers of the raster cell, and the depth of the cell. The developed modeling process allows to automate the design of gravure printing forms with predetermined characteristics that provide the necessary cell volume and ink capacity of the form, since in gravure printing the gradation characteristic of the printing process under stabilized printing conditions is a stable dependence of the optical density of the print from the parameters of the form printing elements, which are taken into account in the generalized parameter of the cell volume or the ink capacity of the form.

Keywords: computer subsystem · software · modeling · parameters · intaglio (gravure) printing

Z. Hu et al. (Eds.): ICCSEEA 2024, LNDECT 242, pp. 329–338, 2025.
https://doi.org/10.1007/978-3-031-84228-3_28

1 Introduction

In the printing field and other industries (pharmaceutical, cable, etc.), the gravure printing method is widely used for the marking process as applying images on various types of products [1, 2]. In many cases, the gravure printing method cannot be replaced by another method due to a number of reasons (technical [3] and technological [4], constructive [5], environmental [6], etc.). The main element in this technological process is the gravure printing form [7, 8]. The printing elements of gravure printing forms, regardless of the type of originals, are very small embedded raster cells [6, 9]. They are separated from each other by thin membranes which are intermediate elements that are located at the level of the forming material surface [10, 11]. Thus, the image different tonality on the printout is ensured by the different ink thickness [12, 13] that is transferred from the form to the printed material [14, 15].

In our research, the gravure printing form is produced with the laser engraving ("computer-to-plate") method using the developed new form polymer materials, which allows to adjust the main parameters of the raster cells of the printing form. This makes it possible to implement the mathematical modeling in terms of the influence of engraving on the volume and ink capacity of gravure printing forms main parameters, and on this basis to optimize the image reproduction process [16, 17].

2 Main Part Representation

Based on the results of experimental studies, there was conducted the calculation of the parameters influence of the polymer gravure printing forms raster image [18] on the ink capacity in the line images printing.

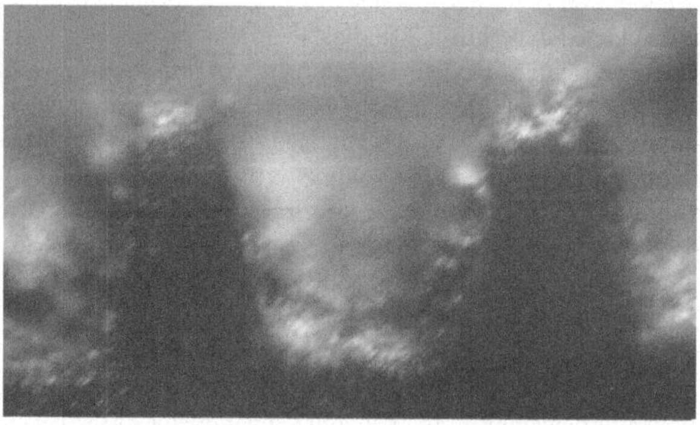

Fig. 1. The microphotograph with the cross-section of a raster cell after the circulation resistance tests

For a mathematical description of the parameters influence of raster forms on the printing form ink capacity (V, cm3/m2) while the line elements printing, the following parameters were selected as the main parameters: the raster ruler equal to 40–100 lin/cm, the depth of printing forms equal to 5…60 μm, a set of cells and membranes (cell parameter) equal to the new range from 1:2 to 1:6. The optimal range of the ink capacity while the line elements printing is 20–23 cm^3/m^2.

For this purpose, the calculations for differences between the cell form produced by laser engraving (Fig. 1), the form taking into account the Gaussian distribution of radiation energy (Fig. 2) in the laser beam (Fig. 3), and the idealized cell form were conducted (Fig. 5).

The energy distribution in a single-mode laser beam was calculated based on the adequate description by the Gaussian distribution [19]:

$$E = \frac{1}{\sqrt{2\pi}\sigma} \exp\left(\frac{-x}{2 \times \sigma^2}\right) \tag{1}$$

where σ is the distance from the center to the inflection point (Fig. 2).

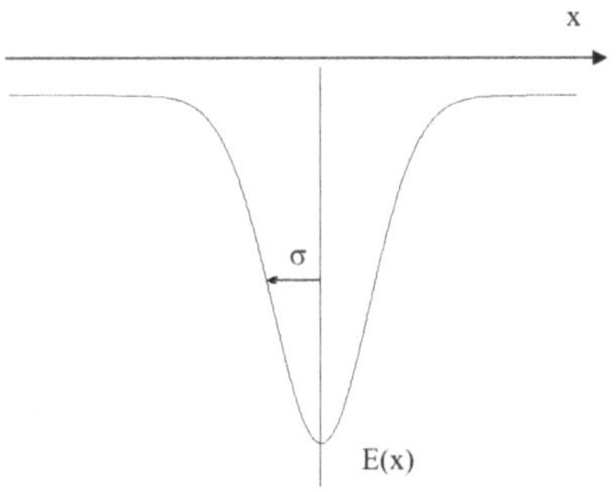

Fig. 2. The Gaussian distribution of energy in a laser beam

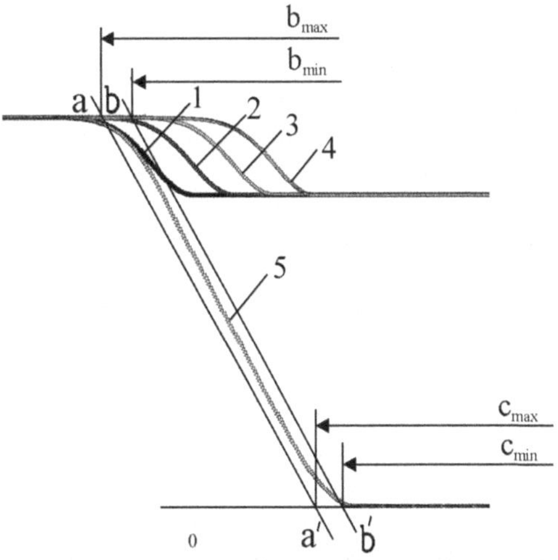

Fig. 3. The formation process modeling of the raster cell side wall

The distribution of the radiation energy in the focus of the laser engraver lens was carried out according to the technical documentation for the optical elements of the laser engraving apparatus.

The Fig. 3 shows the simulation of the formation process of the cell side wall, taking into account the energy distribution in the laser beam and in accordance with the form engraving mode. Lines 1, 2, 3, 4 describe the power distribution profile of each pass, line 5 is the resulting value of the passes sum of the laser beam. Curves 1–4 were obtained by calculating the Gaussian distribution [20] of laser radiation energy using formula 1 and the software algorithm that is presented in the Fig. 4, which uses the following parameters: number of passes (max – 20); the displacement step between passes; for numerical verification is used the Gaussian parameter σ.

It is possible to specify the value of the threshold effect and the possibility of empirically specifying the change in the engraving depth for the case of more than 10 passes, which is related to the change in the specific power density due to defocusing.

With the usage of the graphic-analytical method, there was calculated the accuracy of two types of simulations for the form raster cell wall as a plane and a surface, taking into account the Gaussian distribution of energy in laser radiation (Fig. 3). By drawing auxiliary lines aa' and bb', it is possible to determine the maximum (bmax, bmin) and minimum (cmax, cmin) size of the truncated pyramid bases. Accordingly, there was calculated the difference between the idealized cell shape and the Gaussian model according to formula 3. The difference in the ink capacity is less than 5%. A similar method was used to evaluate the difference between the idealized and real cell form (Fig. 1).

The calculations showed that the deviation of the form ink capacity with idealized cells from the form that was calculated taking into account the Gaussian distribution of energy in the laser beam is less than 5%, and from the real one the difference is equal to $\pm 2.4\%$. Therefore, for the investigated materials, the calculation can be carried out using the model of an idealized cell.

The calculation was carried out taking into account the following restrictions:

- the walls of the raster cell are planes, that means the features of the energy distribution in the laser beam and the features of the burning process were not taken into account;
- the formed pyramid is regular, meaning that the height passes through the base center;
- the base of the raster cell is a square.

Depending on the engraving conditions, the pyramid may be straight or truncated. The cross-section of a cell in the form of a regular truncated pyramid is shown in Fig. 5.

The algorithm was developed (Fig. 6) and a software was written in JavaScript for the cell volume calculations taking into account the raster ruler, the ratio of cells and membranes of the raster cell and the depth of the cell.

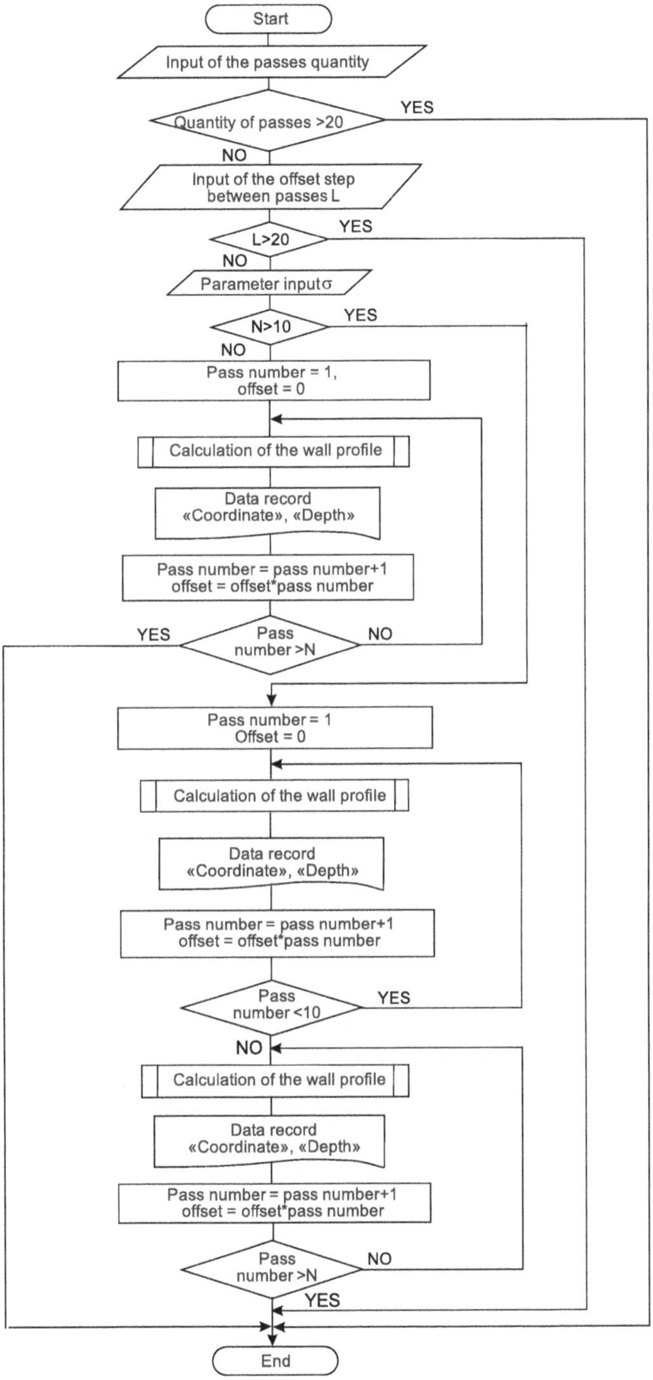

Fig. 4. The software algorithm for the modeling process of the side wall of the gravure printing form cell formation

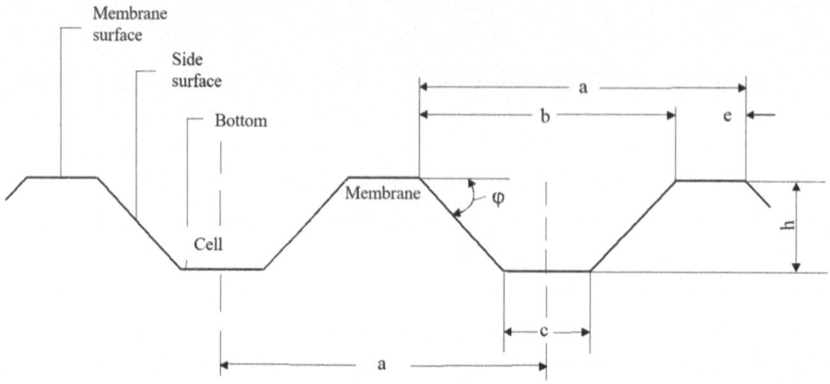

Fig. 5. The cross-section of an idealized raster cell on the graver printing form: a – raster step; b – cell width; c - the width of the bottom; e – the membrane width; h – the cell depth; φ - the angle of the raster cell wall inclination.

The calculation results are recorded in a file in the form of a table (Table 1).

The calculation was carried out under the conditions of changing the cell size, which corresponds to the raster ruler-frame equal to 40–100 lin/cm with a step of 1 lin/cm for the wall inclination angle of the 78° printing element. The input parameter for calculations is the depth of the raster cell.

The results processing and presentation were carried out using the Mi-crocal Origin 6.0 program. The results can be presented in two versions: 3-dimensional (Fig. 7) and 2-dimensional (Fig. 8).

In the 3-dimensional representation, the surfaces formed by a set of points corresponding to the ink capacity for a given depth are depicted (for example, 30 μm - surface 1 and 40 μm - surface 2) while changing the raster ruler for ratios 1:2, 1:3, 1:4, 1:5, 1:6.

In the 2-dimensional representation, the lines of the nomograms are the projections of surfaces 1 and 2 from Fig. 7 per plane. These lines are lines of the same ink capacity (a – for a depth of 30 μm; b – for a depth of 40 μm).

In the gray zone (Fig. 8, a), the ink capacity is more than 20 cm^3/m^2 and less than 23 cm^3/cm^2 for rulers of 40–60 lin/cm and, as can be seen from the nomogram, it can be achieved only with a ratio of cells and membranes of 1:6.

The shaded area in Fig. 8, b) is marked with the ink capacity of 20–23 cm^3/m^2. It can be seen from the nomogram that the required ink capacity for rulers of 40–60 lin/cm is achieved at a ratio of 1:3 and 1:4, but a ratio of 1:3 will provide the required ink capacity for rulers of 40–50 lin/cm, and a ratio of 1:4 will provide for rulers of 50–60 lines/cm.

For a ruler of 50 lin/cm with a ratio of 1:3, the ink capacity is 20.046 cm^3/m^2, and for a ratio of 1:4, the ink capacity is 22.976 cm^3/m^2.

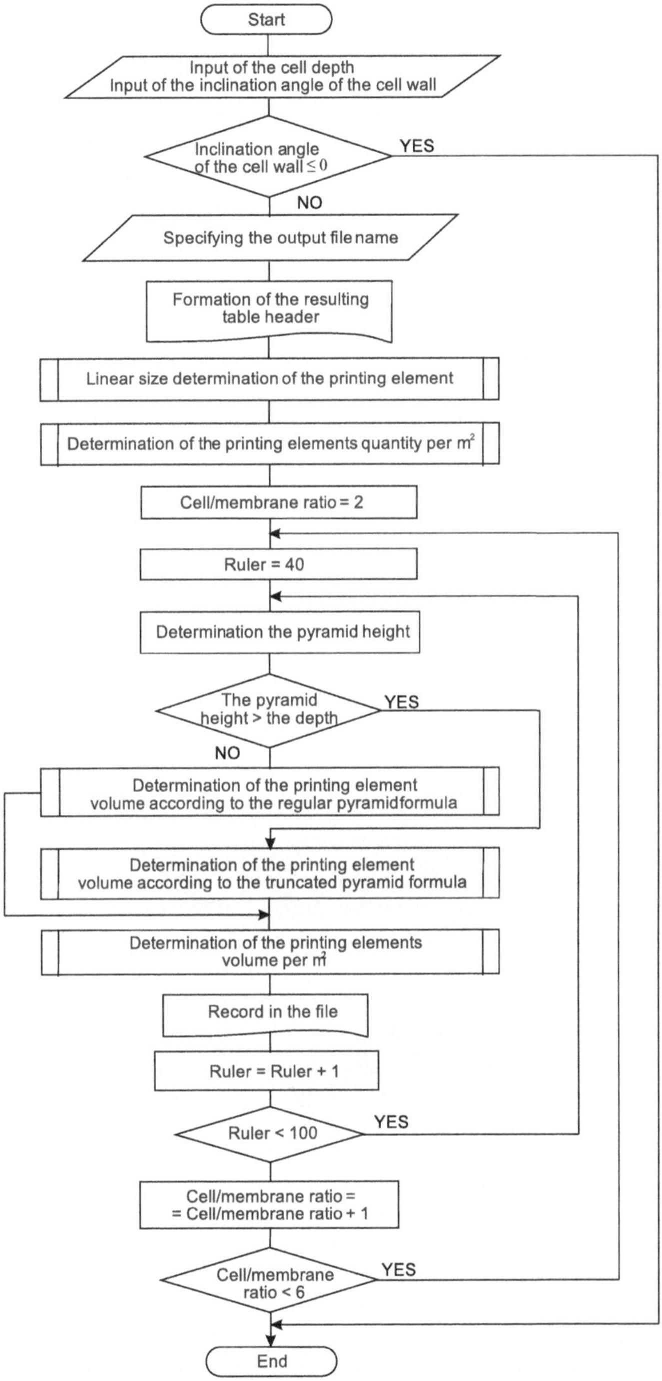

Fig. 6. Algorithm of the program for calculating the cell volume and the ink capacity of the gravure printing form

Table 1. The calculation table fragment for the cell volume and form ink capacity using the developed program

Depth	=	40.000	mkm	Kut = 78	V*(10^-12) cm^3	
Ruler	40 cm^{-1}	41 cm^{-1}	42 cm^{-1}	43 cm^{-1}	44 cm^{-1}	45 cm^{-1}
Cell						
1-do-2	166.67 mkm	162.60 mkm	158.73 mkm	155.04 mkm	151.52 mkm	148.15 mkm
1-do-3	187.50 mkm	182.93 mkm	178.57 mkm	174.42 mkm	170.45 mkm	166.67 mkm
1-do-4	200.00 mkm	195.12 mkm	190.48 mkm	186.05 mkm	181.82 mkm	177.78 mkm
1-do-5	208.33 mkm	203.25 mkm	198.41 mkm	193.80 mkm	189.39 mkm	185.19 mkm
1-do-6	214.29 mkm	209.06 mkm	204.08 mkm	199.34 mkm	194.81 mkm	190.48 mkm
Na_m2	16000000	16810000	17640000	18490000	19360000	20250000
	V	V	V	V	V	V
1-do-2	16.026	15.983	15.941	15.899	15.857	15.815
1-do-3	20.521	20.473	20.425	20.378	20.330	20.282
1-do-4	23.485	23.434	23.383	23.331	23.280	23.229
1-do-5	25.572	25.519	25.465	25.412	25.358	25.305
1-do-6	27.117	27.062	27.007	26.952	26.897	26.842

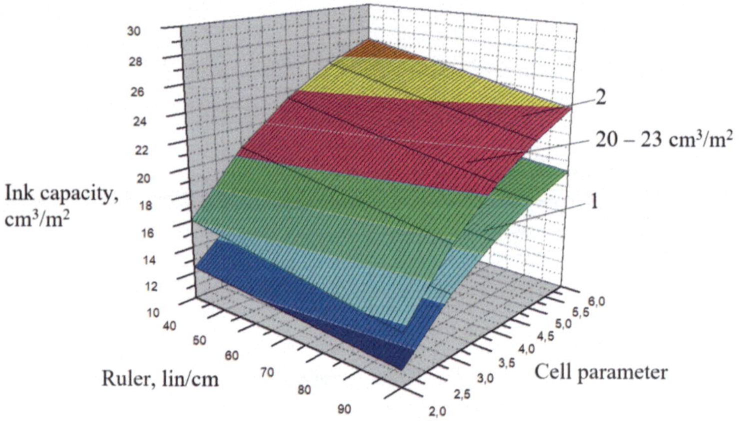

Fig. 7. Dependence of the ink capacity on the ruler and the ratio of cells and membranes: 1 – the depth of cells is 30 μm; 2 – the depth of the cells is 40 μm.

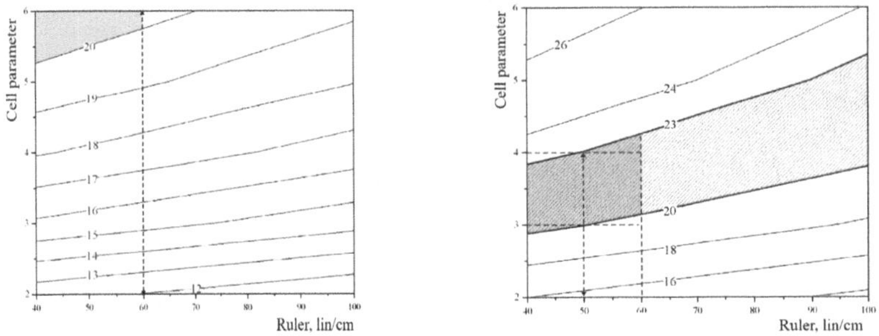

a) b)

Fig. 8. Nomograms for determining the optimal ratio of cells and membranes for different rulers at a given depth (for example, for the case of V = 20–23 cm^3/m^2, $\varphi = 78°$): a – depth of 30 μm; b – depth of 40 μm.

3 Conclusions

The development of computer technologies and the widespread usage of lasers in the printing industry led to the "Computer-to-plate" system creation, which is a new technology for the direct production of polymer gravure forms without the usage of expensive and scarce metals, complex processing of form cylinders, and long-lasting etching solutions, which can be based on the usage of YAG:Nd lasers with a wavelength of 1.064 μm.

The usage of laser technologies allows to adjust the main parameters of the raster cells on the printing form, that enables to carry out mathematical modeling of the effect of the main engraving parameters on the volume and ink capacity of the gravure printing forms.

There was implemented the modeling subsystem as a software that enables the automation of the designing polymer printing forms process with specified characteristics.

The influence of the raster parameters and conditions of the engraving process on the ink capacity of the printing form has been revealed and investigated. The recommended ink capacity of the printing form for the line elements reproducing with the required optical density is within 20–23 cm^3/m^2.

On the basis of studies regarding the main parameters of the raster image influence, a mathematical description of their influence on the ink capacity of gravure printing forms was carried out, and optimized modes of their production and operation were proposed.

The created method of the formation process modeling of the graver printing form raster cell can be used for the investigation of similar technological processes.

References

1. Szentgyörgyvölgyi, R.: Gravure Printing, pp. 199–215. William Andrew Publishing, Printing on Polymers (2016)

2. Lee, J., Kim, J., Noh, Y., Jeong, H., Lee, D.: Advanced technique for achieving 10-μm-width fine lines in roll-to-roll continuous gravure printing. Precis. Eng. **69**, 1–7 (2021)
3. Nguyen, H.A.D., Lee, C., Shin, K.-H.: Approach to optimizing printed conductive lines in high-resolution roll-to-roll gravure printing. Robot. Comput. Integr. Manuf. **46**, 122–129 (2017)
4. Zhang, C., et al.: Roll-to-roll micro-gravure printed large-area zinc oxide thin film as the electron transport layer for solution-processed polymer solar cells. Org. Electron. **45**, 190–197 (2017)
5. Sico, G., Montanino, M., Prontera, C.T., De Girolamo Del Mauro, A., Minarini, C.: Gravure printing for thin film ceramics manufacturing from nanoparticles. Ceram. Int. **44**(16), 19526–19534 (2018)
6. Jyun-Ting, W., Carvalho, M.S., Kumar, S.: Emptying of gravure cavities containing shear-thinning and shear-thickening liquids. J. Nonnewton. Fluid Mech. **268**, 46–55 (2019)
7. Schäfer, J., Roisman, I.V., Sauer, H.M., Dörsam, E.: Millisecond fluid pattern formation in the nip of a gravure printing machine. Colloids Surf. A **575**, 222–229 (2019)
8. Brumm, P., Ciotta, N., Sauer, H.M., et al.: Deep learning study of induced stochastic pattern formation in the gravure printing fluid splitting process. J. Coat. Technol. Res. **20**, 51–72 (2023)
9. Zhang, X., Liu, K., Sunappan, V., Shan, X.: Diamond micro engraving of gravure roller mould for roll-to-roll printing of fine line electronics. J. Mater. Process. Technol. **225**, 337–346 (2015)
10. Brinksmeier, E., Schönemann, L.: Generation of discontinuous microstructures by diamond micro chiseling. CIRP Ann. **63**, 49–52 (2014)
11. Khandavalli, S., Lee, J.A., Pasquali, M., Rothstein, J.P.: The effect of shear-thickening on liquid transfer from an idealized gravure cell. J. Non Newton. Fluid Mech. **221**, 55–65 (2015)
12. Khandavalli, S., Rothstein, J.P.: Ink transfer of non-Newtonian fluids from an idealized gravure cell: The effect of shear and extensional deformation. J. Nonnewton. Fluid Mech. **243**, 16–26 (2017)
13. Chung, C., Kumar, S.: Emptying of viscoelastic liquids from model gravure cells. J. Nonnewton. Fluid Mech. **221**, 1–8 (2015)
14. Dutta, P., Jana, G.G., Paul, S., Pal, S., Dey, S., Sadhu, A.: AI-based smart prediction of liquid flow system using machine learning approach. Int. J. Eng. Manuf. (IJEM) **14**(1), 53–65 (2024)
15. Nazimuddin, A.K.M., Ali, Md. S.: The study of slow manifolds in the Lorenz-Haken model using differential geometry. Int. J. Math. Sci. Comput. (IJMSC) **9**(4), 1–9 (2023)
16. Ahmad, F., Faisal, M.: Assessing similarity between software requirements: a semantic approach. Int. J. Inf. Eng. Electron. Bus. (IJIEEB) **15**(2), 38–53 (2023)
17. Nwandu, I.C., Odii, J.N., Nwokorie, E.C., Okolie, S.A.: Evaluation of software quality in test-driven development: a perspective of measurement and metrics. Int. J. Inf. Technol. Comput. Sci. (IJITCS) **14**(6), 13–22 (2022)
18. Shandilya, S., et al.: YOLO-based segmented dataset for drone vs. bird detection for deep and machine learning algorithms. Data Brief 50, 109355 (2023)
19. Izonin, I., et al.: SGD-based cascade scheme for higher degrees wiener polynomial approximation of large biomedical datasets. Mach. Learn. Knowl. Extr. **4**(4), 1088–1106 (2022)
20. Havryliuk, M.: Investigation of PNN optimization methods to improve classification performance in transplantation medicine. CEUR Workshop Proc. **3609**, 338–345 (2023)

Tree-Based Classification of the Technical Ukrainian Texts

Sergii V. Mashtalir[1](✉) and Oleksandr V. Nikolenko[2]

[1] Kharkiv National University of Radio Electronics, 14, Nauky Avenue, Kharkiv 61166, Ukraine
sergii.mashtalir@nure.ua
[2] Uzhhorod National University, 14, University Street, Uzhhorod 88000, Ukraine
oleksandr.nikolenko@uzhnu.edu.ua

Abstract. Our research leverages anonymous data from an online garage management system. This vast dataset, enriched by transcribed customer calls, forms the basis for our tree-based classification model development, aimed at categorizing automotive repairs and parts. Addressing the challenges of analyzing technical texts, particularly in Ukrainian, we adapted classical NLP methods for our dataset. We explored the most popular data classification algorithms. Despite their potential, none provided sufficient accuracy for classifying parts, leading us to develop and train a custom algorithm tailored to our hierarchical data structure. Our Python library facilitates real-time classification, demonstrating significant speed and accuracy improvements across various datasets. This work not only advances the application of tree-based classification in the automotive sector but also sets the stage for future research on more complex tasks, including extracting and classifying automotive information from large text corpora. In conclusion, our study underscores the utility of tree-based classification for technical text analysis within the automotive industry, promising enhanced operational efficiencies and valuable insights for various stakeholders.

Keywords: NLP · tree-based classification · machine learning

1 Introduction and Literature Review

The growing field of tree-based data classification is increasingly recognized for its capacity to manage and categorize data with inherent hierarchical structures, proving essential in areas such as biology for phylogenetic tree analysis [1], computer science for file system management [2], and information retrieval for document categorization [3]. Its main strength lies in efficiently organizing data according to hierarchical relationships, facilitating effective retrieval, analysis, and modeling.

In tree-based classification, each tree node represents a category or class, with branches to child nodes denoting subcategories. Classification involves assigning data points to nodes based on attributes, following specific rules or models that define their placement.

Z. Hu et al. (Eds.): ICCSEEA 2024, LNDECT 242, pp. 339–348, 2025.
https://doi.org/10.1007/978-3-031-84228-3_29

Despite the comprehensive development and optimization of tree-based classification methodologies for datasets featuring numerical or formal categorical attributes, including Decision Trees [4], Random Forests [5], Gradient Boosted Trees [6], and Hierarchical Clustering [7], the domain of Natural Language Processing (NLP) reveals a significant gap. Specifically, a robust framework for the textual strings categorization into predefined hierarchical classes remains elusive, more so for local languages and specialized technical terminologies.

This paper investigates tree-based classification for textual data, detailing its theory, algorithms, applications, particularly focusing on classifying texts from an automotive garage management system (GMS) into predefined categories of repairs and parts.

Tree-based classification methodologies are instrumental in automating a myriad of processes within the automotive industry, spanning the spectrum from quality monitoring of call center operations at car service centers to the centralized data collation concerning specific vehicle issues for stakeholders such as insurance companies and automobile manufacturers.

Like other NLP tasks, this endeavor falls under the umbrella of unstructured data processing and is marked by its complexity, necessitating a blend of expertise in natural language processing and specialized data science methodologies [8]. The presence of a hierarchical directory of tasks categorizes this problem as a supervised machine learning task, thereby facilitating the construction and training of models and algorithms.

Typically, data science tasks are segmented into several phases: Problem Definition, Data Collection, Data Preprocessing, Data Modeling (encompassing Model Selection, Model Training, Model Evaluation, Model Tuning), and Deployment. These phases are further elaborated upon in the subsequent sections of this paper.

2 Problem Definition

The general objective is to develop an algorithm for extracting and further classifying texts using given tree-based classes sets. For this purpose, we have 4 types of datasets:

1. Classes – two distinct trees for works (180 entries) and parts (1,600 entries).
2. Training – the class trees, augmented with manually labeled data (6,000 entries).
3. Test – additional labeled dataset (11,000 entries).
4. Operational or Input Data – tens of millions of entries.

Car repair works are organized in a three-level tree structure.

Car components are classified in a four-level tree structure, starting with the components most general types (mechanical parts, electric and electronic components, body parts and optics, oils and fluids, wheels and tires, etc.). Further classes are divided into corresponding systems (braking, chassis, steering, transmission, engine, etc.), and then into specific components (for example, the braking system – into pads and shoes, discs and drums, hydraulics, pneumatics, etc.). Finally, components last tier consists of a list of specific spare parts and their subparts (brake pads, brake shoes, parking brake pads, etc.). The extract of car parts data tree is shown in Table 1.

The training and test datasets consist of works and parts combined lists, for instance, "dismantling and disposal of non-asbestos disc brake pads for trucks." These lists have been manually labeled specifically for this study purposes.

Operational or Input data comprises arbitrary text, which may include data from GMS, phone calls transcriptions, messages from messengers, emails, etc. The goal is to determine whether the input text contains information related to car repairs and then correctly assign it to one of the works and parts classes.

Table 1. Car parts tree data extract

ID	Parent	UA	EN
1030000	1000000	Деталі Гальмівної Системи	Brake Parts
1030100	1030000	Колодки гальмівні	Brake pads & shoes
1030101	1030100	Колодки гальмівні дискові	Brake pads
1030102	1030100	Колодки гальмівні барабанні	Brake shoes
1030103	1030100	Колодки стояночного гальма	Parking pads & shoes
1030104	1030100	Колодки гальмівні спортивні	Sport brake pads
1030200	1030000	Диски гальмівні, щити дисків	Brake discs, brake discs shields
1030201	1030200	Диски гальмівні	Brake discs

A classification is deemed successful if: 1) No less than 90% of car repair works are identified and extracted from the incoming unlabeled texts **AND** 2) Of these, no less than 90% of works and parts are correctly assigned to their appropriate classes. For example, the text "inspection of handbrake pads" should be accurately classified both into class 1104 in works and 1030103 in parts.

However, for the scope of this study, we consider a simplified task where we precisely know that the input text contains information about works and components and only the second criterion of successful classification is considered. The task of extracting relevant information about repair works and automotive parts from arbitrary text will be addressed in future studies.

3 Model Selection

The three simplest yet most popular algorithms for text data classification are Naive Bayes, k-Nearest Neighbors (kNN), and Logistic Regression [9]. Each of these algorithms possesses a unique set of strengths and weaknesses, especially when applied to Natural Language Processing (NLP) tasks.

3.1 Naive Bayes

Naive Bayes is an algorithm based on Bayes' theorem with the assumption of independence among input features. The probability that a text with given features x_1, x_2, …, x_n belongs to class C_k is calculated using the formula [10]:

$$P(C_k|x_1, x_2, \ldots, x_n) = \frac{P(C_k)P(x_1, x_2, \ldots, x_n|C_k)}{P(x_1, x_2, \ldots, x_n)}$$

where:

- $P(C_k|x_1, x_2, \ldots, x_n)$ is the conditional probability that the sample belongs to class C_k,
- $P(C_k)$ is the prior probability of class C_k,
- $P(x_1, x_2, \ldots, x_n|C_k)$ is the likelihood of observing features $x_1, x_2, \ldots, x_n$ given that the sample belongs to class C_k,
- $P(x_1, x_2, \ldots, x_n)$ is the probability of observing features $x_1, x_2, \ldots, x_n$ irrespective of the class.

It then assumes that all features x_i are independent of each other, significantly simplifying the probability calculation as the product of individual probabilities.

$$P(x_1, x_2, \ldots, x_n|C_k) = P(x_1|C_k)P(x_2|C_k)\ldots P(x_n|C_k)$$

The algorithm advantages include simplicity, efficiency, and the ability to compute multiple classes simultaneously. It is straightforward to understand and implement, performs reliably even with small training datasets (if the independence assumption holds), and is suitable for large datasets. However, the independence assumption often fails in real-world NLP tasks where words are contextually dependent.

3.2 K-Nearest Neighbors (kNN)

The kNN is one of the simplest yet effective classification algorithms. Its essence lies in calculating the distance from a given object to all others, with the object being classified as belonging to the most frequent class among its k neighbors. This method advantages include that it requires no training, making kNN a lazy learning algorithm. It can handle nonlinear data and easily adapts to various types of NLP tasks. However, kNN fully exhibits the curse of dimensionality in high-dimensional spaces, where distances between points become less meaningful. Moreover, its speed drastically decreases as the dataset size grows, and the algorithm performs poorly with noisy data.

3.3 Logistic Regression

This classification algorithm is based on the logistic function. The logistic or sigmoid function, which is S-shaped, is calculated using the formula [11]:

$$f(x) = \frac{L}{1 + e^{-k(x-x_0)}}$$

where:

- L is the maximum curve value,
- x is the input value,
- x_0 is the value of x at which $f(x) = L/2$,
- k is the steepness of the curve.

This function transforms the linear combination of input parameters into a probability in the interval from 0 to 1. Unlike classical regression, logistic regression is used not for prediction tasks but for classification and probability estimation. In addition to calculating probabilities that can be useful in estimating the classification results confidence, logistic regression advantages include easy scaling to large datasets and the possibility of regularization to avoid overfitting. On the other hand, logistic regression operates under the assumption of a linear relationship between input parameters, which mainly is far from true in NLP tasks. Moreover, it does not determine the importance of features as effectively as tree-based models.

In the practical resolution of our task, the k-Nearest Neighbors (kNN) algorithm was selected for the works classification. Regrettably, none of the implemented algorithms demonstrated adequate accuracy in classifying parts. The highest recorded accuracy was 83%, achieved by the kNN algorithm. This outcome necessitated the development, training, and evaluation of a bespoke algorithm tailored to the tree-like architecture of the input data.

4 Data Collection and Preprocessing

Our research utilized anonymous data from an online garage management system, covering five years and 500 service stations, yielding over 1 million work records and 1.2 million spare parts records annually. Additionally, customer calls recorded via IP telephony and transcribed into text data contributed to a substantial dataset with hundreds of thousands of texts and tens of millions of data rows. Training and test sets were randomly drawn and manually labeled.

In our previous work we explored the challenges of analyzing technical texts in Ukrainian, Russian, and their mix ("surzhyk") [12]. That work summarizes key aspects: the data includes many errors and technical terms, often in bilingual mix, with limited standard cleaning tools for Ukrainian.

Adapting classical NLP methods, we implemented language identification, translated texts into Ukrainian, and performed lemmatization using Ukrainian online dictionaries. This process reduced the original 10,288 sentences to 6,062 tokens and halved the dictionary size by 59%, enhancing efficiency in all processing stages.

5 Works Classification

Considering the fact that the works directory is relatively small, with each record consisting of 1–2 words (during preprocessing 180 initial rows / 358 initial tokens are transformed into 166 tokens), a slightly modified k-Nearest Neighbors (kNN) algorithm demonstrates good results.

This algorithm does not have a training phase, so it belongs to lazy algorithms. Instead of training, works directory initialization is provided, which becomes a set of classes.

Initialization includes two steps:

1. Synonyms are added to each class name based on input data processing (during review, markup, classification error analysis).

2. Key words are found for each name (initial and synonyms). Key words are those with minimal document frequency (DF) values in the corpus of all class names. In other words, these are words that occur least frequently among all others (usually only once).

The modified algorithm consists of the following steps:

- During classification, a list of preprocessed text strings of work names from the online GMS is input.
- All text strings are transformed by a vectorizer into binary vectors set (1 - token present in the string, 0 - token absent).
- The classifier finds the distances between the input names vectors and the nearest vectors of work class names using cosine similarity metric.
- For each of the nearest work classes, an additional check is made to see if there is a match by keyword; if not, the option is discarded.
- Among the classes with the minimum distance, the one whose token is closest to the beginning of the input name is selected.

Cosine similarity was chosen because it is a widely used similarity measure for real-valued vectors, which is especially important for parts classification. Additionally, cosine has the nice property that it is 1.0 for identical and 0.0 for orthogonal vectors [13].

The cosine similarity between vectors A and B is calculated by the formula [14]:

$S_C(A, B) = \cos(\theta) = \frac{AB}{\|A\|\|B\|} = \frac{\sum_{i=1}^{n} A_i B_i}{\sqrt{\sum_{i=1}^{n} A_i^2}\sqrt{\sum_{i=1}^{n} B_i^2}}$

At the algorithm end, tokens corresponding to the selected master-work class name are removed from the input name to leave only tokens corresponding to the spare part name. The method returns a list of pairs ID – work name corresponding to the input names list, as well as a list of cleaned input names for further parts classification.

6 Parts Classification

6.1 Parts Classes and Training Sets Vectorization

The parts catalog is significantly more complex than the master-works directory and encompasses approximately 1,600 entries across all levels. Unlike in works, where we did not engage with a tree-like directory structure, the tree plays a crucial role in the parts classification.

In preparation for training and classification, a class vectors tree is constructed. The foundation for this is the parts directory, wherein each tree vertex undergoes the following stages.

Child nodes initialization, "full names" construction

For each node, a set of names is constructed, which includes both the parent's and all the child names – the so-called "full name":

full_name = [node's own name] + [own name + full_name of every child node].

Classes tree vectorization

The classes tree nodes vectorization occurs through the TF-IDF method use, widely employed in text classification. TF (Term Frequency) represents the ratio of the number of occurrences of a chosen word to the total word count of the document, indicating the word's importance within the document. IDF (Inverse Document Frequency) inversely quantifies the frequency with which a word appears across a collection of documents. Utilizing IDF diminishes the weight of commonly used words.

The TF-IDF metric is calculated using the formula [15]:

$$TF - IDF = TF \times IDF$$

$$TF(t, d) = \frac{f_{t,d}}{\sum_{t' \in d} f_{t',d}},$$

where:

$f_{t,d}$ – raw count of a term t in document d
$\sum_{t' \in d} f_{t',d}$ – number of words in document d

$$IDF(t, D) = log \frac{N}{|\{d \in D : t \in d\}|},$$

where:

- N – total number of documents in the corpus, $N = | D |$
- $|\{d \in D : t \in d\}|$ – number of documents, where term t appears, i.e. $TF(t, d) \neq 0$

For each node in the class tree, a unique matrix is constructed. The matrix rows represent the corresponding TF-IDF values of each token from the child nodes of the given node. Herein, the text for TF calculation is the full name of the parent node, while the corpus for IDF calculation consists of the child nodes full names set. Tokens are considered to be unigrams, direct and inverted bigrams – as the simplest options for the augmented dataset.

Selecting keywords and super keywords

Subsequent steps involve selecting keywords and super keywords. During the training, these receive additional weight, significantly aiding in classification accuracy.

Keywords are identified based on the document frequency values of tokens in each of the child names. Keywords are those with a DF value of 1, meaning they occur in only one of the neighboring child classes.

If a node's name consists of only two words, it automatically becomes a super keyword, receiving even greater weight.

Training data vectorization

At this stage, for each input training string, a matrix is created using the same algorithm as for the class vectorizer.

The matrix rows are the corresponding TF-IDF values of the input string on the specific class node text corpus.

The augmented dataset set is created by tokenizing not only unigrams and bigrams but also bigram permutations. Such bigram permutations of the input string tokens enrich the feature sets, as different word orders can have the same meaning but be found in different classes. For example, from "brake pads and discs," bigram permutations like "brake pads," "brake discs," "pads brake," "discs brake," "pads discs," "discs pads" are formed.

Now, everything is prepared for initiating the training process.

6.2 Parts Classification Algorithm

Prior to commencing the model training, let us examine the classification algorithm itself. If the model construction occurred in a bottom-up fashion, with each parent node encompassing information about all its children, the classification process unfolds in a top-down manner. Initially, an appropriate class at the tree first level is selected, followed by selection among child nodes at the second level, and so forth.

The classification foundation rests on finding the node with the nearest vector. The vectors compared are the weighted rows of the class node matrix and the corresponding input string matrix row. It is reminded that the rows of the class node matrix correspond to the weighted TF-IDF values of tokens from child nodes. Hence, the nearest row in the matrix corresponds to a specific child node, which will determine the next step in the classification.

Similar to the works classification, the distance between vectors is determined using cosine similarity. This metric's capability to work with real numbers becomes particularly useful here. Moreover, the value calculated through cosine similarity can be interpreted as the probability of classifying the input string into that specific class, enabling a confidence score computation.

The confidence score is calculated as the ratio between the most likely class probability, chosen as the classification result, and that of the second most probable class. If the second class probability is 0, meaning no matches were found in any of the child classes and they are all zeros, or a match was found only in one class and all others are zeros, then a confidence is assigned an arbitrarily high value, for example, 1000.

6.3 Model Training and Evaluation

This article's scope limits a detailed exploration of our model's training and evaluation. We intend to address this in a forthcoming separate work, which will detail our training approach and confidence score evaluation comprehensively. Currently, we provide a summary of the training algorithms.

After building the tree and initializing the vectorizer and classifier for each node, we begin a parameter tuning process that operates across several threads, optimizing node matrix and training vector weights. Once all nodes are tuned, the tree is ready for classification.

A critical aspect post-classification is the confidence in its accuracy, especially for unlabeled data. We developed a confidence assessment model using Scikit-learn's ML algorithms [16] to integrate confidence scores into a single estimate, employing boosting trees [17] for iterative hyperparameter recalibration. This approach ensures sequential model improvements and error correction.

The trained data enables calculating a unified confidence score for unlabeled data, facilitating classification verification and immediate algorithm optimization for extensive datasets. The model's effectiveness is demonstrated through classified parts' confidence scores, showing our method precision.

Table 2 shows the top three and bottom three parts classification results, confidence scores by tree levels, and the final unified confidence.

Table 2. Parts classification results

Data	Class		Re-sult	Confidence Scores			Confi-dence
	label.	pred.		lev 1	lev 2	lev 3	
Заміна зовн. ручки і приводу замка чи двері	20113	20113	**true**	35,7	119,3	97,1	**99,9%**
Зняття і установка консолі склоочисника в зб.	12005	12005	**true**	65,9	145,7	73,4	**99,8%**
Замена сцепного шкворня	10509	10509	**true**	31,1	669,8	100,0	**99,8%**
…							
Заміна обох газонаповнених амортизат. капота	20103	20111	**false**	2,2	17,8	3,4	**7,5%**
Ремонт клапана привода передней двери	20113	11404	**false**	6,0	5,4	15,3	**5,5%**
Ремонт ПЖД	11609	11200	**false**	156,2	1,3	1,3	**5,3%**

7 Results Achieved and Conclusions

A Python library was developed based on our approach, achieving classification times of up to 125 ms for a single row and 5 ms for a ten thousand rows sample, enabling real-time application for various problems. Accelerations by two orders of magnitude were noted for large datasets, attributed to Python's efficient algorithms for matrix operations and optimizations for large data processing.

Classification accuracy ranged from 94% to 98% for works and 89% to 95% for parts names, with mechanical parts classification being more accurate than specialized tasks like transmission or truck repairs, likely due to an incomplete class directory for these tasks. Enhancing directories with missing terms emerged as a valuable outcome of this algorithm.

This paper demonstrates the effective use of tree-based classification methods for analyzing Ukrainian technical texts in the automotive sector. By developing a Python library, we efficiently classified texts on automotive repairs and parts, achieving speeds suitable for real-time use. This paves the way for applications ranging from improving call center operations to developing chatbots for service advisors.

We plan to tackle more complex tasks next, including extracting and classifying automotive information from large text corpora and evaluating confidence levels for unlabeled data to enhance our algorithm's accuracy.

In summary, our research highlights the significant potential of tree-based classification for technical text analysis in the automotive industry, offering advanced solutions that could benefit insurers, manufacturers, and vehicle owners. This contributes to a better understanding of vehicle maintenance costs and reliability, offering valuable insights for both consumers and industry stakeholders.

References

1. Zhang, Z., Guo, K., Pan, G.: Improvement of phylogenetic method to analyze compositional heterogeneity. BMC Syst. Biol. **11**(Suppl 4), 79 (2017)
2. Albadri N., Dekeyser S.: A novel file system supporting rich file classification. Computers and Electrical Engineering, vol. 103 (2022)
3. Kim, S.W., Gil, J.M.: Research paper classification systems based on TF-IDF and LDA schemes. Hum. Cent. Comput. Inf. Sci. **9**, 30 (2019)
4. Quinlan, J.R.: C4.5: Programs for Machine Learning. Morgan Kaufmann Publishers, Inc. (1993)
5. Breiman, L.: Random forests. Mach. Learn. **45**, 5–32 (2001)
6. Friedman, J.H.: Greedy function approximation: a gradient boosting machine. Ann. Statist. **29**(5), 1189–1232 (2001)
7. Aggarwal, C.C., Reddy, C.K.: Data Clustering: Algorithms and Applications. Chapman and Hall/CRC (2013)
8. Cielen, D., Meysman, A., Ali, M.: Introducing Data Science: Big Data, Machine Learning, and more, using Python tools. Manning Publications (2016)
9. Jurafsky, D., Martin, J.: Speech and Language Processing, 3rd edn. draft. https://web.stanford.edu/~jurafsky/slp3/ed3book_jan72023.pdf. Accessed 8 Feb 2024
10. Hand, D.J., Yu, K.: Idiot's Bayes – not so stupid after all?. Int. Stat. Rev. **69**(part 3), 385–399 (2001)
11. Hosmer, D. W., Lemeshow, S.: Applied Logistic Regression. 2nd edn. John Wiley & Sons, Inc. (2000)
12. Mashtalir, S.V., Nikolenko, O.V.: Data preprocessing and tokenization techniques for technical Ukrainian texts. Appl. Aspects Inf. Technol. **6**(3), 318–326 (2023)
13. Singhal, A.: Modern information retrieval: a brief overview. Bull. IEEE Comput. Soc. Tech. Committee Data Eng. **24**(4), 35–43 (2001)
14. Tan, P.-N., Steinbach, M., Kumar, V.: Introduction to Data Mining. Addison-Wesley (2005)
15. Vajjala, S., Majumder, B., Gupta, A., Surana, H.: Practical Natural Language Processing: A Comprehensive Guide to Building Real-World NLP Systems. O'Reilly Media, Inc. (2020)
16. Müller, A., Guido, S.: Introduction to Machine Learning with Python: A Guide for Data Scientists. 1st edn. O'Reilly Media, Inc. (2016)
17. Géron, A.: Hands-On Machine Learning with Scikit-Learn, Keras, and TensorFlow: Concepts, Tools, and Techniques to Build Intelligent Systems. 3rd edn. O'Reilly Media, Inc. (2022)

Ensemble Classification Methods of Machine Learning for Analyzing News Texts for Falsity

Liubov Oleshchenko(✉) and Oleksii Melnychuk

National Technical University of Ukraine "Igor Sikorsky Kyiv Polytechnic Institute", Kyiv 03056, Ukraine
oleshchenkoliubov@gmail.com

Abstract. The article introduces ensemble methods to enhance prediction accuracy and evaluates various classifiers using separate datasets. The Naive Bayesian, Passive-aggressive, Linear Support Vector, Logistic Regression, k-nearest neighbors, and Random forest classifiers are assessed. Ensembles comprising combinations of these classifiers are also tested. Results indicate high accuracy for individual classifiers, with ensembles further improving performance. Research results present classification outcomes, showcasing the effectiveness of different methods.

The research emphasizes the significance of ensemble methods in fake news detection, providing insights for future research. For TF-IDF Vectorization, the Support Vector Machine (SVM) classifier stands out with the highest mean accuracy of 95.74%. This indicates that SVM achieved the highest proportion of correct predictions compared to other classifiers when trained on TF-IDF transformed data. When utilizing Hashing Vectorization, the SVM classifier maintained its superior performance, achieving the highest mean accuracy of 97.26%. Ensemble methods, specifically Voting Ensemble 3 (Ens3-SVM+PA+LR), exhibited noteworthy performance, particularly with Hashing Vectorization. This ensemble achieved a mean accuracy of 96.93%.

The main idea of the method is to analyze a purely news text, without extraneous information (dates of publication, website names, additional media). The text is analyzed according to three metrics separately: truthfulness of the news, satire, or hate speech.

Datasets of the Kaggle service were used to train the models according to metrics, and arbitrarily selected texts of news and comments were used for testing in "real conditions". The structure of these datasets is text and a binary label in another column corresponding to the criterion. The dataset included 6,335 lines of news texts and true or false labels. The satire dataset is a combination of two separate datasets, one from the BBC news service and the other from the humor Onion.

Keywords: software method · big data · classification algorithms · machine learning · ensemble methods · analyzing news texts for falsity · TF-IDF

Z. Hu et al. (Eds.): ICCSEEA 2024, LNDECT 242, pp. 349–361, 2025.
https://doi.org/10.1007/978-3-031-84228-3_30

1 Introduction

Fake news has emerged as a pervasive issue in the digital age, propagated through social media platforms and online news outlets. The rapid dissemination of false or misleading information poses serious threats to public discourse, political processes, and societal trust. Traditional methods of news verification are often inadequate in addressing the scale and sophistication of modern misinformation campaigns. Consequently, there is a pressing need for advanced computational techniques to combat the spread of fake news and preserve the integrity of information channels.

The widespread availability of the Internet and social media platforms accelerated the dissemination of information at an unprecedented pace, catching society off guard. This rapid dissemination posed challenges to maintaining the impartiality and accuracy of news content, particularly within social media environments.

Various online platforms employ diverse strategies to categorize and assess similar content. While traditional news agencies rely on manual verification processes, social media platforms like Facebook utilize automated systems to detect hate speech in comments. Platforms like X (previously known as Twitter) employ "community notes," collaborative annotations appended to posts offering additional context and sometimes debunking misinformation. Amidst these efforts, machine learning techniques have emerged as a popular approach for verification. By analyzing vast amounts of text data, machine learning models can make predictions and attempt to discern the veracity of new information.

1.1 Analysis of Recent Research and Related Work

The research [1] explores how technology has led to a shift in news consumption from print media to social media platforms, presenting a challenge of fake news dissemination. To address this, machine learning algorithms like K-Nearest Neighbor, Support Vector Machine, Decision Tree, Naïve Bayes, and Logistic Regression were employed to distinguish fake news from real ones in a dataset. Data preprocessing techniques were utilized to improve algorithm efficiency, particularly in handling imbalanced data. Results indicate high accuracy rates: 89.98% for KNN, 90.46% for Logistic Regression, 86.89% for Naïve Bayes, 73.33% for Decision Tree, and 89.33% for SVM. Over time, the volume of text data has surged, paralleled by advancements in AI and computing power. Fake news has become a global concern with far-reaching impacts. The paper [2] explores how natural language processing and machine learning techniques can address this issue. Methods like bag-of-words, n-grams, and count vectorizers are employed, along with TF-IDF, to train data on five classifiers. Evaluation metrics such as precision, recall, and F1 scores are used to assess the effectiveness of each model on labeled news statements. Scikit-learn is a Python library that offers a wide range of modern machine learning algorithms for medium-scale supervised and unsupervised tasks. The work [3] aims to make machine learning accessible to non-experts through a user-friendly, high-level language. The focus is on usability, performance, thorough documentation, and consistent APIs. It has few dependencies and is freely available under the simplified BSD license, promoting its use in both academic and commercial environments. Ensemble methods combine multiple classifiers to make predictions by aggregating their outputs. They include techniques like Bayesian averaging, error-correcting output coding, Bagging, and boosting. The paper [4] examines these methods and discusses why ensembles often outperform individual classifiers. It reviews previous studies comparing ensemble methods and presents new experiments to understand why Adaboost doesn't overfit quickly. In recent years, there has been extensive research in natural language processing, focusing on text processing stages from preparation to comprehension. Vector space modeling aims to represent words in a language corpus as vectors, grouping similar meanings close together. Two common approaches to vectorization are word-based, considering the entire linguistic context, and document-based, focusing on the corpus of analyzed texts.

The paper [5] compares various text vectorization methods in natural language processing, particularly in Text Mining, using simple classification methods like NBC and k-NN. By assessing classification accuracy, the study lays groundwork for further research in automated text analysis. The paper [6] discusses the rise of fake news on social media, noting its potential negative impacts. It highlights challenges in detecting fake news due to intentional misleading content and the complex nature of user engagements. The survey provides an overview of existing research on fake news detection, covering psychological and social theories, data mining algorithms, evaluation metrics, and datasets. It also identifies open problems and future research directions in this area. The study [7] delves into detecting fake news articles, creators, and subjects on online social networks, evaluating their performance. It tackles challenges arising from the varied nature of fake news and interconnectedness among articles, creators, and subjects. Through thorough data analysis, the paper introduces FakeDetector, an innovative model for automatic fake news credibility inference. FakeDetector utilizes explicit and latent features from textual information to construct a deep diffusive network model, simultaneously learning representations of news articles, creators, and subjects. Extensive experiments on real-world fake news datasets compare FakeDetector with several state-of-the-art models, demonstrating its effectiveness. The paper [8] explores satire as a unique form of deception, contrasting it with legitimate news. It introduces an SVM-based algorithm enriched with five predictive features and tests their combinations on 360 news articles. The best feature combination (Absurdity, Grammar, and Punctuation) achieves 90% precision and 84% recall in detecting satirical news. The algorithmic identification of satire can help mitigate its deceptive impact.

The study [9] examines methods for detecting fake news on social media by analyzing various features extracted from news stories, including sources and social media posts. It introduces a new set of features and evaluates the performance of existing approaches for automatic fake news detection. Results uncover insights into the effectiveness of different features in identifying false news. The paper concludes with a discussion on the practical application of fake news detection methods, outlining challenges and opportunities. The study [10] introduces a machine learning approach for detecting fake news, utilizing a novel classification algorithm based on Bayesian networks. By applying this algorithm to two datasets of news articles, the authors achieve high accuracy, surpassing other existing models in fake news detection. Emphasizing the growing issue of fake news and its detrimental impact on society, the paper advocates for continued research to develop more effective techniques for identifying and curbing its dissemination.

Despite the advancements in methods for analyzing news texts to detect falsity, several unresolved questions persist in this domain. One key challenge revolves around the dynamic and evolving nature of fake news, which necessitates continuous adaptation and improvement of detection techniques. Additionally, the inherent complexity of language and the ever-changing landscape of online communication pose difficulties in accurately discerning deceptive content from genuine news. Furthermore, issues such as bias in training data, the rapid spread of misinformation through social media, and the need for robust evaluation metrics present ongoing hurdles in the development of effective fake news detection methods. Addressing these unresolved questions is essential for enhancing the reliability and efficiency of text analysis methods in combating the proliferation of fake news.

1.2 Research Classification Methods

We selected seven classification algorithms for comparison: Naive Bayesian, passive-aggressive, linear support vectors, logistic regression, k-nearest neighbors, decision trees, and random forest.

Naive Bayes is a probabilistic method used for data classification, employing probability as the likelihood that the data belongs to a certain class. Its multinomial model offers reduced complexity, making it suitable for small training sets without constant retraining.

Passive-aggressive classifiers are interactive learning algorithms that adapt to new data in real time, making them effective for dynamic datasets like those found in social media. They adjust their model incrementally, reacting to changes and making corrections when predictions are incorrect.

Support vector machine (SVM) algorithm aims to find the optimal hyperplane that best divides data into classes. It utilizes linear hyperplanes for classification tasks, effectively separating data points in multidimensional space.

Logistic regression is ideal for binary classification tasks, predicting the probability of a binary outcome using a logistic function. Its simplicity makes it suitable for simplified binary classification datasets.

K-nearest neighbors (KNN) algorithm assigns a class label to a query point based on the majority class of its nearest neighbors, making it versatile for various data and regression problems.

Random forest classifier, an ensemble method, creates decision trees for subsets of data and aggregates their predictions for the final classification decision.

Ensemble methods combine multiple models to achieve better performance than individual models. The ensemble voting method combines forecasts from several models, either through hard voting (choosing the most popular prediction) or soft voting (averaging predictions), providing a more accurate prediction, especially in real-world scenarios like news falsity analysis.

1.3 Objective

The objective of the research is solving the task of developing a classification algorithm for analyzing news texts in English for falsehood, satire, and hate speech. The main idea is to analyze a purely news text, without extraneous information (dates of publication, website names, additional media). The text is analyzed according to three metrics separately: truthfulness of the news, satire, or hate speech.

1.4 Research Data

Datasets of the Kaggle service were used to train the models according to these metrics, and arbitrarily selected texts of news and comments were used for testing in "real conditions". The structure of these datasets very simple - text and binary mark in another columns that responds criterion. The dataset for plagiarism included 6,335 lines of news texts and true or false labels. The dataset for checking for satire is a combination of two separate datasets, one from the news service BBC, another from the humorous Onion. Combined dataset totals 8341 lines. Last dataset consisted of comments network X and included >24,000 lines.

Python programming technologies (sklearn, pandas, numpy) were used for the research, processor AMD Ryzen 5 4500U 6 cores, 16 gigabytes of RAM.

The greatest difficulty in such a combined analysis of metrics is the variety of data for individual metrics, since the data for falsehood are long, formal texts, and user comments are used to check for hate speech.

Before training the models, the texts of the datasets need to be changed in a certain way (data preparation). Data preparation is the process of preparing "raw" data and transforming it into a format suitable for machine learning models. This is the first and mandatory step in creating models. This especially applies to the analysis of textual data of human speech. Such textual data has "noise" manifested in the punctuation, emotions and case of the text, as well as in the declension of words.

Stop words are the most common words in the text that do not provide any information. Since the language of the datasets is English, the stop words include articles, "this", "where" and so on. The NLTK library was used to remove such words. It includes approximately 180 stop words.

Abbreviations of words, or stemming, consists in separating and removing auxiliary parts of the word from the root, or the main form. The NLTK library is used to shorten words. In particular, lemmatization was used in this study. Unlike the usual stemming, where it is possible to lose the original context and meaning of the word (universal – universe) during shortening, during lemmatization a morphological analysis of the word takes place before shortening, although it takes more time (Fig. 1).

Depending on the datasets, numbers are also removed or substituted for words in the data, or abbreviated words are "expanded" (do n't - do not). Machine learning algorithms most often accept numerical values for training, so we need to convert all text documents into their numerical representation. An example of such a numerical representation is a vector, the indices of which are the indices of unique words from the documents, and the values are the number of occurrences of these words in the documents. Stop words can also be removed again during vectorization.

In the research two vectorizers were selected and used: TD-IDF and a vectorizer with hashing. TF-IDF (Term Frequency – Inverse Document Frequency) is an indicator of the importance of words in documents. The importance of a word is the number of

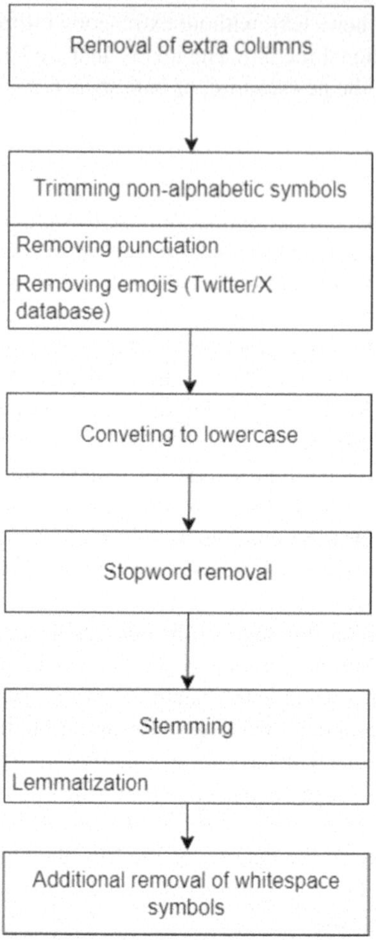

Fig. 1. The data preparation algorithm used

occurrences/uses of a word in documents. Inverse frequency refers to the uniqueness of a word in a collection of documents.

$$TFIDF(t.d) = TF(t.d) * IDF(t) \quad (1)$$

where *TF(t.d)* is number of word usage in documents. Inverse frequency refers to the uniqueness of a word in a collection of documents.

Big value TF-IDF is reached when the value of TF is large and when the weight is in the denominator in the IDF formula is of little importance.

One of the vectorizers that is built into the scikit-learn library and is the easiest to implement. It creates a matrix from all the documents, where each unique word is a column and the documents are rows. The value in each cell is numerical and means the number of times this word appears in the document. In the scikit-learn implementation words are not stored as text, but rather as indexes.

This vectorizer is a modification of the quantitative. The biggest difference from the quantized vectorizer is that the hashed vectorizer does not store a dictionary of word indices. Each word/ token hashed using the Murmurhash 3 function and displayed directly in the column of the matrix. This is an advantage because the available datasets are quite large.

2 Research Method

2.1 Justification of the Use of Classifiers

When compared to deep learning, speech models, or convolutional /recurrent neural networks, conventional classifiers are usually inferior in their prediction accuracy. The use of classifiers was chosen for several reasons.

First, testing ensemble methods. It has been hypothesized that if single classifiers are inferior in accuracy to more sophisticated and complex models, then one can try to use multiple classifiers and select the best predictions to maximize accuracy.

Second, classifiers are much simpler to implement and faster to train, and are therefore ideal for working under time constraints or when other settings or parameters need to be tested.

Third, classifiers work much better with small datasets than more complex models. All used datasets are relatively small in volume and were taken in ready form from separate sources, without the use of additional media or web scrapers and do not have the scale to train complex models of the ChatGPT level, and therefore are well suited for a more compact study. Language models, on the other hand, require quite a large amount of data for optimal results, and this is an additional hassle.

In the study, the texts are divided into 3 categories: fake news, satire and hate speech. In order to simplify the training and use of classifiers, it was decided to train each metric separately, that is, to simplify the classification problem to a binary one. Trained classifiers are stored for further use in ensemble methods.

One of the goals of the research is to find the probability that the text belongs to the category. The scikit - learn library has a built-in function `predict proba ()`, but it does not work with all the classifiers used, so some of them have been modified accordingly.

The Naive Bayesian classifier was chosen primarily due to its simple classification work, and as a consequence due to its popularity. It is suitable for the classification of short texts, and considers that the features of the text are not related to each other, but such an assumption is not always true. One of the datasets is filled with comments from network X (Twitter), and this factor also influenced the choice of the classifier.

Bayes multinomial variant chosen for the study can effectively model the joint distribution of features and their associated classes using the multinomial distribution. The frequency distribution for each term can be improved by including a measure of variance, such as term frequency inverse of document frequency (TF-IDF), which accounts for the number of documents in which each occurs. This can significantly improve performance by giving more weight to terms that appear in fewer documents, and thus improve their discriminating abilities.

Although the multinomial distribution performs well when used directly with term frequencies, it also performs well for fractional values such as TF-IDF values.

Passive-aggressive classifier is considered one of the best. It is often used to divide text into two groups, and is therefore very well suited for use in this study. In addition, it is often used to filter spam emails and detect fraud.

The passive-aggressive classifier does not have a probability method, so it was chosen to create a kind of wrapper for it:

```
    def predict_proba (self , x_test):
  arr1= 1-(1. / (1. + np.exp (- self.decision_function
(x_test))))
  arr2= -(arr1-1)
     return np.stack ((arr2, arr1), axis =1)
```

The built-in function `decision _ function ()` shows how the algorithm makes a decision about the classification and returns a numpy array, where each element is provided for the x values of the sample x _ test (data for testing) shows the distance from the hyperplane from its different sides, with values from -1 to 1, with 0 exactly in the middle.

SVM, although slower than the Bayesian classifier, is considered one of the best and is used for both classification and regression.

In scikit-learn, the SVM module is a wrapper for the libsvm library and has support for various kernels, not just linear ones. The study uses the LinearSVC () function, which only supports a linear kernel but is faster than the SVM module. This function is then calibrated using CalibratedClassifierCV. This was done in order to obtain the probabilities of matching the text to the type.

Logistic regression is ideal for binary classification, and immediately returns the probability. In scikit - learn, the function LogisticRegression () implements regularized logistic regression using the liblinear library.

The k-nearest neighbors classifier (KNN) is one of the simplest and most common algorithms used for text classification. In some specific cases, KNN can outperform cumbersome BERT- type models. In 2023 a group of researchers at the University of Waterloo published sample code where the algorithm was used with the gzip library to compress distance metrics. It was done in order to prove that similarity between by two text documents closely connected with them compressibility into files smaller size.

Random forest classifier is a popular choice for text classification because it handles large amounts of text data well and can handle complex relationships between words in the text and the categories to which they belong. One of the main advantages of this classifier is that it is able to avoid overtraining, when the model is too complex and fits the training data too closely, resulting in poor performance on new data.

The following combined ensemble methods were used in the research:

1. naive Bayesian classifier + k-nearest neighbors classifier + logistic regression;
2. k-nearest neighbors classifier + passive-aggressive classifier;
3. linear support vector classifier + passive-aggressive classifier + logistic regression.

These combinations were chosen according to different, but not rigid, principles. Naive Bayesian, KNN and logistic regression are the most popular algorithms among the selected, so it was interesting to check their performance in relation to each other. SVM, passive-aggressive classifier and logistic regression performed best in preliminary tests on available datasets.

2.2 Classifiers Training

In Python, training a classifier involves several steps. First, we import the necessary libraries, such as scikit-learn, which provides a wide range of machine learning algorithms. Then, we prepare data by splitting it into training and testing sets using functions like `train_test_split()`. Next, we initialize our classifier and fit it to the training data using the `fit()` method. During this process, the classifier learns patterns and relationships in the data. After training, we evaluate the performance of the classifier on the testing data using metrics like accuracy, precision, recall, or F1-score. Finally, we fine-tune the model's hyperparameters using techniques like grid search or cross-validation to optimize its performance.

The classifier vectorization process involves transforming raw text data into numerical features that can be used by machine learning algorithms. This is typically done using techniques like `CountVectorizer` or `TF-IDFVectorizer` from the scikit-learn library for vectorizing text data:

```
from sklearn.feature_extraction.text import
TfidfVectorizer
# Assuming corpus is a list of text documents
corpus = ["This is the first document.", "This document
is the second document.", "And this is the third one.", "Is
this the first document?"]
# Initialize the TF-IDF vectorizer
vectorizer = TfidfVectorizer()
# Fit the vectorizer to the corpus and transform the
documents into TF-IDF features
X = vectorizer.fit_transform(corpus)
# Now X contains the TF-IDF features of the documents,
which can be used for training classifiers
```

Classifier training is performed in cycles, separately for each classifier/combination, separately for each dataset, and separately using vectorizers.

After training the classifiers, they are tested on examples of each text category. Classifiers should give the probability that the text belongs or does not belong to the chosen category. For this, the scikit - learn function is used `predict()` and `predict proba()`. `Predict` method guesses the class of the data, and the `predict proba` method guesses the probability that the data belongs to some class, for each of them, and returns an array of those values.

The program code is organized in such a way that we can simply create combination of classifiers and immediately start testing/training it (Fig. 2).

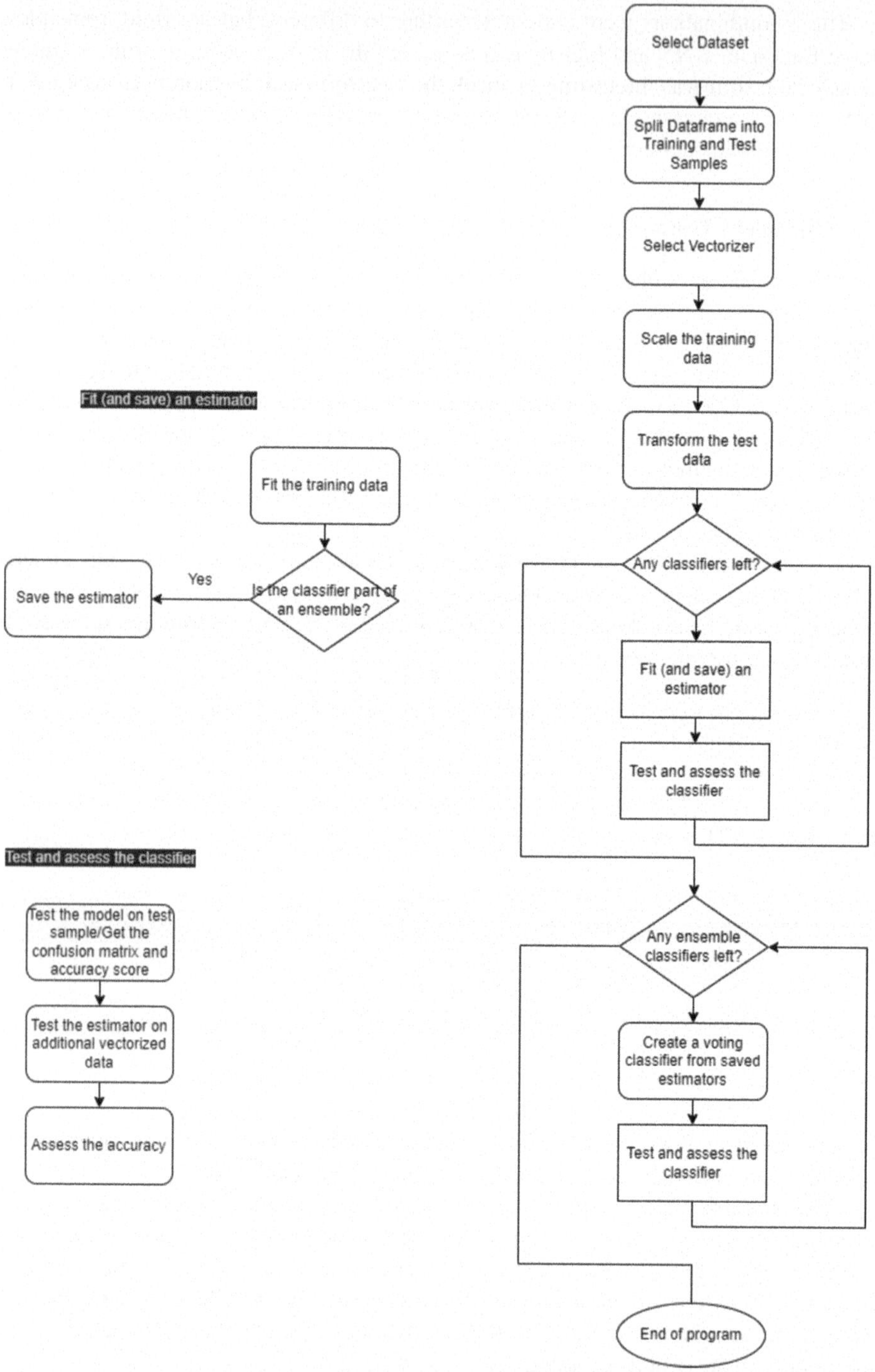

Fig. 2. Training classifiers algorithm

3 Research Results

The analysis of the results presents a comprehensive view of the performance of various classifiers and ensemble methods in the classification of news text data into categories of fake news, satire, and hate speech.

The mean accuracy and mean correct predictions obtained from the provided table offer crucial insights into the efficacy of different approaches (Fig. 3).

TFIDF	Fakes	0.902	0.9254	0.9297	0.9091	0.8431	0.8317	0.9111	0.726	0.9201	by samp	by dataset				
	expecte	Naïve B	Passive	SVM	LR	KNN	Forest	Ens1	Ens2	Ens3		SVM	Ens1	NB	KNN	LR
real1	0	1	1	1	0	0	1	1	0	0	LR	PA	Ens2	KNN	PA	
real2	0	1	1	1	1	1	1	1	1	1	KNN	LR	Ens3	SVM	PA	LR
joke1	1	1	1	1	1	1	1	1	1	1	TREE					
joke2	1	1	1	1	1	1	1	1	1	1						
hate1	1	1	1	1	1	1	1	1	1	1						
hate2	1	0	1	1	1	1	1	1	1	1						
		3	4	4	5	5	4	4	5	5						
hashing	Fakes	0.8967	0.9072	0.9173	0.8977	0.8283	0.8006	0.9015	0.6858	0.9024						
	expecte	Naïve B	Passive	SVM	LR	KNN	Forest	Ens1	Ens2							
real1	0	1	1	1	0	0	1	0	0	0						
real2	0	1	1	1	1	0	1	1	0	1	LR					
joke1	1	1	1	1	1	1	1	1	1	1	KNN					
joke2	1	1	1	1	1	1	1	1	1	1						
hate1	1	1	1	1	1	1	1	1	1	1						
hate2	1	1	1	1	1	0	1	1	0	1						
		4	4	4	5	5	4	5	5	5						
tfidf	sarcasm	0.9855	0.9938	0.9942	0.9695	0.9419	0.8645	0.9847	0.7792	0.9902		SVM				
	expecte	Naïve B	Passive	SVM	LR	KNN	Forest	Ens1	Ens2			PA				
real1	0	1	0	0	1	0	1	1	1	0		NB				
real2	0	1	1	1	1	1	1	1	1	1						
joke1	1	1	1	1	1	1	1	1	1	1	NB					
joke2	1	1	1	1	1	1	1	1	1	1	PA					
hate1	1	1	1	1	1	1	1	1	1	1	KNN					
hate2	1	1	1	1	1	1	1	1	1	1						
		4	5	5	4	5	4	4	4	5						
hashing	sarcasm	0.9241	0.9942	0.9956	0.9844	0.9437	0.8155	0.9815	0.7846	0.9909		SVM				
	expecte	Naïve B	Passive	SVM	LR	KNN	Forest	Ens1	Ens2	Ens3		PA				
real1	0	0	0	0	1	0	1	0	0	0		LR				
real2	0	1	1	1	1	1	1	1	1	1	NB					
joke1	1	1	1	1	1	1	1	1	1	1	PA					
joke2	1	1	1	1	1	1	1	1	1	1	SVM					
hate1	1	1	1	1	1	1	1	1	1	1	KNN					
hate2	1	1	1	1	1	1	1	1	1	1	TREE					
		5	5	5	4	5	4	5	5	5						
tfidf	hate	0.9397	0.9307	0.9484	0.9474	0.9457	0.9462	0.9466	0.8704	0.9473		SVM				
	expecte	Naïve B	Passive	SVM	LR	KNN	Forest	Ens1	Ens2	Ens3	NB	LR				
real1	0	0	0	0	0	0	0	0	0	0	PA	FOREST				
real2	0	0	0	0	0	0	0	0	0	0	KNN					
joke1	0	0	0	0	0	0	0	0	1	0	LR					
joke2	0	0	0	0	0	0	0	0	0	0	SVN					
hate1	1	0	0	0	0	0	0	0	0	0						
hate2	1	0	1	1	1	1	0	0	0	1						
		4	5	5	5	5	4	4	3	5						
hashing	hate	0.9458	0.9364	0.9495	0.9482	0.9451	0.9462	0.9468	0.8777	0.9477		SVM				
	expecte	Naïve B	Passive	SVM	LR	KNN	Forest	Ens1	Ens2	Ens3		LR				
real1	0	0	0	0	0	0	0	0	1	0		FOREST				
real2	0	0	0	0	0	0	0	0	0	0	PA					
joke1	0	0	0	0	0	0	0	0	1	0	SVM					
joke2	0	0	0	0	0	0	0	0	0	0						
hate1	1	0	0	0	0	0	0	0	0	0						
hate2	1	0	1	1	0	0	0	0	0	0						
		4	5	5	4	4	4	4	2	4						

Fig. 3. Classification results

For TF-IDF Vectorization, the Support Vector Machine (SVM) classifier stands out with the highest mean accuracy of 95.74%. This indicates that SVM achieved the highest proportion of correct predictions compared to other classifiers when trained on TF-IDF transformed data. Additionally, the SVM classifier also achieved the highest mean correct prediction of 4.67, signifying its robustness in accurately categorizing news text data.

On the other hand, when utilizing Hashing Vectorization, the SVM classifier maintained its superior performance, achieving the highest mean accuracy of 97.26% and the highest mean correct prediction of 4.67. These results reinforce the effectiveness of SVM in handling the high-dimensional feature space generated by the hashing vectorization technique.

Moreover, ensemble methods, specifically Voting Ensemble 3 (Ens3 - SVM+PA+LR), exhibited noteworthy performance, particularly with Hashing Vectorization. This ensemble achieved a mean accuracy of 96.93% and a mean correct prediction of 4.67, further emphasizing the advantages of combining multiple classifier strategies to enhance classification accuracy and correctness (Table 1).

Table 1. Classification results comparison

Classifier	Average accuracy	Average number of successful predictions
Naive Bayesiv (NB)	93.23%	4
Passive Aggressive (PA)	94.8%	4.67
Support Vector Machines (SVM)	**95.58%**	4.67
Logistic Regression (LR)	94.27%	4.5
k-nearest neighbors (KNN)	90.79%	**4.83**
Random forest (FOREST)	86.74%	4
NB + KNN + LR	94.54%	4.33
KNN + PA	78.72%	4
SVM + PA + LR	94.98%	**4.83**

4 Conclusion

The article introduces ensemble methods to enhance prediction accuracy and evaluates various classifiers using separate datasets. The Naive Bayesian, Passive-aggressive, Linear Support Vector, Logistic Regression, k-nearest neighbors, and Random forest classifiers are assessed.

The research examines classifiers and vectorizers used to analyze news texts for false information. The use of ensemble methods is a common practice, and the proposed combined method can be an alternative. It successfully classified the selected texts, but showed a lower accuracy than one of the standard classifiers.

References

1. Pandey, S., Prabhakaran, S., Subba Reddy, N.V., Acharya, D.: Fake news detection from online media using machine learning classifiers. In: Journal of Physics: Conference Series, 1st International Conference on Artificial Intelligence, Computational Electronics and Communication System (AICECS 2021) 28–30 October 2021, vol. 2161, no. 27, pp. 2–9 (2021). https://doi.org/10.1088/1742-6596/2161/1/012027
2. Vasu Agarwal, H., Sultana, P., Malhotra, S., Sarkar, A.: Analysis of classifiers for fake news detection. Procedia Comput. Sci. **165**(2019), 377–383 (2019). https://doi.org/10.1016/j.procs.2020.01.035
3. Chary, D.P., Singh, R.P.: Review on advanced machine learning model: scikit-learn. Int. J. Sci. Res. Eng. Dev. (IJSRED) **3**(4), 526–529 (2020)
4. Dietterich, T.G.: Ensemble methods in machine learning. In: Multiple Classifier Systems. MCS 2000. Lecture Notes in Computer Science, vol. 1857, pp. 1–15. Springer, Heidelberg (2000). https://doi.org/10.1007/3-540-45014-9_1
5. Krzeszewska, U., Poniszewska-Maranda, A., Ochelska-Mierzejewska, J.: Systematic comparison of vectorization methods in classification context. Appl. Sci. **12**, 5119 (2022). https://doi.org/10.3390/app12105119
6. Shu, K., Sliva, A., Wang, S., Tang, J., Liu, H.: Fake news detection on social media: a data mining perspective. ACM SIGKDD Explor. Newsl. **19**(1), 22–36 (2017). https://doi.org/10.1145/3137597.3137600
7. Wang, W., Cui, P., Zhu, W., Yang, S.: Fake news detection with deep diffusive neural network. In: Proceedings of the 2018 World Wide Web Conference on World Wide Web, pp. 797–806 (2018)
8. Rubin, V.L., Conroy, N.J., Chen, Y.: Fake news or truth? Using satirical cues to detect potentially misleading news. Proc. Assoc. Inf. Sci. Technol. **52**(1), 1–4 (2015). https://doi.org/10.18653/v1/W16-0802
9. Reis, J.C., Correia, A., Murai, F., Veloso, A., Benevenuto, F., Cambria, E.: Supervised learning for fake news detection. IEEE Intell. Syst. **34**(2), 76–81 (2019). https://doi.org/10.1109/MIS.2019.2899143
10. Pal, S., Kumar, T.S., Pal, S.: Applying machine learning to detect fake news. Indian J. Comput. Sci. **4**(1), 7–12 (2019). https://doi.org/10.17010/ijcs/2019/v4/i1/142411

Wuhan Highway Logistics Demand Forecast Based on Gray Prediction Model

Xinrui Du, Xudong Deng, and Yong Wang(✉)

School of Management, Wuhan University of Science and Technology, Wuhan 430081, China
{dengxudong,wang-yong}@wust.edu.cn

Abstract. This article conducted a study on the demand for road logistics in Wuhan city based on the grey prediction model. Firstly, data analysis was carried out on the current situation of road logistics demand in Wuhan, including freight volume of road, railway and air cargo. Then, the GM (1,1) model in grey prediction was introduced. Finally, combining with the economic development situation and statistical yearbook data of Wuhan city, as well as the advantages of the GM (1,1) model, the model was used to predict the freight volume in Wuhan city and put forward measures and strategies for the development of Wuhan's logistics industry.

Keywords: highway logistics · grey forecasting · logistics demand forecasting

1 Research Background

As an essential part of the modern logistics system, road logistics plays a crucial role in global trade and economic development. Especially in Wuhan, a large city known as the "Gateway to Nine Provinces," which claims to reach anywhere across two provinces, roads are essential in modern logistics. As a traditional mode of transportation, roads play an irreplaceable role in the contemporary logistics system.

Road transportation has the characteristics of flexibility and adaptability, which can realize point-to-point and door-to-door transportation and can flexibly adjust the route and mode of transportation according to demand to improve the efficiency of cargo transportation. The road network covers a wide area, connecting cities and villages, and can reach remote areas to meet the transportation needs of all types of goods. Road transport is relatively simple, does not require complex loading and unloading processes, and can achieve fast transportation, especially suitable for time-critical goods, such as fresh food and fast-moving consumer goods. Road transport is more economical than other modes of transport. Its operating costs are relatively small for short and medium-distance transportation, and it is more economical for small-batch, high-frequency cargo transportation. As an end-of-line mode of transportation, road complements other modes of transportation such as railway, aviation, water transport, and others, supports multimodal transport, and can realize the diversified transportation needs of goods.

As an open Chinese mainland city, Wuhan has a significant position as an essential transportation hub and logistics center. In this context, the forecast research on the demand for highway logistics in Wuhan has essential significance and purpose:

Z. Hu et al. (Eds.): ICCSEEA 2024, LNDECT 242, pp. 362–370, 2025.
https://doi.org/10.1007/978-3-031-84228-3_31

1. The forecast research of road logistics demand can help enterprises and logistics practitioners better understand the future market demand and cargo flow to rationally allocate logistics resources, improve transportation efficiency, and reduce logistics costs.
2. The forecast research of road logistics demand can provide scientific data support for the Wuhan municipal government and provide a basis for policy formulation and planning.
3. The study of predicting the demand for road logistics in Wuhan can promote the healthy development of the logistics industry, increase employment opportunities, and promote economic growth.

Therefore, this study aims to deeply explore the development trend and characteristics of highway logistics in Wuhan through the forecast and research of Wuhan's highway logistics demand, put forward scientific and reasonable logistics planning and management measures, and provide theoretical support and practical guidance for the healthy development of Wuhan's highway logistics industry.

2 Literature Review

Yin Yue and Chen Yuting (2023) predicted the demand for agricultural product logistics and scientifically analyzed the demand for agricultural product logistics in the Beijing-Tianjin-Hebei region. The results show that while this phenomenon is likely to be related to the policy of evacuating non-capital functions, practical suggestions for developing agricultural product logistics are proposed for Tianjin and Hebei under the premise of implementing this policy. [1] Zeng Qingguang, Yao Hongfa, et al. (2022) used a gray GM(1,1) model to predict the production of citrus fruits in the next ten years, indicating that citrus fruits will develop rapidly in the next ten years, with a rapid increase in production, and the overall demand for cold chain logistics is on a steady upward trend. Finally, suggestions are given for developing cold chain logistics of citrus fruits in Liuzhou City. [2] Zeng Hao and Zhu Wenjuan (2022) predicted that the demand for cold chain logistics of vegetables in Hunan Province's agricultural products is on the rise, and the demand for other cold chain logistics is on a steady development trend. Countermeasures for developing cold chain logistics of agricultural products in Hunan Province are given to the government and enterprises. [3] Lei Lei, Wu Jie, and Chen Jiaxing (2022) predicted the logistics demand of Guizhou Province in the next five years by applying the gray GM(1,1) model, and the results showed that the logistics demand would be on the rise. In this context, the relevant departments should seize the opportunity to adapt to the new situation and changes, cultivate logistics professionals, integrate the logistics network, and promote the construction of digital and intelligent logistics to achieve sustainable logistics development. The above suggestions can provide a basis for decision-making for the relevant logistics departments to formulate logistics industrial policies, optimize the allocation of logistics resources, and promote the modernization of logistics construction during the "14th Five-Year Plan" period. [4] Chai Lei (2017) used the GM(1,1) model to predict that the demand for logistics in Wuhan is rising. However, Wuhan's logistics industry is still facing some bottlenecks. Therefore, this paper puts forward several recommendations to promote the development of Wuhan's logistics industry. These suggestions include increasing investment in logistics infrastructure, enhancing the integration of logistics networks, promoting digital and intelligent logistics, and achieving sustainable development. The future city of Wuhan will become a national logistics node city, integrating a transportation hub, a commodity processing and distribution hub, and a material distribution center. [5] Huiming Duan et al. established a multivariate gray prediction model with energy structure (EPMGM(1, M, N) model) to analyze multiple principal factor variables. The model was effectively applied to power forecasting in China [6].

3 GM(1,1) Gray Prediction Model

The road transport industry has been hit hard by the global outbreak of COVID-19, and countries and regions worldwide have adopted strict epidemic prevention and control measures, including traffic restrictions, lockdowns, and other measures, which have led to a significant reduction in road transport industry activity. Therefore, during the epidemic, road transport data in many regions has been anomalous and is not suitable for analysis and research on the development trend of the road transport industry in 2023 and beyond. Therefore, the road transport data from 2010 to 2019 are used here to predict that there is no impact of the epidemic, and appropriate data adjustments and corrections are made to ensure the accuracy and reliability of the road transport forecast results in Wuhan.

The Grey Prediction Model is a forecasting method based on the theory of grey systems to deal with problems with a small sample of data, lack of sufficient statistical information, or low data quality. Its models fit and predict existing data based on the history and present development law of objective things, describe and analyze future development trends and conditions through scientific methods, and put forward scientific hypotheses and judgments. It is widely used in many fields, including economy, finance, transportation, environment, and others.

3.1 Construction of Gray GM(1,1)

Let the primitive number be listed as:

$$X^{(0)} = \left\{X^{(0)}(k), k = 1,2,3,\ldots,n\right\} \tag{1}$$

X(0) = (21120, 22442, 24354, 25023, 28083, 28495.59, 28891.92, 34981.98, 38633.9, 41778.2).

First, X(0) is summed once, generating an accumulation sequence:

Thereinto:

$$X^{(1)}(k) = X^{(1)}(k-1) + X^{(0)}(k) \tag{2}$$

$$X^{(1)} = \left\{X^{(1)}(k), k = 1,2,3,\ldots,n\right\} \tag{3}$$

That is,

X(1) = (21120,43562,67916,92939,121022,149517.59,178409.51,213391.49,252 025.39,293803.59).

Build a gray GM(1,1) model. For X (1), with the help of the albino differential equation

$$\frac{dX^{(1)}}{dt} + aX^{(1)} = u \tag{4}$$

Build a gray model

a is the development gray number, and u is the endogenous control gray number

Using the least squares method, let

$$A = [a, u]^T \tag{5}$$

Thereinto

$$B = \begin{bmatrix} -\frac{1}{2}\left[X^{(1)}(1) + X^{(1)}(2)\right] & 1 \\ -\frac{1}{2}\left[X^{(1)}(2) + X^{(1)}(3)\right] & 1 \\ \vdots & \vdots \\ -\frac{1}{2}\left[X^{(1)}(n) + X^{(1)}(n-1)\right] & 1 \end{bmatrix} \tag{6}$$

$$Y = \begin{bmatrix} X^{(0)}(2) \\ X^{(0)}(3) \\ \vdots \\ X^{(0)}(10) \end{bmatrix} \tag{7}$$

$$\text{namely } B = \begin{bmatrix} -32341 & 1 \\ -55739 & 1 \\ -80427.5 & 1 \\ -106980.5 & 1 \\ -135269.795 & 1 \\ -163963.55 & 1 \\ -195900.5 & 1 \\ -232708.44 & 1 \\ -272914.49 & 1 \end{bmatrix}$$

$$Y = \begin{bmatrix} 22442 \\ 24354 \\ 25023 \\ 28083 \\ 28495.59 \\ 28891.92 \\ 34981.98 \\ 38633.9 \\ 41778.2 \end{bmatrix}$$

The resulting a = −0.080126329, u = 18935.86461 will be solved. Substitution, we get the prediction model of X(0), that is:

$$\widehat{X}^{(1)}(k) = \left[X^{(0)}(1) - \tfrac{u}{a}\right]e^{-a(k-1)} + \tfrac{u}{a}(k = 1,2,3,\ldots,n) \tag{8}$$

Reduction is made to calculate the predicted value $\widehat{X}^{(0)}(k)$ Namely:

$$\widehat{X}^{(0)}(k) = \widehat{X}^{(1)}(k) - \widehat{X}^{(1)}(k-1) \tag{9}$$

$$\widehat{X}^{(0)}(k) = (21120{,}21477.08{,}23268.79{,}25209.96$$

$$27313.07, 29591.64, 32060.29, 34734.88, 37632.6, 40772.06)$$

3.2 Accuracy Test of Wuhan Highway Logistics Demand Forecasting Model

3.2.1 Residual Test

Calculate the absolute error:

$$\Delta(k) = \left|\widehat{X}^{(0)}(k) - X^{(0)}(k)\right| \tag{10}$$

Calculate the relative error:

$$\varnothing(k) = \frac{\Delta(k)}{X^{(0)}(k)} \times 100\% \tag{11}$$

Calculate the average relative error (Table 1):

$$\varnothing = \frac{1}{n-1}\sum_{k=1}^{n}\frac{\Delta(k)}{X^{(0)}(k)} \tag{12}$$

3.2.2 Post-Inspection Inspection

The mean and mean square deviation of the original sequence

$$S_1 = \sqrt{\frac{\sum_{k=1}^{n}\left[X^{(0)}(k) - \overline{X^{(0)}}\right]^2}{n}} \tag{13}$$

The mean difference and mean square deviation of the absolute error sequence Δ(0).

$$S_2 = \sqrt{\frac{\sum_{k=1}^{n}\left[\Delta(k) - \overline{\Delta}\right]^2}{n}} \tag{14}$$

Mean square deviation ratio.

$$C = \frac{S_2}{S_1} \tag{15}$$

Table 1. Table of model fit results

year	Original value	Predicted value	Absolute error value	Relative error (%)
2010	21120	21120	0	0
2011	22442	21477.08	964.9168	4.30
2012	24354	23268.79	1085.214	4.456
2013	25023	25209.96	186.959	0.747
2014	28083	27313.07	769.9269	2.742
2015	28495.59	29591.64	1096.05	3.846
2016	28891.92	32060.29	3168.37	10.966
2017	34981.98	34734.88	247.0975	0.706
2018	38633.9	37632.6	1001.297	2.592
2019	41778.2	40772.06	1006.138	2.408

The average relative error of the model is 3.2%, which means that the model fits well.

Table 2. Table of model predictions

Prediction order	Predicted value
2020	44173.43
2021	47858.55
2022	51851.1
2023	56176.72
2024	60863.2
2025	65940.65
2026	71441.68

$C = 0.126 < 0.35$, the model is excellent (Table 2).

According to the forecast results, the total logistics demand in Wuhan will increase by about 30 million tons by the end of 2026. The total logistics demand in Wuhan is expected to continue to increase in the next few years. This trend shows that Wuhan, as an essential transportation and logistics hub, will continue to attract more enterprises and logistics practitioners to settle in and accelerate the improvement and development of the logistics system. At the same time, it also provides new opportunities and challenges for Wuhan's economic development and urban construction. In order to better grasp this opportunity, Wuhan should further improve its road logistics system, improve the quality and efficiency of logistics services, strengthen the innovation and development of the logistics industry, and promote the sustained growth and prosperity of Wuhan's economy.

3.3 Data Projections During the Pandemic

Similarly, a gray prediction model is established and calculated according to the method of Sect. 3.2 (Tables 3 and 4).

Table 3. Parameter table

Development factor A	Gray action amount b
−0.349	11635.295

Table 4. Table of model fit results

Time	Original value	Predicted value	Residuals	Relative error (%)
2019	41778.2	41778.2	0	0
2020	31724.91	31341.463	383.447	1.21
2021	45115.6	44410.412	705.188	1.56

The average relative error of the model is 2.77%, which means that the model fits well.

Mean square deviation ratio $C = \frac{S_2}{S_1}$, C = 0.051 < 0.35, the model is excellent (Table 5).

Table 5. Model Prediction Results Table

Prediction order	Predicted value
2022	62928.930
2023	89169.410
2024	126351.800
2025	179038.726
2026	253695.358

It is essential to understand that Wuhan's road logistics data fluctuates significantly during the pandemic because this situation will significantly impact logistics forecasting and analysis. However, the forecast results for the next six years show that Wuhan's road logistics transportation volume is proliferating, reflecting that the new trends and opportunities of Wuhan's road logistics growth are significant for future road logistics planning.

4 Conclusion and Outlook

4.1 Analysis of Forecast Results

According to the analysis of the forecast results, first of all, Wuhan's unique geographical location and convenient transportation conditions have spawned the close integration of society and the logistics industry, which has promoted Wuhan's economic development. Wuhan is located in the middle reaches of the Yangtze River and is connected to many provinces and cities in China. At the same time, Wuhan is also a critical node city of the "Belt and Road" strategy and has an essential position as an international logistics and transportation hub. These factors will provide strong support for Wuhan's logistics needs.

Secondly, with the urbanization rate and population increase, Wuhan's logistics demand will continue to grow. According to statistics, in recent years, the urbanization rate of Wuhan has been rising, and the number of people is also increasing rapidly, which will further drive the growth of logistics demand in Wuhan. At the same time, consumption upgrading will also promote the continuous expansion of logistics demand in Wuhan.

Third, with the continuous improvement of logistics technology, the logistics demand in Wuhan will also be further improved. At present, Wuhan is vigorously developing the logistics industry and constantly introducing new technologies and models, such as logistics big data, Internet of Things, and others; these new technologies and new models will provide more intelligent and efficient services for Wuhan's logistics needs and further improve the logistics demand in Wuhan.

Finally, the Wuhan Municipal Government is actively promoting the development of the logistics economy, promoting the rapid development of the logistics industry by introducing relevant policies and planning new logistics parks, which will further promote the growth of road logistics demand. For example, the Wuhan High-tech Industrial Development Zone has become a national-level logistics park. Wuhan also plans to build several new logistics parks, which will provide more diversified and rich services for Wuhan's logistics needs.

In short, the combined effect of the above factors will make the demand for road logistics in Wuhan show an overall trend of improvement in the next six years, which provides a broader space and opportunity for Wuhan's logistics industry and economic development.

Given the future growth trend of logistics demand, it is recommended that Wuhan should strengthen the construction of the following aspects in the development of the logistics industry. First of all, to ensure the smooth implementation of policy measures, the government and other relevant departments need to guide and publicize to ensure the effective implementation of policies, followed by increasing the construction of a multimodal logistics system, improving its infrastructure, and finally building a logistics information sharing platform to integrate logistics information such as warehouses, vehicles, and equipment to achieve an effective and reasonable allocation of resources. These construction measures will help improve the level of road logistics services in Wuhan, promote the healthy development of Wuhan's road logistics industry, and promote the steady growth of Wuhan's economy.

4.2 Conclusion and Outlook for the Future

According to the logistics demand forecasting model established in this paper, whether optimistic or pessimistic, forecast results show that the logistics demand in Wuhan will show a steady growth trend in the next six years. The model's predictive accuracy has been validated, and these analysis results also have important guiding significance for logistics planning and decision-making in Wuhan, helping to ensure that the road logistics system can cope with potential adverse situations, find coping strategies, and improve overall resilience. It provides a robust data basis for the formulation of relevant government policies.

This paper believes that the Wuhan Municipal Government should seize the opportunity for logistics development, formulate targeted policies and regulations, and further enhance Wuhan's comprehensive competitiveness in the logistics field. In addition, it is recommended that the government and enterprises increase investment in logistics infrastructure construction, improve logistics efficiency, reduce logistics costs, and promote the rapid development of the logistics industry.

The prediction model established in this paper is relatively single, and there may still be some errors. Therefore, future research must optimize the model further and comprehensively consider more relevant factors to improve prediction accuracy.

References

1. Yin, Y., Chen, Y.: Analysis of logistics demand of agricultural products in the Beijing-Tianjin-Hebei region based on gray prediction. Hubei Agric. Sci. **62**(01), 214–218+223 (2023)
2. Li, S.-C., Ye, J.: Demand analysis and prediction of cold chain logistics of agricultural products based on gray regression model. Highway Transport. Sci. Technol. **39**(5), 166–174 (2022)
3. Zeng, Q., Yao, H., Zeng, J., Nie, J.: Demand forecast analysis of cold chain logistics of citrus fruits in Liuzhou city based on gray model. J. Guilin Inst. Aerospace Indust. **27**(04), 520–526 (2022)
4. Zeng, H., Zhu, W.: Demand forecasting analysis of cold chain logistics of fresh agricultural products in Hunan Province based on the gray GM(1,1) model. J. Xinyang College Agric. Forestry **32**(04), 40–46 (2022)
5. Lei, L., Wu, J., Chen, J.: Forecasting analysis of logistics demand in Guizhou in the 14th Five-Year Plan based on gray GM(1,1). J. Zaozhuang College **39**(05), 78–84 (2022)
6. Huiming, D., Xinyu, P.: A novel gray prediction model with system structure based on energy. J. Clean. Product. (2023)
7. Lei, C.: Research on logistics demand forecasting in Wuhan based on GM(1,1). Guangdong Econ. **8**, 173–174 (2017)
8. Gao, X., Zhang, Z.: Analysis of logistics demands forecasting in post-epidemic Wuhan based on gray prediction model. Logist. Sci. Technol. **45**(7), 12–16 (2022)
9. Xi, W., Zheng, Y.: Application of gray system forecasting model in logistics demand forecasting in Nanjing. Logist. Eng. Manage. **44**(3), 33–35 (2022)

Comparative Analysis of Flow-Based Models for Fast ReRouting in Softwarized Networks

Oleksandra Yeremenko[1](✉), Daniil Nedostup[1], Amal Mersni[2], and Mykhailo Persikov[1]

[1] Kharkiv National University of Radio Electronics, Nauky Avenue 14, Kharkiv 61166, Ukraine
oleksandra.yeremenko.ua@ieee.org, mykhailo.persikov@nure.ua

[2] International University of Sarajevo, Hrasnička cesta br. 15, 71210 Ilidža - Sarajevo, Bosnia and Herzegovina
amersni@ius.edu.ba

Abstract. This work presents the research and comparative analysis of flow-based models for Fast ReRouting proposed for use in softwarized networks. A linear optimization model was chosen among the existing promising flow-based models used in conjunction with link and node protection schemes. Numerical studies were carried out on the structure of the network core fragment selected for the study . The results proved the model's efficiency and confirmed its prospects for use in calculating primary and backup routes by programmable network controllers. The research also made it possible to formulate recommendations for the further development of the approach used to implement Traffic Engineering and advanced protection schemes under Quality of Service guarantees over the primary and backup routes.

Keywords: Multipath routing · Fast ReRoute · Bandwidth · QoS · Software-Defined Network

1 Introduction

The growing need to develop and deploy robust and resilient future communication networks arises from several factors. On the one side, network operators emphasize robustness and resilience in responding to the growing reliance on communication infrastructure by users, service providers, and various institutions, making network reliability a pivotal consideration in their operational strategies [1].

On the other hand, as communication networks rapidly evolve, it is essential to continuously refine new design and management techniques to maintain reliable network operations. One of the most significant technological advances for improving network resilience is Software-Defined Networking (SDN) [2–6]. SDN introduces network centralized management for effective reconfigurability, allowing dynamic adjustment to network conditions. This feature is vital for maintaining uninterrupted communications, supporting critical services, and addressing network challenges. Researchers and practitioners are diligently exploring innovative strategies to construct the next generation

Z. Hu et al. (Eds.): ICCSEEA 2024, LNDECT 242, pp. 371–384, 2025.
https://doi.org/10.1007/978-3-031-84228-3_32

of resilient networks, fully leveraging the transformative potential of SDN in reshaping network architecture and management [7–12].

Despite the extensive exploration of SDN's capabilities and functionalities, achieving consistent resilience against connectivity failures remains a key challenge [1]. SDN-based architectures have become prevalent in environments such as data centers, enterprise networks, and wide-area networks, which are particularly susceptible to communication link failures. These failures, often caused by damage to network interfaces, data plane elements, or cables, lead to significant disruptions in ongoing services [9].

The recovery process within softwarized networking frameworks is designed to address these failures efficiently, involving initial failure reporting to the controller, then the computation of a new routing path, and culminating in the update of switches' flow tables. Despite the potential for rapid flow table updates, the critical phases of path computation and the necessary communication between switches and controllers contribute to extended recovery times [8]. As a result, the recovery duration in SDN environments may exceed that of traditional networks, largely due to the complexity of these procedural steps and the complex interactions inherent in SDN's dynamic management capabilities [8, 11].

Recovery paths in communication networks can be established either reactively after a failure or proactively before one occurs, with reactive methods often facing delays due to route updates. Fast ReRoute (FRR) offers a solution by enabling the pre-computation of recovery paths, allowing immediate rerouting when failures happen, and reducing disruption and recovery time [1, 7–12]. As a crucial fault-tolerant routing tool, FRR is designed to swiftly address service interruptions that may arise from congestion in network links and router queues. Employed primarily at the network's core, FRR ensures continuous packet delivery. It operates by concurrently establishing a primary path for network traffic while also preparing a backup path. This backup path is strategically calculated to circumvent any network elements identified as potential points of failure, such as nodes, links, or segments, thus ensuring a seamless flow of communication even in the face of network disruptions [7].

FRR is crucial in enhancing network resilience and minimizing recovery times post-failures, a critical research focus as networks evolve toward SDN. Traditional IP and MPLS networks benefit from FRR through methods like Loop-Free Alternate (LFA) and RSVP-TE [1, 8, 13, 14], showcasing its foundational role in network recovery. Adapting FRR to SDN's dynamic and programmable nature, using strategies like source routing [7–11], represents a significant shift towards optimizing network resilience in more complex environments.

In this context, [7] introduces the Local Fast Reroute (LFR) algorithm for SDN, which optimizes the rerouting process by aggregating affected flows. This innovation reduces controller-switch operations and decreases the total number of flow entries, enhancing network resilience. Further advancing FRR's application within SDNs, [8] presents a novel algorithm that expedites the computation of recovery paths, improving both cost efficiency and network resilience. The effectiveness of this approach is confirmed through comprehensive evaluations and the development of a prototype. Additionally, [9] elaborates on SD-FAST, a rerouting mechanism tailored for link failures in SDN that targets only affected traffic, thereby maintaining uninterrupted flow for unaffected traffic.

The investigation into flow-based models for fast rerouting [12] represents a relevant direction in adapting FRR strategies within SDN [7–11]. These models' focused analysis and optimization toward network resilience and efficiency represent a promising area for future research.

2 Flow-Based Models for Fast ReRouting

The following initial data should be noted in the description of the flow-based model of fast rerouting in a network:

- number of network links (n);
- number of network nodes (m);
- source node (s);
- destination node (d);
- link capacity ($c_{i,j}$);
- link metrics for the primary (f^p) and backup (f^b) routes;
- flow rate incoming to the network (r, packets per second, pps);
- protection scheme for the network element to be implemented.

The number of network links n determines the dimension of the vector $\overrightarrow{x}$ whose coordinates $x_{i,j}$ characterize the flow share in the link between the ith and jth nodes. In turn, the dimension of the metrics vector $\overrightarrow{f}$ corresponds to the number of links n, whose coordinates $f_{i,j}$ specify the metric of the associated link between the ith and jth nodes.

In the course of the study, to implement multipath routing, the following constraints are imposed on the coordinates of the primary and backup routing vectors [11, 12]:

for the primary route

$$0 \leq x_{i,j} \leq 1; \tag{1}$$

for the backup route

$$0 \leq \overline{x}_{i,j} \leq 1. \tag{2}$$

The physical sense of variables (1) and (2) determines the possibility of flow splitting over network paths, i.e., packets can be transmitted via one or multiple paths.

When solving the routing problem, it is necessary to ensure that the flow conservation conditions for the primary and backup routes in each of the network nodes and the network as a whole are fulfilled [12]:

for the primary route

$$\begin{cases} \sum_{j:(i,j)} x_{i,j} - \sum_{j:(j,i)} x_{j,i} = 1 - \textit{for the source node}; \\ \sum_{j:(i,j)} x_{i,j} - \sum_{j:(j,i)} x_{j,i} = 0 - \textit{for the transit nodes}; \\ \sum_{j:(i,j)} x_{i,j} - \sum_{j:(j,i)} x_{j,i} = -1 - \textit{for the destination node}; \end{cases} \tag{3}$$

for the backup route

$$\begin{cases} \sum_{j:(i,j)} \overline{x}_{i,j} - \sum_{j:(j,i)} \overline{x}_{j,i} = 1 - \textit{for the source node}; \\ \sum_{j:(i,j)} \overline{x}_{i,j} - \sum_{j:(j,i)} \overline{x}_{j,i} = 0 - \textit{for the transit nodes}; \\ \sum_{j:(i,j)} \overline{x}_{i,j} - \sum_{j:(j,i)} \overline{x}_{j,i} = -1 - \textit{for the destination node}. \end{cases} \quad (4)$$

It is also important to avoid overloading the network links under the following conditions, the number of which corresponds to the number of network links:

for the primary route

$$r \cdot x_{i,j} \le c_{i,j}\left(i,j = \overline{1,n}, i \ne j\right) \quad (5)$$

for the backup route

$$r \cdot \overline{x}_{i,j} \le c_{i,j}(i,j = \overline{1,n}, i \ne j) \quad (6)$$

As mentioned above, in the fast rerouting process, several basic protection schemes for network elements can be supported: node, link, path, and bandwidth. Therefore, additional constraints should be considered for implementing link and node protection schemes on the routing variables $\overline{x}_{i,j}$ responsible for determining the backup path [11, 12].

In the case of multipath routing, additional constraints for the link protection scheme are as follows [12]:

$$0 \le \overline{x}_{i,j} \le \delta_{i,j} \quad (7)$$

where

$$\delta_{i,j} = \begin{cases} 0 \ \textit{under the link } (i,j) \ \textit{protection}; \\ 1 \ \textit{otherwise}. \end{cases} \quad (8)$$

The fulfillment of conditions (7) and (8) ensures that the link (i,j), which is considered protected, will not be included and used by the backup route.

For the node protection scheme in the case of multipath routing, additional restrictions are associated with the fact that the links adjacent to the protected node will be covered [12]:

$$0 \le \overline{x}_{i,j} \le \delta_{i,j} \ \textit{for links adjacent to the protected node} \quad (9)$$

where $\delta_{i,j}$ values, the choice of which is subject to condition (8).

The following objective function must be minimized to determine the values of the routing variables $x_{i,j}$ and $\overline{x}_{i,j}$ when solving the fault-tolerant routing problem in the network:

$$F = \sum_{(i,j)} f_{i,j}^{p} x_{i,j} + \sum_{(i,j)} \overline{f}_{i,j}^{b} \overline{x}_{i,j} \quad (10)$$

where $f_{i,j}^{p}$ is the link (i,j) metric for the primary route; $\overline{f}_{i,j}^{b}$ is the link (i,j) metric for the backup route.

3 Numerical Research

The structure of the studied network fragment consists of 9 nodes and 12 links. The first router was considered the source node, and the eighth router was selected as the destination node. Figure 1 shows the network's topology, taking into account the network links' conditional bandwidth (Table 1).

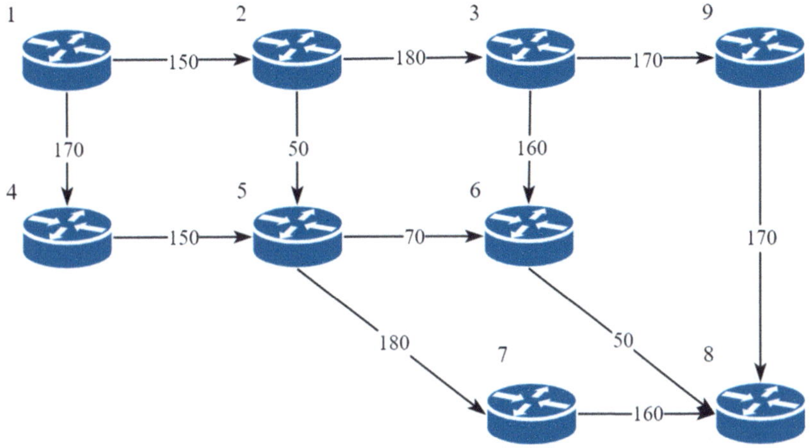

Fig. 1. Network fragment topology under study.

Table 1. Network links bandwidths.

Link #	Link	Link Bandwidth $c_{i,j}$, pps	Link #	Link	Link Bandwidth $c_{i,j}$, pps
1	(1,2)	150	7	(4,5)	150
2	(1,4)	170	8	(5,6)	70
3	(2,3)	180	9	(5,7)	180
4	(2,5)	50	10	(6,8)	50
5	(3,6)	160	11	(7,8)	160
6	(3,9)	170	12	(9,8)	170

The desired vector $\vec{x}$ for the network structure and its links capacities, shown in Fig. 1, contains both types of routing variables for primary and backup paths. For further investigation we will use the following link metrics:

- $f_{i,j}^{RIP}$ – metrics by analogy with the RIP protocol;
- $f_{i,j}^{OSPF}$ – metrics by analogy with the OSPF protocol.

Then the $f_{i,j}^{RIP}$ will be calculated using the rule:

$$f_{i,j}^{RIP} = 1. \tag{11}$$

Using metric (11) enables the calculation of the primary and backup routes containing the minimum number of communication links.

Also, for calculations and comparative analysis, a metric by analogy with the OSPF protocol will be used to calculate the best routes in terms of bandwidth. Therefore, the $f_{i,j}^{OSPF}$ metric will be obtained as shown below:

$$f_{i,j}^{OSPF} = 10^8/c_{i,j}. \tag{12}$$

Let's formulate the flow conservation conditions (3) and (4) on the nodes of the network fragment under study, which are the same for both the primary and backup routes:

$$\begin{cases} x_{1,2} + x_{1,4} = 1; \\ -x_{1,2} + x_{2,3} + x_{2,5} = 0; \\ -x_{2,3} + x_{3,6} + x_{3,9} = 0; \\ -x_{1,4} + x_{4,5} = 0; \\ -x_{2,5} - x_{4,5} + x_{5,6} + x_{5,7} = 0; \\ -x_{3,6} - x_{5,6} + x_{6,8} = 0; \\ -x_{5,7} + x_{7,8} = 0; \\ -x_{3,9} + x_{9,8} = 0; \\ -x_{6,8} - x_{7,8} - x_{9,8} = -1; \end{cases} \begin{cases} \overline{x}_{1,2} + \overline{x}_{1,4} = 1; \\ -\overline{x}_{1,2} + \overline{x}_{2,3} + \overline{x}_{2,5} = 0; \\ -\overline{x}_{2,3} + \overline{x}_{3,6} + \overline{x}_{3,9} = 0; \\ -\overline{x}_{1,4} + \overline{x}_{4,5} = 0; \\ -\overline{x}_{2,5} - \overline{x}_{4,5} + \overline{x}_{5,6} + \overline{x}_{5,7} = 0; \\ -\overline{x}_{3,6} - \overline{x}_{5,6} + \overline{x}_{6,8} = 0; \\ -\overline{x}_{5,7} + \overline{x}_{7,8} = 0; \\ -\overline{x}_{3,9} + \overline{x}_{9,8} = 0 \\ -\overline{x}_{6,8} - \overline{x}_{7,8} - \overline{x}_{9,8} = -1. \end{cases} \tag{13}$$

In the vector-matrix representation, the system of equations will take the next form:

$$Aeq * \vec{x} = \vec{b}\,eq \tag{14}$$

where according to (13)

$$Aeq = \begin{bmatrix} Aeq1 & 0 \\ 0 & Aeq1 \end{bmatrix}, \quad \vec{b}\,eq = \begin{bmatrix} \vec{b}\,eq1 \\ \vec{b}\,eq1 \end{bmatrix}$$

under

$$Aeq1 = \begin{bmatrix} 1 & 1 & 0 & 0 & 0 & 0 & 0 & 0 & 0 & 0 & 0 & 0 \\ -1 & 0 & 1 & 1 & 0 & 0 & 0 & 0 & 0 & 0 & 0 & 0 \\ 0 & 0 & -1 & 0 & 1 & 1 & 0 & 0 & 0 & 0 & 0 & 0 \\ 0 & -1 & 0 & 0 & 0 & 0 & 1 & 0 & 0 & 0 & 0 & 0 \\ 0 & 0 & 0 & -1 & 0 & 0 & -1 & 1 & 1 & 0 & 0 & 0 \\ 0 & 0 & 0 & 0 & -1 & 0 & 0 & -1 & 0 & 1 & 0 & 0 \\ 0 & 0 & 0 & 0 & 0 & 0 & 0 & 0 & -1 & 0 & 1 & 0 \\ 0 & 0 & 0 & 0 & 0 & -1 & 0 & 0 & 0 & 0 & 0 & 1 \\ 0 & 0 & 0 & 0 & 0 & 0 & 0 & 0 & 0 & -1 & -1 & -1 \end{bmatrix};$$

$$beq1 = \begin{bmatrix} 1 \\ 0 \\ 0 \\ 0 \\ 0 \\ 0 \\ 0 \\ 0 \\ -1 \end{bmatrix}. \tag{15}$$

The conditions for preventing the overloading of network links for the primary route are represented by a system of inequalities and supplemented by conditions for the backup route:

$$\begin{cases} r * x_{1,2} \le c_{1,2}; \\ r * x_{1,4} \le c_{1,4}; \\ r * x_{2,3} \le c_{2,3}; \\ r * x_{2,5} \le c_{2,5}; \\ r * x_{3,6} \le c_{3,6}; \\ r * x_{3,9} \le c_{3,9}; \\ r * x_{4,5} \le c_{4,5}; \\ r * x_{5,6} \le c_{5,6}; \\ r * x_{5,7} \le c_{5,7}; \\ r * x_{6,8} \le c_{6,8}; \\ r * x_{7,8} \le c_{7,8}; \\ r * x_{9,8} \le c_{9,8}; \end{cases} \begin{cases} r * \overline{x}_{1,2} \le c_{1,2}; \\ r * \overline{x}_{1,4} \le c_{1,4}; \\ r * \overline{x}_{2,3} \le c_{2,3}; \\ r * \overline{x}_{2,5} \le c_{2,5}; \\ r * \overline{x}_{3,6} \le c_{3,6}; \\ r * \overline{x}_{3,9} \le c_{3,9}; \\ r * \overline{x}_{4,5} \le c_{4,5}; \\ r * \overline{x}_{5,6} \le c_{5,6}; \\ r * \overline{x}_{5,7} \le c_{5,7}; \\ r * \overline{x}_{6,8} \le c_{6,8}; \\ r * \overline{x}_{7,8} \le c_{7,8}; \\ r * \overline{x}_{9,8} \le c_{9,8}. \end{cases} \tag{16}$$

The system of inequalities (16) in vector-matrix form will be as described under the following:

$$A * \vec{x} \le \vec{b} \tag{17}$$

where

$$A = \begin{bmatrix} A1 & 0 \\ 0 & A1 \end{bmatrix}, \ \vec{b}\,eq = \begin{bmatrix} \vec{b}\,1 \\ \vec{b}\,1 \end{bmatrix}$$

under

$$A1 = \begin{bmatrix} r\,0\,0\,0\,0\,0\,0\,0\,0\,0\,0\,0 \\ 0\,r\,0\,0\,0\,0\,0\,0\,0\,0\,0\,0 \\ 0\,0\,r\,0\,0\,0\,0\,0\,0\,0\,0\,0 \\ 0\,0\,0\,r\,0\,0\,0\,0\,0\,0\,0\,0 \\ 0\,0\,0\,0\,r\,0\,0\,0\,0\,0\,0\,0 \\ 0\,0\,0\,0\,0\,r\,0\,0\,0\,0\,0\,0 \\ 0\,0\,0\,0\,0\,0\,r\,0\,0\,0\,0\,0 \\ 0\,0\,0\,0\,0\,0\,0\,r\,0\,0\,0\,0 \\ 0\,0\,0\,0\,0\,0\,0\,0\,r\,0\,0\,0 \\ 0\,0\,0\,0\,0\,0\,0\,0\,0\,r\,0\,0 \\ 0\,0\,0\,0\,0\,0\,0\,0\,0\,0\,r\,0 \\ 0\,0\,0\,0\,0\,0\,0\,0\,0\,0\,0\,r \end{bmatrix}; \quad \vec{b}1 = \begin{bmatrix} c_{1,2} \\ c_{1,4} \\ c_{2,3} \\ c_{2,5} \\ c_{3,6} \\ c_{3,9} \\ c_{4,5} \\ c_{5,6} \\ c_{5,7} \\ c_{6,8} \\ c_{7,8} \\ c_{9,8} \end{bmatrix}. \tag{18}$$

Let link (2,3) of the network fragment shown in Fig. 1 be protected. Also, a flow with an intensity of $r = 50$ pps will be transmitted from the first to the eighth node. Consequently, the following routing solutions were obtained as a result of solving the linear programming problem using the code implemented using Python and the functionality of the NumPy and Scipy libraries.

Using metric (11), the routing order for the primary and backup routes was obtained, as shown in Fig. 2.

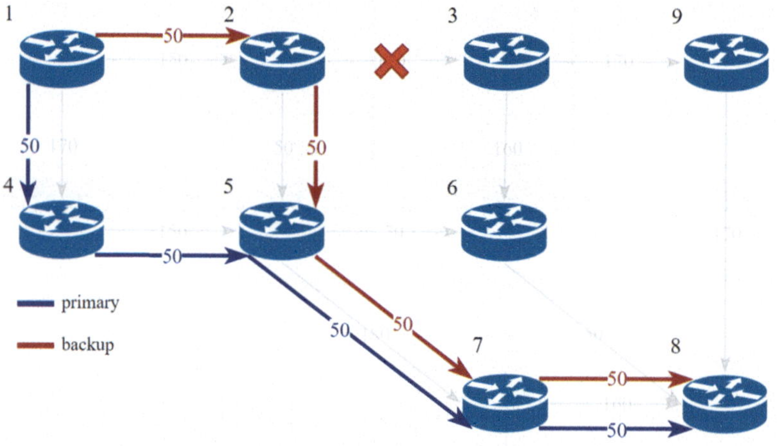

Fig. 2. Primary and backup routes with metric (11) at flow rate $r = 50$ pps and link (2,3) protection.

At the same time, using metric (12), we obtained the routing solutions shown in Fig. 3.

The analysis of the obtained results revealed the outcomes as follows. Metric (11), which focuses on choosing the shortest route regarding the number of communication links, facilitates the formation of the primary and backup routes to bypass the protected

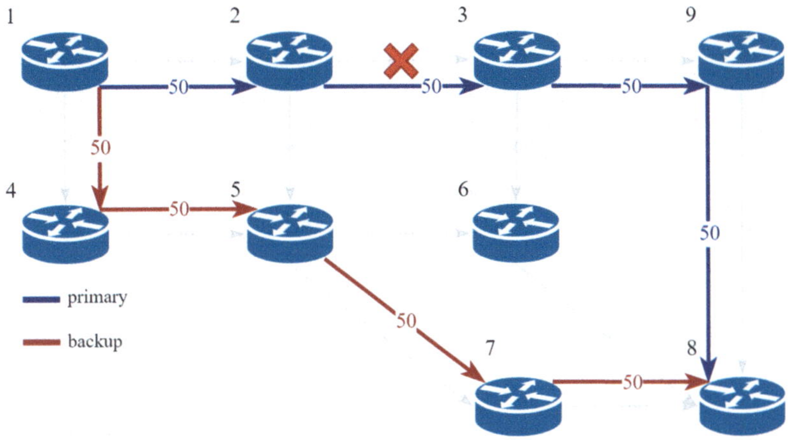

Fig. 3. Primary and backup routes with metric (12) at flow rate $r = 50$ pps and link (2.3) protection.

link. We can see that the primary and backup routes differ only by two links between them. For its part, metric (12) allows us to calculate the primary and backup routes that do not overlap and, moreover, are chosen so that the bandwidth (BW) of the primary and backup routes is maximized (Table 2).

Table 2. Bandwidths and communication links of the primary and backup routes at a flow rate of $r = 50$ pps and link (2,3) protection.

Route type	Route			
	f^{RIP} metric	BW, pps	f^{OSPF} metric	BW, pps
Primary	$1 \to 4 \to 5 \to 7 \to 8$	150	$1 \to 2 \to 3 \to 9 \to 8$	150
Backup	$1 \to 2 \to 5 \to 7 \to 8$	50	$1 \to 4 \to 5 \to 7 \to 8$	150

For further analysis, we increase the flow rate to $r = 160$ pps and link to be protected is still (2,3). The resulting routing order for the primary and backup routes using metric (11) is shown in Fig. 4.

From the results of calculating the routes according to metrics (11) and (12), it can be seen that when the flow rate increases to $r = 160$ pps, both metrics use load balancing, and the packet flow is distributed over several paths. In the case of metric (11), load balancing for the primary and backup routes is performed on the same paths and bypasses the security link. Meanwhile, metric (12) has the same route for the backup path as metric (11), but the primary route is calculated considering the maximum bandwidth. It passes through the link (2,3), which is subject to protection. The characteristics of the resulting primary and backup multipaths are shown in Table 3.

While using metric (12), the resulting routing order for the primary and backup routes is shown in Fig. 5. It should be noted that the total bandwidth of the primary

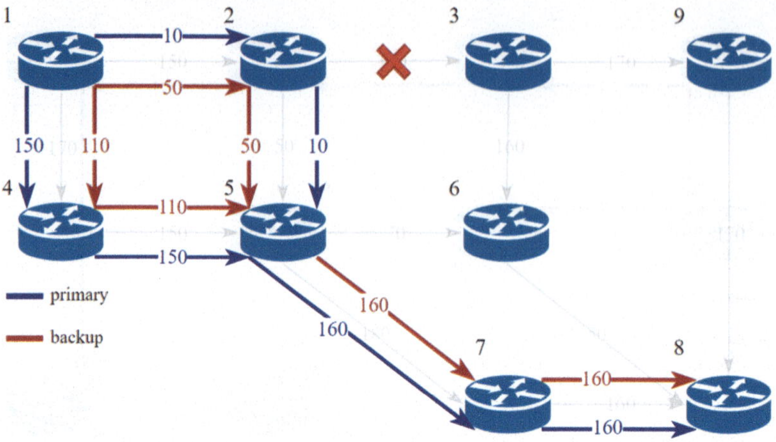

Fig. 4. Primary and backup routes with metric (11) at flow rate $r = 160$ pps and link (2.3) protection.

route obtained based on (12) is 300 pps, which is one and a half times higher than the bandwidth of the primary route with metric (11) – 200 pps (Table 3).

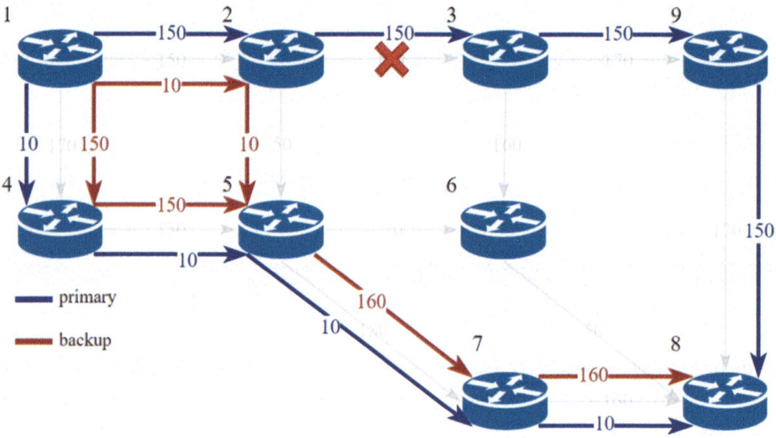

Fig. 5. Primary and backup routes with metric (12) at flow rate $r = 160$ pps and link (2.3) protection.

For the next example, we increase the flow rate to $r = 200$ pps, and the link to be protected remains (2,3). The resulting routing solutions for the primary and backup routes using metrics (11) and (12) coincide and are shown in Fig. 6 and Table 4.

Therefore, the routing solutions for both metrics coincide concerning the primary and backup paths because the flow rate of 200 pps is the maximum possible for this network fragment and the optimization problem statement.

Table 3. Bandwidths and communication links of the primary and backup routes at a flow rate of $r = 160$ pps and link (2,3) protection.

Route type	Route			
	f^{RIP} metric	BW, pps	f^{OSPF} metric	BW, pps
Primary	$1 \rightarrow 4 \rightarrow 5 \rightarrow 7 \rightarrow 8$ $1 \rightarrow 2 \rightarrow 5 \rightarrow 7 \rightarrow 8$	150 50	$1 \rightarrow 2 \rightarrow 3 \rightarrow 9 \rightarrow 8$ $1 \rightarrow 4 \rightarrow 5 \rightarrow 7 \rightarrow 8$	150 150
Backup	$1 \rightarrow 4 \rightarrow 5 \rightarrow 7 \rightarrow 8$ $1 \rightarrow 2 \rightarrow 5 \rightarrow 7 \rightarrow 8$	150 50	$1 \rightarrow 4 \rightarrow 5 \rightarrow 7 \rightarrow 8$ $1 \rightarrow 2 \rightarrow 5 \rightarrow 7 \rightarrow 8$	150 50

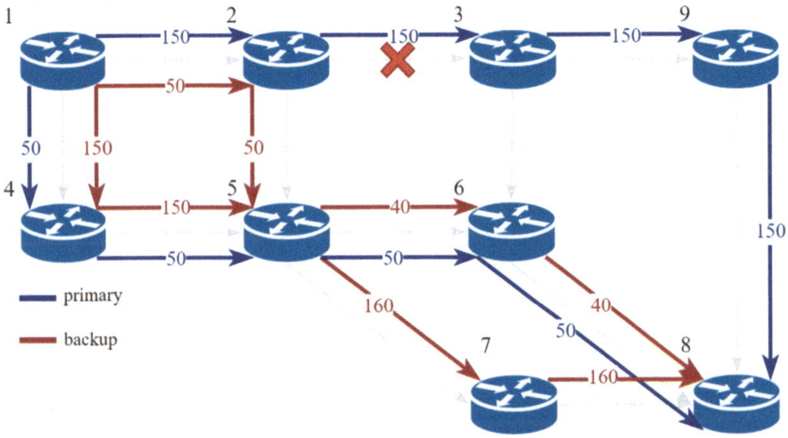

Fig. 6. Primary and backup routes with metrics (11) and (12) at flow rate $r = 200$ pps and link (2.3) protection.

Table 4. Characteristics of the primary and backup routes at a flow rate of $r = 200$ pps and link (2,3) protection.

Route type	Route			
	f^{RIP} metric	Flow share, pps	f^{OSPF} metric	Flow share, pps
Primary	$1 \rightarrow 2 \rightarrow 3 \rightarrow 9 \rightarrow 8$ $1 \rightarrow 4 \rightarrow 5 \rightarrow 6 \rightarrow 8$	150 50	$1 \rightarrow 2 \rightarrow 3 \rightarrow 9 \rightarrow 8$ $1 \rightarrow 4 \rightarrow 5 \rightarrow 6 \rightarrow 8$	150 50
Backup	$1 \rightarrow 4 \rightarrow 5 \rightarrow 7 \rightarrow 8$ $1 \rightarrow 2 \rightarrow 5 \rightarrow 6 \rightarrow 8$ $1 \rightarrow 2 \rightarrow 5 \rightarrow 7 \rightarrow 8$	150 40 10	$1 \rightarrow 4 \rightarrow 5 \rightarrow 7 \rightarrow 8$ $1 \rightarrow 2 \rightarrow 5 \rightarrow 6 \rightarrow 8$ $1 \rightarrow 2 \rightarrow 5 \rightarrow 7 \rightarrow 8$	150 40 10

4 Discussion

The study showed that at an intensity of less than 150 pps, a single path can be used for any metric, and at an intensity of more than 150 pps, the formation of a multipath begins. However, the primary and backup multipaths are formed in different ways. The dependence of the number of routes in the primary and backup multipaths for metric (12) on the intensity of the transmitted flow is shown in Fig. 7. The f^{OSPF} metric (12) was chosen for analysis because it allows obtaining the most productive routing solutions.

These dependencies show that the primary multipath is formed from the most productive communication links, and the backup is formed based on the network resource that remains after the primary one is formed. However, the high-performance link is not used (3,6). This suggests that to improve load balancing in the network, it is also necessary to use the Traffic Engineering (TE) concept when the formation of multipath takes into account the utilization of communication links.

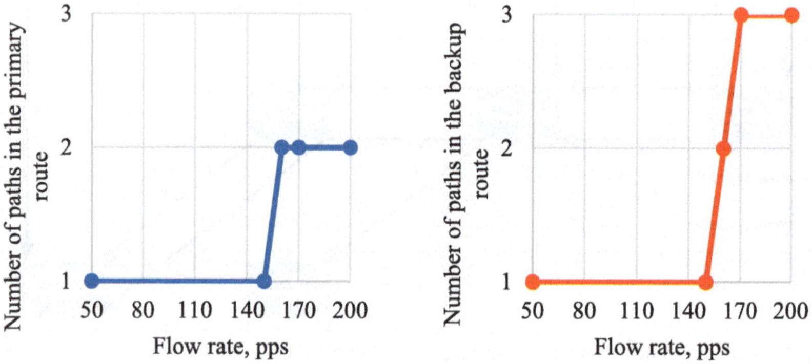

Fig. 7. Dependence of the number of routes in the primary and backup multipaths for metric (12) on the intensity of the transmitted flow.

5 Conclusion

The work solves the problem of analyzing and studying promising methods of ensuring fault tolerance through routing in softwarized networks, and choosing an effective mathematical model of fast rerouting. The current state and prospects for developing fault tolerance technologies employing routing in SDN are investigated.

The experimental part of the work is devoted to the study and comparative analysis of flow-based models of fast rerouting. Within the optimization problem statement, mathematical models of fast rerouting with different metrics are presented and described, provided that the link and node protection (redundancy) schemes are implemented in the network. In the example of the studied network fragment, the modeling was carried out using the Python libraries NumPy and Scipy.

The technological task of multipath fast rerouting was formulated in an optimization form. Expressions (1)–(10) represent the corresponding conditions and constraints for calculating the primary and backup routes, as well as the link and node protection conditions (7)–(9). The study was carried out on the selected topology (Fig. 1), which was used for further modeling in the Google Colab environment and software implementation in Python. We used metrics by analogy with the RIP and OSPF protocols (11) and (12) to calculate the primary and backup routes.

Considering various route formation metrics, a study was conducted to build routing solutions for the primary and backup paths when the flow intensity value changes and the network element (link) to be protected is selected. The research allowed us to conclude the choice of metrics and load-balancing strategy.

The analysis of the results showed the following at low flow intensity and link protection mechanism. Metric (11), which is focused on selecting the shortest route by the number of communication links, contributes to forming the primary and backup routes to bypass the protected link. For its part, metric (12) allows us to calculate the disjoint primary and backup routes, which are selected so that the bandwidth of the primary and backup routes is maximized. With the increase in the intensity of the flow transmitted between the source and destination nodes, both metrics contributed to the formation of multipaths when the packet flow was distributed over several paths. Consequently, the chosen multipath routing strategy also contributed to the implementation of load balancing during fast rerouting. The research also showed that the primary and backup multipaths were formed differently. The corresponding dependences of the number of routes in the primary and backup multipaths for metric (12) on the intensity of the transmitted flow were obtained.

The results of additional studies of constructing routing solutions for the primary and backup paths using metrics (11) and (12) in the node protection scheme confirmed the dynamics of the solutions obtained for implementing the link protection scheme.

The general conclusion is that in softwarized networks, it is recommended to use multipath fault-tolerant routing strategies in conjunction with load balancing based on TE. In addition, it is necessary to provide protection (redundancy) not only for links, nodes, and paths to increase network resilience but also redundancy with Quality of Service guarantee over the primary and backup routes (e.g., protection of bandwidth and other QoS indicators).

Acknowledgment. This chapter was published due to work on the Erasmus+ Project Jean Monnet module "The European experience for enhancing the resilience of critical entities in Ukraine" Project No.: 101085825 – ERASMUS-JMO-2022-MODULE.

References

1. Chiesa, M., Kamisiński, A., Rak, J., Rétvári, G., Schmid, S.: A survey of fast-recovery mechanisms in packet-switched networks. IEEE Commun. Surv. Tutorials **23**(2), 1253–1301 (2021). https://doi.org/10.1109/COMST.2021.3063980
2. Ahmad, S., Mir, A.H.: SDN Interfaces: Protocols, Taxonomy and Challenges. International Journal of Wireless and Microwave Technologies (IJWMT) **12**(2), 11–32 (2022). https://doi.org/10.5815/ijwmt.2022.02.02

3. Segeč, P., Moravčik, M., Uratmová, J., Papán, J., Yeremenko, O.: SD-WAN-architecture, functions and benefits. In: 2020 18th International Conference on Emerging eLearning Technologies and Applications (ICETA) Proceedings, pp. 593–599. IEEE (2020). https://doi.org/10.1109/ICETA51985.2020.9379257
4. Barabash, O., Kravchenko, Y., Mukhin, V., Kornaga, Y., Leshchenko, O.: Optimization of parameters at SDN technologie networks. Int. J. Intell. Syst. Appl. **9**(9), 1–9 (2017). https://doi.org/10.5815/ijisa.2017.09.01
5. Al Mtawa, Y., Haque, A., Lutfiyya, H.: Migrating from legacy to software defined networks: a network reliability perspective. IEEE Trans. Reliab. **70**(4), 1525–1541 (2021). https://doi.org/10.1109/TR.2021.3066526
6. Rohitaksha, K., Rajendra, A.B., Mohan, J.: Routing in hybrid software defined networking using hop-count approach. Int. J. Wireless Microwave Technol. **12**(3), 54–60 (2022). https://doi.org/10.5815/ijwmt.2022.03.04
7. Zhang, X., Cheng, Z., Lin, R., He, L., Yu, S., Luo, H.: Local fast reroute with flow aggregation in software defined networks. IEEE Commun. Lett. **21**(4), 785–788 (2016). https://doi.org/10.1109/LCOMM.2016.2638430
8. Qiu, K., Zhao, J., Wang, X., Fu, X., Secci, S.: Efficient recovery path computation for fast reroute in large-scale software-defined networks. IEEE J. Sel. Areas Commun. **37**(8), 1755–1768 (2019). https://doi.org/10.1109/JSAC.2019.2927098
9. Moyeen, M.A., Tang, F., Saha, D., Haque, I.: SD-fast: a packet rerouting architecture in SDN. In: 2019 15th International Conference on Network and Service Management (CNSM) Proceedings, pp. 1–7. IEEE (2019). https://doi.org/10.23919/CNSM46954.2019.9012703
10. Subramanian, K., Abhashkumar, A., D'Antoni, L., Akella, A.: D2R: policy-compliant fast reroute. In: ACM SIGCOMM Symposium on SDN Research (SOSR) Proceedings, pp. 148–161. ACM (2021). https://doi.org/10.1145/3482898.3483360
11. Lemeshko, O., Yevdokymenko, M., Yeremenko, O., Hailan, A.M., Segeč, P., Papán, J.: Design of the fast reroute QoS protection scheme for bandwidth and probability of packet LOS in software-defined WAN. In: 2019 IEEE 15th International Conference on the Experience of Designing and Application of CAD Systems (CADSM) Proceedings, pp. 1–5. IEEE (2019). https://doi.org/10.1109/CADSM.2019.8779321
12. Yeremenko, O.S., Lemeshko, O.V., Tariki, N.: Fast ReRoute scalable solution with protection schemes of network elements. In: 2017 IEEE First Ukraine Conference on Electrical and Computer Engineering (UKRCON) Proceedings, pp. 783–788. IEEE (2017). https://doi.org/10.1109/UKRCON.2017.8100353
13. Previdi, S., Filsfils, C., Decraene, B., Litkowski, S., Horneffer, M., Shakir, R.: Source packet routing in networking (SPRING) problem statement and requirements, RFC 7855 (2016). https://www.rfc-editor.org/rfc/rfc7855.txt
14. Farrel, A.: Overview and principles of internet traffic engineering, RFC 9522 (2024). https://www.rfc-editor.org/rfc/rfc9522.html

An Approach Towards Performance Improvement of Design Processes Based on Associative Rules

Solomiya Liaskovska[1(✉)] and Yevgen Martyn[2]

[1] Department of Artificial Intelligence, Lviv Polytechnic National University, Lviv 79905, Ukraine
solomiya.y.lyaskovska@lpnu.ua

[2] Department of Project Management, Information Technologies and Telecommunications, Lviv State University of Life Safety, Lviv 79007, Ukraine

Abstract. The proposed development involving the utilization of information technologies in the form of a recommendation system for designers based on associative rules. The development is aimed at providing designers with free access to information related to their professional activities in the processes of designing industrial machinery and mechanisms. Proposed methods for organizing the study of the Apriori algorithm, illustrations of its functioning, identified shortcomings, and ways to improve it in terms of achieving expected results in increasing designers' awareness, particularly by improving the efficiency of the algorithm. The object of the conducted research was the process of designing industrial machinery and mechanisms using the example of designing Maltese mechanisms. A program has been proposed that allows designers to receive recommendations to improve the performance of design processes based on associative rules. The scientific results obtained in the research emphasize the effectiveness of the proposed approach and open up the possibility of recommending the application of the system we have created for the quick and comfortable identification of associative design tools that contribute to the optimization and organization of work in company.

Keywords: recommendation system · algorithm Apriori · associative rules · design · Maltese mechanism

1 Instruction

Fuzzy association rule mining (FARM) with the Apriori algorithm can capture the selection of scientific developments based on historical transaction data. It identifies fuzzy association rules to reflect similar products. Experimental results show that the proposed FARM approach provides a wealth of information for decision-makers. In research [2], the aim is to explore market associations between products and offer alternatives. Apriori is used along with the natural language processing methods Word2Vec and FastText. Article [3] presents the integration of educational resources and courses to provide students, future designers, with an independent learning platform and information-oriented

Z. Hu et al. (Eds.): ICCSEEA 2024, LNDECT 242, pp. 385–394, 2025.
https://doi.org/10.1007/978-3-031-84228-3_33

learning environment. According to the authors, association rule mining is the most researched and widely used method of data mining. The proposed method of multimedia network integration of educational resources is based on association rule algorithms. Experiment results show that with an increase in data volume, the average testing time of Apriori is reduced by 56.49 s compared to MapReduce. In articles [4, 5], the authors propose an intelligent recommendation system aimed at facilitating the academic advising process and thus interaction among designers. This system is built on a combination of association rule mining and content-based filtering approach. In work [6], an approach to characterizing user context parameters and determining the optimal subset of parameters to achieve the best results from CARS algorithms is investigated. This work proposes a new method of determining the optimal subset of user-context-dependent parameters in any domain, allowing CARS algorithms to generate the most relevant recommendations for each user. Article [7] proposes association rule theory and the Apriori algorithm for completing basket analysis and product recommendations, including engineering requests. It is shown that the XGBoost model achieved better performance and accuracy with an F1 score of approximately 0.789. The proposed recommendation model provides good recommendation results. In research [8], the RFM model is used to create behavioral features, TF-IDF method for product descriptions to create product categories; K-means clustering algorithm for consumer grouping, association rules algorithm by Apriori. Research [9] aims to incorporate intelligent data analysis to improve services. The research results include apriori recommendations in the form of associative rules or consumer transaction templates. In work [10], the feasibility of using one of the included algorithms depending on the parameters of the database being studied is justified. These and other results obtained in this scientific work can be used to facilitate the task of selecting an association rule mining algorithm after determining the studied database. Works such as [1–3] and [7] can be considered more theoretical, leaning towards more decisive applied scientific work rather than purely problem-oriented applied works focusing on how to solve the problem. It is worth mentioning works [4, 8–10] which are more problem-oriented, focusing on problematics rather than method, delving mostly into more intricate algorithm variations or complements, such as article [8] complicating existing systems, to some extent provided in [11–18], with complex genetic algorithm dimensionality reduction T-SNE.

2 Information and Mathematical Support

The input consists of a file with training records using the Apriori algorithm, as well as data in string format where specific products and the quantity of items they will be associated with will be specified.

Input data for the Apriori algorithm are provided in the form of a table or database containing information about transactions. Each transaction must be represented as a set of elements, for example, a list of products, details purchased in one set. Therefore, the input data for the Apriori algorithm consists of a table or database containing information about transactions, unique identifiers, or names of transaction elements. As for the output data, the program will output lists of details that fit the parameters of the input data for designers.

2.1 Apriori Algorithm for Engineearing Tasks

The Apriori algorithm is an algorithm that attempts to operate on database records, including transaction records, or records containing a certain number of fields or elements and uses a "bottom-up" approach to gradually contrast complex records. Using this algorithm, rules of item co-occurrence are found within each transaction, which are then sorted by their strength. To output associative rules, we will use a series of key concepts useful in engineering practice.

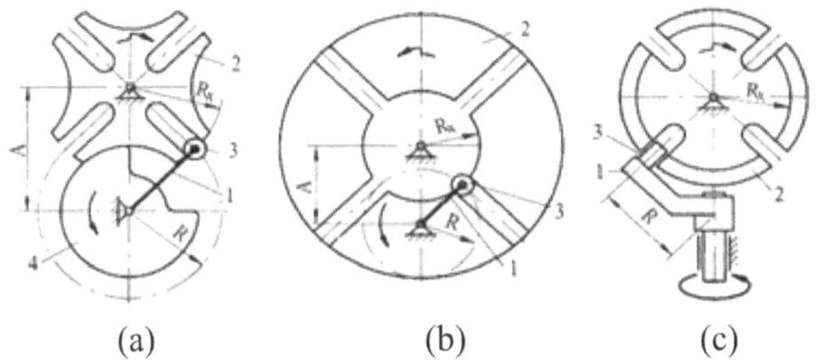

Fig. 1. Transaction Example: Types of Maltese Mechanisms a - with external engagement; b - with internal engagement; c - spherical; 1 - camshaft; 2 - slider; 3 - pin; 4 - locking sector

It may turn out that the details for a certain type of Maltese mechanism in Fig. 1a, b exceed a certain proportion. This means that they typically have a significant influence on the design process. Therefore, the possibility of using this proportion as a threshold support value can be considered. Then sets of details with support values above this threshold can be identified as important detail sets for assembly units.

It should be noted that designers use technical information in three forms in their practice.

At the beginning of the design process, information about mechanism diagrams is useful. Figure 2a, b. provides schematic diagrams of Maltese mechanisms, which are useful for designers of similar mechanisms in the process of setting design tasks.

The designer have to gain an understanding of both the working principle in general and the differences in individual design solutions. Therefore, information technology should provide both a detailed description of the mechanism and the interaction of its parts in the technological process. For example, carousel 1 and Maltese cross 5 are typically rigidly mounted on the same shaft. Continuous rotation of the cam 4 is achieved from the electric motor 7 through the belt 8 and worm gear b transmission. For precise positioning of the carousel relative to the working elements mounted on the frame, a latch 2 is applied. During continuous rotation of shaft I, the cam 3 disengages the latch 2 from the carousel socket, overcoming the force of the spring. Additional elements may be introduced into the schemes that information technology should provide: a one-way clutch 9, a position sensor 11 for shaft I (correspondingly for cam 4), and the clutch actuator - electromagnetic 10. The torque transmission from the worm wheel to the shaft is carried out by clutch 9. The scheme operates in such a way that during continuous rotation of the motor shaft after turning the carousel and its fixation, shaft I continues

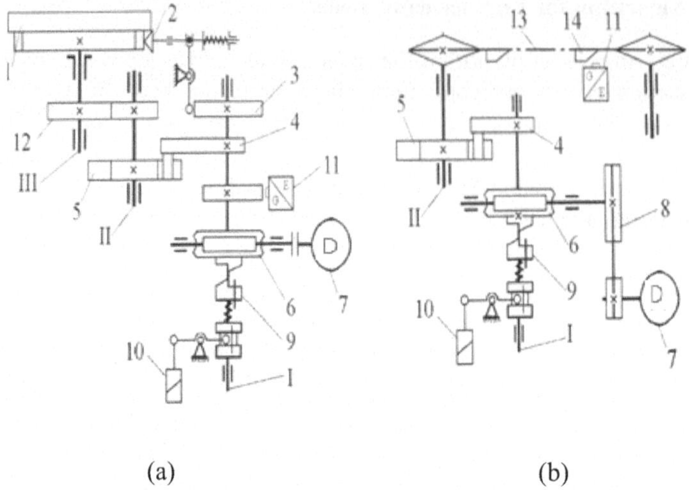

Fig. 2. (a) The carousel drive scheme with the Maltese mechanism. (b) The conveyor drive scheme with the Maltese mechanism, the camshaft of which rotates periodically.

to rotate until the position at which the sensor 11 triggers. At this point, a command is issued to switch off the electromagnetic 10. The clutch disconnects the worm wheel from the shaft, and its movement stops.

The next step is the analysis of processes within the mechanism. To do this, it is necessary to provide in the information technology the ability to obtain dependencies of parameters. They can accompany the scheme (Fig. 3a) with static characteristics in Fig. 3b, or dynamic characteristics taking into account the dimensionality of the parameter space of the mechanism. Free components of transient processes (Fig. 3c) are obtained based on solving the generalized differential Eq. (1). The characteristic parameter is taken as the defining one, for example, u - the force in the latch. The graphs of the free components of transient processes (Fig. 3b) are obtained based on solving the generalized differential Eq. (1). The characteristic parameter is taken as the determining one, for example, the force in fixture 2. Solutions are obtained in a mathematical processing environment, such as Matlab, in four three-dimensional coordinate systems with dimensions *y(1), y(2), y(3), y(4)* as components of the determining force u. The obtained dependencies *y(1) = y(y(2), y(3)), y(2) = y(y(3), y(4)), y(1) = y(y(2), y(4)), y(1) = y(y(3), y(4))* allow the designer to evaluate the mutual influence of parameters on individual details and nodes of the mechanism being designed. The availability of the designer to such scientific resources through the proposed technology allows him to make comparisons, avoid mistakes, and significantly increase the effectiveness of the work.

$$\frac{d^4y}{dt^4} + \frac{d^3y}{dt^3} + \frac{d^2y}{dt^2} + \frac{dy}{dt} = 0 \quad (1)$$

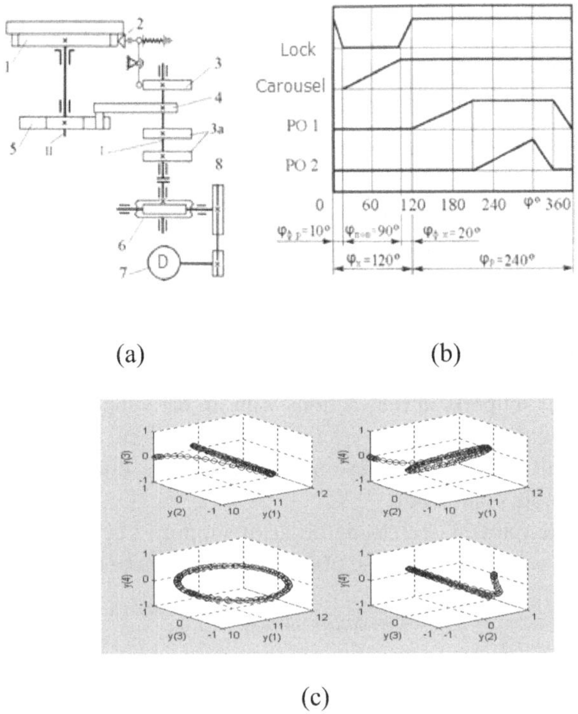

Fig. 3. (a) Schematic diagram of the continuously rotating camshaft drive (b) Cyclogram of the carousel drive with a Maltese mechanism, the camshaft of which rotates continuously. Phase three-dimensional projection planes of system (1).

Confidence refers to how likely it is that detail Y will be found when detail X is found, expressed as {X - > Y}. This is measured by the frequency of operations with detail X, in which detail Y also appears. {cam - > cross} is 3 out of 4, or 75%.

$$\begin{gathered} Support(\text{slider, cross}) \\ Confidence(\text{slider} \rightarrow \text{cross}) \\ Support(\text{slider}) \end{gathered}$$

One of the drawbacks of the confidence measure is that it may inaccurately reflect the importance of the association. This is explained by how important the cam is, but not the cross. If the cross is also important in the mechanism, there is a greater likelihood that a transaction containing the cam will also contain the cross, thus increasing the confidence score. To account for the basic popularity of both component parts, we use a third indicator called lift. 2.2.3. Lift demonstrates how likely it is to find detail Y when detail X is sought and controls how demanded detail Y is. In Table 1, the lift {cam - > cross} equals one, indicating no association between the details. A lift value greater than one means that detail Y is more likely to be purchased if detail X is purchased, while a value less than one means that detail Y is unlikely to be purchased if detail X is purchased.

$$
\begin{aligned}
&Support(\text{slider, cross})\\
&Lift(\text{slider} \rightarrow \text{cross})\\
&Support(\text{slider}) \times Support(\text{cross})
\end{aligned}
$$

2.2 Conviction, the Error Rate of the Rule

The higher the result above one, the better. Steps for the Apriori algorithm:

Step 1: Determine the support of itemsets in the transaction database and select the minimum support and confidence.

Step 2. Take all supports in transactions with higher support than the minimum or selected support value.

Step 3. Find all the rules of these subsets that have higher confidence than the threshold or minimum confidence.

Step 4. Sort the rules in decreasing order of support. Let's consider the following example. Suppose we have a database of engineering documentation:

$$
\begin{aligned}
&1 - Support(x_2)\\
&Conv(x_{1,}x_2)\\
&1 - confidence(x_1, x_2)
\end{aligned}
$$

Let's consider the following example. Suppose we have a database of engineering documentation:

Table 1. The database of library

Transaction	Detail 1	Detail 2	Detail 3
	1	2	3
1	Slider	Cross	Finger
2	Slider	–	–
3	Cross	Finger	–
4	Finger	Bracket	–
5	Slider	Cross	–
6	Cross	Slider	Shaft
7	Shaft	Slider	Cross
8	Finger	Bracket	Slider
9	Cross	Slider	
10	Bracket	Shaft	Finger
11	Slider	Cross	Shaft

Table 2. First block of Sets

Set	Support
Slider	7
Cross	6
Shaft	4
Pin	5
Bracket	3

Table 3. Second Block of Sets

Set	Support
{slider, cross}	6
{slider, shaft}	3
{slider, pin}	2
{slider,bracket} bracket}	1
{cross, shaft}	3
{cross, pin}	2
{cross, bracket}	0
{shaft, pin}	1
{shaft, bracket}	1

Table 4. Final Block of Sets

Set	Support
{slider, cross, shaft}	3

Let the occurrence of a detail be considered frequent if it appears in at least three transactions of the database. The first step of Apriori is to calculate the occurrence count (support) for each detail separately. During the initial scan of the database, we obtain the following result (Table 2):

The next step is to create a list of all pairs of frequent items. For example, for the pair {slider, cross}: the first table shows that the products First aid kit slider and Gloves cross appear together in five sets. Therefore, we say that the item {slider, cross} has a support of five (Table 3).

All pairs of items whose support is greater than three are considered frequent. Now, because some pairs of items are not frequent, any larger set containing it cannot be frequent. Thus, we can reduce the item sets: now we will look for frequent triples of items in the database, but we can already exclude all triples containing one of these two pairs of items (Table 4):

In this example, the only frequent itemset of three details is {slider, cross, shaft}. Other triplets were excluded because they were subsets of pairs that were already below the support threshold.

Thus, we have identified frequent itemsets of details in the database and demonstrated how some details were not counted because one of their subsets was already known to be below the threshold.

This Python script utilizes the Apriori algorithm to discover associative rules in a dataset of details.

The step-by-step execution is as follows:

1. The function **apriori_algo(df)** takes a DataFrame **df** as an argument. The first step in the function is to fill all missing values (NaN) in **df** with zeros.
2. Next, an empty list **transactions** is created. In a loop, for each row of data in **df**, a list containing the values of all columns (except zeros) is created. Each such list is added to **transactions**.
3. The **TransactionEncoder** is used to encode our transactions into a Boolean format required for the Apriori algorithm. The result is then transformed back into a DataFrame.
4. Using the **apriori** function from the mlxtend library, elements that frequently occur with a minimum support of 0.001 are computed.
5. A new column **length** is added, representing the length of each set of details.
6. Sets of details with lengths of two (**n2**) and three (**n3**) are selected, and they are sorted by support (Fig. 4).

```
n2 = freq_items.loc[(freq_items['length'] == 2)]
n2.sort_values("support", ascending=False)
n3 = freq_items.loc[(freq_items['length'] == 3)]
n3.sort_values("support", ascending=False)
```

Fig. 4. Selection and sorting of sets

The function returns two lists: the first contains all sets of details with a length of two, and the second contains sets with a length of three (Fig. 5).

```
return n2['itemsets'].tolist(), n3['itemsets'].tolist()
```

Fig. 5. Returning the results

Summary and Conclusion

The integration of information technologies in the form of a recommendation system for designers, indicating a forward-looking approach to enhancing design processes.

1) By providing designers with free access to information relevant to their professional activities, the development aims to streamline and enrich the design process for industrial machinery and mechanisms.
2) Comprehensive Analysis of Apriori Algorithm: the article conducts a thorough examination of the Apriori algorithm, elucidating its functioning, identifying shortcomings, and proposing methods for improvement, thereby ensuring the algorithm's efficacy in increasing designers' awareness.
3) A case study on the design of Maltese mechanisms, the research underscores its practical relevance and applicability in real-world industrial settings, ensuring its potential to address industry-specific challenges.
4) The proposed program offers designers actionable recommendations based on associative rules, demonstrating tangible outcomes that can optimize design processes and organizational efficiency within companies.

References

1. Dogan, O., Kem, F.C., Oztaysi, B: Fuzzy association rule mining approach to identify e-commerce product association considering sales amount [J]. Complex Intell. Syst. **8**(2), 1551–1560 (2022)
2. Bayrak, A.T., Ceren, O., Gencer, S.M.: Using word embedding methods for product recommendation. IEEE, Safranbolu, Turkey (2022)
3. Mochurad, L., Shchur, G.: Parallelization of cryptographic algorithm based on different parallel computing technologies. In: Proceedings of the Symposium on Information Technologies & Applied Sciences (IT&AS 2021), vol. 2824, pp. 20–29 (2021)
4. Mechery, S., Preethi, N., Iyer, B, Crick, T., Peng, S.-L.: An intelligent recommendation system using market segmentation. Applied Computational Technologies: Springer Nature Singapore, pp. 468–476 (2022)
5. Zomchak, L., Starchevska, I.: Macroeconomic determinants of economic development and growth in ukraine: logistic regression analysis. In: Proceedings of the CSDEIS. Advances in Intelligent Systems, Computer Science and Digital Economics IV, pp. 358–368. Springer, Cham (2023)
6. Alghamdi, S., Sheta, O.M. Adrees, M.A.: framework of prompting intelligent system for academic advising using recommendation system based on association rules. IEEE (2022)
7. Izonin, I., Tkachenko, R., Mitoulis, S.A., Faramarzi, A., Tsmots, I., Mashtalir, D.: Machine learning for predicting energy efficiency of buildings: a small data approach. Procedia Comput. Sci. **231**, 72–77 (2024)
8. Kavu, T.D.: Characterisation of user-contexts for context-aware recommendation systems. In: Proceedings of the Future Technologies Conference (FTC), pp. 89–106 (2021)
9. Zhao, X., Keikhosrokiani, P.: Sales prediction and product recommendation model through user behavior analytics. Comput. Mater. Continua **70**(2), 3855–3874 (2022)
10. Mochurad, L., Hladun, Y., Zasoba, Y., Gregus, M.: An approach for opening doors with a mobile robot using machine learning methods. Big Data Cogn. Comput. **7**(2), 69 (2023)
11. Shen, B.: E-commerce customer segmentation via unsupervised machine. ACM, Stanford CA USA (2021)

12. Izonin, I., Tkachenko, R., Holoven, R., Yemets, K., Havryliuk, M., Shandilya, S.K.: SGD-based cascade scheme for higher degrees wiener polynomial approximation of large biomedical datasets. Mach. Learn. Knowl. Extr. **4**(4), 1088–1106 (2022)
13. Laurentinus, O., Data mining using apriori algorithm and linear regression in product recommendations. IEEE, Yogyakarta, Indonesia (2021)
14. Zomchak, L., Lapinkova, A.: Key interest rate as a central banks tool of the monetary policy influence on inflation: the case of Ukraine. In: Proceedings of the CSDEIS. Advances in Intelligent Systems, Computer Science and Digital Economics IV, pp. 369–379. Springer, Cham (2023)
15. Han, J., Pei, J., Yin, Y.: Mining frequent patterns without candidate generation. ACM SIGMOD Rec. **29**(2), 1–12 (2000)
16. Fielding, R.T.: Architectural Styles and the Design of Network-based Software Architectures. Doctoral dissertation, University of California, Irvinem (2000)
17. Richardson, L., Ruby, S.: RESTful web services. O'Reilly Media (2007)
18. Pautasso, C., Zimmermann, O., Leymann, F.: Restful web services vs. "big" 'web services: making the right architectural decision. In: Proceedings of the 17th International Conference on World Wide Web, pp. 805–814 (2008)

Procedural Recursive Algorithm of Generation Based on Triangles

Andrii Kudriavtsev[1] and Oleksandra Bulgakova[2](✉)

[1] Black Sea National University, Mykolaiv 54000, Ukraine
[2] Taras Shevchenko National University of Kyiv, Kyiv 04116, Ukraine
sashabulgakova2@gmail.com

Abstract. The article presents a comprehensive investigation into the procedural generation of digital content, with a particular focus on the application of fractals and noise functions. It evaluates traditional methods such as the Diamond-Square algorithm and Perlin Noise, and proposes the Triangular Mesh Fractal (TMF) algorithm, a novel approach based on stochastic fractals that advances previous techniques by employing triangular tessellation. This research the TMF algorithm's efficiency, scalability, and its capacity to render content with a more natural appearance. The experimental results underscore the TMF algorithm's enhancements in detail and texture over conventional methods, showcasing its potential in 3D graphics and virtual environment simulations.

Keywords: Procedural generation · Algorithms · Fractals · Stochastic fractals · Noises · Diamond-square algorithm · TMF algorithm · Triangular mesh fractal algorithm

1 Introduction

Until recently, geometric models of various natural landscapes were traditionally constructed based on relatively simple geometric figures or entirely manually. It is evident that the method, while sufficient for describing elementary structures, has certain shortcomings when applied to creating complex objects such as the outlines of continental coastlines, mountains, etc. In the last 5–10 years, new geometric concepts, particularly fractals like the Diamond-Square Algorithm [1], and noises such as Perlin Noise [2], have been increasingly used to describe landscapes and similar structures, replacing many traditional practices. The role of fractals in computer graphics is quite significant today. They come to aid, for instance, when it is necessary to define lines and surfaces of a very complex form with the involvement of several coefficients. Noises, in turn, are used in computer graphics to enhance the realism or graphical complexity of the surfaces of geometric objects. They can also be applied to generate effects like smoke, fog, etc.

Algorithmic procedural generation is a type of computer algorithm that automatically generates content in real-time according to established rules, coefficients, and constraints. In contrast to manual-made content, where an artist or designer determines each object and all parameters of the created content by hands, procedural generation can

Z. Hu et al. (Eds.): ICCSEEA 2024, LNDECT 242, pp. 395–405, 2025.
https://doi.org/10.1007/978-3-031-84228-3_34

result in somewhat unpredictable outcomes, generating content that may differ with each run. One of the key advantages of procedural generation is the reduction in development costs for various commercial products. In the gaming industry, procedural generation elevates design to a new level and enhances key game systems.

However, it is easy to make mistakes in understanding and applying procedural generation: it is crucial to recognize that it is not a tool to solve all problems. It can be used to obtain a myriad of digital content or introduce an element of randomness into objects that would be time-consuming to create manually. Developing and testing algorithms can be time-consuming, especially when they interact with other application's systems. Moreover, procedural generation is not just a means of creating content. Procedurally generated content can serve various purposes, such as adaptive animations, infinite game levels, seamless textures, endless replayability, intuitive systems, adaptive music, or anything else.

The research and developments, especially regarding terrain generation, were conducted by variety developers. Key contributions were made by Yin and Zheng, who developed the parametrically controlled terrain generation (PCTG) method [3]. Additionally, in [4] focused on procedural content generation methods in video games. The Dungeon template library (DTL) and THREE.Terrain engine, tools for procedural generation, represent collaborative efforts from developers aiming to advance the field further [5, 6]. Specific studies on genetic algorithm-based approaches for terrain generation were discussed in [7], highlighting the potential of these methods in creating customized terrains based on user preferences and aesthetic considerations.

The proposed algorithm aims to address and improve several key aspects of procedural generation. Firstly, it seeks to enhance the realism and complexity of generated landscapes, providing more natural and varied terrains that can better mimic the intricacies of real-world environments. Additionally, it focuses on improving computational efficiency, reducing the time and resources required to generate terrains without sacrificing quality, aims to offer greater versatility and adaptability, allowing for its application across a broader range of digital content creation, from gaming to virtual simulations [8].

2 The Ways to Generate Surfaces

2.1 Perlin Noise

Perlin Noise is a procedural texture primitive, used in computer graphics to generate natural-looking textures and terrains. Developed by Ken Perlin, it is a gradient noise function that produces a more natural and organic appearance compared to purely random noise [2]. This algorithm has become a fundamental tool in graphics for creating complexity in visual effects and environments, such as clouds, landscapes, and textures on surfaces, by simulating the randomness found in nature. Its ability to produce a wide range of effects with simple adjustments makes it highly versatile for both 2D and 3D graphics.

Perlin Noise is typically implemented as a function of two, three, or four dimensions but can be defined for any number of dimensions. The implementation usually involves three steps: defining a grid of random gradient vectors, calculating the dot product between the gradient vectors and their offsets, and interpolating between these values [9].

In the first step, an n-dimensional grid is defined, where each intersection of the grid is associated with a fixed random n-dimensional gradient vector of unit length. The exception is in the case of one-dimensional space, where gradients are random scalars between -1 and 1.

The next step involves calculating the scalar product. To compute the value at any potential point, it is necessary to identify the unique grid cell in which the point is located. Then, identify the $2n$ vertices of this cell and their associated gradient vectors. Next, for each vertex, compute the displacement vector. The displacement vector is the vector from that vertex to the potential point.

For each vertex, it is necessary to find the scalar product between its gradient vector and the displacement vector to the potential point. This scalar product will be zero if the potential point is exactly at the grid vertex.

It is important to note that the influence of the gradient vector increases with distance, which can be mitigated by normalizing the displacement vector to a length of 1. This may lead to noticeable abrupt changes unless the distance is considered in the next step of interpolation. However, normalizing the displacement vector is not a universally adopted practice.

For a point in a two-dimensional grid, this requires computing four displacement vectors and scalar products, whereas in three dimensions, it requires eight displacement vectors and eight scalar products. Overall, the algorithm has a complexity of $O(2n)$ where n is the number of dimensions.

The final stage is interpolation between the $2n$ scalar products. Interpolation is performed using a function that has a zero first derivative (and possibly also a zero second derivative) at all $2n$ grid nodes. Thus, in points close to the grid nodes, the result will approximate the scalar product of the gradient vector at the node and the displacement vector to the node. This means that the noise function will pass through 0 at each node, giving Perlin noise its characteristic appearance.

If $n = 1$, an example of a function that interpolates between the value a_0 at grid node 0 and the value a1 at grid node 1 is the smooth(x) function (1).

$$(x) = a_0 + smooth(x) \cdot (a_1 - a_0), \quad 0 \le x \le 1 \tag{1}$$

Noise functions used in computer graphics typically generate values in the range of $[-1.0, 1.0]$ and can be scaled accordingly.

2.2 Diamond-Square Algorithm

The Diamond-Square algorithm is a method used to generate fractal surfaces that approximate natural terrains. It is particularly useful in computer graphics for creating landscapes and other irregular, natural-looking surfaces in 3D models and environments. The algorithm operates on a two-dimensional square grid and proceeds through a series of iterations, each of which refines the grid's detail, adding randomness to simulate the complexity of natural landscapes.

The fractal surface it creates is a surface, a fractal object that mimics the appearance of natural terrain. In other words, the fractal surface emerges not from a rigidly defined procedure but is rather a pseudo-random object with fractal properties. Many natural objects exhibit some form of statically self-similarity, which can be modeled using fractal surfaces. Furthermore, changes in surface texture contain important information about surface orientation and inclination. The use of nearly self-similar objects can help simulate natural visual effects [2].

The algorithm starts with a 2D grid and then, from four initial values, randomly generates a heightmap arranged in a grid of points so that the entire plane is covered with squares, Fig. 1. The 2D grid should have dimensions of n by n, where $n = 2x + 1$, where $x \in \mathbb{Z}$. Initial values generated by pseudo-random numbers are set at the four corner points of the net. The "diamond" step. For a square with four initial values, the central point is found. It is set to the average value of the four corner points plus a pseudo-random value. The "square" step. For each rhombus in the net, a central point is set, assigned the average of the four corner points plus a pseudo-random value. In the square steps, points located on the edges of the overall net will have only three neighboring values instead of four (the fourth point would exceed the dimensions). There are several ways to deal with this, the simplest – is to take the average of three edge points instead of four.

Next, for the obtained four squares, the "diamond" and "square" steps are repeated until the values of all points are found. The diamond-square algorithm is recursive, and the exit condition is when the difference in x coordinates of two adjacent points is equal to one, where $x_1 - x_0 = 1$, with $x_1 > x_0$. When running this algorithm, cases of double determination of point values may occur. To ensure correct operation, already found values should be marked to avoid recalculating them, because this can lead to visual artifacts. Also, the pseudo-random value added at each step should be kept in a variable range, considering the distance between anchor points, Fig. 1.

Pearson correlation coefficient is a non-dimensional way to describe the correlation or similarity between series. However, continuity function needs to be described by inner product. Completing curve alignment of heterogeneous data by building the following function (2):oordinatesoftheanchorpoi

$$f(a_c) = \frac{a_0 + a_1 + a_2 + a_3}{4} + rand(x_1 - x_0) - \frac{(x_1 - x_0)}{2} \tag{2}$$

where a_c is the value of the current point, $a_0, \ldots, a_3$ are the values of the anchor points, $x_0 - x_1$ are the x coordinates of the anchor points, and rand(max) is a function returning a pseudo-random value in the range $[0, max]$.

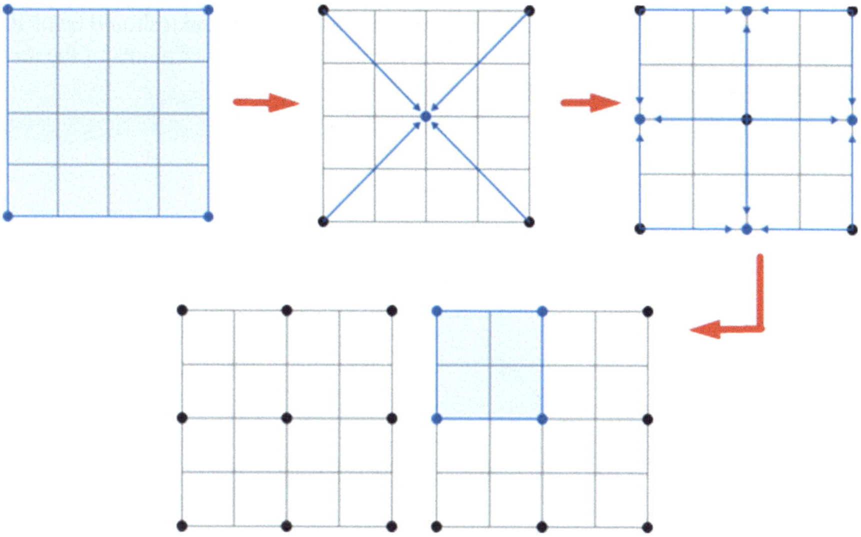

Fig. 1. Result of the first iteration

3 Triangular Mesh Fractal Algorithm (TMF)

The developed TMF algorithm is based on the principles of the "diamond-square" algorithm, meaning it is also a variant of a stochastic fractal, representing a stochastic surface. Similar to the mentioned algorithm, it mimics the appearance of a natural surface, has a plasma effect, and can be used for generating complex landscapes, clouds, etc.

The algorithm also starts with a 2D grid, but unlike the base algorithm, the new algorithm does not require all four initial points to begin generation. It is sufficient to have three points to fill half of the grid with values and then replicate the same process for the other half to complete the grid. Additionally, the new procedure involves fewer steps to achieve a result similar to the base algorithm (Fig. 2).

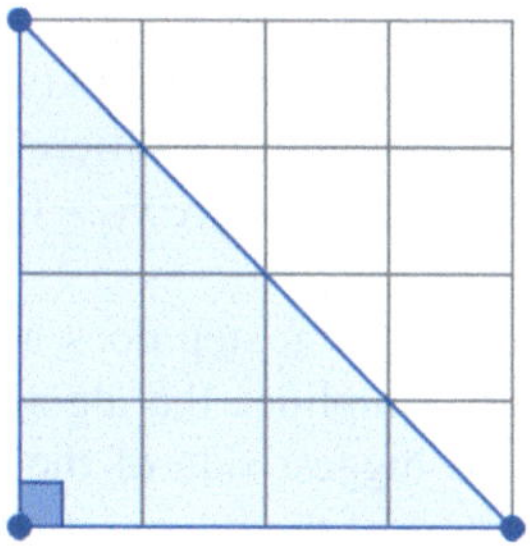

Fig. 2. Initial grid with three initial points

The dependency on the size of the 2D grid remains the same, and it should be of size n by n, where $n = 2x + 1$, $x \in \mathbb{Z}$. Initial values are set in three corner points of the array and are generated as pseudo-random numbers, Fig. 3.

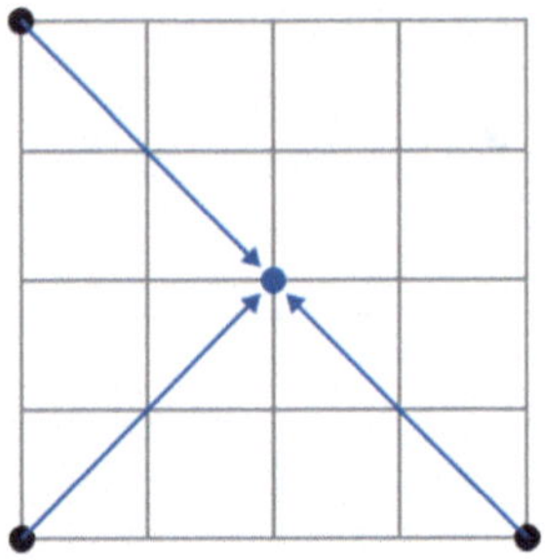

Fig. 3. Initial grid with three initial points

Next, the central point of the longest side of the triangle is calculated. In this case, it is the hypotenuse of an isosceles right triangle. The determined point is set to the average value of the three vertices of the triangle plus a pseudo-random value, Fig. 4.

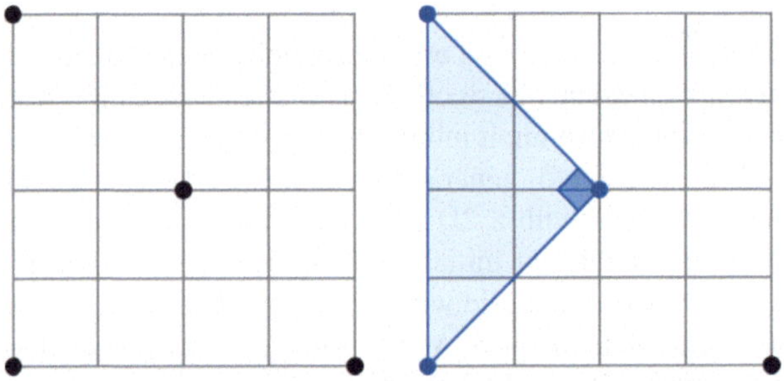

Fig. 4. The result after the first iteration of the algorithm

The found point divides the previous triangle into two new isosceles right triangles to which the same operation is applied until the difference between the coordinates of two adjacent points equals one: $x_1 - x_0 = 1$, where $x_1 > x_0$. In other words, this algorithm is also recursive.

The formal definition of the algorithm step does not require having a square 2D grid, but such a grid significantly simplifies the algorithm's execution. In this case, there is no need to search for the biggest side of the triangle, as it is known – the hypotenuse. Additionally, the centers of these longest sides will coincide with the grid points. Therefore, to find the correct values, interpolation methods will not be necessary.

In this case, when working with this algorithm, cases of double determination of point values will arise, similar to the "diamond-square" algorithm. Therefore, to ensure proper operation, it is essential to remember already found values for points to avoid recalculating them. Additionally, the pseudo-random value added at each step should be kept within a variable range, considering the distance between the vertices of the biggest side of the triangle, for example:

$$f(a_c) = \frac{a_0 + a_1 + a_2}{3} + rand(dist((x_0, y_0)(x_1, y_1))) - \frac{dist((x_0, y_0)(x_1, y_1))}{2} \quad (3)$$

a_c is the value of the current point, $a_0 - a_2$ are the values of the vertices, x_0, y_0, x_1, y_1 are the coordinates of the vertices, given as pairs *(x, y)*, where $x, y \in \mathbb{Z}$, *rand(max)* is a function that returns a pseudo-random value in the interval [0, *max*], and *dist*((*x*, *y*), (*x*, *y*)) is a function that returns the distance between two pairs of coordinates.

The TMF algorithm has fewer steps compared to "diamond-square", and, most importantly, it operates on triangles, considering the possibility of interpolating values – any triangle shape can be used. Taking this into account, along with the fact that the new approach possesses the same property as the original, namely, when used sequentially with specified common initial values, this method allows "stitching" together the generated fractal surfaces. Now, this feature of the algorithm extends to the triangle shape. For instance, it becomes possible to reconstruct a geometric space divided into triangles using triangulation methods.

4 Simulation Case

The Figs. 5, 6 and 7 depict height maps generated by three different algorithms: the Diamond-Square algorithm, Perlin Noise, and a Triangular Mesh Fractal (TMF) algorithm. Each algorithm produces a distinct pattern and complexity in the terrain's elevation data.

The height map generated by the Diamond-Square algorithm shows a balanced mix of high and low areas, creating a natural and smooth gradient, Fig. 5. This suggests that the algorithm is effective in simulating the variance seen in natural terrains. However, the transitions between heights can be somewhat uniform, which might not capture more rugged features accurately.

The Perlin Noise algorithm generates a height map with a more stochastic pattern, which is evident in the speckled distribution of heights, Fig. 6. This pattern can be excellent for simulating certain types of terrain or surface details such as rock, soil, or other granular materials. Perlin noise often produces a more 'cloud-like' structure which may be less suitable for terrains that require sharp elevation changes.

The TMF algorithm, a triangular-based modification of the Diamond-Square algorithm, produce a height map with features that are less grid-aligned compared to the original Diamond-Square algorithm, Fig. 7. This could potentially lead to more natural-looking terrains as natural landscapes do not follow a strict grid. The TMF algorithm's use of triangles also allow for more intricate detail in the terrain's elevation since the triangle is a more flexible shape for modeling surfaces than the square.

In Table 1 compared algorithms. TMF algorithm handles the highest resolution effectively, managing a large-scale map of 2048 × 2048 pixels, which is indicative of its robustness in creating detailed terrains. The time efficiency of the TMF algorithm is competitive, taking only 3.2 s to generate the map, which is slightly slower than Perlin Noise but faster than the Diamond-Square algorithm. Despite its slower speed, the TMF

Fig. 5. The generated height map by Diamond-Square algorithm

Fig. 6. The generated height map by Perlin nose

algorithm excels in the level of detail it can produce, with 300 unique points per area, surpassing the other two algorithms. This attention to detail does come with a higher memory cost, requiring 80 MB, which is significantly more than the other algorithms.

The scalability of the TMF algorithm is particularly notable, as it exhibits minimal increase in time for doubling the map resolution, making it highly efficient for scaling up projects. Moreover, it features stitching capability, allowing it to integrate additional terrain data seamlessly with existing maps, an essential attribute for ongoing or expansive

Fig. 7. The generated height map by TMF algorithm

Table 1. Comparative analysis of generation algorithms

Feature	Diamond-Square algorithm	Perlin Noise	TMF algorithm
Resolution	1024 × 1024	512 × 512	2048 × 2048
Time spent (seconds)	2.5	1.8	3.2
Number of iterations	8	N/A	7
Memory usage (MB)	50	45	80
Detail level	200 unique points/area	250 unique points/area	300 unique points/area
Scalability	1.2× time increase for double resolution	1.5× time increase for double resolution	1.1× time increase for double resolution
Stitching capability	No	No	Yes

terrain projects. This feature is absent in the other algorithms, highlighting the TMF algorithm's superiority in adaptability and expansion.

In Fig. 8 presented a three-dimensional visualization of terrain generated by TMF algorithm. This visualization showcases the algorithm's capacity to produce varied elevations and intricate landscape features, offering a rich tapestry of peaks and valleys. The nuanced shades of green imply a gradation of altitude, while the white areas suggest the presence of cloud cover or snow-capped summits, adding to the realistic depiction of a natural environment.

Fig. 8. TMF algorithm terrain visualization

5 Summary and Conclusion

A new algorithm, the Triangular Mesh Fractal (TMF) algorithm, was proposed, which improves upon the visual complexity and natural appearance of terrains compared to traditional methods like the Diamond-Square algorithm. Experimental results demonstrated that the TMF algorithm is highly efficient in handling high-resolution maps and exhibits significant advancements in detail, with the ability to produce 300 unique elevation points per area unit. Furthermore, the TMF algorithm showcased excellent scalability, with minimal time increase for map resolution doubling, and unique stitching capabilities for seamless integration with existing terrain data.

Future research will delve deeper into optimizing the TMF algorithm for even greater efficiency and reduced memory usage. Research will also explore stitching techniques to enhance the algorithm's capability to merge images without visible seams, further improving its application in digital worlds. The adaptability of TMF algorithm in real-time will be key areas of development, aiming to push the boundaries of dynamic and interactive environments in gaming, simulation, and virtual reality applications.

References

1. Jakob, W., Haan, M., et al.: Discrete stochastic microfacet models. ACM Transactions on Graphics. Proceedings of SIGGRAPH, vol. **33**(4) (2014)
2. Parberry, I.: Designer worlds: procedural generation of infinite terrain from real-world elevation data. J. Comput. Graph. Tech. **3**, 74–85 (2014)

3. Yin, H., Zheng, C.: A parametrically controlled terrain generation method. In: 7th International Conference on Computer Science & Education (ICCSE), pp. 775–778. Melbourne, VIC, Australia (2012)
4. Frade, M., de Vega, F., et al.: Automatic evolution of programs for procedural generation of terrains for video games. Soft. Comput. **16**, 1893–1914 (2012)
5. Dungeon template library. https://github.com/AsPJT/DungeonTemplateLibrary
6. THREE.Terrain engine. https://github.com/IceCreamYou/THREE.Terrain
7. Walsh, P., Gade, P.: Terrain generation using an Interactive Genetic Algorithm Conference. In: 2010 IEEE Congress on Evolutionary Computation (CEC). https://doi.org/10.1109/CEC.2010.5585913
8. Bulgakova, O., Zosimov, V.: Gamified active learning and 3d technologies for enhancing students' academic achievements. ITLT **96**(4), 46–57 (2023)
9. Fangxia, Z., Ahmad, K.: 3D animation simulation of computer fractal and fractal technology combined with diamond-square algorithm. Appl. Math. Nonlinear Sci. **8**(1), 467–474 (2023)

Investing in People, Investing in Growth: the Regional Model for Human Capital and Economic Development

Larysa Zomchak[1(✉)], Maryna Nehrey[2,3], and Yuliia Perevertailo[1]

[1] Ivan Franko National University of Lviv, Lviv 79000, Ukraine
lzomchak@gmail.com, perevertayilo.yuliia@lnu.edu.ua

[2] National University of Life and Environmental Sciences of Ukrain, Kyiv 03041, Ukraine
marina.nehrey@gmail.co

[3] Collegium Helveticum, CH-8006, Zurich, Switzerland

Abstract. The study investigates the reciprocal relationship between human capital and economic development at the regional level, focusing on the Lviv region of Ukraine. The authors argue that while human capital is considered crucial for economic growth, the impact of economic development on human capital formation also deserves attention. A system of simultaneous equations is employed to analyze the mutual influence between human capital (measured by the number of employed population) and economic development (measured by gross regional product, EPD). The model incorporates various factors affecting both sides, including capital investments, consumer price index, scientific research expenditures, average salary, and number of graduates from higher education institutions. The model explains 99.4% of the variance in gross regional product by capital investments, consumer price index, scientific research expenditures, and human capital (employed population). The model explains 86.4% of the variance in employed population by average salary and number of graduates from higher education institutions. These results suggest that the model adequately captures the interdependence between human capital and economic development in the Lviv region. The developed simultaneous equation model allows for quantitative assessment of the mutual influence between human capital and economic development factors, providing valuable insights for regional policymakers. This approach goes beyond simply interpreting individual factors' impact and offers a deeper understanding of their complex interactions.

Keywords: human capital · economic development · human resources · capital investment · gross regional product · employed population · region · regional economic development · economic growth · inflation · average wage · model · econometric model

Z. Hu et al. (Eds.): ICCSEEA 2024, LNDECT 242, pp. 406–416, 2025.
https://doi.org/10.1007/978-3-031-84228-3_35

1 Instruction

One of the important factors of economic development at various levels - from micro to macro - is human capital, which determines competitiveness, opportunities for socio-economic and technological growth. The impact of human capital on economic development is no less than that of physical capital. Working with human capital involves not only the transfer of knowledge, but also the production of new knowledge, which is a source of innovation and technological change. The formation of human capital, professional training and advanced training takes place within the region. Also, the income level of residents and their standard of living are determined within the region, which, in turn, determines the economic conditions of development. The impact of human capital on economic development has been described by economists, for example, in the Mankiw-Romer-Weil model [1], but there is also an inverse relationship, when economically developed regions attract more qualified human resources.

Human capital is identified as one of the important factors of economic growth. However, there is an inverse effect of the economic development of territories on the formation and improvement of human capital, which is less researched. To identify the mutual influence between human capital and economic development, it is advisable to apply the method of the system of simultaneous equations (simultative modeling). Another problem in the study of interdependencies between economic development and human capital is that the laws that are fulfilled at the level of the country will not necessarily be fulfilled at the level of the region, therefore, separate studies should be conducted for the region, because the formation, accumulation and use of human capital of capital takes place within the territorial limits of the region.

The intricate link between human resources (HR) and economic development has long been recognized. Human capital, encompassing knowledge, skills, and abilities of individuals, plays a crucial role in driving economic growth, productivity, and innovation. A central theme in the literature emphasizes the role of human capital, defined as the knowledge, skills, and abilities of individuals, as a critical driver of economic growth. Studies point to the positive impacts of education, training, and healthcare investments in enhancing worker productivity, fostering innovation, and promoting entrepreneurship. For instance, authors of [2] finds that a one-year increase in average education leads to a 5–10% increase in GDP per capita. While human capital development is crucial for economic growth, the literature also acknowledges the dynamic nature of this relationship. Economic growth, in turn, generates resources that can be invested back into human capital development, creating a virtuous cycle. For example, in [3] authors argue that economic growth allows for increased public spending on education and healthcare, further propelling human development, similar ideas can be find in [4].

If we talk about Ukraine, the war has a huge impact on human resources [5–9]. This articles specifically focuses on the negative consequences of the war on Ukraine's workforce, including deaths and injuries reducing the size of the labor force; displacements and disruptions to education hindering skill development; mental and physical health impacts affecting productivity. The articles highlight the long-term potential for these effects to hinder Ukraine's economic recovery.

Different methods are used by scientists in modeling the economy and other related fields: nonlinear modeling [10], self-organizing algorithms [11], simultaneous equations [12], spatial panel [13], optimization [14, 15], machine learning [16–19], logistic dynamics [20], cryptographic algorithm [21, 22], multilevel model [23], neural networks [24–26], network theory [27] and fuzzy logic [28] and others.

A review of educational practices in Ukraine highlights its centrality to human capital development. Authors of articles [29, 30] emphasize aligning education with industry needs and fostering social responsibility, both crucial for a skilled and adaptable workforce. Another authors [31] analyze the current state of education, including digitalization's role, a key skill for the modern world; in [32] authors focus on entrepreneurship education, demonstrating how education can cultivate specific skillsets that drive economic growth. While authors of [33] address assessment methods, effective evaluation ensures graduates possess valuable knowledge, another pillar of human capital. In conclusion, these studies showcase how education in Ukraine, by equipping its citizens with relevant skills, fostering social responsibility, and promoting adaptability, plays a critical role in developing a strong human capital base for the nation.

The relationship between human resources and economic development is a central concern for policymakers and researchers alike. This review has highlighted the key themes and debates surrounding this complex topic, emphasizing the crucial role of human capital, the dynamic interplay between economic growth and human development, and the importance of institutional frameworks, governance, and effective human resource management practices. As we move forward, addressing the challenges and inequalities that exist will be crucial in harnessing the full potential of human resources for sustainable and inclusive economic development.

2 Model of the Mutual Influence of Economic Development and Human Capital

The link between human capital and economic development has been extensively explored in economic literature, establishing human capital as a fundamental driver of economic growth e.g., Lucas [34], Mankiw [1]. However, the literature has primarily focused on the unidirectional impact of human capital on economic development, neglecting the potential reciprocal influence.

Recent studies acknowledge the complex interplay between these factors, highlighting the possibility that economic development can also impact human capital formation and improvement [35, 36]. For example, economic prosperity might facilitate increased investments in education and healthcare, leading to a more skilled and healthy workforce [37].

Recognizing this bidirectional relationship, the present study employs a system of simultaneous equations to model the interdependence between human capital and economic development in the Lviv region of Ukraine. This approach aligns with recent advancements in economic analysis, enabling the quantification of both direct and indirect effects between variables [38].

The chosen indicators (capital investments, consumer price index, scientific research expenditures, average salary, and higher education graduates) represent established factors influencing both human capital and economic development [39, 40]. By incorporating these factors into the simultaneous equations model, the study aims to provide a more comprehensive understanding of the complex dynamics at play in the Lviv region.

To model the mutual influence of human capital and economic development of the Lviv region, the following indicators were chosen:

Endogenous variables:

y_1 - volume of gross regional product, million UAH;

y_2 – number of employed population in the region, thousands of people;

Exogenous variables:

x_1 – capital investments, UAH million;

x_2 - consumer price index in the region, %;

x_3 – scientific expenses of enterprises for carrying out scientific research, UAH million;

x_4 - average salary in the region, UAH;

x_5 – number of graduates of higher educational institutions, persons;

Independent variables can be divided into those that characterize the economic development of the Lviv region:

- consumer price index in the region;
- expenses of enterprises for carrying out scientific research;
- Capital Investments;

And also those that affect the human capital of the Lviv region:

- average salary in the region;
- number of university graduates;

The input statistics for the construction and implementation of the model for all indicators were collected from the official page of the Main Department of Statistics in the Lviv region [41] and the State Statistics Service of Ukraine [42] for the period 2004–2021.

The simulative model describing the structure of relationships between economic variables and quantitative characteristics of human capital of the Lviv region consists of two equations:

1. Dependence of the volume of the gross regional product as a quantitative characteristic of the economic condition of the Lviv region on the number of the employed population of the region, capital investments, the consumer price index, and scientific expenses of enterprises for conducting scientific research (million hryvnias):

$$y_{1t} = a_{10} + a_{11}x_{1t} + a_{12}x_{2t} + a_{13}x_{3t} + b_{12}y_{2t} + e_{1t}$$

where a_{10} is the free term of the equation;

$a_{11}, a_{12}, a_{13}, b_{12}$– parameters of the equation;

e_{1t} is a random variable.

2. The dependence of the number of the employed population as a quantitative characteristic of human capital of the Lviv region on the volume of the gross regional product, the average salary and the number of university graduates:

$$y_{2t} = a_{20} + a_{24}x_{4t} + a_{25}x_{5t} + b_{21}y_{1t} + e_{2t}$$

where a_{20} is the free term of the equation;
a_{24}, a_{25}, b_{21} are parameters of the equation,
e_{2t} is a random variable.
In general, the model will look like this:

$$\begin{cases} y_{1t} = a_{10} + a_{11}x_{1t} + a_{12}x_{2t} + a_{13}x_{3t} + b_{12}y_{2t} + e_{1t} \\ y_{2t} = a_{20} + a_{24}x_{4t} + a_{25}x_{5t} + a_{26}x_{6t} + b_{21}y_{1t} + e_{2t} \end{cases} \quad (1)$$

The next step is to check the model for identification (Table 1). For this, it is necessary to determine the number of excluded exogenous variables and included endogenous variables in our model:

Table 1. The condition of the order of identification of the simulative model

Equation	Number of exogenous variables not included	Number of included endogenous variables	Conclusion
1	3	1	Overidentified
2	2	1	Overidentified

Therefore, based on the results of checking the equations of the simulative model for identification, it is advisable to estimate the parameters of the model by the two-step method of least squares, the idea of which is to replace the endogenous features of the model with auxiliary variables that will be highly correlated with the features they replace, but will not be correlated with random deviations ε.

3 Econometric Analysis of the Economy and Human Resource Impact on the Regional Level of Ukraine

As a result of estimating the unknown parameters of the system of Eq. (1) by the two-step method of least squares, the model takes the form:

$$\begin{cases} y_{1t} = 1940779 + 0{,}98x_{1t} - 172355x_{2t} + 544{,}63x_{3t} - 1717{,}09y_{2t} \\ y_{2t} = 1052{,}15 + 0{,}11x_{4t} + 0{,}001x_{5t} - 0{,}005y_{1t} \end{cases}$$

The results of the econometric analysis of the simulative model of mutual dependence of GRP and the number of employed population of Lviv region are presented in Table 1. Estimates of parameters for the GRP equation are shown in Table 2, and for the human capital equation - in Table 3.

Table 2. Econometric analysis of the simulative model

Equation 1		Equation 2	
Multiple R	0.997	Multiple R	0.93
R-squared	0.994	R-squared	0.864
Adjusted R-squared	0.993	Adjusted R-squared	0.835
Standard error	7173.25	Standard error	9,921
Observation	18	Observation	18

The value of the determination ratio for the first equation is approximately 0.994. This means that approximately 99.4% of the variability of the gross regional product of the Lviv region can be explained by the independent variables included in the model, for 99.4% the behavior of the dependent variable (gross regional product) is described by the independent variables given in it, for the remaining 0.6% GRP is determined by other factors not described.

The R-squared for the second equation is 0.86. This means that the model explains 86.43% of the variation in the dependent variable with the help of the independent variables. Economic interpretation for the number of the employed population - that is, 86.4% is described by the independent variables given in it, the remaining 13.6% of the employed population is determined by other factors not described. So, in addition to the economic data included in the model, there are also political, sociological and other factors that are not included in this model.

Table 3. Parameter estimates for the GRP equation

	Coefficients	Standard error	t-statistics	P-value
Free member	1940779	116339,449	16,682	0.00000000037
X 1	0.98	0.478	2,054	0.046
X 2	−172355	18610,153	−9,261	0.000000435
X 3	544.63	36,589	14,885	0.0000000015
In 2	−1717.09	96,484	−17.8	0.00000000017

Table 4. Parameter estimates for the employed population equation

	Coefficients	Standard error	t-statistics	P-value
Free member	1052,151	22,648	46,456	0.0000000000000001
X 4	0.114	0.018	6,328	0.000019
X 5	0.001	0.0006	1,608	0.0013
In 1	−0.005	0.0008	−6,428	0.000016

Since the p-value < 0.05 for all parameters of the equation, we reject the null hypothesis of statistical insignificance of the parameters (Table 4).

The dynamics of the actual and forecast values of the model's endogenous features (GDP and employed population) are shown in Figs. 1 and 2, respectively. As can be seen from the graphs, the model quite accurately describes the development trends of both processes during the studied period.

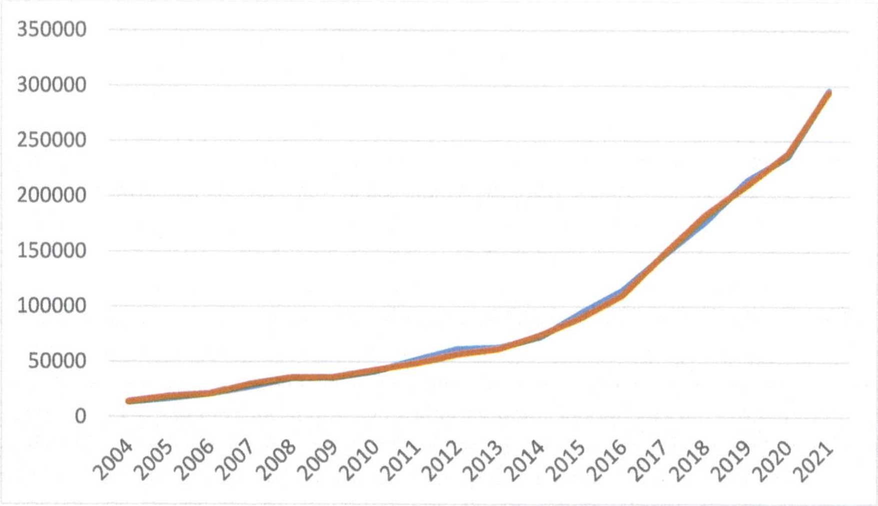

Fig. 1. Actual and theoretical values of GRP of Lviv region

As can be seen from the graph (Fig. 1), the simulative equation of the gross regional product is quite accurate, which is also evidenced by the high correlation value. While there is some discrepancy for the number of employed population, this is also explained by the correlation value of 86%, meaning there is another 14% of indicators that complement the model.

Regarding the economic interpretation of the gross regional product equation, the following conclusions can be drawn:

- With a change in capital investments by UAH 1 million, the indicator of the gross regional product will increase by UAH 0.98265 million;
- If the consumer price index changes by 1%, the indicator of the gross regional product will decrease by UAH 172,355 million;
- With a change in the scientific expenses of enterprises for conducting scientific research by UAH 1 million, the indicator of the gross regional product will increase by UAH 544.627 million
- If the number of the employed population changes by 1,000 people, the indicator of the gross regional product will decrease by UAH 1,717.09 million.

Similarly, regarding the economic interpretation of the equation of the number of the employed population:

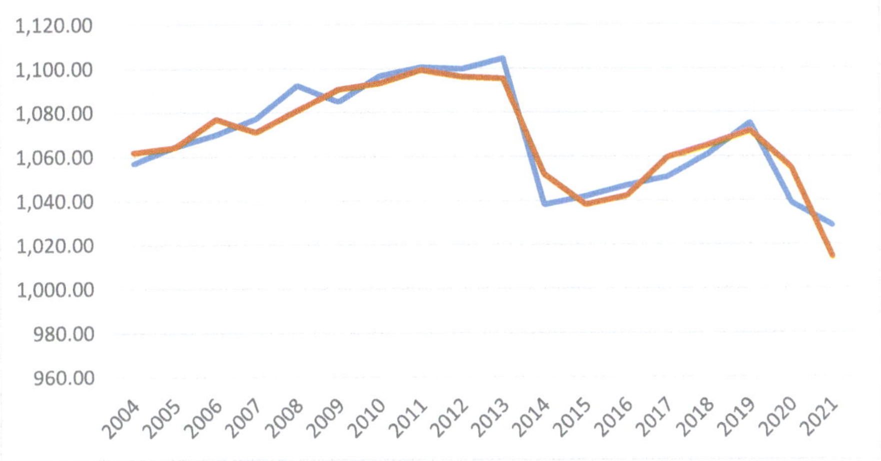

Fig. 2. Actual and theoretical values of the employed population of Lviv region

- If the average salary in the region changes by UAH 1, the number of employed people will increase by 0.11422 thousand people;
- If the number of released specialists in the region changes by 1 person, the number of employed population will increase by 0.00097 thousand people.

If the gross regional product changes by UAH 1 million, the number of employed people will decrease by 0.00505 thousand people.

4 Summary and Conclusion

To study the dependence of economic development (EPD) and human capital (number of employed population) of Lviv region, such indicators as capital investments, consumer price index, scientific expenses of enterprises for carrying out scientific research, average salary per employee, number of specialists graduated from higher education institutions were used educational institutions of the Lviv region. The simulative model of the influence of human capital on regional development is described by two equations: GRP and the number of the employed population.

As shown by the results of calculations based on statistical data for 2004–2021, 99.4% of the gross regional product is explained by the influence of such basic indicators as capital investments, the consumer price index, scientific expenses of enterprises for carrying out scientific research and human capital - employed population of Lviv region. At the same time, 86.4% of the number of employed population is explained by the influence of the following main indicators: the average salary per employee, the number of graduates from higher education in the Lviv region.

Therefore, analyzing the obtained results, we can conclude that the model adequately reflects the influence of human capital on the regional development of the Lviv region, but the work can be expanded by taking into account other indicators of human capital, which in the future can be compared with the obtained model.

References

1. Mankiw, N.G., Romer, D., Weil, D.N.: A contribution to the empirics of economic growth. Q. J. Econ. **107**(2), 407–437 (1993)
2. Patrinos, H.A., Psacharopoulos, G.: Education: past, present and future global challenges. World Bank Policy Research Working Paper (5616) (2011)
3. Filmer, D., Pritchett, L.: The impact of public spending on health: does money matter? Soc. Sci. Med. **49**(10), 1309–1323 (1999)
4. Bingshui, L., Hong-wei, H., Ya, Y.: The impact of human capital on income—based on the perspective of education. Int. J. Educ. Manage. Eng. **2**(12), 16 (2012)
5. Libanova, E., Pozniak, O.: War-driven wave of Ukrainian emigration to Europe: an attempt to evaluate the scale and consequences (the view of Ukrainian researchers). Stat. Trans. New Ser. **24**(1), 259–276 (2023)
6. Libanova, E., Pozniak, O., Tsymbal, O.: Scale and consequences of forced migration of the population of Ukraine as a result of armed aggression of the Russian Federation. Demography Soc. Econ. **2**(48), 37–57 (2022)
7. Gorodnichenko, Y., Kudlyak, M., Şahin, A.: The effect of the war on human capital in Ukraine and the path for rebuilding. Centre for Economic Policy Research (2022)
8. Strzelecki, P., Growiec, J., Wyszyński, R.: The contribution of immigration from Ukraine to economic growth in Poland. Rev. World Econ. **158**(2), 365–399 (2022)
9. Hrynkevych, O., Levytska, O., Baranyak, I.: Human resources for regional development in Ukraine: a roadmap for forecasting and determining a regional training request. Reg. Sci. Policy Pract. **15**(1), 95–107 (2023)
10. Oliskevych, M., Lukianenko, I.: Labor force participation in Eastern European countries: nonlinear modeling. J. Econ. Stud. **46**(6), 1258–1279 (2019)
11. Lytvynenko, V., Kryvoruchko, P., Lurie, I., Savina, N., Naumov, P., Voronenko, M.: Comparative studies of for forecasting economic parameters. Int. J. Mod. Educ. Comput. Sci. **12**, 1–15 (2020)
12. Zomchak, L.M.: Sustainable development of Ukraine as a combination of social, economic and environmental components: structural econometric model of three-pillar approach. IOP Conf. Ser. Earth Environ. Sci. **1254**(1), 012125 (2023)
13. Sahai, A., Sankat, C.K., Khan, K.: Decision-making using efficient confidence-intervals with meta-analysis of spatial panel data for socioeconomic development project-managers. Int. J. Intell. Syst. Appl. **4**, 92–103 (2012)
14. Martyn, Y., Liaskovska, S., Gregus, M., Izonin, I., Velyka, O.: Optimization of technological's processes Industry 4.0 parameters for details manufacturing via stamping: rules of queuing systems. Procedia Comput. Sci. **191**, 290–295 (2021)
15. Ljaskovska, S., Martyn, Y., Malets, I., Velyka, O.: Optimization of parameters of technological processes means of the flexsim simulation simulation program. In: Proceedings of the 2020 IEEE 3rd International Conference on Data Stream Mining and Processing, DSMP 2020, pp. 391–397 (2020)
16. Zomchak, L., Melnychuk, V.: Creditworthiness of individual borrowers forecasting with machine learning methods. In: International Conference of Artificial Intelligence, Medical Engineering, Education, pp. 553–561 (2023)
17. Izonin, I., Tkachenko, R., Holoven, R., Yemets, K., Havryliuk, M., Shandilya, S.K.: SGD-based cascade scheme for higher degrees wiener polynomial approximation of large biomedical datasets. Mach Learn Knowl Extr. **4**(4), 1088–1106 (2022)
18. Mochurad, L., Hladun, Y., Zasoba, Y., Gregus, M.: An approach for opening doors with a mobile robot using machine learning methods. Big Data Cogn. Comput. **7**(2), 69 (2023)

19. Kiv, A.E. et al.: Machine learning for prediction of emergent economy dynamics. In: CEUR Workshop Proc., vol. 3048, pp. 1–31 (2021)
20. Hrytsiuk, P., Parfeniuk, O., Shevchenko, I.: The logistic dynamics of population growth as a prerequisite for global sustainable development. IOP Conf. Ser. Earth Environ. Sci. **1126**, 012030 (2023)
21. Mochurad, L., Shchur, G.: Parallelization of cryptographic algorithm based on different parallel computing technologies. In: Proceedings of the Symposium on Information Technologies & Applied Sciences (IT&AS 2021), vol. 2824, pp. 20–29 (2021)
22. Mochurad, L., Panto, R.: A parallel algorithm for the detection of eye disease. In: Hu, Z., Wang, Y., He, M. (eds.) Advances in Intelligent Systems, Computer Science and Digital Economics IV. CSDEIS 2022. Lecture Notes on Data Engineering and Communications Technologies, vol. 158, pp. 11–125. Springer, Cham (2023). https://doi.org/10.1007/978-3-031-24475-9_10
23. Yu, M., Bai, H.: Study on influencing factors of regional economy based on multilevel model. Int. J. Math. Sci. Comput. (IJMSC) **9**, 29–43 (2023)
24. Izonin, I., Tkachenko, R., Holoven, R., Yemets, K., Havryliuk, M., Shandilya, S.K.: SGD-based cascade scheme for higher degrees Wiener polynomial approximation of large biomedical datasets. Mach. Learn. Knowl. Extr. **4**(4), 1088–1106 (2022)
25. Tkachenko, R., Izonin, I., Gregušml, M., Tkachenko, P., Dronyuk, I.: Committee of the SGTM neural-like structures with extended inputs for predictive analytics in insurance. In: Younas, M., Awan, I., Benbernou, S. (eds.) Big Data Innovations and Applications. Innovate-Data 2019. Communications in Computer and Information Science, vol. 1054, pp. 121–132. Springer, Cham (2019). https://doi.org/10.1007/978-3-030-27355-2_9
26. Koroliuk, Y., Hryhorenko, V.: ANN model of border regions development: approach of closed systems. Int. J. Intell. Syst. Appl. (IJISA) **11**, 1–8 (2019)
27. Auzinger, W., Obelovska, K., Dronyuk, I., Pelekh, K., Stolyarchuk, R.: A continuous model for states in CSMA/CA-based wireless local networks derived from state transition diagrams. In: Saraswat, M., Roy, S., Chowdhury, C., Gandomi, A.H. (eds.) Proceedings of International Conference on Data Science and Applications. Lecture Notes in Networks and Systems, vol. 287, pp. 571–579. Springer, Singapore (2022). https://doi.org/10.1007/978-981-16-5348-3_45
28. Pryimak, V., Melnyk, B., Holubnyk, O., Kostyshyna, T., Brych, V.: A fuzzy assessment of the development of the national labor market of Ukraine. In: 10th International Conference on Advanced Computer Information Technologies (ACIT), pp. 682–686 (2020)
29. Oleksiv, I., Izonin, I., Kharchuk, V., Tkachenko, R., Doroshenko, A.: Identification of IT sector stakeholder's requirements to masters program in information system in Lviv region. In: Proceedings of ICTERI, pp. 429–437 (2018)
30. Oleksiv, I., Lema, H., Kharchuk, V., Lisovych, T., Dluhopolskyi, O., Dluhopolska, T.: Identification of stakeholders importance for the company's social responsibility using the analytic hierarchy process. In: Proceedings of the 10th International Conference on Advanced Computer Information Technologies (ACIT), pp. 1–4 (2020)
31. Nehrey, M., Klymenko, N., Kostenko, I.: Formal and non-formal education of Ukraine: analysis of the current state and the role of digitalization. In: Hu, Z., Dychka, I., He, M. (eds.) Advances in Computer Science for Engineering and Education VI. ICCSEEA 2023. Lecture Notes on Data Engineering and Communications Technologies, vol. 181, pp. 1085–1098. Springer, Cham (2023). https://doi.org/10.1007/978-3-031-36118-0_92
32. Dimitrov, I., Dimitrova, A., Davydenko, N.: Entrepreneurial intentions research of business students in the Ukrainian context. In: 2020 III International Conference on High Technology for Sustainable Development (HiTech), pp. 1–4. IEEE (2020)
33. Guryanova, L., Yatsenko, R., Zarzhetskyi, V., Lytovchenko, I.: Set of models for assessing knowledge in distance learning systems. In: CEUR Workshop Proceedingsthis, vol. 2927, pp. 66–82 (2021)

34. Lucas Jr., R.E.: On the mechanics of economic development. J. Monet. Econ. **22**(1), 3–42 (1988)
35. Psacharopoulos, G., Patrinos, H.A.: Returns to investment in education: a further update. Educ. Econ. **12**(2), 111–134 (2004)
36. Hanushek, E.A., Woessmann, L.: The economics of international differences in educational achievement. In: Handbook of the Economics of Education, vol. 3, pp. 89–200 (2011)
37. Marimuthu, M., Arokiasamy, L., Ismail, M.: Human capital development and its impact on firm performance: evidence from developmental economics (2009)
38. Guisan, M.C., Neira, I.: Direct and indirect effects of human capital on world development, 1960–2004. Appl. Econometrics Int. Develop. **6**(1) (2006)
39. Faggian, A., Modrego, F., McCann, P.: Human capital and regional development. In: Handbook of Regional Growth and Development Theories, pp. 149–171 (2019)
40. Mačerinskienė, I., Vaikšnoraitė, B.: The role of higher education to economic development. Vadyba [Vilniaus universitetas] **2**, 82–90 (2006)
41. Main Statistical Office in Lviv Region. https://www.lv.ukrstat.gov.ua. Accessed 10 Jan 2024
42. State Statistics Service of Ukraine. http://www.ukrstat.gov.ua/. Accessed 10 Jan 2024

Analysis of the Relationship Between CSR Elements Applying Isomorphic Distances

Ihor Oleksiv[1,2](✉), Halyna Lema[1], Liubomyr Vankovych[1], and Serhii Chapran[1]

[1] Lviv Polytechnic National University, Lviv 79000, Ukraine
I.Oleksiv@gold.ac.uk, {halyna.v.mykhailiak,liubomyr.y.vankovych, serhii.p.chapran}@lpnu.ua

[2] University of London, Goldsmiths, Institute of Management Studies, London 146NW, UK

Abstract. We analyzed the relationships between various Corporate Social Responsibility (CSR) elements. For this purpose, methods of multivariate analysis were utilized. By applying the matrix of isomorphic distances, we revealed the shortest paths between CSR elements. As a result, the interaction and interdependence between CSR elements were analyzed. In addition, applying cluster analysis based on the calculation of similarities, we combined the CSR elements into groups. The companies will allocate efforts and resources for CSR more effectively using the study results. Furthermore, we identified future research directions focused on studying relationships between various CSR elements.

Keywords: CSR elements · expert evaluation · isomorphic distances · cluster analysis

1 Introduction

Corporate and social orientation of domestic enterprises is an important manifestation of the social and economic relations interaction. Among Ukrainian enterprises, the publication of a non-financial report on their own corporate and social interaction is increasingly popular. Such a report provides an opportunity for interested groups to familiarize themselves with the company's interaction with social activities. One of the disadvantages of creating a report of an enterprise corporate social responsibility is the lack of a clear structure. In addition, there are currently no indicators or provisions for assessing of enterprises corporate social responsibility. Thus, most enterprises prepare reports in an arbitrary form. Accordingly, there are complications in analyzing social reports, as they are not mandatory.

Thus, the main goal of the article is the selection of CSR elements that can be evaluated by the density of connections with each other, which will further minimize the company's costs. This information is crucial for understanding how management decisions aimed at regulating the impact of one element can have repercussions on others. By considering the interplay between different Corporate Social Responsibility

Z. Hu et al. (Eds.): ICCSEEA 2024, LNDECT 242, pp. 417–426, 2025.
https://doi.org/10.1007/978-3-031-84228-3_36

(CSR) elements, it becomes possible to determine which cluster of elements is most suitable for maximizing results under conditions of limited resources.

In essence, knowing the value of relationships between CSR elements allows managers to strategically allocate resources to achieve the most significant overall impact. This involves understanding how changes in one aspect of CSR may affect others and leveraging this knowledge to make informed decisions that align with organizational goals and values. By understanding which CSR elements are closely related and how they interact within clusters, managers can prioritize their efforts and allocate resources efficiently.

Using these CSR elements, it is important to understand by what indicators they can be evaluated in natural measurement. Therefore, a prerequisite for the formation of social and corporate responsibility is the consideration by enterprises of such CSR elements as: Work Results (financial indexes); Key sustainability indicators; Innovation support; Security; Human Rights; Sustainable Professional Development; Resources Efficiency; Teaching to Treat Wastes and Preventing from Climate Changes; Product safety and quality; IT Efficiency; Interaction with Stakeholders; Investment into Society (Social Partnership) [1–3].

The modern business environment is characterized by high complexity and uncertainty. Thus, there is a need to quickly respond to changes in production conditions and implement the necessary measures. The use of information technologies will contribute to the automation of many production processes, which will lead to the preservation of financial and labor resources. The main participants in the context of stakeholder relations are customers, employees, shareholders and investors, creditors, communities, government institutions and other. Surveys, internal communications, participation in conferences and discussions with investors, public discussions, presentations and reports on the company's activities are important methods of interaction between the company and stakeholders.

Therefore, an important task in the process of researching the corporate social responsibility of domestic enterprises is to determine the interrelationships and dependencies of each CSR elements among themselves and to study their impact on the internal and external environment of the organization. It is expedient to display the interrelationships of CSR elements with the help of cluster analysis and a matrix of isomorphic curves between each proposed elements.

By understanding which CSR elements are closely related and how they interact within clusters, managers can prioritize their efforts and allocate resources effectively. The latter improves strategic decision-making, and maximize the positive impact on the organization's overall CSR performance while managing potential risks or conflicts between elements. Therefore, utilizing the insights obtained from cluster analysis empowers managers to make informed decisions that align with the company's goals and values [4].

2 Literature Review

Academic literature on corporate social responsibility (CSR) in different countries studies common meaning of CSR and the impact of enterprise social and ecological factors on the environment: Feng [5] studied bibliometric analysis of CSR, citation analysis, publication trend of CSR. Rodrigues summarizes the most cited articles from CSR by country's, according to the author's analysis, Australia (32%), Canada (25%) and England (14%) have the highest percentage of CSR publications [6]. The scientist also analyzed the number of publications by one author on CSR issues. Rogowski analyzed CSR-related activities of Polish companies [7]; Vlastelica et al. investigated the influence of CSR on corporate reputation in Serbian companies [8]; Strouhal et al. investigated the link between CSR reporting and company performance in Estonia and the Czech Republic [9]. Mura explored the evolution of Sustainability measurement research and formed a cluster analysis of CSR publications [10]. It is noted that CSR is often considered as a construct with four aspects: Market place, Working place, Environment, and Community [1].

Additionally, it is highlighted that there is only one publication that focused on studying separate aspects of the CSR construct at companies operating in emergent markets, specifically in Serbian companies [8]. According to references [11, 12], corporate social responsibility involves responsibility to all spheres with which the company faces (personnel, partner organizations, society). According to the study [13], corporate social responsibility is expressed in the form of mutual relations between the enterprise and the social environment.

The exploration of methods to assess the impact of CSR elements on a company's operations and the environment is crucial [14–16]. Such methods provide insights into the level of interaction of social responsibility in tangible terms. Weighting factors for measuring corporate and social responsibility can be determined through expert opinions and analytical tools. By employing the matrix of isomorphic distances, we can uncover the interactions and interdependencies among each CSR element, while cluster analysis can then group them based on shared characteristics [17–21].

3 Data and Method

Structuring the relationship between the elements of CSR at work, using the following: the concept of CSR and company sustainability; a structural design approach that combines natural intelligence and artificial intelligence, allowing: to make all CSR processes intelligentization; conduct multi-criteria analysis of big data intelligent analysis and decision-making; make overall CSR organizational management intelligent; consider the possibility of making management at all levels intelligent and establish intelligent corporate social responsibility (ICSR) accordingly.

The theory of Markov stochastic processes and the graph theory of applied isomorphism curves are based on dynamic and static mathematical models solved by computer technology and are designed to support decision-making regarding the development status and selection of corporate social responsibility or any element in its hierarchy [22]. This method performs better compared to other expert methods, such as Analytic Hierarchy Process (AHP) and Cross-impact analysis [3]. The theory of Markov facilitates informed decision-making by highlighting areas where strategic interventions can yield the most significant positive outcomes while considering resource constraints and environmental sustainability. Ultimately, this holistic perspective enables companies to enhance their CSR practices effectively, aligning with societal expectations and long-term business sustainability goals.

A cognitive approach is adopted to develop Intelligent corporate social responsibility (ICSR), which proposes treating ICSR as a logical cognitive model of social agents and integrating ICSR into a multi-agent system based on comprehensive artificial intelligence and taking into account the appropriate formalization of the decision-making process. In order to understand the sequence of the experiment with analysis of the relationship between CSR elements applying isomorphic curves, it is necessary to propose a methodological scheme of the study (Fig. 1).

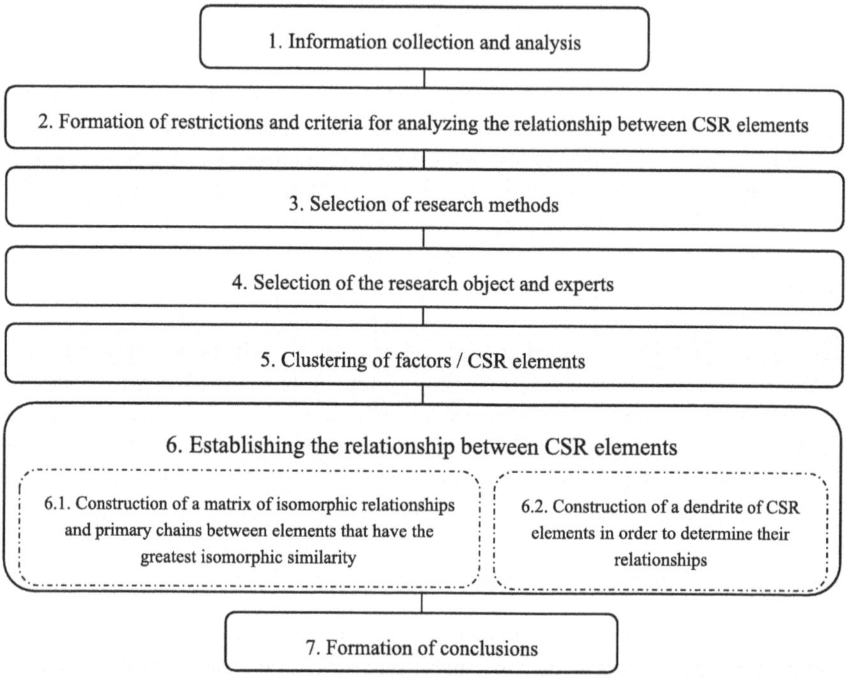

Fig. 1. Methodological scheme of the analysis of the relationship between CSR elements

We used survey results conducted at "Concern Galnaftogaz" in 2022–2023. The process of selecting a sample of experts for the interview consisted of two stages: identifying all possible experts and selecting the final panel of experts considering their competencies such as willingness to participate in the survey, experience in implementing CSR projects, holding a management position in the implementation of CSR projects, etc.

To identify the minimal number of experts who should participate in the study we used formula (1) with an acceptable level of significance "0.12" [23].

$$k = \frac{\frac{40}{260}\left(1 - \frac{40}{260}\right)2^2}{0,12^2} \approx 36 \tag{1}$$

In order to obtain reliable results, we had to interview at least 36 experts. From 260 possible experts, we selected a panel of 40 experts who had the most relevant competences for the study.

When determining the validity of a factor, we recommend using one of two methods: the direct assessment method and the paired equation method. According to the prescribed methodology, experts are provided to determine the importance and significance of each factor within a certain score range, which is the same for all experts. The mean score of *i*-factor and the validity coefficient of *i*-factor are determined and can be calculated based on the following correlations (formula (2)):

$$W_i = \frac{\sum_{j=1}^{N} z_{ij}}{\sum_{j=1}^{N} \sum_{j=1}^{N} z'_{ij}} \tag{2}$$

Where: W_i - coefficient of validity of i-factor, Z_{ij} - *i*-factor, evaluated by *j*-expert, N - the quantity of experts.

There are two traditional calculation methods for the paired comparison method: 1) based on specific paired comparisons; 2) based on complete paired comparisons.

The method of cluster analysis was used to form the degree of density of connections between CSR elements.

Clustering of factors using the bullet method involves the following algorithm:

1) formation of the initial data matrix;
2) construction of a matrix of isomorphic distances, i.e. grouping of factors based on structural similarity. Isomorphic grouping is performed according to the formula (3) [23]. The isomorphic distance between objects is determined by formula (4) [24].

$$Z_{ij} = \frac{X_{ij} / \sum_{i=1}^{n} X_{ij}}{\sum_{j=1}^{m} \frac{X_{ij}}{\sum_{i=1}^{n} X_{ij}}} \tag{3}$$

where: X_{ij} – the value of the *j*-feature for the *i*-object.

$$d_{ik} = \sqrt{\sum_{j=1}^{m} \left(Z_{ij} - Z_{kj}\right)^2} \tag{4}$$

As a result of applying the mathematical apparatus (1)–(4), we obtain a matrix of isomorphic distances (5).

$$\begin{pmatrix} 0 & k_{12} & k_{13} & k_{14} & k_{15} & k_{16} & k_{17} & k_{18} & k_{19} & k_{110} \\ k_{21} & 0 & k_{23} & k_{24} & k_{25} & k_{26} & k_{27} & k_{28} & k_{29} & k_{210} \\ k_{31} & k_{32} & 0 & k_{34} & k_{35} & k_{36} & k_{37} & k_{38} & k_{39} & k_{310} \\ k_{41} & k_{42} & k_{43} & 0 & k_{45} & k_{46} & k_{47} & k_{48} & k_{49} & k_{410} \\ k_{51} & k_{52} & k_{53} & k_{54} & 0 & k_{56} & k_{57} & k_{58} & k_{59} & k_{510} \\ k_{61} & k_{62} & k_{63} & k_{64} & k_{65} & 0 & k_{67} & k_{68} & k_{69} & k_{610} \\ k_{71} & k_{72} & k_{73} & k_{74} & k_{75} & k_{76} & 0 & k_{78} & k_{79} & k_{710} \\ k_{81} & k_{82} & k_{83} & k_{84} & k_{85} & k_{86} & k_{87} & 0 & k_{89} & k_{810} \\ k_{91} & k_{92} & k_{93} & k_{94} & k_{95} & k_{96} & k_{97} & k_{98} & 0 & k_{910} \\ k_{101} & k_{102} & k_{103} & k_{104} & k_{105} & k_{106} & k_{107} & k_{108} & k_{109} & 0 \end{pmatrix} \tag{5}$$

In accordance with the above algorithm of the method of balls, using formula (6), it is possible to determine the minimal distances between the factors and the critical point.

3) on the basis of the matrix of isomorphic distances, a critical point is selected, which divides the studied factors into separate groups. The critical point is determined by the formula [23]:

$$C_l(p) = \frac{1}{k}\sum\nolimits_{l=1}^{G}\sum\nolimits_{p=1}^{p_1} C_l(p), \tag{6}$$

where: $C_l(p) = \min_{q=g_1} C_{ll}(p,q), p = 1, 2..P_l, K = \sum_{l=1}^{G} P_l$; $C_{ll}(p,q)$ – the distance between the p- and q-factors, which belong to the l-cluster; $C_l(p)$ – is the distance from p-factor to the neighboring factor in l-cluster; P_l – is the number of factors in l-cluster; G – is the number of clusters.
4) identification of clusters;
5) on the basis of isomorphic distances, the formation of chains of factors and the construction of a matrix of interchain distances;
6) building a dendrite based on the data obtained in step 5.

4 Results and Discussion

The practical demonstration of interrelationships and dependencies among CSR elements is exemplified through an expert-analytical method, using "Concern Galnaftogaz" as a case study. The study's findings will elucidate the extent of CSR elements' influence on the company's operations, while also predicting the grouping of elements with significant influence into one cluster group, and identifying those requiring enhanced efficiency in another influence group.

The initial data for the construction of the interdependence matrix of CSR elements and the subsequent construction of isomorphic distances between them are given below and presented in the Table 1: 1- Work Results(financial indexes); 2- Key sustainability indicators; 3- Innovation support; 4- Security; 5- Human Rights; 6- Sustainable Professional Development; 7- Resources Efficiency; 8- Teaching to Treat Wastes and Preventing from Climate Changes; 9- Product safety and quality; 10 - IT Efficiency; 11- Interaction with Stakeholders; 12- Investment into Society (Social Partnership) [3]. Further use of CSR elements in a mathematical application will involve only entering the numerical value of the element. Based on the grouping of evaluation results the initial data, based on the signs of similarity, we will get a matrix of distances between CSR elements (Table 1).

According to the presented data, the minimum distance are between the CSR elements as Work Results (financial indexes) and Innovation support (15,3); Key sustainability indicators and Teaching to Treat Wastes and Preventing from Climate Changes (18,2); Innovation support and Interaction with Stakeholders (14,0); Security and Teaching to Treat Wastes and Preventing from Climate Changes (15,2); Human Rights and IT Efficiency (16,2); Sustainable Professional Development and IT Efficiency (15,7); Human Rights and Product safety and quality (16,1).

To study the relationships between CSR elements, it is advisable to use the tools of cluster analysis, in particular the sphere method, which involves grouping elements based on establishing isomorphic similarity between them (Fig. 2).

Table 1. Matrix of distances between CSR elements of "Concern Galnaftogaz"

	1	2	3	4	5	6	7	8	9	10	11	12
1	0,0	21,1	15,3	26,7	18,4	21,9	22,8	22,4	27,9	18,3	18,7	25,6
2	21,1	0,0	24,5	20,7	31,1	27,6	23,8	18,2	19,8	21,8	29,4	30,5
3	15,3	24,5	0,0	23,1	22,0	23,1	22,2	20,9	27,2	16,9	14,0	25,1
4	26,7	20,7	23,1	0,0	29,0	24,2	21,8	15,2	16,1	17,8	26,4	23,5
5	18,4	31,1	22,0	29,0	0,0	17,2	20,9	25,3	27,4	16,2	18,0	25,5
6	21,9	27,6	23,1	24,2	17,2	0,0	24,2	25,0	24,5	15,7	21,0	29,4
7	22,8	23,8	22,2	21,8	20,9	24,2	0,0	19,6	23,9	16,1	20,1	15,8
8	22,4	18,2	20,9	15,2	25,3	25,0	19,6	0,0	20,1	16,2	23,2	20,5
9	27,9	19,8	27,2	16,1	27,4	24,5	23,9	20,1	0,0	19,0	31,1	32,6
10	18,3	21,8	16,9	17,8	16,2	15,7	16,1	16,2	19,0	0,0	17,9	21,5
11	18,7	29,4	14,0	26,4	18,0	21,0	20,1	23,2	31,1	17,9	0,0	20,5
12	25,6	30,5	25,1	23,5	25,5	29,4	15,8	20,5	32,6	21,5	20,5	0,0

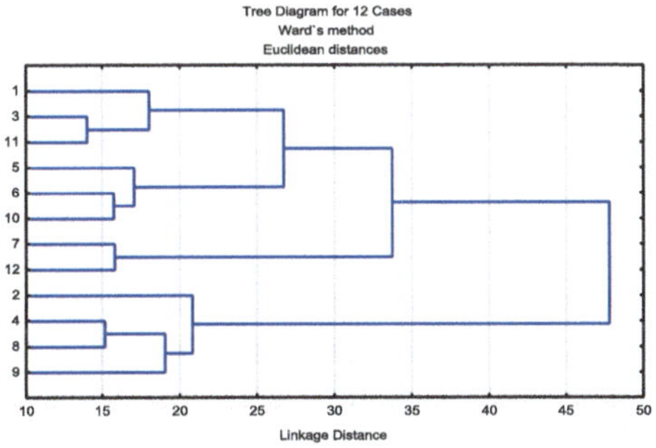

Fig. 2. Cluster analysis of Euclidean distances between the CSR elements of Galnaftogaz

Utilizing the data presented in Table 1 and Fig. 1, we will construct a matrix of isomorphic relationships and primary chains between elements exhibiting the highest isomorphic similarity. This process will result in the formation of chains based on the signs of similarity and interrelationships among the elements, as outlined in Table 2.

The chains data of isomorphic relationships make it possible to build a dendrite of CSR elements, which shows their relationships, presented in Fig. 3.

Constructed chains of the matrix of isomorphic distances show the interdependence between CSR elements, which are grouped by corresponding similar features into cluster groups. If the elements are close to each other, the closer are the connection between them. Thus, chains 1 counts the elements Work Results (financial indexes), Innovation support and Interaction with Stakeholders (interdependence between factors are 1,7329;

Table 2. Isomorphic distances between CSR elements of "Concern Galnaftogaz"

Chain	Element	Distance	Element	Distance	Element	Distance	Element	Distance
Chain 1	1	1,732	3	1,255	11	1,631	-	-
Chain 2	5	2,085	6	2,509	10	2,429	-	-
Chain 3	7	2,926	12	2,752	-	-	-	-
Chain 4	2	1,516	4	1,338	8	1,516	9	1,338

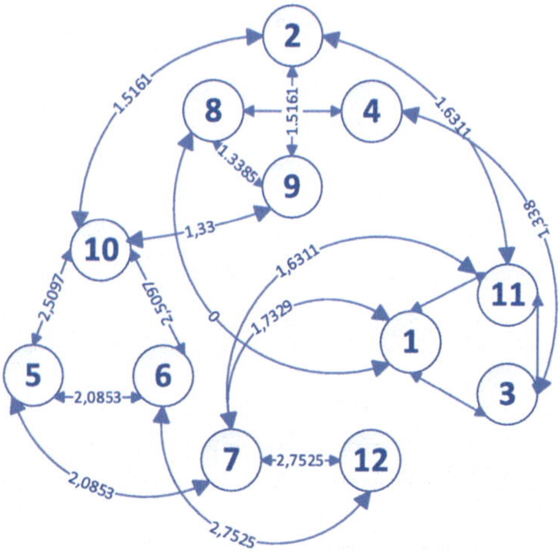

Fig. 3. Dendrite definition of relationships between CSR elements

1,2558 and 1,6311). This means that innovative support in the enterprise's activities will primarily depend on its financial results. If the enterprise is financially stable, the more actively it will involve innovations in its activities, as a result of which the activity of interested parties to such an enterprise will increase.

Chain 2 includes the following elements: Human Rights, Sustainable Professional Development and IT Efficiency (interdependence between factors are 2,0853; 2,5097 and 2,4294). There is a close relationship between each element, as Sustainable Professional Development must interact with Human Rights and IT Efficiency. Chain 3 talks about the interaction of the elements as Resources Efficiency and Investment into Society (Social Partnership), since efficient resources use is possible with the right social partnership and investment in it. Chain 4 presents elements of Key sustainability indicators, Security, Teaching to Treat Wastes and Preventing from Climate Changes, Product safety and quality. This cluster says that only with the correct management of waste, safe products that do not affect changes in the climate, security will be felt and indicators of sustainable development will increase.

5 Summary and Conclusion

In order to determine the relationships between CSR elements, an expert study was conducted. The object of the study was the "Concern Galnaftogaz". The object was investigated twice (in 2022 and 2023). Based on the use of cluster analysis and the construction of a matrix of isomorphic distances, groups of CSR elements that are similar in common features were determined.

In the process of forming an enterprise's CSR, it is important to have information not only about the list, content of factors, their relative importance, but also about the extent to which these elements are related to each other. This information is important from the point of view of taking into account how the management decisions made to regulate the effect of one or another element will be reflected on other element.

That is, having the value of the relationships between CSR elements, it is possible to influence which group (cluster) of elements is appropriate in conditions of limited resources to ensure the maximum result.

Understanding which CSR elements are closely related and how they interact within clusters empowers managers to prioritize efforts and allocate resources efficiently. This knowledge facilitates strategic decision-making, maximizing the positive impact on the organization's overall CSR performance while managing potential risks or conflicts between elements. The insights obtained from cluster analysis will enable managers to make informed decisions that align with the company's goals and values, fostering sustainable business practices and societal impact.

The study paves the way for further research focused on the interconnections between various CSR elements within the identified chain. The scholars might target factors that influence the relations between financial indexes, innovation support, interaction with stakeholders, etc.

References

1. Carroll, A.B.: The pyramid of corporate social responsibility: toward the moral management of organizational stakeholders. Bus. Horizons 39–48 (1991)
2. Odrekhivsky, M., Vankovych, L., Pshyk-Kovalska, O.: Problems of construction of smart innovative enterprises. COLINS, pp. 1015–1029 (2022)
3. Oleksiv, I., Lema, H., Kharchuk, V., Lisovych, T., et al.: Identification of stakeholders importance for the company's social responsibility using the analytic hierarchy process. Adv. Comput. Inform. Technol. 573–576 (2020)
4. Izonin, I., Tkachenko, R., Holoven, R., et al.: SK shandilya SGD-based cascade scheme for higher degrees wiener polynomial approximation of large biomedical datasets. Mach. Learn. Knowl. Extract. **4**(4), 1088–1106 (2022)
5. Feng, Y., Zhu, Q., Lai, K.-H.: Corporate social responsibility for supply chain management: a literature review and bibliometric analysis. J. Clean. Prod. **158**, 297–307 (2017). https://doi.org/10.1016/j.jclepro.2017.05.018
6. Rodrigues, M., Mendes, L.: Mapping of the literature on social responsibility in the mining industry: a systematic literature review[J]. J. Clean. Prod. **181**, 88–101 (2018). https://doi.org/10.1016/j.jclepro.2018.01.163
7. Rogowski, W.: Organization and management examples of CSR related activities at polish companies. Organ. Manage. 65–72 (2013)

8. Vlastelica, T., Cicvaric Kostic, S., Okanovic, M., et al.: How corporate social responsibility affects corporate reputation: evidence from an emerging market. JEEMS 10–29 (2018). https://doi.org/10.5771/0949-6181-2018-1-10
9. Strouhal, J., Gurvitš, N., Nikitina-Kalamäe, M., Startseva, E.: Finding the Link between CSR reporting and corporate financial performance: evidence on czech and estonian listed companies. Central Eur. Bus. Rev. 48–59 (2015)
10. Mura, M., Longo, M., Micheli, P., Bolzani, D.: The evolution of sustainability measurement research. Int. J. Manag. Rev. **20**, 661–695 (2018)
11. Izonin, I., Tkachenko, R., Yemets, K., Havryliuk, M.: An interpretable ensemble structure with a non-iterative training algorithm to improve the predictive accuracy of healthcare data analysis. Scientific Reports (2024)
12. Berjanir, A.L.: Corporate social responsibility as a factor in increasing the efficiency of modern business. Econ. Manage. Organ. 43–47 (2014)
13. Chen, S., Fan, J.: Measuring corporate social responsibility based on a fuzzy analytical hierarchy process. IJCNIS **3**(5), 13–22 (2011)
14. Fahim, A.: Homogeneous densities clustering algorithm. IJITCS **10**(10), 1 (2018). https://doi.org/10.5815/ijitcs.2018.10.01
15. Sumei, Z.: Comprehensive evaluation of teaching quality based on cluster analysis and factor analysis. IJEME **1**(3), 16–23 (2011)
16. Nahirna, O., Viunyk, O., Vysotska, I., Vankovych, L., et al.: Legal aspects of the formation, development and use of human capital in financial activities. Cuestiones Políticas, pp. 50–62 (2023). https://doi.org/10.46398/cuestpol.4178.03
17. Wahyu, C.F., Gayuh, U.V., Prathivi, R., Dewi, C.: The impact of financial statement integration in machine learning for stock price prediction. IJITCS **16** (2024)
18. Izonin, I., Tkachenko, R., Gregus, M., Zub, K., Tkachenko, P.: A GRNN-based approach towards prediction from small datasets in medical application. Procedia Comput. Sci. **184**, 242–249 (2021)
19. Zomchak, L., Melnychuk, V.: Creditworthiness of individual borrowers forecasting with machine learning methods. In: Hu, Z., Ye, Z., He, M. (eds.) Advances in Artificial Systems for Medicine and Education VI. AIMEE 2022. Lecture Notes on Data Engineering and Communications Technologies, vol. 159. Springer, Cham (2023). https://doi.org/10.1007/978-3-031-24468-1_50
20. Mochurad, L., Panto, R.: A parallel algorithm for the detection of eye disease. Lect. Notes Data Eng. Commun. Technol. **158**, 111–125 (2023)
21. Pyroh, O., Prokopenko, M., Chernobay, L., Kovalenko, R., et al.: Management of business processes and export-import activity of industrial enterprises in the digital economy. Estudios de Economia Aplicada **39**(5) (2021)
22. De la Fuente, A.: Mathematical methods and models for economists. Cambridge University Press (2000)
23. Lapach, S., Chubenko, A., Babych, P.: Statistics in science and business. Moryon, Kyiv (2002) (in Ukrainian)
24. Shchebel, A.: Cluster analysis of factors influencing the rationality of enterprise potential management. Int. Sci. e-J. ΛΌΓΟΣ (2020). (in Ukrainian)

Influence of External and Internal Factors on Company Activity in War Conditions Applying Expert Integral Evaluation System: A Case Study from Ukraine

Halyna Lema[1], Taras Lisovych[1], Ihor Oleksiv[1,2](✉), Viktoriya Kharchuk[1], Roman Shulyar[1], and Oleh Lema[1]

[1] Lviv Polytechnic National University, Lviv 79000, Ukraine
{halyna.v.mykhailiak,taras.y.lisovych,viktoriia.y.kharchuk, roman.v.shuliar,oleh.i.lema}@lpnu.ua, I.Oleksiv@gold.ac.uk
[2] University of London, Goldsmiths, Institute of Management Studies, London 146NW, UK

Abstract. The paper objective is to study the problem of analyzing the enterprise's activities in different Ukrainian regions in war conditions. The basis of the information analysis was the results of expert studies conducted at enterprises. The study contributes to understanding the impact of the war on the enterprise's activities in different Ukrainian regions. The latter emphasizes the need to analyze internal and external factors that affect the success of enterprises in such conditions. On the basis of the obtained data, we developed the integral indicator, which allows clear distinguishing and understanding of the influence of factors on the enterprise's activities. In particular, the research aimed to elaborate company strategies considering the prevailing balance of factors affecting each other. The development of adaptive strategic management at enterprises in war conditions received further development. In general, this research is an important contribution to the understanding and management of company activities in the conditions of war. It is useful both for academic debate and practical activity in business management, investment, and innovations. Moreover, it allows a better understanding of the need to consider the change in the individual weights of the factors that influence the company's activities in war conditions, particularly in regions where hostilities are happening.

Keywords: business management · integral analysis · innovative development

1 Introduction

War conditions create an extremely difficult and unpredictable business environment. In such conditions, the analysis and research of factors become critically important for the successful functioning and survival of the enterprise. It is important for the economy of Ukraine to support and develop business, in particular in the territories that are close to military operations. Thus, there is a significant increase in risk and uncertainty.

Z. Hu et al. (Eds.): ICCSEEA 2024, LNDECT 242, pp. 427–437, 2025.
https://doi.org/10.1007/978-3-031-84228-3_37

Analyzing the external environment helps to understand the geopolitical, economic and social factors that can affect business activities. Regarding the study of the internal environment, here, first of all, the available resources and their effective use are evaluated, as well as a SWOT analysis of the enterprise is carried out. Thus, the application of the integral evaluation method will allow analyzing the change in the weighting structure of the influencing factors, as a result of the change in the focus of the company's resources.

The dynamism of the external environment during the war period can complicate the adaptation processes of the enterprise. By scrutinizing these changes and making predictions, enterprises can swiftly adjust and mitigate adverse effects. Insight into both external and internal factors enables the formulation of strategic plans, facilitating the identification of opportunities, avenues for growth, and risk mitigation strategies. Ultimately, this comprehension empowers businesses to navigate uncertainties effectively, ensuring resilience and sustainable development.

The war's conditions can affect the reputation of a business, especially if its actions or participation in the conflict may be perceived negatively. This is a big threat for Ukrainian companies, because most companies actively help the Ukrainian army to defend the independence of Ukraine in the fight against the invaders. [1]. Taking into account the conditions of the external environment and the constant missile attacks on Ukrainian territories, the business still managed to adapt to these conditions and showed growth at the beginning of 2023. It is also worth noting that a significant number of enterprises changed their geolocation completely or partially, having lost their assets in the territories occupied by the russian federation. Despite significant difficulties, they were able to resume production in the western, central and northern Ukrainian regions.

Amidst the challenges posed by the aftermath of the war, Ukrainian manufacturers have found reorientation towards export markets to be a pivotal strategy. Many companies, notably within the agro-industrial complex, have amplified their export volumes while concurrently scaling down domestic sales. However, despite this shift, impediments to expanding exports persist, largely stemming from the external conditions shaped by the hostilities. This is a limitation of transportation through the ports of the southern regions, as well as a complication of transport connections due to the overloading of western routes by road transport. This is also accompanied by restrictions on transportation due to the blocking of the western borders by Polish protesters [1]. Maintaining resilience, adapting to change, and increasing competitiveness are paramount to successfully managing a business in wartime.

2 Literature Review

Zacharias T. [2] studied «organization environment for employee performance». Scientist Ghaleb E. [3] was engaged in Environment and Organizational Factors in Big Data Readiness. Hao J. [4] built research models that show relations technical, organizational and environmental perspectives in the technology organization environment. Dys S. [5] researched pairwise correlations among different factors ("organizational environment, staff outcomes and staff assessment of person-directed care"). Zhang Y. [6] studied green innovation in the business environment through the main factors of the organization such as employees, technology, resources, etc. The current nature of the working environment of employees and organizational support for such employees was researched by Rasool S. [7].

Ukrainian researchers were also engaged in the study of factors that have an impact on the company's activity. In particular, Sabadash L. [8] evaluated factors influencing the economic sustainability of the enterprise. Fedotova I. [9] studied the impact of exogenous and endogenous factors on transport and logistics companies. Petkov O. [10] studied the structure of factors of the external and internal environment and their influence on the enterprise's economic efficiency. The works [11–17] describes the influence of stakeholders (elements of the external environment) on the organization's CSR, investigated the identification of factors for the formation of the enterprises economic potential.

The impact of external and internal factors of Ukrainian enterprises requires further study and identification of strong factors and those that need to be improved in various regions of the country where hostilities are taking place, caused by the full-scale russian invasion, and in safer areas of the country. Amidst such circumstances, the application of integral assessment emerges as a valuable tool for evaluating these factors comprehensively. Integral assessment enables the determination of standardized values for both external and internal factors, along with their respective weighting factors. Such an assessment will make it possible to determine the standardized values of external and internal factors and their weighting factors, which help to see their strong and weak influence [11, 18–22].

3 Data and Method

The integral evaluation system will include the formation of indicators standardized values and their weighting factors chosen by experts. According to [12], it is possible to objectively assess the impact of external and internal factors on the company's activities using the R-model. The probability of evaluating the model is 81%:

$$Rint^{/} = \sqrt{\frac{k_1}{\sum k}x_1^2 + \frac{k_2}{\sum k}x_2^2 + \ldots \frac{k_n}{\sum k}x_n^2}, \tag{1}$$

where *Rint*- is an integral estimate; $x_1, x_2, \ldots x_n$- standardized values of indicators; k_1, $k_2, \ldots k_n$- weight coefficients of indicators chosen by experts.

Determination of weighting coefficients will be carried out using L. Thurstone's matrix of pairwise comparisons. The matrix of co-paircomparisons enables expert analysts to make a sequential comparison of all external and internal factors with each other in order to identify their pairwise priority. The mathematical apparatus for applying the expert procedure of pairwise comparisons according to the Thurstone's matrix is presented below. The matrix of the comparative importance of indicators that characterize the factors influence is calculated as follows:

$$i\begin{pmatrix} 0 & k_{12} & k_{13} & k_{14} & k_{15} & k_{16} & k_{17} \\ - & 0 & k_{23} & k_{24} & k_{25} & k_{26} & k_{27} \\ - & - & 0 & k_{34} & k_{35} & k_{36} & k_{37} \\ - & - & - & 0 & k_{45} & k_{46} & k_{47} \\ - & - & - & - & 0 & k_{56} & k_{57} \\ - & - & - & - & - & 0 & k_{67} \\ - & - & - & - & - & - & 0 \\ & & & j & & & \end{pmatrix}, \tag{2}$$

where k_{ij} - the indicator number of the external and internal factors impact assessment on the enterprise activity the, which has an advantage; i - the number of the matrix rows for assessing indicators values the external and internal factors impact on the enterprise's activities; j - the number of the matrix columns for assessing the level indicators values for external and internal factors influence on the enterprise's activities.

The cumulative frequency of preferences is determined by the sum of the indicators of the level assessment for external and internal factors influence on the enterprise activity with the greatest frequency by i and the level indicators values for external and internal factors influence on the enterprise's activities with the greatest frequency by j:

$$\begin{pmatrix} e_1^s \\ e_2^s \\ \dots \\ e_n^s \end{pmatrix} = \begin{pmatrix} e_{1j}^i \\ e_{2j}^i \\ \dots \\ e_{ij}^i \end{pmatrix} + \begin{pmatrix} e_{1i}^j \\ e_{2i}^j \\ \dots \\ e_{ij}^j \end{pmatrix}, \tag{3}$$

where e_{ij}^i - assessment indicators of the level for external and internal factors influence on the enterprise activity with the highest frequency according to i;. - assessment indicators of the level for external and internal factors influence on the enterprise activity with the highest frequency according to j; e_n^s is the cumulative frequency of preference indicators for assessing the level for external and internal factors influence on the enterprise activity.

The total sum of the advantages cumulative frequency for evaluation indicators:

$$m = \sum_{l=1}^{n} e_n^s \tag{4}$$

where n - the indicators number for assessing the level of external and internal factors influence on the enterprise activity; m - the frequencies sum of preference indicators for assessing the impact of external and internal factors influence on the enterprise activity; l - the indicator number for assessing the level of external and internal factors influence on the enterprise activity.

The importance of indicators for assessing the level of influence of environmental factors on enterprise activity is determined by the ratio of the cumulative frequency of the advantages of such indicators to their total sum:

$$\begin{pmatrix} e_1 \\ e_2 \\ \dots \\ e_a \end{pmatrix} = \begin{pmatrix} e_1^s/m \\ e_2^s/m \\ \dots \\ e_n^s/m \end{pmatrix}, \tag{5}$$

where e_a - the indicators weight for assessing the level of external and internal factors influence on the enterprise activity.

The model evaluation results suggest applying the rule that the best weighting factor is 1. When the weighting factor is equal to 1, it means that the corresponding factor has the maximum influence. The worst weight factor is 0, which indicates the absence of influence of this factor.

4 Results and Discussion

The calculation of the integral assessment for external and internal factors influence on the enterprise in the conditions of war considered on the example of analyzing the agro-industrial sector in three regions of Ukraine (west region, east region and south region). The choice of such regions is explained by determining the influence of external and internal factors on the organization's activities in the conditions where hostilities are conducted and in safer areas of the country. The results of the integrated assessment will show the level of external and internal factors influence on the company's activities and the identification of strong factors and those that need to improve their effectiveness in different regions of the country, where hostilities are taking place, caused by a full-scale russian invasion, and in safer areas of the country (Table 1).

Qf - the number of tasks actually completed; Qt - the total number of assigned tasks; Rf - the number of effectively used resources; Rt - the total number of resources; Ntech - the cost of new fixed assets raised by the enterprise; Ttech - the total value of the company's fixed assets; Dcvt - the number of creative tasks performed by the employee; Dt - the total number of tasks completed by the employee; Rc - regular customers; Tc - the total number of consumers; *Tcreg* – total number region competitors; *Ts* - total number of competitors; Se - efficient suppliers; St - the total number of suppliers.

According to the results of the data analysis obtained by experts, the standardized values of factors that affect enterprises in war conditions were calculated (Table 2) (Tables 3 and 4).

In the Table 2, indicators between different regions are significantly different. If the enterprise is located at a small distance from the combat zone, its assets are destroyed under shelling, accordingly, the performance of the assigned tasks is not at a high level. The high risk of doing business in such regions causes a low level of technological renewal of production assets and investment, which is expressed in the corresponding indicator. Destroyed roads and railways, the high risk of damage to moving vehicles greatly complicates logistics, and as a result, the level of reliability of the resources supply, as well as finished products to the final consumer.

The integral indicator calculation of the factors influence on the enterprises activities of the agro-industrial sector of the western Ukrainian region ($Rint_1$) (Tables 5 and 6):

$$Rint_1 = \sqrt{\frac{0{,}08}{1} * 0{,}85^2 + \frac{0{,}12}{1} * 0{,}81^2 + \frac{0{,}2}{1} * 0{,}89^2 + \frac{0{,}12}{1} * 0{,}81^2 + \frac{0{,}2}{1} * 0{,}89^2 + \frac{0{,}12}{1} * 0{,}95^2 + \frac{0{,}16}{1} * 0{,}92^2}$$

$Rint_1 = 0{,}74.$

The integral indicator calculation of the external and internal factors influence on the enterprises activities of the agro-industrial sector of the eastern Ukrainian region will have the form ($Rint_2$) (Tables 7 and 8):

Table 1. Indicators for assessing the external and internal factors influence of enterprises in war conditions

External and internal factors	Indicators	Formula for calculation
Task	Task performance level (Lt)	$Lt = Qf / Qt$
Resource	The level of effectively used resources (Lr)	$Lr = \frac{Rf}{Rt}$
Technology	The level of fixed assets technological upgrading (Ltech)	$Ltech = \frac{Ntech}{Ttech}$
Workers	The level of creative performance of assigned tasks (Lc)	$Lcvt = Dcvt / Dt$
Consumers	The level of stability of consumption (Nc)	$Nc = Rc / Tc$
Competitors	Level of competition (Lc)	$Lc = Tcreg / Ts$
Suppliers	Level of reliability of supply (Ls)	$Ls = Se / St$

Table 2. The results of determining the standardized values of the enterprises external and internal factors in the conditions of war

Indicators	Indicator of Western region in Ukraine	Indicator of Eastern region in Ukraine	Indicator of Southern region in Ukraine
Task performance level (Lt)	0,85	0,52	0,72
The level of effectively used resources (Lr)	0,81	0,49	0,68
The level of fixed assets technological upgrading (Ltech)	0,89	0,23	0,86
The level of creative performance of assigned tasks (Lc)	0,81	0,75	0,65
The level of stability of consumption (Nc)	0,89	0,23	0,82
Level of competition (Lc)	0,95	0,35	0,79
Level of reliability of supply (Ls)	0,92	0,39	0,76

Note: the authors' calculations are based on observations of enterprise activity

$$Rint_2 = \sqrt{\frac{0{,}16}{1} * 0{,}52^2 + \frac{0{,}12}{1} * 0{,}49^2 + \frac{0{,}16}{1} * 0{,}23^2 + \frac{0{,}08}{1} * 0{,}75^2 + \frac{0{,}2}{1} * 0{,}23^2 + \frac{0{,}12}{1} * 0{,}35^2 + \frac{0{,}16}{1} * 0{,}39^2}$$

Table 3. The matrix of co-pair comparisons the importance of enterprise's external and internal factors of the agro-industrial industry of the western region of Ukraine

Indicators	1	2	3	4	5	6	7	e^1_{ij}
Task performance level (Lt)		2	3	1	5	1	7	2
The level of effectively used resources (Lr)	-		2	2	5	6	7	2
The level of fixed assets technological upgrading (Ltech)	-	-		3	3	3	7	3
The level of creative performance of assigned tasks (Lc)	-	-	-		5	6	5	2
The level of stability of consumption (Nc)	-	-	-	-		5	5	2
Level of competition (Lc)	-	-	-	-	-		7	1
Level of reliability of supply (Ls)	-	-	-	-	-	-		-
e^2_{ij}	-	0	1	1	3	2	4	

* *Note: formed based on the results of expert evaluation (* explanation of indicators paragraph 3)*

Table 4. The results of determining the importance of enterprise's factors on the example of the agro-industrial industry of the western region of Ukraine

Indicators	e^1_{ij}	e^2_{ij}	e^s_{ij}– total prevalence frequency	e_{ij}- importance of indices ($e^s_{ij}/\sum e^s_{ij}$)
Task performance level (Lt)	2	–	2	0,08
The level of effectively used resources (Lr)	3	0	3	0,12
The level of fixed assets technological upgrading (Ltech)	4	1	5	0,2
The level of creative performance of assigned tasks (Lc)	2	1	3	0,12
The level of stability of consumption (Nc)	2	3	5	0,2
Level of competition (Lc)	1	2	3	0,12
Level of reliability of supply (Ls)	–	4	4	0,16
Total	14	11	25	1

* *Note: calculated by the author (* explanation of indicators paragraph 3)*

$R_{int2} = 0{,}39.$

The integral indicator calculation of the factors influence on the enterprises activities of the agro-industrial sector of the southern Ukrainian region ($Rint_3$):

Table 5. The matrix of co-pair comparisons the importance of enterprise's external and internal factors of the agro-industrial industry of the eastern region of Ukraine

Indicators	1	2	3	4	5	6	7	e_{ij}^{1}
Task performance level (Lt)		1	3	1	5	1	7	3
The level of effectively used resources (Lr)	-		2	2	5	2	7	3
The level of fixed assets technological upgrading (Ltech)	-	-		3	5	3	3	3
The level of creative performance of assigned tasks (Lc)	-	-	-		5	6	7	1
The level of stability of consumption (Nc)	-	-	-	-		5	7	1
Level of competition (Lc)	-	-	-	-	-		6	1
Level of reliability of supply (Ls)	-	-	-	-	-	-		-
e_{ij}^{2}	-	0	1	1	4	1	4	

* *Note: formed based on the results of expert evaluation (* explanation of indicators paragraph 3)*

Table 6. The results of determining the importance of enterprise's external and internal factors of the agro-industrial industry of the eastern region of Ukraine

Indicators	e_{ij}^{1}	e_{ij}^{2}	e_{ij}^{s}– total prevalence frequency	e_{ij}- importance of indices ($e_{ij}^{s}/\sum e_{ij}^{s}$)
Task performance level (Lt)	3	–	3	0,16
The level of effectively used resources (Lr)	3	0	3	0,12
The level of fixed assets technological upgrading (Ltech)	3	1	4	0,16
The level of creative performance of assigned tasks (Lc)	1	1	2	0,08
The level of stability of consumption (Nc)	1	4	5	0,2
Level of competition (Lc)	1	1	3	0,12
Level of reliability of supply (Ls)	–	4	4	0,16
Total	14	11	25	1

* *Note: calculated by the author (* explanation of indicators paragraph 3)*

$$Rint_3 = \sqrt{\frac{0{,}045}{1} * 0{,}72^2 + \frac{0{,}045}{1} * 0{,}68^2 + \frac{0{,}18}{1} * 0{,}86^2 + \frac{0{,}09}{1} * 0{,}65^2 + \frac{0{,}23}{1} * 0{,}82^2 + \frac{0{,}23}{1} * 0{,}79^2 + \frac{0{,}18}{1} * 0{,}76^2}$$

Table 7. The matrix of co-pair comparisons the importance of enterprise's external and internal factors on the example of the agro-industrial industry of the southern region of Ukraine

Indicators	1	2	3	4	5	6	7	e_{ij}^1
Task performance level (Lt)		2	3	1	5	6	7	1
The level of effectively used resources (Lr)	-		3	2	5	6	7	1
The level of fixed assets technological upgrading (Ltech)	-	-		3	3	6	7	2
The level of creative performance of assigned tasks (Lc)	-	-	-		5	6	7	1
The level of stability of consumption (Nc)	-	-	-	-		5	5	2
Level of competition (Lc)	-	-	-	-	-		6	1
Level of reliability of supply (Ls)	-	-	-	-	-	-		-
e_{ij}^2	-	0	2	1	3	4	4	

* *Note: formed based on the results of expert evaluation (* explanation of indicators paragraph 3)*

Table 8. The results of determining the importance of enterprise's external and internal factors on the example of the agro-industrial industry of the southern region of Ukraine

Indicators	e_{ij}^1	e_{ij}^2	e_{ij}^s– total prevalence frequency	e_{ij}- importance of indices ($e_{ij}^s / \sum e_{ij}^s$)
Task performance level (Lt)	1	-	1	0,045
The level of effectively used resources (Lr)	1	0	1	0,045
The level of fixed assets technological upgrading (Ltech)	2	2	4	0,18
The level of creative performance of assigned tasks (Lc)	1	1	2	0,09
The level of stability of consumption (Nc)	2	3	5	0,23
Level of competition (Lc)	1	4	5	0,23
Level of reliability of supply (Ls)	–	4	4	0,18
Total	8	14	22	1

* *Note: calculated by the author (* explanation of indicators paragraph 3)*

$R_{int3} = 0{,}45.$

5 Summary and Conclusion

The integrated assessment results of the external and internal factors influence on the organization's activities showed an average value of influence in the eastern and southern regions (0.39 and 0.45, respectively), while in the western region there is a high (0.74) influence of each of the proposed factors. In this case, military actions change the fluctuation of calculations for the external and internal factors influence on the average values observed before their beginning. The level of the proposed indicators implementation decreases, which in turn leads to a decrease in the integral value. Based on the calculation of the integral indicator, it is possible to draw a conclusion regarding the influence of specific factors in dynamic conditions. Taking these calculations into account, it is possible to form a short-term business strategy. Therefore, the integral assessment of factors provides a quantitative assessment of their impact on the company's activities and is a methodical analysis tool.

References

1. The state and needs of business in wartime: the results of the survey in September (2022). https://business.diia.gov.ua/cases/novini/stan-ta-potrebi-biznesu-v-umovah-vijni-rezultati-opituvanna-u-veresni
2. Zacharias, T., Rahawarin, M., Yusriadi, Y.: Cultural reconstruction and organization environment for employee performance. J. Ethnic Cult. Stud. **8**(2), 296–315 (2021)
3. Ghaleb, E.A.A., Dominic, P.D.D., Fati, S.M., Muneer, A., Ali, R.F.: The assessment of big data adoption readiness with a technology organization-environment framework: a perspective towards healthcare employees. Sustainability **13**(15), 8379 (2021). https://doi.org/10.3390/su13158379
4. Hao, J., Shi, H., Shi, V., Yang, C.: Adoption of automatic warehousing systems in logistics firms: a technology organization environment framework. Sustainability **12**(12), 5185 (2020). https://doi.org/10.3390/su12125185
5. Dys, S., Tunalilar, O., Hasworth, S., Winfree, J., White, D.L.: Person-centered care practices in nursing homes: Staff perceptions and the organizational environment. Geriatr. Nurs. **43**, 188–196 (2022)
6. Zhang, Y., Sun, J., Yang, Z., Wang, Y.: Critical success factors of green innovation: technology, organization and environment readiness. J. Clean. Product. **264** (2020)
7. Rasool, S.F., Wang, M., Tang, M., Saeed, A., Iqbal, J.: How toxic workplace environment effects the employee engagement: the mediating role of organizational support and employee. Wellbeing. Int. J. Environ. Res. Public Health **18**(5), 2294 (2021). https://doi.org/10.3390/ijerph18052294
8. Sabadash, L.O.: Assessment of factors influencing the economic stability of the enterprise in crisis conditions. Innov. Econ. 1–2, [82], 117–124 (2020)
9. Fedotova, I.V.: Exogenous and endogenous factors influencing the viability of enterprises in the transport and logistics sphere. Econ. Transp. Complex **35**, 94–121 (2020)
10. Petkov, O.I.: Economic efficiency of enterprises and factors affecting it. Ukrainian J. Appl. Econ. **6**(1), 392–399 (2021)
11. Oleksiv, I., Lisovych, T., Shulyar, R.: Formation of organizational change realization ability within the system of enterprise control. Actual Probl. Econ. **171**(9), 235–241 (2015)
12. Kuzmin, O., Oleksiv, I., Mykhailyak, G.: Integral evaluation of company employees' competence system. Actual Probl. Econ. **155**(5), 506–513 (2014)

13. Peleshko, D., Rak, T., Izonin, I.: Image superresolution via divergence matrix and automatic detection of crossover. IJISA. **8**(9), 1–8 (2016)
14. Yu, M., Bai, H.: Study on influencing factors of regional economy based on multilevel model. IJMSC. **9**(4), 29–43 (2023)
15. Tkachenko, R., Izonin, I., Greguš, M., Tkachenko, P., Dronyuk, I.: Committee of the SGTM neural-like structures with extended inputs for predictive analytics in insurance. Big Data Innov. Appl. Innovate-Data. Commun. Comput. Inform. Sci. **1054**, 121–132 (2019)
16. Zomchak, L.M.: Sustainable development of Ukraine as a combination of social, economic and environmental components: structural econometric model of three-pillar approach. In: Proceedings of the 4th International Conference on Sustainable Futures: Environmental, Technological, Social and Economic Matters (ICSF-2023). Kryvyi Rih. IOP Conference Series: Earth and Environmental Science, vol. 11. IOP Publishing (2023)
17. Mochurad, L., Shchur, G.: Parallelization of cryptographic algorithm based on different parallel computing technologies. In: Proceedings of the Symposium on Information Technologies & Applied Sciences, vol. 2824, pp. 20–29 (2021)
18. Izonin, I., Trostianchyn, A., Duriagina, Z., Tkachenko, R., Tepla, T., Lotoshynska, N.: Image superresolution via divergence matrix and automatic detection of crossover **10**(9), 1–8 (2018)
19. Omri, M.-N., Mribah, W.: Towards an intelligent machine learning-based business approach. IJISA **14**(1), 1–23 (2022)
20. Tang, P., Tian, H.: Research on the layout of national economic mobilization logistics centers. IJISA **2**(1), 44–50 (2010)
21. Vitynskyi, P., Tkachenko, R., Izonin, I., Kutucu, H.: Hybridization of the sgtm neural-like structure through inputs polynomial extension. In: 2nd International Conference on Data Stream Mining and Processing, pp. 386 – 391 (2018)
22. Auzinger, W., Obelovska, K., Dronyuk, I., Pelekh, K., Stolyarchuk, R.: A continuous model for states in CSMA/CA-based wireless local networks derived from state transition diagrams. In: International Conference on Data Science and Applications, vol. 287 (2022)

Aspect-Based Sentiment Classification Using Enhanced CNN-BiLSTM Networks and Attention

Ousmane Sangary[1(✉)] and Umar Muhammad Mustapha kumshe[2]

[1] School of Computer Science, Hubei University of Technology, Wuhan 430070, China
sangaryousmane@gmail.com

[2] College of Computer and Information, Hohai University, Nanjing 210098, China

Abstract. In the realm of sentiment classification using deep learning models like Convolutional Neural Network (CNN), current research is striving to enhance the models' ability to capture contextual relationships and long-range dependencies in textual data. While CNNs excel at extracting local features, their simplicity in network structure may hinder their capacity to incorporate nuanced contextual relationships necessary for sentiment analysis. This limitation underscores the importance of models that can comprehend and utilize contextual dependencies for more accurate sentiment analysis, highlighting the need for solutions beyond traditional CNN architectures. To address these limitations, specialized variants of Recurrent Neural Networks (RNNs) such as Long Short-Term Memory (LSTM) and Gated Recurrent Unit (GRU) have been introduced to mitigate issues like gradient vanishing and explosion. The vanishing gradient problems can severely impact the training and performance of RNNs by hindering the network's ability to capture long-term dependencies and stabilize the training process. In response, LSTM and its bidirectional counterpart, BiLSTM, offer enhanced capabilities in capturing contextual information bidirectionally along the input sequence, aiding in more effective sentiment analysis in NLP tasks. To handle these problems, a hybrid CNN-BiLSTM model with an attention mechanism is proposed to enhance sentiment classification by combining the strengths of CNN and BiLSTM architectures with an added attention layer. The model leverages CNN for high-level feature extraction and local pattern capture, complemented by BiLSTM for contextual feature representation processing. The integration of an attention mechanism further enhances the model's ability to prioritize and interpret different parts of the input sequence, refining its sentiment analysis capabilities. By incorporating these advanced layers, the hybrid CNN-BiLSTM model aims to improve contextual features and sentiment classification performance in handling the intricacies of textual data, paving the way for more accurate and nuanced sentiment analysis outcomes. We evaluate the model on four benchmark datasets, and it outperformed previous models with state-of-the-art results.

Keywords: Sentiment classification · CNN · BiLSTM · Attention

Z. Hu et al. (Eds.): ICCSEEA 2024, LNDECT 242, pp. 438–447, 2025.
https://doi.org/10.1007/978-3-031-84228-3_38

1 Introduction

Aspect-based Sentiment Analysis (ABSA) delves deeper than traditional sentiment analysis methodologies by pinpointing specific aspects, features, and topics within a text and assigning sentiment to each. This granular approach provides a more nuanced understanding of the data, uncovering the specific elements that drive positive or negative sentiment. In contrast, traditional sentiment analysis often yields a more general overview, lacking the precision to identify the specific drivers of sentiment. The attention vector provides a representation focused on the relevant aspect, leading to better sentiment classification. This focused representation is more informative compared to the average pooling used in CNN-BiLSTM models without attention. The current challenges within aspect-based sentiment analysis (ABSA) encompass aspect sentiment analysis, extraction of opinion targets, and classification of aspect sentiment polarity [2]. The major role of the three is polarity classification in sentiment, which helps stakeholders make decisions. The presence of numerous aspects in an opinionated text indicates that there are comparable sentiment polarities for each aspect [1]. This work focuses on sentiment polarity classification by extending the prior two-way classification to include the neutral polarity class (negative, positive, and neutral), which makes it more efficient for decision-making procedures.

This paper applies bidirectional LSTM to the CNN network output to capture past and future contextual information inherent in opinionated writings. Section 3 provides a more extensive discussion of the model.

The contributions of this paper are twofold:

1. The paper introduces an attention-based CNN-BiLSTM that learns in both forward-backward direction with semantics and contextual processing with added attention. This study examined two distinct model architectures. The original architecture used a CNN layer to obtain detailed information from input data. The collected data was then input into an output layer with the SoftMax activation function for sentiment polarity classification. The second architecture added a BiLSTM layer directly to the input data, capturing both forward and backward contextual dependencies. The hidden representation obtained from the BiLSTM layer was then input into the same output layer for sentiment classification.
2. Our model improves the results of previous models by 10% – 15%, respectively. This is because most of the previous models used standard LSTM that learns in one direction (forward), which is unable to detect the important textual contents for fine-grained sentiment classification. Comparisons of our model with previous works were also done and the results show state-of-the-art improvement over baseline.

2 Model Description

Figure 1 includes a description of the architecture of the base model. Sub-Sects. 2.3 and 2.4, covers the CNN-BiLSTM model and their interactions as well as related problems.

2.1 Problem Formation

In a given sentence $s = (w_1, w_2, \ldots, w_m)$ containing n words, and an opinion target present in the sentence $a = (a_1, a_2, \ldots, a_m)$ as a subsequence of m consecutive words from s, aspect-level sentiment classification aims to assess the sentiment expressed in sentence s towards the opinion target a. To initiate text corpus analysis, the first step involves associating each word w with a continuous feature vector $e_w \in R_d$, with d representing the embedding dimension. These word vectors are then linked to the rows of a word embedding matrix $ERVd$ where V signifies the vocabulary size [13].

2.2 Bidirectional Neural Networks

The difficulty of overcoming the vanishing gradient problem in traditional RNN (recurrent neural network) led to the development of LSTM (long short-term memory), an RNN architecture engineered to effectively tackle the vanishing gradient challenge. Within the LSTM architecture, there exist three gates (input denoted as i, forget denoted as f, and output denoted as o), along with a cell memory activation vector c. The vector formulas for the recurrent hidden layer function H of the LSTM network are realized in the following manner:

$$i_t = \sigma\left(W_{x_i x_t} + W_{h_i h_t} - 1 + b_i\right) \tag{1}$$

$$f_t = \sigma\left(W_{x_f x_t} + W_{h_f h_t} - 1 + b_f\right) \tag{2}$$

b_i, b_f and bo are the to be trained biases. σ is a logistic Sigmoid function.

$$c_t = f_t c_t - 1 + i t_\tau\left(W_{x_c x_t} + W_{h_c h_t} - 1 + b_c\right) \tag{3}$$

The τ and θ are the input and output of the cell's activation functions which are stated as *tanh*.

$$o_t = \sigma\left(W_{x_o x_t} + W_{h_o h_t} - 1 + b_o\right) \tag{4}$$

$$h_t = o_t \theta(c_t) \tag{5}$$

Regular LSTMs can only analyse data in a forward direction. They employ prior time steps to provide context for the present step. However, in many circumstances, future context can be beneficial. BiLSTMs fix this by running two independent LSTM layers on the input sequence at the same time, forward and backward. The outputs of each layers are concatenated and utilized to produce predictions. The reverse LSTM focuses on incorporating past data within its context, while the forward LSTM integrates forthcoming contextual details. For our model, input sequences are produced by either extracting the next sequence from the convolution layer or utilizing the embeddings of subjective sentences. Consequently, the ultimate unseen state output is determined by the combination of these contextual indicators (Eq. 6).

$$h_t = \mu\left(\underset{h_t,}{\rightarrow}\ \underset{h_t}{\leftarrow}\right) \tag{6}$$

In this context, μ symbolizes the mean of both concealed states.

2.3 Convolutional Neural Networks

CNN are a type of neural networks mostly use for image classification, processing and face recognition. Convolutional neural networks have found application in solving sequence modelling problems like sentiment analysis, predictions problems et al. CNN, unlike classic feature extraction algorithms, does not require human feature extraction. Our CNN module employs a one-dimensional convolutional layer with m filters and k kernels. The embedding vector e_w is derived from an embedding matrix $E \in R_{V \times d}$, V represents the lexicon size of the matrix of the embedding and d denotes the length. This vector is then input into the convolutional layer and undergoes activation using the ReLU activation function for individual word (w) in the sentence s [1]. The result is subsequently passed on to the BiLSTM layer.

2.4 Attention Mechanism and BiLSTM

Attention mechanism is one of the cardinal advances in deep learning model in recent years, and it is widely used in Natural Language Processing tasks. Introducing Attention Mechanism to BiLSTM yields Attention-based BiLSTM which produces high accuracy in sentiment Analysis. Based on the work of [3], every text which comprises of the words w_i are identify by their word embeddings e_{w_i} still from $E \in R_{V \times d}$, where V represents the embedding matrix's vocabulary size and d denotes the dimensions. BiLSTM with trainable parameter $\theta bilstm$ captures sequential alongside relevant contextual details and the ultimate result comprises the concealed representations generated from the provided input, as delineated in Eq. (7).

$$[h_1, \ldots, h_n] = Bilstm\left(\left[e_{w_1}, \ldots, e_{w_n}\right], \theta bilstm\right) \tag{7}$$

Z provides the output of the target representation for a sentence s of length n as in Eq. (8).

$$z_s = \sum\nolimits_{i=1}^{n} a_i h_i \tag{8}$$

In Eq. (9), α represents the layer responsible for assigning weights to each word in the document, while h denotes the unseen representation of the phrase.

$$\alpha = \frac{exp(b_i)}{\sum_{i=1}^{n} exp(b_j)} \tag{9}$$

fscore is a method that calculates the weight in every word w_i in sentence s given the meaning and context between h_i and t_s.

$$b_i = fscore(h_i, t_s) \tag{10}$$

t_s is the mean of the target's hidden view.

$$t_s = \frac{1}{m} \sum_{i=1}^{n} h_i a_i \tag{11}$$

The input represented as z in the output layer is subject to processing using the SoftMax activation function to estimate the distribution among emotion labels, as illustrated in Eq. (12).

$$P = softmax\left(w_{o_z} + b_o\right) \tag{12}$$

The output of the weight and bias, represented by *wo* and *bo* respectively, optimizes the model by minimizing the cross-entropy loss function, detailed in Eq. (13).

$$J = -\sum\nolimits_{i \in D} log_{P_i}(C_i) \tag{13}$$

D represents the training dataset, C_i denotes the actual label for the *ith* sample, and $Pi(Ci)$ signifies the probability of the true label given the sample.

3 Experiments and Datasets

We conducted experiments on four benchmark aspect-level datasets and three document-level datasets. Our aspect-level dataset included restaurant and laptop domains from SemEval-2014 Task 4 [2], SemEval-2015 restaurant domain [4], and SemEval-2016 restaurant domain task 5. In general, the datasets used are Rest14, Lap14, Rest15 and Rest16. These datasets are used for aspect-level sentiment analysis, which involves identifying the sentiment of specific aspects or features of a product or service, such as the food quality or service quality of a restaurant. The Restaurant 14 dataset contains reviews of restaurants in English, while the Restaurant 15 and 16 datasets contain reviews in multiple languages, including English, Spanish, and Dutch. The laptop dataset contains reviews of laptops in English and is used to evaluate systems that can identify the sentiment of specific aspects of laptops, such as the battery life or screen quality. The Laptop 14 dataset is similar to the Restaurant datasets in terms of its format and purpose.

Table 3 depicts the model's results on the document-level datasets, where we use the YELP2014 dataset, the Amazon review dataset covering the electronics domain, and the IMDB movie review datasets. The Yelp dataset contains reviews of restaurants and other businesses on the Yelp platform. The Amazon Review Electronic Domain dataset contains reviews of electronic products sold on Amazon. The dataset includes reviews, ratings, and product metadata, such as descriptions, category information, price, brand, and image features. The IMDB dataset contains reviews of movies and TV shows on the IMDB platform. The dataset includes over 50,000 reviews and is commonly used for sentiment analysis research.

4 Discussions and Results

The Table 1 depicts the result of our models. Our study compares the performance of the model across multiple versions. CNN-I refers to the version that only uses CNN, whereas BiLSTM-I refers to the variant that uses both. CNN-BiLSTM-I represents the combine networks (CNN-BiLSTM) variation, while CNN-BiLSTM-A represents the attention-based form. Our results demonstrated that the combined CNN-BiLSTM variant

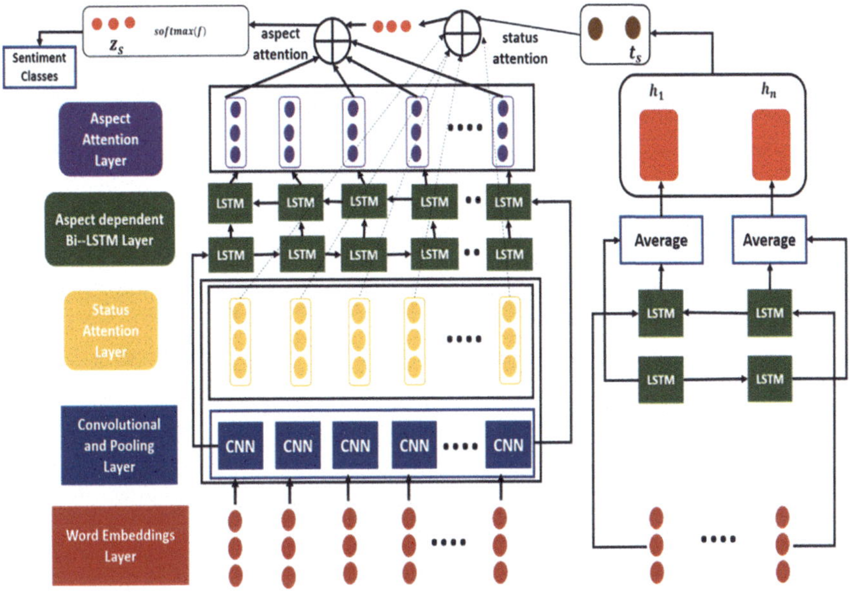

Fig. 1. Model Architecture

performed better on the Restaurant14 and Restaurant15 datasets, respectively. The Fig. 2 depicts our model's performance results in terms of validation accuracy, it illustrates the validation accuracy per epoch throughout the training procedure. Figure 2 also depicts the performance of the model on the validation dataset, which provides insights into how effectively the model generalizes to unseen data.

Figure 3 depicts the changes of validation loss over training epochs. The horizontal axis depicts the number of iterations across the full dataset throughout the training phase(epochs). The validation loss is represented on the y-axis, and it represents the difference between the actual validation dataset results and the model's prediction. This demonstrates that our approach accurately predicts sentiment with minimum error. The ultimate goal is to minimize validation loss, which represents the model's capacity to accurately capture and identify sentiment in text input.

The results of the experiments conducted on the Restaurant14 and 15 datasets and the Laptop domain of the aspect level dataset show that the CNN-BiLSTM-A model had the best performance. The experiments were conducted using our model, which was implemented using the Python and Keras deep learning framework. The optimal parameter settings of the model are shown Table 2, which shows that the CNN-BiLSTM-A model outperformed the CNN-BiLSTM models' accuracy and Micro-F1 score. Table 3 depicts our model with other document related methods and the results shows that our model performs well on the Rest14, Rest15 and the Lap15 datasets respectively as compare to baselines. The majority of results were generated from open source software, but others were obtained as published. Best outcomes are in bold. A hyphen (–) indicates that no results were obtained for that dataset. A quick overview of the compared relevant methodologies is provided below.

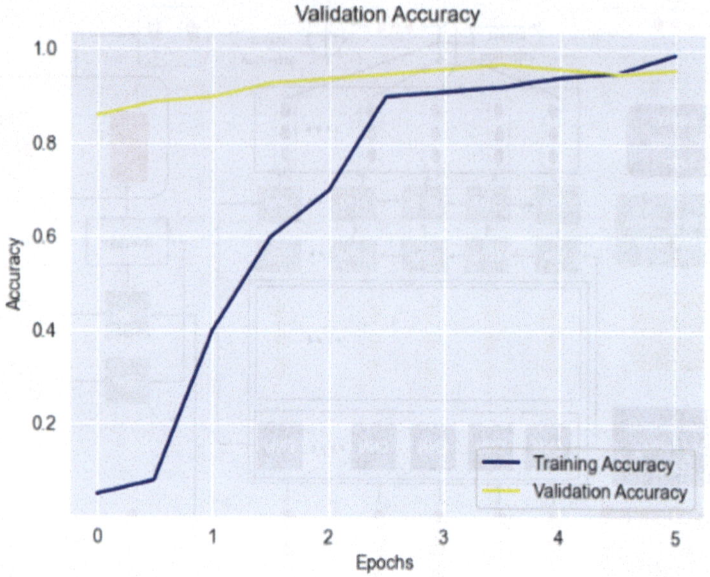

Fig. 2. Performance result on validation accuracy.

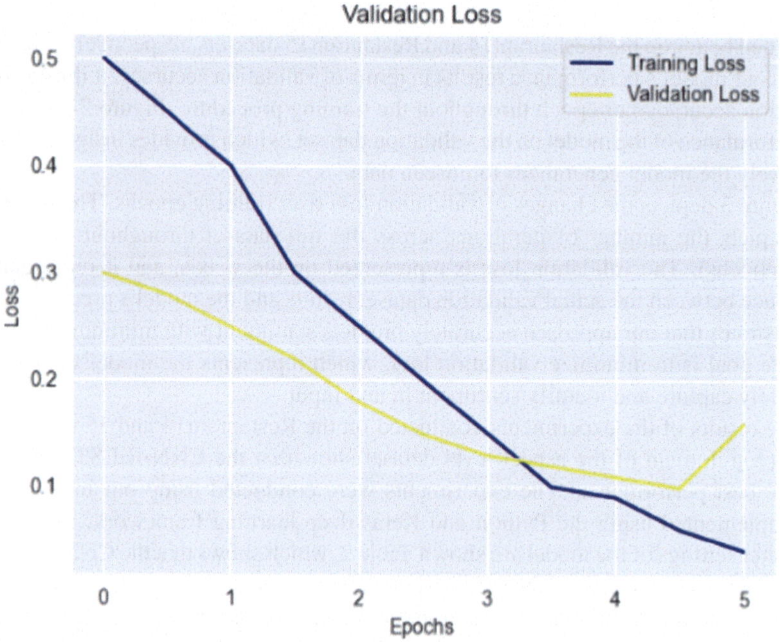

Fig. 3. This graph visually illustrates how the validation loss changes as the model is trained over multiple epochs.

i. EMGF: EMGF [10] propose a novel Extensible Multi-Granularity Fusion Network for Aspect-based Sentiment Analysis.
ii. PABS-BiGRU: PABS-BIGRU [5] Using bidirectional recurrent gated units, proposed a position-less attention-based sentiment categorization method.
iii. MGAN: MGAN [5] introduces a novel approach to sentiment analysis by employing a multiple-attention network to effectively detect sentiments within review texts.
iv. IACAPSNET: IACAPSNET [11] was developed to address the issue of overlapping feature representations arising from sentences containing multiple aspects by introducing an interactive attention capsule network.
v. BiSyn-GAT+: BiSyn-GAT introduced a technique that maximizes the utilization of syntax information within a sentence's constituent tree. This method effectively models the sentiment-oriented context for each aspect and discerns the sentiment interactions among different aspects within the sentence.

Table 1. Results comparison with baseline models.

Method	Yelps		Amazon Review		IMDB	
	Acc%	Micro-F1	Acc%	Micro-F1	Acc%	Micro-F1
BiLSTM-I	87.50	**80.64**	69.22	63.63	78.91	65.43
CNN-I	79.52	67.61	81.81	64.81	80.83	72.34
CNN-BiLSTM-I	88.53	80.30	**86.33**	**78.45**	83.10	75.46
CNN-BiLSTM-A	**94.50**	79.21	86.10	77.54	**87.93**	**80.35**

Table 2. The performance of BILSTM-CNN on the aspect-level dataset

Method	Rest14		Lap14		Rest15		Rest16	
	Acc (%)	Micro-F1	Acc%	Micro-F1	Acc%	Micro-F1	Acc%	Micro-F1
BiLSTM-I	78.50	70.64	72.22	65.63	68.91	59.43	80.4	74.21
CNN-I	80.52	75.61	83.81	77.81	78.83	72.34	**81.3**	76.34
CNN-BiLSTM-I	88.53	**80.30**	85.33	**78.45**	83.10	75.46	79.2	65.45
CNN-BiLSTM-A	**90.50**	79.52	**88.10**	77.54	**87.93**	**80.35**	78.88	**77.65**

Table 3. Experiment with various document-level datasets

Method	Rest14		Lap14		Rest15		Rest16	
	Acc%	Micro-F1	Acc%	Micro-F1	Acc%	Micro-F1	Acc%	Micro-F1
EMGF	82.11	79.24	88.42	83.20	78.87	78.06	85.48	**84.73**
PABS-BiGRU	81.37	72.06	75.39	70.50	80.88	80.88	**89.30**	66.93
MGAN	81.25	71.94	75.39	72.47	79.36	57.26	87.06	62.14
AOA	79.97	70.42	72.62	67.52	78.17	57.02	87.50	66.21
BiSyn-GAT +	87.94	82.43	82.91	79.38	85.85	85.49	–	–
IACAPSNET	79.27	70.85	73.87	70.10	80.13	74.14	–	–
Our Model	**95.50**	**85.3**	**98.38**	**82.34**	**87.9**	**85.00**	78.88	69.45

5 Conclusion

Unquestionably, one of the main issues influencing sentiment classification accuracy is still a lack of training data. As a result, we demonstrate in this study that data augmentation transfer and the implementation of a combined neural network (CNN-BiLSTM) on document and aspect-based sentiment analysis can greatly enhance aspect sentiment classification. Moreover, we conduct experiments and demonstrate how the attention mechanism with DNN might benefit aspect sentiment classification by efficiently representing the intricate semantic elements in opinionated texts. Results from experiments, such as ablation studies and the implementation of various variants of our model, validate the efficacy of our approach.

Acknowledgement. I express deep gratitude to my supervisor for her invaluable guidance and unwavering support. The unwavering love and belief of my parents have been a constant source of motivation, for which I am truly thankful to have them in my life.

References

1. Eniafe Festus Ayetiran, Attention-based aspect sentiment classification using enhanced learning through cnn-Bilstm networks, Knowledge-Based Systems, vol. 252, p. 109409, ISSN 0950–7051 (2022). https://www.sciencedirect.com/science/article/pii/S0950705122007067
2. Pontiki, M. et al.: Aspect based sentiment analysis. In: Bethard, S., Cer, D.M., Carpuat, M., Jurgens, D., Nakov, P., Zesch, T. (eds.) Proceedings of the 10th International Workshop on Semantic Evaluation, SemEval@NAACL-HL, June 16–17, pp. 19–30. The Association for Computer Linguistics, San Diego, CA, USA
3. Pontiki, M., Galanis, D., Pavlopoulos, J., Papageorgiou, H., Androutsopoulos, I., Manandhar, S.: SemEval-2014 Task 4: aspect based sentiment analysis. In: Proceedings of the 8th International Workshop on Semantic Evaluation (SemEval), pp. 27–35. Association for Computational Linguistics, Dublin, Ireland
4. Fan, F., Feng, Y., Zhao, D.: Multi-grained attention network for aspect-level sentiment classification. In: Proceedings of the 2018 Conference on Empirical Methods in Natural Language

Processing, pp. 3433–3442. Association for Computational Linguistics, Brussels, Belgium (2018)

5. Yadav, R.K., Jiao, L., Goodwin, M., Granmo, O.-C.: Positionless aspect-based sentiment analysis using attention mechanism. Knowledge. -Based Syst. **226**, 107136 (2021). https://www.sciencedirect.com/science/article/pii/S0950705121003993
6. Chollet, F., et al.: Keras https://keras.io
7. Ayetiran, E.F., Sojka, P., Novotný, V.: EDS-MEMBED: multi-sense embeddings based on enhanced distributional semantic structures via a graph walk over word senses. Knowl.-Based Syst. **219**, 106902 (2021)
8. Chen, Z., Qian, T.: Transfer capsule network for aspect level sentiment classification. In: Korhonen, A., Traum, D.R., Màrquez, L. (eds.), Proceedings of the 57th Conference of the Association for Computational Linguistics
9. Interactive attention for aspect-level sentiment classification. In: Inui, K., Jiang, J., Ng, V., Wan, X. (eds.) Proceedings of the 2019 Conference on Empirical Methods in Natural Language Processing and the 9th International Joint Conference on Natural Language Processing, EMNLP-IJCNLP 2019, 3–7 November 2019, pp. 5488–5497, Association for Computational linguistics, Hong Kong, China (2019)
10. Zhao, X., Zhou, Y., Xu, X., Liu, Y.: Extensible multi-granularity fusion network for aspect-based sentiment analysis. 12 Feb 2024 16:52:26 GMT
11. Liang, S., Wei,W., Mao, X.-L., Wang, F., He, Z.: 2022. BiSyn-GAT+: bi-syntax aware graph attention network for aspect-based sentiment analysis. In: Findings of the Association for Computational Linguistics: ACL 2022, pp. 1835–1848. Association for Computational Linguistics, Dublin, Ireland

Time Series Anomaly Detection Algorithm Incorporating Data Augmentation in Data Scarcity Scenarios

Qi Huang(✉)

Scholl of Computer Science, Hubei University of Technology, Wuhan 430070, China
huangcs@hbut.edu.cn

Abstract. In recent years, with the development of various industries, we have access to an increasing amount of data with timestamps, and people are increasingly valuing the industrial value brought by the temporal features within this data. Research on temporal features can extract patterns from historical data, which can then be utilized for the analysis and modeling of time series. Among the studies of time series, anomaly detection is an important branch. Detecting anomalies in time series enables the monitoring of data generated in real-time, thereby reducing data management costs, which is a meaningful practical problem. The industrial time series data nowadays exhibits characteristics such as high dimensionality and large volume, posing greater challenges for anomaly detection in time series. This is primarily manifested in traditional models performing reasonably well on low-dimensional and small-scale data but struggling to model a large number of high-dimensional temporal features. Additionally, anomaly definitions vary across different industrial scenarios, and data labels are often limited. Consequently, one of the challenges faced by anomaly detection in high-dimensional time series is how to achieve good detection results in various scenarios with limited data labels, where samples are restricted. Therefore, this paper proposes a Transformer model combined with data augmentation methods to investigate algorithms for anomaly detection in small-sample time series.

Keywords: Time series anomaly detection · Transformer · Data Augmentation

1 Instruction

Due to the continuously increasing number of sensors and devices in modern Internet of Things (IoT) platforms, sensor data is highly correlated in complex topological and nonlinear ways. The volatility of data and the variability of patterns are becoming increasingly complex, making unsupervised time series anomaly detection highly challenging in practice. Models not only need to learn information representations from complex temporal changes through unsupervised learning but also require a distinguishable criterion to detect rare anomalies from a large number of normal time points. In recent years, the success of machine learning algorithms has largely relied on the availability of large

Z. Hu et al. (Eds.): ICCSEEA 2024, LNDECT 242, pp. 448–457, 2025.
https://doi.org/10.1007/978-3-031-84228-3_39

amounts of data and powerful computational capabilities, but in practice, such conditions are not always available. Time series observations can be seen as the realization of a latent dynamic system controlled by unknown rules. In real-world applications, it is often not possible to sample data indefinitely, and model training with limited data can often lead to severe overfitting, hindering generalization. Therefore, it is a meaningful practical challenge to expect machine learning models to make correct decisions with a limited number of samples. To address these issues, additional information such as structure or invariance in the data can be utilized to help machine learning algorithms learn more efficiently and focus on solving the most important features of the task. Therefore, this paper proposes a lightweight model based on Transformer and incorporating the RIM [1] data augmentation algorithm. The main contributions can be summarized as follows:

1. Proposing a Transformer-based time series model and learning time trends through adversarial training.
2. Incorporating the RIM data augmentation algorithm during the training phase of the model to expand the dataset, thereby improving the model's generalization ability.
3. Experimental results on real datasets demonstrate that the proposed mode

2 Related Work

Current mainstream anomaly detection methods, usually include autoregressive These methods include autoregressive-based techniques such as VAR [2] and LSTM-NDT [3]; clustering-based methods such as Deep-SVDD [4]and THOC [5]; and probability- and density-estimation-based methods such as LOF [6] and DAGMM [7].Also covered are reconstruction-based techniques, including LSTM-VAE[8], OmniAnomaly [9], MAD-GAN [10], USAD [11], and MTAD-GAT [12].

Many of the early enhancement techniques for time series data drew inspiration from image data enhancement, such as operations like cropping, flipping, and adding noise. These enhancement strategies mainly rely on performing random transformations on the training data, which are categorized into three main domains: magnitude domain, time domain, and frequency domain. Time-series decomposition-based data enhancement methods are an innovative strategy that centers on exploiting the intrinsic properties and characteristics of time series to generate new time series with similar characteristics. The core idea of generative model-based data enhancement methods is to utilize generative models [13](e.g., generative adversarial networks) to generate new time series samples.

3 Methodology

3.1 RIM (Recursive Interpolation Method)

The RIM [1] data enhancement algorithm is a time series generation method with rigorous theoretical proofs to ensure the validity of the enhanced data, and we apply the RIM algorithm to the field of anomaly detection for the first time.

Firstly, let the sample dimension of the time series be $\mathrm{d}+1$ where the label set for each sample is $\{0,1,\cdots,\mathrm{n}\}$. Thus, each sample belongs to $\mathbb{R}^{\mathrm{d}+1}\times 0,1,\ldots,\mathrm{k}$. We define the set of time series samples as $S = \{s_0, s_1, \ldots, s_N\}$, and specify a distribution $\mathcal{D}$ with support interval [0,1). For each time $i \in [1 : d]$, we independently draw λ_i from $\mathcal{D}$, denoted as $\lambda_i \sim \mathcal{D}$. Furthermore, we define an interpolation vector $\overrightarrow{\lambda} = (\lambda_1, \ldots, \lambda_d)$, and for simplicity, we use $\overrightarrow{\lambda} \triangleq \lambda$ and $\lambda \sim D$ to denote that each component λ_i of λ is independently sampled from $\mathcal{D}$.

For a given vector $\lambda \in [0,1)^d$ and a time series sample $s \in S$, we can generate an augmented time series sample as follows.

Let $s = (x_0, x_1, \ldots, x_d, y)$ be the original time series sample, where $(x_0, x_1, \ldots, x_d) \in \mathbb{R}^{d+1}$ are the sample features and $y \in 0,1, \ldots, k$ is the corresponding sample label. Let each $\lambda = (\lambda_1, \ldots, \lambda_d) \in [0,1)^d$, and λ_0 is defined as a virtual variable. We define the augmented sample as $s_\lambda = \left(x_{0,\lambda_0}, x_{1,\lambda_1}, \ldots, x_{d,\lambda_d}, y\right)$, where:

$$x_{0,\lambda_0} = x_0,\ x_{i,\lambda_i} = (1 - \lambda_i)x_i + \lambda_i x_{i-1,\lambda_{i-1}} \tag{1}$$

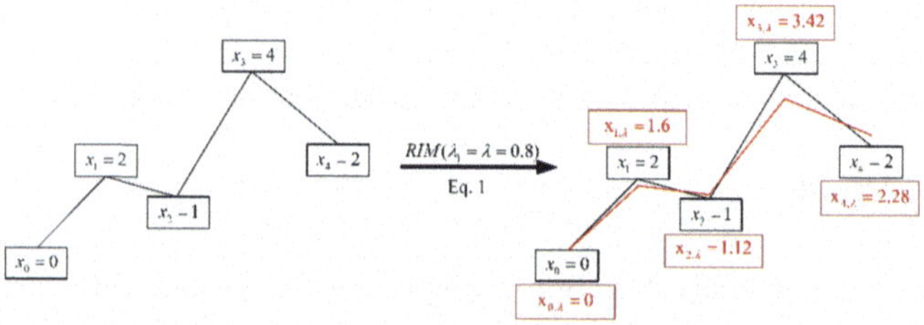

Fig. 1. Illustration of RIM.

As illustrated in Fig. 1, the newly generated augmented sample s_λ shares the same label y as the original sample s. Through this recursive definition, a new realization of the time series can be generated, preserving the trajectory of the original data within a certain range.

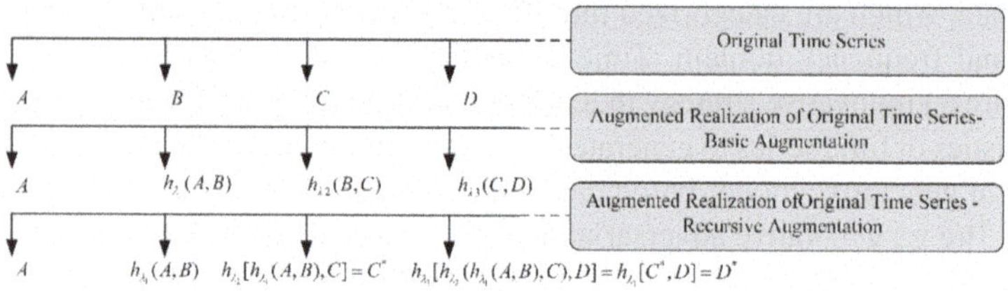

Fig. 2. The RIM Process

In Fig. 2, points A, B, C, and D define the original time series, where the core principle of RIM can be represented by $h_\lambda(A, B) = (1 - \lambda)A + \lambda B$.

3.2 Problem Formulation

Assuming observation of a continuous system composed of d measurement values, and recording equidistant observations over a period of time. The observed time series T is represented by a set of timestamps $\{x_1, x_2, \ldots, x_T\}$, where $x_t \in \mathbb{R}^d$ denotes the observation result at time t.When $d = 1$, T represents a univariate time series, and when $d > 1$, T represents a multivariate time series. The unsupervised time series anomaly detection problem involves determining whether x_t is an unlabeled anomalous data point.

We are given a training input time series T. For any unknown testing time series $\hat{T}$ of length $\hat{T}$ and with the same pattern as the training sequence, we need to predict $Y = \left\{y_1, y_2, \ldots, y_{\hat{T}}\right\}$. Here, we use $y_t \in \{0,1\}^d$ to indicate whether the data at the t-th timestamp in the testing set is anomalous (1 indicates an anomalous data point).

To establish the dependency between the current and previous time points, we define a local context window with a time step of K as:

$$W_t = \{x_{t-K+1}, \ldots, x_t\} \tag{2}$$

We convert the input time series T into a sliding window sequence $\{W_1, \ldots, W_T\}$. When $t < K$, we use replication padding to maintain the window length of each time point t as K. We do not use T as training data, but use W for model training, and take unseen windows $\widehat{W}$ as testing data.

3.3 Overall Architecture

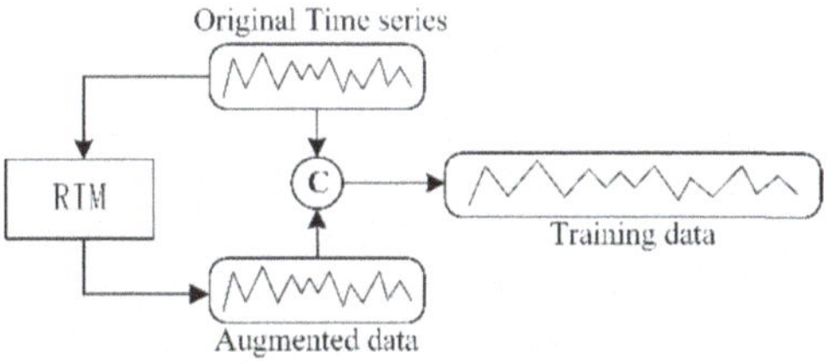

Fig. 3. Model Architecture

Figure 3 illustrates the process of obtaining training data using data augmentation (Fig. 4).

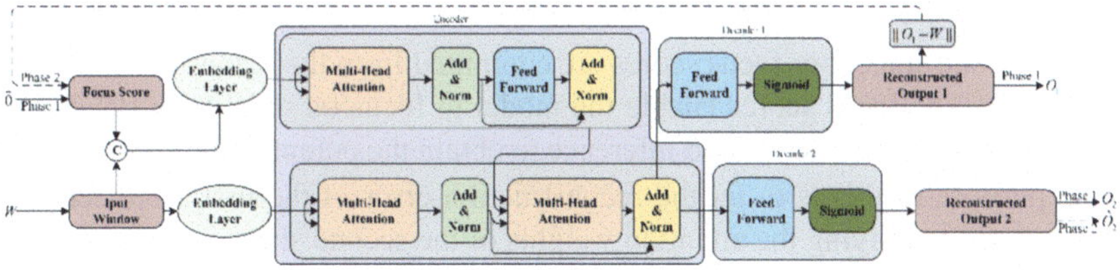

Fig. 4. Model Architecture

The model uses an efficient method of GAN adversarial training. The model consists of an encoder and two decoders, which are composed of Transformer [14] blocks. Before the model training, the data is first augmented with data using the RIM algorithm to expand the original dataset. Then, we divided the model inference into two stages, first, we take W as input and obtain a focus score F (initially a zero matrix with the same dimension as F). We broadcast F to match the dimension of W with the appropriate zero padding, and then join the two. We then apply positional coding to get the input to the Encoder. The Encoder operation is:

$$\begin{aligned} X_1^1 &= \mathrm{LayerNorm}(X_1 + \mathrm{MultiHeadAtt}(X_1, X_1, X_1)) \\ X_1^2 &= \mathrm{LayerNorm}\left(X_1^1 + \mathrm{FeedForward}\left(X_1^1\right)\right) \\ X_2^1 &= \mathrm{LayerNorm}(X_2 + \mathrm{Mask}(\mathrm{MultiHeadAtt}(X_2, X_2, X_2))) \\ X_2^2 &= \mathrm{LayerNorm}\left(X_2^1 + \mathrm{MultiHeadAtt}\left(X_2^1, X_1^2, X_1^2\right)\right) \end{aligned} \tag{3}$$

In the encoder, X_1^2 is made cross-attentive to X_2^1 with the goal of better capturing short-term time trends. Finally, we decode the results of the encoder using two identical decoders, which operate as required below

$$O_i = \mathrm{Sigmoid}\left(\mathrm{FeedForWard}\left(X_2^2\right)\right) \tag{4}$$

To match the normalized input window, the decoder generates outputs in the range [0,1] using the Sigmoid function. The input window W generates two outputs O_1 and O_2.

3.4 Two-Phase Adversarial Training

In the first stage, the objective of the model is to generate an approximate reconstruction of the input window and use the reconstruction error as the focus score, so that the model focuses on sequences with large deviations. Therefore, the bias in the first stage will affect the output of the second stage. In the first stage, the input window $W \in \mathbb{R}^{K \times d_{model}}$ (with Focus Scores [15] $F = [0]_{K \times m}$) is transformed into a compressed latent representation X_2^2. Then, this latent representation is transformed into generated outputs O_1 and O_2 through $O_i = \mathrm{Sigmoid}\left(\mathrm{FeedForWard}\left(X_2^2\right)\right)$, and the training objective is defined by the L2-norm of the reconstruction losses of two decoders:

$$\begin{aligned} \mathrm{Loss}_1 &= \|O_1 - W\|_2 \\ \mathrm{Loss}_2 &= \|O_2 - W\|_2 \end{aligned} \tag{5}$$

In the second stage, our focus scores are computed from the reconstruction error of the first stage. The focus scores enable the model to focus on short-term trends in the time series, so we perform a second inference to obtain the output of the second decoder $\widehat{O}_2$. The purpose of this is to let the reconstruction error of the first stage serve as a prerequisite for modifying the attention weights of the second stage.

$$\min_{\mathrm{Decoder1}} \max_{\mathrm{Decoder2}} \left\|\widehat{O}_2 - W\right\|_2 \tag{6}$$

Here, the second decoder acts as a discriminator at this point, aiming to distinguish between candidate reconstructions (focus scores) generated by the first decoder in the first stage and the input window by maximizing the L2 distance $\left\| \widehat{O}_2 - W_2 \right\|$. On the other hand, the first decoder acts as a generator to fool the discriminator (the second decoder) by minimizing the reconstruction loss to achieve perfect reconstruction (i.e., $O_1 = W$), causing the focus scores to degrade into a zero vector. This results in the second decoder generating the same output as the first stage in the second stage, unable to distinguish between real and generated data. Their loss functions can be expressed as:

$$\begin{aligned} Loss_1 &= +\left\| \widehat{O}_2 - W \right\|_2 \\ Loss_2 &= -\left\| \widehat{O}_2 - W \right\|_2 \end{aligned} \quad (7)$$

In our model architecture, each decoder has different objectives in different stages. Now, we combine the losses of the two decoders in different stages to obtain an evolved loss function:

$$\begin{aligned} Loss_1 &= \eta^{-n}\|O_1 - W\|_2 + \left(1 - \eta^{-n}\right)\left\| \widehat{O}_2 - W \right\|_2 \\ Loss_2 &= \eta^{-n}\|O_2 - W\|_2 - \left(1 - \eta^{-n}\right)\left\| \widehat{O}_2 - W \right\|_2 \end{aligned} \quad (8)$$

Where n represents the training period, and η is a training parameter close to 1 but greater than 1.

3.5 Anomaly Detection

We now use the trained model for anomaly detection. For an unseen observation we define $\widehat{W}$ and the anomaly score is denoted as:

$$s = \frac{1}{2}\left\| O_1 - \widehat{W} \right\|_2 + \frac{1}{2}\left\| \widehat{O}_2 - \widehat{W} \right\|_2 \quad (9)$$

We use the same two-stage inference in testing, where in the first stage inference yields a reconstruction of O_1, and in the second stage inference yields a reconstruction of $\widehat{O}_2$. If an observation follows the normal pattern of the time series, then it can be reconstructed well and the anomaly score will be low. On the contrary, when an observation cannot be reconstructed well, the anomaly score will be high. Once the timestamp anomaly scores for each dimension s_i are obtained, we distinguish normal data from anomalous data by setting a threshold, and when the anomaly score of a timestamp is higher than the set threshold, we judge this timestamp as anomalous data. However, it is difficult to apply the same criterion to different types of faults due to the variable operating environments and diverse operating conditions of complex devices. Therefore, we use the Peak Over Threshold [16] (POT) to automatically find a suitable threshold for the current data.

4 Experiments

4.1 Datasets and Evaluation Indicators

The experiments in this section evaluate the algorithm on three publicly available datasets. The time series data are normalized in order to make the model more robust. The evaluation metric is the F1 score, which is the reconciled mean of precision and recall, indicating the combined evaluation metrics of precision and recall.A higher F1 score indicates a better overall performance of the model. F1 score calculation process:

$$\text{Precision} = \frac{TP}{TP+FP} \quad (10)$$

$$\text{Recall} = \frac{TP}{TP+FN} \quad (11)$$

$$F1\ \text{Score} = \frac{2\times\text{Precision}\times\text{Recall}}{\text{Precision}+\text{Recall}} \quad (12)$$

Where TP (True Positive) is True Positive, which indicates that the model correctly judges a sample that is actually abnormal as the number of instances that are abnormal; FP (False Positive) is False Positive, which indicates that the model incorrectly judges a sample that is actually normal as the number of instances that are abnormal; and FN (False Negative) is False Negative, which indicates that the model incorrectly judges a sample that is actually abnormal as the number of instances that are normal.

4.2 Data Augmentation Analytics

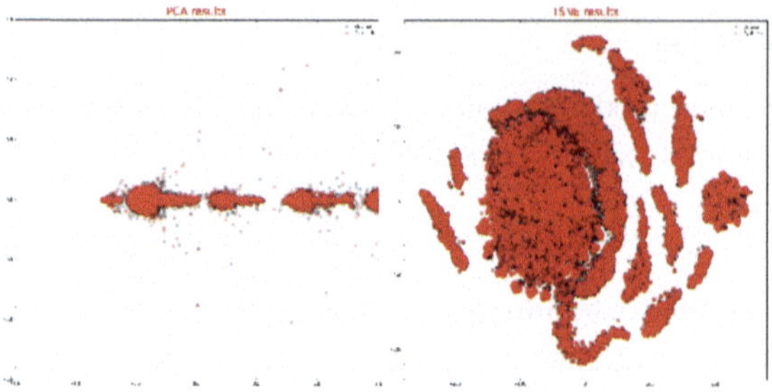

Fig. 5. PCA, T-SEN visualization of the Augmented data

In Fig. 5, the 2D features of the real MSDS data and the RIM generated data are demonstrated. Where the PCA results do not seem to draw a valid conclusion, this is due to the fact that the temporal features of the MSDS dataset are not a linear correlation, whereas the PCA dimensionality reduction is linear, and PCA is not able to deal with linearly indistinguishable data in the dimensionality reduction process. Whereas T-SEN is a nonlinear dimensionality reduction method, in the T-SEN results, it can be seen that the original data (black) and the generated data (red) follow a similar distribution.

4.3 Results

In real industrial scenarios, the cost of obtaining data is usually high, and therefore the data collected is often limited. In order to simulate this real situation, we choose to use only 20% of the data as the training set (and the same 20% of the data for the test set) to simulate the data scarcity situation. The specific implementation is to split the complete data into five groups, train and test them separately, and finally the average performance of the five groups will be counted. Doing so can effectively test the performance of the model when the training data is scarce. Among the experimental benchmarks are autoregressive-based techniques such as LSTM-NDT [3]; clustering-based methods such as THOC [5]; and probabilistic and density-estimation-based methods such DAGMM [7]. Also covered are reconstruction-based techniques including LSTM-VAE [8], OmniAnomaly [9], MAD-GAN [10], USAD [11], MTAD-GAT [12] (Table 1).

Table 1. Dataset Statistics

Dataset	Train	Test	Dimensions	Anomalies(%)
SMAP	135183	427617	25	13.13
MSL	58317	73729	55	10.72
SMD	708405	708420	38	4.16
MSDS	146430	146430	10	5.37

Tables 2 show the model performance using 20% training set. The model proposed in this paper has the highest AUC and F1 compared to all baseline models. On average, the average F1 score of Our without data augmentation is 0.8945. Not only that, when data augmentation is used, the performance of Our* is substantially improved to 0.9273, which is a performance enhancement of 3.6%, demonstrating that the performance of the model in data-scarce scenarios can be greatly improved after applying the RIM data augmentation algorithm.

Table 2. Experimental Results

Method	SMAP		MSL	
	AUC	F1	AUC	F1
LSTM-NDT	0.7007	0.5418	0.9520	0.7608
THOC	0.9273	0.8654	0.9376	0.8735
DAGMM	0.9881	0.8369	0.9606	0.8010
LSTM-VAE	0.9567	0.7627	0.9324	0.7954
OmniAnomaly	0.9879	0.8131	0.9703	0.8424
MAD-GAN	0.9877	0.8468	0.9649	0.8190
USAD	0.9883	0.8379	0.9649	0.8190
MTAD-GAT	0.9814	0.8225	0.9782	0.8024
Our	0.9899	0.8984	0.9824	0.9197
Our*	0.9937	0.9176	0.9873	0.9327

(continued)

Table 2. (*continued*)

Method	MSDS		SMD	
	AUC	F1	AUC	F1
LSTM-NDT	0.7813	0.7912	0.9563	0.6754
THOC	0.8027	0.8032	0.9297	0.8039
DAGMM	0.7763	0.8389	0.9845	0.8986
LSTM-VAE	0.8237	0.7985	0.9570	0.7856
OmniAnomaly	0.5613	0.8389	0.9859	0.9352
MAD-GAN	0.5002	0.7390	0.8635	0.9318
USAD	0.7613	0.8389	0.9854	0.9213
MTAD-GAT	0.6122	0.8248	0.9798	0.6661
Our	0.8795	0.8408	0.9827	0.9431
Our*	0.9341	0.9232	0.9901	0.9497

5 Summary and Conclusion

In this paper, a time series anomaly detection model incorporating the RIM [1] data enhancement algorithm is proposed to capture the time series trend through adversarial training, and the distinction between normal and anomalous data is performed by reconstruction error. In the experiments, only 20% of the training data is used to simulate real data scarcity scenarios, and the experimental results show that the model has excellent anomaly detection performance with less training data, and the performance of this model outperforms all the baseline model performance. The expansion of RIM augmented data on training data achieves consistent enhancement results across different datasets and tasks, demonstrating its broad applicability and robustness.

References

1. Aboussalah, A.M., Kwon, M., Patel, R.G., et al.: Recursive time series data augmentation. In: The Eleventh International Conference on Learning Representations (2022)
2. Melnyk, I., Banerjee, A., Matthews, B., Oza, N.: Semi-markov switching vector autoregressive model-based anomaly detection in aviation systems. In: Proceedings of the 22nd ACM SIGKDD International Conference on Knowledge Discovery and Data Mining, in KDD '16. Association for Computing Machinery, pp. 1065–1074. New York, NY, USA (2016). https://doi.org/10.1145/2939672.2939789
3. Hundman, K., Constantinou, V., Laporte, C., Colwell, I., Soderstrom, T.: Detecting spacecraft anomalies using LSTMs and nonparametric dynamic thresholding. In: Proceedings of the 24th ACM SIGKDD International Conference on Knowledge Discovery & Data Mining, pp. 387–395 (2018). https://doi.org/10.1145/3219819.3219845
4. Ruff, L., et al.: Deep one-class classification. In: Proceedings of the 35th International Conference on Machine Learning, PMLR, pp. 4393–4402 (2018). Accessed 18 Nov 2022

5. Shen, L., Li, Z., Kwok, J.: Timeseries anomaly detection using temporal hierarchical one-class network. In: Advances in Neural Information Processing Systems, pp. 13016–13026. Curran Associates, Inc. (2020). Accessed 18 Nov 2022
6. Breunig, M.M., Kriegel, H.-P., Ng, R.T., Sander, J.: LOF: identifying density-based local outliers. In: Proceedings of the 2000 ACM SIGMOD international conference on Management of data, in SIGMOD '00, pp. 93–104. Association for Computing Machinery, New York, NY, USA (2000). https://doi.org/10.1145/342009.335388
7. Zong, B., et al.: Deep autoencoding gaussian mixture model for unsupervised anomaly detection. Presented at the International Conference on Learning Representations (2022). Accessed 18 Nov 2022
8. Park, D., Hoshi, Y., Kemp, C.C.: A Multimodal Anomaly Detector for Robot-Assisted Feeding Using an LSTM-based Variational Autoencoder. arXiv:1711.00614 [cs] (2017). Accessed 14 Mar 2022
9. Su, Y., Zhao, Y., Niu, C., Liu, R., Sun, W., Pei, D.: Robust anomaly detection for multivariate time series through stochastic recurrent neural network. In: Proceedings of the 25th ACM SIGKDD International Conference on Knowledge Discovery & Data Mining, pp. 2828–2837 ACM, Anchorage AK USA (2019). https://doi.org/10.1145/3292500.3330672
10. Li, D., Chen, D., Jin, B., Shi, L., Goh, J., Ng, S.-K.: MAD-GAN: multivariate anomaly detection for time series data with generative adversarial networks. In: Tetko, I.V., Kůrková, V., Karpov, P., Theis, F. (eds.) Artificial Neural Networks and Machine Learning – ICANN 2019: Text and Time Series, LNCS, vol. 11730, pp. 703–716. Springer International Publishing, Cham (2019). https://doi.org/10.1007/978-3-030-30490-4_56
11. Audibert, J., Michiardi, P., Guyard, F., Marti, S., Zuluaga, M.A.: USAD: unsupervised anomaly detection on multivariate time series. In: Proceedings of the 26th ACM SIGKDD International Conference on Knowledge Discovery & Data Mining, pp. 3395–3404. ACM, Virtual Event CA USA (2020). https://doi.org/10.1145/3394486.3403392
12. Zhao, H., et al.: Multivariate time-series anomaly detection via graph attention network. In: 2020 IEEE International Conference on Data Mining (ICDM), pp. 841–850 (2020). https://doi.org/10.1109/ICDM50108.2020.00093
13. Yoon, J., Jarrett, D., van der Schaar, M.: Time-series generative adversarial networks. In: Advances in Neural Information Processing Systems, Curran Associates, Inc. (2019). Accessed 4 Jul 2023
14. Vaswani , A., et al.: Attention is all you need. Adv. Neural Inform. Process. Syst. **30** (2017)
15. Tuli, S., Casale, G., Jennings, N.R.: TranAD: deep transformer networks for anomaly detection in multivariate time series data. Proc. VLDB Endow. **15**(6), 1201–1214 (2022). https://doi.org/10.14778/3514061.3514067
16. Siffer, A., Fouque, P.-A., Termier, A., Largouet, C.: Anomaly detection in streams with extreme value theory. In: Proceedings of the 23rd ACM SIGKDD International Conference on Knowledge Discovery and Data Mining, in KDD '17, pp. 1067–1075. Association for Computing Machinery, New York, NY, USA (2017). https://doi.org/10.1145/3097983.3098144

Engineering and Systematical Technologies

The Water Transport Configuration of Nanning City Under the Background of the Western Land Sea New Corridor Pinglu Canal

Lanfang Zhang[1], Ning Yu[2], and Jie Shang[1(✉)]

[1] Guangxi Key Laboratory of International Join for China-ASEAN Comprehensive Transportation, Nanning University, Nanning 530000, Guangxi, China
zhanglanfang@unn.edu.cn

[2] Guangxi Beigang Planning and Design Institute Co, Nanning 530000, Guangxi, China

Abstract. August 28, 2022, the western land sea new corridor backbone project - Pinglu Canal officially started construction, the western land sea new corridor is located in the hinterland of China's western region, north of the Silk Road Economic Belt, south of the 21st Century Maritime Silk Road, synergistic convergence of the Yangtze River Economic Belt, in the pattern of coordinated development of the region has an important strategic position, and the Pinglu Canal will be opened up between China and the ASEAN The Pinglu Canal will open up a convenient water transportation channel for trade between China and ASEAN, extend the strategy of a strong nation of the sea to the cities along the middle and upper reaches of the Xijiang River, open up a new channel for the shortest distance of the Xijiang River mainstem into the sea, realize the direct connection between the 5, 873-km inland waterway network of Guangxi, part of the region of Yunnan and Guizhou, and the ocean transportation, effectively alleviate the pressure on the lower reaches of the Xijiang River, and in the long term, connect with Xianggui Canal which is being studied and demonstrated, and vertically connect with the Yangtze River, the Pearl River, and the Gulf of Tonkin. In this context, as the location of the Pinglu Canal, Nanning City, the waterway transportation pattern will change greatly with the construction of the Pinglu Canal. In this paper, we analyze the water transport pattern of Nanning City after the construction of the Pinglu Canal by quantitative argumentation and analysis of the choice of waterway transport routes by shippers, and through the comparison of transport economy and technology.

Keywords: Western Land Sea New Corridor · Pinglu Canal · Water transportation · Transport routes · Transport economy

1 Introductory

Nanning City has the characteristics of "two along and two near" along the river, along the railroad line, near the sea and near the border, Nanning City's waterway connects to Longzhou on the Left River, Baise and Yunnan on the Right River, Guizhou on the Red

Z. Hu et al. (Eds.): ICCSEEA 2024, LNDECT 242, pp. 461–474, 2025.
https://doi.org/10.1007/978-3-031-84228-3_40

River, and Guigang, Wuzhou, Guangzhou, Hong Kong, and Macao on the lower part of the city. 2022, Nanning City completed the waterway freight transport volume of *4, 550, 000* tons, a decrease of *1.15%* compared with the 2021; completed the waterway freight transport turnover of *38.772* billion ton kilometers, an increase of *10.14%* year on year. *1.15%;* completed waterway freight turnover 38.772 billion tons of kilometers, an increase of *10.14%*, with the improvement of waterway conditions, the average distance of waterway freight is increasing, water transport in Nanning City, accounting for *11.1%* of the total external freight, in the country vigorously promote the "public to water", Under the background of the country vigorously promoting the construction of "public to water", "iron to water" and Pinglu Canal, water transportation in Nanning City has greater development potential [1].

After the completion of the Pinglu Canal, Nanning City, the waterway transport pattern will undergo significant changes, specifically reflected in the following aspects: First, Nanning Port and Beibu Gulf Port to achieve the waterway through the waterway, Nanning City, Nanning City, inland waterway transport to increase the southward channel to the sea to enhance the waterway transport in the city of Nanning City, the role of the integrated three-dimensional transport network, bringing the "land-to-water" transport increment; The second is to change the direction of some of the freight flows, with some of the goods transported via the Xijiang shipping trunk line to the Pearl River Delta ports to be transported via the Pinglu Canal to the Beibu Gulf port; Third, strengthening the position of Nanning Port as the hub of inland water transport in the southwest region, which will bring the incremental volume of "land-to-water" transportation; Fourth, in the future, such as the opening of the Xianggui Canal, the Pinglu Canal can be directly linked to areas along the Xiangjiang River, and the goods in Hunan Province can be directly connected with the Xiangjiang River, so that the goods in the Hunan Province can be transferred to the other areas. Fourthly, if Xianggui Canal is opened in the future, Pinglu Canal can directly connect to the areas along Xiangjiang River, and the goods in Hunan Province can be conveniently exported to the sea through Beibu Gulf Port.

In this context, to promote the efficient and green development of waterway transportation in Nanning City, considering the choice of waterway transportation routes by cargo owners, quantitatively analyzing and demonstrating the waterway transportation pattern in Nanning City after the construction of Pinglu Canal can be used as a reference for relevant decision-making.

2 Scope of Water Transportation Argumentation in Nanning

Nanning City, the waterway transportation pattern analysis fully consider the completion of the Pinglu Canal, its impact on the region's cargo through the Nanning port transit economy, considering the different flow of goods, specifically subdivided into two major sources of goods. First, for the southward exchange of goods with ASEAN, Hainan and other regions, will be compared to the Pinglu Canal influence region of the source of goods directly transported to the Pinglu Canal - Beibu Gulf port to the sea and through the Nanning port transit and then through the Pinglu Canal - Beibu Gulf port to the sea in two cases of transport economics, because the different programs after the sea path is consistent, therefore, the scope of the argument for the source of the port - Beibu Gulf port, in addition, for Hunan goods may be through the Yangtze River port to the sea, the scope of the argument for the source of goods to the port, the scope of the argument is to the port. In addition, for Hunan cargoes that may go to sea via Yangtze River ports, the scope of the demonstration is from the port of origin to the port of destination; secondly, for cargoes exchanging with northern ports of China in the north direction, the transportation economy of cargoes in the Pinglu Canal influence area will be compared in two scenarios: southward to the Beibu Gulf port via the Pinglu Canal, and eastward to the ports of the Pearl River Delta via the Xijiang River Mainline, and the scope of the demonstration is from the port of origin to the ports of the north, because the two ports of departure are involved.

Select 11 typical cargo nodes in Guangxi, Yunnan, Guizhou and Hunan. Combined with the planning and construction of the Pearl River waterway network in Guigang and above, the cargo source nodes in Guangxi are Nanning and Guigang in Guigang and above, Chongzuo in the Zuojiang basin, Baise in the Right River basin, Liuzhou in the Liujiang-Qianjiang basin, Guilin in the Guijiang basin, and Hezhou in the Hejiang basin; in Yunnan Province, the generation of large quantities of materials is mainly concentrated in Kunming, Qujing, Wenshan and other cities, mainly relying on the Beibu Gulf port for access, and Kunming and Wenshan are selected as typical cargo sources; in Guizhou Province, the distance from the high level of high quality is selected, and Kunming and Wenshan are selected as typical cargo sources. In Yunnan Province, bulk material generation is mainly concentrated in Kunming, Qujing, Wenshan and other cities, mainly relying on Beibu Gulf Port to enter and exit, and Kunming and Wenshan are selected as the typical cargo sources; Guizhou Province selects Qiandongnan Prefecture, which is closer to the high-level waterways, as the typical cargo source; in Hunan Province, under the premise of opening the Xianggui Canal, the Pinglu Canal can be directly connected to the areas along the Xiangjiang River, and Yongzhou City is selected as the typical cargo source [2].

3 Analyzing and Justifying Parameter Values

3.1 Water Transportation

Water transport tariffs refer to the Yangtze River Shipping Index Analysis Report and Pearl River Water Transportation Economic Performance published on the official website of the Ministry of Transportation and Communications (MOTC). Among them, the average value of tariffs in the last 6 months for the mainline of the Yangtze River and the average value of tariffs in the last 3 years for the mainline of the Xijiang River are taken into account, and the influence of the inland waterway dry season is considered simultaneously in the values [3] (Table 1).

Calculation of the cost of transportation by water is based on the following formula:

$$\mathrm{W_{tc}} = \mathrm{C_{ih}} + \mathrm{R} \times \mathrm{L} + \mathrm{F_{rt}} + \mathrm{T_{iw}} + \mathrm{C_{sr}} \quad (1)$$

Among them: W_{tc} is water transportation cost, port to port cost; C_{ih} is inland river port loading and unloading fee, calculated according to RMB 7/ton; R is water transportation tariff rate; L is billing transportation distance; F_{rt} is inland river transit fee, inland river transit is considered according to water barge, taking RMB 8/ton; T_{iw} is inland river gate fee, taking RMB 0.6/gross ton; C_{sr} sea and river dressing fee, taking RMB 28/ton with reference to Beibu Gulf port fee standard [4].

3.2 Railroads

The following formula is used to calculate the cost of railroad transportation with reference to the benchmark tariff rate for railroad transportation published by the NDRC:

$$C_{rt} = P_{rb1} + P_{rb2} \times L + C_{rl} + C_h \tag{2}$$

$$P_{rb2} = 0.098 + F_{rc} \tag{3}$$

Among them, C_{rt} is the cost of railroad transportation for station to station; P_{rb1} is the base price 1, taking 16.3 yuan; P_{rb2} is the base price 2, the calculation method is as follows; L is the billing transportation distance, referring to the railroad network of each railroad bureau, selecting the shortest path to be considered; C_{rl} is the cost of the railroad special line, taking this cost into account in the case of needing to be charged; C_h is the cost of unloading and handling of the two ends, sea-rail intermodal handling charges are taken as RMB 36/ton with reference to Beibu Gulf Port charges. F_{rc} is the railroad construction fund, taking 0.033.

3.3 Highways

With reference to the price of short-haul road transport in a typical area, the cost of road transport within a distance of 200 km is taken as RMB 0.5/ton-km.

4 Transportation Options and Cost Estimation Related to Waterway Transportation in Nanning City

In order to analyse the change of waterway transport pattern in Nanning after the construction of the Pinglu Canal, on the basis of giving full consideration to the economic hinterland of the Pinglu Canal, we will focus on comparing the transport schemes of goods from different sources within the hinterland to the Pinglu Canal - Beibu Gulf Port directly by waterway and the transport schemes of the water-to-water transshipment in the port of Nanning to the Pinglu Canal - Beibu Gulf Port and also compare the transport schemes with the existing sea-railway corridor of the port of Beibu Gulf, sea-roadway corridor of the port of Beibu Gulf, and the Pearl River Deltalt will also be compared with the existing Beibuwan Port Sea-Rail Intermodal Transport Corridor, Beibuwan Port Sea-Public Intermodal Transport Corridor and Pearl River Delta (PRD) River-Sea Intermodal Transport Corridor, among which Hunan Province will be compared with Yangtze River Delta (YRD) River-Sea Intermodal Transport Corridor [5]; and the areas west of the upper reaches of Nanning in the PRD water system (Left and Right river basins and Yunnan) will be added to the transport scheme of transshipment to the Xijiang Mainline to the PRD ports through Nanning Port, as well as the transport scheme of direct transport to the PRD ports [6].

At the same time, in compliance with the objective reality and the principle of nearby transit, considering that the cargoes in Gui River, He River Basin and Hunan area may transit in Wuzhou Guijiang River Estuary, the Wuzhou Guijiang River Estuary Transit Program has been set up as part of the cargo transportation program for Gui River, He River Basin and Hunan area [7]; considering that the cargoes in Liujiang River Basin and Guizhou area may transit in Shilong Sanjiangkou and Guiping River Estuary, the Shilong Sanjiangkou and Guiping River Estuary Transit Program has been set up as part of the cargo transportation program for Liujiang River Basin and Guizhou area [8]. Considering the possibility of cargoes from Liujiang basin and Guizhou area to transit in Shilong Sanjiangkou and Guipingkou, the cargo transportation program also establishes Shilong Sanjiangkou and Guipingkou transit program [9].

When setting up the transportation scheme and cost estimation, the transportation schemes of Xijiang River trunk line, Zuo and Ruijiang River Basin, Liujiang-Qianjiang River Basin, Duliujiang-Rongjiang River Basin, Guijiang River, Hejiang River, Xianggui Canal Basin to the south and to the north to the sea are considered respectively, and the specific scenarios and transportation costs per ton are shown in Figs. 1, 2, 3, 4, 5, 6, 7 and Tables 2, 3, where A - I in the Figures and Tables represent the number of each transportation scenario [10].

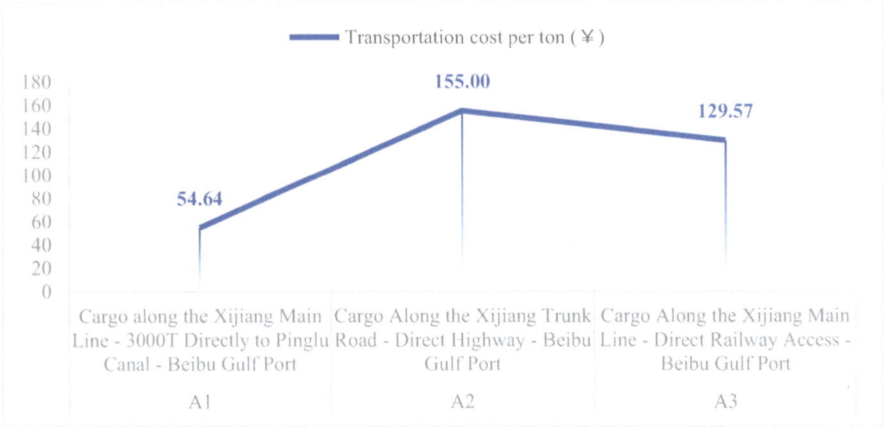

Fig. 1. Schematic illustration of transportation cost per ton of cargo along the Xijiang trunk line going out to sea via Beibu Gulf port

Table 1. Water transport tariff rates for ships of different tonnage classes (bulk cargo) Unit: yuan/ton-kilometer

class in tons (of a passenger ship)	1000 tons	2000 tons	3000 tons	5000 tons	coast
Xijiang tariff rate	0.095	0.06	0.04	0.034	0.02

Note: The tariff rate for 5000-ton ships is the value taken after considering the impact of dry water period, and the dry water period is taken as 5 months

Table 2. Transportation options for typical cargoes from the Left and Right River Basins, including Yunnan Province, to the sea via Beibu Gulf Port

programmatic	Transportation routes	Transportation cost per ton (rmb)	note
B_1	Left and Right River Basin Cargo - 1000T Directly to Pinglu Canal-Beibu Gulf Port	87.62	At present, the Zuo River and the Right River are navigable by ships of 1, 000 tons
B_2	Left and Right River Basin Cargo - 1000T - Nanning Port Transit - 3000 T - Pinglu Canal - Beibu Gulf Port	76.62	
B_3	Left and Right River Basin Cargo - Railway - Beibu Gulf Port	115.81	
B_4	Left and Right River Basin Cargo - Direct Highway - Beibu Gulf Port	125.00	
B_5	Left and Right River Basin Cargo - 2000T - Pinglu Canal - Beibu Gulf Port	70.22	In the long term, in the case of 2000-ton ships on the Left and Right Rivers and 5000-ton ships on the Pinglu Canal
B_6	Left and Right River Basin Cargo - 2000T - Nanning Port Transit - 5000T Pinglu Canal - Beibu Gulf Port	70.99	
B_7	Left and Right River Basin Cargo - Railway - Beibu Gulf Port	115.81	
B_8	Left and Right River Basin Cargo - Direct Highway - Beibu Gulf Port	125.00	

5 Analysis of Waterway Transportation Pattern in Nanning

Through the comparison of the above transportation schemes, we quantitatively demonstrate the waterway transportation pattern of Nanning City after the opening of Pinglu Canal from the perspective of demand side, i.e., based on the viewpoint of cargo owner's choice, and get the following conclusions from the perspective of transportation economy:

Table 3. Transportation schemes for the exchange of cargoes between left and right river basin sources, including typical sources in Yunnan Province, and domestic ports in the north of the country

programmatic	Transportation routes	Transportation cost per ton (¥)	note
C_1	Left and Right River Basin Cargo - 1000T Directly to Pinglu Canal - Beibu Gulf Port - Yingkou Port	124.02	Recently, the Left and Right Rivers have been navigated by 1000-ton ships, including cargoes from typical sources in Yunnan Province
C_2	Left and Right River Basin Cargo - 1000T - Nanning Port Transit - 3000T Pinglu Canal - Beibu Gulf Port - Yingkou Port	113.02	
C_3	Left and Right River Basin Cargo-1000T Direct-Pearl River Delta Port - Yingkou Port	169.30	
C_4	Left and Right River Basin Cargo - 2000T Direct-Pearl River Delta Port - Yingkou Port	132.30	
C_5	Left and Right River Basin Cargo - 2000T - Pinglu Canal - Beibu Gulf Port - Yingkou Port	106.62	In the long term, if the Left and Right Rivers are open to 2, 000-ton vessels and the Pinglu Canal is open to 5, 000-ton vessels
C_6	Left and Right River Basin Cargo - 2000T - Nanning Port Transit - 5000T Pinglu Canal - Beibu Gulf Port - Yingkou Port	107.01	
C_7	Left and Right River Basin Cargo - 2000T - Nanning Port Transshipment - 5000T Xijiang Mainline - Pearl River Delta Ports - Yingkou Port	117.10	

(continued)

Table 3. (*continued*)

programmatic	Transportation routes	Transportation cost per ton (¥)	note
C_8	Left and Right River Basin Cargo - 1000T - Nanning Port Transshipment - 5000T Xijiang Mainline - Pearl River Delta Ports - Yingkou Port	127.50	

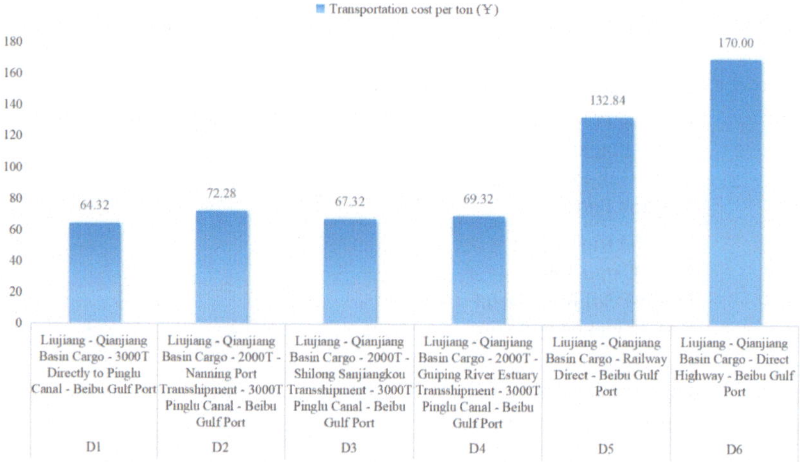

Fig. 2. Transportation cost per ton for the transportation scheme of cargoesLiujiang - Qianjiang Basin to the sea via Beibu Gulf port

(1) For local cargo sources in Nanning, the construction of the Pinglu Canal has opened up a new channel for southward water transportation to the sea, which is more economical than highway and railroad transportation to the Beibu Gulf Port and to the sea via the Xijiang River Main Line to Zhuhai Port. For other areas along the main line of the Xijiang River, the direct transport through the Pinglu Canal to the Beibu Gulf Port is more economical than the transit in Nanning Port [11].

(2) For the cargo sources in the Left and Right River Basin (including Yunnan cargo sources), whether it is southbound sea cargo or northbound cargo exchanging with northern ports, under the circumstance that the Left and Right Rivers are open to 1, 000-ton ships, it is more economical to transit through the Pinglu Canal to the port of Qinzhou at the port of Nanning rather than direct transportation to the Pinglu Canal to the port of Qinzhou [12]; for the domestic trade exchanges between the cargo sources in the Left and Right River Basin and the Pearl River Delta (PRD), it is more economical to transit through the Xijiang Mainline to the PRD port than

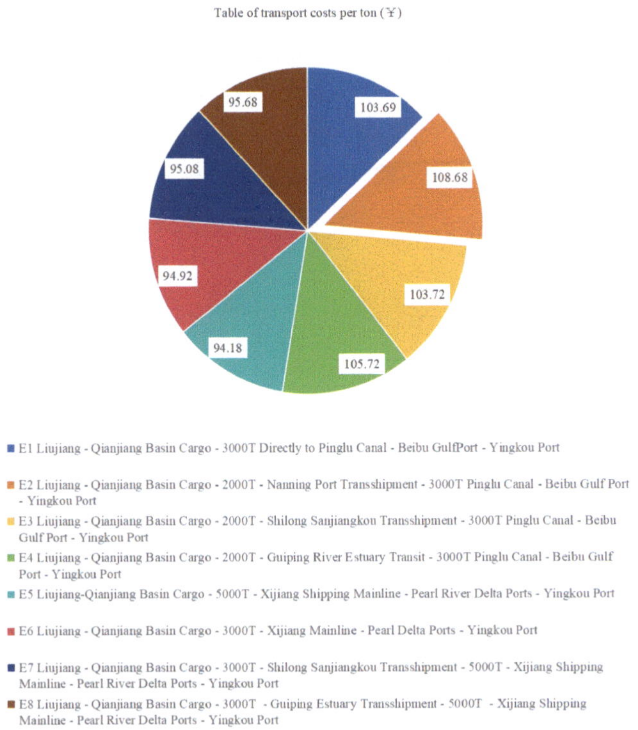

Fig. 3. Schematic diagram of transportation costs per ton of the transportation scheme for the exchange of cargoes between the cargo sources in the Liujiang - Qianjiang basin and the ports in the north of China

to direct transportation through the Xijiang Mainline to the PRD port. For domestic trade exchanges between the left and right river basins and the PRD region, it is more economical to transit through the Nanning Port via the Xijiang Mainline to the PRD ports than through the Xijiang Mainline to the PRD ports. At the same time, for the domestic trade exchanges between the left and right river basins and the PRD region, taking into account the waiting time for crossing the locks caused by the serious congestion at the Changzhou locks, it is a new alternative to transit through the Nanning Port via the Pinlue Canal to be transported by sea from Beibu Gulf Port to the PRD region. Before the left and right rivers can be navigated by 2, 000-ton ships, transshipment at Nanning Port will be one of the most important ways of organizing the transportation of cargo from the left and right river basins (including Yunnan), whether it is southbound cargo going to the sea, or northbound domestic cargo exchanging with northern ports and the Pearl River Delta region [13].

(3) For the southward seaward cargoes of Hongshui River and Liujiang-Qianjiang River Basin (including Guizhou cargoes), transshipment in Nanning Port is not as economical as direct transportation by the same tonnage ship, and not as economical as transshipment in Shilong Sanjiangkou. Among them, for the cargoes in Liuzhou and Laibin, direct transportation is more economical [14]; for the cargoes along other

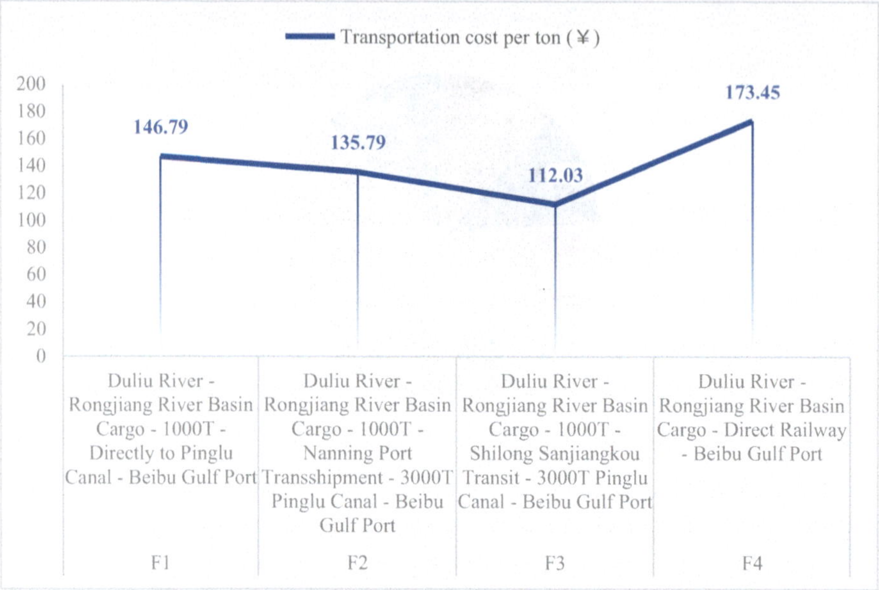

Fig. 4. Transportation cost per ton of cargo from theDuliuRiver - Rongjiang River Basin going to sea via Beibu Gulf port

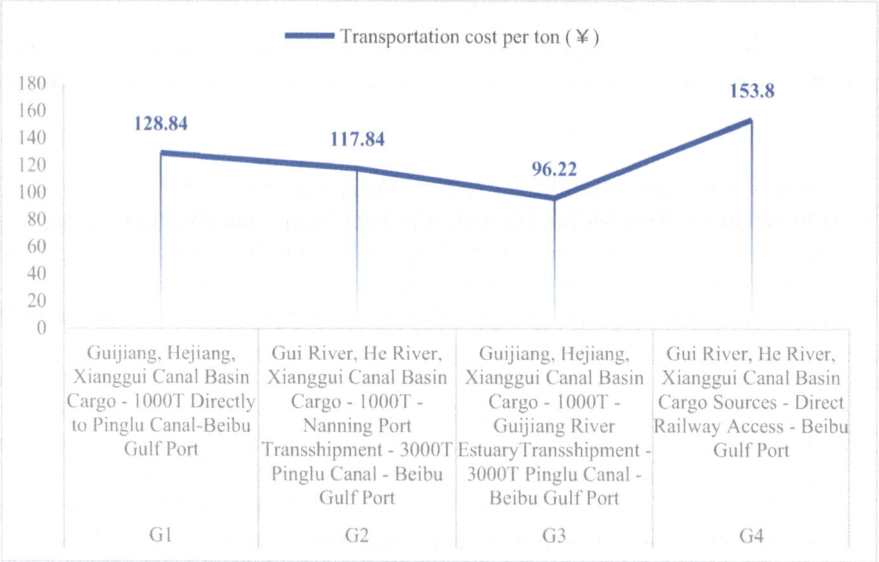

Fig. 5. Schematic diagram of transportation cost per ton for the southward seaborne transportation scheme for cargoes from the Gui River, He River and Xianggui Canal Basin

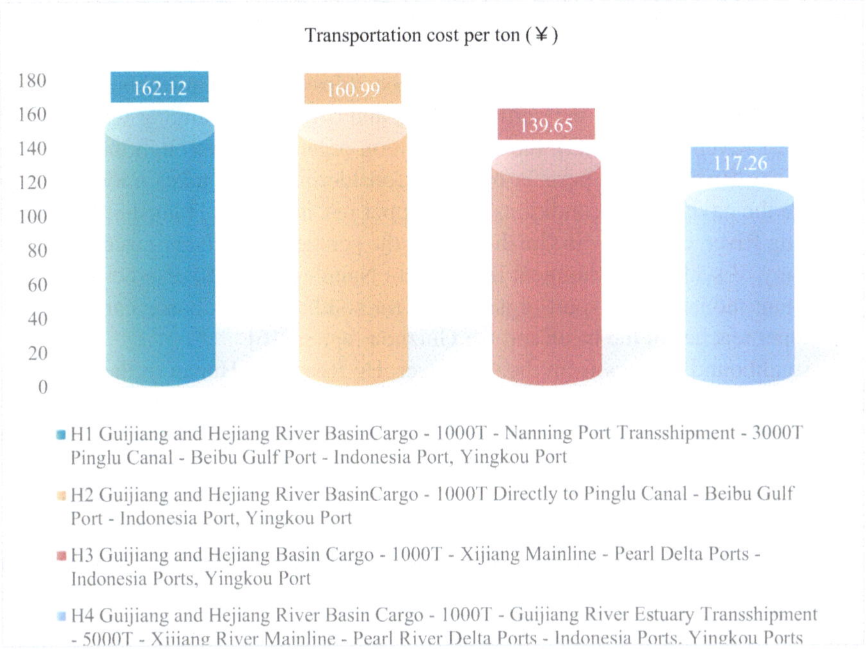

Fig. 6. Schematic diagram of transportation cost per ton for the northward seaborne transportation scheme for cargoes from the Gui and He River basins

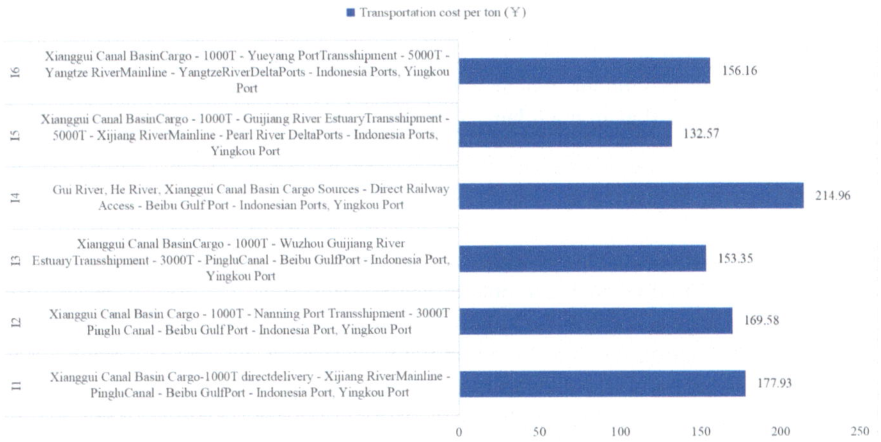

Fig.7. Schematic diagram of transportation cost per ton for the northward seaborne transportation scheme of Xianggui Canal Basin cargoes Transportation Scheme

tributaries, including Guizhou cargoes, it is more economical to use 1000-ton ships to transit 3000-ton ships in Liuzhou Jiangkou operation area or Shilong Sanjiangkou, followed by 1000-ton ships for direct transportation, and 1000-ton ships to transit

3000-ton ships in Nanning port via Pinglu Canal. The economy of the transportation organization of Beibu Gulf port is worse, but it is not much different from the economy of direct transportation by 1000-ton ships [15]. For northbound cargoes exchanging with ports in the north of China and cargoes exchanging with ports in the Pearl River Delta (PRD), the transportation organization through the mainline of Xijiang River ~ PRD ports is optimal. Considering that it takes a long time to change the navigational conditions in the upper reaches of the Hongshui River and Liujiang River, especially in Guizhou, from the perspective of large-sized ships and economy of scale, transshipment transport in Nanning port will also be one of the important modes of transport organization for southbound seaborne cargoes from the upper reaches of the basin and the Guizhou region [16].

(4) For southbound cargoes from the Gui River, He River and Hunan in the Xianggui Canal Basin, the transportation mode of 1000-ton ships transiting the port of Nanning and 3000-ton ships going through the Pinglu Canal to the port of Beibu Gulf is more economical than that of the same tonnage ships transporting to the port of Pinglu Canal to the port of Beibu Gulf, but not as economical as that of the same tonnage ships transiting through the port of Wuzhou at the mouth of the Gui River and the port of Beibu Gulf [17]; among them, for cargoes generated from Wuzhou at the mouth of the Gui River, it is more economical to transport cargo directly because of the capacity of 3000 to 5000-ton ships. In particular, for cargoes generated in Wuzhou at the mouth of the Guijiang River, direct transportation is more economical due to the capacity of 3, 000-5, 000-ton ships. For goods exchanging with ports in the north of China and with ports in the Pearl River Delta (PRD), the best way of transportation is through the Xijiang Main Line to the PRD ports [18].

In the long term, with the upgrading of the grade of Pinglu Canal waterway, with the condition of passing 5000-ton river vessels, the transportation advantage of Pinglu Canal will be more prominent, which will further magnify the water-to-water transit advantage of Nanning Port in the inland waterway network in the southwest region.

6 Concluding Remarks

This paper first analyzes the cargo hinterland of Nanning waterway transport after the completion of the Pinglu Canal, and puts forward a collection of various transportation schemes for the cargo sources in the hinterland, and on this basis, analyzes the economy of various transportation schemes based on the viewpoint of the cargo owner, so as to come to the conclusion of the analysis of the pattern of waterway transport in Nanning City after the completion of the Pinglu Canal [19]. For the cargo sources of the Xijiang River shipping trunk line, Pinglu Canal is the best channel for its southward to the sea; for the cargo sources of Zuo and Ruijiang River Basin (including the cargo sources of Yunnan), under the condition that the left and right rivers are open to 1, 000-ton ships, its southward and northward to the sea is the best choice for the cargo sources to be transited to take the Pinglu Canal in the port of Nanning, and under the condition of the open to 2, 000-ton ships, the cargo sources to go directly to the Pinglu Canal to the sea is the best choice; for the Red River, Liujiang-Qianjiang River Basin, Liujiang-Qianjiang River Basin, the Pinglu Canal is the best choice for the cargo sources to go straight to the sea. Liujiang-Qianjiang basin, Guijiang, Hejiang and Hunan in Xianggui Canal basin (including Guizhou), it is more economical for 3000-ton ships to go directly to the sea, and it is more economical for 1000-ton ships to transit 3000-ton ships at Sanjiangkou of Shilong (Wuzhou port at the mouth of the Guijiang River), and for goods exchanged with ports of the northern part of China in the northern part of China as well as those with ports of the Pearl River Delta (PRD), the transportation of cargoes through the Xijiang shipping route is the most economical choice [20]. The best way to organize transportation is through the Xijiang shipping line to the PRD ports. In general, after the completion of the Pinglu Canal, the cargo sources in the hinterland of the Pinglu Canal within the scope of Guangxi, Yunnan, Guizhou and Hunan, except for the Hongshui River, Liujiang-Qianjiang River Basin, Guijiang River, Hejiang River and Xianggui Canal Basin of Hunan, etc., which are more economical to go out to the sea through the Xijiang River and the Pearl River Delta, the rest of the hinterland is more economical to go out to the sea through the Pinglu Canal. Undertake part of the southwest region transit cargo transportation to undertake the whole of Guangxi and Yunnan, Guizhou and Hunan Pinglu Canal hinterland within the scope of the function of waterway transportation, will bring the rapid growth of Nanning City waterborne freight volume, Nanning City, the need for corresponding waterborne transportation supporting facilities, such as the construction of ports, upgrading the level of navigation, anchorage, etc., for the future of the city of Nanning's waterborne transportation volume of rapid growth. Provide guarantee for the future rapid growth of waterway transportation volume in Nanning.

Acknowledgment. This project is supported by Nanning University Model Courses (No.2022SZSFK08).

References

1. Liu, X., Wang, T., Wu, X.: Impact of the construction of Pinglu Canal on the pattern of freight transportation to the sea in the southwestern region. Water Transport. Eng. **11**, 15–22 (2023). (in Chinese)
2. Wang, T., Liu, X., Wu, X.: Freight demand analysis of Pinglu Canal based on the construction of comprehensive three-dimensional transportation network. Water Transport. Eng. **11**, 88–93 (2023). (in Chinese)
3. Hao, L.: Research on economic development of waterway transportation under low carbon economy. China Ship. Weekly **40**, 41–43 (2023). (in Chinese)
4. Natter, A.: Senate Committee expected to focus on transportation, water infrastructure. Environ. Report. **43**(44), 2835–2836 (2012)
5. Grzesiak, M.: Application of transportation model for rationalization of water supply in the region. Res. Papers Wrocaw Univ. Econ. **238**, 11 (2011)
6. Baatz, E.: Price trends. Logist. Manag. **51**(3), 12 (2012)
7. Pramraksa, S., Suthiwartnarueput, K.: Evaluation of transportation alternatives for automotive products. WMS J. Manage. **3**, 1–16 (2016)
8. Batarlien, N., Akalys, R.: Criteria impacting synchronization of transport flows along international transport corridor. Promet - Traffic - Traffico **32**(3), 399–408 (2020)

9. Osowska, A.: Point infrastructure of inland waterway transport of the Lower Vistula in the position of the transport corridor Baltic Sea- Adriatic Sea. AUTOBUSY – Technika Eksploatacja Systemy Transportowe **19**(12), 947–950 (2018)
10. Ghiba, M.G., Olei, B.A.: The technique of using ris corridor data for the logistic integration of inland water transportation. Adv. Eng. Forum **34**, 241–246 (2019)
11. Bray, L., Dager, C., Burton, M.: Willingness to pay for water transportation in the Ohio River Basin. Transport. Res. Record J. Transport. Res. Board **1871**, 5–12 (2004)
12. Fu, X.: Opportunities and challenges of waterway transportation in the new period. Pearl River Water Transport. **11**, 21–23 (2023). (in Chinese)
13. Li, S.: Research on water transportation development strategy of Nanning City in the context of the construction of the Millennium Project Pinglu Canal. Pearl River Water Transport. **10**, 26–28 (2023). (in Chinese)
14. Suwannarongsri, S., Bunnag, T., Klinbun, W.: Traveling transportation problem optimization by adaptive current search method. Int. J. Mod. Educ. Comput. Sci. **6**(5), 33–45 (2014)
15. Narayanamoorthy, S., Saranya, S., Maheswari, S.: A method for solving fuzzy transportation problem (FTP) using fuzzy russell's method. Int. J. Intell. Syst. Appl. **5**(2), 71–75 (2013)
16. Su, S.: Low-carbon policy design of coastal transportation system based on route optimization. Dalian Maritime University (2022)
17. Xu, H.: Research on the evaluation of waterway transportation efficiency in China. Zhejiang Ocean University, Zhoushan (2022)
18. Chakraborty, R., Mandal, U.K., Barman, R.N.: A comparative study of ANN and GEP Model to predict the pressure drop in the water transportation system. Int. J. Inform. Eng. Electron. Bus. **12**(5), 47–57 (2020)
19. Jian, G.: Comprehensive evaluation of energy transportation corridor in China. Int. J. Inform. Technol. Comput. Sci. **3**(4), 36–42 (2011)
20. Su, H.: Study on effect theory of price changes with MathCAD. Int. J. Educ. Manage. Eng. **2**(7), 9–15 (2012)

Prediction Model for Throughput and Induced Traffic Volume of Newly Built Inland River Ports

Hongyu Wu[1,2], Lulu Li[3], and Xunran Yu[1](✉)

[1] Water Transport Office, Transport Planning and Research Institute Ministry of Transport, Beijing 100028, China
13611058357@163.com

[2] Port Resources Monitoring Laboratory, Beijing 100028, China

[3] Department of Transport, Guangxi Zhuang Autonomous Region, Nanning 530012, China

Abstract. To predict the throughput of newly developed ports and scientifically determine the planning capacity of ports, the induced traffic volume is identified as the core of this paper based on existing studies and current engineering applications. The induced traffic volume prediction model of newly developed ports is constructed by considering the constraints of port land and shoreline resources, the type and scale of port-related industries, the input and output situation of port-related industries, the generation of freight volume and the proportion of water transportation sharing, and so on. Port A, developed using Canal B, is selected as a case study. The induced traffic volume of Port A in 2035 is predicted to be 20 million tons, which is consistent with the actual demand for developing port-backed industries; the predictive model is well-adapted to the needs of port-backed industries.

Keywords: Newly developed · Throughput prediction · Industry guidance model · Resource constrained model · Induced traffic volume

1 Introduction

In 2022, China's ports completed a cargo throughput of 15.68 billion tons and a container throughput of 296 million TEUs, and the scale of port throughput ranked first in the world for many consecutive years. With the development of port shoreline resources for many years and the rapid growth of port transportation demand, there has been a gradual increase in new ports or newly developed port areas in many places in recent years. Accurately grasping the types of cargo transportation demand and forecasting the scale of demand of newly developed ports is of great significance to scientific planning of port scale and determining the pace of port development.

Port throughput prediction is essential for the government to carry out port planning and for enterprises to make investment decisions. The related research is relatively abundant, and the engineering application is relatively mature. At present, port throughput forecasting methods at home and abroad are mainly divided into two categories; one is

Z. Hu et al. (Eds.): ICCSEEA 2024, LNDECT 242, pp. 475–484, 2025.
https://doi.org/10.1007/978-3-031-84228-3_41

focusing on modelling research, relying on time series forecasting models, causal forecasting models, modern forecasting models and combination forecasting models [1–5]; the other is focusing on economic and industrial research, focusing on macroeconomic operation and industrial development trend research and analysis [6–8]. Meanwhile, prediction methods and models in other fields also provide new ideas for predicting port cargo throughput. Febrian Wahyu Christanto et al. use Machine Learning (ML) to make prediction, including stock price prediction, optimal playing position prediction and meteorological changes. Pijush Dutta establish the AI-Based Smart Prediction Model to predict the liquid flow rate [9–12]. AI-Based smart prediction of liquid flow system using machine learning approach [13]. All types of studies need to rely on historical statistical data to predict the port throughput level for the coming year.

For newly developed ports or port areas, due to the lack of historical statistical data, the forecast of throughput scale relies more on the future transportation demand feedback from port-related enterprises, which is highly uncertain, lacks science and is unable to scientifically guide the planning of ports and support the investment decision-making of enterprises. This paper focuses on newly developed inland ports, constructs a throughput prediction model for newly developed inland ports and conducts an empirical study to provide a reference for the government and enterprises to make decisions.

2 Newly Developed Port Throughput Prediction Forecasting Ideas

2.1 Research Scope

For newly developed ports, throughput prediction needs to focus on two factors. One is based on the newly developed port industry-driven role, predicting port construction investment, investment in port industry construction and port industry development generated by the "induced traffic volume"; two is based on the completion of the port for the region's existing demand for cargo transportation, regional highways, railroads, pipelines and other modes of transport of goods transferred to the waterway transportation of the "Transfer cargo volume". The throughput of newly developed ports generally consists of the above two components. Considering that the application of the "transfer cargo volume" is more mature in the field of research and engineering, the study in this thesis focuses on predicting the induced incremental transportation volume of newly developed ports.

2.2 Forecasting Ideas

Considering that there is no historical throughput data for the newly developed port, the causal class analysis, time series forecasting method and modern forecasting methods, which are widely used in port throughput forecasting, are not applicable. After the completion of the new port, the scale of port throughput not only depends on the local economic and social operation and industrial development needs but also is closely related to the regional industrial layout, changes in the comprehensive transportation system, the organization mode of water logistics and transportation, as well as the development of neighboring ports. The induced traffic volume prediction of newly developed ports, with the service of port industrial parks or port industrial zones as the core, needs to comprehensively consider many factors, such as the situation of land and shoreline resources, the types and scales of port industries, the transportation needs of raw materials and products of different types of industries, and the flow and direction of cargo.

3 Newly Developed Port Throughput Prediction Model Construction

3.1 Port Resource Constraint Model

Analyzing and predicting the controllable resource scale of a newly developed port is the starting point of throughput prediction. Resource scale is the most essential input data for throughput prediction of newly developed ports. Considering the resource demand characteristics of port development, the initial input data is usually:

$$\begin{cases} Y_0 \\ X_0 \end{cases} \tag{1}$$

Y_0 means the scale of initial planning land resources (km^2), X_0 means the initial planned shoreline size.

Calculate the effective resource size of newly developed ports:

$$Y = \sum_i \beta_i \sum_k \alpha_k Y_{ik} \tag{2}$$

Y_{ik} is the total land resource size in square kilometers consistent with the shoreline attribute k and the land attribute i.

α_k is the resource utility evaluation parameter for the shoreline attribute k, which takes the value of [0–1].

β_i is the resource utility evaluation parameter for the land attribute k, , which takes the value of [0–1].

3.2 Model for Guiding the Port Industry

Determining the direction, type and scale of port-related industries is the most basic part of the throughput prediction of newly developed ports, and is also the basis for determining the scale of investment in fixed assets in port-related industries in newly developed ports. Combined with economic and social, economic zone, industrial park and other types of planning, to determine the industrial orientation program of newly developed ports:

$$P_1, P_2, \cdots P_t \tag{3}$$

t means the type of industry. Considering the different resource demand characteristics of typical industries, the typical industry-resource demand nature parameters are established to clarify the effective resource land scale of each industry Y_t.

$$Y = \sum_t Y_t \tag{4}$$

Clarify the parameters of investment intensity per unit of resources for typical industries I_t, and measure the effective scale of investment in newly developed port-related industries.

$$I = \sum_t Y_t \times I_t \tag{5}$$

3.3 Input-Output Model of Pro-port Industry

Forecasting the gross domestic product (GDP) of newly developed port-served port-related industries by modeling the input-output analysis of leading industries is a key component of the study on the mechanism of generating transportation demand for port-related industries.

I_t define the scale of fixed asset investment in various types of port-related industries P_t,establish input-output parameters of typical industries based on international and domestic experience analysis γ_t, and then find out the industrial value-added formed by the leading industries. Where γ_t is the new value-added of industry formed by every 100 million yuan of new fixed asset investment in P_t industry.

Then there are, corresponding to the newly developed ports, industrial value-added.

$$I_p = \sum_t \gamma_t I_t \tag{6}$$

Gross Domestic Product (GDP) of newly developed ports:

$$GDP = I_p / m \tag{7}$$

where m is the share of industry in the industrial structure of the new port district.

3.4 Port Throughput Generation Models

Analyzing and determining the amount of transportation demand generated in the new port area and reasonably analyzing and forecasting port demand are the last stages of the throughput forecast of the newly developed port. According to the results of input-output measurement, the "freight generation density" analysis parameter δ_t is established to determine the total freight demand generated by the corresponding port industry in the newly developed port area under a specific gross generation value. In this paper, the freight generation density parameter analysis method mainly adopts the regional analogy method, i.e., analyzing the current data of output value and freight volume of similar industrial zones at home and abroad that have been built and analogically analyzing and calculating the δ value.

The transportation mode share model is used to finally determine the total throughput scale of the newly developed port. The transportation mode-sharing model is more mature and commonly used in transportation planning theory and methods. More practical mathematical models include the Logit model, the sharing rate curve method, etc. This paper adopts the Logit model. Then there are:

$$G_n = \exp(X_n) \Big/ \sum_{n=1}^{v} \exp(X_n) \tag{8}$$

where G_n is the sharing rate of transportation mode n,v is the total number of transportation modes, and X_n is the utility function of transportation mode n.

4 Case Studies on Throughput Prediction in Newly Developed Ports

4.1 Basic Information

Port A is a port to be newly developed, with several port areas based on the newly constructed Canal B. In 2021, the Party Central Committee and the State Council issued the "Outline of the National Comprehensive Three-dimensional Transportation Network Plan," including Canal B in the "four longitudinal, four transverse and two networks" national high-level waterway layout. The planned scope of Canal B's economic belt includes five counties in City C, with an area of about 12,440 square kilometres and a population of about 4.64 million people. The planning area of Canal B Economic Zone includes five counties in City C, with an area of about 12,440 square kilometres and a population of about 4.64 million. Currently, the status of the development of Canal B Economic Zone is as follows.

First, the geographical location advantage is unique. B Canal economic zone is located in the western land and sea new channel main channel, is located in the hinterland of the Beibu Gulf city cluster centre, has unique location advantage of the river to the sea and gives full play to the green and low-carbon features of the transport advantages of the water transport, is Guangxi to build a full range of open development of the new pattern of the essential links.

Secondly, the economic strength continues to strengthen. In 2022, the gross regional product along Canal B reached 191.7 billion yuan, with a year-on-year growth rate of 8.2%, ranking first in Guangxi Autonomous region. The seaward economy has been vigorously promoted in recent years, and new economic growth points continue to emerge.

Third, the characteristics of the industry are beginning to take shape. B Canal on both sides of the distribution of several industrial parks and agricultural industrial parks has formed a green chemical industry, equipment manufacturing, new energy, new building materials, modern agriculture, biomedicine and other leading industrial systems along the river, with the canal to build a demonstration zone for the development of exceptional agricultural development and scale of industrial aggregation of the development of a good foundation.

4.2 Resource Constraints

The development of Port A mainly serves the area along Canal B, specifically the Canal B economic zone and the industrial parks adjacent to the port. In terms of land resources, considering the route direction of Canal B and the distribution of Port A's port area, the "scale of planned land resources" to support Port A's throughput forecast includes two aspects: first, the planned land resources of industrial parks immediately adjacent to the port, which is the scope of the traditional concept of port-related industries; second, the planned land resources of industrial parks along the route that are not adjacent to the port but closely connected to the port through the land-based collection and evacuation system, and the raw materials and product transportation of this industrial park. The second is the planning of land resources for the industrial park along the route, which is not adjacent to the port but is closely connected to the port through the land transportation system. The transportation of raw materials and products in this industrial park is also highly dependent on Port A, which is the " pan-port" conceptual scope. Regarding shoreline resources, according to the planning concept of Port A, five operation zones and many port points are proposed, and the length of the planned port shoreline reaches 12 km. Considering that Port A is a new port planning, there is sufficient space for port area arrangement, shoreline planning and supporting industrial park planning, so it is possible to thoroughly select a high-quality shoreline for inclusion in the port master plan and land suitable for industrial land for the planning of the industrial park. The land area of the port is ample, and the utility of the land and shoreline resources is good. The port's land area is abundant, and the resource utility of land and shoreline is good. In summary, it can be obtained:

$$\begin{cases} Y_0 = 333\,km^2 \\ X_0 = 12\,km \end{cases}$$

4.3 Industry Guidance

According to the regional economic and industrial development plan and the industrial plan of Canal Economic Belt B, the leading industries of several port industrial parks relying on Port A in the future will be petrochemical and chemical industry, equipment manufacturing, new building materials, processing of agricultural products (including processing of forest and wood products, processing of cereals, oil and food products, and processing of paper products, etc.), manufacturing of equipment (ship repairing and building, wind power equipment and mechanical equipment, etc.), port logistics, and ship repairing and building. The scale of industrial investment can be measured according to the "Control Indicators for Land Use for Construction of Industrial Projects in Guangxi (Revised in 2021)", "Control Indicators for Investment and Output Intensity of Land Use for Construction of Industrial Projects in Industrial Parks in Nanning City (Revised in 2019)", as well as the parameters of the investment intensity of other cities and typical industrial parks in China.

Since the scale of investment mainly serves the measurement of input and output, the scale of the output value of the relevant industrial parks relying on Port A has already been the planning target. Hence, the practical scale of investment in the port industry is no longer measured in this paper.

4.4 Inputs and Outputs

Considering the characteristics of the development of significant industries, input and output rates, and referring to the relevant indicators of similar functional zones in China and the world, the output value of the industrial parks can be estimated. Combined with the analysis in the "Industrial Guidance" section, the output value of the relevant industrial parks in 2030 can reach 94 billion yuan in total. Combined with the planning and construction pace of each industrial park, and considering the industrialization process of China and Guangxi, it is expected that the output value of the above industrial parks will further increase to about 133.5 billion yuan in 2035. See Table 1 for details.

Table 1. Planning targets for industrial park output value based on Port A

Serial number	Park Name	Projected size of production value in 2035
1	Industrial Park 1	13.0 billion dollars
2	Industrial Park 2	4.0 billion dollars
3	Industrial Park 3	2.0 billion dollars
4	Industrial Park 4	51.0 billion dollars
5	Industrial Park 5	29.0 billion dollars
6	Industrial Park 6	9.0 billion dollars
7	Industrial Park 7	2.0 billion dollars
8	Industrial Park 8	35 billion dollars
9	Industrial Park 9	2.0 billion dollars
Add up the total		145 billion dollars

Combined with the evolution of the relationship between gross industrial output value and gross national product in China and the Guangxi Autonomous Region, the calculation obtained that the regional GDP in 2035 amounted to 120 billion yuan. Figure 1 illustrates the correlation between industrial value-added and GDP in both China and the Guangxi Autonomous Region.

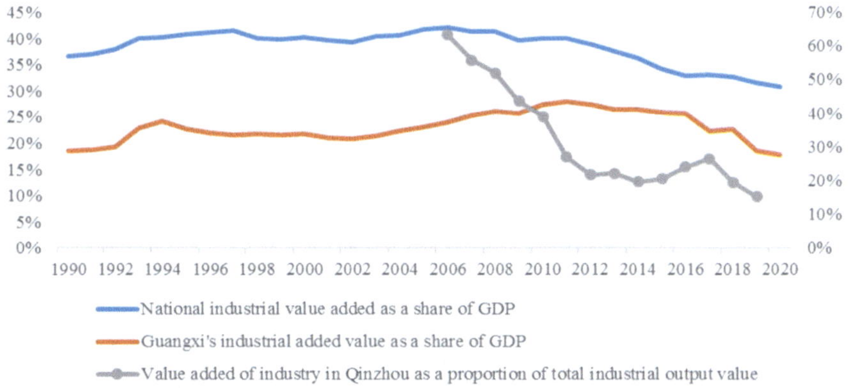

Fig. 1. Relationship between industrial value-added and GDP in the China and Guangxi Autonomous Region

4.5 Carryover Generation

4.5.1 Analysis of Freight Generation Intensity

According to the statistical data, the freight generation intensity of China and Guangxi Autonomous Region in 2021 will be 4.6 tons/million yuan and 8.7 tons/million yuan, respectively. From the development trend, with China's economic restructuring, industrial transformation, and upgrading, freight intensity shows a downward trend yearly. From the stage of development, the developed provinces with large economic scale and industrial structures have freight generation intensity lower than the national average. From the regional differences, the current situation of freight generation intensity in Guangxi is much higher than the national average level and some more developed provinces; one is because the port of Beibu Gulf in Guangxi bears a large number of the hinterland goods of the upper and lower water, the second is relying on the port layout of the many "big into the big out of the" type of port industry, the overall greatly enhanced the freight generation intensity of Guangxi Autonomous Region. This phenomenon is also evident in other coastal regions of China.

Combining the overall evolution trend of freight generation intensity in China and typical provinces, the current situation of freight generation intensity in Guangxi and the analysis above of regional industrial development, it is predicted that the regional freight generation intensity in 2035 will be 3.6 tons per ten thousand yuan, which corresponds to the generation of 44 million tons of regional total social freight volume. Figure 2 demonstrates the changes in freight generation intensity throughout the country and in typical provinces since 2002.

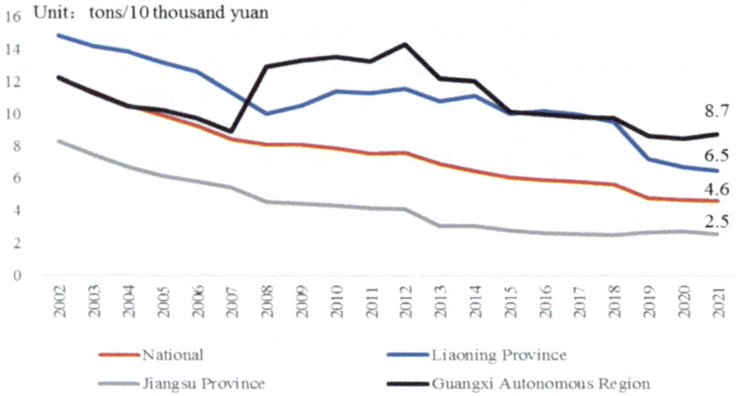

Fig. 2. Changes in freight generation intensity across the country and typical provinces

4.5.2 Measurement of the Share of Waterborne Transport

Many factors affect the proportion of waterborne freight transport, including the situation of waterborne resources, industrial development structure, transportation economy, differences between the place of registration and operation of ships, etc. By comparing the proportion of freight transported by waterways in typical provinces in China, it can be seen that in the areas with developed inland waterway networks, the competitive advantage of water transportation is further enhanced, and the proportion of freight transported by waterway transportation is much higher than the national average.

In the current situation, the scale of water transport volume of goods in Guangxi Autonomous Region continues to grow; the continuous improvement of waterway conditions, port service capacity and service level of the continuous improvement of the inland waterway network in Guangxi is further optimized, the advantages of water transport is gradually played, in recent years, the scale of water transport freight volume is proliferating, accounting for the proportion of the region's freight volume continues to increase. In 2020, the water transport freight volume in Guangxi was 328,520,000 tons, accounting for a proportion of the region's freight volume of 17.5%, accounting 11 percentage points higher than in 2000 and 6 percentage points higher than in 2010.

Combined with the development trend of water transportation in Guangxi concerning the proportion of water transportation in coastal and riverine areas of the whole society, and focusing on the consideration of the layout of the industry along the B canal should mainly rely on water transportation for raw materials transfer in or out, it is predicted that the proportion of regional freight water transportation of Port A in 2035 will be 45%, and it is predicted that the throughput generated by Port A in 2035 in the industrial area will be about 20 million tons. Figure 3 shows the share of national and typical provinces' waterborne cargo volume since 2002.

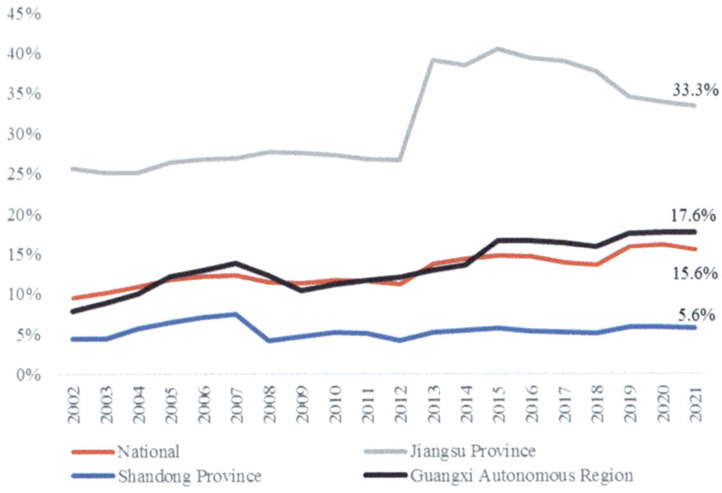

Fig. 3. Share of national and typical provinces' waterborne cargo volume

5 Conclusions of the Study

In this paper, the throughput prediction model and its application to newly developed ports are investigated, and the main conclusions are as follows:

1) A prediction model for the throughput of newly developed ports was constructed by comprehensively considering factors such as land and shoreline resource constraints of ports, the type and scale of port-related industries, the input and output situation of port-related industries, the generation of freight volume and the proportion of water transportation sharing.
2) Port A, to be newly developed, is selected as a case study, and the induced traffic volume of Port A in a typical year is predicted based on the constructed prediction model. It is predicted that the induced traffic volume of Port A in 2035 will be 20 million tons/year, which is consistent with the actual demand of the back-end industrial development of the port and the adaptability of the prediction idea and the prediction model is verified.

References

1. Yager, R.R., Filev, D.P.: Induced ordered weighted averaging operators. IEEE Trans. Syst. Man Cybern. B Cybern. **29**(2), 141–150 (1999)
2. Yager, R.R.: Induced aggregation operators. Fuzzy Sets Syst. **137**(1), 59–69 (2003)
3. Wang, Z., Chang, D., Zhu, Z., et al.: Analysis and prediction of port quarterly container throughput with ES-Markov model. Navigation of China **42**(4), 125–130 (2019). (in Chinese)
4. Huang, Y., Chen, X.: Optimized GM (1,1) model for predicting port throughput. Navigation of China **42**(4), 136–140 (2019). (in Chinese)
5. Yasmine, R., Hilde, M., Christa, S., et al.: A combined approach to forecast container throughput demand: Scenarios for the Hamburg-Le Havre range of ports. Transportation Research Part A: Policy and Practice **117**, 127–141 (2018)
6. Liu, C., Zhong, M., Zhou, Q., et al.: Structured trend prediction of cargo throughput of coastal ports in the Peral River Delta during 14th five year plan. Port & Waterway Eng. **08**, 14–20 (2023). (in Chinese)
7. Wang, R., Ding, W., Feng, Y., et al.: Prediction and development suggestions of port container throughput in Guangdong-Hong Kong-Macao Greater Bay Area. Port & Waterway Engineering **01**, 1–6 (2023). (in Chinese)
8. Liu, C., Gao, T., Chen, Z., et al.: Structural forecast of cargo throughput in China's ports during 14th Five-Year Plan period. Port & Waterway Engineering **12**, 1–6 (2022). (in Chinese)
9. Christanto, F.W., Utomo, V.G., Prathivi, R., et al.: The impact of financial statement integration in machine learning for stock price prediction. International Journal of Information Technology and Computer Science (IJITCS), (16), 35–42 (2024)
10. Salehpour, A.: Predicting automobile stock prices index in the Tehran stock exchange using machine learning models. International Journal of Intelligent Systems and Applications (IJISA) **15**, 12–27 (2023)
11. Utomo, K.S., Wiradinata, T.: Optimal playing position prediction in football matches: a machine learning approach. International Journal of Information Engineering and Electronic Business (IJIEEB), (15), 30–47 (2023)
12. Patil, H.: Cyclone prediction from remote sensing images using hybrid deep learning approach based on alexnet. International Journal of Image, Graphics and Signal Processing (IJIGSP), (16), 96–107 (2024)
13. Dutta, P., Jana, G.G., Paul, S., et al.: AI-Based smart prediction of liquid flow system using machine learning approach. International Journal of Engineering and Manufacturing (IJEM), (14), 53–65 (2024)

Determination of the Parameters of the Device's Piezoelectric Actuator for Collagen-Containing Meat Raw Material Shredding

Olexandr Batrachenko[1], Constantine Bazilo[1](✉), Sergey Filimonov[1], Nadiia Filimonova[1], and Volodymyr Shadkhin[2]

[1] Cherkasy State Technological University, Cherkasy 18006, Ukraine
{o.batrachenko,k.bazilo,s.filimonov,n.filimonova}@chdtu.edu.ua
[2] Department of Information Technology, Media Infrastructure in Schools and Network Projects, Department of Schools Administration, Stuttgart, Germany
Volodymyr.Shadkhin@Stuttgart.de

Abstract. The preservation and support of robust health and immunity for military personnel in extreme conditions during combat are among the priority tasks of the country's military-defense complex. One of the ways to address this task is the production and consumption of high-quality functional food products. However, known technologies for processing collagen-containing meat raw materials involve a prolonged and energy-intensive preparation, which hinders the production of functional meat products in mobile equipment complexes. The use of modern ultrasonic technologies will not only successfully address these issues but also streamline and significantly reduce the cost of manufacturing most types of meat products in industrial conditions.

The rational parameters of the device's piezoelectric actuator for collagen-containing meat raw material fragmentation are substantiated in the work through numerical modelling of its operation using the COMSOL Multiphysics software complex. The maximum amplitude of the piezoelectric actuator's vibrations corresponds to frequencies of 28 kHz, 41.2 kHz, 42.9 kHz, and 45.9 kHz. The most rational frequency is 42.9 kHz, at which more volumetric deformations of the knife occur, distributed practically over the entire blade surface. In this case, the acceleration is 1.22×10^6 m/s^2, and the maximum amplitude of the piezoelectric actuator's knife oscillations is 18.7 μm at a frequency of 41.2 kHz with a force of 15 kN. The adopted design scheme of the piezoelectric actuator is capable of ensuring effective intensification of meat collagen-containing raw material fragmentation.

Keywords: piezoelectric element · ultrasonic system · collagen · fragmentation · actuator · modelling

1 Introduction

High physical and emotional loads on military personnel during the performance of their duties impose specific requirements on their nutrition ration. Preventing poisoning of combatants is a particularly important task.

Z. Hu et al. (Eds.): ICCSEEA 2024, LNDECT 242, pp. 485–493, 2025.
https://doi.org/10.1007/978-3-031-84228-3_42

Functional meat products with collagen content can provide detoxification of the body, prevent cardiovascular and joint diseases, as well as increase the energy level of military personnel and stimulate muscle development. However, existing technologies for processing collagen-containing raw materials have significant drawbacks, such as high energy consumption and low process productivity.

The connective tissue of the raw material, which contains the majority of collagen, has a strength that exceeds the strength of meat muscle tissue by 200 times. Before fragmentation, the raw material must undergo a preparation stage: either boiling for 6–8 h or soaking in special acidic solutions for 18–24 h at low temperatures of 0–4 °C.

According to previous research by the authors, the use of ultrasonic vibrations will intensify the fragmentation process of the raw material and thus significantly reduce energy consumption in the production of functional meat products by excluding the raw material preparation stage. This approach aims to enhance the quality of the products and enable the equipment for collagen-containing raw material fragmentation to be compact and highly productive.

2 Problem Statement

Preserving and supporting the strong health and immunity of military personnel in extreme conditions during combat is one of the priority tasks of the country's military-defense complex. One way to address this task is the production and consumption of high-quality functional food products.

Therefore, it is relevant to create and develop highly efficient multi-component ultrasonic oscillatory systems for the food industry, public catering establishments, and individual use, including a mobile, energy-efficient ultrasonic complex for intensifying the process of meat raw material fragmentation when preparing functional meat products for combatants. These products are aimed at detoxifying their bodies, preventing cardiovascular diseases, joint and bone disorders, as well as increasing their energy levels and stimulating muscle development.

In [1], the exceptional role of a proper specific diet for rapid wound healing is demonstrated. The article in [2] provides an assessment of the impact of specialized oral food supplements on wound healing. The research in [3] investigates the wound healing process, tissue restoration, and regeneration in individuals with diabetes, which is crucial in modern mobilization conditions. The authors of [4] found that specialized functional nutrition has a highly positive effect on the performance and energy levels of Canadian military personnel.

The feasibility and possibility of collagen extraction from untreated waste of a leather manufacturing enterprise were studied in [5]. The authors in [6] demonstrated that ultrasonic irradiation increased the degree of hydrolysis and the speed of protein recovery. Ultrasonication suppressed the formation of hydrogen bonds, reduced collagen molecule interweaving, trans-formed part of the complex structure into a spiral structure, and decreased the thermal stability of collagen molecules. Atomic force microscopy was utilized in [7] to study the impact of ultra-sonic vibrations on the ultrastructure of collagen fibrils in cattle meat tendons.

The study in [8] evaluated the influence of ultrasound power and processing time on the oxidation and protein structure of beef. The results indicate that ultrasound treatment leads to changes in the structure and oxidation of beef proteins caused by mechanical effects of cavitation and the formation of free radicals. In [9], the effect of ultrasound power on the kinetic model of collagen extraction and the physical, chemical and structural characteristics of collagen were investigated. It was found that increasing ultrasound power significantly increased the ex-traction rate and equilibrium concentration of collagen.

However, in these works, as well as in other known scientific publications and intellectual property patents, the question of intensifying the process of collagen-containing raw material fragmentation using ultrasound has not been investigated.

The results of studies on the efficiency of meat raw material fragmentation using an improved cutter knife head model are presented in [10]. The authors in [11] outlined modern trends in the development of cutting tools for meat raw material fragmentation machines. How-ever, these works, like other well-known scientific publications and patents, do not provide descriptions of the designs of ultrasonic cutting tools for meat cutting machines and do not investigate the efficiency of their operation.

The works [12–14] are dedicated to the research, design, and development of highly efficient ultrasonic systems based on piezoceramics. They note that in designing highly efficient ultrasonic systems for intensifying biotechnological processes, piezoelectric transducers, widely used in various fields of science and technology, are employed. However, research on the operation of piezoelectric actuators to ensure complex spatial movement of a cutting tool has not been conducted.

The aim of the study is to justify the rational parameters of the device's piezoelectric actuator for collagen-containing meat raw material fragmentation through numerical modelling of its operation in the COMSOL Multiphysics software complex.

To achieve this aim, the following tasks were addressed: creating a mathematical model of the device's piezoelectric actuator for collagen-containing meat raw material fragmentation using the COMSOL Multiphysics software complex; determining the amplitude-frequency characteristics and vibration forms of this piezoelectric actuator; providing recommendations for the structural design of the device for collagen-containing meat raw material fragmentation.

3 Materials and Methods

Taking into account the technical features of piezoelectric ultrasonic systems and certain complexities arising in their manufacturing, it is advisable to use numerical modelling methods implemented in specialized computer-aided design systems (CAD) [15].

Figure 1 depicts the investigated 3D model of the device's piezoelectric actuator (hereinafter referred to as PA) for collagen-containing meat raw material fragmentation, designed for the production of functional meat products.

Information on the features of constructing mathematical models that take into account multiphysical processes described by a combination of different partial differential equations, and some key computational aspects using the Piezoelectric Effects module and COMSOL's Solid Mechanics and Electrostatics tools, is detailed in [16, 17]. The analysis of the PA's operation was carried out in Frequency response mode. Among other things, the acceleration of the blade edge during cutting was determined.

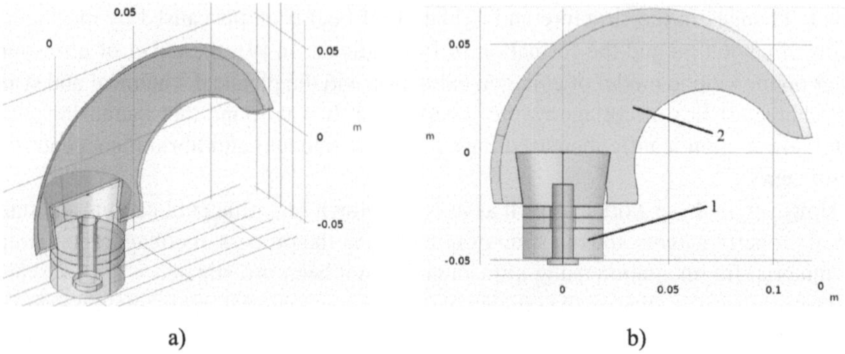

Fig. 1. 3D model of PA: a) general view; b) side view: 1 – ultrasonic emitter; 2 – blade

The Direct solver was used, in which the SPOOLES numerical method was selected to solve systems of linear equations with sparse matrices [18].

The dimensions of the blade in the model correspond to a standard cutter knife for a 6-L cutter: length 123 mm, width 119 mm, thickness 2.4 mm. Dimensions of the ultrasonic actuator in the model: maximum waveguide diameter 55 mm, minimum waveguide diameter 45 mm, waveguide length 24 mm, reflector thickness 12 mm, piezoelectric element's diameter 45 mm, overall length 46 mm. PZT-4 piezoceramic material was used for modelling the piezoelectric motor. The material grade of the waveguide and reflector of the ultrasonic emitter corresponds to AISI 4340 steel, and the blade corresponds to AISI 420 steel.

In modelling the operation of the PA, Fixed boundary conditions were applied to the external end face of the upper piezoelectric element in the piezoelement package, tightened with a bolt. An electric potential of 300 V was applied to the external sides of the piezoelectric elements, and the common "ground" was applied to the sides of the piezoelectric elements that are connected to each other. The polarization of the piezoelectric elements is opposite.

The principle of operation of the PA is as follows. When an alternating electrical voltage is applied to the piezoelectric elements, mechanical vibrations occur, which are transmitted to the blade. As a result, the edge of the blade begins to oscillate, leading to the intensification of the raw material fragmentation process.

A crucial step in obtaining reliable results in numerical modelling is the construction of a finite element model of the PA's structure by introducing a Mesh grid, which significantly influences the computation results. The computational finite element mesh was selected under the following conditions. Lagrangian finite elements with second order elementary basis functions (Lagrange-Quadratic) were used. The number of Lagrangian elements is 6 per local wavelength that propagates in the structure. The mesh was constructed using a tetrahedral division, and the investigated three-dimensional models were represented by a total of more than twenty thousand elements for each component of the structure (see Fig. 2).

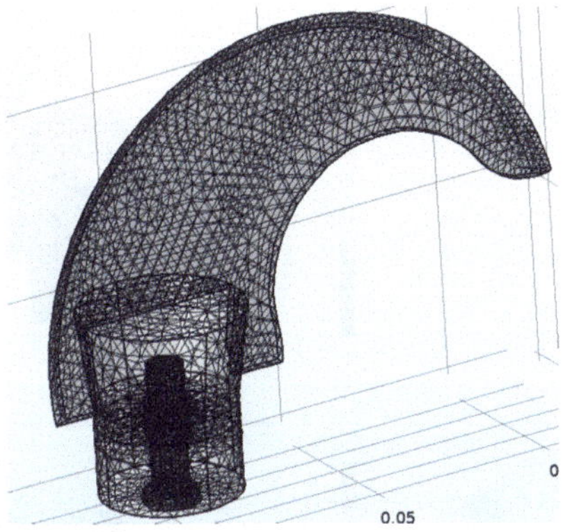

Fig. 2. Finite element model of PA

4 Numerical Experiments and Research Results

Since the efficiency of the PA directly depends on the choice of its operating (resonant) frequency [19], the first step was to determine the maximum amplitude of oscillations at frequencies ranging from 19 kHz to 60 kHz (the used ultrasonic emitter is designed for a frequency of 40 kHz). The simulation results are presented in Fig. 3.

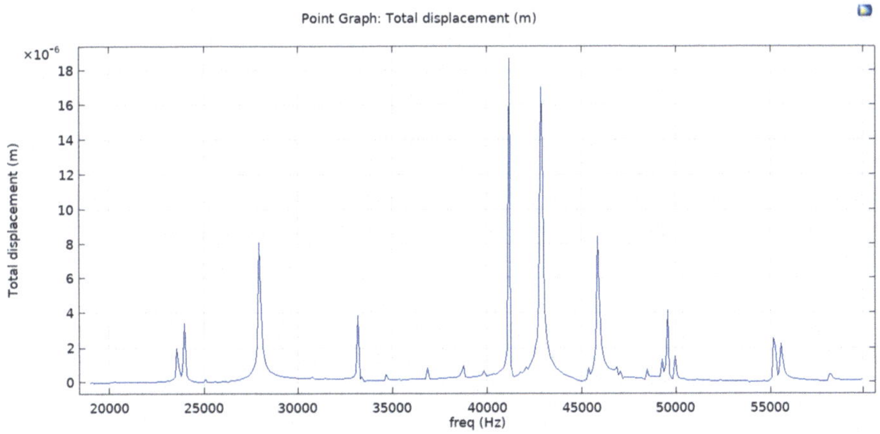

Fig. 3. Amplitude-frequency characteristic of PA

From Fig. 3 it can be seen that the maximum amplitude of oscillations corresponds to frequencies of 28 kHz, 41.2 kHz, 42.9 kHz, and 45.9 kHz. In Fig. 4, based on the results from Fig. 3, a visual representation of the deformation of the PA corresponding to the maximum amplitudes of oscillations is presented.

From Fig. 4 it can be seen that the most rational frequency is 42.9 kHz (Fig. 4, b). This figure illustrates that at this frequency, more extensive deformations of the blade occur, distributed practically over the entire surface of the edge.

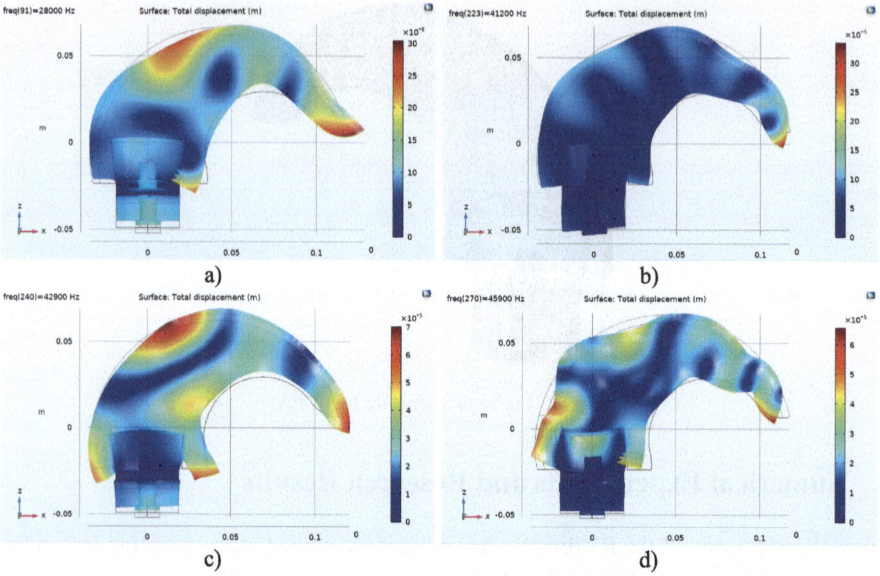

Fig. 4. The results of determining the forms of PA's oscillations in the Frequency response mode at frequencies: a) 28 kHz; b) 41.2 kHz; c) 42.9 kHz; d) 45.9 kHz

The next step was to determine the acceleration of the blade edge during cutting. Using the built-in function in the COMSOL Multiphysics program, solid.accZ, the acceleration of the PA's blade edge can be determined. The calculation results are presented in the graph (Fig. 5). It should be noted that the values of the oscillation amplitude (Fig. 5 and Fig. 6) above zero correspond to stretching, and below zero correspond to compression of the PA's blade.

From Fig. 5 it can be observed that at the rational frequency the acceleration is 1.22×10^6 m/s^2. Similarly, using the built-in function solid.RFtotalz, the force that the PA's blade will exert during its operation can be determined. The obtained results are presented in Fig. 6.

From Fig. 6 it can be seen that the maximum force generated by the PA's blade is 121.5 kN at a frequency of 28 kHz. At this frequency, the amplitude of the oscillations of the cutting edge of the blade corresponds to 8.1 μm. It is worth noting that significant amplitude of vibrations is necessary when cutting with the PA's blade, and this is achieved in the resonance mode of the entire ultrasonic system. Therefore, the maximum amplitude of the PA's knife oscillations is 18.7 μm at a frequency of 41.2 kHz, generating a force of 15 kN.

A direction for further research by the authors could be the experimental determination of the amplitude-frequency characteristics of the PA and its efficiency in shredding collagen-containing meat raw materials.

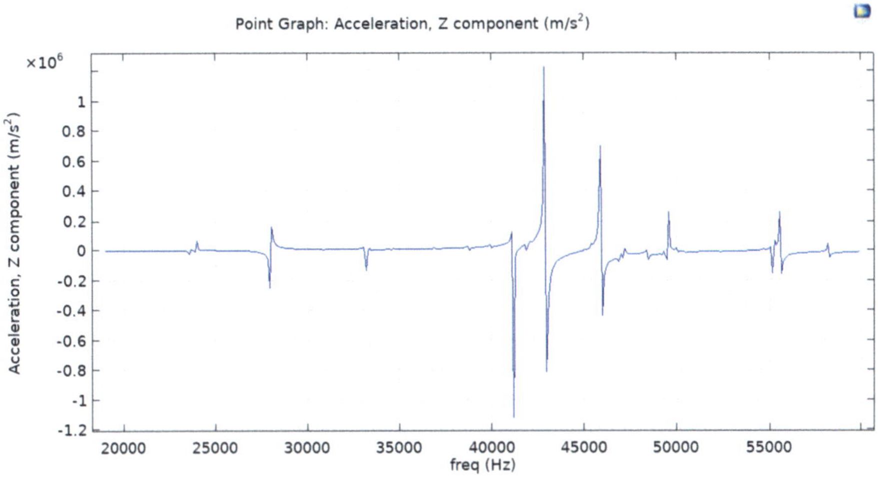

Fig. 5. Amplitude-frequency dependence of PA's blade acceleration

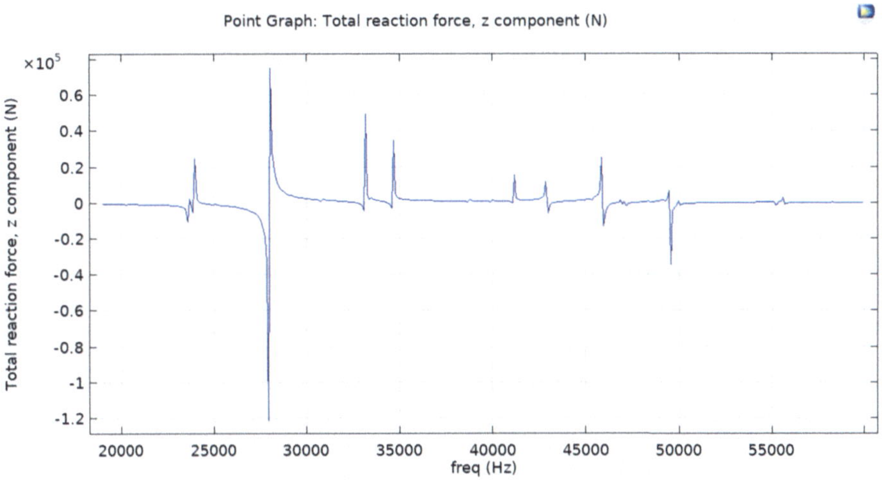

Fig. 6. Amplitude-frequency dependence of PA's blade reaction force

5 Conclusions

The preservation and support of strong health and immunity for military personnel in extreme conditions during combat are among the top priorities of the country's defence complex. One way to address this challenge is the production and consumption of high-quality functional food products.

Existing technologies for processing collagen-containing meat raw materials involve prolonged (up to 2 days) energy- and resource-intensive preparation, making it impractical for manufacturing functional meat products in mobile equipment complexes. The use of modern ultrasonic technologies can successfully address these challenges, simplifying and significantly reducing the cost of producing various types of meat products in industrial settings.

The paper justifies the rational parameters of the device's piezoelectric actuator for collagen-containing meat raw material shredding through numerical modelling of its operation using the COMSOL Multiphysics software.

The scientific novelty of the results lies in establishing dependencies between the structural design of the piezoelectric actuator and its amplitude-frequency characteristics, as well as the forms of oscillations of the piezoelectric actuator's blade. The maximum amplitude of piezoelectric actuator's oscillations corresponds to frequencies of 28 kHz, 41.2 kHz, 42.9 kHz, and 45.9 kHz. The most rational frequency is 42.9 kHz, where more voluminous deformations of the blade occur, distributed practically over the entire surface of the blade. At this frequency, the acceleration is 1.22×10^6 m/s^2, and the maximum amplitude of the piezoelectric actuator's blade oscillations is 18.7 μm at a force of 15 kN.

The practical value of the work lies in the development of the piezoelectric actuator's design to intensify the process of shredding collagen-containing meat raw materials in the production of functional meat products and substantiating the parameters of this design. Moreover, the adopted structural scheme of the piezoelectric actuator is capable of ensuring effective intensification of meat collagen-containing raw material shreddings.

References

1. Grada, A., Phillips, T.J.: Nutrition and cutaneous wound healing. Clin. Dermatol. **40**(2), 103–113 (2022). https://doi.org/10.1016/j.clindermatol.2021.10.002
2. Kerr, K., Clark, R., Stampas, A., et al.: Assessing the impact of specialized oral nutrition supplementation use on wound healing. Arch. Phys. Med. Rehabil. **103**(3), e36 (2022). https://doi.org/10.1016/j.apmr.2022.01.101
3. Madeshiya, A.K., et al.: Nutrition and diabetic wound healing. In: Wound Healing, Tissue Repair, and Regeneration in Diabetes, pp. 403–413 (2020)
4. Ahmed, M., et al.: Dietary intakes from ad libitum consumption of canadian armed forces field rations compared with usual home dietary intakes and military dietary reference intakes. Mil. Med. **188**(1–2), e205–e213 (2023). https://doi.org/10.1093/milmed/usab222
5. Maistrenko, L., Iungin, O., Lastovetska L.: Extraction of collagen from non-tanned wastes of leather production. Herald Khmelnytskyi National Univ. **5**(289), 269–273 (2020) http://journals.khnu.km.ua/vestnik/?p=4650
6. He, L., et al.: Ultrasonication promotes extraction of antioxidant peptides from oxhide gelatin by modifying collagen molecule structure. Ultrason. Sonochem. **78**, 105738 (2021). https://doi.org/10.1016/j.ultsonch.2021.105738
7. Li, G., et al.: Effects of various processing methods on the ultrastructure of tendon collagen fibrils from qinchuan beef cattle observed with atomic force microscopy. J. Food Qual. **2018**, 9090831 (2018). https://doi.org/10.1155/2018/9090831
8. Kang, D.-C., et al.: Effects of power ultrasound on oxidation and structure of beef proteins during curing processing. Ultrason. Sonochem. **33**, 47–53 (2016). https://doi.org/10.1016/j.ultsonch.2016.04.024
9. Zou, Y., Yang, H., Zhang, X., et al.: Effect of ultrasound power on extraction kinetic model, and physicochemical and structural characteristics of collagen from chicken lung. Food Prod Process and Nutr **2**, 3 (2020). https://doi.org/10.1186/s43014-019-0016-1

10. Batrachenko, O.: Simulation of Influence of Constructive Parameters of Meat Bowl Cutterknives on Their Endurance at Alternative Oscillations. In: Information Technology for Education, Science, and Technics. ITEST 2022. Lecture Notes on Data Engineering and Communications Technologies **178**, 363–385 (2023) https://doi.org/10.1007/978-3-031-35467-0_23
11. Haack, E., Schnäckel, W.: Der 3D-Druck eröffnet Möglichkeiten - Gedruckte Werkzeuge für die Fleischbearbeitung bringen technisch-technologische Vorteile. In: Fleischwirtschaft **99**(4), 136–139 (2019)
12. Panda, D., Manickam, S.: Cavitation technology—the future of greener extraction method: a review on the extraction of natural products and process intensification mechanism and perspectives. Appl. Sci. **9**(4), 766 (2019). https://doi.org/10.3390/app9040766
13. Chen-Glasser, M., et al.: Piezoelectric materials for medical applications. In: Piezoelectricity – Organic and Inorganic Materials and Applications, IntechOpen (2018) https://doi.org/10.5772/intechopen.76963
14. Xu, J., Huanhuan, R.: Design and finite element simulation of an ultrasonic transducer of two piezoelectric discs. J. Measurements in Eng. **5**(4), 266–272 (2017) https://doi.org/10.21595/jme.2017.19396
15. Wang, L., et al.: High accuracy Comsol simulation method of bimorph cantilever for piezoelectric vibration energy harvesting. AIP Adv. **9**(9), 095067 (2019). https://doi.org/10.1063/1.5119328
16. Halchenko, V.Y., Bondarenko, Y.Y., Filimonov, S.A., Filimonova, N.V.: Determination of influence of geometric parameters of piezoceramic plate on amplitude characteristics of linear piezomotor. Electrical Engineering & Electromechanics **1**, 17–22 (2019) https://doi.org/10.20998/2074-272X.2019.1.03
17. Halchenko, V.Y., Filimonov, S.A., Batrachenko, A.V., Filimonova N.V.: Increase the Efficiency of the Linear Piezoelectric Motor. J. Nano- Electron. Phys. **10**(4), 04025 (2018) https://doi.org/10.21272/jnep.10(4).04025
18. Behera, B., Nemade, H.B.: Investigating translational motion of a dual friction-drive surface acoustic wave motor through modeling and finite element simulation. SIMULATION **95**(2), 117–125 (2019). https://doi.org/10.1177/0037549718778770
19. Aladwan, I.M., Bazilo, C., Faure, E.: Modelling and development of multisectional disk piezoelectric transducers for critical application systems. Jordan J. Mech. Ind. Eng. **16**(2), 275–282 (2022). https://jjmie.hu.edu.jo/vol16-2/09-48-21.pdf

Hardware Implementation of Two-Level Digital Frequency Synthesizer with Poisson Distribution of Output Pulses

Roman Stakhiv[1](✉) and Joanna Michalowska[2]

[1] Department of Computerized Automatic Systems, Lviv Polytechnic National University, Lviv 79013, Ukraine
roman.i.stakhiv@lpnu.ua

[2] The Institute of Technical Sciences and Aviation, The University College of Applied Science in Chelm, Pocztowa 54, 22-100 Chełm, Poland
jmichalowska@panschelm.edu.pl

Abstract. The paper presents a new structure of a two-level digital frequency synthesizer, which enables the formation of an output pulse sequence with a Poisson distribution law, allowing for the dynamic adjustment of the average output frequency value. Using the proposed methodology, statistical characteristics of the output sequence of the synthesizer are investigated, and ranges of control code values that meet specified requirements are determined. The study demonstrates the possibility and convenience of using the synthesizer to simulate the output signals of dosimetric detectors, facilitating the design and calibration of dosimetric devices without the use of hazardous sources of radiation. In the future, there are plans to develop a frequency synthesizer with a Poisson distribution of output pulses based on other pseudo-random number generators, including shift registers with linear feedback connections, which will simplify the structure of synthesizers while ensuring specified statistical characteristics.

Keywords: Two-level digital frequency synthesizers · Poisson Pulse Sequence · Statistical Characteristics · the Dosimetric Detector

1 Introduction: State of Affairs, Motivation, Relevance of the Work

Two-level digital frequency synthesizers (TLDFS) are widely used in measurement devices, communication tools, and various other technical fields [1–18]. Their main advantages compared to analog frequency synthesizers include the accuracy of the output signal reproducing, resistance to external influences, reliability, and high technological sophistication. Among all TLDFS, a separate group can be distinguished, where the capability of controlling the frequency change step is implemented, thus enhancing their operational characteristics [19, 20].

Poisson pulse sequence generators (PPSG) are utilized to simulate various natural and technical processes, including the emulation of output signals of dosimetric detectors

Z. Hu et al. (Eds.): ICCSEEA 2024, LNDECT 242, pp. 494–506, 2025.
https://doi.org/10.1007/978-3-031-84228-3_43

[9–32]. Typically, they are based on pseudo-random number generators, the enhanced structures of which have been proposed in previous works [31, 32].

In the study [28], we proposed a modified TLDFS (MTLDFS), in which the capability of controlling the step change of the output frequency's mean value and forming the output pulse sequence according to the Poisson distribution law are implemented. However, the presence of a code multiplier significantly complicates its hardware implementation.

Several questions has been arisen, connected to the introduction of the proposed MTLDFS [28], including:

1) the creation of the MTLDFS structure, with Poisson distribution law of output pulses and the capability of controlling the step change of the output frequency's mean value, for its structural implementation;
2) determination of the characteristics of such MTLDFS;
3) utilization of MTLDFS for simulating the output signals of dosimetric detectors (DD).

This work aims to address the aforementioned issues, which underscores its relevance.

The objective of the study is the hardware implementation of MTLDFS, investigation of its characteristics, and adaptation of its parameters for simulating the output signals of dosimetric detectors (DD).

2 Structural Scheme and Operating Principle

The structural scheme of MTLDFS intended for hardware implementation is illustrated in Fig. 1.

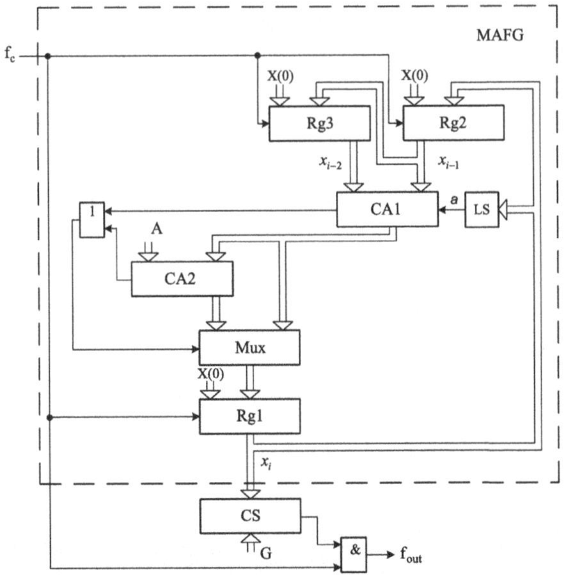

Fig. 1. Structural Scheme of MTLDFS

The synthesizer comprises a Modified Additive Fibonacci Generator (MAFG), a Comparator Scheme (CS), and a logical multiplication. In turn, MAFG consists of registers Rg1-Rg3, combinational adders CA1 and CA2, a multiplexer Mux, a logical scheme LS, and a logical multiplication.

The MAFG operates according to the equation:

$$x_i = (x_{i-2} + x_{i-1} + a) mod(m) \tag{1}$$

where x_i, x_{i-1}, x_{i-2}- numbers in registers Rg1, Rg2, and Rg3 respectively. The number of binary digits of the s structural elements of the scheme (Rg1 - Rg3, CA1, CA2) is selected based on the need to satisfy the condition $2^s > m$. At the same time, the smallest value of s is chosen at which this condition is met. The number A applied to one of the groups of inputs CA2 is determined by the equation $A = 2^s - m$. In the absence of carry signals at the outputs of CA1 and CA2, the number from the output of CA1 passes to the information inputs of Rg1 through multiplexer Mux, and in the presence of any of these signals - the number from the output of CA2. At the output of logical scheme LS, the signal a is formed, the value of which is determined by the logical equation.

$$a = a_0 \oplus a_1 \oplus ...a_{s-1}, \tag{2}$$

where a_i $(i = 0, 1, 2, ..., s - 1)$ - the value of the binary digits of the number in Rg1. The number of terms in Eq. (2) may vary to ensure high speed with satisfactory statistical characteristics of the output signal. The initial number $X(0)$ is recorded in registers Rg1 - Rg3. Clock pulses are applied to the clock inputs of registers Rg1 - Rg3.

The described operating mode of the generator ensures a change in the numbers in registers Rg1 - Rg3 within the range of values $0 \div m - 1$. As a result, and also because clock pulses pass to the output of the synthesizer through the logical multiplication under the condition $x_i < G$, the average value of the output frequency of MTLDFS is determined by the equation:

$$f_{out} \cong \frac{G}{N_s - A} f_c \tag{3}$$

where $N_s = 2^s$, G - is the control code of the synthesizer, f_c - is the frequency of clock pulses.

3 Assessment of the Conformity of the Output Pulse Sequence to the Poisson Distribution Law

Conformity of the pulse sequence with the Poisson distribution law depends on the statistical characteristics of the pseudo-random number sequence x_i at the output of the MAFG.

Below are individual results of the MTLDFS study, constructed according to the structure shown in Fig. 1.

Figure 2 shows the dependencies of the average values of the MTLDFS output frequency f_{out} on the control code value G for three different fixed values of the control code A. Dependencies were obtained for $N_s = 1024$ ($s = 10$) and $X(0) = 111$. The values of $X(0)$, as well as the number of terms in Eq. (2), were selected as a result of simulation modeling aimed at improving the quality of the synthesizer's output signal - matching its output pulse sequence to the Poisson distribution law in as wide a range of control code values G and A as possible.

The output frequency was determined by the number of output pulses per second at the clock frequency value $f_c = 1\,kHz$. Here, solid lines represent theoretical dependencies according to Eq. (3). Dashed lines denote boundaries in accordance with equations.

$$f_{out\_h} = f_{out} + 2 \cdot \sqrt{f_{out}}, \tag{4}$$

$$f_{out\_l} = f_{out} - 2 \cdot \sqrt{f_{out}}, \tag{5}$$

this allows to assess the conformity of the output pulse sequence to the Poisson distribution law [33].

In [28], a methodology for assessing the quality of MTLDFS is provided, whose output signal follows a Poisson distribution law. The methodology utilizes interval estimates and Pearson's chi-squared test of mathematical statistics [33]. It is quite versatile, yet it takes into account the specifics of the construction and operation of TLDFS structures.

The division of the input pulse stream of MDLFS into n equal groups, each consisting of $i_{\max}$ pulses, lies at the core of this methodology. The maximum number of groups is denoted as $n_{\max}$. The groups of input pulses correspond to groups of output pulses with the number of pulses $k_1, k_2, \ldots, k_{n_{\max}}$.

Next, for each control code value G, using the Pearson criterion (criterion χ^2), the following are determined:

- k_e *(x_v)* - the average value of $k_1, k_2, \ldots, k_{n_{\max}}$;
- χ_c^2 *(SS_n)* - the value of the Pearson criterion;
- χ_{cp}^2 *(SS_n_pot)* - the current average value of the last five values of χ^2;
- K_L *(level)* - the number of values of χ_p^2 greater than χ^2.

Figures 3, 4, 5 and 6 present the results of investigating the output signal of MTLDFS (Fig. 1) using the proposed methodology – Fig. 3 for $A = 0$, Fig. 4 for $A = 100$, Fig. 5 for $A = 200$, and Fig. 6 for $A = 300$. Dependencies were obtained at $N_s = 1024$ ($s = 10$), $X(0) = 111$, and clock frequency $f_c = 1kHz$. It was assumed that $n_{\max} = 1000$, $k = 10$ and $\chi^2 = 25$. Figures 3, 4, 5 and 6 display the graphs of k *(x_v)*, χ^2 *(SS_n)*, χ_p^2 *(SS_n_pot)*, and K_L *(level)* values as a function of the control code G.

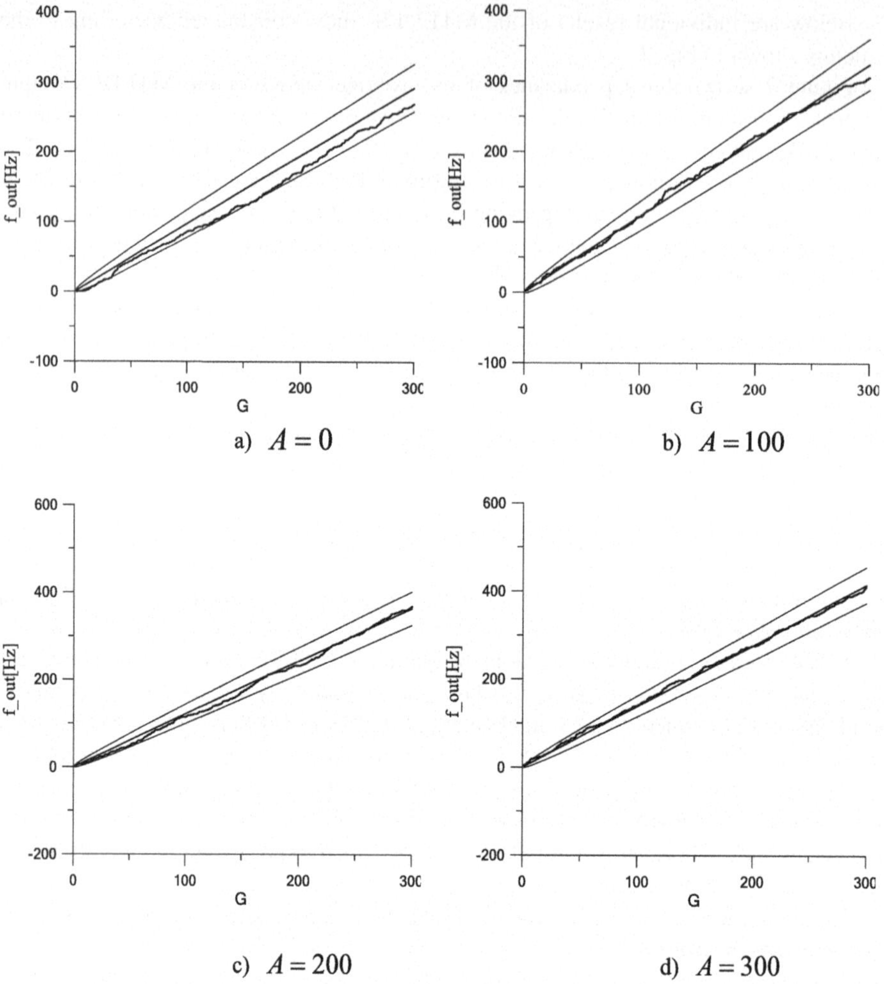

Fig. 2. Dependencies of the output frequency of MTLDFS f_{out} on the control code value G at various values of the control code A.

The results provided allow determining the range of control code values G for which the output signal of the generator corresponds to the Poisson distribution law. For $A = 0$, this range is from $0 \div 110$, for $A = 100$ - from $0 \div 76$, for $A = 200$ - from $0 \div 75$, and for $A = 300$ - from $0 \div 38$.

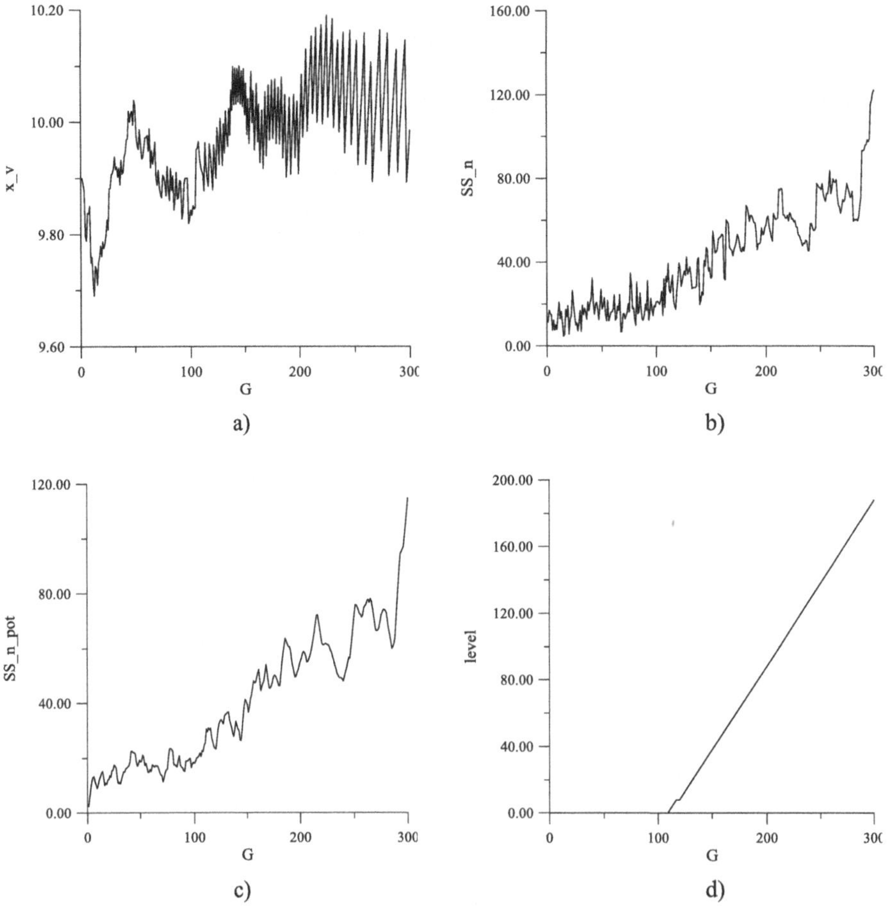

Fig. 3. Statistical Characteristics of MTLDFS at $A = 0$

4 The Use of MTLDFS for Simulating the Output Signals of Dosimetric Detectors (DD)

Let's provide an example of using MTLDFS to simulate the output signals of DD.

The average frequency of the pulse stream f_{out} at the output of DD, without considering its dead time, can be determined by the equation [27].

$$f_{out} = \lambda \cdot \gamma, \tag{6}$$

where λ- represents the exposure dose rate, and γ- represents the sensitivity of the DD.

From (3) and (6) follows

$$\lambda = \frac{G \cdot f_c}{(N_s - A) \cdot \gamma} \tag{7}$$

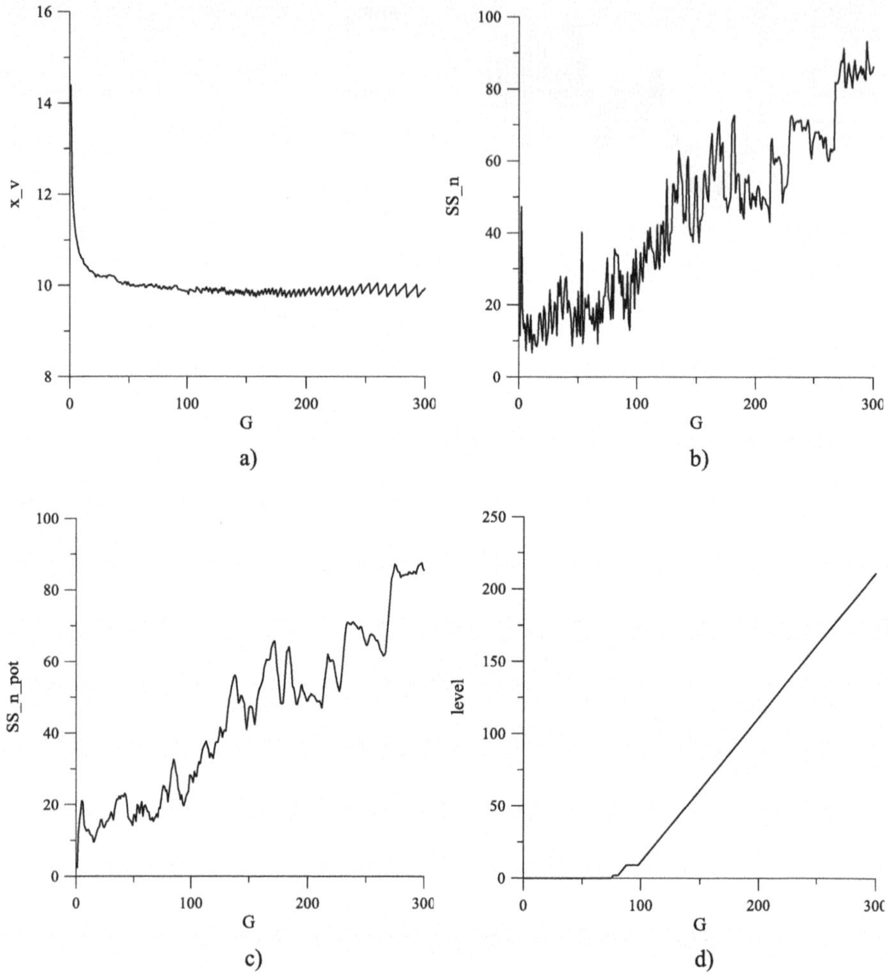

Fig. 4. Statistical Characteristics of MTLDFS at $A = 100$

Let's assume that $\gamma = 0,02\left[\frac{Hz}{\mu R/h}\right]$. Then, when $f_c = 10^6 Hz$, $N_s = 2^{20}$ $(s = 20)$, and $s = 20$ equals

$$\lambda = 50 \cdot G[\mu R/h]. \tag{8}$$

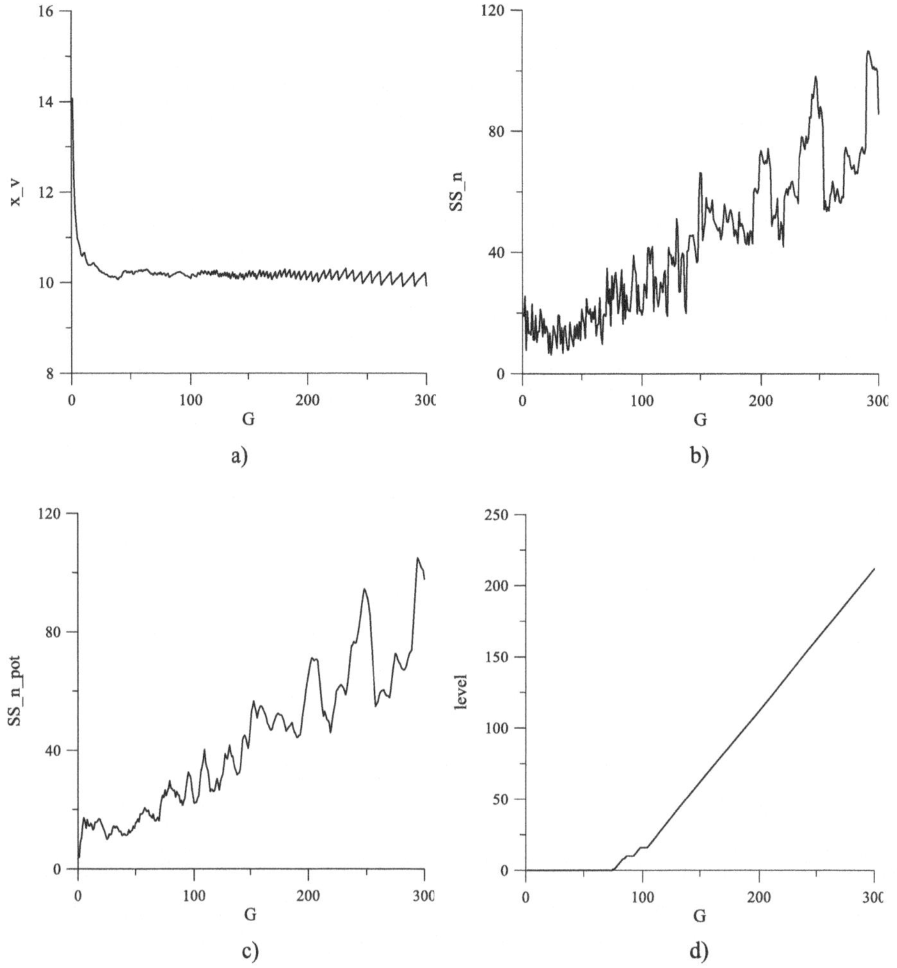

Fig. 5. Statistical Characteristics of MTLDFS at $A = 200$

The dependency (8) allows simulating the exposure dose rate of radiation in a convenient form. In this regard, for any sensitivity value of the DD, by appropriately adjusting the control code value A, the relationship between λ and G can be obtained with an integer proportionality coefficient. This significantly simplifies the calibration and testing of dosimetric devices.

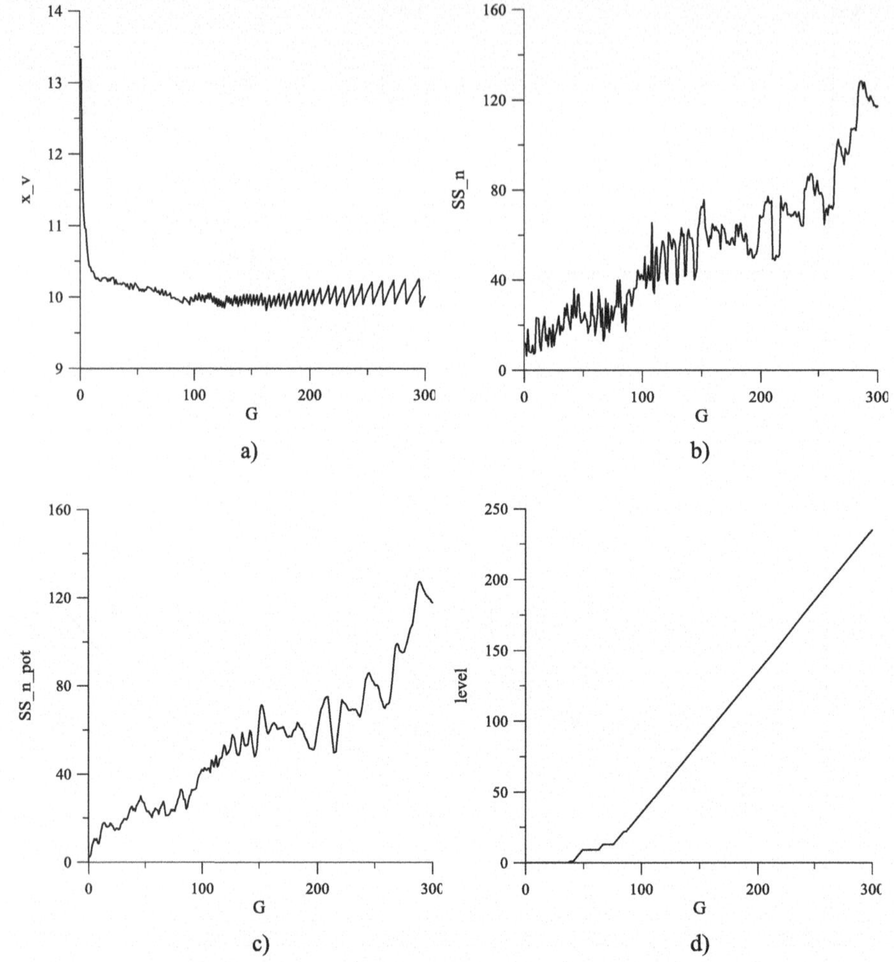

Fig. 6. Statistical Characteristics of MTLDFS at $A = 300$

On Fig. 7, the relationship between the average output frequency of MTLDFS f_{out} and the control code G is depicted for $f_c = 10^6 Hz$, $N_s = 2^{20}$ $(s = 20)$, $X(0) = 111111$ and $A = 48576$. The values of $X(0)$, as well as the number of terms in Eq. (2), were chosen through simulation modeling to ensure better quality of the synthesizer's output signal – adhering to the Poisson distribution law over a wider range of control code values G. The solid straight line represents the theoretical dependency according to Eq. (3). Dotted lines denote the limits according to Eqs. (4) and (5), allowing for the assessment of conformity of the output pulse sequence to the Poisson distribution law.

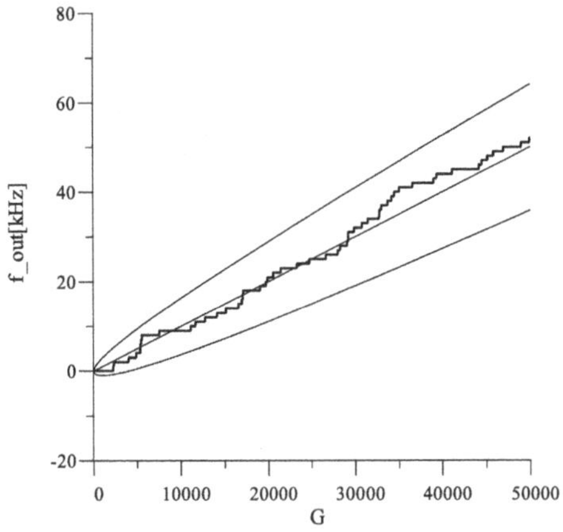

Fig. 7. Dependence of the output frequency of MTLDFS f_{out} on the control code G for $f_c = 10^6 Hz$, $N_s = 2^{20}$ ($s = 20$), $X(0) = 111111$ and $A = 48576$

Figure 8 shows the results of research of the MTLDFS output signal quality based on the proposed methodology for $f_c = 10^6 Hz$, $N_s = 2^{20}$ ($s = 20$),$X(0) = 111111$ and $A = 48576$. Here, we adopted $n_{\max} = 1000$, $k = 10$, $\chi^2 = 25$. The graphs in Fig. 8 represent the values of $k(x\_v)$, χ^2 (SS_n), χ_p^2 (SS_n_pot), and K_L $(level)$ as functions of the control code G.

The presented results allow us to assert that within the range of $G = 1 \div 50000$ values of the control code, the output signal of the generator conforms to the Poisson distribution law. This enables the simulation of the output pulse flow of the DD, corresponding to the range of $50 \div 2,5 \cdot 10^6[\mu R/h]$ values of the exposure dose power with a step size of $\Delta\lambda = 50[\mu R/h]$.

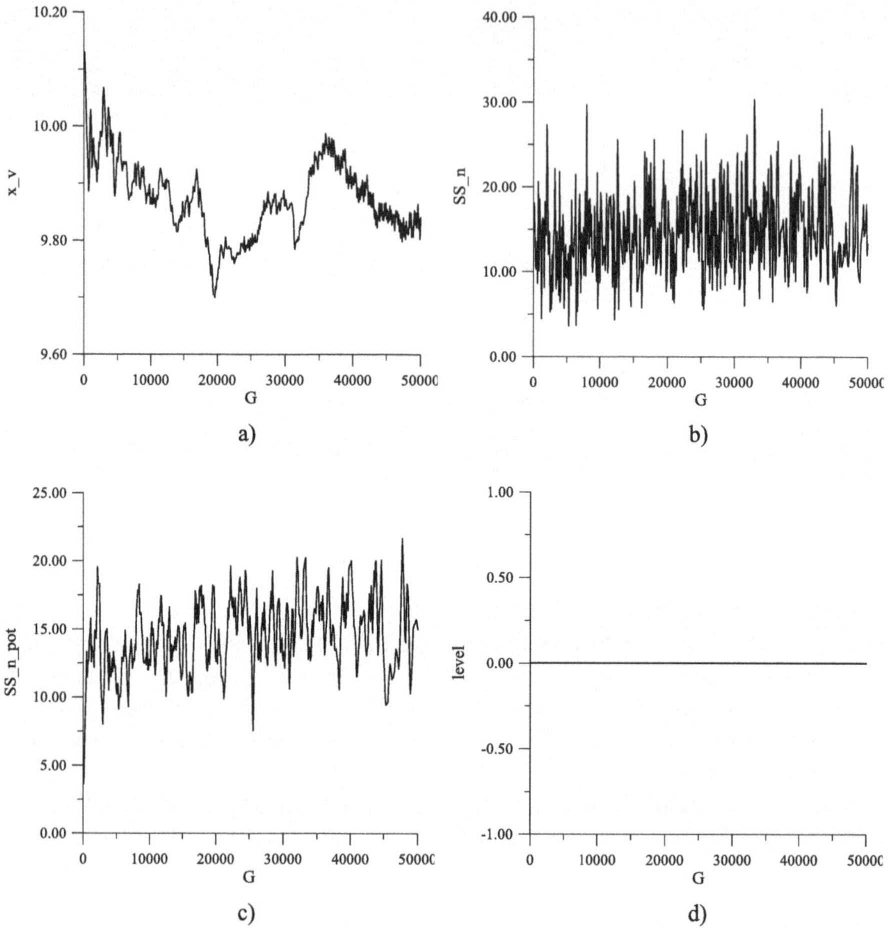

Fig. 8. Statistical Characteristics of MTLDFS at $f_c = 10^6 Hz, N_s = 2^{20}$ $(s = 20), X(0) = 111111$ and $A = 48576$

5 Conclusions and Future Research Directions

The paper introduces a new structure of a two-level digital frequency synthesizer capable of generating an output pulse sequence following a Poisson distribution law while allowing for the rapid adjustment of the mean output frequency value. Unlike the previous development [28], the proposed device eliminates the need for a code multiplication scheme, significantly simplifying its hardware implementation. Thus, the proposed structure is primarily designed for implementation based on programmable logic integrated circuits.

The statistical characteristics of the synthesizer's output sequence have been investigated based on the proposed methodology, and ranges of control code values that meet specified requirements have been identified.

The study demonstrates the practicality and convenience of using the synthesizer for simulating the output signals of dosimetric detectors, enabling the design and testing of dosimetric devices without the use of hazardous radiation sources.

Future plans include the development of frequency synthesizers with Poisson distribution of output pulses based on other pseudorandom number generators, particularly those utilizing shift registers with linear feedback, aiming to simplify synthesizer structures while maintaining desired statistical characteristics.

Acknowledgment. The author expresses gratitude to the team of the Department of Computerized Control Systems at Lviv Polytechnic National University, Ukraine, for their assistance and universal support in preparing this article.

References

1. Kroupa, V.F.: A Digital Frequency Synthesizer. Wiley-IEEE Press (1999)
2. Vankka, J., Halonen, K.A.: Direct Digital Synthesizers: Theory. Springer, Design and Applications (2013)
3. Stakhiv, R., Maksymovych, V.: Using two-level digital frequency synthesizers to simulate the output signals of dosimetric Detectors. Collection of Scientific Papers "Computer Printing Technologies". 27, 176–180 (2011)
4. Phadke, A.G., Pickett, B., Adamiak M., et al.: Ukrainian Academy of Printing. (in Ukrainian). Synchronized sampling and phasor measurements for relay and control. IEEE Trans. on Power Delivery **9**(1), 442–452 (2014)
5. Maksymovych, V., Przystupa, K., Harasymchuk, O., Shabatura, M., Stakhiv, R., Kuts, V.: Hardware modified additive Fibonacci generators using prime numbers. Lecture Notes on Data Engineering and Communications Technologies. – 2023. – Vol. 181: Advances in Computer Science for Engineering and Education VI. Proceedings of the 6th International Conference on Computer Science, Engineering, and Education Applications (ICCSEEA2023), Warsaw, Poland, March17–19, pp. 486–498 (2023)
6. Stakhiv, R.I., Maksymovych, V.M.: Evaluation of the performance of a two-level digital frequency synthesizer on accumulators with code multiplexing. Computer Printing Technologies: Collection of Scientific Papers of the Ukrainian Academy of Printing **11**, 59–67 (2004). (in Ukrainian)
7. Maksymovych, V.M., Stakhiv, R.I.: Pipeline two-level frequency synthesizer. Computer Printing Technologies. Collection of Scientific Papers, 19, 129–138 (2008). (in Ukrainian)
8. Liu, J., Wang, S.: Survey on applications of wide area measurement system in power system analysis. High Voltage Eng. **33**(7), 182–185 (2015)
9. Hu, Z., Dychka, I., Sulema, Y., Radchenko, Y.: Graphical data steganographic protection method based on bits correspondence scheme. International Journal of Intelligent Systems and Applications (IJISA) **9**(8), 34–40 (2017). https://doi.org/10.5815/ijisa.2017.08.04
10. Hu, Z., Khokhlachova, Y., Sydorenko, V., Opirskyy, I.: Method for optimization of information security systems behavior under conditions of influences. International Journal of Intelligent Systems and Applications (IJISA) **9**(12), 46–58 (2017). https://doi.org/10.5815/ijisa.2017.12.05
11. Hu, Z., Gnatyuk, S., Okhrimenko, T., Tynymbayev, S., Iavich, M.: High-speed and secure prng for cryptographic applications. International Journal of Computer Network and Information Security (IJCNIS) **12**(3), 1–10 (2020). https://doi.org/10.5815/ijcnis.2020.03.01
12. Tereikovskyi, O., Wang, D.: Procedure for processing biometric parameters based on wavelet transformations. International Journal of Modern Education and Computer Science (IJMECS) **13**(2), 11–22 (2021). https://doi.org/10.5815/ijmecs.2021.02.02
13. Zhengbing, H., et al.: Data science applications to improve accuracy of thermocouples. In: 2016 IEEE 8th International Conference on Intelligent Systems (IS), pp. 180–188 (2016)
14. Beshley, M., Kryvinska, N., Beshley, H., Kochan, O., Barolli, L.: Measuring end-to-end delay in low energy SDN IoT platform. Computers, Materials & Continua **70**(1), 19–41 (2022)

15. Bubela, T., Kochan, R., Więcław, Ł., Kuts, V., Yatsuk, J.: Disassembly-free metrological control of analog-to-digital converter parameters. Metrology and Measurement Systems. This link is disabled. **29**(4), pp. 669–684 (2022)
16. Xiong, G., Przystupa, K., Teng, Y., et al.: Online measurement error detection for the electronictransformer in a smart grid. Energies **14**(12), 3551 (2021)
17. Hu, Z., Mashtalir, S.V., Tyshchenko, O.K., Stolbovyi, M.I.: Video shots' matching via various length of multidimensional time sequences. International Journal of Intelligent Systems and Applications (IJISA) **9**(11), 10–16 (2017). https://doi.org/10.5815/ijisa.2017.11.02
18. Hu, Z., et al.: Statistical techniques for detecting cyberattacks on computer networks based on an analysis of abnormal traffic behavior. International Journal of Computer Network and Information Security (IJCNIS) **12**(6), 1–13 (2020). https://doi.org/10.5815/ijcnis.2020.06.01
19. Stakhiv, R., Maksymovych, V.: Two-level frequency synthesizer. Bulletin of the Lviv Polytechnic National University: Computer Engineering and Information Technologies **468**, 29–34 (2002). (in Ukrainian)
20. Stakhiv, R.I., Maksymovych, V.M., Stakhiv, M.Y.: Modified structure of two-level digital frequency synthesizer. [Bulletin of Lviv Polytechnic National University "Measurement Technology and Metrology" **80**(1), 17–20 (2019). (in Ukrainian)
21. Adelfio, G., Chiodi, M., D'Alessandro, A., Luzio, D., D'Anna, G., Mangano, G.: Simultaneous seismic wave clustering and registration. Comput. Geosci. **44**, 60–69 (2012)
22. Mandrona, M.N., Maksymovych, V.M., Kostiv, Y.M., Harasymchuk, O.I.: Investigating the statistical characteristics of poisson pulse sequences generators constructed in different ways. J. Automation and Information Sciences **49**(10), 11–19 (2017)
23. Liu, X., Yang, M.C.K.: Simultaneous curve registration and clustering for functional data. Computational Statistics & Data Analysis **53**(4), 1361–1376 (2009)
24. Maksymovych, V.M., Harasymchuk, O.I., Mandrona, M.N.: Designing generators of Poisson pulse sequences based on the additive Fibonacci generators. J. Automation and Information Sciences **49**(12), 1–12 (2017)
25. State Grid Corporation of China. Technical specification for WAMS. State Grid Corporation of China, Beijing (2006). (in Chinese)
26. Maksymovych, V., Harasymchuk, O., Opirskyy, I.: The designing and research of generators of poisson pulse sequences on base of fibonacci modified additive generator. Advances in Intelligent Systems and Computing - Springer Publisher, (ICCSEEA 2018, Kiev, Ukraine 18–20 January, 2018), **754**, pp. 43–53 (2018)
27. Maksymovych, V., Mandrona, M., Harasymchuk, O.: Dosimetric detector hardware simulation model based on modified additive fibonacci generator. In: Advances in Intelligent Systems and Computing , **938**, pp. 162–171 (2020). Springer Verlag. https://doi.org/10.1007/978-3-030-16621-2_15
28. Wang, J., Przystupa, K., Maksymovych, V., Stakhiv, R., Kochan, O.: Computer modelling of two-level digital frequency synthesizer with poisson probability distribution of output pulses. Measurement Science Review, **2**, pp. 65–72 (2020)
29. Maksymovych, V., Harasymchuk, O., Shabatura, M.: Modified generators of poisson pulse sequences based on linear feedback shift registers. Advances in Intelligent Systems and Computing **1247**, 317–326 (2020)
30. Liu, D., Song, D., Wang, H.: Voltage stability online evaluation system based on WAMS and EMS. Power System Technology **07**, 1934–1938 (2014)
31. Maksymovych, V., Harasymchuk, O., Karpinski, M., Shabatura, M., Jancarczyk, D., Kajstura, K.: A new approach to the development of additive fibonacci generators based on prime numbers. Electronics **10**(23), 2912 (2021). https://doi.org/10.3390/electronics10232912
32. Maksymovych, V., Shabatura, M., Harasymchuk, O., Karpinski, M., Jancarczyk, D., Sawicki, P.: Development of additive fibonacci generators with improved characteristics for cybersecurity needs. Applied Sciences (Switzerland), **12**(3), 1519 (2022)
33. DeCoursey, W.: Statistics and Probability for Engineering Applications. Elsevier (2003)

Innovative Process for Granulation of Heterogeneous Liquid Systems in a Fluidized Bed

Bogdan Korniyenko(✉), Yaroslav Kornienko, Serhii Haidai, and Andrii Liubeka

National Technical University of Ukraine «Igor Sikorsky Kyiv Polytechnic Institute», Kyiv 03056, Ukraine
bogdanko@gmx.net, ynk@kpi.ua

Abstract. The results of experimental determination of temperature fields in the vicinity and in bowl of a rotary disperser during granulation of multicomponent heterogeneous liquid systems in a fluidized bed are presented. These results confirm the provisions of the physical model regarding a significant increase in the geometric dimensions of the irrigation zone. Based on the results of experimental studies, it was confirmed that the use of a rotary bowl disperser with a perforated lateral surface, due to the increased spray zone and the presence of intensive circulation of moistened particles, allowed the implementation of a layered-structure granulation mechanism during dehydration of liquid multicomponent heterogeneous systems with a stable process kinetics with a granulation coefficient of more than 90% and an increase in productivity of the granulated product.

Keywords: Granulation · Fluidization · Rotary Disperser · Liquid Heterogeneous System

1 Introduction

Application of the fluidization technique for dehydration and granulation of liquid systems ensures efficient use of thermal energy with a thermal coefficient of more than 50%. It also has a high level of environmental safety and high technological flexibility for obtaining a product with specified physicomechanical and morphological properties. Two mechanisms of granule formation are possible: agglomeration and layer-by-layer. Agglomeration is typically implemented in the dehydration of single-component solutions and used in the food and pharmaceutical industry. The layer-by-layer mechanism is used in the dehydration of multicomponent solutions to obtain a product with a uniform distribution of components throughout the volume of the granule, specified morphological structure and specified physicomechanical properties.

The main factors for the implementation of stable kinetics of the process of dehydration of liquid systems in a fluidized bed include the hydrodynamics of the fluidized bed and the method of introducing the liquid phase. In the existing ways of organizing the process, the supply of heat to the bed of solids, the absorption of vapors and the renewal of interfacial contact between components in the zone of irrigation and intensive heat

Z. Hu et al. (Eds.): ICCSEEA 2024, LNDECT 242, pp. 507–518, 2025.
https://doi.org/10.1007/978-3-031-84228-3_44

and mass exchange is ensured by the supply of a gas fluidizing agent [1, 2]. The injection of the liquid phase is provided by pneumatic nozzles in the form of torches, which has minimal impact on the intensity of mixing of the solids in the bed and the intensification of the renewal of the interfacial contact surface.

The aim of the article is to enhance the efficiency of dehydration and granulation processes for highly concentrated heterogeneous liquid systems in a fluidized bed.

2 Analysis of Scientific Papers

In the "Spray granulation" technology, there are two ways of injection the liquid phase: above the bed (*Glatt*, *GEA* «*Niro*», *Glatt*, *Vector* and *Vanguard*) and directly in the bed (*Glatt* type *ProCell*) [1, 2]. However, when using such methods of injection of the liquid phase an active direct influence on the mixing of the solids is not provided. In this case, the agglomeration mechanism of granulation is mainly realized. To activate the dispersed medium in *Glatt* devices [1, 2] the *Bottom-Spray* («*Wurster*») technique was proposed, which has certain advantages over the previous ones. However, these methods are unacceptable for spraying highly concentrated heterogeneous liquid systems with suspended solid particles.

The authors of [3] proposed to create a directed vertical circulation of granules in a parallelepiped-shaped apparatus by inserting a disk disperser and two vertical plates in the center. The presence of a distributor allows to create a directed circulation of the dispersion movement through the central channel (where the liquid solution is sprayed by the disk disperser) to the side channels with a downward movement of the solids. This method of granulation can also be attributed to the «*Wurster*»- technology, but the advantage is the possibility of injecting heterogeneous liquid systems into the apparatus when using a rotary disk disperser. However, the authors [4, 5] experimentally confirmed the formation of material adhesion zones on the inner surface of the plates at a distance of 10 mm from the edge of the disk disperser.

3 Physical Model of a Layered-Structure Granulation of Liquid Heterogeneous Systems in a Fluidized Bed

In the ideal case, after the irrigation zone heat is supplied to the liquid phase film which is held by adhesive and sorption forces on the surface of the granule, mainly from the granule. As a result, moisture evaporates and micro crystals are formed from the dissolved component on the surface of the granules and simultaneously precipitate macro impurities of insoluble components and colloidal particles of humic substances. The time during which the nucleation and formation of a new microcrystal on the surface of the granule occurs is τ_{min} and is the minimum relaxation time [4]. There are many granules in the irrigation zone, for which it is necessary to provide conditions under which the risk of agglomerate formation is reduced. To realize such a granulation mechanism, the size of the liquid phase droplets and the method of distribution of moistened granules after the irrigation zone among the heated dry granules in the fluidized bed are very important. The main provisions of the physical model and the conditions for preventing the formation of agglomerates are described by the authors [4].

To implement the process was chosen a device with jet-pulsating fluidization in self-oscillatory mode in which three technological zones *I*, *II* and *III* with the same width are formed in the granulator chamber [2]. With this, in zone *I* the solids moves only downward, and has an almost constant porosity of layer-by-layer granulation $\varepsilon = 0.4$. In zone *III*, under the action of the coolant, an upward movement of the solids is cyclically formed, which subsequently moves to zone *I*. The bed porosity in zone *III* changes cyclically from $\varepsilon = 0.5$ to 0.8 and then again to $\varepsilon = 0.5$. In zone *II* with intermediate movement the porosity change is similar with a range of porosity change of $0.512 \leq \varepsilon \leq 0.76$. One cycle of pulsations for a bed of solids with a size of $D_e = 2.5$ mm lasts 0.4 s (frequency 2.5 Hz). In the middle of zone *II*, at a height of $H_{1d} = 0.68\ H_0$ a rotary disperser is located. The regularities of changes in the average porosity of bed in time under heterogeneous fluidization are given by the author [4].

To increase the size of the irrigation zone, the rotary bowl disperser has the shape of a cut cone with a perforated side surface. The main purpose of using a rotary bowl disperser is to spray a heterogeneous liquid phase into the middle of a fluidized bed, create an intense directed circulation of solids through the internal volume of the disperser bowl, rationally distribute moistened and dry granules, and provide a layer-by-layer granulation mechanism, Fig. 1.

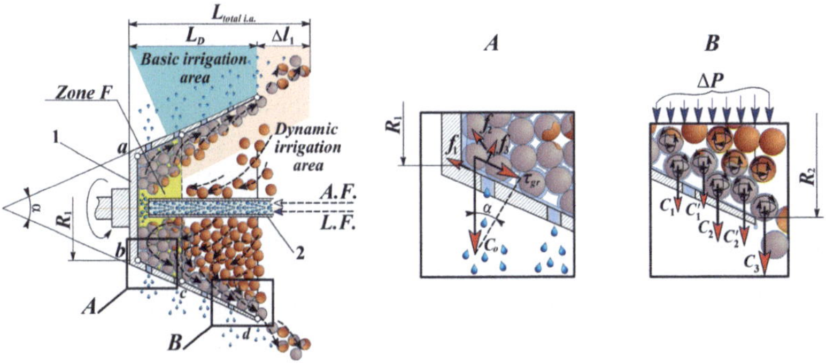

A.F., *L.F.* – aerated liquid phase supply; 1 – rotary disperser; 2 – liquid phase supply pipe; 3 – dry granules with bed temperature; 4 – moistened granules
A) scheme of the forces acting on the granules;
B) scheme of the interaction of granules in the disperser bowl

Fig. 1. Physical model of the rotary disperser operation

The liquid fed through the feed tube 2, Fig. 1, under the action of centrifugal force moves to the inner side surface of the bowl rotary disperser and forms a film of liquid that moves along the inner perforated bowl of the rotary disperser. At the same time, a part of the liquid is dispersed outward in the form of droplets, creating a basic irrigation zone L_D. The rest of the liquid falls on the surface of the granules, which, under the action of centrifugal force, are pressed against the inner surface of the bowl and begin to move along the inner conical surface from the smaller radius R_1 of the bowl to the larger edge of the dispersant with a radius R_2.

At the same time, a driving force acts on the bed of moistened solids by analogy with a filter centrifuge, which removes an excessive amount of the liquid phase that may be on the surface of the granules $\Delta P_{(1,2)}$:

$$\Delta P_{(1,2)} = \frac{\rho_{liquid}\omega^2}{2}(R^2_{(1,2)} - d^2_{gr}) \quad (1)$$

It follows from Eq. (1) that if $R_2 = 2\ R_1$ then $\Delta P_2 = 2\ \Delta P_1$. Consequently, the driving force for removing liquid from the surface of the moistened solids increases at the edge of the bowl. The accelerated movement of wet granules in the disperser bowl is achieved by the presence of a pushing force τ_{gr}, which depends on the value of the centrifugal force C_o for the minimum bowl radius R_1, Fig. 1 *A*. Thus, at point *b*, located at the junction of the lateral inner surface of dispersant and the bottom, there is a granule with a diameter of $\boldsymbol{d_{gr}}$ ($d_{gr} = 2.5{\cdot}10^{-3}$ m), which is subject to the centrifugal force C_o (2), which forms a pushing force $\boldsymbol{\tau_{gr}}$ (3):

$$C = m_{gr}\omega^2(R_1 - \frac{d_{gr}}{2}) \quad (2)$$

$$\tau_{gr} =_o \sin(\frac{\alpha}{2}) \quad (3)$$

where ω is the angular velocity of the dispersant, 1/s; α is the opening angle of the conical dispersant bowl (taken α = 50° – sin(α / 2) = 0.42).

Thus, the proposed dispersant design provides the following dependence: $\tau_{gr(i)} = 0.42\ C_{O(i)}$.

As a result, the intensive removal of moistened solids by a fan-shaped flare from the dispersant bowl to the fluidized bed with a frequency of 2.0 … 2.5 Hz, occurs, which causes the formation of a dynamic zone with a size of Δl_1 and leads to an increase in the size of the basic irrigation zone ΔL_D to the total – $\Delta L_{total\ i.a.}$. A twofold increase in the radius of the larger edge increases the likelihood of ensuring at least two dry granules in contact with moistened ones at the outlet of the dispersant bowl and systematic removal of excess moisture from their surface.

This method of supplying the liquid phase will ensure a high uniformity of the distribution of the liquid heterogeneous phase in the fluidized bed during the dehydration process and create a volumetric circulation of the granular material along the axis of rotation of the disperser.

4 Experimental Setup

Checking the physical efficiency of using a rotary bowl disperser, Fig. 2 *a*, on the kinetics of the granulation process [6–13] of saturated solutions of ammonium sulfate with humic substance impurities was carried out on a pilot plant with the dimensions of the granulator chamber $A \times B \times H = 0.3 \times 0.11 \times 1.5$ m, Fig. 2 *b*.

On the pilot plant, Fig. 2 *b*, was used a bowl rotary disperser, Fig. 2 *a*, made of stainless steel with a polished surface. The average perforation coefficient on the side surface was 7.7%.

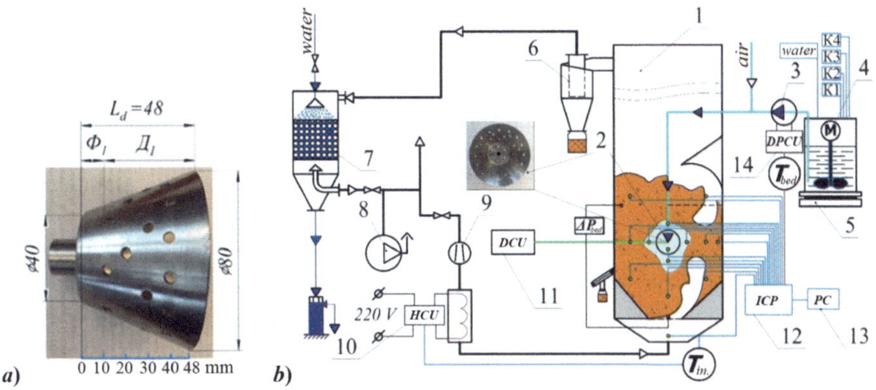

1 – granulator; 2 – rotary disperser; 3 – dosing pump; 4 – a container with a mixer; 5 – weights; 6 – cyclone; 8 – vacuum pump; 9 – chamber diaphragm; 10 – heater with control unit; 11 – unit for adjusting the rotation of the disperser; 12 – thermocouple controller; 13 – computer; 14 – dosing pump control unit

Fig. 2. Scheme of the pilot plant

In previous studies, it was found that at certain values of the centrifugal Froude number at a smaller radius of the dispersant bowl R_1 and the film number at the maximum radius $\mathrm{Re}_{film(max)}$ the movement of a second layer of liquid on top of the hydrodynamic viscous sublayer is observed on the inner surface of the dispersant bowl. This second layer of liquid is little affected by the centrifugal force, which leads to the formation of liquid droplets larger than 300 μm in the outer base irrigation area, Fig. 1.

It is proposed to evaluate the hydrodynamic mode of the liquid phase movement by the factor of film formation H_g under the action of centrifugal force [4]: $H_g = \mathrm{Re}_{film(max)}\ \mathrm{Fr}_{(min)}$, where $\mathrm{Fr}_{(min)}$ is the Froude number at a smaller dispersant bowl radius R_1.

When the linear velocity of the dispersant's larger edge $V_{l(\max)}$ varies from 12.6 to 22.6 m/s and the mass flow rate $G_1 = 12.6$ kg/hour (the volume density of irrigation of the bowl surface $\Gamma_V = 3.71 \times 10^{-4}$ m^3/(m^2 s)) a single-layer mode of fluid movement was recorded in the range of parameter values $H_g = (2.24 \ldots 7.25) \times 10^4$ [4].

5 Materials and Methods

The starting centers of granulation were ammonium sulfate solids containing 2% humic substances with an initial diameter $D_e = 1.82$ mm. The height of the fixed bed in the granulator chamber was $H_0 = 0.32$ m. The liquid phase containing 38% ammonium sulfate, 2% humic substances and 60% water was fed into the fluidized bed using a bowl rotary disperser.

The disperser rotary speed was maintained within 4200 rpm (70 Hz), $V_{l(max)} = 17.6$ m/s. The process was carried out under heterogeneous jet-pulsation fluidization [4].

In the apparatus maintained a constant mass of the bed of solids, which was controlled by unloading a portion of the granulated product from the granulator chamber.

6 Results and Discussion

The stability of the process kinetics was determined by the dynamics of the coolant temperature change at the inlet to the apparatus and in the bed. In the transient process A at $0 \le \tau \le 0.51$ h., the inlet temperature increased within $T_{ent.} = 150 \ldots 180$ °C. In the steady-state process B at $0.51 \le \tau \le 2.38$ h. the inlet temperature was maintained at $T_{ent.} = 180 \pm 5$ °C and the temperature in the bed of solids was $T_{bed} = 94 \ldots 96$ °C.

The average value of the linear growth rate of the equivalent diameter in the bed in the steady-state process (zone ***B***), $\Lambda = dD/d\tau = 0.44$ mm/h, Fig. 3 and Fig. 4.

The dynamics of changes in the mass percentages of individual fractions, shown in Fig. 4, is characterized by a sequential transition of particles from the 1.0 mm fraction to the 2.0 mm fraction, and from the 2.0 mm fraction to the 3.0 mm fraction, which confirms the layer-by-layer mechanism of solid composite formation.

The determination of rational technological parameters for the granulation process with a layered structure is confirmed by the high values of the granulation coefficient in the steady-state process B ($\psi \ge 93\%$).

Based on the results of the analysis of experimental studies, it was found that in the jet-pulsating mode of fluidization with driving force values $\Delta T = T_{\text{ВХ}} - T_{\text{Ш=}} = 95 \pm 3$ °C the granulation coefficient is $\psi = 92 \div 98\%$ at values of $H_g = (2.95 \ldots 3.2) \times 10^4$.

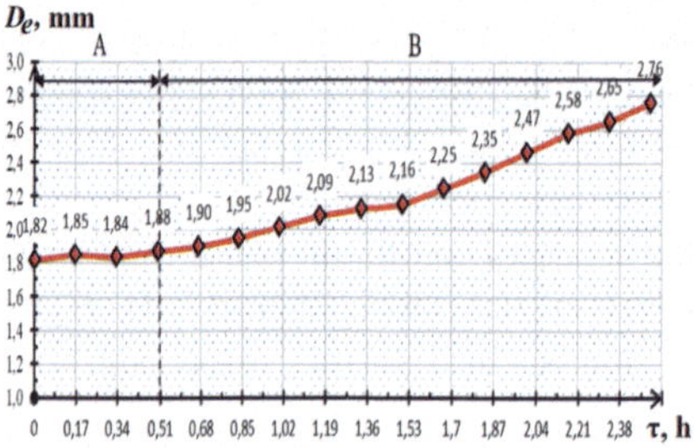

Fig. 3. Dynamics of changes in the equivalent diameter of granules

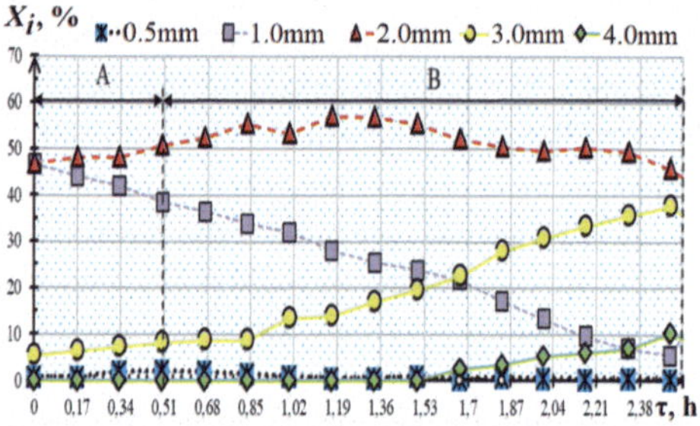

Fig. 4. Dynamics of changes in mass percentages of individual fractions

The temperature fields in different zones of the granulator chamber were measured according to the scheme of thermocouple tracks, Fig. 5 *a*, which was superimposed on the working hydrodynamics of jet-pulsating fluidization, Fig. 5 *b*, simultaneously with the determination of the kinetic characteristics of the granulation process [2]. A more detailed scheme for measuring temperatures in the vicinity of the dispersant is shown in Fig. 6 *a*.

The temperature field inside the bowl of the rotary disperser is determined by the readings of the thermocouple of the track *T*5 which is located along the vertical axis of the device 20 mm below the supply pipe. This zone will be primarily filled with granular material from the fluidized bed, Fig. 7. Thus, the minimum temperature value in this track at a distance of 20 mm from the bottom of the rotary disperser is 57 °C, which is almost equal to the temperature of the wet thermometer, and at the outlet of the disperser – 85 °C. That is, the temperature gradient along the length of the disperser is $dT_s/dY = 0.675$ deg/mm, which indicates the presence of intensive movement of dry

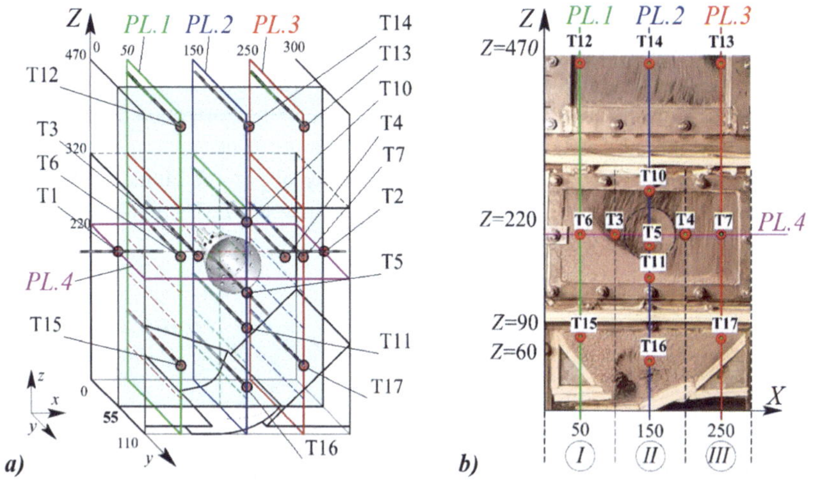

a) scheme of the 3D arrangement of thermocouple tracks in the granulator chamber;
b) overlay of thermocouple tracks on the granulator working chamber;
Conditional zones of granule movement: *I* – downward, *II* – intermediate; *III* – upward

Fig. 5. Scheme of the location of thermocouple tracks in the granulator chamber and of conditional planes *PL*.1; *PL*.2; *PL*.3; *PL*.4

granules with a temperature of 95 °C in the lower sector of the disperser bowl. This is achieved due to the effective distribution of wet granules among dry granules with a coordination number greater than 4.

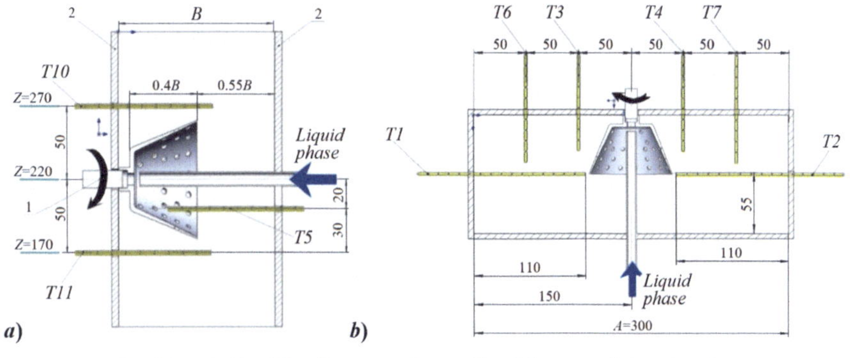

1– drive shaft of the disperser; 2 – walls of the granulator chamber

Fig. 6. Scheme of thermocouple tracks arrangement: *a*) *T5*, *T10*, *T11* in the vertical plane of *PL*.2 ($X = 150$ mm) in the vicinity of the dispersant; *b*) *T1*, *T2*, *T3*, *T4*, *T6*, *T7* in the horizontal plane of *PL*.4 ($Z = 220$ mm)

A further increase in the temperature of the dispersion outside the dispersant in the jet (which moves from the bowl of the rotary disperser along the track of the thermocouple *T5* with a temperature from 85 °C to 95 °C) occurs at a distance of $\Delta Y_5 = \Delta l_1 =$

55 mm. In this case, the mixture of wet and dry granules interacts with the gas coolant with a temperature gradient $dT_s/dY = 0.182$ deg/mm, which is 3.7 times less than the temperature gradient in the dispersant bowl. This confirms the statement of the physical model about the presence of a dynamic irrigation zone in the horizontal direction, which leads to a 2-fold increase in the size of the total irrigation zone.

Temperature measurements in *PL.2* in the middle of zone *II* along tracks *T10* and *T11*, located along the vertical axis of the apparatus in zone *II* 10 mm above and below the larger edge of the dispersant, are shown in Fig. 8.

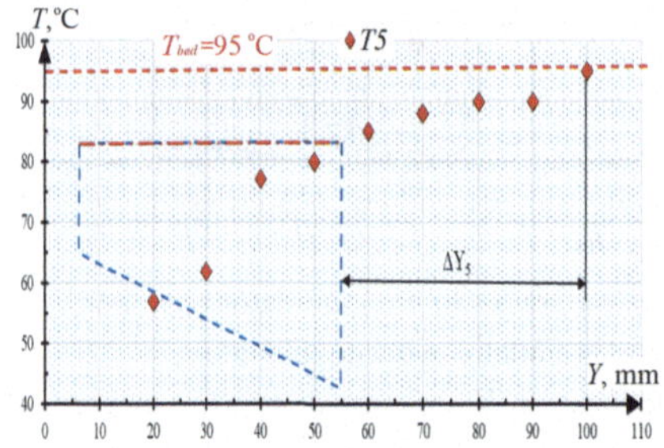

PL.2, T5, X = 150 mm,
30 ≤ *Y* ≤ 110 mm, *Z* = 200 mm,

Fig. 7. Temperature in track *T5*, *PL.2*

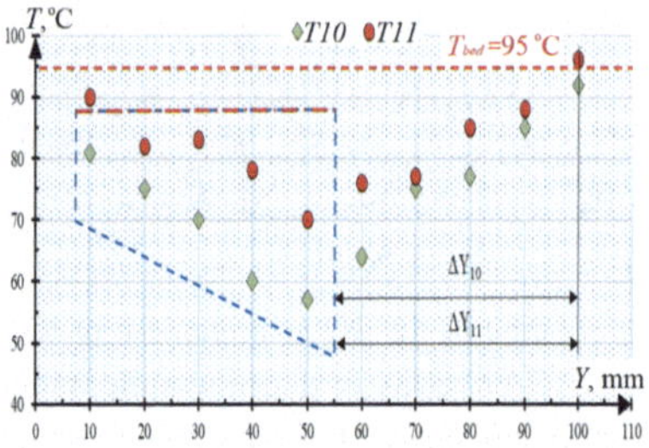

PL.2, T10, X = 150, 0 ≤ *Y* ≤ 110 mm, *Z* = 170 mm
PL.2, T11, X = 150, 0 ≤ *Y* ≤ 110 mm, *Z* = 270 mm

Fig. 8. Temperature in tracks *T10* and *T11*

Considering that the distance of the thermocouple tracks *T10*, *T11* from the disperser bowl along the length decreases from 30 to 10 mm when changing $10 \leq Y \leq 50$ mm, therefore, for the *T10* track (located below the disperser bowl) a temperature decrease is observed from 82 °C to 57 °C with a negative gradient $dT10 / dY = -0.52$ deg/mm. And for the *T11* track (located above the cup disperser), with a change of $50 \leq Y \leq 100$ mm the temperature changes from 90 °C to 70 °C with a negative gradient $dT11 / dY = -0.42$ deg/mm. Outside the dispersant at $50 \leq Y \leq 100$ mm in both tracks, the temperature increases from 57 °C to 92 °C (track *T10* with a gradient $dT10 / dY = +0.70$ deg/mm) and, accordingly, for track *T11* from *T11* 70 °C to 96 °C with a gradient $dT11 / dY = +0.52$ deg/mm. That is, the dynamic irrigation zone in these tracks is also $\Delta l_1 = \Delta Y_{10} \approx \Delta Y_{11} \approx 50$ mm and confirms a 2-fold increase in the total irrigation zone. By analogy with the previous case, in this dynamic irrigation zone, a conditional moistened granule comes into active contact with a conditional set of dry granules (heated to a bed temperature of 95 °C) in the presence of a gas coolant. As a result, there is a dynamic redistribution of moisture between the granules and their heating, and the risk of liquid bridges is completely eliminated, which provides a layer-by-layer granulation mechanism.

To determine the size of the dispersant action in the horizontal planes, were used the thermocouple tracks shown in Fig. 6 *b*. The tracks of thermocouples *T3* and *T4*, located on the border of hydrodynamic zones *I-II* and *II-III*, Fig. 9, recorded a decrease in temperature within the basic irrigation zone for *T3* from 85 °C to 72 °C, and for *T4* – from 95 °C to 60 °C. Outside the basic irrigation zone at $55 \leq Y \leq 110$ mm an intense increase in temperature to 90 … 96°C was observed at a distance of $\Delta Y_3 =$

45 mm, and $\Delta Y_4 = 55$ mm. This also confirms the existence of a dynamic irrigation zone $\Delta l_1 \approx 45 \ldots 55$ mm in the horizontal plane of *PL4*.

Tracks *T6* and *T7*, Fig. 6 *b*, are located in the same plane as the previous tracks *T3* and *T4* inside zone *I* and zone *III*, respectively. The temperature inside zone *I* (track *T6*), Fig. 10, during the baseline and dynamic irrigation zones at $0 \leq Y \leq 110$ mm is within 86 ± 2 °C, the porosity is ε = 0.4 … 0.45 (directed downward movement of the granular material).

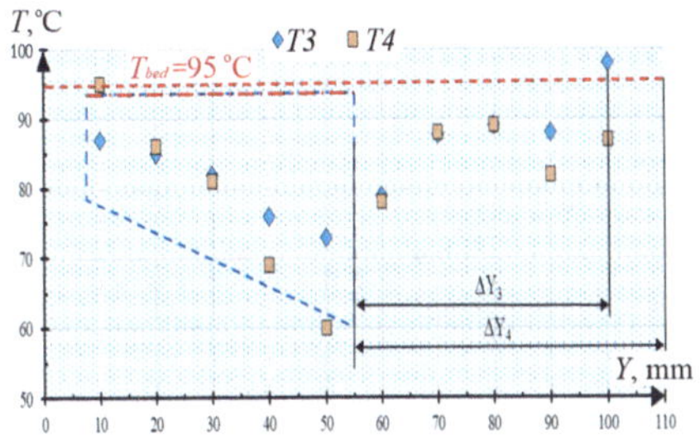

Fig. 9. Temperature in tracks *T3* (X = 100 mm, $0 \leq Y \leq 110$ mm, $Z = 220$ mm) and *T4* ($X = 200$, $0 \leq Y \leq 110$ mm, $Z = 220$ mm)

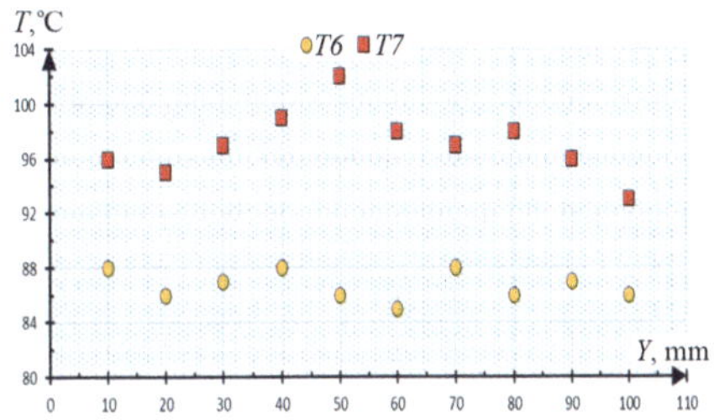

Fig. 10. Temperatures in tracks *T6* (X = 50 mm, $0 \leq Y \leq 110$ mm, $Z = 220$ mm) and *T7* ($X = 250$ mm, $0 \leq Y \leq 110$ mm, Z = 220 mm)

The temperature in the middle of zone *III*, track *T7* is 98 ± 2 °C. At the same time, at $50 \leq Y \leq 110$ mm the temperature in track *T7* decreased from 97 to 92 °C, and in track T6 it slightly increased by 2 °C from 85 to 87 °C.

Measurements of the temperature field in the vicinity of the bowl disperser determined the existence of a dynamic irrigation zone in the form of a truncated cone with a height.

$\Delta l_1 \approx 45 - 55$ mm with a horizontal axis of rotation and a diameter close to 200 mm. The nature of the temperature change in track *T5* (located 20 mm below the axis of rotation inside the dispersant) indicates the presence of a powerful movement of granules from dispersant with a linear size of about 80 mm. That is, the volume of the total irrigation zone is about 29%, in half of which there is an intense fan-shaped circulation movement of moistened granules, which provides a layer-by-layer granulation mechanism and eliminates the risk of agglomeration.

The liquid phase fed for granulation (according to the technical specifications from industry) was a mixture containing nutrients of mineral and organic origin to create granular humic-organic-mineral fertilizers. To test the efficiency of the bowl rotary disperser, a heterogeneous multicomponent liquid phase with a concentration of solids 1.5 times higher than that of the existing granulation process (40%) was fed to the granulation. It consisted of 27% ammonium sulfate, 29% bone meal, 2% humates, 2% potassium chloride and 40% water. Preliminary experiments [5] have established that to ensure a mechanism with a layered structure with a granulation coefficient $\psi > 65\%$ the suspended bone meal particles should not exceed 30 μm.

Ammonium sulfate granules with humate admixtures $D_e = 1.8$ mm were used as the initial granulation centers.

To spray the liquid phase, a rotary bowl disperser was used, placed inside the fluidized bed in the same way as when determining the temperature field. The height of the fixed bed $H_0 = 320$ mm. During the experiments, the disperser operation parameters were kept constant.

The dynamics of changes in the equivalent diameter of the granules shown in Fig. 11., shows the stable kinetics of granulation in experiments №1 and №2. The average growth rate of the equivalent diameter in experiment №1 is $dD_e/d\tau = 0.44$ mm/h, and in experiment №2 is $dD_e/d\tau = 0.47$ mm/h, that confirms the stable kinetics of granulation.

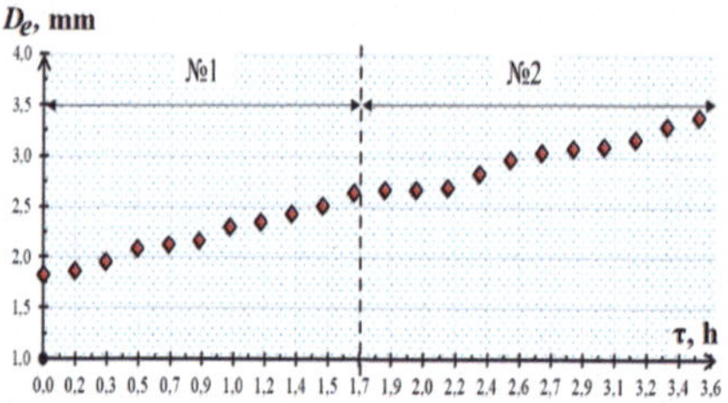

Fig. 11. Dynamics of changes in the equivalent diameter of granules

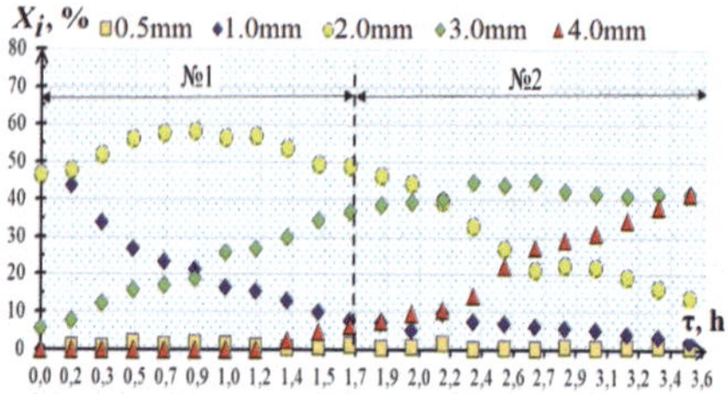

Fig. 12. Dynamics of changes in mass percentages of individual fractions

The dynamics of changes in the mass percentages of individual granule fractions during dehydration and granulation of highly concentrated liquid systems, Fig. 12, confirms the intensive layer-by-layer mechanism of granulation by sequential transition of granules as their size increases due to the layer-by-layer mechanism of granulation.

Comparison of the specific productivity of the granular product when using disk and bowl rotary dispersers, Fig. 13, shows an increase in productivity from 3 to 4 times. This is explained by a 1.5-fold increase in the concentration of solids in the liquid phase, a multiple increase in the irrigation zone when using a bowl rotary disperser and a jet-pulsating mode of fluidization.

The dynamics of changes in the calculated factor of film formation H_g in experiments with a steady-state granulation process is shown in Fig. 14. Generalization of the experimental results made it possible to determine the value of the parameter $H_g = (3.5 \ldots 4.2) \times 10^4$, at which the granulation coefficient $\psi = (82 \ldots 97)$ % is achieved.

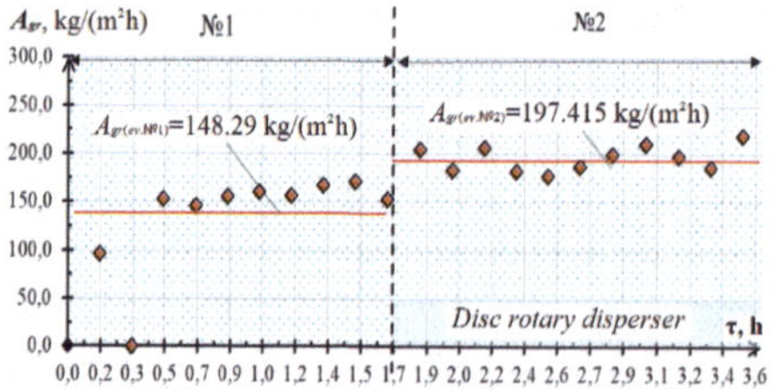

Fig. 13. Dynamics of changes in specific productivity for a granular product in comparison with the use of a rotary disk disperser

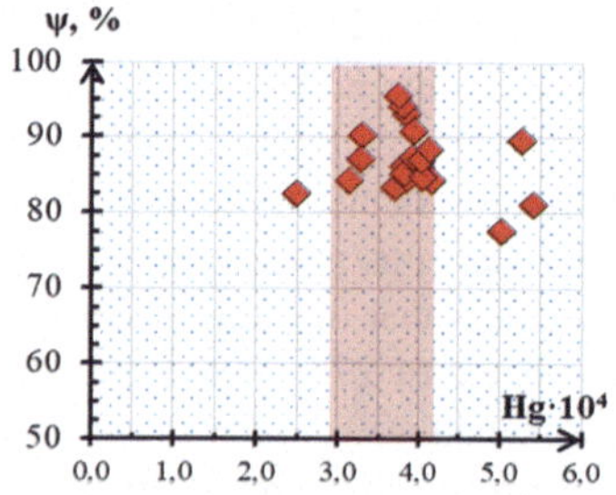

Fig. 14. Dependence of the granulation coefficient on the film formation factor H_g

The general view of humic-organic-mineral (Humic-Phosphorus-Calcium-Nitrogen-Potassium-Sulfur-containing fertilizers) with the composition H:P:Ca:N:K:S =

1:10:19:11:2:12.5 is shown in Fig. 15. The commercial fraction of + 4.0 mm has a spherical shape with a complete absence of agglomerates. The photographs of the granule cross-section clearly show a new structure of the initial center of granulation - granular ammonium sulfate 1, Fig. 16.

Fig. 15. Photo of humic-organic-mineral fertilizers with a stimulating effect: H:P:Ca:N:K:S = 1:10:19:11:2:12.5

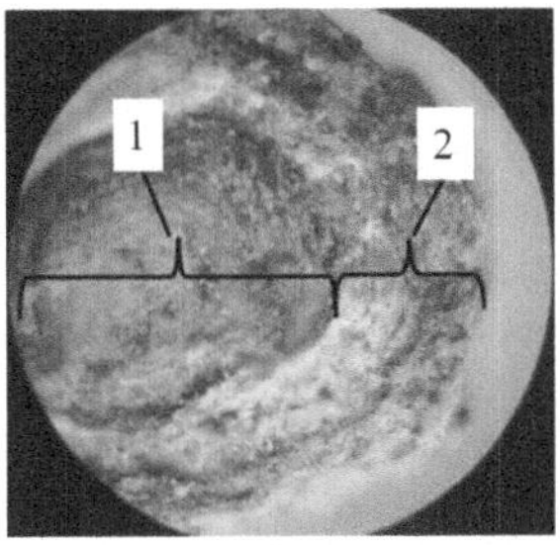

1 – initial granulation center, 2 – layer structure

Fig. 16. Cross-section of a granule with *De* = 4.0 mm

The photo, Fig. 16, shows the newly formed layer structure with a new chemical composition, which confirms the creation of an innovative technology that allowed us to produce granular humic-organic-mineral fertilizers containing phosphorus and calcium of organic origin (bone meal) and ammonium sulfate of mineral origin for the first time.

7 Summary and Conclusion

The experimentally determined temperature fields in the vicinity and in the bowl of the rotary disperser confirmed the physical model, regarding the presence of intensive circulation of granules through the internal space of the disperser bowl, which leads to a powerful removal of moistened granules into the surrounding space of the fluidized bed. As a result, the geometric size of the irrigation zone has significantly increased, which is at least 10 times larger than the irrigation zone of the disk rotary disperser.

Thus, the use of a rotary bowl disperser with a perforated side surface due to the increased spray area and the presence of intensive circulation of moistened particles made it possible to implement a layered-structure granulation mechanism for dehydrating liquid multicomponent heterogeneous systems, with stable process kinetics with a granulation coefficient $\psi \geq 90\%$ with an increase in granulation productivity by at least 3 times compared to the use of disk rotary disperser.

References

1. Kornienko, Y.M., Haidai, S.S., Sachok, R.V., Liubeka, A.M., Korniyenko, B.Y.: Increasing of the heat and mass transfer processes efficiency with the application of non-uniform fluidization. ARPN J. Eng. Appl. Sci. **15**(7), 890–900 (2020)
2. Korniyenko, B., Kornienko, Y., Haidai, S., Liubeka, A., Huliienko, S.: Conditions of non-uniform fluidization in an auto-oscillating mode. Advances in Computer Science for Engineering and Manufacturing. ISEM 2021. Lecture Notes in Networks and Systems 463, 14–27 (2022). https://doi.org/10.1007/978-3-031-03877-8_2

3. Shevchenko, Y.M., Kornienko, Y.M., Haidai, S.S., Denisenko, V.R.: Gas distributing device of the apparatus of fluidized bed. Patent UA 136196 U Ukraine, IPC B01J 8/44, 18 February 2019, Bulletin № 15 (Ukr.)
4. Korniyenko, B., Kornienko, Y., Haidai, S., Liubeka, A.: The heat exchange in the process of granulation with non-uniform fluidization. Advances in Computer Science for Engineering and Manufacturing. ISEM 2021. Lecture Notes in Networks and Systems **463**, 28–37 (2022). https://doi.org/10.1007/978-3-031-03877-8_3
5. Kornienko, Y.M., Liubeka, A.M., Sachok, R.V., Korniyenko, B.Y.: Modeling of heat exchangement in fluidized bed with mechanical liquid distribution. ARPN J. Eng. Applied Sciences **14**(12), 2203–2210 (2019)
6. Kornienko, Y.N., Podmogilnyi, N.V., Silvestrov, A.N., Khotyachuk, R.F. Current control of product granulometric composition in apparatus with fluidized layer. J. Automation and Information Sciences **31**(12), 97–106 (1999)
7. Korniyenko, B., Ladieva, L. Mathematical modeling dynamics of the process dehydration and granulation in the fluidized bed. Advances in Intelligent Systems and Computing, 1247 AISC, pp. 18–30 (2021). https://doi.org/10.1007/978-3-030-55506-1_2
8. Korniyenko, B., Ladieva, L., Galata, L.: Control system for the production of mineral fertilizers in a granulator with a fluidized bed. ATIT 2020 - Proceedings: 2020 2nd IEEE International Conference on Advanced Trends in Information Theory, № 9349344, pp. 307–310 (2020). https://doi.org/10.1109/ATIT50783.2020.9349344
9. Korniyenko, B., Galata, L., Ladieva, L. Research of information protection system of corporate network based on GNS3. 2019 IEEE International Conference on Advanced Trends in Information Theory, ATIT 2019 - Proceedings, № 9030472, pp. 244–248 (2019). https://doi.org/10.1109/ATIT49449.2019.9030472
10. Korniyenko, B., Galata, L., Ladieva, L.: Mathematical model of threats resistance in the critical information resources protection system. CEUR Workshop Proceedings **2577**, 281–291 (2019)
11. Korniyenko, B.Y., Ladieva, L.R., Galata, L.P.: Mathematical model of heat transfer process of production of granulated fertilizers in fluidized bed. ARPN J. Eng. Applied Sciences **16**(20), 2126–2131 (2021)
12. Korniyenko, B., Ladieva, L., Galata, L., Nesteruk, A., Matviichuk-Yudina, O. Information control system for the production of mineral fertilizers in the granulator with a fluidized bed. 2021 IEEE 3rd International Conference on Advanced Trends in Information Theory, ATIT 2021 - Proceedings, pp. 250–255 (2021)
13. Korniyenko, B., Zabolotnyi, V., Galata, L. The optimization of the critical resource protection system of a mineral fertilizers manufacturing facility. Proceedings of the 11th IEEE International Conference on Intelligent Data Acquisition and Advanced Computing Systems: Technology and Applications, IDAACS 2021, **1**, pp. 172–178 (2021)
14. Malekzadeh, M., Khosravi, A., Noei, A.R., Ghaderi, R.: Application of adaptive neural network observer in chaotic systems. International Journal of Intelligent Systems and Applications (IJISA) **6**(2), 37–43 (2014). https://doi.org/10.5815/ijisa.2014.02.05
15. Bhagawati, K., Bhagawati, R., Jini, D.: Intelligence and its application in agriculture: techniques to deal with variations and uncertainties. International Journal of Intelligent Systems and Applications (IJISA) **8**(9), 56–61 (2016). https://doi.org/10.5815/ijisa.2016.09.07
16. Wang, W., Cui, L., Li, Z.: Theoretical design and computational fluid dynamic analysis of projectile intake. International Journal of Intelligent Systems and Applications (IJISA) **3**(5), 56–63 (2011). https://doi.org/10.5815/ijisa.2011.05.08
17. Patnaik, P., Das, D.P., Mishra, S.K.: Adaptive inverse model of nonlinear systems. International Journal of Intelligent Systems and Applications (IJISA) **7**(5), 40–47 (2015). https://doi.org/10.5815/ijisa.2015.05.06

A Two-Level Model of a Discrete Manufacturing System. Management Methods and Software Architecture

Alexander Pavlov[1](✉), Elena Khalus[1], Kateryna Lishchuk[1], Mykhailo Medvediev[1], and Cennuo Hu[2]

[1] National Technical University of Ukraine "Igor Sikorsky Kyiv Polytechnic Institute", Kyiv 03056, Ukraine
pavlov.fiot@gmail.com, lishchuk_kpi@ukr.net

[2] Department of Computer Science, College of Science, Purdue University, West Lafayette, IN 47907, USA
hu945@purdue.edu

Abstract. We consider a discrete manufacturing system with a mathematical representation in the form of a two-level model. The model defines the structure of a technological process in the form of a special two-stage scheduling problem. For the proposed new model of a discrete production system, we solve a management problem consisting of two interconnected subproblems of operational scheduling and operative operational planning. We present an original solution to the formulated problem of a discrete manufacturing system management. We substantiate an efficient software architecture, which includes a model for efficient static management of asynchronous computing processes. The management model is an organic consequence of the proposed method of managing a two-level discrete manufacturing system. We show that the efficiency of the proposed method and of the corresponding algorithms depends significantly on the efficiency of solving several single-stage scheduling problems, which were the object of the authors' previous theoretical and experimental research. Finally, we give an illustrative example that confirms the practical value and efficiency of the proposed models and methods.

Keywords: discrete manufacturing system · optimization · scheduling and operative operational planning · asynchronous parallel computing

1 Introduction

1.1 Related Works Review Substantiating the Theoretical and Practical Relevance of the Results that We Present

The problem of formalization and practical implementation of efficient management of discrete manufacturing systems (DMS) is still relevant today, since the systemic complexity of such objects still does not allow creating a universal mathematical theory, which practical implementation would fully satisfy an administrative management. The application field of such models is extremely large. As an example, only small-scale production takes about 70% of the volume of the world manufacturing [1].

Z. Hu et al. (Eds.): ICCSEEA 2024, LNDECT 242, pp. 519–532, 2025.
https://doi.org/10.1007/978-3-031-84228-3_45

The problems of creating efficient models and methods of DMS management are the subject of research presented, e.g., in [2–16]. The most common goal of DMS management is the profit maximization that in turn is reached only under the condition of such organization of the manufacturing process that minimizes the uncertainty of the final production result. In this case, the most efficient methods of DMS management are interconnected models and methods of scheduling and operative operational planning from the class of multi-stage scheduling problems, and the models define the network representation of technological processes [16]. For such problems, the use of universal directed enumeration algorithms, such as the branch and bound method, dynamic programming, the "Kyiv broom" (the method of sequential analysis of options developed by V. S. Mikhalevich and N. Z. Shor [17]), genetic algorithms, etc., leads to exponential complexity. These methods become implementable only when additional heuristic constraints are used, but this practically leads to the impossibility of obtaining optimal and, in most cases, efficient solutions. The most promising direction is to achieve a compromise between the simplicity of multi-stage scheduling models and their adequacy to real types of DMSs. This would allow creating special optimization methods for such models primarily on the base of the theoretical properties of such partial models and allowing to efficiently use algorithms created for simpler single-stage scheduling problems. In particular, this approach is implemented in the scientific research of Prof. Alexander Pavlov and his scholars [16, 18–21]. This work belongs right to this class of scientific works. We consider a new partial case of a universal two-level model of DMS that can have a wide practical application, and the efficiency of the proposed original algorithms for scheduling and operative operational planning is based on the use of efficient solution methods for four single-stage scheduling problems [20–23].

1.2 The General Methodology that Formed the Basis of the Proposed Methods

The proposed methods are the implementation of the following universal methodology for multi-level scheduling models solving.

The stages sequence of the problem solving is inverse (from the end to the beginning), and each stage is a special single-stage scheduling problem. For each of the problems, we use only theoretically or statistically substantiated efficient solving methods. Thus, the efficiency of the solving method for the complex problem directly follows from the efficiency of the systemically connected set of solutions of intractable, in the general case, but simpler combinatorial optimization problems.

1.3 The Structure of the Paper

In Sect. 2, we present a formal problem statement, namely, a formal model of a two-level system. We also give the efficiency criteria for the operational scheduling and operative operational planning problems. In Sect. 3, we present an original locally optimal iterative method for solving the operational scheduling problem. In Sect. 4, we formulate a heuristic algorithm for operative operational planning. Its practical efficiency is based on implementation of three interconnected operational procedures. In Sect. 5, we present a model and the corresponding method for static management of parallel asynchronous computing processes. Its implementation significantly increases the efficiency of a software that implements the proposed locally optimal method of operational scheduling that simultaneously optimizes two criteria given in lexicographical order. Section 6 provides an illustrative example that confirms the practical significance and efficiency of the proposed models and methods. Section 7 contains the conclusions of the work.

2 A Multi-level DMS Management Problem Statement

We show the structure of a two-level DMS model in Fig. 1.

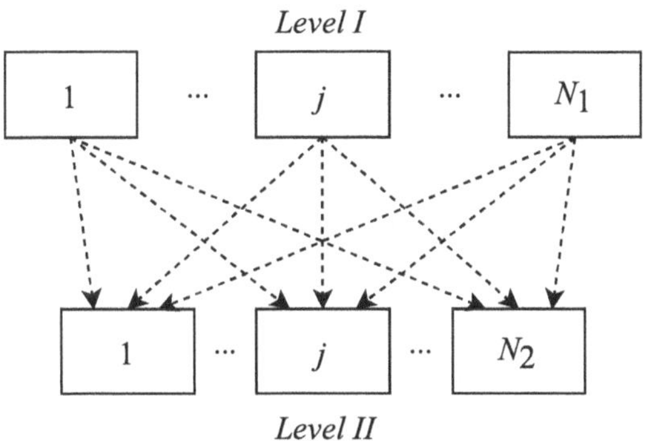

Fig. 1. The structure of a two-level DMS model

The two-level DMS model consists of N_1 subsystems of the first level and N_2 subsystems of the second level. The formal model of each subsystem of the first or second level is a machine that sequentially executes a given set of operations. For each of the operations we know its processing time on a given machine of the first or second level. A set of jobs $J_1, J_2, \ldots, J_n$ is given, each of the jobs consists of two sequential operations. The first operation can be executed on a given subset of subsystems of the first level, the second one on a given subset of subsystems of the second level.

We introduce the following formal notation.

The first level of the two-level DMS. For each j-th subsystem of the first level, where $j = \overline{1, N_1}$, we are given a set of operations

$$\left\{J_{i_1}^{1j}, \ldots, J_{i_{l_j}}^{1j}\right\}_t \subset \left\{J_{p_1}^{1j}, \ldots, J_{p_{L_j^1}}^{1j}\right\} \tag{1}$$

where $\left\{J_{p_1}^{1j}, \ldots, J_{p_{L_j^1}}^{1j}\right\}$ are all possible operations, which the j-th subsystem of the first level can process; $\left\{J_{i_1}^{1j}, \ldots, J_{i_{l_j}}^{1j}\right\}_t$ is the set of operations assigned to the j-th subsystem of the first level at the t-th iteration of the operational scheduling method; for each operation $J_{p_f}^{1j}, f = \overline{1, L_j^1}$, we are given its processing time $l_{p_f}^{1j}$.

The second level of the two-level DMS. For each j-th subsystem of the second level, $j = \overline{1, N_2}$, we are given, in a similar way,

$$\left\{J_{i_1}^{2j}, \ldots, J_{i_{l_j}}^{2j}\right\}_t \subset \left\{J_{p_1}^{2j}, \ldots, J_{p_{L_j^2}}^{2j}\right\}, l_{p_f}^{2j}, f = \overline{1, L_j^2} . \quad (2)$$

For each job J_i, $i = \overline{1, n}$, a deadline d_i for its completion is given.

We need to build such a *feasible* (that does not violate the deadlines) operational schedule of the two-level DMS jobs processing that would correspond to the optimum according to two criteria given in lexicographic order:

1. To achieve the fulfilment of the condition

$$\max\left\{\min_{j=\overline{1,N_1}} r_j\right\} \quad (3)$$

where r_j is the processing start time of the j-th subsystem of the first level provided that the obtained operational schedule of the DMS is feasible.

2. If (3) is met, then fulfill

$$\min_{\forall\sigma} \sum\nolimits_{i=1}^{n} \omega_i(d_i - C_i(\sigma)), C_i(\sigma) \le d_i \quad (4)$$

where $C_i(\sigma)$ is the completion time of job J_i; $\omega_i > 0$ are expert weight coefficients, σ is the operational schedule of all the subsystems of the two-level DMS.

3 Locally Optimal (in Two Versions) Iterative Method of Operational Scheduling

3.1 Locally Optimal Iterative Operational Scheduling Method (the First Version)

The method consists of T iterations, T is the number of all possible distinct variations of assigning subsets of operations (1), (2) to the subsystems of the first and second levels of the DMS.

A locally optimal method of building an operational schedule that satisfies two optimality criteria (3), (4) given in lexicographic order for fixed sets (1), (2) (the method of performing each iteration) is given below.

1. For each j-th subsystem of the second level, $j = \overline{1, N_2}$, we solve the following single-stage scheduling problem: find such a schedule σ_j of operations assigned to be processed at t-th iteration that achieves

$$\min_{\forall\sigma_j} \sum\nolimits_{m=1}^{l_j} \omega_{i_m}^j \left(d_{i_m}^j - C_{i_m}^j(\sigma_j)\right), C_{i_m}^j(\sigma_j) \le d_{i_m}^j \quad (5)$$

where (5) is met on the set of operations $J_{i_m}^{2j} \in \left\{J_{i_1}^{2j}, \ldots, J_{i_{l_j}}^{2j}\right\}_t$; $d_{i_m}^j$ is the deadline; $C_{i_m}^j(\sigma_j)$ is the completion time of the operation $J_{i_m}^{2j}$, $\omega_{i_m}^j > 0$ is the corresponding expert coefficient. (5) is achieved under the condition that the start time of the subsystem of the second level is as large as possible.

Remark 1. The functional of the form (5) is used when fulfilling the condition $\forall C^j_{i_m}(\sigma_j) = d^j_{i_m}$ is required. But in discrete optimization problems, almost never exists a feasible schedule σ_j that satisfies this condition. Then this condition is replaced by the functional (5), and the values of $\forall \omega^j_{i_m} > 0$ for each value of i_m correspond to the degree of importance of fulfilling the condition $C^j_{i_m}(\sigma_j) = d^j_{i_m}$. That is, the larger the value of $\omega^j_{i_m}$, the smaller the difference $d^j_{i_m} - C^j_{i_m}(\sigma_j)$ should be in the schedule that minimizes (5).

To solve this problem, we propose to use the modified branch and bound method [20]. The methodology of its implementation for the problem given in the general form is the following.

$$\min_{\forall \sigma} \sum\nolimits_{i=1}^{n} \omega_i(d_i - C_i(\sigma)),\ C_i(\sigma) \le d_i,\ i = \overline{1, n},\ r = r_{\max}, \tag{6}$$

Step 1. Using the algorithm from [21], find r_{max}, the maximum starting time of the machine at which a feasible schedule exists.
Step 2. According to the algorithm given in [21], find a set of all vertices, each of which can be the last in the feasible schedule.
Step 3. For each such vertex, build in the inverse order a complete tree of feasible schedules.

The paper [20] gives two theoretically substantiated rules for pruning branches of a complete tree of feasible schedules built in the inverse order. The first rule follows from the fact that for two immediately preceding jobs i_1, i_2 in the optimal schedule:

- if $l_{i_2} < l_{i_1}$, $\omega_{i_2} > \omega_{i_1}$, then job i_1 must precede job i_2;
- if $l_{i_2} < l_{i_1}$, $\omega_{i_2} < \omega_{i_1}$, then the processing order of jobs i_1 and i_2 in the optimal schedule depends on the sign of the inequality: $\omega_{i_1} \cdot l_{i_2} \gtrless \omega_{i_2} \cdot l_{i_1}$.

The second pruning rule is based on the well-known result [24]: if the schedule built in deadline increasing order is not feasible, then no feasible schedule exists.

The experimental studies carried out by the authors showed the high efficiency of the exact method of solving problem (6), and therefore, any problem (5). The experimental studies of the efficiency of the exact method are explained by the fact that the start time is the maximum possible. This leads to the fact that the second pruning rule is quite strict and that the set of possible operations that are candidates to be the next branch in the first steps of the inverse building of the schedules tree are significantly smaller than the total number of operations.

Step 4. From all constructed feasible schedules, we choose the one corresponding to the minimum of (6).

2. For each j th subsystem of the second level, the start times of each second operation in the built optimal feasible schedule become deadlines for the first operations of each job J_i, $i = \overline{1, n}$, in the corresponding subsystem of the first level. Next, for each subsystem of the first level, the algorithm [21] finds a feasible schedule of operations in their deadlines increasing order, with the maximum processing start time of each subsystem of the first level.

The local optimality of the proposed method consists in the following:

- The general problem is decomposed into $N_1 + N_2$ single-stage scheduling problems and finding a strictly optimal schedule for each subproblem. The solutions of these problems in the inverse procedure of finding the operational schedule for a two-level DMS unambiguously define the input data (deadlines) of each subsequent single-stage scheduling problem.
- For each subsystem, the maximum start time of the processing is found, and only when this condition is met, then a schedule for subsystems of the second level that satisfies the condition (5) is found by the exact method.

3.2 Locally Optimal Iterative Operational Scheduling Method (the Second Version)

If T (the number of iterations of the method described in Subsect. 3.1) is unacceptably large, then we propose to use the experience of managers who are capable to efficiently adjust the sets (1), (2) for the next iteration by analyzing the schedule obtained at each iteration and to decide as well which iteration is the last.

Corollary. In an obvious way, the method of building an operational schedule can be applied to the case of a multi-level DMS, namely:

1. For each subsection of the last level, independently solve problems (5) for the maximum machine start time $r_{\max}$. The start times of the last operations of each job become the deadlines for the penultimate operations of the jobs for the subsystems of the penultimate level.
2. For each subsystem of the penultimate level, find $r_{\max}$ for the schedule in which the jobs are arranged in deadline increasing order. The start times of the operations become deadlines for the operations of subsystems of the next (in the inverse order) level; repeat steps 1 and 2 until we have a complete operational schedule.

4 The Method of an Operative Operational Planning for a Two-Level DMS

The first step is implemented for each j th subsystem of the second level, $j = \overline{1, N_2}$. Let t be the adjustment time of the operational plan. Exclude all completed operations and the operation that was processing at time t from the operation set of the j th subsystem of the second level. If necessary, adjust the input data (the start time of operations at the first and second levels). For the remaining operations of the j th subsystem of the second level, if the machine start time is equal to $\max(t_1, r_{\max})$, and $r_{\max} > t_1$ where t_1 is the completion time of the operation that was processing by the j th subsystem at time t, then solve the problem of the total weighted earliness of jobs minimization on a single machine. If $r_{\max} < t_1$, then solve for this subsystem at time t_1 the problem of the total weighted tardiness of jobs minimization on this set of operations. Find such operational plans for all $j = \overline{1, N_2}$ subsystems of the second level.

Remark 2. The formal model of the total weighted tardiness minimization problem and the PSC-algorithm for its solution only for the set of operations of an arbitrary subsystem of the first level are given below.

The second step. For each j th subsystem of the first level, $j = \overline{1, N_1}$, exclude all completed operations and the operation that was processing at time t. The release time of the j th subsystem of the first level is the completion time of the operation that was processing at time t. The start times of operations at the second level, according to the operational plan built in the first step of operative operational planning, are the deadlines for the remaining operations for all subsystems of the first level. For each subsystem of the first level, solve the problem of minimizing the total weighted tardiness of operations regarding the deadlines on a single machine. The formal problem statement is as follows:

$$\min_{\forall\sigma} \sum_{m=1}^{\hat{l}_j} \omega_m^j \max\left(0, c_{i_m}^{1j} - d_{i_m}^{1j}(\sigma)\right) \tag{7}$$

where $\omega_m^j > 0$ is the expert weight from expression (5), it refers in expression (7) to the first operation of the corresponding job; $d_{i_m}^j$ is the deadline for the corresponding first operation obtained in the first step of the algorithm; $\left\{i_1, \ldots, i_{\hat{l}_j}\right\}$ is the set of the first operations processed by the j th subsystem of the first level in the operative operational planning algorithm.

Let us rewrite problem (7) in general form:

$$\min_{\forall\sigma} \sum_{j=1}^{m} \omega_j \max\left(0, c_j - d_j(\sigma)\right). \tag{8}$$

To solve problem (8) and therefore all problems of the form (7) we will use the simplified version [19] of the PSC-algorithm [22]: at the first stage, we build the schedule σ^{ord} [22] (the jobs are ordered by non-increasing priorities $p_j = \omega_j / l_j$ where $\omega_j > 0$ is the expert weight, l_j is the processing time of the jth job). At the second stage, we build the schedule σ^{fp} [22] for each machine. The algorithm for building σ^{fp} by σ^{ord} consists in the sequential execution of all possible free permutations [22] in σ^{ord}. A free permutation of a job in the schedule σ^{ord} means its moving to the position with the maximum number where this job does not violate its deadline and at the same time the functional value (8) decreases. In [22], we formulated five sufficient signs of optimality of the obtained solution for the schedule σ^{fp}. We find such operational plans for all subsystems of the first level.

The third step. Adjust the operational plans of all subsystems of the second level built in the first step according to the operational plans of the subsystems of the first level built in the second step (the first level). For the j th subsystem of the second level, $j = \overline{1, N_2}$, sequentially assign the operations ready to be processed, which positions in the operational plan of the j th subsystem of the second level found in the first step are the smallest.

We show schematically the operative operational planning method in Fig. 2.

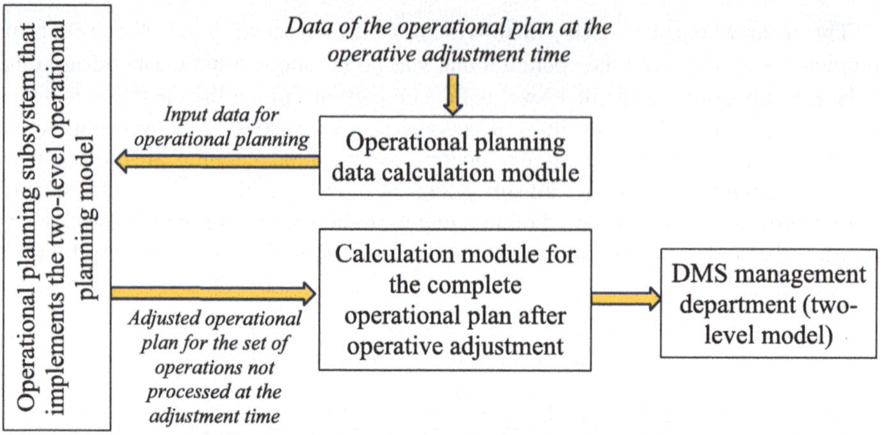

Fig. 2. Scheme of operative operational planning of a DMS (two-level model)

5 Elements of Optimization of the Software that Implements the Proposed Two-Level DMS Management Method

It is obvious that the most convenient software architecture is monolithic, and a significant increase in its efficiency is associated with the use of parallel computing when implementing the method of operational scheduling. Indeed, the logic of the method (its first version) shows that both the execution of each iteration can be carried out independently of each other, and during the execution of each individual iteration of the method (both versions), the trees of the forest can also be built independently of each other. Thus, the efficient organization of parallel computing can be reduced to the following universal static model of managing asynchronous computing processes.

A single meta-program consists of n subprograms that can independently start execution at the same time or at arbitrary fixed times. The number of independent identical processors is $m < n$. We know the average statistical duration l_i of the i th subprogram. In our case, the input information is the average statistical dependence of the number of building operations of a single tree as a function of the number of operations (5).

We need to construct a schedule σ that would achieve

$$\min_{\sigma}\{C_{\max}(\sigma) - C_{\min}(\sigma)\} \tag{9}$$

where $C_{\max}(\sigma)$ is the completion time of all subprograms in the schedule σ and $C_{\min}(\sigma) = \min_{i=\overline{1,n}} C_i(\sigma)$, where $C_i(\sigma)$ is the processing completion time by i th processor.

It is obvious that criterion (9) is the most adequate, in particular, when m processors execute not a single meta-program but their sequence.

To efficiently solve the problem according to criterion (9), the authors suggest using the algorithm proposed in [23].

6 An Illustrative Example to the Execution of a Single Iteration of the Operational Scheduling Method

We give the initial data of the problem in Table 1 and Table 2. Figure 3 shows the corresponding scheme of the two-level DMS.

Table 1. Input data of the machines of the first level

Machine 1-1		Machine 1-2		Machine 1-3	
Operation	Processing time	Operation	Processing time	Operation	Processing time
1-1	5	1-5	14	1-9	23
1-2	9	1-6	3	1-10	11
1-3	4	1-7	8	1-11	8
1-4	12	1-8	34	1-12	3

Table 2. Input data of the machines of the second level

Machine 2-1				Machine 2-2			
Operation	Processing time	Deadline	Weight	Operation	Processing time	Deadline	Weight
2-1	9	96	1	2-7	5	56	1
2-2	12	45	1	2-8	6	79	7
2-3	11	85	8	2-9	3	89	3
2-4	7	90	2	2-10	2	64	2
2-5	14	66	1	2-11	7	84	15
2-6	10	90	4	2-12	8	70	4

The problem solving. At the first stage, we apply the total weighted earliness minimization algorithm for each machine at the second (last) level at the maximum start times of the machines ($r_{\max}$) that guarantee the existence of feasible solutions.

We give below a description of the steps of the algorithm that builds inversely (from the end to the beginning) a tree of feasible schedules.

1. Only operation 2-1 is suitable for the last place in the schedule for Machine 2-1, since only its deadline (96) is greater than or equal to the sum of $r_{\max} = 33$ and the total processing time of operations ($33 + 63 = 96$). Accordingly, operations 2-9, 2–11 are suitable for Machine 2-2 ($r_{\max} = 51$, the sum of $r_{\max}$ and the total processing time of operations is 82).
2. At each level, check for all available operations the conditions of preceding the operations added in previous iterations. The conditions are:

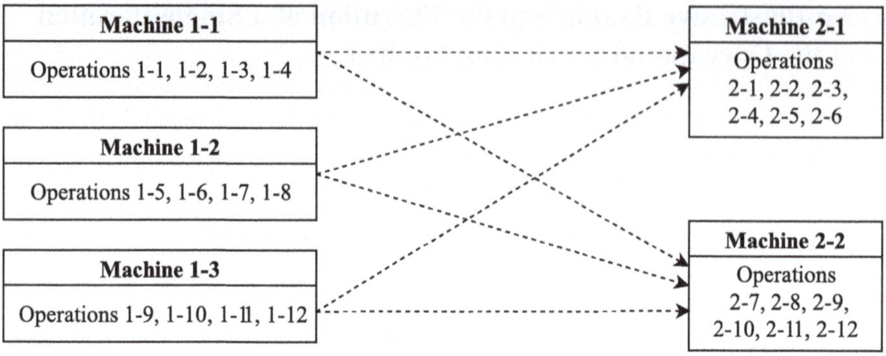

Fig. 3. Structure of the two-level DMS

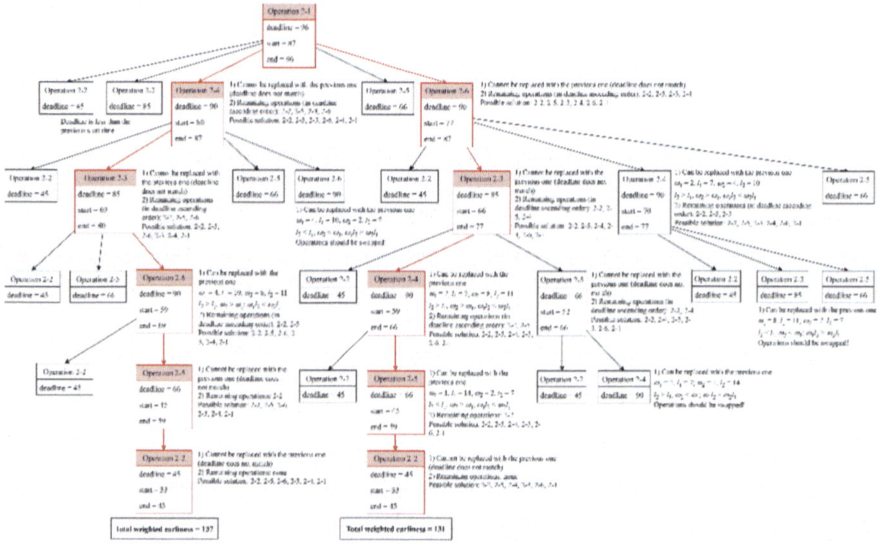

Fig. 4. The forest tree for Machine 2–1

- The deadline of the current operation is greater than or equal to the start time of the next operation.
- The first pruning rule: swapping the current and the next operations makes the schedule non-feasible according to the deadlines or increases the total earliness.
- Fulfillment of the opposite condition to the second pruning rule: adding (in the inverse order, from the end to the beginning) all the remaining operations, starting from the first position, in the deadlines increasing order, before the operations of the current branch makes the schedule non-feasible according to the deadlines.

  If all the conditions are met, add the operation to the end of the current branch, building the schedule in the inverse order, and execute the algorithm recursively for this operation.

3. After adding all available operations, save the obtained result until the algorithm ends.
4. Choose the schedule with the minimum weighted earliness from all the obtained complete feasible schedules. The end of the algorithm.

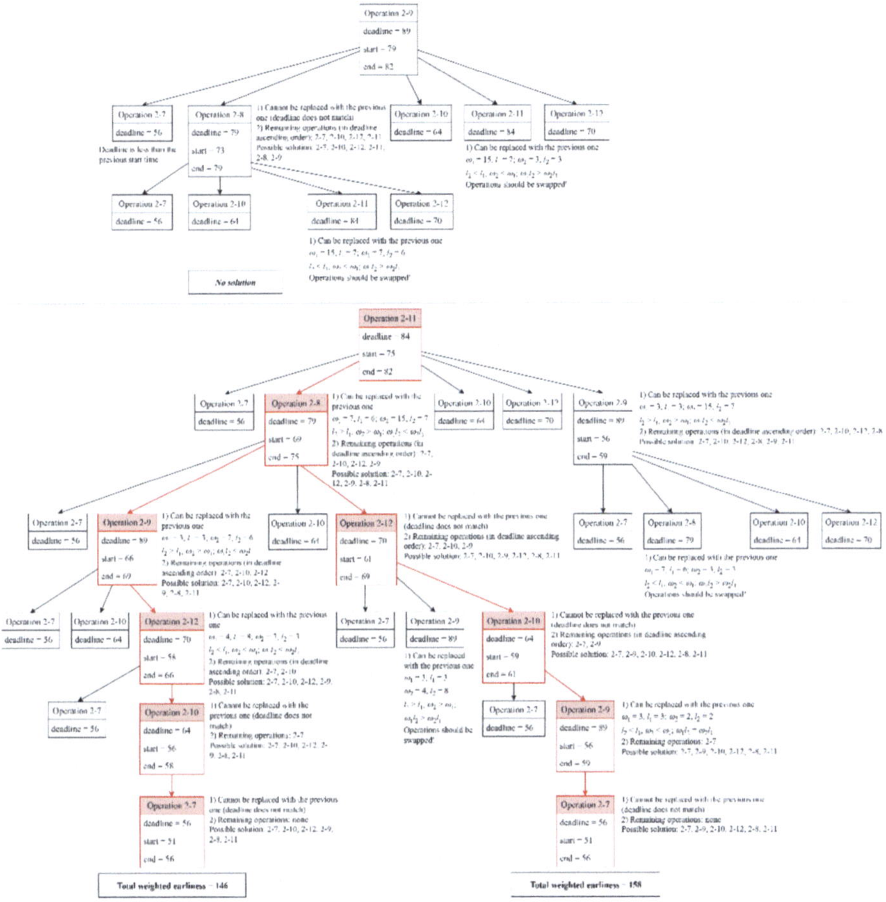

Fig. 5. The forest trees for Machine 2–2

Figure 4 and Fig. 5 show the forest trees for Machine 2–1 and Machine 2–2, respectively.

Complete feasible schedules for Machine 2–1 are the branches of operations highlighted in red in Fig. 4. The optimal solution is the branch of operations 2–2, 2–5, 2–4, 2–3, 2–6, 2–1 that corresponds to the smallest functional value. For Machine 2–2, the optimal solution is the branch of operations 2–7, 2–10, 2–12, 2–9, 2–8, 2–11 (Fig. 5). The start times of jobs in the optimal solutions of Machines 2–1, 2–2 are deadlines for the first operations of the corresponding subsystems of the first level. For each of the subsystems we find a feasible solution (see Remark 1) for the maximum start time of each machine of the first level. The tables in Fig. 6 show the problem solution and the parameters of the obtained operational plan for the two-level DMS.

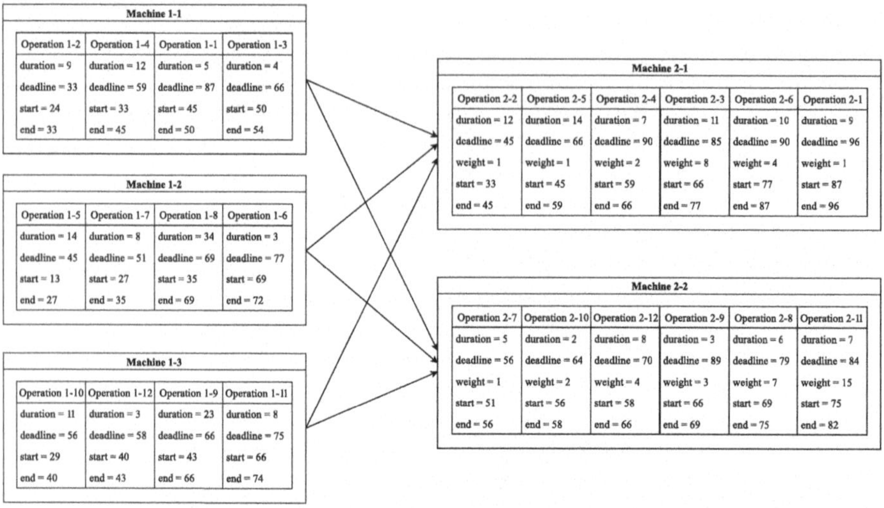

Fig. 6. The problem solution

7 Conclusions

1. We have presented an original universal two-level model of a discrete manufacturing system for which we have formulated a management problem consisting of building an operational schedule and, if necessary, an operative operational plan.
2. We have proposed an efficient iterative (in two versions) method of building an operational schedule. We have shown that it can be generalized for a multi-level model of a discrete manufacturing system in an obvious way.
3. We have proposed a method of an operative operational plan building.
4. The analysis of the given methods has shown that their efficiency directly depends on the efficiency of the proposed solutions to several single-stage scheduling problems.
5. We have presented a model and a method of implementing a static management of asynchronous parallel computing processes. Their use significantly increases the efficiency of the software implementation for the described method of a two-level discrete manufacturing system management.

References

1. Zgurovsky, M.Z., Pavlov, A.A.: Introduction. In: Combinatorial Optimization Problems in Planning and Decision Making: Theory and Applications. 1st edn. Studies in Systems, Decision and Control **173**, pp. 1–14. Springer, Cham (2019). https://doi.org/10.1007/978-3-319-98977-8_1

2. Leng, J., Jiang, P.: Dynamic scheduling in RFID-driven discrete manufacturing system by using multi-layer network metrics as heuristic information. J. Intell. Manuf. **30**(3), 979–994 (2019). https://doi.org/10.1007/s10845-017-1301-y
3. Li, J.: Overlapping decomposition: a system-theoretic method for modeling and analysis of complex manufacturing systems. IEEE Trans. Autom. Sci. Eng. **2**(1), 40–53 (2005). https://doi.org/10.1109/tase.2004.835576
4. Wang, C., Jiang, P.: Manifold learning based rescheduling decision mechanism for recessive disturbances in RFID-driven job shops. J. Intell. Manuf. **29**(7), 1485–1500 (2016). https://doi.org/10.1007/s10845-016-1194-1
5. Sabuncuoglu, I., Bayiz, M.: Analysis of reactive scheduling problems in a job shop environment. Eur. J. Oper. Res. **126**(3), 567–586 (2000). https://doi.org/10.1016/s0377-2217(99)00311-2
6. Ouelhadj, D., Petrovic, S.: A survey of dynamic scheduling in manufacturing systems. J. Sched. **12**(4), 417–431 (2009). https://doi.org/10.1007/s10951-008-0090-8
7. Khodke, P.M., Bhongade, A.S.: Real-time scheduling in manufacturing system with machining and assembly operations: a state of art. Int. J. Prod. Res. **51**(16), 4966–4978 (2013). https://doi.org/10.1080/00207543.2013.784414
8. Buzacott, J.A., Shanthikumar, J.G.: Stochastic Models of Manufacturing Systems. Prentice-Hall, Englewood Cliffs, New Jersey, p. 553 (1993)
9. Neumann, A., Hajji, A., Rekik, M., Pellerin, R.: Genetic algorithms for planning and scheduling engineer-to-order production: a systematic review. International Journal of Production Research, pp. 1–30 (2023). https://doi.org/10.1080/00207543.2023.2237122
10. Neumann, A., Hajji, A., Rekik, M., Pellerin, R.: A didactic review on genetic algorithms for industrial planning and scheduling problems. IFAC-PapersOnLine **55**(10), 2593–2598 (2022). https://doi.org/10.1016/j.ifacol.2022.10.100
11. Aliyev, A.G.: Development of models of manufacturing processes of innovative products at different levels of management. International Journal of Information Technology and Computer Science (IJITCS) **11**(5), 23–29 (2019). https://doi.org/10.5815/ijitcs.2019.05.03
12. Aliyev, A.G., Shahverdiyeva, R.O.: Application of mathematical methods and models in product – service manufacturing processes in scientific innovative technoparks. International Journal of Mathematical Sciences and Computing (IJMSC) **4**(3), 1–12 (2018). https://doi.org/10.5815/ijmsc.2018.03.01
13. Mekidiche, M., Belmokaddem, M., Djemmaa, Z.: Weighted additive fuzzy goal programming approach to aggregate production planning. International Journal of Intelligent Systems and Applications (IJISA) **5**(4), 20–29 (2013). https://doi.org/10.5815/ijisa.2013.04.02
14. Qin, G., Sun, S., Ye, H., Lu, D.: Strategies on teaching reformation for mechanical manufacturing technology. International Journal of Education and Management Engineering (IJEME) **1**(6), 70–75 (2011). https://doi.org/10.5815/ijeme.2011.06.11
15. Wang, L., Tang, D.: Optimization for manufacturing system based on Pheromone. International Journal of Information Technology and Computer Science (IJITCS) **3**(3), 15–21 (2011). https://doi.org/10.5815/ijitcs.2011.03.03
16. Zgurovsky, M.Z., Pavlov, A.A.: The four-level model of planning and decision making. In: Combinatorial Optimization Problems in Planning and Decision Making: Theory and Applications. 1st edn. Studies in Systems, Decision and Control **173**, pp. 347–406. Springer, Cham (2019). https://doi.org/10.1007/978-3-319-98977-8_8
17. Mihalevitch, V.S., Shore, N.Z.: Metod posledovatel'nogo analiza variantov pri reshenii variatsionnykh zadach upravleniya, planirovaniya i proektirovaniya [Method of sequential analysis for solving variational problems in management, planning and design]. Materials of the IV All-Union Mathematical Congress, 91 (1961). (in Russian)

18. Pavlov, A.A., Khalus, E.A., Borysenko, I.V.: Planning automation in discrete systems with a given structure of technological processes. In: Hu Z., Petoukhov S., Dychka I., He M. (eds) Advances in Computer Science for Engineering and Education. ICCSEEA 2018. Advances in Intelligent Systems and Computing **754**, pp. 177–185 (2019). https://doi.org/10.1007/978-3-319-91008-6_18
19. Pavlov, A.A., Khalus, E.A., Misura, E.B., Melnikov, O.V., Medvediev M.Ye.: PDS-alhorytmy dlya dvoetapnoyi zadachi kalendarnoho planuvannya v determinovaniy postanovtsi ta v umovakh nevyznachenosti [PSC-algorithms for a two-stage scheduling problem in the deterministic formulation and under uncertainty conditions]. Adaptive Systems of Automatic Control **1**(42), pp. 184–196 (2023). https://doi.org/10.20535/1560-8956.42.2023.279170 (in Ukrainian)
20. Pavlov, A.A., Khalus, E.A., Medvediev, M.Y.: Minimizatsiya sumarnoho zvazhenoho momentu vyperedzhennya vykonannya zavdan' na odnomu pryladi. Physico-Mathematical Modelling and Informational Technologies **37**, pp. 57–61 (2023). https://doi.org/10.15407/fmmit2023.37.057 (in Ukrainian)
21. Zgurovsky, M.Z., Pavlov, A.A., Khalus, E.A.: Zadacha postroenija dopustimogo raspisanija s maksimal'no pozdnim momentom zapuska i minimal'nym summarnym operezheniem [The problem of constructing a feasible schedule with maximum startup time and minimum total earliness]. System research and information technologies **2015**(2), pp. 7–15 (2015). URL: http://journal.iasa.kpi.ua/article/view/51990. Accessed 13.02.2024 (in Russian)
22. Zgurovsky, M.Z., Pavlov, A.A.: The total weighted tardiness of tasks minimization on a single machine. In: Combinatorial Optimization Problems in Planning and Decision Making: Theory and Applications. 1st edn. Studies in Systems, Decision and Control **173**, pp. 107–217. Springer, Cham (2019). https://doi.org/10.1007/978-3-319-98977-8_4
23. Telenyk, S.F., Nowakowski, G., Pavlov, A.A., Misura, E.B., Melnikov, O.V., Khalus, E.A.: Highly efficient scheduling algorithms for identical parallel machines with sufficient conditions for optimality of the solutions obtained. Bulletin of the Polish Academy of Sciences: Technical Sciences **72**(1), e148939 (2024). https://doi.org/10.24425/bpasts.2024.148939
24. Tanaev, V.S., Shkurba, V.V.: Vvedenie v Teoriju Raspisanij [Introduction to Scheduling Theory]. Nauka, Moscow, p. 280 (1975) (in Russian)

Improving the Energy Efficiency of Industrial Enterprise's Power Supply System by Relocating Movable Local MicroGrid Based on Mobile Power Plant

Natalia Savchenko[1](✉), Oksana Dovgalyuk[2], and Andrii Tretiak[3]

[1] National Aerospace University named after M. E. Zhukovsky "Kharkiv Aviation Institute", Kharkiv 61070, Ukraine
natali_a_savchenko@ukr.net

[2] National Technical University "Kharkiv Polytechnic Institute", Kharkiv 61002, Ukraine
Oksana.Dovhaliuk@khpi.edu.ua

[3] National University "Poltava Polytechnic named after Yury Kondratyuk", Poltava 36011, Ukraine
tretjak_a.v@ukr.net

Abstract. The ability to regulate power flows in the power supply system of an industrial enterprise is an important factor in its reliable operation in the conditions of the existing electricity market and the critical state of the power system of a regime type. An innovative and promising method of improving the power supply system of an industrial enterprise is the use of mobile power plants with renewable energy sources to ensure reliability, quality, energy efficiency and controllability of processes in internal local power grids, taking into account environmental safety issues. The paper proposes the design of a universal mobile power plant with an average power of 15 kW, developed by the authors, in order to solve the problems of regulating the imbalances of electricity at an industrial enterprise by controlling power flows. Also, the mobile power plant can additionally perform the functions of a backup power source and a regulator consumer. A scheme for connecting a mobile power plant and an algorithm for selecting its operating modes in the enterprise MicroGrid system have been developed. A schematic implementation of a local MicroGrid with a mobile power plant is presented. The modelling of the proposed solutions is carried out, the results of which confirm the effectiveness of the proposed method of power control in the enterprise power supply system.

Keywords: Power supply system · Universal mobile power plant · Renewable energy sources · Operating mode

1 Instruction

Alternative energy is developing at a rapid pace. Many researchers and scientists around the world are working on the latest technologies in this field, as renewable energy is the solution to all critical issues of the energy crisis. The possibility of using renewable

Z. Hu et al. (Eds.): ICCSEEA 2024, LNDECT 242, pp. 533–544, 2025.
https://doi.org/10.1007/978-3-031-84228-3_46

energy sources (RESs) together with energy storage systems at all levels of energy supply will completely reformat the structure of power systems in the near future. At the same time, we will have a complete rejection of centralised control of power system modes, and the basis will be localisation and intellectualisation of the control process [1–4]. Today, the implementation of a system based on Smart Grid technology and its variant, the MicroGrid system, is already the basis for building autonomous systems and distributed generation systems with intelligent management of energy resources [5–8]. For Ukraine, this issue is becoming particularly relevant today, providing the possibility of autonomous power supply to consumers of various types and purposes in the event of significant damage to the centralised power supply system.

It should be noted that the introduction of the MicroGrid system as a component of industrial power supply systems is becoming increasingly widespread. It is the introduction of micro-generation that allows solving a number of problems related to the reliability and quality of the power supply itself and imbalances in energy consumption [9–11]. The greatest effect of such systems is obtained when they are integrated with the enterprise's SCADA systems.

Most industrial enterprises use stationary photovoltaic installations to implement a number of local MicroGrid networks, due to the possibility of installing them on the roof and walls of buildings. But such installations are inefficient because they do not have solar tracking systems and their generation depends on the time of day. In addition, the operation and maintenance of photovoltaic installations installed in this way causes a number of problems.

A progressive solution proposed by the authors is the introduction of mobile electrical installations with hybrid RESs into local MicroGrid, which will allow changing the structure of the power grid to effectively manage power flows and increase the energy efficiency of the enterprise's power supply system as a whole. The ability to quickly move, deploy and connect mobile power plants at different points of an industrial enterprise's power supply system is their feature and advantage over stationary installations. It should also be noted that such mobile power plants are environmentally friendly, which is an important issue for the energy sector today.

2 Design of the Developed Mobile Power Plant with Hybrid Sources of Renewable Energy

The developed design of a universal mobile wind-solar power plant, shown in Fig. 1, is essentially a hybrid of two types of wind turbines and two types of photovoltaic installations. Such hybridisation significantly expands the capabilities of the power plant for use as a source of microgeneration as part of the local MicroGrid of an industrial enterprise and increases its energy efficiency. The developed design can also act as a backup power source and additionally perform the functions of regulating the load schedule of an industrial enterprise in order to reduce imbalances in energy consumption. The versatility, environmental friendliness and energy efficiency of the proposed mobile power plant does not limit its implementation to industrial enterprises only; this design solution makes it possible to use it in various industries.

The universal mobile wind-solar power plant includes a wind generator 1, which is constructed from a combination of two types of blades. The air blades of the lower cylindrical part of the wind turbine are designed as a Savonius rotor, and the air blades of the upper conical part are structurally a blade-type rotary wind turbine, with a blade angle of 45° relative to the wind turbine axis. The design of the combined wind turbine is attached to a telescopic stand 2, which, together with a screw brace 3, ensures complete folding of the wind turbine for transport on a trailer 4 (Fig. 1b) and full deployment in the operating state (Fig. 1a). Such a wind turbine design has improved technical characteristics.

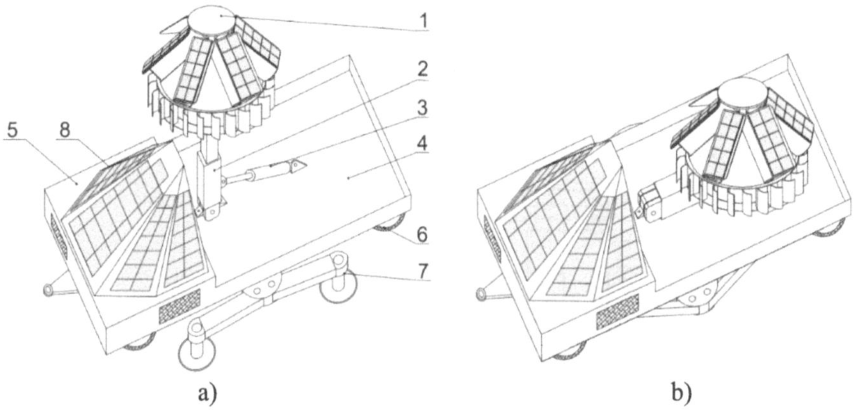

Fig. 1. Universal mobile wind-solar power plant

To increase the energy efficiency of the mobile power plant, two types of solar panels are used as a second source of energy, and their installation also has design features. The first type is flexible thin-film photovoltaic panels located on the blades of the upper cone part of the wind turbine. The use of thin-film photovoltaic panels has a number of significant advantages in terms of technical and operational characteristics compared to other types, and the design solution of blade mounting improves their energy efficiency due to additional cooling during the wind turbine rotation. This arrangement also stabilises the amount of electricity generated regardless of the hours of daylight and without the use of expensive solar tracking systems.

The main advantages of thin-film panels include the following: high efficiency, which reaches about 68.9%; low operating temperature; the ability to generate electricity even in low light conditions; relatively small value of the temperature coefficient, which reflects the change in the electrical parameters of the module when its temperature fluctuates.

The second type of photovoltaic installation as part of a universal mobile power plant (UMPP) is a permanently installed installation on a trailer 4 with monocrystalline panels located on a trapezoidal semi-pyramid structure 8. This structure is located on the closed part of the trailer in the form of a block 5, which houses the control system of the installation and accumulating elements. The main purpose of this photovoltaic system is to directly charge the battery packs, and during transportation, it improves the aerodynamics of the entire system by reducing the resistance to wind load [12].

The flatbed trailer is mounted on a wheelbase 6, and to increase the stability of the structure during deployment, it is proposed to attach a reinforced steel frame with sliding outriggers 7 to the trailer.

It should also be noted that the position of the wind turbine is fixed during transport.

The block diagram of the developed UMPP, which demonstrates the principles of organisation and operation of the installation, is shown in Fig. 2.

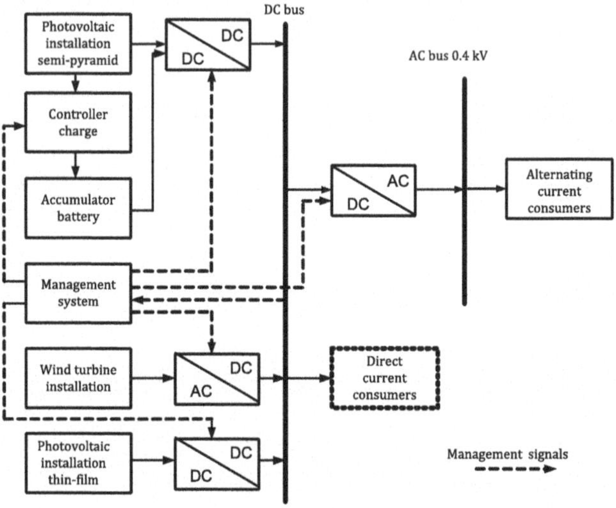

Fig. 2. Block diagram of a UMPP

A UMPP can provide several modes of operation, namely: a mode of joint generation of electricity, which corresponds to the maximum generation mode, a mode of separate generation with a certain capacity, and a mode of using purely reserve power from the accumulator battery (AB). In the maximum joint generation mode, which is actually the main mode of operation, all sources simultaneously generate and supply electricity to the grid when the batteries are fully charged. When there is a need to charge the batteries, power is taken from the PV modules with monocrystalline panels. In case of insufficient power from all power plant energy sources, additional power from the AB can be added. All generation, distribution and consumption management processes rely on an intelligent control system and smart DC/DC and AC/DC inverters.

When the UMPP operates in the mode of separate generation with a certain capacity, some of the sources can be disconnected if necessary.

In case of impossibility of generation or low-power consumption, consumers can be supplied with power from the battery.

The algorithm for selecting the operating modes of a universal mobile power plant is shown in Fig. 3.

Mathematically, the operating modes of a UMPP can be represented as follows:

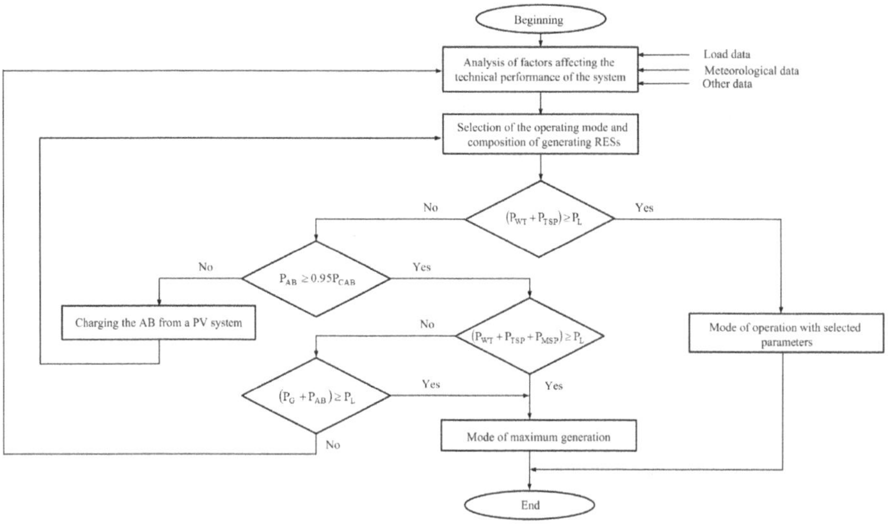

Fig. 3. Algorithm for selecting operating modes of a UMPP

– maximum generation mode without battery power:

$$\begin{cases} P_G = P_{WT} + P_{TSP} + P_{MSP} - \Delta P \\ P_{AB} \geq 0,95P_{CAB} \\ P_G \geq P_L \end{cases}, \tag{1}$$

where P_G is the total generation capacity; P_{WT} is the power generated by the wind turbine; P_{TSP} is the power generated by thin-film solar panels; P_{MSP} is the power generated by monocrystalline solar panels; ΔP is the power losses in power plant equipment; P_{AB} is the AB capacity; P_{AB} is the AB charge capacity; P_L is the load power;

– maximum generation mode from the battery:

$$\begin{cases} P_G = P_{WT} + P_{TSP} + P_{MSP} + P_{AB} - \Delta P \\ 0,3P_{CAB} \leq P_{AB} \leq 0,95P_{CAB} \\ P_G \geq P_L \end{cases}; \tag{2}$$

– mode of operation with the selected parameters:

$$\begin{cases} P_G = P_{WT} + P_{TSP} - \Delta P \\ P_{MSP} = P_{AB} \\ P_G \geq P_L \end{cases}; \tag{3}$$

– operating mode using the battery backup power:

$$\begin{cases} P_G = 0 \\ P_{AB} \geq P_L \\ P_{AB} \geq P_{CAB} \end{cases}. \tag{4}$$

The ability to implement various operating modes of the developed UMPP allows it to be organically integrated into the enterprise's power system with a change in its structure, which leads to a reduction in energy losses and efficient energy management. This is ensured by creating an appropriate scheme for connecting the UMPP to the power supply system of an industrial enterprise in combination with the use of an intelligent mode control system for the developed installation.

3 Development of a Scheme for Connecting a UMPP to the Power Supply System of an Industrial Enterprise and Management Intelligence

Energy efficient management of energy consumption at an industrial enterprise is an important task today that can only be solved through automation, the latest technologies and sources of distributed generation based on alternative energy. Automation using the integration of automated systems of automated commercial electricity metering system (ACEMS) and SCADA systems is actually a new level of enterprise energy balance management, and decentralisation of power supply with the construction of local MicroGrid is ensuring the reliability and quality of the processes of generation, distribution and consumption of electricity. The possibility of using the developed universal mobile power plant as a source of distributed generation will allow for the rapid construction of a local MicroGrid to power energy-intensive electrical equipment or to reconfigure technological processes with additional energy consumption, which cannot be provided by the general power grid. Thanks to automation and control of electricity consumption, the ACEMS and SCADA systems will allow timely detection of electricity losses at the controlled facility or electrical equipment and, if possible, eliminate them by organisational methods or compensate for them by connecting a universal mobile power plant to maintain the enterprise's electricity balance and regulate electricity consumption imbalances to reduce financial losses in settlements with energy supply organisations. A block diagram of a possible connection of a mobile power plant for the implementation of a local MicroGrid is shown in Fig. 4.

As can be seen from the structural diagram of the local MicroGrid as part of the industrial enterprise power supply system shown in Fig. 4, the structural diagram of the local MicroGrid as part of the power supply system of an industrial enterprise, the mobile power plant can be connected both directly to power individual electrical equipment and to the busbars of switchgear and busbars of workshop transformer substations. The ability to connect at different points of the UMPP circuit will also allow you to control the power flows of the power supply system.

The intellectualisation of power grids poses new challenges to the devices for collecting and transmitting information from the ACEMS and SCADA systems:

- ensuring the stability of the enterprise's power supply system;
- greater volume of operational information;
- remote monitoring and control of electricity generation from sources of dispersed generation, in this case from a mobile power plant;
- load management;

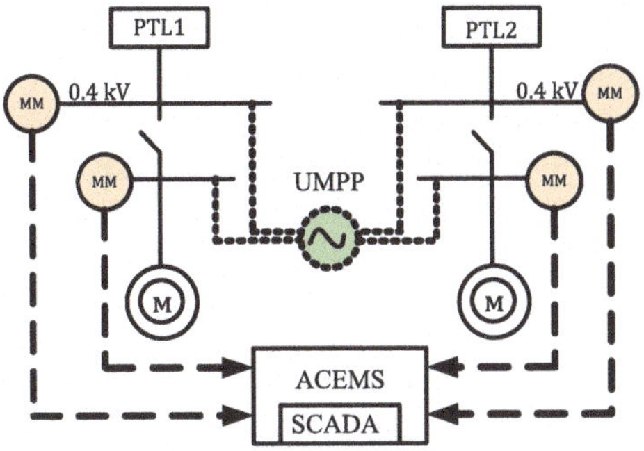

Fig. 4. Schematic diagram of a local MicroGrid with a mobile power plant: PTL - power transmission line, M - industrial electrical installation with asynchronous motors, MM - microprocessor-based meters, ACEMS - automated commercial electricity metering system

- detection of emergency currents, isolation of damaged areas and restoration of power supply, automatic switching.

Thus, the schematic implementation of local MicroGrid with a UMPPs and with intelligent control and the possibility of redeployment movable MicroGrid in the enterprise's power system is a new approach to building energy-efficient power supply systems. For the justified use and redeployment of movable local UMPPs, it is important to predict and analyse the modes of operation of the power grid by these units, which should be performed using mathematical modelling.

4 Analysis of Modes of the Power Supply System of an Industrial Enterprise with Microgeneration

The use of mobile power plants in electrical installations for various purposes requires justification and quantitative analysis of the conditions of further operation, since the impact of these installations on the operation of the power grid is multifactorial and very diverse, which can lead to both positive and undesirable effects, as evidenced by a significant number of publications [13, 14]. In order to meet the regulated requirements for the modes of power grids with mobile power plants, important issues include taking into account the nature of generation [15, 16], current monitoring of the mode parameters and indicators of the quality of electricity in the grid [17–19], and the reasonable use of appropriate technical means to normalise the state and optimize the mode of the power grid [20–22].

To test the efficiency of the developed universal mobile power plant, we analysed the modes of the power supply system of an industrial enterprise with microgeneration. This task was solved by modelling the electrical network of an industrial enterprise using the DigSILENT PowerFactory software and calculation complex.

The medium-sized industrial enterprise under investigation is located in the north-eastern part of Ukraine. It belongs to the woodworking industry and supplies products to the domestic market of Ukraine. The production activities of the enterprise involve interaction with many enterprises of other sectors of the Ukrainian economy and several

European countries, which determine both the sustainable development of the region's infrastructure and the country's foreign economic relations. The power supply system of the investigated woodworking enterprise includes a 110/10 kV main step-down transformer substation (TSS) with two 16 MVA power transformers, which provides power from the centralised power supply system. The 7 shop TSSs of 10/0.4 kV and consumers are connected using cable power lines.

The characteristics of solar insolation and wind load indicate the feasibility of using the developed UMPP for this region. According to the proposed approach, it is envisaged to meet the needs of an industrial enterprise to cover power unbalances relative to the volumes declared to the system operator by connecting the proposed 15 kW UMPP to the power grid in accordance with the scheme presented in Fig. 4.

The model was used to calculate the modes of operation of the power grid when the proposed universal mobile power plant was connected to it, which, according to the developed algorithm (Fig. 3), provided electricity in different generation modes. The sources of data for the calculation were the power grid diagram and the load characteristics of the shops' consumers. The results of modelling the operation of a part of the power grid in the mode of maximum UMPP generation are given in Fig. 5. The calculation results for typical steady-state modes of operation of the power grid under different climatic conditions and the corresponding parameters of the universal mobile power plant are presented in Table 1.

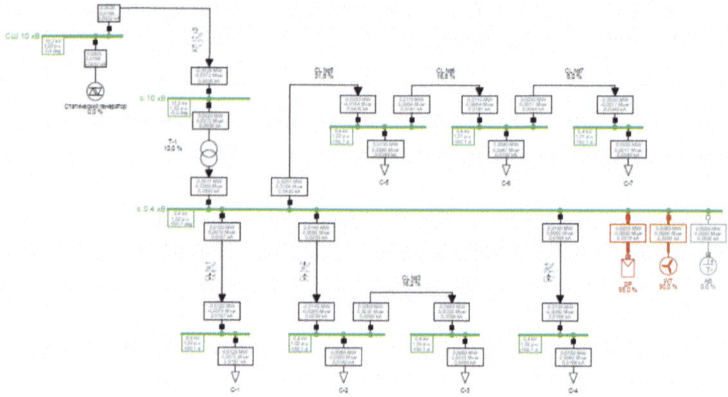

Fig. 5. Calculation of a part of the power grid mode of an industrial enterprise with a UMPP for the maximum generation mode

The analysis of the results of calculating the modes of the power grid of the industrial enterprise under consideration shows that the connection of a UMPP provides a significant reduction in power losses, which, depending on the characteristics of weather

Table 1 Results of calculation of the power grid modes of an industrial enterprise with a UMPP

Mode	UMPP parameters			Power losses in the grid, MW	Power from the energy system, MW
	P_{SPP}, kW	P_{WPP}, kW	P_{ESS}, kW		
Basic	0	0	0	1.546	64.046
Maximum generation	5.5	6	0	1.5237	64.0239
	5.5	6	3.5	1.5187	64.0179
	5.5	6	−3.5	1.5301	64.0302
Separate generation with a defined capacity	3.5	5	0	1.52917	64.0292
	3.5	5	−3.5	1.53487	64.0359
	3.5	5	3.5	1.52288	64.023
	1.1	5	0	1.53372	64.0338
	1.1	5	3.5	1.52715	64.0273
	5.5	1	0	1.5329	64.0331
	5.5	1	3.5	1.5264	64.0265
	0	6	0	1.5339	64.0340
Power output from the AB	0	6	3.5	1.5273	64.0274
	2	0	3.5	1.5315	64.0368
	0	0	3.5	1.5338	64.0385

conditions, are on average 1.2%. At the same time, compared to the basic mode of operation of the power grid without UMPP, in the maximum generation mode, power losses in the power grid are reduced by 1.77%, and at the worst wind load indicators (at $P_{WPP} = 1$ kW, when power losses in the grid are 1.5329 MW), the reduction in power losses is 0.7%. According to the modelling results, even in the absence of electricity generation by one of the RESs types used in the UMPP, the presence of the battery allows to cover consumption needs and ensure the required mode of operation of the power grid. Such performance indicators of the power grid indicate an increase in efficiency.

In addition, it should be noted that the power output of the UMPP provides a reduction in the volume of electricity supply from the central power system, reducing unbalances. Compared to the baseline mode, for various parameters of the UMPP, the reduction in power consumption is $6 \div 22.1$ kW for the studied shop, which allows reducing the total unbalances of the enterprise by up to 80 MW per month.

The algorithm for selecting the operating modes of the UMPP ensures the adjustment of the generated power in such a way as to achieve maximum efficiency in compensating for the current imbalance in accordance with the consumer load schedule (Fig. 6). The use of modern intelligent control systems to ensure maximum adaptation of the generation mode to the needs for imbalance compensation ensures high efficiency of the developed UMPP. It should also be noted that it is important in the technical implementation of such an approach to generation to take into account the peculiarities of the operation of technical means to ensure the level of reliability and quality of electricity in the power grid in the general algorithm [23–26].

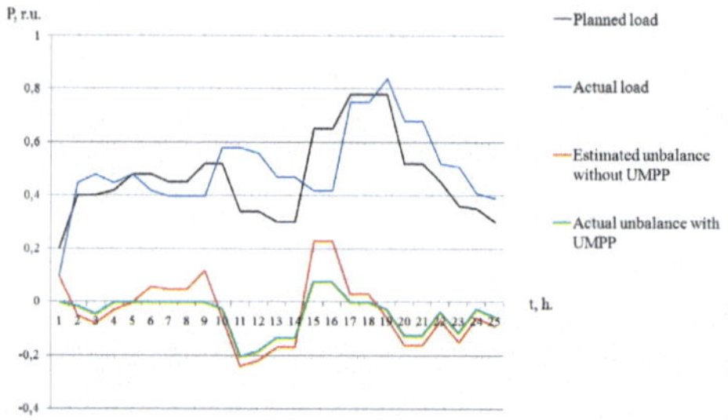

Fig. 6. Graph of consumer load of an industrial enterprise and power imbalances

The analysis of the obtained results shows that, compared to the basic mode of operation of the power grid, when microgeneration is not used, the use of UMPP provides a reduction in the level of unbalances and power losses in the studied network. All three modes, in which the control system provides for the operation of the UMPP, are characterised by voltage and current levels in the transmission lines within the permissible limits. The adaptation of the UMPP to the actual change in the load schedule helps to reduce the financial costs of the industrial consumer to pay for power imbalances, which indicates the high efficiency of the developed installation.

Thus, the analysis of the operation modes of the power supply system of an industrial enterprise with microgeneration from a UMPP proves the feasibility of using the proposed control method to solve the problems of ensuring the optimal power balance of consumers and increasing their financial stability in a liberalised energy market.

5 Summary and Conclusion

The complete intellectualisation of an industrial enterprise's power supply system is a rather complex and multifactorial task. However, more modern enterprises are already widely implementing ACEMS and SCADA systems, so changing the approach to the principles of building power supply systems based on centralisation by introducing local MicroGrid systems based on mobile power plants as sources of distributed generation will significantly expand the possibilities of managing the processes of generation, distribution and consumption of electricity. This article presents the following research results:

1) A UMPP plant based on alternative energy sources has been developed and proposed, with a special feature of its design with hybrid renewable energy sources and high environmental friendliness.
2) A structural diagram of a UMPP has been developed, which fully reveals the principle of its functioning and the links between its components.
3) Possible modes of operation of a UMPP are considered and an algorithm for their selection is developed, a mathematical description of each mode is given.
4) The developed structural diagram of the intellectualisation of the power supply system through the introduction of ACEMS and SCADA and local MicroGrid systems is presented.
5) A new method for implementing the variable structure of the power system of an industrial enterprise with movable local MicroGrid based on a mobile power plant with intelligent control is proposed.
6) The analysis of the modes of operation of the power supply system of an industrial enterprise, performed by means of modelling, has proved the high efficiency of the proposed method of power control of a UMPP used for microgeneration. Adaptation of the generation process to the current needs of covering imbalances allows to increase the financial stability and profitability of industrial enterprises in a liberalised energy market.

References

1. Idries, A., Krogstie, J., Rajasekharan, J.: Challenges in platforming and digitizing decentralized energy services. Energy Inform. **5**, 8 (2022)
2. Dokk, S.I., Kirkegaard, J.K., Szulecki, K.: A functional approach to decentralization in the electricity sector: learning from community choice aggregation in California. J. Environ. Plann. Manage. **66**(6), 1305–1335 (2023)
3. Al Shibli, M., Marques, P.: Artificial intelligent nonlinear auto-regressive external input neural network modeling, design and control of a sea wave electro-mechanical power generating system. Int. J. Intell. Syst. Appl. (IJISA) **6**(11), 1–12 (2019)
4. Yacouba, Y.H., Diabagaté, A., Maiga, A., Coulibaly, A.: Multi-agent system for management of data from electrical smart meters. Int. J. Inf. Technol. Comput. Sci. (IJITCS) **1**(13), 18–43 (2021)
5. Maduranga, M.W.P., Wijerathna, A.S.B.: NB-IoT based status measurement system for 33 kV power distribution networks in smart grids. Int. J. Eng. Manuf. (IJEM) **4**(12), 37–45 (2022)
6. Denysiuk, S., Hilevich, K.: Optimization of the use of distributed energy resources in local electrical power systems according to the criterion of minimum loss of electricity. Power Eng. Econ. Tech. Ecol. **4**, 7–21 (2023). (in Ukrainian)
7. Bin, L., Bingzhao, Z., Ziyou, G., et al.: Energy router interconnection system: a solution for new distribution network architecture toward future carbon neutrality. Energy Convers. Econ. **3**, 181–200 (2022)
8. Uddin, M., Mo, H., Dong, D., et al.: Microgrids: a review, outstanding issues and future trends. Energy Strategy Rev. **49**(4), 101127 (2023)

9. Ndeke, C.B., Adonis, M., Almaktoof, A.: Energy management strategy for a hybrid micro-grid system using renewable energy. Discov Energy **4**, 1 (2024)
10. Ginevičius, R., Trishch, R., Bilan, Y., et al.: Assessment of the economic efficiency of energy development in the industrial sector of the European Union area countries. Energies **15**(9), 3322 (2022)
11. Shevchenko, S., Savchenko, N., Dovgalyuk, O., Danylchenko, D., Tretiak, A.: Comparative analysis of energy efficiency of geometric constructions solar low-power plants. In: Proceedings of the 4th KhPIWeek International Conference on Advanced Technology, pp. 1–6. IEEE, Kharkiv (2023)
12. Savchenko, N., Tretiaк, A., Dovgalyuk, O., et al.: Mobile solar power plant "Pyramid" with a kinetic energy storage. In: Proceedings of the International Science and Technology Conference on "Integrated Computer Technologies in Mechanical Engineering – Synergetic Engineering", pp. 1–15. Springer, Kharkiv (2023)
13. Shahzad, S., Abbasi, M.A., Ali, H., et al.: Possibilities, challenges, and future opportunities of microgrids: a review. Sustainability **15**(8), 6366 (2023)
14. Sinha, P., et al.: Comprehensive review based on the impact of integrating electric vehicle and renewable energy sources to the grid. Energies **16**, 2924 (2023)
15. Saukh, S.Y., Dzhyhun, E.N.: Retrospective models of wind power plants and solar power plants in the problems of planning the models of electric power systems. Èlektronnoe modelirovanie **42**, 18–33 (2020)
16. Dovgalyuk, O., Omelianenko, H., Bondarenko, R., et al.: Analysis of operation modes of electric networks with solar power plants. In: Proceedings of the 4th IEPS International Conference on Intelligent Energy and Power Systems, pp. 196–201. IEEE, Istanbul (2020)
17. Nerubatskyi, V., Plakhtii, O., Hordiienko, D.: Control and accounting of parameters of electricity consumption in distribution networks. In: Proceedings of the XXXI MMA International Science Symposium on Metrology and Metrology Assurance, pp. 114–117. IEEE, Sozopol (2021)
18. Stanimirović, A., Bogdanović, M., Frtunić, M., Stoimenov, L.: Low-voltage electricity network monitoring system: design and production experience. Int. J. Distrib. Sensor Netw. **16**(1), 1–12 (2020)
19. Malek, R., Yang, Q.: Analyzing interrelationships and prioritizing performance indicators in global product development: application in the Chinese renewable energy sector. Sustainability **15**, 11212 (2023)
20. Plakhtii, O., Nerubatskyi, V., Karpenko, N., et al.: Studying a voltage stabilization algorithm in the cells of a modular six-level inverter. Eastern-Eur. J. Enterp. Technol. **6/8**(102), 19–27 (2019)
21. Dovgalyuk, O., Savchenko, N., Batalin, V., et al.: Optimisation of operating modes of distribution electric networks in Ukraine. In: Proceedings of the 4th KhPIWeek International Conference on Advanced Technology, pp. 1–6. IEEE, Kharkiv (2023)
22. Das, N.R., Rai, S.C., Nayak, A.: Intelligent scheduling of demand side energy usage in smart grid using a metaheuristic approach. Int. J. Intell. Syst. Appl. (IJISA) **6**(10), 30–39 (2018)
23. Momeni, M.R., Haghighat, F., Haghighat, M.: An efficient and cloud based architecture for smart grid security. Int. J. Wirel. Microwave Technol. (IJWMT) **5**(11), 35–42 (2021)
24. Hovorov, P., Kindinova, A., Hovorov, V.: Mode control of urban electrical networks based on the smart grid concept. In: Proceedings of the 2nd KhPI Week International Conference on Advanced Technology, pp. 88–93. IEEE, Kharkiv (2021)
25. Zhang, X.-P., Yan, Z.: Energy quality: a definition. Power Energy **7**, 430–440 (2020)
26. Agrawal, M., et al.: Priority based strategic modes of operation and control for renewable energy sources micro-grid. Mater. Today Proc. **46**, 5585–5593 (2021)

Education Reform and Innovation

Construction and Practice of Logistics Engineering Education and Teaching Ecosystem Based on "Integration of Industry Chain Driven by Digital Intelligence"

Yanfang Pan(✉)

College of Traffic and Transportation, Nanning University, Nanning 530200, China
281340537@qq.com

Abstract. With the upgrading of industries and the upgrading and replacement of logistics fields such as intelligent warehousing, intelligent sorting, and intelligent distribution, the job requirements have also been raised to higher levels, and the professional knowledge system and quality requirements have also put forward higher requirements for professional talents. To seek development in a profession, it is necessary to delve into industries and fields, and adapt to the accelerating transformation of the logistics industry. The article takes "intelligent transportation, intelligent logistics, and intelligent vehicles" as the central characteristics of the times, explores the intersection and entry point of the integration and development of logistics engineering and industrial chains, and uses the standardized job matrix of "five horizontal and five vertical" logistics engineering as a reference for abilities. It constructs a "four chain and three circle" logistics engineering education and teaching ecosystem, promoting the cultivation of professional talents to keep up with the development of the times and the needs of enterprises.

Keywords: Digital and smart logistics · Logistics industry chain · Education and teaching ecology · Logistics Engineering major

1 Introduction

With the explosive trend of technology applications such as virtual reality and big data analysis, industrial chains, supply chains, and digital operations in different fields are undergoing changes [1–3]. The emergence of new terms such as intelligent supply chain graph, intelligent supply chain, national chain supply chain, and supply chain digitization not only profoundly transforms the appearance of the logistics industry, but also profoundly affects the means and methods of logistics engineering professional education and teaching [4–6].

1.1 Characteristics and Development Trends of Digital and Smart Logistics

In the era of big data, lean production, intelligent manufacturing, and precise consumption prediction greatly promote the digitalization and intelligent upgrading and transformation of China's supply chain [7, 8]. The logistics industry, as the lifeline of the national

Z. Hu et al. (Eds.): ICCSEEA 2024, LNDECT 242, pp. 547–558, 2025.
https://doi.org/10.1007/978-3-031-84228-3_47

economy, runs through the entire development of the industrial chain. The intelligent supply chain empowered by digital technology and artificial intelligence will become an important pole of China's economic growth. The content shown in Fig. 1 shows the revenue scale of digital services in China's supply chain from 2017 to 2026.

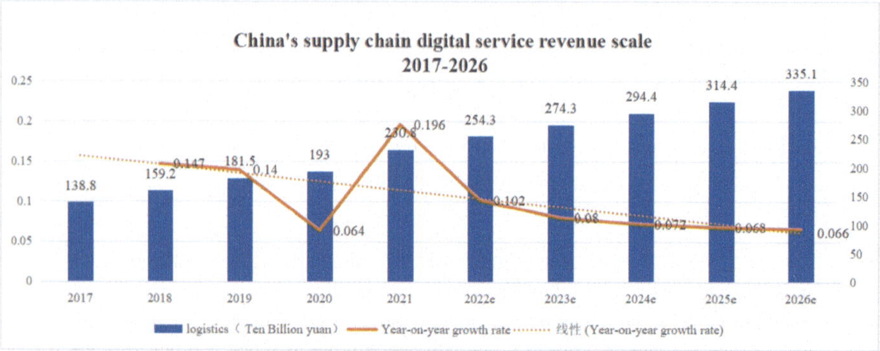

Fig.1. China's supply chain digital service revenue scale, 2017–2026

"Digital and smart logistics" refers to the use of advanced Internet of Things technologies such as barcodes, electronic radio frequency technology, sensor applications, GPS positioning technology, etc., to apply information processing and network communication in logistics activities such as warehousing, transportation, distribution, packaging, handling, and loading and unloading, achieving automated operation, efficient optimization, and low-cost operation in the process of material flow [9, 10]. The prerequisite for the transformation and upgrading of the logistics industry is to utilize advanced technologies such as information collection, data transmission, data processing, data analysis, and data updates for integrated processing and collaborative office work, to achieve optimized integrated management of logistics practice [11].

1.2 The Core of the Transformation in the Logistics Industry is Integrated Development

In the new logistics industry environment, undergraduate education in logistics engineering should focus on knowledge updating, technological innovation, and industry integration to enhance students' comprehensive abilities. Facing the characteristics of cross industry and cross domain logistics industry, it is an urgent need for the entire industry to cultivate composite logistics talents who possess both basic knowledge and digital and intelligent knowledge [12–14].

Facing the future development trend of digital logistics, driven by solving practical logistics problems, with real enterprise applications as the core, based on the integration of industry and education, school enterprise cooperation, and assisted by school cooperation and school communication, integrating internal and external resources, complementing each other's advantages, and forming a joint force, we can find a suitable model and mechanism for the development of the logistics industry and logistics education in line with the changing era [15].

1.3 The Supply-Demand Contradiction of Digital and Intelligent Talents

The overall recruitment demand in the current logistics industry has significantly increased, especially in the four categories of marketing, business operations, overseas specialists, and customer service, accounting for over 60% of the job demand. There is a significant increase in talent demand in emerging fields such as cross-border e-commerce, multimodal transportation, and logistics informatization [16]. On the other hand, the development of the logistics industry is gradually shifting from labor-intensive to technology, knowledge, and service intensive. There are not many high-quality talents working in emerging fields such as supply chain optimization, intelligent logistics system design, and logistics planning.

Faced with the challenges and requirements of new environments, new technologies, and new professions, it is necessary to break free from boundaries and constraints, break free from the traditional teaching system for cultivating logistics professionals, and at the same time, build a logistics education and teaching ecosystem [17, 18]. The task of schools is to cultivate students who possess both profound professional knowledge, innovative thinking, practical experience, and broad perspectives and abilities [19].

2 The Construction Goals of the Education and Teaching Student Circle

Creating a good educational ecosystem is conducive to the cultivation of talents in logistics engineering.The construction of logistics engineering education and student circle needs to focus on the following goals.

2.1 Taking Moral Education and Talent Cultivation as the Fundamental Task

Taking the logistics teaching and research room as the physical teaching and research room, and the logistics engineering major as the construction object, we deeply explore the integration of the development needs of logistics enterprises and the ideological and political elements of courses, and do a good job in correct value orientation and value guidance [23, 24].

2.2 Driven by Serving Local Industrial Economy

Focusing on the development of regional economy and logistics in ASEAN, with the goal of meeting the requirements of the era and professional positions for the development of "digital + intelligent" logistics enterprises, we seize the opportunities for the construction of "New Energy Vehicle Industry College" and "Digital Economy Industry College", integrate the intersection and common characteristics of logistics engineering, transportation, automotive service engineering, new energy vehicle engineering, and digital economy, and strive to achieve the integration of professional talent training and industry enterprise application [25].

2.3 Targeting Innovative Forms of Teaching and Research

It is needed to form a new type of teaching organization that spans multiple colleges and majors, strengthen professional construction and curriculum development, and strive to achieve the goal of a curriculum construction system that can be shared at the grassroots level, integrated at the middle level, and mutually selected at the senior level. We will promote the construction of an industry group collaborative education platform that includes "textbook co compilation, teacher training, resource sharing, project planning, management co governance, quality evaluation, and employment assistance".

3 Preliminary Construction of an Education Ecosystem Based on Digital and Intelligent Driven Industrial Chain

The construction of educational circles for students can be carried out from the following three aspects.

3.1 Building a Fusion Platform Based on the Concept of Sharing

Based on the commonalities and development trends of industrial development, we will sort out the knowledge structure and application level of professional groups and subject development, build a digital transportation technology and logistics technology teaching and research platform, to solve the problems of industrial chain demand and talent cultivation connection. Against the background and knowledge system of "modern big transportation and big logistics", we will break down the professional wall, enable various secondary colleges to cooperate and win in talent cultivation, teaching communication, innovation and entrepreneurship, subject competitions, ideological and political education, and integrate online and offline, on campus and off campus, and school government and enterprise cooperation services, cultivate students to establish a correct career development concept, and do a good job in the responsibility and responsibility of "transportation personnel".

3.2 Deepen Industry Enterprises and Construct Job Matrices

By conducting in-depth investigations into enterprises and industries, conducting a thorough analysis and sorting of professions, a standardized job matrix of "five horizontal and five vertical" positions in the logistics engineering profession is constructed, as shown in Fig. 2.

Analyze job responsibilities in the transportation, logistics, and automotive industries, divide them according to job content and professional functions, remove positions that are weakly related to logistics, and horizontally classify them based on logistics business models and operational logic. Divide them into five directions: operational operations, logistics planning, big data operations, information design and operation, and intelligent equipment design and operation. At the same time, vertically classify them into five levels: primary, intermediate, advanced, comprehensive, and industrial according to the requirements of job professional abilities,

	Logistics planning and design position	Logistics Operation Post	Logistics Information Post	Logistics Intelligent Equipment Post	Logistics Big-Data Operations Post
Primary	Warehouse network planning Transportation capacity planning	Warehouse allocation operation Express delivery	Basic research and development	material management Installatio n testing	data acquisition Data cleaning
Interme -diate	Warehouse interior planning and design	Warehouse distribution operation	Front-end research and development	Research and development quality control	Operational analysis Data Mining and Applications
Senior	Efficiency planning improvement	Business optimization management	Technical project management	Business architectu re design	Data analysis management optimization
Compre hensive level	Business operation management decisions, logistics technology consultants, planning management decisions, big data operation management decisions				
Industri al level	Special agricultural products, whole vehicles, multimodal transportation services, cold chain aquatic products and other import and export businesses				

Fig. 2. The standardized job matrix of "five horizontal and five vertical"

Starting from the needs of industrial positions, relying on industry standards, industry business processes, real projects and other industry best practices, we aim to create high-quality courses for the entire supply chain from end to end, including supply chain planning, planning, operation, and technology at all levels. We will optimize the structure of intelligent supply chain professional group courses and build a professional group course system that can be shared at the grassroots level, integrated at the middle level, and mutually selected at the senior level.

3.3 Precise Positioning and Building an Ecosystem

On the base of clear positioning, resources were integrated to construct the "four chains and three circles" logistics engineering professional education and teaching ecosystem shown in Fig. 3.

The so-called "three four chains and three circles ecosystem" is a form of "sharing and integration + open cooperation", which opens up an information exchange channel for talent cultivation in logistics engineering, and does a good job of professional docking, docking with banks and enterprises, docking with society, and docking with technology.

"Logistics full chain" emphasizes the smoothness and connection of the entire chain, attaches importance to the flow of materials, information, and funds, and focuses on the enhancement of product value. The supply chain encompasses communication and business processing from the source planning, planning, production, and sales of products, considering the overall cost reduction and efficiency enhancement empowerment of the chain. The "professional chain" strengthens the construction of professional clusters in universities, fully leverages the advantages of clusters, and realizes the complementarity and sharing of resources. The "industrial chain" is the core module of logistics engineering education and teaching, focusing on the needs and requirements of modern factories, modern logistics, and new retail industries for logistics engineering. It also needs to be based on digital technology, emphasizing practice and application, integration and solution, virtual and reality.

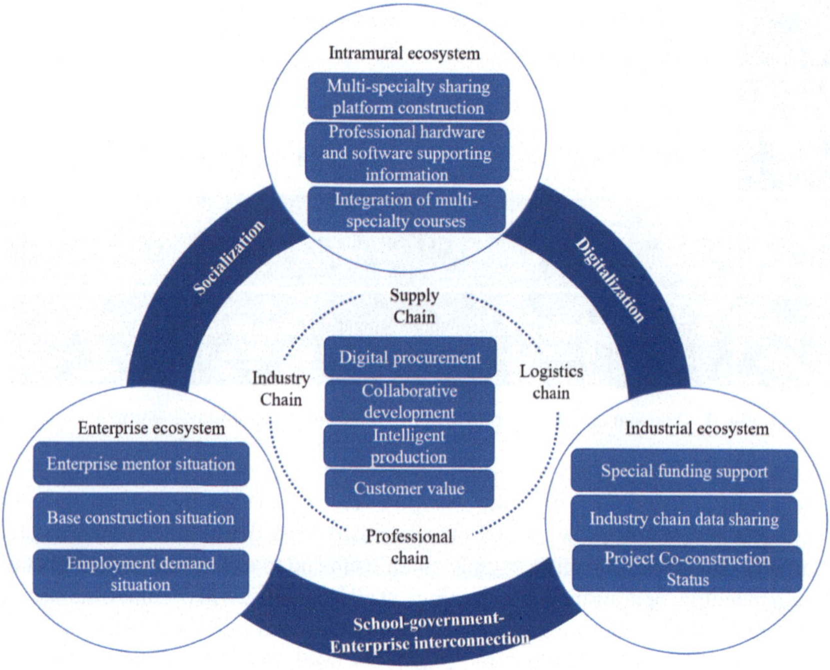

Fig. 3. Logistics engineering education and teaching ecosystem

Socialization and digitization as well as intelligent are the cornerstone, with one side being economy and the other side being technology. Socialization requires that the training of logistics engineering professionals must be based on regional economic and social development needs, and comprehensively promote the construction of cross-border e-commerce, cold chain collaboration, rural revitalization, transportation, green and dual carbon. "Digitalization" requires that the training of logistics engineering professionals must meet the innovative requirements of current new technologies, methods, and tools, and do a good job in updating educational knowledge, professional knowledge, and teaching methods.

4 Establishment of Key Indicators for the Education and Teaching of Logistics Engineering Majors in Student Circles

The construction of a student ecosystem in logistics engineering education requires the joint efforts of schools, governments, industries, and enterprises. The joint efforts of multiple departments require communication and coordination among each other. It is necessary to identify key indicators and optimize the hierarchical structure of ecosystem indicators, as shown in Table 1.

Table 1. Indicator factors of the student attitude circle in logistics engineering education

Overall goal	Sub objectives	Criterion layer
The logistics engineering professional education and teaching ecosystem of "digitalization + industrial chain" (X)	On campus education ecosystem (Y_1)	Multi professional sharing platform (Z_{11})
		Software and hardware support (Z_{12})
		Integration of multiple professional courses (Z_{13})
	Social Ecosphere (Y_2)	Special funding support (Z_{21})
		Industry Chain Data Sharing (Z_{22})
		Project Co-construction (Z_{23})
	Enterprise Ecosystem (Y_3)	Corporate Mentor (Z_{31})
		Base construction (Z_{32})
		Employment needs (Z_{33})

The first step is to build a judgment matrix. According to the research method, combined with expert scoring, the judgment matrices of each criterion layer and each scheme layer are obtained through pairwise comparison. The weight table is obtained by column normalization of the judgment matrices of each criterion layer and each scheme layer, and consistency testing is carried out. The formula used for consistency testing is as follows:

Maximum eigenvalue: $\lambda = \frac{1}{n}\sum_{i=1}^{1}\frac{(A\omega)_i}{\omega_i}$, (where, n is the order)
Consistency indicator: $CI = \frac{\lambda-1}{n-1}$
Consistency ratio: CR = CI/RI

According to the random consistency index RI, Table 2 is obtained.

Table 2. Random consistency index RI values

Order	1	2	3	4	5
RI	0	0	0.539	0.916	1.143

Based on the comparison of various indicator factors provided by experts, the weight index of the establishment of the student ecosystem in school education (Y_1) and the social ecosystem (Y_2) is 1:3, the weight index of the establishment of the student ecosystem in school education (Y_1) and the enterprise ecosystem (Y_3) is 1:7, and the weight index of the establishment of the social ecosystem (Y_2) and the enterprise ecosystem (Y_3) is 1:5. The Analytic Hierarchy Process table is constructed, as shown in Table 3.

Table 3. Judgment matrix of criterion layer

A	Y_1	Y_2	Y_3
Y_1	1	1/3	1/7
Y_2	3	1	1/5
Y_3	7	5	1

For the weight values of each indicator, a simple weighted average was used to calculate the consistency test CR, and Table 4 was obtained. It can be found that the establishment of the enterprise ecosystem (Y3) has the highest weight in the criterion layer.

Table 4. Criteria layer weight

A	Y_1	Y_2	Y_3	w	AW	Consistency check CR
Y_1	0.091	0.053	0.106	0.083	0.251	0.061<0.1
Y_2	0.273	0.158	0.149	0.193	0.588	
Y_3	0.636	0.789	0.745	0.724	2.273	

Similarly, based on expert ratings, a product matrix of indicator factors for each stage is obtained, as shown in Table 5.

Table 5. Judgment Matrix of Student Attitude Circle in Campus Education

A	Z_{11}	Z_{12}	Z_{13}
Z_{11}	1	3	7
Z_{12}	1/3	1	1
Z_{13}	1/7	1	1

According to the consistency test CR and Table 6, it can be seen that the social ecosystem (Y_1) with the highest weight is the establishment of a multi professional sharing platform.

Table 6. Weights of the campus education and teaching ecosystem

A	Z_{11}	Z_{12}	Z_{13}	w	AW	Consistency check CR
Z_{11}	0.677	0.600	0.778	0.685	2.174	0.076<0.1
Z_{12}	0.226	0.200	0.111	0.179	0.543	
Z_{13}	0.097	0.200	0.111	0.136	0.413	

Through calculation, it presents that the construction of multi professional sharing platforms has the highest weight in the school education and teaching ecosystem (Y_1) (Z_{11}).

Based on expert ratings, construct an Analytic Hierarchy Process table to obtain a judgment matrix for the social ecosystem, as shown in Table 7.

Table 7. Social ecosystem judgment matrix

A	Z_{21}	Z_{22}	Z_{23}
Z_{21}	1	1/4	1/6
Z_{22}	4	1	1/2
Z_{23}	6	2	1

According to the consistency test CR, the table below shows that the project co construction situation has the highest weight in the social ecosystem (Y_2), as shown in Table 8.

Table 8. Weight of social ecosphere

A	Z_{21}	Z_{22}	Z_{23}	w	AW	Consistency check CR
Z_{21}	0.091	0.077	0.100	0.089	0.268	0.009<0.1
Z_{22}	0.364	0.308	0.300	0.324	0.974	
Z_{23}	0.545	0.615	0.600	0.587	1.770	

Based on expert ratings, construct an Analytic Hierarchy Process table to obtain a judgment matrix for the enterprise ecosystem, as shown in Table 9.

Through the consistency test of CR values, it can be seen that the most weighted factor in the enterprise ecosystem (Y_3) is the employment demand situation (Z_{33}).

In summary, if all matrices pass the consistency test, the comprehensive weight of the scheme layer factors relative to the target layer factors can be calculated. Sort them to obtain the most important indicators of feasibility.

The calculation formula is:

Table 9. Matrix for judging the corporate ecosystem

A	Z_{31}	Z_{32}	Z_{33}
Z_{31}	1	1/3	1/6
Z_{32}	3	1	1/5
Z_{33}	6	5	1

comprehensive weight of feasibility indicators = weight of criterion layer x weight of scheme layer.

The calculation results and comprehensive ranking are shown in Table 10.

Table 10. Comprehensive weights of the feasibility of teaching student circle in logistics engineering education

Criterion layer			Plan layer			Compre-hensive weigh	Overall sorting
Factor	Weight	Factor ranking	Factor	Weight	Factor ranking		
Y_1	0.083	3	Z_{11}	0.685	1	0.057	6
			Z_{12}	0.179	2	0.015	8
			Z_{13}	0.136	3	0.011	9
Y_2	0.193	2	Z_{21}	0.089	3	0.017	7
			Z_{22}	0.324	2	0.063	5
			Z_{23}	0.587	1	0.113	3
Y3	0.724	1	Z_{31}	0.092	3	0.067	4
			Z_{32}	0.201	2	0.146	2
			Z_{33}	0.707	1	0.512	1

From the Table 10, it can be got that the key to building a student circle for logistics engineering education is for universities and enterprises to plan and communicate their employment needs, strengthen the construction of multi professional sharing platforms and projects on campus, promote education through industry, and promote professional connotation construction through integration.

5 Conclusion

"Digital and Smart Logistics" is an inevitable trend in the development of modern logistics, and it also puts forward corresponding requirements for the cultivation of logistics professionals today. In the current employment situation where multiple contradictions are equally important, to deepen the implementation of the national government's decisions and deployments on "stabilizing employment" and "ensuring employment", various measures should be taken to deepen higher education reform, promote the organic linkage between employment education and talent cultivation, effectively connect talent supply and demand, and adapt to economic and technological development.

The logistics engineering major needs to integrate the cooperation needs in fields such as smart logistics and supply chain, connect enterprise job positions and employment needs with the talent training system of universities, timely integrate cutting-edge industry technologies into classroom teaching, actively promote the inclusion of university teachers and enterprise mentors, university resources and enterprise resources in the education and teaching ecosystem, empower each other, and serve students in high-quality employment.

Acknowledgment. This project is supported by (1) The second batch of the Ministry of Education's supply and demand docking employment and education project "Targeted talent cultivation and training" (202301111597); (2) The Guangxi Higher Education Undergraduate Teaching Reform Project "Research and Application of Logistics Engineering Major Connection Professional Group Construction under the Background of Industrial College" (2023JGA392); (3) Nanning University Virtual Teaching and Research Office Project "Regional Logistics and Digital Intelligent Logistics Virtual Teaching and Research Office" (2023XNJYS01).

References

1. Abhishek, P.: Making digital empowerment a fundamental right. Dataquest **34**(11), 91–92 (2016)
2. Mithas, S., Mcfarlan, F.W.: What is digital intelligence? IT Professional **19**(4), 3–6 (2017)
3. Bughin, J., Zeebroeck, N.V.: Artificial intelligence: why a digital base is critical. McKinsey Quarterly **2018**(3), 13–16 (2018)
4. Park, K.O.: A Study on sustainable usage intention of blockchain in the big data era: logistics and supply chain management companies. Sustainability **12**(24), 10670 (2020)
5. Vikaliana, R.: Big data on digital logistics in supply chain risk perspective. J. Logistik Indonesia **2**(1), 21–30 (2018)
6. Kumar, A., Kumar, P.: Smart logistics and efficient supply chain for discrete industry. International, Organization of Scientific Res. **11**, 11 (2021)
7. Casiano, M.: The supply chain industry hits a crossroad for sustainability. Supply Chain Brain **2**, 26 (2022)
8. Bhowan, T.: Supply chain solutions provider fills industry gaps. Creamer Media s Engineering News & Mining Weekly **35**, 41 (2021)
9. Cranswick, E.: Stockland explains what's driving the logistics boom. MHD Supply Chain Solutions **14**, 52 (2022)
10. Wahab, S.N., Rajendran, S.D., Yeap, S.P.: Upskilling and reskilling requirement in logistics and supply chain industry for the fourth industrial revolution. Logforum **17**(3), 399–410 (2021)
11. Rao, A.V.N.: Exploring the factors for enhancing Supply chain responsiveness in Logistics Industry: a case study of select logistics companies in Delhi NCR Region. International Journal of Indian Culture and Business Manage. **1**(1), 1 (2020)
12. Chen, G.: Exploration and practice of teaching reform on "multi cross integration" of digital intelligence empowered logistics management course under the background of new liberal arts. Foreign Economic and Trade Journal **01**, 112–115 (2024). (in Chinese)
13. Gao, S.: Exploration of the rise of digital supply chain on the training of logistics talents in universities. Times Automotive **22**, 93–95 (2023). (in Chinese)

14. Wang, H.: Research on the training path of new professional talents in intelligent logistics technology under the transformation and upgrading of digital intelligence. Logistics Technology **46**(18), 167–170 (2023). (in Chinese)
15. Chan, V., Malik, N., Chun, C.C.E.: Educating Gen-Z: developing a teaching and learning ecosystem in a 21st century learning. Int. J. Comput. Math. **1**, 168–182 (2020)
16. Li, Z.: Opportunities and challenges in cultivating logistics professionals in the era of digital logistics. China Logistics and Purchasing **18**, 90–91 (2023). (in Chinese)
17. Llinares, A., Nashaat-Sobhy, N.: What is an ecosystem?: defining science in primary school CLIL contexts. Language Teaching for Young Learners **3**(2), 337–362 (2021)
18. Jun, L.: Study on the transition of teachers' ecological role in college English teaching in new media age. Overseas English **13**, 2 (2019)
19. Zakiah, A., Prihatmanto, A.S., Mahayana, D., et al.: The philosophy of smart learning using the approach Thomas Kuhn Paradigm Shift. International Journal of Information Engineering and Electronic Business (IJIEEB), **16**(1), 54–62 (2024)
20. Porkodi, S., Tabash, B.K.H.: A comprehensive meta-analysis of blended learning adoption and technological acceptance in higher education. International Journal of Modern Education and Computer Science (IJMECS), **16**(1), 47–71 (2024)
21. Mhina, J.R.A., Lashayo, D.M.: Implementing enterprise resource planning systems in Tanzanian higher education institutions: the influence of task-technology fit on staff performance. International Journal of Information Technology and Computer Science (IJITCS) **15**(5), 12–24 (2023)
22. Eden, A.: Ecosystem explorers: utilizing inaturalist to promote inquiry in the classroom. The American Biology Teacher: Journal of the National Association of Biology Teachers **4**, 85 (2023)
23. Ujwal, U.J., Saleem, M.: A hybrid weight based feature selection algorithm for predicting students' academic advancement by employing data science approaches. International Journal of Education and Management Engineering (IJEME) **13**(5), 1–22 (2023)
24. Karsenti, T., Parent, S., Cuerrier, M., et al.: Creation of an interactive mapping of artificial intelligence (IA) in education. Formation et Profession **28**(3), 122 (2020)
25. Li, X., Feng, L., Zhou, L., et al.: Learning in an ambient intelligent world: enabling technologies and practices. IEEE Trans. Knowl. Data Eng.Knowl. Data Eng. **21**(6), 910–924 (2009)

Exploration of the Integration Model of Professional Course Teaching Under the Background of Emerging Engineering Education-Taking Mechanical Engineering Testing Technology as an Example

Dan Meng(✉)

College of Intelligent Manufacturing, Nanning University, Nanning 530200, China
mengdan508@163.com

Abstract. In order to meet the emerging engineering education background, cultivate future diversified and innovative excellent engineering talents. Therefore, this paper puts forward the exploration of the integration of science and education and the integration of curriculum ideology and politics in the teaching of professional course-taking "Mechanical Engineering Testing Technology" as an example. With the help of "multi-dimensional mind map", the relationship between the knowledge points of "Mechanical Engineering Testing Technology" course is demonstrated, and science education is integrated with curriculum ideology and politics. Finally, through the closed-loop control and feedback adjustment of "teach-exam-evaluation", the students' achievement of curriculum goals is evaluated comprehensively through multiple, multi-level and multi-faceted. It lays a foundation for further promoting the development of mechanical specialty under the background of emerging engineering education.

Keywords: Science and education · Curriculum ideology and politics · Professional course · Multi-dimensional mind map

1 Introduction

In recent years, with the advancement of emerging engineering education construction, domestic colleges and universities pay more and more attention to the construction of integrated curriculum [1, 2]. In order to meet the needs of the training of engineering education professional certification personnel Ruisheng Xiong analyzed the shortcomings of the traditional teaching methods and course design schemes of steel structure course, put forward the idea of integrating course design and unit practice, and gave the specific implementation scheme and quality assurance measures of "one student one problem" [3]. "One student one question" is based on the situation of each student to get the same question type and different parameters of the question. This shows that students can not only grasp the knowledge system of the course as a whole, but also apply the knowledge of each chapter to practical problems. In addition, it also allows students to master

Z. Hu et al. (Eds.): ICCSEEA 2024, LNDECT 242, pp. 559–569, 2025.
https://doi.org/10.1007/978-3-031-84228-3_48

the basic ideas of structural design, students' independent thinking ability and analysis, the ability to solve practical problems, and the quality of students' training have also been improved. In the course of "Engineering cost Management", Huizhong Yu et al. adopted the practical integration teaching method to closely link the theoretical course, the experimental course and the practical training course, so as to realize the integration and integration of professional course [4]. These also cultivate students' practical and professional skills, and effectively mobilize and stimulate students' interest in learning. Shuangyan Chen et al. integrated the teaching reform of "microcomputer principle" and "single-chip microcomputer technology" into the two courses, taking the task as the lead, and closely linked theory with practice, so as to improve the teaching quality and provide useful ideas and methods for personnel training and course teaching of electrical majors in applied undergraduate colleges [5].

These studies show that the integrated construction of specialized basic course focuses on the integration of content, while the integrated construction of specialized course with strong application focuses more on the combination of theory and practice. These are not enough to meet the "emerging engineering education" to be guided by virtue and cultivate people, to respond to changes and shape the future as the construction concept, to inherit and innovation, cross and integration, coordination and sharing as the main way, to cultivate diversified and innovative engineering talents in the future. Therefore, this paper puts forward the research and discussion on the integration of science and education and curriculum ideological and political integration in the teaching of specialized course - taking mechanical engineering testing technology as an example.

2 Science and Education Integration and Curriculum Ideological and Political Integration of Education Ideas

With the development of students as the center, pay attention to theory and practice, results-oriented education, science education and curriculum ideological and political integration in teaching. In the face of the changing environment at home and abroad and the global innovation transformation driven by a new round of scientific and technological revolution and industrial revolution, the development of science and technology is the primary productive force, and the historical mission of colleges and universities to cultivate application-oriented cross-compound talents with both virtue and ability has become the development trend of college education. The integration of science and education and the integration of curriculum, ideology and politics, teachers should take moral education as the foundation of education, not only to study hard in professional knowledge, but also to absorb diversified knowledge, professional knowledge and cultural literacy in one, to preach and enlighten students [6]. Based on the "trinity" talent training goal of value shaping, knowledge imparting and ability training, curriculum as the core element of talent training, value shaping is the first priority for course teaching [7–9].

3 Analysis of Teaching Design of "Mechanical Engineering Testing Technology" Course

The integration of science and education is to integrate scientific thinking into teaching, realize knowledge transfer, inheritance and innovation in the process of research-teaching-learning, and cultivate students' scientific thinking and innovation ability. Curriculum thought and politics is the combination of moral education thought and professional knowledge point, so that students in the professional learning of moral education thought edification and guidance. The integration of science and education and curriculum ideology and politics complement each other and support each other in order to train the builders of all-round development of morality, intelligence, physical fitness, the United States and labor. Specialized course is the important technical basis for engineering students to work, and the applied and comprehensive specialized course is taken as the research object. Taking "Mechanical engineering testing Technology" of mechanical design, manufacturing and automation specialty as an example, this paper studies how to apply the integration of science and education and curriculum ideology and politics into professional course.

"Mechanical Engineering Test Technology" is a compulsory course of the university. The research object is the working principle of sensors, signal conditioning circuits and recording instruments commonly used in dynamic testing of mechanical engineering, the evaluation method of static and dynamic characteristics, and the analysis and processing of test signals. The teaching content of the course focuses on the basic theory, basic method and application of test technology. Focus on training students to use the knowledge to solve practical measurement problems. It plays a connecting role in the whole teaching plan of mechanical majors, and is a bridge from the basic course to the professional course. At present, the total class hours of this course are 32, of which 24 are theoretical and 8 are experimental. To learn this course, taking courses such as "Advanced Mathematics", "Mechanical Precision Design and testing", "Principle and Application of single chip Microcomputer", and "Foundation of Control Engineering" first. This course has the characteristics of strong comprehensiveness, multi-disciplinary interlacing, various theories, difficult learning, emphasis on practice and wide range of applications. Based on students' development as the center, pay attention to theory and practice, results-oriented education, integrate science and education with curriculum ideology and politics in teaching, explore new teaching models, display abstract and difficult theoretical knowledge in the way of "multidimensional mind mapping", facilitate students to understand and memorize, connect knowledge with real life, and apply what they learn.

4 Teaching Design Process of "Mechanical Engineering Test Technology" Course

(1) Setting of course teaching objectives. The course "Mechanical Engineering Testing Technology" formulates the course teaching objectives according to the development trend of the machinery industry, the talents training opinions of the Ministry of Education and the orientation of the school, divides the teaching objectives into

the course content, designs corresponding teaching activities, and finally evaluates students' learning results through a multi-dimensional evaluation mode, and finally feeds back into the formulated course objectives. Formed the teaching process management of teaching-exam-evaluation curriculum. The corresponding relationship between the teaching objectives of the course "Mechanical Engineering Testing Technology" and the graduation requirements of the major of mechanical design, manufacturing and automation is shown in Table 1.

Table 1. The corresponding relationship between the teaching objectives of the course "Mechanical Testing Technology" and the graduation requirements of the mechanical major

Course teaching objective	Graduation requires two index points	Secondary index point support strength
Objective 1	Application analysis and design, processing and manufacturing, testing and control, system and integration capabilities and other professional abilities, responsible for the design, development, manufacturing and management of complex engineering systems under the multidisciplinary background of mechanical engineering and related fields	*M*
Objective 2	Be able to develop or select modern tools that meet specific needs for specific objects of complex engineering problems in the field of mechanical engineering, simulate and predict professional problems, and be able to analyze their limitations	*M*
Objective 3	Be able to set up the experimental system correctly, operate the experimental equipment correctly, determine the test parameters correctly, collect and organize the experimental data correctly	*H*

Remarks: *H*- High correlation; *M*- Medium association; *L*- Low correlation

(2) Course teaching design process. Taking the course "Mechanical Engineering Testing Technology" as an example, using the method of "multi-dimensional mind mapping", the content of the course is divided rationally to form several traditional mind maps. Then establish hyperlinks between related knowledge points, especially when a certain knowledge point appears in multiple different mind maps, the hyperlinks

will remind students what other concepts the knowledge point is related to in different chapters, and how the mind maps are related to each other. "Multidimensional mind map" is adopted to visually cover the knowledge points of "Mechanical Engineering Testing Technology" course and form a knowledge system. Micro-level, it is committed to making the knowledge points explanation vivid, intuitive, easy to read and understand -relying on the knowledge backbone provided by multidimensional mind map. Use multimedia technology (such as pictures, animations, videos, web links, etc.) to make abstract concepts concrete [10]. The basic content of the introduction to "Mechanical Engineering Testing Techniques" can be summarized by the mind map in Fig. 1. The main line of the course content can be expressed in Fig. 2. The contents of part 1 and 2 can be summarized in Fig. 3. From Fig. 1 to Fig. 3, the relationship of the mind map is shown in Fig. 4, where the two boxes connected by each arrow line represent interrelated concept groups. The effect of "multidimensional mind mapping" is shown in Fig. 4. By adding hyperlinks to the related concepts between different mind maps, the content of the whole course is formed as a whole, and the relationship between various knowledge points can be displayed at the same time.

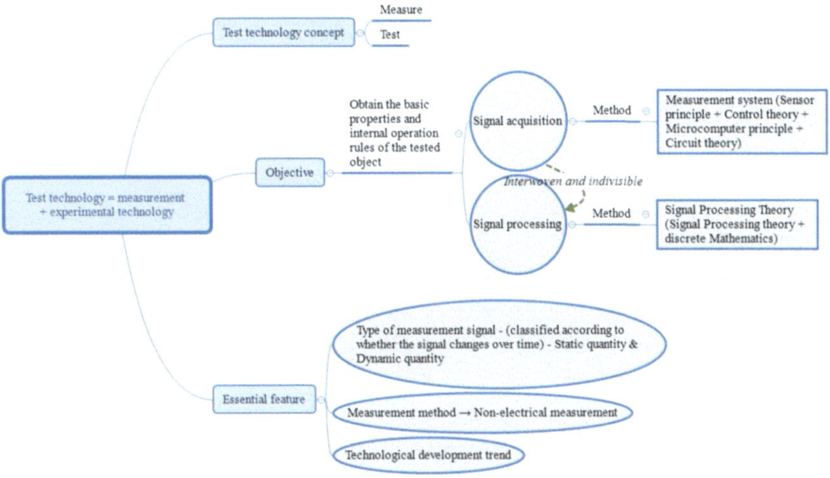

Fig. 1. Mind map of "Mechanical Engineering Test Technology Course Overview"

With the help of "multi-dimensional mind map", the relationship between knowledge points is well displayed, which is convenient for students to understand and remember. Application of science and education integration and curriculum ideological and political integration in the course chapter of "Mechanical Engineering Testing Technology". Taking the common sensors in Chapter 4 as an example, the common sensors are first linked by "multidimensional mind map", and then the abstract concept knowledge points are concretized, as shown in Fig. 5 below. The knowledge of piezoelectric sensor includes the working principle and piezoelectric

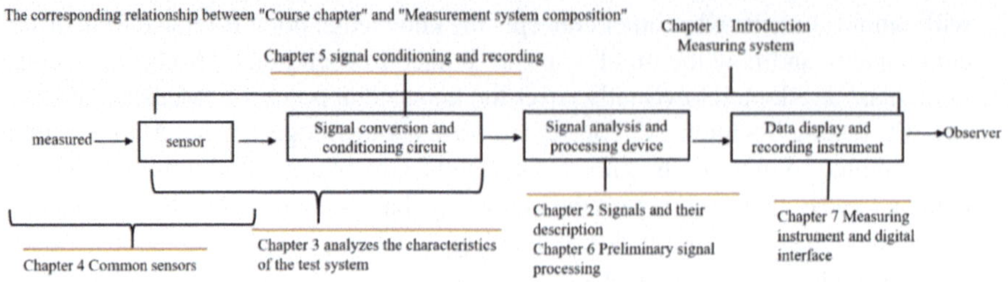

Fig. 2. Mind map of "Main content of Mechanical Engineering Test Technology Course"

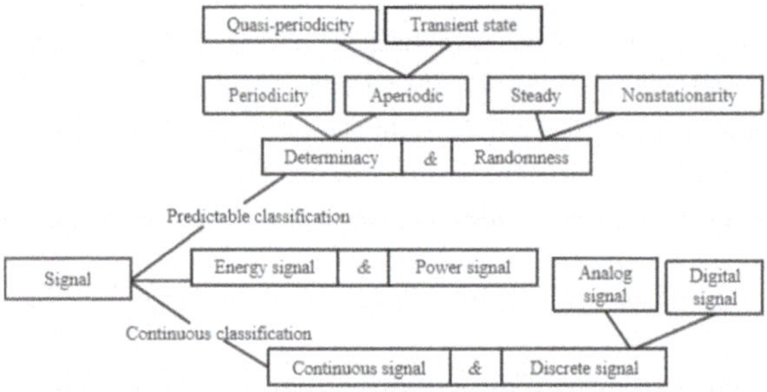

Fig. 3. Mind map of "Signal Processing and Analysis" (some screenshots)

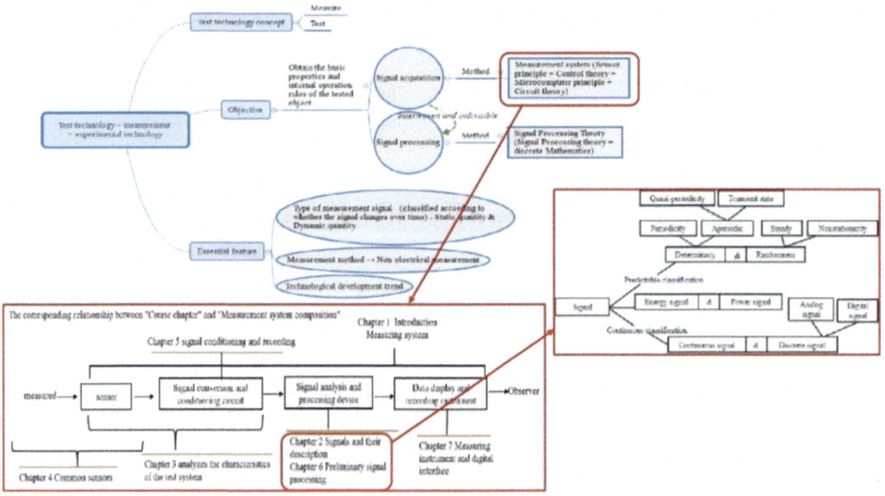

Fig. 4. Add hyperlinks between different mind maps

materials. Piezoelectric materials are mainly divided into piezoelectric monocrystals, piezoelectric ceramics and organic piezoelectric films. With the continuous development of science and technology, developers can mainly change the properties of piezoelectric materials and then develop new types of sensors. For example, the earliest barium titanate ceramic ($BaTiO_3$) materials can be doped to change the microstructure and properties of the material. If $CaHfO_3$ is doped, the piezoelectric coefficient can be effectively increased to *400 pC/N* [11]. The doping of $Sr(Zn_{1/3}Nb_{2/3})O_3$ changed its microstructure, as shown in Fig. 6, and also makes it become a new type of high-density energy storage device. The change process of energy storage is shown in Fig. 7 [12]. Small doping amounts of materials can produce different reactions, developing new products with high performance. In class, students not only learn theoretical knowledge, but also combine theory with practice and scientific research to stimulate students' enthusiasm for learning.

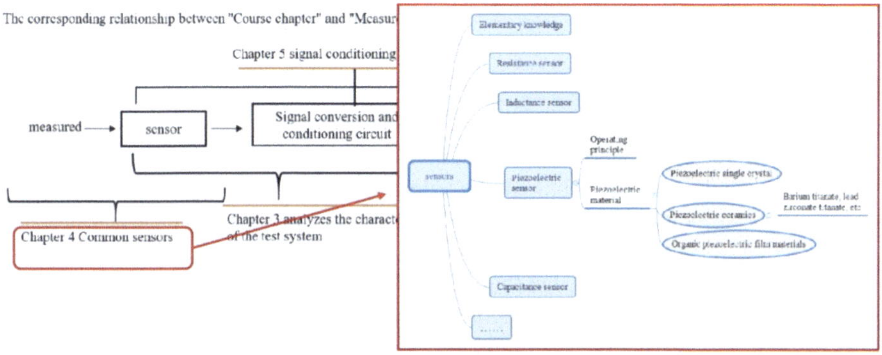

Fig. 5. Mind map of "Common sensors" (partial screenshots)

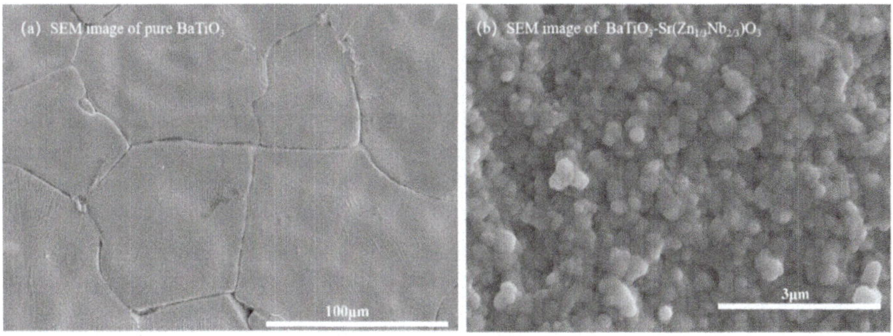

Fig. 6. SEM images of Barium titanate -based ceramics (a) Pure BaTiO3 ceramics, (b) BaTiO3 ceramics modified by Sr (Zn1/3Nb2/3)O3

It is not only limited to the change of the ratio of materials that will affect the performance of its products, but also leads students to review the previous course "Engineering Materials and Mechanical Manufacturing Fundamentals" and think that the preparation process of materials will also affect the performance of products. Similarly, other types of sensors and chapter knowledge points can also be taught in a similar way. The teaching content integrates the latest knowledge points of machinery, materials, electronics, information and control according to the content

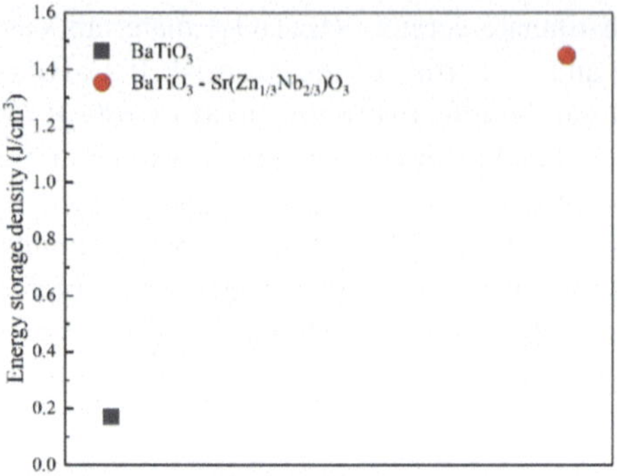

Fig. 7. The energy storage density between pure barium titanate and doping

of each chapter, so that students can feel the changes of computer, network technology, communication technology and other technologies in class. This includes the standardized description of basic knowledge, the exercise of comprehensive knowledge, the exploration of innovative knowledge, and the guidance of cutting-edge knowledge, forming an organic integration of certain functional modules or system knowledge, laying a knowledge foundation for improving students' engineering literacy, enabling students to have engineering system thinking ability and engineering practice ability, and effectively stimulating students' innovative spirit [13]. The latest research progress of various disciplines is integrated into the classroom teaching, which is not only the integration of science and technology, but also the scientific spirit, craftsman spirit and innovation spirit to carry out curriculum thinking and politics in the way of "touching things quietly", so as to achieve the integration of science and education in professional course teaching and the integration of curriculum thinking and politics.

(3) Course teaching process management. "Mechanical Engineering Testing Technology" course assessment to check students' mastery of knowledge points, covering three main course teaching objectives (Table 1). Course assessment is mainly divided into four parts, which are homework (*15%*), classroom performance (*10%*), experiment (*15%*), and final score (*60%*). The specific evaluation methods are as follows: 1) Daily work. The main test is whether the students' homework is completed and whether it is completed on time; 2) Classroom performance. Examine students' class discussion, attendance, quizzes and innovation ability (the ability to change a question, solve multiple problems, and win prizes in competitions); 3) Experiment. The practice and innovation ability of students are tested by the course training report; 4) Final exam result. It mainly tests students' mastery of theoretical knowledge and innovative ability (the ability of changeable and multi-solution of a problem).

The process of "teaching, examination and evaluation" of the curriculum is the key element to achieve the curriculum objectives. In the traditional teaching process, each link is isolated or information sharing is blocked, which is not conducive to the continuous improvement of teaching quality. Therefore, "teaching-exam-evaluation" should be a process of mutual correlation, interaction and dynamic adjustment. According to the feedback control idea of the course "Foundation of

Control Engineering", teaching management (" teaching") is regarded as the input link (with students as the control object), teaching assessment (" examination") is the output link, and teaching evaluation (" evaluation") is the feedback link to compare the input and output and correct the results. In this way, the curriculum teaching process management system has formed a closed-loop control system with the elements of "teaching, examination and evaluation" as shown in Fig. 8 (a), which realizes the course teaching process management with timely information sharing and feedback of all links. As shown in Fig. 8 (b), the teacher implemented teaching organization management for the course "Mechanical Engineering Testing Technology" in class. The management process of "teaching, examination and evaluation" firstly takes the form of the integration of science and education with the integration of curriculum ideology and politics, the integration of theory and practice, and the combination of virtual simulation and practice. Then add the classroom's project-based, heuristic, interactive guided classroom management mode to the students (i.e., "control object"). Then, students through effective classroom management through the assessment of the course, the assessment by multiple, multi-level, multi-faceted comprehensive evaluation, can effectively and comprehensively assess the comprehensive ability of students. Finally, the output of teachers' teaching results and students' learning results, and then feedback to "teaching" and "management" in the way of "evaluation", forming a closed-loop feedback adjustment. In this way to achieve dynamic curriculum teaching management adjustment.

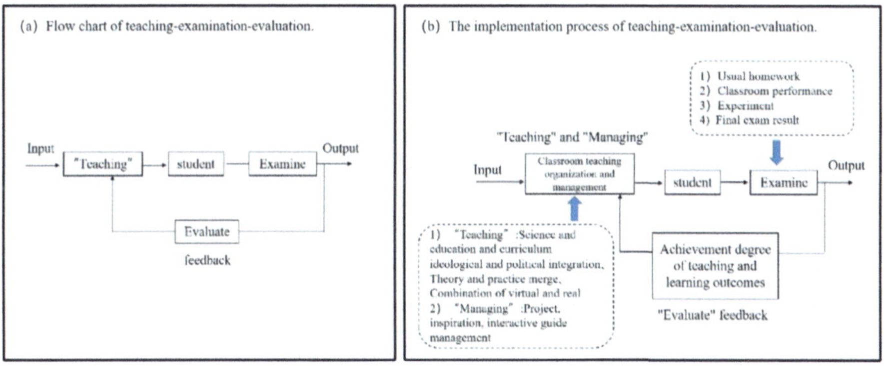

Fig. 8. The course teaching process management of "Mechanical Engineering Test Technology" teaching - examination - evaluation

5 Conclusion

Based on the concept of science and education integration and curriculum ideological and political integration, this paper demonstrates the relationship between the knowledge points of "Mechanical Engineering Testing Technology" with the help of "multi-dimensional mind mapping", integrates science and education with curriculum ideological and political integration.

1) Through the closed-loop control feedback of "teaching-exam-assessment", students can adjust the degree of achieving the course objectives.
2) Through multiple, multi-level and multifaceted evaluation, it not only examines students' mastery of theoretical knowledge points, but also further examines the integration of science and education and curriculum ideological and political integration.
3) It provides a good reference for the teaching reform of mechanical design, manufacturing and automation major in applied undergraduate colleges in promoting the further development of emerging engineering education.

Acknowledgment. This project is supported by the Project to improve the basic scientific research ability of young and middle-aged teachers in Guangxi universities (Grants No. 2023KY1855).

References

1. Lv, Y., Su, Y., Zhao, W.: Research on mechanical engineering teaching system based on integration of science, education, industry and perspective of new engineering. Advances in Educational Technology and Psychology **7**(9), 87–91 (2023)
2. Wang, Y., Luo, Y.: Research on the integration of ideological and political elements in the teaching of professional courses-taking the course of "automobile consumption psychology" as an example. OALib **08**(02), 1–5 (2021)
3. Xiong, R.: The practice and application of the concept of "integration" in the teaching of steel structure. J. Henan Radio and Television University **32**(04), 85–89 (2019). (in Chinese)
4. Yu, H., Liu, X.: The breakthrough of education principle and practice treatment of integration of theory and practice-taking "engineering cost management" course as an example. Contemporary Vocational Education **08**, 43–45 (2013). (in Chinese)
5. Chen, S., Tang, Q.: Research on integrated teaching reform of microcomputer principle and single-chip microcomputer technology course. Equipment Manufacturing Technology **06**, 260–262 (2016). (in Chinese)
6. Li, Y., Ji, N., Yuan, S.: Probe into probability and statistics teaching under the new development concept. J. Educational Research and Policies **5**(5) (2023)
7. Chen, X., Ge, M., Teng, Y., et al.: Exploration on ideological and political teaching reform of microeconomics from the perspective of applied talents training. International Journal of New Developments in Education **5**(15), 89–94 (2023)
8. Li, Y.F., Zhao, Q.F.: Research and practice of curriculum of network engineering in application-oriented undergraduate teaching. International Journal of Education and Management Engineering (IJEME) **1**(5), 43–48 (2011)
9. Gong, Y., Lu, X., Cui, D.: Research on the construction strategies of "curriculum ideology and politics" for professional courses. International J. Frontiers in Sociology **3**(8), 118–122 (2021)
10. Denggao, G., Jinghui, L., Jianping, L., et al.: Material Physics & Chemistry quality network curriculum construction and teaching practice. International Journal of Education and Management Engineering (IJEME) **2**(5), 24–30 (2012)
11. Yang, Y., Zhou, Y., Ren, J., et al.: Coexistence of three ferroelectric phases and enhanced piezoelectric properties in $BaTiO_3$–$CaHfO_3$ lead-free ceramics. J. Eur. Ceram. Soc. **38**(2), 557–566 (2018)

12. Meng, D., Feng, Q., Luo, N., et al.: Effect of $Sr(Zn_{1/3}Nb_{2/3})O_3$ modification on the energy storage performance of $BaTiO_3$ ceramics. Ceram. Int. **47**(9), 12450–12458 (2021)
13. Zhao, C., Wang, Y., Ma, J., et al.: Study on the examination pattern of the engineering specialized courses. International Journal of Education and Management Engineering (IJEME) **2**(2), 59–65 (2012)

The Impact of Art Competitions on the Enhancement of Social Skills Among Art Students in Science and Engineering University

Fang Li[1](✉), Peipei Wang[2](✉), and Xingyu Zhang[3]

[1] Quality Evaluation Department, Wuhan University of Technology, Wuhan 430070, China
fangfangdancing@163.com

[2] Graduate School, Wuhan University of Technology, Wuhan 430070, China

[3] School of Marxism, Wuhan University of Technology, Wuhan 430070, China

Abstract. With the aim of exploring the impact of art competitions on the social skills of art majors at Wuhan University of Science and Technology, this article conducted a survey of 280 questionnaires and conducted differential analysis and multiple linear regression analysis on the data obtained. The results indicate that there are gender differences in the influence of college students on art competitions held by schools. However, the research findings indicate that all college students are generally satisfied with art competitions and recognize the role of art competitions in improving their social practice performance. We can continuously improve the impact of art competitions on the artistic characteristics and comprehensive quality of science and engineering universities through internal and external joint driving, establishing a scientific evaluation system, paying attention to student differences, implementing classroom teaching, and creating unique brands.

Keywords: Art Competition · Social Practice Activities · University Students

1 Introduction

The Art major is of great practicality. Sustainable development of art major entails that universities promote its competitiveness and response to the reform and development of teaching method so as to cultivate more high quality talents for social development and break new path for talent-cultivation. In recent years, the art competition has become one of the important entry points for the cultivation of university students' practical ability [1]. Educating students through social practice is the key issue facing the universities. It calls for both reformed art competitions and social practice to improve the practical ability and professional capability [2]. Cui Yan liu et al. analyzed the development of innovation and entrepreneurship teaching for art majors in colleges and universities and found new problems. Based on this, they defined the orientation of the education mode of innovation and entrepreneurship competition for art majors in colleges and universities from the two levels of goals and concepts, and built a "hybrid, multi-dimensional and three-dimensional" education mode of innovation and entrepreneurship competition for

Z. Hu et al. (Eds.): ICCSEEA 2024, LNDECT 242, pp. 570–579, 2025.
https://doi.org/10.1007/978-3-031-84228-3_49

art majors. It provides new ideas for the cultivation of innovative talents in art in colleges and universities [3]. On the base of the international PBL education concept, Huo Kai et al. conducted a reform of the teaching of art and design competition. They studied and explored the teaching reform of art and designed competition based on the international PBL education concept from four aspects: teaching reform orientation, teaching reform content, art and design innovation, and art and design innovation competition reform, providing new ideas for art and design education [4]. It can be seen that art competition has been paid more and more attention by colleges and universities, which plays an indispensable role in improving their own competitiveness. In addition, in 2022, Huo Kai et al. studied the current situation of discipline competition under the background of first-class university construction, as well as the problems and shortcomings encountered in the development process, and pointed out the positioning of talent training construction, teacher team construction and education environment construction of discipline competition under the background of first-class university construction. It also puts forward the characteristic practice trend of the construction of art disciplines, excavates the regional cultural characteristics to deeply integrate with discipline competitions, integrates multi-disciplinary resources to concentrate advantages in the competition, and innovates with the help of new media, new means and new technologies in the future development [5].Taking the School of Art and Design of Wuhan University of Technology as a statistical sample, this paper analyzes the data differences in terms of gender, age, social practice evaluation and other dimensions to evaluate the current situation of university art competitions. This paper discusses the necessity and requirement of establishing innovation-driven talent training system, implementing academic competition and social practice, and start a new development path for the development of art major. At present, the development of art major tigers a series of relevant academic competitions in various categories, creating a vibrant atmosphere in this regard. Participating in academic competitions is an important way to nurture university students' ability of organizing, innovating and self-learning and to improve the teamwork spirit of students majored in art. The result of academic competitions is an indicator of the effect of education in art major and the rank of universities in academic competitions will impact on their reputation. As more competitions come out, more frequently the competitions are held and more high-level competitions spring up. It is a vital task of art major to deliver professional and innovative education by using art competitions to improve the performance in social practice, upgrading teaching quality and improving the quality of innovative talents cultivated, creating quality art talents that are needed by the society [6].

The competition education model has accelerated the development of high-quality professionals and unique talents, and fully implemented the competition-driven innovative human resource development model. It can improve students ' professional learning and comprehensive research ability [7]. Driven by high-level competition, students continue to improve and enhance their competition ability and professional level. This process can quickly improve the academic understanding of professional knowledge, accelerate the application of theoretical knowledge, and develop into a talent that needed by schools and even the country [8]. The improvement of students' comprehensive abilities has always been a concern for universities around the world [9–11]. Adopting various methods to strengthen student training and tap into their potential abilities and qualities is worth the continuous efforts of the education community [12].

2 Research Design

2.1 Questionnaire Design

This paper adopted questionnaire to carry out the survey and collected and analyzed the data, presenting the overall evaluation of the status quo and effect of university art competition; difference analysis was conducted on the evaluation of art competitions and social practice based on genders, age and majors; analysis and conclusion were made on the status quo and influence factor of art academic competitions in China's universities.

The questionnaire consists of three parts:

1) evaluation of art competition in terms of the content, coaching team and importance placed by the university;
2) evaluation of social practice in terms of the cooperation between companies and the school, social practice information exchanging platform, and the coaching teams of the school and company;
3) evaluation of the influence of art competitions on social practice, including its positive impact on enhancing the creativity and organizing ability of students and hence on students' performance in social practice. All single-choice questions were rated by 5-piont Likert scale and each indicator was divided into 5 scales.

Art majored students in Wuhan University of Technology has been the objection for the survey. 280 questionnaires were distributed with the valid callback rate of 88.43%. 247 students participated the survey, in which 40.92% were males and 49.08% were females; Sample distribution is shown in the Fig. 1.

The data on the dimension of student grade is freshmen accounted for 14.33%, sophomores 45.64%, juniors 21.67%, seniors 8.33%, and postgraduates 10%, Sample distribution is shown in the Fig. 2.

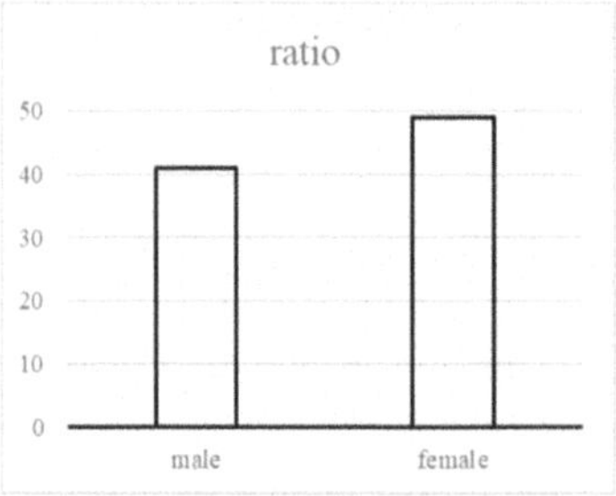

Fig. 1. Sample of genders

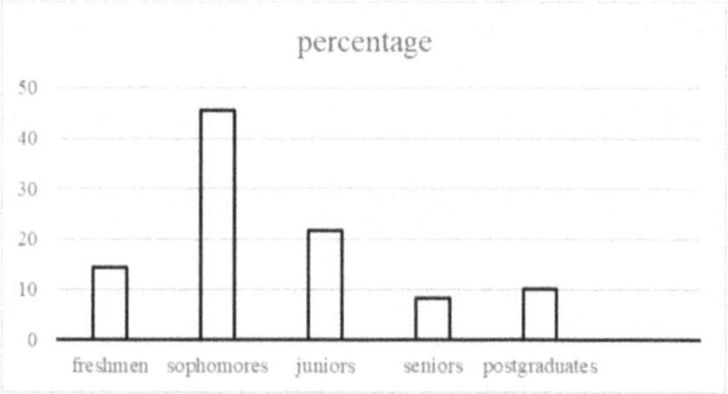

Fig. 2. Sample of ages distribution

2.2 Reliability and Validity Test

2.2.1 Reliability Test

In order to ensure the credibility of the conclusion, the reliability and validity of the returned questionnaires need to be tested. Reliability indicates the consistency of the questionnaire. Cronbach α was one of the major assets of classical test theory, which is the indicator of the internal reliability of the questionnaire [13]. The higher the result is, the more consistent the measures in the questionnaire are. The Cronbach α can be calculated as Eq. (1):

$$\alpha = \frac{K}{K-1}\left(1 - \frac{\sum_{i=1}^{K}\sigma_{Y_I}^2}{\sigma_X^2}\right) \quad ((1))$$

where, "K" is the number of samples, "$\sigma_{Y_I}^2$" is the variance of the total sample.

Usually, the Cronbach α coefficient has a value between 0 and 1. If the X coefficient does not exceed 0.6, it is generally considered that the internal reliability is insufficient; When it reaches 0.7–0.8, it indicates that the scale has considerable reliability, and when it reaches 0.8–0.9, it indicates that the reliability of the scale is very good. An important feature of Cronbach α coefficients is that their values increase with the increase of scale items. Therefore, Cronbach α coefficients may be artificially and inappropriately increased due to the inclusion of redundant measurement items in the scale. There is also a coefficient that can be used in conjunction with the Cronbach α coefficient. The coefficient can help to evaluate whether the calculation of the mean conceals some irrelevant measurement items in the calculation of the Cronbach α coefficient. Different researchers have different views on the limit value of the reliability coefficient. Some scholars believe that the Cronbach α coefficient should be at least 0.8 in basic research, 0.7 in exploratory research, and only 0.6 in practical research.

2.2.2 Validity Test

Validity, namely the effectiveness, refers to how well an instrument is as measures when it is designed to measure. Validity is the most important factor in evaluating the questionnaire and plays a decisive role in the value of the survey. This paper used KMO test to measure the questionnaire. Through factor analysis, it measured how suitable the questionnaire was for the hypothesis model came up by the researcher when designing it. The KMO can be estimated by (2).

$$KMO = \frac{\sum\sum_{i\neq j} r_{ij}^2}{\sum\sum_{i\neq j} r_{ij}^2 + \sum\sum_{i=j} p_{ij}^2} \quad ((2))$$

Kaiser gives a commonly used KMO metric: above 0.9 indicates a good fit; 0.8 indicates suitable; 0.7 indicates general; 0.6 indicates that it is not suitable. Below 0.5 indicates very unsuitable. The KMO statistic is between 0 and 1. When the sum of squares of the simple correlation coefficients among all variables is much larger than the sum of squares of the partial correlation coefficients, the KMO value is close to 1. The closer the KMO value is to 1, the stronger the correlation between variables is, and the original variables are more suitable for factor analysis. When the sum of squares of the simple correlation coefficients among all variables is close to 0, the KMO value is close to 0. The closer the KMO value is to 0, the weaker the correlation between variables, and the less suitable the original variables are for factor analysis.

The Cronbach α of the questionnaire was *0.845*, indicating a high reliability. KMO was *0.838*, suggesting a sound validity.

3 Empirical Analysis

3.1 Difference Analysis

This research adopted a questionnaire with 12 questions to collect needed data and then analyzed the overall evaluation of all participants, and evaluations of male university students and their female counterparts of university art competition's influence on social practice of students. Then, the evaluations of male and female students were analyzed by T-test to show if any significant differences existed. The T-test can be represented by (3):

$$t = \frac{\overline{X_1} - \overline{X_2}}{\sqrt{\frac{\sigma_{\overline{X}_1}^2 + \sigma_{\overline{X}_2}^2 - 2\gamma\sigma_{X_1}\sigma_{X_2}}{n-1}}} \qquad ((3))$$

where, "$\overline{x}$" is the average of sample X, "$\sigma^2 X$" is the variance of the total sample, and "Y" is the Correlation coefficients of relevant samples. The results are shown in Table 1.

Results are shown in Table 1, in general, the average score is higher than assumed, suggesting that university students speak highly of the art competition and the performance in social practice of the school and agree on the positive impact of art competitions on the performance in social practice. However, certain scores, including creativity of art competition and inter-disciplinary development, are lower than the average, showing that there is room for improvement.

The results of T-test show that, with the confident level of 10%, female students are general more satisfied with the positive impact of art competition on the performance in social practice than male students. The results of the T-test ($p<0.1$)of the professional level and inter-disciplinary development show that female students are more satisfied than their male counterparts while male students are more content than female ones in terms of publicity of art competitions.

In conclusion, universities should further promote the creativity of art competitions and broaden the disciplines evolving in art competitions to foster a sound atmosphere of social practice. In addition, universities should also make targeted adjustments in terms of professional level, inter-disciplinary development and publicity of art competitions to cater to students' need, boosting their satisfaction of art education.

3.2 Regression Analysis

In order to analyze the correlation between university art competitions and the performance in social practice, this paper adopted multiple linear regression method. Based on gender difference, the regression relationships were established by having the total score

Table 1. Analysis on the Results of the Questionnaire

Measures		Objects				
		Assumed average score	Overall	Male students	Female Students	T-test
			Average score	Average score	Average score	Significance
	Overall evaluation	30	32.26	31.356	33.167	
Art competition	Internationalization	3	3.175	3.079	3.271	0.140
	Professional level	3	3.14	2.967	3.313	0.078
	Importance	3	3.939	3.899	3.979	0.173
	Creativity	3	2.859	2.761	2.958	0.070
	Inter-disciplinary development	3	2.947	2.79	3.104	0.014
	Publicity	3	3.785	3.8	3.771	0.177
Influence of art competition on social practice	Organization	3	3.125	3.021	3.229	0.139
	Social practice base	3	3.165	3.101	3.229	0.572
	Information exchange platform	3	3.054	2.921	3.188	0.637
	Cooperation between the school and companies	3	3.071	3.017	3.125	0.199

of performance in social practice as the dependent variable and the score of art competitions as the independent variable. The respective regression models can be represented by (4) and (5), respectively:

$$Male\ Social\ practice\ level = \alpha_0 + \alpha \sum Art\ competition + \varepsilon \tag{4}$$

$$Female\ Social\ practice\ level = \alpha_0 + \alpha \sum Art\ competition + \varepsilon \tag{5}$$

The regression models reflect the correlation between dependent variable and independent variable as shown in the Table 2:

Model (1), from the perspective of male students majored in art, reflects the relation between art competitions and their impact on the performance in social practice. The results show significant positive correlation among the professional level, importance and publicity of art competitions which can improve the performance in social practice. In terms of the regression coefficient, publicity of art competition is the highest, meaning that it has the largest impact on the performance in social practice. Therefore, to better students' performance in social practice, universities should make targeted plans for male students majored in art, including promoting the publicity, professional level and importance for art competitions.

Table 2. The result of empirical analysis

	Model (1)	Model (2)
Internationalization	*0.563*	*0.421*
Professional level	*0.712***	*1.687****
Importance	*0.896****	*0.712*
Creativity	*1.266*	*0.456***
Inter-disciplinary development	*1.643*	*1.243***
Publicity	*1.702****	*1.875*
R^2	*0.542*	*0.762*
F	*46.60*	*104.69*

Model (2), from the perspective of female students majored in art, reflects the relation between art competitions and their impact on the performance in social practice. The results show significant positive correlation among the professional level, creativity and inter-disciplinary development of art competitions which can improve the performance in social practice. Professional level has the highest regression coefficient, meaning that it has the largest impact on the performance in social practice. Therefore, from the view of female students, universities can better the performance in social practice by strengthening the professional level, creativity and inter-disciplinary development of art competitions.

4 Optimistic Path

4.1 On-campus and Off-campus Circulation Drive

To promote students' social practice activities through art competitions, it is necessary to promote the combination of teaching design and work practice to maximize the overall effectiveness [13]. Therefore, the school should build a social practice base, combine the actual needs of the industry and enterprises, and combine the content of the competition with the specific course content and professional training objectives. It integrates competition, learning and practice, and takes competition as an opportunity to improve the professional function and innovation consciousness of participating students, so to improve the professional level and practical ability [14].

4.2 Establishing Scientific Evaluation System

To measure the quality of the art competition, the key is to see the actual effect of the competition on the practical activities of university students. The scientific evaluation system can not only help students sublimate individual experience into rational knowledge, but also find out the problems in practice. Quality evaluation can be improved only by combining qualitative analysis with quantitative analysis effectively and combining empirical methods with humanistic methods [15]. Do a good job of data statistics after the art competition in time, and carry out structured analysis through scientific statistical techniques to ensure the creativity and practicality of the art competition.

4.3 Focus on Student Differences to Achieve Personalization

The art competition should be designed according to the characteristics of the school and the needs of the students. We should adhere to the unity of theory and practice, and strengthen the organic combination of teaching and practice [16]. Adhere to the student-centered personalized teaching, pay attention to the differences in men and women, personal personality, educational background, academic performance and other aspects, create different types of art competition, change the past education mode focusing on theoretical teaching and basic knowledge teaching. Through competition, students' social practice ability can be trained, team cooperation consciousness can be cultivated, knowledge in different fields can be integrated, and interdisciplinary development can be improved [17]. While improving the professional ability, through the accumulation of practical experience to improve the overall quality of students, to ensure that every student free and comprehensive development.

4.4 Create a Brand of Art Competition

According to the above statistics, it can be concluded that the current students have a high evaluation of the art competition and social practice in universities. However, there are still deficiencies in professional level and publicity, which reflects that university students' participation in art competitions is not enough, and the influence of art competitions needs to be improved. Therefore, universities should increase publicity efforts, according to the characteristics of regional and school discipline construction, to create a brand with its own characteristics, and then to mobilize the enthusiasm and initiative of university students to participate in the competition, to achieve the broad participation of students.

5 Conclusion

Art competition in universities plays a vital role in improving the students' performance in social practice.It helps to improve students' working ability and cultivate students' awareness of innovation and entrepreneurship. Universities should be in line with the specific need of certain artistic fields to carry out art competitions with a targeted aims. Being adjusted according to the advices from teachers and students, art competitions can be held in a more reasonable manner. The goal of competition should be set in accordance with students' need and their development levels and characteristics. The social practice arrangement should give prominence to both theoretical and practical capabilities to create all-round talents.

According to salient features of different regions and locations of campus, targeted market surveys should be conducted to set competition goals with unique characteristics and to design creative competition models rather than following the trend blindly.

Creativity and originality are the essence of art. Universities should combine art competitions with social practice and chart a better course for art major by adopting an education system that gives equal attention to academic competitions and social practice, delivering creative art talents to the society. The improvement of the organization and management mechanism of art competition is a long-term and continuous process. Only when the system is perfect, the division of labor is clear, the departments cooperate with each other, and the teachers and students actively participate, can the ideal competition results be obtained, and the art competition can be used as the traction to promote the continuous improvement of art education in university and universities.

Universities should improve the publicity through various channels in order to increase the popularity of art competitions among students in relevant or other majors. Such effort will also add weight to the influence of art competitions, motivating students to participate in and thus boosting their performance in social practice.

In the research process, these research methods are more likely to lead to biased results due to the probability problem of random sampling, insufficient sample size, limited references related to the research topic and other factors.

References

1. Wezhi, Y., LiuCheng, Z.: Research on the improvement of aractical qbility of economic and management student by subject competition. Laboratory Science **22**(2), 213–217 (2019)
2. Li, P.: Research on the application of experience theory in the social practice activities of university students. J. Social Science and Humanities **4**(4), 24–28 (2022)
3. Cui, Y., Huo, K.: Research and practice of "mixed, multi-dimensional and three-dimensional" innovation and entrepreneurship competition education model for art majors in universities. Theoretical research and Practice of Innovation and Entrepreneurship **6**(06), 108–112 (2023)
4. Huo, K., Zhou, J.: Research on teaching reform of art design competition based on international PBL education concept. Theoretical Research and Practice of Innovation and Entrepreneurship **6**(02), 156–160 (2023)
5. Huo, K., Lv, Y.: Practice research of art discipline competition under the background of first-class university construction. Art and Design (Theory) **2**(06), 140–142 (2022)
6. Li, J, Lin, J.: Promoting the cultivation of innovative talents through combination of practical teaching and academic competition. Experimental Technology and Management **28**(11), 1–3,16 (2011)
7. Chen, K.: Reflection and Prospect: rethinking the social practice of postgraduates in the new era. Degree and Postgraduate Education **08**, 73–77 (2021)
8. Zhang, H., Tang, Z.: The historical experience of social practice education for young students since the founding of New China. Ideological and Theoretical Education Guide **11**, 127–133 (2022). (in Chinese)
9. Liu, Z., Fei, J., Wang, F., et al.: Study on higher education service quality based on student perception. International Journal of Education and Management Engineering (IJEME) **2**(4), 22–27 (2012)
10. Oyetoke, O.O.: Leveraging information technology in automating school management and student activities by successfully integrating a java- based school management application software. International Journal of Information Technology and Computer Science (IJITCS) **8**(1), 22–32 (2016)

11. Dada, E.G., Alkali, A.H., David, O.: An investigation into the effectiveness of asynchronous and synchronous elearning mode on students' academic performance in national open university (NOUN), Maiduguri Centre. International Journal of Modern Education and Computer Science (IJMECS) **11**(5), 54–64 (2019)
12. Al-Hagery, M.A., Alzaid, M.A., Alharbi, T.S., et al.: Data mining methods for detecting the most significant factors affecting students' performance. International Journal of Information Technology and Computer Science (IJITCS) **5**, 1–13 (2020)
13. Jin, J.: Countermeasures for Deepening University Students' Social Practice. Chinese Youth Studies (07), 72–73+86 (2007). (in Chinese)
14. Chen, Y., Liu, J., Tang, X., et al.: Exploring and practicing the teaching mode of combining competition with teaching for engineering university students. China Management Informatization **18**(13), 246–248 (2015). (in Chinese)
15. He, Y., Yang, M.: Study on quality evaluation system for university students. International Journal of Wireless and Microwave Technologies (IJWMT) **1**(3), 29–34 (2011)
16. Zou, H., Liu, Y., Liu, M.: Discussion on the cultivation of practical ability of science and engineering students in application-oriented university. Computer Knowledge and Technology **16**(22), 97–98,101 (2020). (in Chinese)
17. Huo, K., Wei, X.: Research on the effectiveness of art discipline competition in the context of double first-class. Scientific Consulting (Educational Research) **02**, 54–56 (2023)

Exploring the Mediating Role of Learning Behavior and Parental Participation in Minors Based on PLS-SEM Model

Jing Zuo[1], Yujie Huang[1], and Yanxin Ye[2](✉)

[1] Nanning University, Guangxi 530200, China
[2] Guangxi Modern Polytechnic College, Guangxi 547000, China
dariayyx@126.com

Abstract. This study used PLS-SEM to explore how parents' educational beliefs, parental involvement, and minors' learning behaviors jointly impact minors' academic performance. Based on the data from the China Family Panel Studies (CFPS) in 2020, we analyzed 3631 samples. The research results show that parents' educational beliefs do not have a significant direct impact on minors' academic performance, but their indirect impact through parental involvement and minors' learning behaviors significantly and positively influences minors' academic performance. Specifically, parents' educational beliefs positively influence parental involvement and minors' learning behaviors, further promoting minors' academic achievement. This study enriches the theoretical framework of the impact of family education on minors' academic achievement.

Keywords: The learning behavior of minors · Parental educational philosophy · PLS-SEM model · Parental involvement

1 Introduction

High-quality education can cultivate talents with innovative abilities and critical thinking, providing individuals with personal development opportunities and laying the foundation for national economic development [1]. In the education process, parents and minors play indispensable roles. Parents' educational beliefs and involvement, as well as minors' learning behaviors, together form a complex network that influences students' academic performance. Research has shown that parents' educational beliefs have profound effects on minors' educational outcomes. These beliefs not only directly affect minors' learning motivation but also indirectly impact through influencing parents' behavior [2–4]. For example, parental involvement, including monitoring study, assisting with homework, and participating in school activities, has been shown to have a positive effect on improving minors' academic performance [4]. In addition, students' learning behaviors, such as studying hard, completing and reviewing homework, and maintaining concentration, are also key factors influencing academic achievement [5]. Scholars have conducted some research by focusing on the influence and role of parents in children's growth, and trying to involve parents in their children's education [6, 7]. However, despite valuable insights

Z. Hu et al. (Eds.): ICCSEEA 2024, LNDECT 242, pp. 580–590, 2025.
https://doi.org/10.1007/978-3-031-84228-3_50

provided by existing literature, research on how parental educational beliefs jointly influence academic performance through parents' and minors' behaviors is relatively scarce. Particularly, how parental educational beliefs translate into parental involvement and students' learning behaviors and subsequently influence academic performance is still in need of further exploration. Through quantitative research methods, this study will evaluate the key pathways and mechanisms in this process, aiming to provide empirical evidence for education policy makers and practitioners to promote the improvement of students' academic performance.

2 Literature Review

2.1 The Relationship Between Parental Educational Beliefs and Minors' Academic Performance

When exploring the impact of parental educational beliefs on minors' academic performance, the Expectancy-Value Theory provides a theoretical framework. This theory emphasizes how individuals' expectations of success and the value of tasks jointly influence their achievement motivation and behavior [8]. In an educational context, parents' educational beliefs may affect their expectations of educational outcomes, which, in turn, influence minors' learning motivation and behavior, thus impacting academic performance. According to Kaplan et al. (2001), the relationship between parents' own educational achievement and their expectations for minors' academic performance is influenced by parents' self-perceptions. These self-perceptions further affect parents' expectations for minors' current academic achievement [9]. Furthermore, McGrath and Repetti (2000) found a positive correlation between mothers' satisfaction with minors' academic performance and minors' perception of their academic abilities [10]. Another study on the relationship between family educational practices and minors' intelligence and academic performance further confirms the impact of parents' educational beliefs on minors' academic performance [2]. Finally, Campbell et al. (1991) found that minors whose mothers received early childhood education interventions scored lower on traditional educational beliefs, which were negatively correlated with minors' reading achievement. This finding reveals the potential impact of parents' educational beliefs on minors' academic performance [3]. Based on these, the following hypothesis is proposed:

Hypothesis 1 (*H1*): Parental educational beliefs have a positive effect on minors' academic performance.

2.2 The Relationship Between Parental Educational Beliefs, Parental Involvement, and Children's Academic Performance

The Theory of Planned Behavior (TPB) provides a theoretical basis for how parental educational beliefs influence minors' academic performance through parental involvement. This theory emphasizes three key determinants of individual behavior: attitude (positive or negative evaluation of the behavior), subjective norm (the pressure individuals feel from important others to perform or not perform certain behaviors), and perceived behavioral control (individuals' perceptions of the ease or difficulty of performing the behavior) [10]. According to TPB, parental educational beliefs form the basis for the attitude toward educational involvement. When parents believe that being involved in their minors' education significantly improves their academic performance, this positive attitude can enhance their intent to participate in educational activities. Georgiou and Tourva's (2007) study supported this point, finding that parents' positive educational beliefs can positively influence actual parental educational involvement by enhancing the belief in "involvement" [11]. Additionally, Reed et al.'s (2000) study found that parents' construction of their role in their minors' education and the invitation and expectation of involvement from the school are key factors influencing parental educational involvement decision-making. This research provides empirical support for understanding how parental educational beliefs influence their educational involvement [12].

Similarly, subjective norm and perceived behavioral control in TPB also play an important role in the relationship between parental educational involvement and minors' academic performance. If parents perceive the social expectation for them to play a positive role in their minors' education and believe they have the ability to participate effectively (e.g., helping with homework, communicating with teachers), they are more likely to engage in these educational involvement behaviors. Khajehpour and Ghazvini's (2011) study showed that various forms of involvement in the family (such as checking minors' homework, discussing school topics at home) are associated with minors' better academic performance [4]. A review of 42 studies conducted between 2003 and 2021 also confirmed that there is a positive correlation between parental involvement and minors' academic achievement [13]. Therefore, the following hypotheses are proposed:

Hypothesis 2 (H2): Parental educational beliefs have a positive effect on parental involvement.
Hypothesis 3 (H3): Parental involvement has a positive effect on minors' academic performance.
Hypothesis 4 (H4): Parental involvement mediates the relationship between parental educational beliefs and minors' academic performance.

2.3 The Relationship Between Parental Educational Beliefs, Minors's Learning Behavior, and Children's Academic Performance

Understanding how parental educational beliefs influence minors' academic performance through their impact on minors' learning behavior can also be analyzed further using the Theory of Planned Behavior (TPB). Parents' educational beliefs also subtly influence their minors' attitudes toward learning, which then affects their learning behavior, ultimately leading to learning outcomes. Bubić and Tošić's (2016) study showed that parents' self-efficacy and beliefs in their minors' abilities predicted their educational involvement behavior, which, through providing learning models and reinforcing positive learning, influenced minors' learning behavior [14]. Chi and Rao (2003) found in their study of rural Chinese families that parents generally believe that effort is crucial to improving minors' academic performance. This belief then affects minors' learning behavior and academic achievement [15]. And students' learning behavior, including attitudes, habits, and strategies, are key factors influencing academic performance. Tus (2020) found that students' actual learning behaviors, such as regular reviewing and active participation in classroom discussions, directly impact academic performance

[5]. Cobb (2003) also found that students' self-regulated learning behaviors, while varying across different courses, are significantly related to academic performance [16]. Therefore, the following hypotheses are proposed:

Hypothesis 5 (H5): Parental educational beliefs have a positive effect on minors' learning behavior.
Hypothesis 6 (H6): Children's learning behavior has a positive impact on their academic performance.
Hypothesis 7 (H7): Children's learning behavior mediates the relationship between parental educational beliefs and minors' academic performance.

3 Methodology

3.1 Sample Description

The data used in this study is from the *2020* China Family Panel Studies (CFPS) conducted by the China Social Science Survey Center at Peking University. This survey is a large-scale comprehensive social tracking survey covering 25 provinces, municipalities, and autonomous regions nationwide, known for the representativeness and universality of its data. The CFPS survey contains four main questionnaires: the household member questionnaire, the household economy questionnaire, the individual self-response questionnaire, and the parent proxy questionnaire for minors aged *0* to *15*. This study focuses on the data from the parent proxy questionnaire collected in *2020*, with the screening criteria that participants are in primary school or higher educational stages. After excluding missing data, the final effective sample size is *3631*. Among them, there are *1681* females (*46.3%*) and *1950* males (*53.7%*). Specifically, there are *1275* minors aged 6–9 (35.11%), 1646 minors aged *10–13* (*45.33%*), and *710* minors aged *14–16* (*19.55%*).

3.2 Variables

3.2.1 Dependent Variable

Academic performance is the dependent variable in this study, measured by the average scores of minors in Chinese and mathematics.

3.2.2 Independent Variables

The independent variable is parental educational attitudes, using items from the questionnaire about parents' attitudes towards raising minors, including two items: "If necessary, parents should cut down on living expenses to pay for their minors' education" and "I have a great responsibility for my child's academic performance", both of which are on the Likert 5-point scale.

3.2.3 Mediating Variables

The first mediating variable is parental educational involvement, using items from the questionnaire about parents' involvement in their minors' education, including 6 items: such as giving up watching TV in conflict with the child's study, discussing school matters with the child, requiring the child to complete homework, checking the child's homework, preventing or stopping the child from watching TV or videos, and restricting the types of TV or video programs the child watches, all of which are on the Likert 5-point scale.

The second mediating variable is minors' study behavior, using items from the questionnaire about minors' study behaviors, including 4 items: studying hard, checking the work several times after completing homework, only playing after completing homework, and focusing when doing tasks, all of which are on the Likert 5-point scale.

3.2.4 Control Variables

Gender (male or female), educational stage (primary school, junior high school, or senior high school), participation in coaching classes (yes or no), and household registration location (urban or rural).

3.3 Data Analysis

This study uses Partial Least Squares Structural Equation Modeling (PLS-SEM) for analysis, mainly for two reasons. First, PLS-SEM is more suitable for exploratory research [17], and the dynamic relationship between parental educational attitudes, parental educational involvement, minors' study behavior, and academic performance explored in this study has not been fully revealed in the literature, also making it an exploratory study. Second, PLS-SEM is suitable for testing mediating effects by estimating indirect effects of latent variables, effectively revealing the complex mediating mechanisms between variables [17].

4 Result

4.1 Measurement Model Analysis

Before conducting Structural Equation Modeling (SEM) analysis, the reliability and validity of the measurement model were evaluated to ensure that the scales used accurately reflect the latent variables. Through preliminary analysis, it was found that some item factor loadings of parental educational attitudes and parental educational involvement were below 0.5, and these low-loading items were deleted to improve the model's reliability. After adjustments, both parental educational attitudes and academic achievement are single-item measurements, so the reliability analysis of these constructs is no longer displayed separately. For the remaining variables, their reliability and validity are presented in Tables 1 and 2.

From the analysis results in Table 1, the adjusted measurement model shows good reliability and validity overall in parental educational involvement and minors' study behavior. Although the Cronbach's Alpha (CA) value for parental educational involvement is *0.515*, slightly lower than the usually recommended threshold, indicating weak internal consistency of this variable, the Composite Reliability (CR) and Average Variance Extracted (AVE) both exceed the recommended thresholds of *0.7* and *0.5*, indicating that despite slight reliability issues, the overall reliability and validity of this construct are still credible. Data from Table 2 demonstrates that the analysis of the correlations between variables verifies the discriminant validity of the model, with each variable showing lower correlations with other variables compared to their own square root AVE values, clearly indicating good differentiation between the variables.

Table 1. Construct reliability, validity, and factor loadings

Constructs	CA	CR	AVE	Item	Loadings
Parental Involvement	*0.515*	*0.805*	*0.673*	PI2	*0.819*
				PI4	*0.823*
Children's Learning Behavior	*0.779*	*0.855*	*0.596*	CLB1	*0.838*
				CLB2	*0.777*
				CLB3	*0.712*
				CLB4	*0.756*

Table 2. Correlation of constructs

Constructs	AP	PEB	PI	CLB
Academic Performance	*1.000*			
Parental Educational Beliefs	*0.025* *(0.025)*	*1.000*		
Parental Involvement	*0.190* *(0.265)*	*0.070* *(0.097)*	*0.821*	
Children's Learning Behavior	*0.270* *(0.289)*	*0.117* *(0.135)*	*0.015* *(0.062)*	*0.772*

Note: AP = Academic Performance; PEB = Parental Educational Beliefs; PI = Parental Involvement; CLB = Children's Learning Behavior; square root of AVE is presented in diagonal; value within bracket is the value of HTMT ratio

4.2 Structural Model Analysis

Analysis of the path coefficients, R^2, f^2, and Q^2 of the model shows from Fig. 1 that parental educational attitudes have a significant positive impact on parental involvement in education ($\beta = 0.070, p < 0.001$) and on minors' learning behavior ($\beta = 0.117, p < 0.001$), indicating a positive effect on both parental involvement and minors' learning behavior. However, the direct impact of parental educational attitudes on academic performance is not significant (*β = -0.020, p = 0.206*), indicating that parental educational attitudes do not significantly influence academic performance. The influence of parental involvement on academic performance is significant (*β = 0.189, p < 0.001*), as is the influence of minors' learning behavior on academic performance (*β = 0.269, p < 0.001*), further emphasizing the important roles of parental involvement and minors' learning behavior in improving academic performance.

In terms of the model's explanatory power, the R^2 value for academic performance is *0.107*, indicating that the model variables can explain *10.7%* of the total variance in academic performance, suggesting a relatively weak explanatory level (*0.190* is weak). The R^2 value for parental involvement is *0.005*, and for minors' learning behavior is *0.014*, indicating limited explanatory power of the model in these areas, requiring further research to explore other factors influencing these variables. In terms of effect size (f^2), the impact of parental educational attitudes on parental involvement ($f^2 = 0.005$) and minors' learning behavior ($f^2 = 0.014$) is considered low (*0.020* is low). The impact of parental involvement on academic performance ($f^2 = 0.039$) and of minors' learning behavior on academic performance ($f^2 = 0.080$) also demonstrate low effects. The results of predictive relevance (Q^2) for academic performance is *0.105*, for parental involvement is *0.003*, and for minors' learning behavior is *0.008*, all exceeding *0*, indicating a certain predictive power of the model for these variables.

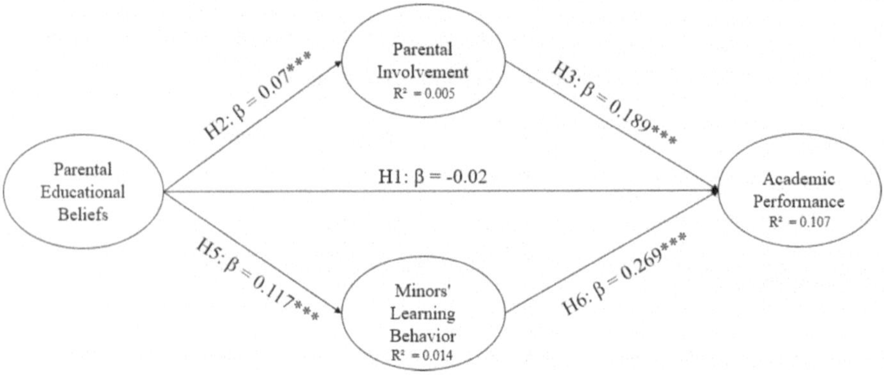

Fig. 1. The results of the structural model. Note: ***p < .001.

4.3 Mediating Effects

The analysis of mediating effects (see Table 3) reveals several key findings. Firstly, the direct impact of parental educational attitudes on academic performance is not significant (*β = -0.02, p = 0.21*), indicating that parents' educational attitudes do not directly affect minors' academic performance. However, through the indirect path analysis of parental involvement and minors' learning behavior, it is shown that parental educational attitudes have a significant positive indirect impact on minors' academic performance. These findings align with the model of Indirect-only mediation, indicating that the relationship between parental educational attitudes and minors' academic performance is fully realized through the mediating variables of parental involvement and minors' learning behavior.

Table 3. Mediating effects on the structural model paths

Path	Effects	Estimate	Bootstrap 5000 Times			Percentile 95%	
			STDEV	T-stat	P values	Low	Upper
PEB → PI → AP	Direct	*−0.02*	*0.02*	*1.26*	*0.21*	*−0.05*	*0.01*
	Indirect	*0.01*	*0.00*	*3.70*	*< 0.001*	*0.01*	*0.02*
	Total	*0.02*	*0.02*	*1.46*	*0.14*	*−0.01*	*0.06*
PEB → CLB → AP	Direct	*−0.02*	*0.02*	*1.26*	*0.21*	*−0.05*	*0.01*
	Indirect	*0.03*	*0.01*	*6.19*	*< 0.001*	*0.02*	*0.04*
	Total	*0.02*	*0.02*	*1.46*	*0.14*	*−0.01*	*0.06*

Note: AP = Academic Performance; PEB = Parental Educational Beliefs; PI = Parental Involvement; CLB = Children's Learning Behavior

5 Discussion

This study explores the comprehensive impact of parental educational attitudes, parental involvement, and minors' learning behavior on minors' academic performance. The analysis results indicate that the direct impact of parental educational attitudes on minors' academic performance is not significant, indicating that holding positive educational attitudes alone, without actual parental involvement and promotion of minors' learning behavior, does not have a direct positive impact on minors' academic performance. Although the direct impact of parental educational attitudes on minors' academic performance is not significant, it has a significant positive impact on minors' academic performance through the indirect path of parental involvement and minors' learning behavior. These results further emphasize the importance of the interaction of various factors in the family educational environment and the indispensable mediating role of parental behavior in promoting minors' academic achievement. In addition, in this study, the values of R^2 and f^2 are not high, indicating that parental educational attitudes, parental involvement, and minors' learning behavior are important factors influencing academic performance, but other factors are worth further exploration; and in designing educational interventions, multiple factors need to be considered, and solely relying on increasing parental involvement and improving minors' learning behavior may not be sufficient to achieve significant academic performance improvement.

This study enriches the theoretical framework in the field of family education research, particularly in explaining how parental educational attitudes influence minors' academic performance through the indirect path of parental involvement and minors' learning behavior. This aspect has not been fully explored in previous studies. The results provide insights for education policy makers and practitioners, emphasizing the potential pathways to improve family education quality and promote minors' academic performance, particularly by cultivating positive parental educational attitudes, promoting parental involvement in education, and enhancing minors' learning behavior as effective strategies.

This study underscores the importance of parental engagement, home-school collaboration, and fostering positive learning environments to improve minors' academic achievement. Recommendations include launching educational awareness programs for parents, promoting effective communication channels between schools and homes, and encouraging parents to actively support their minors' learning through discussion, exploration, and setting up a conducive study environment at home. These strategies aim to enhance minors' learning motivation and confidence, highlighting the pivotal role of parental attitudes and behaviors in educational outcomes.

However, this study still has some limitations. Because it uses existing database data, the measurement of the data may not fully meet the operational requirements of the variables, particularly in the measurement composition of the variables. In addition, the internal consistency of parental involvement is relatively low, which may affect the interpretation of the results. Future research can use more refined scales to measure parental involvement and minors' learning behavior to improve the accuracy and reliability of the measurements. At the same time, exploring other possible mediating or moderating variables, such as family socioeconomic status, cultural background, etc., can provide a more comprehensive understanding.

6 Conclusion

This study delves into the impact of parental educational beliefs on the academic performance of minors through a PLS-SEM analysis, exploring the indirect influence mediated by parental involvement in education and the learning behavior of minors. The findings reveal that the direct impact of parental educational beliefs on minors' academic performance is not significant. However, the indirect impact through parental involvement and minors' learning behavior is significantly positive. This underscores the critical importance of focusing on parental educational beliefs and behaviors to enhance the academic performance of minors.

From a theoretical perspective, this research offers a new lens through which to understand the influence of the family educational background on minors' academic achievement. Traditionally, the focus has been on the direct effects of parental educational beliefs. Our study, however, illustrates that these effects are more significantly manifested through the mediating roles of parental educational practices and the learning behaviors of minors. This finding suggests that parental educational beliefs need to be translated into specific behaviors to effectively contribute to beneficial academic outcomes for minors, thereby providing important guidance for family educational practices and policies.

Moreover, the outcomes of this research have practical implications for educational policymakers and practitioners. It emphasizes that efforts to improve minors' academic performance should not only focus on the quality of school education but also consider the role of the home environment and parental behaviors. Specifically, strategies for parental involvement and support for minors' learning behaviors represent effective ways to enhance academic achievements.

Acknowledgment. This project is supported by Project of Nanning University Education and Teaching Reform Project (2022XJJG35).

References

1. Zuo, J., Huang, T.: Research on the influencing factors of students' general competency improvement in application-oriented universities. In: International Conference on Computer Science, Engineering and Education Applications. Cham: Springer Nature Switzerland, pp. 1000–1010 (2023)
2. Kaviyarasi, R., Balasubramanian, T.: Exploring the high potential factors that affects students' academic performance. International Journal of Education and Management Engineering (IJEME) **8**(6), 15–23 (2018)
3. Campbell, F.A., Goldstein, S., Schaefer, E.S., et al.: Parental beliefs and values related to family risk, educational intervention, and child academic competence. Early Childhood Research Quarterly **6**(2), 167–182 (1991)
4. Khajehpour, M., Ghazvini, S.D.: The role of parental involvement affect in children's academic performance. Procedia Soc. Behav. Sci. **15**, 1204–1208 (2011)
5. Tus, J.: The influence of Study Attitudes and Study Habits on the Academic Performance of the Students. Ijarwl ISSN (o)-2582-1008 October **2**(4) (2020)
6. Angon, P.B., Mondal, S., Das, C.R., Bishnu, M.K.: Behavioral changes of children intelligence for the extreme affection of parents. International Journal of Education and Management Engineering (IJEME) **11**(6), 20–28 (2021)
7. Razak, S.F.A., Abdurahim, B., Mashhod, F.: Keeping parents involved using 360-class monitoring application. International Journal of Modern Education and Computer Science (IJMECS) **8**(12), 10–16 (2016)
8. Wigfield, A., Eccles, J.S.: Expectancy–value theory of achievement motivation. Contemp. Educ. Psychol. **25**(1), 68–81 (2000)
9. Kaplan, D.S., Liu, X., Kaplan, H.B.: Influence of parents' self-feelings and expectations on children's academic performance. J. Educ. Res. **94**(6), 360–370 (2001)
10. McGrath, E.P., Repetti, R.L.: Mothers' and fathers' attitudes toward their children's academic performance and children's perceptions of their academic competence. J. Youth Adolesc. **29**(6), 713–723 (2000)
11. Georgiou, S.N., Tourva, A.: Parental attributions and parental involvement. Soc. Psychol. Educ. **10**, 473–482 (2007)
12. Reed, R.P., Jones, K.P., Walker, J.M., et al.: Parents' Motivations for Involvement in Children's Education: Testing a Theoretical Model (2000)
13. Şengönül, T.: A review of the relationship between parental involvement and children's academic achievement and the role of family socioeconomic status in this relationship. Pegem Journal of Education and Instruction **12**(2), 32–57 (2022)
14. Bubić, A., Tošić, A.: The relevance of parents' beliefs for their involvement in children's school life. Educ. Stud. **42**(5), 519–533 (2016)
15. Chi, J., Rao, N.: Parental beliefs about school learning and children's educational attainment: evidence from rural China. Ethos **31**(3), 330–356 (2003)

16. Cobb, R.: The Relationship Between Self-Regulated Learning Behaviors and Academic Performance in Web-Based Courses. Virginia Tech (2003)
17. Hair, J.F., Ringle, C.M., Sarstedt, M.: PLS-SEM: INDEED a silver bullet. J. Marketing theory and Practice **19**(2), 139–152 (2011)

Teaching Effect Evaluation of Management System Engineering Based on Analytic Hierarchy Process and Fuzzy Comprehensive Evaluation Method

Lei Zhang[1,2] and Han Gui[1(✉)]

[1] Guangzhou City University of Technology, Guangzhou 510800, China
569642014@qq.com

[2] Research Center for Hubei Business Service and Development, Wuhan 430065, China

Abstract. Aiming at solving the problems that the traditional teaching effect evaluation method is subjective, the weighted value is difficult to determine, and the evaluation result is inaccurate, this paper actively discusses the need to evaluate the teaching effect of this course from three main aspects including course paper, classroom performance, daily assignment and experiment as the unified consideration and evaluation. Taking the course "Management System Engineering" as the research object, this paper studies the three-level evaluation index system according to its characteristics and its implementation of educational reform, applies the analytic hierarchy process (AHP) and fuzzy comprehensive evaluation method, and obtains the evaluation result of educational reform effect, which can be used as a new pattern to evaluate the effect of curriculum education reform, and has certain popularization and application value and reference value.

Keywords: Analytic Hierarchy Process · Fuzzy Comprehensive Evaluation Method · Management System Engineering

1 Introduction

With systematic thought, viewpoint, methodology and corresponding method providing, Management System Engineering is one of the basic courses of industrial engineering and management science, which is an interdisciplinary engineering technology, that provides new ideas and new methods for the development of modern science and technology. The purpose of this course is to train the students' concept of system and their ability of modeling, analysis, and synthesis of actual system based on systematically introducing the basic theories and methods of Management System Engineering [1]. By learning from this course, students can have the professional ability to correctly and flexibly apply system view and system engineering methods to analyze and deal with problems and provide scientific basis for decision-making. Teaching system theory, system engineering methodology, system engineering method and other knowledge; system theory application and system engineering modeling, simulation, decision analysis and other

Z. Hu et al. (Eds.): ICCSEEA 2024, LNDECT 242, pp. 591–602, 2025.
https://doi.org/10.1007/978-3-031-84228-3_51

methods training; Guide students to pay attention to the trends of discipline research and application and understand the frontier of the discipline; To develop students' ability to apply the knowledge of the subject to practical problems comprehensively, flexibly, and innovatively [2].

By studying Management Systems Engineering, students can understand the role of management systems engineering in business, describe a system by using systems engineering concepts with skilled systematic professional methods and common concepts, apply scientific and systems thinking skills to analyze problems and propose effective solutions. Able to apply system concept to solve practical problems in system engineering management. Aiming at system analysis, evaluation and decision-making, students can also analyze, evaluate, and optimize the enterprise by using systematic methods [3].

The traditional weighted average method of curriculum is difficult to give the weights of all levels of indicators accurately and quantitatively, which affects the objectivity and accuracy of the evaluation of teaching reform effect. To accurately evaluate the effect of teaching reform, this paper takes the course of "Management System Engineering" as the object, studies the evaluation method of teaching effect based on analytic hierarchy process and fuzzy comprehensive evaluation, and provides a feasible method for accurately and reasonably evaluating the effect of teaching reform, which has good popularization and application value [4].

2 Theory and Principle

2.1 The Concept and Application of AHP

Analytical Hierarchy Process (AHP) is an evaluation and decision-making method that combines qualitative and quantitative analysis proposed by American scholar Satie in 1970s [6]. Analytical Hierarchy Process (AHP) was officially proposed by the mathematician Thomas Saaty in 1980, which is a method of analysis greatly used in the context of a multi-criteria analysis; it allows the comparison and the choice between the preset options [7]. The idea of this method is to use the multi-objective comprehensive evaluation method to decompose the various elements related to decision-making into a hierarchical system. Based on the establishment of this hierarchical system, a comprehensive evaluation decision method is carried out according to the quantitative weight calculation [8].

AHP is characterized by clear analytical thinking, which can systematize, mathematize, and model the thinking process of analysts: less quantitative data is required for analysis, but it requires clear analysis of the nature of the problem, the factors included and their internal relations: it can be used for decision-making analysis of multi-criteria, multi-objective problems, and other kinds of problems. AHP is one of the MCDM method that is very good in modeling expert opinion in decision support systems. In developing the model, AHP perform pairwise comparisons variables that become a decisive factor in the decision-making process [9].

The analytic hierarchy process mathematizes the thinking process of decision makers [11]. It provides a method that can synthesize people's different subjective judgments and give quantitative analysis results [12]. Finally, the very complex system research is simplified to the pair comparison and simple calculation among various factors [13].

General steps of AHP modeling:

(1) The hierarchical structure of AHP is established

The element at the top of the hierarchy is the target to solve the problem, so the top layer is called the target layer. Each level below the goal level contains various intermediate links needed to achieve the goal, collectively referred to as the criterion level [14].

(2) Construct judgment matrix

After the hierarchical structure of AHP is established, the membership relationship between different hierarchical elements can be determined. This degree of importance can be measured on a scale of 1–9.

(3) Calculation of relative ordering weights of elements under certain criteria

According to the lower elements u1, u2... For the judgment matrix A of some criterion C in the upper layer, un finds their relative ranking weights w1, w2..., wn. The relative sort weight wi can be calculated by the following formula:

$$w_i = \frac{1}{n}\sum\nolimits_{j=1}^{n}(a_{ij}/\sum\nolimits_{k=1}^{n}a_{kj}) \tag{1}$$

(4) The consistency test of the judgment matrix is carried out

Firstly, the consistency index IC is calculated, and its calculation formula is as follows:

$$I_C = \frac{\lambda_{max} - n}{n - 1} \tag{2}$$

where λ max is the maximum eigenvalue of the judgment matrix and n is the order of the judgment matrix.

(5) Calculate the combined weight of each layer element on the target layer

(6) Overall performance evaluation

The overall performance is determined by the following formula:

$$\mathrm{T} = \sum\nolimits_{\mathrm{i}=1}^{\mathrm{n}} \mathrm{S_i W_i} \tag{3}$$

where, T is the overall performance value of the evaluation, Si is the score value of each secondary indicator, W_i is the combined weight value of each secondary indicator i to the total target layer, and n is the number of all indicators participating in the evaluation.

2.2 Fuzzy Comprehensive Evaluation Method

Fuzzy comprehensive evaluation was first put forward by Wang Peizhuang, a Chinese scholar [15]. The basic principle is as follows: first, determine the factor (index) set and evaluation (grade) set of the evaluated object; Secondly, the weights of each factor and their membership vectors are determined respectively to obtain the fuzzy evaluation matrix. Finally, the fuzzy evaluation matrix and the weight vector of the factors are carried out fuzzy operation and normalization, and the comprehensive result of fuzzy evaluation is obtained [16].

The advantage of this method is that the evaluation model is simple and easy to master, and the quantitative evaluation of multi-factor and multi-level fuzzy problems is relatively good. In the evaluation of many management systems and economic systems, the fuzzy comprehensive evaluation method has obtained good economic and social benefits [17].

General steps of fuzzy comprehensive evaluation method.

(1) Invite experts from interested parties to form evaluation panels;
(2) Through discussion, the system evaluation index set is determined. The evaluation scale set of each evaluation index is determined;
(3) Through analytic hierarchy process and other methods, or according to the experience of experts, determine the weight limit of each evaluation index;
(4) Membership matrix is constructed for each candidate scheme;
(5) According to the comprehensive evaluation concept of chess and paste theory, the comprehensive evaluation vector of each candidate scheme is calculated;
(6) Finally, each candidate scheme is evaluated according to the different evaluation scale [18].

3 Evaluation of the Effect of Curriculum Teaching Reform

According to the characteristics of curriculum evaluation, the effect of teaching reform is evaluated by combining the calculation process of analytic hierarchy process and fuzzy comprehensive evaluation.

3.1 Calculation of Analytic Hierarchy Process

The key parts of the AHP evaluation process are as follows.

(1) Evaluation index system design

    According to the implementation characteristics and construction status of the curriculum, the evaluation index system of the teaching reform effect designed by the curriculum group is a three-level system, and the indicators at all levels are shown in the following table.

(2) Establishment of judgment matrix

    Judgment matrix is the basic information of AHP, and it is also the basis of relative importance calculation and single hierarchy ranking. The judgment matrix is a certain factor C of the above level as the evaluation criterion, and pairwise comparison is made to determine the matrix elements [19] (Table 1 and (Table 2)).

(3) Relative importance calculation

    After the establishment of the judgment matrix, the weight of the relative importance of the elements at this level relative to an element at the next level should be calculated according to the judgment matrix, that is, the hierarchical single ranking is carried out, which is the basis for ranking the importance of all the elements at the level relative to the highest level.

Here, the root method is chosen to calculate the weight coefficient (Table 3).

The weight coefficient matrix is obtained $W = 0.188$.

Table 1. Teaching reform effect evaluation index system

Primary index	Secondary index	Three-level index
Effect of teaching reform C	Course Paper C_1	Contents C_{11}
		Structure C_{12}
		Analytical Methods C_{13}
		Format Specifications C_{14}
	Classroom performance and daily assignments C_2	Attendance Rate C_{21}
		Classroom Interaction Performance C_{22}
		Correct Rate of Classroom Questioning C_{23}
		Assignments Quality C_{24}
	Experimental Lesson C_3	Use of Experimental Software C_{31}
		Experiment Completion Rate C_{32}
		Experimental Performance C_{33}
		Experimental Report Results C_{34}

Table 2. Judgment Matrix C

C	C_1	C2	C_3
C_1	1	5	7
C_2	1/5	1	3
C_3	1/7	1/3	1

Table 3. Judgment Matrix C weight coefficient

C	C_1	C_2	C_3	Row Multiplication	To the power of 1/n	Weight Coefficient Wi
C_1	1	5	7	35	3.27106631	0.730645
C_2	1/5	1	3	0.6	0.843432665	0.188394
C_3	1/7	1/3	1	0.05	0.362460124	0.080961

(4) consistency check

As mentioned above, the judgment matrix obtained by pairwise comparison may be self-contradictory, and it is impossible to have complete consistency. So how much inconsistency exists, that does not affect the evaluation results, so that it can be accepted, which is the content of the consistency discussion.

$CW=0.577,$

Therefore, $0.188\lambda_2=0.577$

$\lambda_1=3.06;\lambda_2=3.06;\lambda_3=3.06.$

Then, $\lambda_{max}=3.06,$

Coincidence Indicator C.I.=0.03<0.1, therefore, the consistency of the results obtained from the calculation of the judgment matrix can be accepted, that is, the relative importance or weight coefficient obtained can be accepted.

$R=0.03/0.52=0.057<0.1,$ Consistency is still acceptable.

The following uses the same method to calculate the weights of indicators in each dimension, respectively (Table 4).

Table 4. Judgment Matrix C_1

C_1	C_{11}	C_{12}	C_{13}	C_{14}	Row Multiplication	To the power of 1/n	Weight Coefficient Wi	C1W	C1W/Wi
C_{11}	1	3	5	6	90	3.08	0.55	2.30	4.20
C_{12}	1/3	1	4	5	6.67	1.60	0.29	1.19	4.19
C_{13}	1/5	1/4	1	3	0.15	0.62	0.11	0.46	4.20
C_{14}	1/6	1/5	1/3	1	0.01	0.32	0.06	0.24	4.21

$\lambda_{max}=4.207;\ CI=0.069;\ CR=0.078<0.1,$ therefore, it meets the consistency requirement (Table 5).

Table 5. Judgment Matrix C_2

C_2	C_{21}	C_{22}	C_{23}	C_{24}	Row Multiplication	To the power of 1/n	Weight Coefficient Wi	C2W	C2W/Wi
C_{21}	1	1/5	1/7	1	0.029	0.41	0.07	0.29	4.03
C_{22}	5	1	1	7	35	2.44	0.43	1.72	4.02
C_{23}	7	1	1	5	35	2.44	0.43	1.72	4.03
C_{24}	1	1/7	1/5	1	0.029	0.41	0.07	0.29	4.03

$\lambda_{max}= 4.028;\ CI = 0.009;\ CR = 0.01<0.1,$ therefore, the consistency requirement is met (Table 6).

$\lambda_{max}= 4.025;\ CI = 0.008;\ CR = 0.009<0.1,$ therefore, it meets the consistency requirement.

Table 6. Judgment Matrix C_3

C_3	C_{31}	C_{32}	C_{33}	C_{34}	Row Multiplication	To the power of 1/n	Weight Coefficient Wi	C3W	C3W/Wi
C_{31}	1	7	5	3	0.11	3.20	0.59	2.36	4.02
C_{32}	1/7	1	1/2	1/3	0.02	0.39	0.07	0.29	4.02
C_{33}	1/5	2	1	1/2	0.2	0.67	0.12	0.49	4.02
C_{34}	1/3	3	2	1	2	1.19	0.22	0.88	4.01

Finally, the indicators in each dimension calculated above are summarized and weighted to calculate the final weight (Table 7).

Table 7. Judgment Matrix C Final weight

	Index Level	Relative Weight	Final weight
C_1 (0.731)	C_{11}	0.55	0.39
	C_{12}	0.29	0.21
	C_{13}	0.11	0.08
	C_{14}	0.06	0.04
C_2 (0.188)	C_{21}	0.07	0.01
	C_{22}	0.43	0.08
	C_{23}	0.43	0.08
	C_{24}	0.07	0.014
C_3 (0.081)	C_{31}	0.59	0.0488
	C_{32}	0.079	0.01
	C_{33}	0.129	0.01
	C_{34}	0.22	0.02

From the relative importance of teaching effect shown above, course papers are an important element of teaching effect. Among the three dimensions of the criterion layer, the weight of the course paper accounts for the largest proportion, reaching 73.1%, followed by the class performance and daily assignment, accounting for 18.8%, and the proportion of experiments only 8.1%, indicating that the course paper is the most important factor in the teaching effect. In addition, in the course paper, the content and structure of the paper are the most critical, accounting for the highest weight, directly related to the quality of the course paper; In classroom performance and daily assignment, classroom interaction performance and the correct rate of questions and answers occupy the highest weight, which affects the score of classroom performance; In the experimental link, the results of the experimental report are the important embodiment of the experimental effect. In summary, based on this information, AHP can also help managers to diagnose the main problems in teaching effectiveness programs to be evaluated and explore possibilities for improvement [20].

3.2 Fuzzy Comprehensive Evaluation Method

Fuzzy comprehensive evaluation method is a method of comprehensive evaluation of a certain evaluation system by applying fuzzy set theory, that is, language variables are first used to describe the evaluation system of teaching effect qualitatively, and then experts are scored according to the importance of each evaluation factor, and quantitative evaluation results are calculated. Finally, membership degree theory of fuzzy mathematics is applied to the overall evaluation of teaching effect. Using this method to evaluate the teaching effect, the result is clear and logical, and can better solve the fuzzy and difficult to quantify problems in the teaching effect evaluation.

(1) The set of comments for establishing an evaluation is: $V = \{V_1, V_2, \cdots, V_m\}$, which represents *{very satisfied, satisfied, average, dissatisfied, very dissatisfied}*, and indicates the degree of satisfaction from highest to lowest.
(2) The weights are determined, and the evaluation results are shown in Table 8.

Table 8. Evaluation index weight and evaluation table

	Index Level	Final Weight	Very Satisfied	Satisfied	Average	Dissatisfied	Very Dissatisfied
C_1 (0.731)	C11	0.39	0.05	0.25	0.4	0.3	0
	C12	0.21	0	0.5	0.25	0.25	0
	C13	0.81	0.1	0.55	0.35	0	0
	C14	0.04	0.2	0.65	0.15	0	0
C_2 (0.188)	C21	0.02	0.1	0.5	0.3	0.1	0
	C22	0.02	0.15	0.55	0.3	0	0
	C23	0.04	0.1	0.45	0.35	0.05	0.05
	C24	0.10	0.15	0.25	0.5	0.1	0
C_3 (0.081)	C31	0.05	0.1	0.7	0.15	0.05	0
	C32	0.01	0.05	0.25	0.2	0.45	0.05
	C33	0.01	0.05	0.2	0.25	0.5	0
	C34	0.02	0.15	0.5	0.35	0	0

(3) The membership matrix of each evaluation result is determined according to the voting proportion of members:

$W = (0.73, 0.19, 0.08)$

W1= (0.399, 0.21, 0.081, 0.04)
W2 = (0.02, 0.02, 0.04, 0.10)
W3 = (0.048, 0.01, 0.01, 0.021)

$$M1 = \begin{bmatrix} 0.05 & 0.25 & 0.4 & 0.3 & 0 \\ 0 & 0.5 & 0.25 & 0.25 & 0 \\ 0.1 & 0.55 & 0.35 & 0 & 0 \\ 0.2 & 0.65 & 0.15 & 0 & 0 \end{bmatrix}$$

$$M2 = \begin{bmatrix} 0.1 & 0.5 & 0.3 & 0.1 & 0 \\ 0.15 & 0.55 & 0.3 & 0 & 0 \\ 0.1 & 0.45 & 0.35 & 0.05 & 0.05 \\ 0.15 & 0.25 & 0.5 & 0.1 & 0 \end{bmatrix}$$

$$M3 = \begin{bmatrix} 0.1 & 0.5 & 0.3 & 0.1 & 0 \\ 0.15 & 0.55 & 0.3 & 0 & 0 \\ 0.1 & 0.45 & 0.35 & 0.05 & 0.05 \\ 0.15 & 0.25 & 0.5 & 0.1 & 0 \end{bmatrix}$$

4) The comprehensive evaluation vector of each evaluation result was calculated: According to V = Wi*Mi. Therefore,

$$V1 = W1 * M1 = (0.39, 0.21, 0.08, 0.04) * \begin{bmatrix} 0.05 & 0.25 & 0.4 & 0.3 & 0 \\ 0 & 0.5 & 0.25 & 0.25 & 0 \\ 0.1 & 0.55 & 0.35 & 0 & 0 \\ 0.2 & 0.65 & 0.15 & 0 & 0 \end{bmatrix}$$

= (0.04, 0.28, 0.25, 0.17, 0)
Similarly, other evaluation results can be obtained, as shown in Table 9.

Table 9. Judgment Matrix V

	Very Satisfied	Satisfied	Average	Dissatisfied	Very Dissatisfied
V_1	0.04	0.28	0.25	0.17	0
V_2	0.02	0.07	0.08	0.01	0.01
V_3	0.01	0.05	0.02	0.01	0.01

5) Calculate the weight of each evaluation result.

The fuzzy evaluation matrix is established according to the data in the above table R.

$$R = \begin{pmatrix} V_1 \\ V_2 \\ V_3 \end{pmatrix} = \begin{pmatrix} 0.04\ 0.28\ 0.25\ 0.17\ 0 \\ 0.02\ 0.07\ 0.08\ 0.01\ 0.01 \\ 0.01\ 0.05\ 0.02\ 0.01\ 0.01 \end{pmatrix}$$

$$V = W * R = \begin{pmatrix} 0.73\ 0.19\ 0.08 \end{pmatrix} * \begin{pmatrix} 0.034 & 0.28 & 0.25 & 0.17 & 0 \\ 0.02 & 0.07 & 0.08 & 0.01 & 0.01 \\ 0.01 & 0.05 & 0.02 & 0.01 & 0.01 \end{pmatrix}$$

= *(0.03 0.22 0.19 0.13 0.01)*

After the normalization of V, then, *V= (0.56, 0.38, 0.34, 0.22, 0.01)*.

According to the processing and analysis of the above data, 56% of the overall evaluation of teaching effect is very satisfied, 38% is relatively satisfied, 22% is not satisfied. In general, after several years of development and improvement, the teaching effect evaluation system is becoming more and more perfect, and the overall satisfaction is getting higher and higher, basically meeting the needs. However, there are still some dissatisfactions, which points out the direction for the next step of development and needs to be further improved and improved.

4 Conclusion

With the teaching constantly iterating and evolving basing on the rapid development of information technology, the evaluation system and evaluation method of curriculum teaching should also be constantly improved to meet the needs of education development. Regard to the design of teaching quality evaluation system and the establishment of evaluation method of online courses, on the one hand, the characteristics and new characteristics of online courses should be fully considered, on the other hand, the student-centered and highlight the main position of student evaluation should be fully reflected. Only by fully considering these two aspects above, objective, and valuable curriculum evaluation results consistent with students' learning satisfaction and learning experience can be obtained finally. Meanwhile, it is necessary for teachers to design the course contents, which requires teachers to master professional knowledge, and adjust the teaching content and methods in time according to the problems encountered by students in the learning process. In addition, it is also necessary for teachers to create a suitable classroom atmosphere and fully respect each student in the teaching class, so that students can be more actively involved in classroom activities, and then have deeper thinking and understanding of learning, to promote the growth of students, which requires teachers to have strong empathy, sensitivity, reflective ability, and leading ability.

To evaluate the effectiveness of course teaching accurately and objectively in recent years, this paper introduces analytic hierarchy process (AHP) and fuzzy comprehensive evaluation to evaluate the effect of teaching reform. With practice, it is found that the combination of analytic hierarchy process and fuzzy comprehensive evaluation method is more accurate and objective than the traditional average or weighted average method, which can reflect the practical effect of teaching reform, and explore a new model for evaluating the effect of curriculum teaching reform, which is practical and has important popularization and application value.

Acknowledgment. This paper is supported by: Hubei Business Service Development Research Center Fund Project: Hubei Intelligent Logistics Development Support System Research (2020Z01).

References

1. Liu, Y., Hu, R.: Research on performance appraisal of university experimental teachers based on fuzzy comprehensive evaluation. J. Ningbo Inst. Technol. **30**(02), 102–107 (2018). (in Chinese)
2. Li, Z., Liu, Z., Nie, J.: Construction of evaluation index system of teaching academic ability of online teachers in colleges and universities. J. Dist. Educ. **5**, 81–89 (2020). (in Chinese)
3. Wang Lianfen, X., Shubai.: Introduction to Analytic Hierarchy Process. China Renmin University Press, Beijing (1990). (in Chinese)
4. Wang, T., Lin, J.: Research on multi-level fuzzy comprehensive evaluation of university teachers' teaching quality. Comput. Products Distrib. (08), 215–245 (2018). (in Chinese)
5. Zhang, Y.: Application Research of Fuzzy Comprehensive Evaluation Model Based on Fuzzy Hierarchical Analysis. Anqing Normal University, Anqing (2022). (in Chinese)
6. Lu, M.: Research on teacher performance appraisal based on fuzzy comprehensive evaluation model. Heilongjiang Sci. **15**(01), 59–61 (2024). (in Chinese)
7. Essaid, E.L., Azmani, A., El Harzli, M.: Using AHP method for educational and vocational guidance. Int. J. Inf. Technol. Comput. Sci. (IJITCS) **9**(1), 9–17 (2017)
8. Zhang, Y.: Research on difficulties and development dynamics of integration of production and education in applied undergraduate universities. J. Xi 'an Aeronaut. Univ. **02**, 79–84 (2023). (in Chinese)
9. Listyaningsih, V., Utami, E.: Decision support system performance-based evaluation of village government using AHP and TOPSIS methods: Secang sub-district of Magelang regency as a case study. Int. J. Intell. Syst. Appl. (IJISA) **10**(4), 18–28 (2018)
10. Jiang, Z.: Construction of quality evaluation system for integration of production and education in vocational education. Vocat. Educ. Forum **05**, 34–39 (2018)
11. Suartini, N.K., Divayana, D.G., Dewi, L.J.: Comparison analysis of AHP-SAW, AHP-WP, AHP-TOPSIS methods in private tutor selection. Int. J. Modern Educ. Comput. Sci. (IJMECS) **15**(1), 28–45 (2023)
12. Zhu, M.: Hierarchical and collaborative research of experimental teaching system - based on comfort evaluation and optimization of Nanjing audit station. J. Graph. **41**(5), 788–795 (2020). (in Chinese)
13. Gao, D., Yin, G., Sun, J.: Research on improving the teaching quality of higher education courses from the perspective of new engineering. Res. High. Educ. Heilongjiang Province **12**, 144–147 (2018). (in Chinese)
14. Wu, L., Qu, Y., Cheng, W.: Evaluation, analysis, and practice of fuzzy comprehensive risk in university laboratories. Exper. Technol. Manage. **30**(8), 129–131 (2013). (in Chinese)
15. Gu, Y.: Application of project management maturity model based on analytic hierarchy process and fuzzy comprehensive evaluation. Sci. Technol. Innov. Appl. **34**, 172–173 (2018). (in Chinese)

16. Li, F.: Research on the application of fuzzy comprehensive evaluation in teaching performance evaluation of university teachers. Packer **44**(S1), 663–669 (2023)
17. Xie, H.: Research on evaluation system of experimental teaching quality of mechanical major based on fuzzy AHP. Exper. Sci. Technol. **16**(6), 120–124, 128 (2018). (in Chinese)
18. Tu, H., Huang, L.: Research on current situation and countermeasures of production-education integration in higher vocational colleges in Jiangxi Province. Vocat. Educ. Res. **04**, 35–41 (2023). (in Chinese)
19. Wang, L., Tian, B., Cai, Z.: Significance, problems, and strategies of multi-subject evaluation in secondary vocational schools under the integration of production and education: a case study of Engineering cost major. J. Sichuan Vocat. Tech. Coll. **02**, 6–12 (2023). (in Chinese)
20. Gao, X., Wang, H., Song, Q., et al.: College experimental teaching reform practice of orchard working platform based on AHP-TOPSIS. Educ. Acad. Monthly **7**, 106–110 (2020). (in Chinese)

Research on Cultivating Smart Logistics Talents in the Digital Economy Era Based on Analytic Hierarchy Process

Mei E. Xie[1], Hui Ye[1], and Lichen Qiao[2](✉)

[1] School of Business Administration, Wuhan Business University, Wuhan, China
[2] College of Life Sciences, Northeastern University, Shenyang, China
253637228@qq.com

Abstract. With the rapid development of information technologies such as big data, the Internet of Things, artificial intelligence, and 5G networks, the logistics industry is undergoing digital transformation. The application of smart logistics technology has become an effective way to improve logistics efficiency and reduce logistics costs. In order to adapt to the progress of logistics information technology and the development of digital transformation, it is urgent to cultivate talents with smart logistics skills. Therefore, researching how to vigorously cultivate smart logistics talents has become an inevitable requirement for promoting the rapid development of China's digital economy and society. Starting from the current development status of smart logistics talents, this article elaborates on the connotation and target positioning of smart logistics talents. The Analytic Hierarchy Process is used to study the ability cultivation of smart logistics talents, explore the path of smart logistics talent cultivation in the digital economy era, and aim to provide high-quality smart logistics talents for the transformation and upgrading of the traditional logistics industry.

Keywords: Analytic Hierarchy Process · Digital Economy · Smart Logistics Talents

1 Introduction

In December 2022, the General Office of the State Council issued the *14th Five Year Plan* for the development of modern logistics, which clearly pointed out that modern logistics plays an important supporting role in extending industrial chains, enhancing value chains, and building supply chains. At the same time, it also plays a leading, fundamental, and strategic role in building a modern circulation system, forming a strong domestic market, promoting high-quality development, and building a modern economic system [1]. With the continuous improvement of the political environment, the gradual formation of the logistics Internet, and the wide application of big data in the logistics field, the conditions for the development of China's smart logistics have been met. *Smart logistics* has become a new trend in the development of the logistics industry [2]. According to statistics from the China Academy of Commerce and Industry, the size of the smart logistics market

Z. Hu et al. (Eds.): ICCSEEA 2024, LNDECT 242, pp. 603–611, 2025.
https://doi.org/10.1007/978-3-031-84228-3_52

reached 699.5 billion yuan in 2022, a year-on-year increase of 8%. It is expected that by 2025, this market size will exceed the trillion yuan mark.

The scale of China's smart logistics market in recent years is shown in Fig. 1.

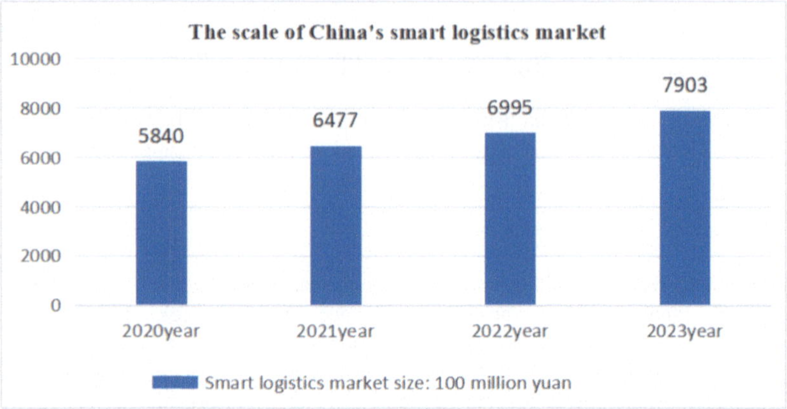

Fig. 1. 2020–2023year the scale of China's smart logistics market

With the application and promotion of new generation information technologies such as the Internet of Things, artificial intelligence, big data, virtual reality, blockchain, and 5G communication, the cultivation of smart logistics talents is urgent. This article adopts research methods such as literature review, field investigation, and Analytic Hierarchy Process (AHP) to start with the current development status of smart logistics talents in the digital economy era. It elaborates on the target positioning of smart logistics talents and uses the Analytic Hierarchy Process (AHP) to study the ability cultivation of smart logistics talents. It explores the path of smart logistics talent cultivation in the digital economy era, aiming to provide high-quality smart logistics talents for the transformation and upgrading of the logistics industry.

2 Literature Review

2.1 Abroad Research Status

Research on smart logistics has been initiated in foreign countries for a long time, and there are many dimensions of research on smart logistics. Starting from the theoretical aspect, empirical research is conducted based on this. Arkadiusz Kawa (2012) proposed a conceptual model of smart logistics based on agent technology and cloud computing. The model mentions the self integration, self-organization, and adaptive characteristics of smart logistics, making information acquisition and flow more convenient and efficient, and making smart logistics management systems better utilized [3]. Henk Zijm (2016) explores the combination of smart logistics and supply chain, and argues that technology integration and the application of new business models pose significant challenges to smart logistics [4]. Karakikes et al. (2017) improved existing smart logistics solutions and validated their influencing factors, providing new ideas and paths for the construction of smart logistics systems [5].

2.2 Domestic Development Status

The development of smart logistics in our country has just begun, but with the rapid development of information technology and artificial intelligence technology, smart logistics will undoubtedly promote the transformation and upgrading of the logistics industry. He Liming conducted a discussion on seven aspects: connectivity, data, mode, experience, intelligence, green, and supply chain upgrading [6]. From the perspective of logistics technology development, Tan Hua et al. believe that informatization, intelligence, intensification, and small batch customization are the future trends in logistics development [7]. Fu Yu pointed out that the development prospects of China's smart logistics are relatively optimistic, but there are still problems such as insufficient logistics data sharing, incomplete logistics information platforms, insufficient professional talents in smart logistics, and weak capabilities in end of line logistics services [8].

Through a review of the existing research achievements and current situation of smart logistics at home and abroad, it is found that many scholars at home and abroad attach great importance to the interpretation of the connotation, development path, and solutions of smart logistics. Scholars have conducted research from different perspectives and adopted diverse research methods, laying a solid foundation for the operation and system platform construction of smart logistics, and providing strong theoretical support for the development of smart logistics and the construction of logistics systems in China [9, 10].

3 The Impact of the Digital Economy Era on Logistics

3.1 Policy Support for Logistics Intelligence

Chinese governments at all levels and relevant functional departments attach great importance to the development of logistics informatization and intelligence, and have successively studied and introduced a series of policy measures to encourage the intelligent development of the logistics industry, creating a good policy environment for the rapid development of logistics intelligence. In 2015, the State Council issued the *Opinions on Promoting the Modernization of Domestic Trade Circulation and Building a Legalized Business Environment*, To vigorously develop third-party logistics and smart logistics, encourage the application of Internet of Things and other technologies in warehousing systems, support the construction of logistics information service platforms, promote efficient matching of vehicle sources, goods sources, and logistics services, support the construction of cold chain logistics systems, and improve the level of logistics socialization, standardization, informatization, and specialization [11, 12].

The application of digital technology in logistics management is showed as Fig. 2:

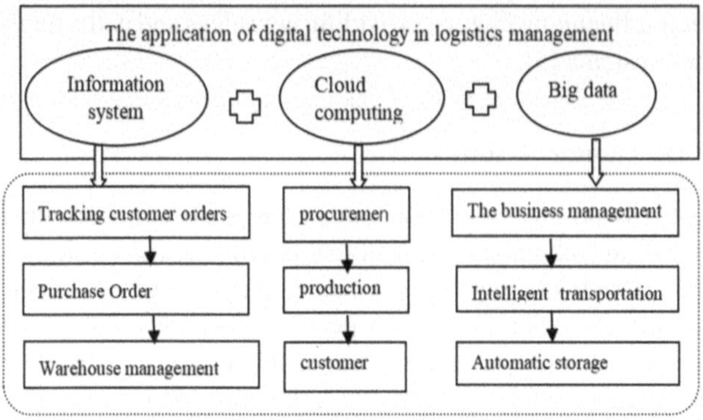

Fig. 2. The application of digital technology in logistics management

3.2 The Requirements of the Digital Economy for New Logistics Talents

The digital economy is an economic form that directly or indirectly uses data to guide the optimal allocation of resources and achieve high-quality development. Its essence is informatization and digitization, which quickly and massively promote the overall transformation and upgrading of enterprises of different scales and formats. The data economy has significant effects on multiple aspects of logistics management, which requires new logistics management talents to have a better understanding of digital technology and data analysis technology. Data organization and advanced analysis will become key abilities for smart logistics management talents. Therefore, logistics management personnel need to significantly improve their knowledge, abilities, and literacy to meet the new requirements of modern logistics [13].

In the current digital economy era, intelligent technologies such as artificial intelligence, big data, and blockchain are rapidly developing, providing new opportunities for the transformation and upgrading of the logistics industry. At the same time, the traditional logistics industry is also facing severe challenges, and smart logistics and logistics big data have become new trends in the development of the logistics industry. Future logistics industry managers must have a basic understanding of logistics business as well as relevant digital knowledge such as computer technology and network technology. Smart logistics management requires high-quality composite and technology applied logistics management talents who are familiar with the operation rules of modern logistics informatization, able to apply emerging digital technologies such as the Internet of Things, cloud computing, big data, and artificial intelligence to assist decision-making, and improve logistics efficiency.

4 The Target Positioning of Intelligent Logistics Talents

4.1 Skilled

Smart logistics talents in the digital economy era must possess advanced technical knowledge and professional skills, including Internet of Things technology, artificial intelligence, big data analysis, cloud computing, and other aspects. With the application of information technology, new requirements have been put forward for the understanding and application capabilities of Internet of Things technology, including sensor technology, communication protocols, data collection and analysis, and other aspects. The large amount of data generated in the era of smart logistics has also enhanced the demand for big data analysis capabilities, including data mining, data visualization, predictive analysis, etc. [14].

4.2 Digital Thinking

Smart logistics talents in the digital economy era need to possess digital thinking, especially the ability to fully understand and apply digital technology, data analysis, and information systems. In terms of data-driven decision-making, smart logistics talents are needed to use data analysis and visualization tools to understand and utilize the data in the logistics process to make decisions and solve problems. At the same time, it is necessary to be familiar with commonly used information systems and technical tools in logistics management, including supply chain management systems, warehouse management systems, transportation management systems, etc. [15].

4.3 Innovation Capability

Smart logistics talents in the digital economy era need to possess innovative thinking and abilities, be able to seek improvement and breakthroughs around their main business, and promote the innovative development of logistics technology and business. Taking technological innovation as an example, it is necessary to be able to apply emerging cutting-edge technologies and tools to the management system of logistics, in order to improve efficiency, reduce costs, or create new service models. At the same time, smart logistics talents should be able to identify shortcomings in traditional logistics and improve them through innovative methods [16].

4.4 Cross Border Cooperation

Smart logistics talents in the digital economy era need to have the ability to engage in cross-border cooperation, collaborate with professionals from different fields, solve complex logistics problems, and promote cross regional, cross departmental, and cross organizational collaboration. In the era of digital economy, smart logistics also involves interdisciplinary knowledge integration and industry exchanges. Therefore, smart logistics talents should have a comprehensive knowledge background in multiple disciplines, and be able to master and integrate knowledge in fields such as logistics, information technology, data analysis, and operations research [17].

5 Cultivating the Ability of Intelligent Logistics Talents Based on Analytic Hierarchy Process

5.1 Determine the Quantitative Evaluation Model

Using methods such as expert scoring and Analytic Hierarchy Process, and comparing 12 elements from 4 aspects of logistics talent competency standards, establish a competency standard element evaluation table (see Table 1) and construct a judgment matrix (see Table 2) to comprehensively evaluate these competency elements.

Table 1. Quantitative evaluation table for elements of logistics talent capability standards

Overall goal	First level evaluation indicators	Weight P	Secondary evaluation indicators	Weight C
Logistics Talent Capability Standards N	Professional competence (A1)	P1	Values (D1)	C1
			Team awareness (D2)	C2
			innovative thinking (D3)	C3
	Logistics knowledge (A2)	P2	Logistics knowledge (D4)	C4
			Information knowledge (D5)	C5
			English knowledge (D6)	C6
	Logistics skills (A3)	P3	Logistics operation (D7)	C7
			Logistics planning (D8)	C8
			Logistics Information Technology (D9)	C9
	Expand knowledge (A4)	P4	computer technology (D10)	C10
			Data analysis (D11)	C11
			market analysis (D12)	C12

5.2 Comparative Analysis of Questionnaire Surveys

Adopting the school enterprise expert evaluation method, the first and second level indicators of smart logistics talents are compared pairwise to obtain their respective matrices, and the coefficients of the first and second level evaluation indicators are determined. Collect and collect 48 questionnaires on the ability indicators of smart

Table 2. Importance judgment matrix for first level indicators of logistics talent capabilities

N	Professional competence	Logistics knowledge	Logistics skills	Expand knowledge
Professional competence	1	1/3	1/5	1
Logistics knowledge	3	1	1/3	3
Logistics skills	5	3	1	5
Expand knowledge	1	1/3	1/5	1

logistics talents from school enterprise experts. Based on the statistical results, establish a judgment matrix for the first level evaluation indicators of professional competence, logistics knowledge, logistics skills, and expanded knowledge of smart logistics talents.

Use formulas (1) for each column to construct a judgment matrix (2):

$$\overline{b_{ij}} = \frac{b_{ij}}{\sum_{k=1}^{n} b_{ij}} \tag{1}$$

$$\begin{pmatrix} 0.1\ 0.07\ 0.12\ 0.1 \\ 0.3\ 0.21\ 0.19\ 0.3 \\ 0.5\ 0.64\ 0.58\ 0.5 \\ 0.1\ 0.07\ 0.12\ 0.1 \end{pmatrix} \tag{2}$$

Adding each row of the matrix yields 0.39, 1, 2.22, and 0.39, respectively.

The result obtained by normalizing the vector using formula (3) is: 4.

$$\sum_{j=1}^{n} \overline{Wij} = 0.39 + 1 + 2.22 + 0.39 = 4 \tag{3}$$

The W_i values are 0.0975, 0.25, 0.555, and 0.975, respectively. The feature vectors are [0.0975, 0.25, 0.555, 0.0975], which are also the weight coefficients corresponding to the four primary indicators of intelligent logistics talent. Finally, consistency check is performed, and by calculating that the CR is less than 0.1, it indicates that the judgment matrix has satisfactory consistency.

Using the same method, the weight coefficients corresponding to the secondary indicators in four aspects can be obtained. Then, using the weighted average method, the coefficients of the primary indicators are multiplied by the corresponding coefficients of the secondary indicators to obtain the importance ranking of the twelve ability elements of logistics talents (see Table 3).

From the table below, it can be got that from the first level indicator of intelligent logistics talent ability, logistics skills have the highest weight (0.555). From the second level indicator of intelligent logistics talent ability, logistics operation ability, logistics planning ability, logistics information technology ability, and logistics knowledge have a higher weight ratio. In the era of digital economy, there is an urgent need for logistics talents to keep up with the development of the times, master corresponding abilities and knowledge.

Table 3. Overall ranking of the importance of intelligent logistics talent capability indicators

Project	D1	D2	D3	D4	D5	D6	D7	D8	D9	D10	D11	D12
First level weight	0.0975			0.25			0.555			0.0975		
Secondary weight	0.26	0.63	0.1	0.6	0.2	0.2	0.6	0.2	0.2	0.14	0.43	0.43
Comprehensive weight	0.025	0.061	0.009	0.15	0.05	0.05	0.33	0.11	0.11	0.014	0.042	0.042

6 Conclusion

The integration of digital technology and logistics has triggered a transformation in the logistics industry, making smart logistics a trend in industry development. Smart logistics talents in the digital age must adapt to industry changes and have the ability to respond to new technologies and models. Strengthening the cultivation of smart logistics talents has become an urgent social demand. Through the joint education of society, schools, enterprises, and individuals, we will highlight the key points of cultivating smart logistics talents, break through regional restrictions, adapt to the development of the digital economy, increase the proportion of high-quality smart logistics talents, and promote the rapid development of China's smart logistics industry. Several thoughts on the cultivation of intelligent logistics talents: (1) Multi party cooperation and gathering of intelligent logistics talent capacity cultivation; (2) Highlighting key areas and cultivating versatile and intelligent logistics talents; (3) Establish sound mechanisms and expand channels for cultivating intelligent logistics talents.

Acknowledgment. This paper is supported by the Cooperative Education Project of the Ministry of Education "Intelligent Logistics Planning and Designer Training" (220600924215841).

References

1. Zheng, Q.: The development model, problems, and countermeasures of smart logistics in China. Bus. Econ. Res. (18), 108–111 (2019). (in Chinese)
2. Wei, Y.: Research on the development of smart logistics industry in the era of sharing economy. Res. Technol. Econ. Manage. 2020283(02), 108–112 (in Chinese)

3. Uckelmann, D.: A definition approach to smart logistics. Next Gen. Teletraffic Wired/Wireless Adv. Network. 273–284 (2008)
4. Kawa, A.: SMART logistics chain. In:. Proceedings of the 4th Asian conference on Intelligent Information and Database Systems, no. 01, pp. 432–438 (2012)
5. Zijm, H., Klumpp, M.: Logistics and supply chain management: developments and trends. Logist. Supply Chain Innov. 1–20 (2016). (in Chinese)
6. Karakikes, I., Nathanail, E.: Simulation techniques for evaluating smart logistics solutions for sustainable urban distribution. Procedia Eng. **178**, 569–578 (2017)
7. He, L.: The current situation and trend of smart logistics development in China. China's Natl. Conditions Strength **12**, 9–12 (2017). (in Chinese)
8. Tan, H., Lin, K., Yang, S.: Analysis of the current development trends of smart logistics technology. Mob. Commun. **40**(21), 45–49, 57 (2016)
9. He, S., Yuan, B.: Modern logistics upgrade and transformation based on smart modernization - taking Panzhihua as an example. Enterpr. Econ. **5**, 121–123 (2011)
10. Zhang, G., Ma, L., Gu, K.: Exploring the development of smart cities and smart logistics in Hebei Province. Coop. Econ. Technol. **10**, 57–58 (2016). (in Chinese)
11. Zhang, L., Zhan, P., Li, X., et al.: Research on strategies and policy support for promoting smart logistics in Liaoning coastal port area. Econ. Res. Ref. (28), 57–61 (2014). (in Chinese)
12. Fu, Y.: The development trends and enhancement strategies of smart logistics in China. Foreign Econ. Trade Pract. **1**, 90–92 (2018)
13. Anitha, P., Malini, M., Patil.: A review on data analytics for supply chain management: a case study. Int. J. Inf. Eng. Electr. Bus. (IJIEEB), **09**(08), 30–39 (2018)
14. Benotmane, Z., Belalem, G., Neki, A.: A cost measurement system of logistics process. Int. J. Inf. Eng. Electr. Bus. (IJIEEB) **09**(08), 23–29 (2018)
15. Ting, P.H.: An efficient and guaranteed cold-chain logistics for temperature-sensitive foods: applications of RFID and sensor networks. Int. J. Inf. Eng. Electr. Bus. (IJIEEB) **12**(18), 1–5 (2013)
16. Kirch, M., Poenicke, O., Richter, K.: RFID in logistics and production –applications, research and visions for smart logistics zones. Procedia Eng. **178**, 526–533 (2017)
17. Jiang, Y.: Exploration of cultivating smart logistics talents in the digital economy era. Logist. Eng. Manage. **12**(45), 170–172 (2023)

Construction and Implementation of a Hierarchical Practical Teaching Model Integrating AI Technology

Ying Zhang and Haoliang Zhu(✉)

Intelligent Manufacturing College of Nanning University, Nanning, China
412363055@qq.com

Abstract. This article takes the practical teaching process of "Principles and Applications of Microcontrollers" as the research object, explores a hierarchical practical teaching mode that integrates artificial intelligence. It integrates validation experiments, improvement layer experiments, design layer experiments, and free play layer experiments into one actual product, and educates the secondary development process of semi-finished products, organically combining isolated knowledge points to form systematic knowledge. Students use artificial intelligence technology (chatGPT, ERNIE Bot) to complete the program design of large workload, and solve the problem that students cannot complete the product program design due to short classroom teaching time. This teaching model can effectively solve various problems in the traditional practical teaching process, such as insufficient personalized teaching space, insufficient teaching hierarchy, and detachment between teaching and production practice. At the same time, it can stimulate students' interest in learning, improve their initiative in learning, and cultivate their skills and abilities in dialogue with AI.

Keywords: Hierarchical practical teaching · AI technology · Semi finished products · AI dialogue

1 Introduction

The Implementation Opinions of the General Office of the State Council on Deepening the Reform of Innovation and Entrepreneurship Education in Higher Education Institutions were released on May 13, 2015, which clearly stated that entrepreneurship and entrepreneurship education should be integrated with professional education. As an important part of professional education, practical teaching is an effective embedding point for integrating entrepreneurship and entrepreneurship education [1, 2]. To effectively extend traditional confirmatory experiments in innovation and entrepreneurship education, many domestic universities (scholars) have conducted a lot of research on hierarchical design of experimental projects and have achieved some results [3–5]. Since the emergence of AI technologies such as ChatGPT in just one year, the application of AI technology in the field of education has been unleashed to a certain extent, with great potential and development value [6, 7]. For example, building personalized

Z. Hu et al. (Eds.): ICCSEEA 2024, LNDECT 242, pp. 612–621, 2025.
https://doi.org/10.1007/978-3-031-84228-3_53

learning paths and teaching plans through AI, dynamically adjusting students based on their learning habits, ability levels, and progress, intelligently recommending learning content, and conducting automatic adaptability tests [8].

AI chatbots can provide instant Q&A services without time constraints, answering students' questions. For example, artificial intelligence assistants such as "Xiaoai Classmates" launched by Xiaomi have been piloted in some schools for teaching evaluation, accurately evaluating students' learning outcomes, providing real-time feedback on learning outcomes, and helping teachers optimize teaching strategies [9, 10]. In addition, AI technology can also simulate the role of teachers in teaching, and monitor and analyze student behavior and emotional states in the classroom through technologies such as facial recognition and speech recognition, assisting teachers in classroom management and improving the intelligence of teaching resources [11–14]; Through AI technology, students can also help screen and recommend the highest quality learning resources, improve learning efficiency, promote personalized learning, and promote educational equity, allowing more students to enjoy educational resources that better meet their own needs and characteristics [15]. However, due to the short time since the emergence of AI technologies such as chatGPT, many traditional classrooms have not yet been effectively integrated and applied, and teacher centered indoctrination teaching is still relatively common. This is the driving force and entry point of this article's research.

2 Problems in Practical Teaching of Microcontrollers

There are two main problems in the teaching of the course "Principles and Applications of Microcontrollers".

2.1 The Unclear and Explicit Implementation Effect of Hierarchical Teaching

The experimental project level of "Principles and Applications of Microcontrollers" can be roughly divided into basic layer, improvement layer, design layer (development layer/exploration layer, etc.), avoiding some drawbacks of previous experiments that mainly focused on basic verification; But many times, increasing the level only increases the difficulty of the experiment, and the design level only integrates more knowledge points on the basis of improving the level to improve its comprehensiveness. There are still some shortcomings: (1) The lack of consideration for differences in student learning abilities and the standardized experimental process and content have led to students with good foundations having "nothing to do" and those with poor foundations having "no confidence"; (2) Insufficient personalization of experimental content and lack of space for innovation. Although the experimental design has a certain level gradient, the tasks completed by students are still the same, with slightly different methods, means, and difficulty, which restricts students' innovative thinking and limits their space for expression. (3) In order to implement differentiation, students are subjectively divided into batch levels (excellent, average, and poor) by teachers, and assigned learning content. This "colored" mode is not conducive to stimulating students' learning interest and developing their learning potential, and to some extent, it also increases the workload of teachers; (4) The technical applicability of the experimental project is not reflected enough, and the experiments are all conducted in experimental boxes, which leads to students having to operate the equipment step by step and unable to explore the content with deep application value behind the experiments.

2.2 The Huge Workload of Microcontroller Project Development

Students are unable to complete a systematic program design for a work in class. For example, to design a "temperature and humidity detection device", even if the teacher has provided a complete design simulation model, students still need to write the drivers for each module, such as LCD display, sensor driver, and other modules that involve various communication protocols. The large workload and time required for program debugging can lead to many issues for beginners in terms of program design standardization and logical structure. During class, students are unable to complete practical tasks perfectly, which undermines their learning enthusiasm. They need to spend a lot of time researching literature; On the other hand, teachers need to constantly help students solve programming problems, which can greatly reduce teaching effectiveness and efficiency.

AI technology plays a significant role in program design. AI can provide students with real-time and accurate code prompts and automatic completion functions by learning a big amount of programming examples and patterns. It can intelligently predict and recommend the code that may need to be written next based on context, greatly improving programming efficiency. It can also check students' code in real-time to discover potential syntax errors, logical errors, or non-standard programming habits; In addition, AI can help students understand the existing code logic. Through semantic analysis, complex code blocks can be decomposed and interpreted to optimize algorithms, making it easier for students to learn and master.

3 Construction and Implementation of a Three-Level Practical Teaching Model

3.1 Construction of a Hierarchical Practical Teaching Model that Integrates AI Technology

3.1.1 Design of Hierarchical Teaching Mode

Design validation experiments, improvement layer experiments, design layer experiments, and innovative experiments on a practical product, rather than conducting experiments on experimental equipment. In the actual teaching process, students do not need to design and produce products, but rather complete the design and verification of the final process (module) on semi-finished products. While completing the basic functions of the work, it is also possible to improve certain performance aspects of the product through the knowledge learned. In addition, new features and application scenarios of the product can be extended based on the first two layers of experiments, so that students can complete the functional design of the product while completing experimental verification.

3.1.2 Classroom Teaching Process Centered on Student Self-Study Based on AI Technology

Design a complete practical teaching process into four processes: "task assignment", "group discussion", "AI dialogue", and "teacher evaluation summary". In class, the teacher first issues experimental requirements and product design indicators, and proposes the problems that need to be solved. Then, students conduct group discussions based on the tasks to come up with the design ideas and task assignments for each module program. Next is the key part of teaching, "AI dialogue". In this part, the teacher's role is not a knowledge transmitter, but a guide for student learning, guiding students to master communication skills with robots; Students are the center of learning, and the acquisition of most knowledge is proposed by students, answered by AI, and then digested by students themselves. In the final stage, teachers evaluate students based on their learning outcomes and provide additional explanations of knowledge points based on actual situations.

3.2 Deconstruction and Reconstruction of Teaching Content

According to the teaching outline requirements of "Principles and Applications of Microcontrollers", the practical teaching content of the course has been deconstructed, and the original 8 experimental projects have been reconstructed using 6 semi-finished products such as cubic lamps, basketball scorers, and pulse meters as carriers, so that isolated knowledge points are organically combined to form a systematic system. Finally, the secondary development process of semi-finished products is processed in a teaching oriented manner. Each of the six experimental works designed can complete all the projects in traditional experiments, with different focuses. Each work also leaves some interfaces for students to freely play. The list of works and corresponding experiments at each level are shown in Table 1.

4 Implementation and Evaluation of Teaching Modes

Taking the 2020 Electrical Engineering and Automation major at Nanning University as an example, a total of 168 students from four classes were selected as the research subjects. Among them, 85 students from Class 1 and Class 2 were selected as the experimental group, and a hierarchical practical teaching model integrating AI technology was adopted. Even if students obtained the required knowledge through AI technology according to their own learning situations, they were able to apply AI technology; A total of 83 students from 3 classes and 4 classes were used as the control group, using the traditional experimental teaching model, where the teacher assigned unified experimental requirements and tasks, and provided unified guidance to students in carrying out practical teaching activities. After investigation, these four classes have comparability in terms of student source structure, gender ratio, class size, learning foundation, and final comprehensive scores. During the research process, in order to objectively evaluate the teaching effectiveness, three models were adopted: horizontal comparison of single teaching effectiveness, vertical comparison of multiple teaching effectiveness, and overall teaching effectiveness for comprehensive evaluation.

Table 1. List of "semi-finished products" in the hierarchical experiment of microcontroller courses

Product project	Traditional experiments	Verification layer	Upgrade layer	Verification layer
1. Cube lamp	I/O port operation	Flow light	Flat flow light	St lights and ors
2. BKB scorer	Keys and Di tubes	Key input	Computer input	Mo Ctl, etc
3. Pulse MEAS	Interrupt and timer	Detecting pulse	P cycle MEAS	U com. Data Rec
4. Smart Ele fan	Serial port expt	speed control	Automatic adjmt	Voice, rem Ctl
5. Di PSW lock	E2ROM expt	Key PSW	DYN PSW	Fingerprints,
6. Intell ES lights	AD/DA, bus exp	Intell light off	M CH Prog Cntrl	Fuzzy Ctl

4.1 Evaluation of Single Teaching Effectiveness

The practical teaching of the "basketball scorer" was selected as the case study object for the single effect, mainly assessing four evaluation indicators: "completion speed", "level ratio", "number of questions", and "project completion degree"; Among them, "experimental speed" refers to the average time required for students to complete basic validation experiments, "hierarchical ratio" refers to the proportion of students who complete the improvement and design layers, "number of questions" refers to the number of times students encounter problems and ask teachers during the learning process, and "project completion" refers to the proportion of students who complete validation layer experimental tasks in the classroom. In Table 2, the experimental group and the control group are respectively labeled as "EP" and "CP".

Table 2. Statistical Analysis of Single Teaching Evaluation

Class	n	Time (min)	Question number	Hierarchy ratio	Project completion
EP 1	42	45.2	9	35.7%	85.7%
EP 2	43	44.4	8	37.3%	86.1%
CP 1	39	62.3	15	25.6%	76.9%
CP 2	44	66.7	18	27.3%	77.3%

From Table 2 (Single Teaching Evaluation), it can be found that the experimental group completed the basic validation layer experiment almost 20 min faster than the control group on average, reducing nearly 30% of the experimental time. Through feedback analysis, it was found that AI technology mainly improved students' clearer and more standardized programming process, reduced some grammar errors, and thus improved programming efficiency. In terms of the number of classroom questions, the experimental group had an average of 8 fewer than the control group. The precise answers provided by AI technology reduced the number of questions asked by students by nearly 50%, which not only reduced the burden on teachers but also enhanced their initiative in learning, reflecting the student-centered teaching philosophy. In addition, with the assistance of AI technology, students have improved their project completion by 8.8%, and the proportion of students completing improvement level experiments (performance level experiments) has increased by nearly 46.9%. From the analysis of the above data, it can be concluded that integrating AI technology into experimental courses can effectively improve the quality of classroom teaching, stimulate student interest, implement problem-based teaching, and implement a student-centered teaching philosophy.

4.2 Vertical Comparison of Multiple Teaching Effects

Multiple vertical comparisons of teaching effects were conducted, with Class 1 and Class 2 of the experimental group as the research objects. All 6 experiments in the practical teaching process were collected, and "completion speed", "level ratio", "number of questions", and "project completion degree" were also used as evaluation indicators. Then, the vertical comparison of teaching effects was carried out based on the order of teaching time, and the data in Fig. 1 was obtained.

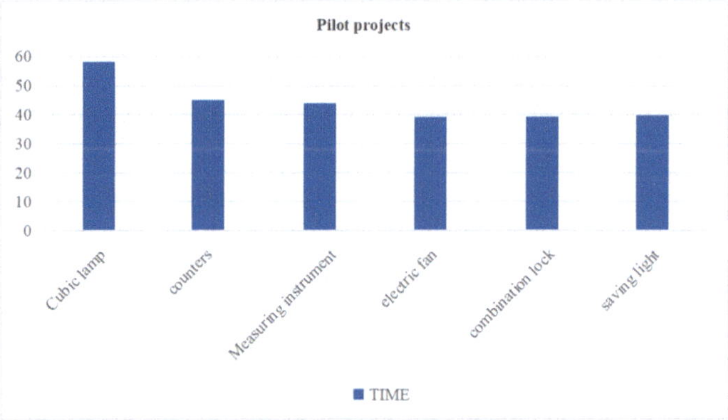

Fig. 1. Completion time of experimental project

From Fig. 1, it can be intuitively observed that in the case where the experimental difficulty of the six projects is equivalent, the speed of students completing the experiment increases significantly as the experiment progresses. The first and second experiments were improved by nearly 13 min, and after three experimental processes, the speed at which students completed the experiment remained stable at around 39 min. It can be analyzed that AI technology has played a significant role in improving the speed of student experiments.

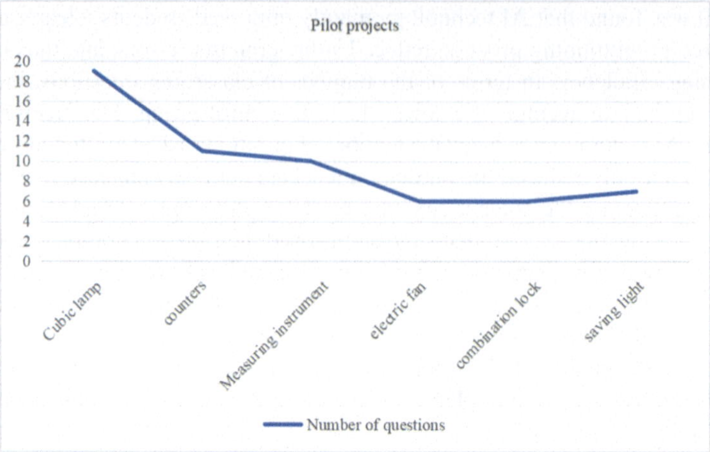

Fig. 2. Number of experimental questions

From Fig. 2, it can be intuitively observed that in the case where the experimental difficulty of the six projects is equivalent, the problems encountered by students in the classroom significantly decrease as the experiment progresses. The first and second experiments showed an average reduction of 8 students in both classes. After three experimental processes, the number of student questions remained stable at 6 to 7. This trend of change was basically mutually verified with the data on completion time. The experimental results show that after about three rounds of experimental process, students' communication skills and abilities with AI have been greatly improved. As a "teacher", AI technology has helped students solve nearly half of their questions, effectively demonstrating self-learning ability and improving certain teaching effects.

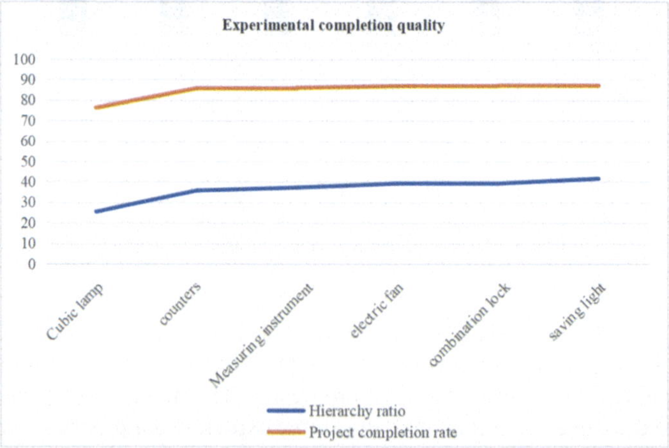

Fig. 3. Quality of experimental completion

The two indicators of "experimental level" and "project completion" in the experimental teaching process collected in Fig. 3 reflect to some extent the quality of student experimental completion. The more students complete basic experiments, the more students will challenge the "improvement level" and "utilization level", and therefore the "level ratio" will correspondingly increase. From the trend of the line in the graph, it can also be analyzed that the quality of students completing the experiment shows a strong negative correlation with the "completion time" and "number of questions" mentioned in Figs. 1 and 2 earlier. Therefore, by comparing the longitudinal data of these four data points, it is not difficult to find that the reasonable application of AI technology in the classroom has indeed achieved good results in improving the quality of learning and teaching, teaching efficiency, and student autonomy.

4.3 Overall Teaching Effectiveness Evaluation

The overall evaluation of teaching effectiveness was conducted based on four dimensions: teaching satisfaction, excellent rate of experimental assessment, test scores, and questionnaire assessment scores after the entire course was completed; The satisfaction survey of the experiment was obtained anonymously by the class counselor through entrusting a third party; The assessment of test scores is based on the same set of experimental questions, and the assessment platform used is a computer-aided teaching system; Questionnaire assessment refers to looking back six months after the course ends to assess the timeliness of students' mastery of course knowledge.

Table 3. Statistical Analysis of Overall Teaching Evaluation

Class	n	Satisfaction	Excellent rate	Exam score	Assessment score
EP 1	42	85.2	14.2%	82.3	67.5
EP 2	43	84.3	15.7%	81.4	66.3
CP 1	39	72.5	6.5%	79.1	47.2
CP 2	44	76.1	8.8%	79.3	49.5

From Table 3, it can be got that the student satisfaction of the experimental group is significantly higher than that of the control group. According to the satisfaction questionnaire analysis, the interest of students in the experiment is significantly higher in the experimental group; From the analysis of the excellence rate of the experiment, the experimental group is also higher (but with a smaller sample size), mainly because "excellent students" in the experiment can have more freedom to play; From the analysis of the final exam scores, the experimental group was slightly higher than the control group, and the main gap was found in the "procedural questions" aspect of the exam questions; Through the analysis of questionnaire assessment data (tested six months after the end of the course), the scores of the experimental group were significantly higher than those of the control group, indicating that the timeliness of the "AI + hierarchical" teaching model is significantly better than traditional teaching, which can help students better understand and remember the knowledge points of the course.

5 Conclusion

By implementing a hierarchical practical teaching model that integrates AI technology, on the one hand, traditional experiments in the course are grafted into a product closely integrated with real life, which can effectively help students combine knowledge and application, stimulate their interest in learning, and answer the question of "what is the use of the course"; At the same time, due to the large space for personalized development of experimental projects, it also improves the challenge, innovation, and higher-order of the course to a certain extent. On the other hand, AI technology is integrated into the classroom, effectively implementing the student-centered teaching philosophy, improving classroom teaching efficiency, and greatly releasing personalized learning space, achieving differentiated experimental teaching, which can effectively stimulate students' creative and innovative thinking, and achieve the organic integration of entrepreneurship education and experimental courses. Therefore, this teaching mode has certain reference value in the design of experimental teaching.

Acknowledgment. This research is supported by 2019 Nanning University Undergraduate Professional Core Curriculum Construction Project (2021BKHXK09).

References

1. El Atmani, Z., Madrane, M.: Reflective practice as a way of developing the professional identity of teachers and professionalizing nursing education. Int. J. Mod. Educ. Comput. Sci. (IJMECS) **15**(4), 57–68 (2023)
2. Simpson, T., Danso, J., Addor, J.A., et al.: Survival analysis computers at a university's computer laboratory and implication on maintainability. Int. J. Educ. Manage. Eng. (IJEME) **14**(2), 17–28 (2024)
3. Philip, A., Neeraj, M., Soopy, K., et al.: Assessment of some famous empirical equation and artificial intelligent model (MLP, ANFIS) to predicting the side weir discharge coefficient. Veh. Technol. IEEE Trans. **2**(4), 75–79 (2014)
4. Ghalkhani, H., Golian, S., Saghafian, B., et al.: Application of surrogate artificial intelligent models for real-time flood routing. Water Environ. J. **27**(4), 535–548 (2013)
5. Guo, J.: Strategies for improving the teaching ability of college teachers from the perspective of artificial intelligence. Nanjing: Nanjing Univ. Aeronaut. Astronaut. **56** (2020). (in Chinese)
6. Gómez, S., Zervas, P., Sampson, D.G., Fabregat, R.: Context-aware adaptive and personalized mobile earning delivery supported by UoLmP. J. King Saud Univ. Comput. Inf. Sci. **26**(1), 47–61 (2014)
7. Wu, L., Cao, Y., Cao, Y.: Framework construction of artificial intelligence empowered classroom teaching evaluation reform and technological implementation. China Electr. Educ. **5**, 94–101 (2021). (in Chinese)
8. Zakiah, A., Prihatmanto, A.S., Mahayana, D., et al.: The philosophy of smart learning using the approach Thomas Kuhn Paradigm shift. Int. J. Inf. Eng. Electr. Bus. (IJIEEB) **16**(1), 54–62 (2024)

9. Yang, Y., Luo, Y.: Construction of a hierarchical independent experimental curriculum system based on research-based learning mode. Exper. Technol. Manage. **5**(31), 7–10 (2014). (in Chinese)
10. Miaomiao, W., De Witt, D., Mohamad Nazry, N.N., et al.: Design and validity of an instrument to measure digital literacy among pre-service teachers involved in inclusive education. Int. J. Mod. Educ. Comput. Sci. (IJMECS) **16**(1), 84–96 (2024)
11. Zhou, J., Wang, H., Nie, J.: Construction and practice of a multi-level innovation experimental platform for college students. Exper. Technol. Manage. **28**(4), 24–26 (2011). (in Chinese)
12. Yang, Y., Luo, Y.: Construction of a hierarchical independent experimental curriculum system based on research-based learning mode. Exper. Technol. Manage. **31**(5), 7–10 (2014). (in Chinese)
13. Porkodi, S., Tabash, B.K.H.: A comprehensive meta-analysis of blended learning adoption and technological acceptance in higher education. Int. J. Modern Educ. Comput. Sci. (IJMECS), **16**(1), 47–71 (2024)
14. Petruk, V., Rudenko, Y., Yurchenko, A., et al.: Analysis of the results of the pedagogical experiment on the integrated analysis of the average and dispersions. Int. J. Modern Educ. Comput. Sci. (IJMECS) **14**(6), 25–34 (2022)
15. Ujwal U.J., Malik, S.: A hybrid weight based feature selection algorithm for predicting students' academic advancement by employing data science approaches. Int. J. Educ. Manage. Eng. (IJEME), **13**(5), 1–22 (2023)

Implementation and Application of Economic Law Course Evaluation and Assessment Utilizing OBE Principles

Ting Song(✉)

School of Management, Guangzhou City University of Technology, Guangzhou 510800, China
1326938849@qq.com

Abstract. Outcome-Based Education (OBE) focuses on achieving specific learning outcomes by emphasizing both the implementation and monitoring processes. Utilizing the principles of outcome-based education and drawing from various evaluation systems, along with hands-on experience in education, we implemented the BOPPPS teaching model to investigate and apply a continuous evaluation system for the Economic Law (CPA perspective) course, resulting in the development of a comprehensive assessment and evaluation framework. Practice shows that implementing continuous assessment can successfully transform students' learning attitudes, boost their motivation and proactivity, ultimately leading to enhanced academic performance.

Keywords: OBE concept · BOPPPS teaching mode · Continuous assessment · Economic Law

1 Introduction

1.1 Research Background and Significance

The evaluation process is a crucial aspect of educational practice, providing a foundational method to assess and enhance teaching methods, ultimately fostering the growth of creative individuals. The goal model and formative evaluation theory were proposed by renowned American educators Taylor and Bloom. They emphasized that assessing teaching objectives should not exclusively rely on final exams, positing that instructors ought to assess and analyze students' learning processes and performance throughout the teaching process. Over the past few years, China has consistently enhanced its focus on assessment methods centered around processes. In 2019, the Ministry of Education released directives aimed at furthering reform in undergraduate education and teaching to enhance the overall development of college students, mandating that all universities nationwide fully adhere to these policies. Over the past few years, multiple colleges and universities have developed and updated their assessment systems in undergraduate education, enhancing and advancing the academic evaluation framework that integrates both process-oriented and outcome-oriented assessments. A multitude of professors in higher education institutions have engaged in studies focusing on process-based evaluation methods, leading to notable outcomes.

Z. Hu et al. (Eds.): ICCSEEA 2024, LNDECT 242, pp. 622–633, 2025.
https://doi.org/10.1007/978-3-031-84228-3_54

The concept of Outcome-Based Education (OBE) focuses on student learning outcomes and places an emphasis on student-centered teaching methodologies [1]. Students engage in purposeful learning, understanding the intended outcomes of their educational journey. The educational philosophy has successfully shifted from a lecturing classroom to an interactive classroom, from a focus on memorization to a focus on skill development, from a competitive atmosphere to a collaborative environment, and from grading based on comparisons to grading based on individual accomplishments [2]. Certainly, merely altering teaching philosophy is insufficient, and it is imperative to also revamp particular teaching modes and methodologies. Henceforth, this research implemented the BOPPPS (Bridge, Outcome, Pre-assessment, Participatory learning, Post-assessment, Summary) instructional approach. The pedagogical approach of this model is rooted in the principles of constructivism theory, emphasizing interactive learning to revolutionize conventional rote instruction. It establishes a teaching cycle consisting of six phases: Bridge in, Objective, Pre-assessment, Participatory Learning, Post-assessment, and Summary. BOPPPS prioritizes active student engagement, leading to improved self-confidence in learning when compared to conventional teaching practices. The student-focused OBE principle aligns perfectly with this trait. Utilizing the OBE approach, we have implemented the BOPPPS teaching framework and documented our findings on the assessment and evaluation processes for the Economic Law (CPA perspective) course through the ChaoXing digital teaching platform. We hope to provide some practical experience for deepening the reform of undergraduate course education evaluation in the new era.

1.2 Literature Review

Continuous evaluation has shifted the evaluation paradigm from "assessment of learning" to "assessment for learning", in line with the trend of higher education teaching evaluation towards a more scientific paradigm [3]. Domestic and foreign scholars have conducted extensive research on continuous evaluation, which not only enriches the theory of continuous evaluation to a certain extent, but also promotes its practical development.

1.2.1 Research on the Connotation of Continuous Evaluation

There are currently two main views on continuous evaluation, each with different perspectives and points of fit. The first viewpoint holds that continuous evaluation is a comprehensive evaluation of students at different learning stages. Representative researchers include Geoffrey N. Masters and John Evans. Their view is that continuous evaluation in teaching practice is to collect and record the learning scores of students at different times in a course, and finally, at the end of the course, weight and integrate to obtain an overall evaluation or cumulative score. Anthony J. Nitko (2004) believes that continuous evaluation involves collecting and providing appropriate explanations for student learning information. Therefore, this evaluation involves teaching content and student learning situations [4]. Yigzaw, Abiy (2013) found that continuous evaluation can be seen as a continuous test, which is more like the detection of dynamic processes in physics. However, measurement can have a certain impact on the original system, which is well known to physicists [5]. In addition, Gao Lingbiao (2004) found that some Chinese scholars believe that continuous evaluation is mainly based on "process observation" to evaluate the cognitive learning effectiveness of students. Some scholars also believe that continuous evaluation should focus on non-intellectual factors such as emotions, attitudes, and values [6]. The second viewpoint believes that continuous evaluation is a mechanism that covers many aspects of students, including cognition, emotion, and spirit. Ojerinde Dibu(2011) believes that continuous evaluation is the systematic weighted accumulation of final evaluations of students in terms of cognition, emotion, and spirit during the semester. Therefore, continuous evaluation not only includes evaluating students' thinking processes, but also evaluating behaviors, personality traits, and agility.

In summary, there are different understandings of the connotation of continuous evaluation, but domestic and foreign scholars agree with the second viewpoint. They believe that continuous evaluation is a way to evaluate students' learning in the process of curriculum implementation, which not only focuses on their cognition, but also on their emotions, values, and other aspects.

1.2.2 Research on the Value of Continuous Evaluation

The first viewpoint is that continuous evaluation is beneficial for increasing student engagement in learning and improving student learning outcomes. Gao Lingbiao (2004) believes that in continuous evaluation, teachers provide timely feedback to students, and students know their own problems, which leads to better engagement in learning and further improvement in learning outcomes. Cole, Jonathan S. (2012) pointed out that continuous evaluation promotes student participation in the curriculum, improves teaching attendance compared to before, and significantly improves student exam scores. Students recognize that evaluation systems can help encourage them to build confidence in learning and continue learning throughout the entire semester [7]. A. Kangaiammal (2013) suggests that students generate internal opinions and evaluate progress towards achieving goals while monitoring their participation in learning activities and tasks [8]. Zhao Jing (2018) introduced continuous evaluation into primary school mathematics classroom learning and conducted a survey through the development of a scale. It was found that the implementation of continuous evaluation mobilized students' enthusiasm, significantly improved their learning effectiveness, and also promoted the improvement of their mathematical literacy [9].

The second viewpoint is that continuous evaluation improves students' innovation ability and is conducive to cultivating innovative talents. Kuihuan (2015) believes that continuous evaluation is not limited to the standardization of teaching objectives and methods, nor is it overly focused on the accuracy of evaluation, nor is it limited to formal evaluation procedures. This is conducive to students fully utilizing their talents, divergent thinking, and stimulating their innovative consciousness [10].

The third viewpoint is that continuous evaluation is beneficial for improving teaching methods and enhancing professional development abilities of teachers. Mohammad Iqbal (2017) believes that continuous evaluation is an evaluation during the teaching process, providing teachers with diagnostic information on the strengths and weaknesses of student learning. This timely feedback also provides teachers with opportunities to identify problems during the teaching process, improve their teaching methods, and continuously promote the improvement of teaching abilities [11].

In short, continuous evaluation not only contributes to the comprehensive development of students, but also strongly supports the teaching process of teachers, promotes the further development of higher education evaluation, and helps to improve the quality of talent cultivation.

1.2.3 Research on Continuous Evaluation Methods and Tools

With the development of network and electronic resources, continuous evaluation has also begun to emerge with network resources and emerging scientific and technological advancements. Alda Pereira and Isolina Oliveira, among others, use electronic records (E-Folio) as an evaluation tool in online courses to enhance students' learning and other related abilities. Xiong Mingfu (2012) designed an electronic portfolio evaluation system, which improved the evaluation efficiency. Yang Weijie (2013) designed a process-based evaluation model for teacher teaching quality based on Assessment Portfolio, which realizes the information management of teacher teaching quality evaluation activities. Morales, Luc í a believes that clickers can positively influence classroom dynamics, liven up the classroom atmosphere, and effectively enhance interaction. Wei Wei (2016) designed an intelligent course inspection system for mobile devices based on the basic theoretical research of continuous evaluation and applied it to conventional classroom teaching [12].

2 The Current Status of Assessment and Evaluation of the Course of Economic Law (CPA Perspective)

2.1 Disproportionate Weight Given to Final Exam Scores

Currently, the Economic Law course (CPA perspective) continues to assess students' comprehension using the traditional evaluation model of combining 30% from regular assessments with 70% from the final exam to judge their learning outcomes. Focusing primarily on final grades rather than ongoing performance can lead students to overlook the learning process during the course, regardless of their exceptional classroom performance, thus impacting their overall assessment and evaluation. As a result, students often overlook the importance of fully engaging in course material, leading them to believe that simply focusing on memorizing key concepts before an exam is sufficient for achieving good grades. The heavy reliance on final exam scores in this assessment model not only promotes a habit of last-minute cramming among students, but also dampens their enthusiasm for learning, ultimately leading to poor teaching outcomes and diminished quality.

2.2 Ambiguous Grading Criteria for Assessments and Evaluation

The typical criteria for assessing student performance in Economic Law include regular attendance, classroom participation, and group presentations related to case studies. The evaluation process lacks a comprehensive approach with general assessment guidelines that do not cater to individual student abilities, resulting in an incomplete evaluation of student capabilities. Initially, when assessing students' performance in the classroom, the constraints posed by large class sizes, limited instructional time, and voluminous course material in Economic Law classes result in inadequate opportunities for students to participate in answering questions. Thus, evaluating students solely based on their participation in classroom question-answering favors certain individuals and creates an unfair advantage. Hence, at present, the measure of class attendance is frequently utilized as the principal factor in assessing classroom performance. However, relying solely on attendance to evaluate students' performance is clearly inadequate given that it is a fundamental responsibility of students. Additionally, in the assessment of group case presentations by students, typically, it is the responsibility of peer leaders to evaluate them and calculate the final average score. Due to the competitive dynamics between our group and other groups, the assessment process lacks complete impartiality and fairness. Therefore, it is necessary to include indicators such as teacher evaluation and intra group evaluation.

2.3 Single Assessment Methods

Following the closed-book examination at the end of the term, teachers evaluate students' performance based on established grading criteria. Typical grading criteria often include three components: presence, academic performance, and collaborative presentation. The absenteeism rate is high in face-to-face classes taught by teachers, due to the large number of students and low effectiveness of in-person attendance monitoring. The evaluation of student academic performance in a classroom heavily relies on the teacher's personal judgment, often lacking standardized criteria. Hence, the evaluation of students solely based on one exam fails to accurately represent their overall learning journey, making the assessment method overly simplistic. In order to address the current situation, it is imperative and essential to revise the evaluation methods employed in this course.

3 Reform and Design of the Continuous Assessment and Evaluation System for the Course of Economic Law

3.1 Reform of Assessment Weights, Increasing the Proportion of Process-Based Assessment Results Reasonably

Given the need to enhance our teaching process through assessment reform, we decided to modify the weight of regular grades from 30% to 40% and decrease the weight of final exam grades from 70% to 60%. The contribution of usual grades and final exams to the course objectives is shown in Table 1. Utilizing the OBE framework, we have formulated three distinct tiers of learning outcomes for the Economic Law course focused on preparing students for the Certified Public Accountant (CPA) profession, which are outlined in Table 1. The daily performance contribution rate of 40% is the sum of the performance assessment at 20%, display assessment at 10%, and stage assessment at 10%, as detailed in Table 2.

Table 1. The contribution of usual grades and final assessments to course objectives

Administrative Levels	Course Objectives	Assessment Weight		Total Contribution
		Usual grades (contribution rate of 40%)	Final exam (contribution rate of 60%)	
1	Knowledge Objectives	10%	10%	20%
2	Skill Objectives	20%	40%	60%
3	Emotional Objectives	10%	10%	20%
Amount	Amount	40%	60%	100%

Table 2. The combination of OBE concept and BOPPS teaching mode in daily performance assessment indicators and weight table

Teaching Model	Teaching Link	Assessment Content	Assessment Indicators	Proportion
B	Bridge in	Performance Assessment	Classroom interaction (including pre-class quizzes of 5%, classroom performance of 10%, and online discussions of 5%)	20%
O	Objective			
P	Pre-assessment			
P	Key and difficult points explanation			
	Participatory Learning	Demonstrative Assessment	Group Case Presentation	10%
P	Post-assessment	Periodic Assessment	Unit Evaluation	5%
S	Summary		Mind Map	5%
	Amount			40%

3.2 Reform of Assessment Content, Integrating the BOPPPS Teaching Model Throughout the Entire Teaching Process

In the specific implementation, we utilize the BOPPPS teaching model to divide the teaching process into six steps: introduction (B) - objectives (O) - pre-assessment (P) - explanation of key and difficult points + participatory learning (P) - post-assessment (P) - summary (S). We create specific evaluation metrics and allocate appropriate weights for each stage. In the realm of evaluating student learning, the assessment content for the

Economic Law course is categorized into performance evaluation, demonstrative evaluation, periodic evaluation, and knowledge evaluation. Performance assessment mainly refers to classroom interaction (including pre-class quizzes of 5%, classroom performance of 10%, and online discussions of 5%), which runs through the introduction of the BOPPPS teaching mode, learning objectives, pre-tests, and explanation of key and difficult points. Demonstrative assessment mainly refers to the presentation of group cases, which runs through the participatory learning process of the BOPPPS teaching model. Periodic assessment mainly refers to unit assessment and mind mapping, which run through the post-test and summary stages of the BOPPPS teaching mode.

3.3 Reform of Assessment Indicators, Refining Grading Rules for Usual Grades

The evaluation criteria for the Economic Law class encompass various components such as participation in class discussions (including 5% for quizzes before each class, 10% for performance in class, and 5% for online discussions), group presentations, unit assessments, and the utilization of mind maps. In accordance with the established grading guidelines, the pre-class quiz carries a 5% weight towards the final grade. The educators will distribute pre-class quiz queries via the ChaoXing instructional tool, then calculate the mean score for this section. Classroom performance accounts for 10% of the total. In the ChaoXing teaching platform, teachers choose students at random to respond to questions. Students will be awarded 1 point for each correct answer, with no points deducted for incorrect responses. The ChaoXing teaching platform facilitates teacher-students interaction by allowing instructors to post online discussion questions, which make up 5% of the overall course grade. Each student will earn one point for every session they actively engage in answering questions. 10% of the grade is attributed to group case presentations, which are evaluated by teachers via the ChaoXing teaching platform. The overall score for group case presentations is calculated by combining scores from teacher evaluations (30%), evaluations within the group (30%), evaluations between groups (30%), and self-evaluation (10%). The assessment of units comprises 5% of the total grade, and educators distribute the questions related to the unit tests via the ChaoXing educational platform. Teachers utilize the ChaoXing teaching platform to distribute mind mapping assignments, which make up 5% of the overall curriculum. Educators manually assess student work and allocate points.

3.4 Reform of Assessment Methods, Constructing a Hybrid Assessment Model of "Online + Offline"

We have created a blended assessment strategy by combining online and traditional methods in our evaluation process. The ChaoXing platform was used to develop an online teaching resource video library, SPOC, online question bank, and student discussion space to address the challenge of recording students during the learning process. Evaluation strategy tools were also constructed based on the ChaoXing teaching platform, including "guidance videos," "learning objectives," "explanations of key and difficult points," "in class testing," "discussion zones," "expanding materials," and "group tasks".

By incorporating tools such as prompt feedback, engaging activities, encouraging self-assessment, fostering teamwork, and displaying abilities, our goal is to successfully meet the evaluation goals of the Economic Law course.

4 Implementation of Continuous Assessment and Evaluation for the Course of Economic Law

Our institution became the inaugural university in China to achieve IACBE international business certification after undergoing a successful formal membership evaluation by IACBE in July 2018.Our School of Management is now fully accredited by the IACBE and aligns with the principles of OBE. The main focus is on outcomes, centered around the needs of students, and constantly striving to enhance. The primary goal is achieving desired results, catering to the individual needs of students, and consistently seeking ways to improve instructional practices. The course of Economic Law (CPA perspective) is a fundamental course in the field of accounting, which has been offered since its establishment. The course is typically provided during the latter half of the academic year and is specifically designed for sophomores studying accounting. Approximately 600 students receive instruction on a yearly basis. Since 2014, we have utilized the official CPA exam textbook created by the Chinese Institute of Certified Public Accountants as the primary instructional resource in this course. The textbook is highly professional, difficult, and has outstanding professional characteristics. In our implementation of the Economic Law course, we have structured a comprehensive formative assessment approach that integrates both continuous assessment and outcome assessment. The BOPPPS pedagogy is employed to deconstruct the instructional procedure into six phases: bridge in(B) - objectives (O) – pre-assessment (P) - explanation of key and difficult points + participatory learning (P) - post-assessment (P) - summary (S). When evaluating student performance in the Economic Law course, we carefully design assessment criteria and allocate proportional weights accordingly to ensure accurate measurement of their progress. The assessment content for the Economic Law course is categorized into performance evaluation, demonstrative evaluation, periodic evaluation, and knowledge evaluation.

4.1 Performance Assessment of the BOPPPS Teaching Model

The initial phase of the BOPPPS instructional framework involves bridging in knowledge through case or videos with students (Bridge), outlining educational goals (Objectives), and evaluating student knowledge through pre-assessment techniques (Pre-assessment). Students view the video on economic laws as a means to spark curiosity in the subject. The instructor establishes educational goals for the section on economic law, empowering students to discern the essential knowledge, principles, and skills required for acquisition. Students complete the pre-test questions, and the teacher moderately adjusts the teaching content based on the pre-test results. The second stage of the BOPPPS teaching model: explanation of key and difficult points + participatory learning (P), where students watch videos explaining the key and difficult points of economic law, incorporating ideological and political elements into specific cases. The educational videos are structured around key concepts, with each video typically kept under 15 min in length to align with principles of microlearning. Learners have the flexibility to engage in educational activities at any location and time, surpassing constraints related to space and time, and resolving challenges associated with ineffective learning due to sluggish internet connections and other factors during live streaming sessions. The performance assessment in the Economic Law course includes three aspects: pre class quizzes, classroom performance, and online discussions. We mainly rely on the ChaoXing teaching platform to record, statistically analyze, and assess performance.

4.2 Demonstrative Assessment of the BOPPPS Teaching Mode

Engaging in participatory learning is the focus of the third stage in the BOPPPS teaching model. The demonstrative assessment in the assessment of the Economic Law course refers to the group case presentation task. The textbook consists of 6 chapters, each with a legal topic selected by team drawing. Students can freely form teams across classes, and teachers require that each group must have at least 8 participants, with a time limit of around 15–20 min. Legal cases, videos, and simulation of performance scenes can be discussed, and students can independently complete the search, writing of cases, role-playing in the early stage, and video editing in the later stage. Through the facilitation of critical thinking opportunities, teachers empower students to cultivate their own ideas and perspectives, fostering holistic growth. The group case presentation task adopts teacher evaluation (30%) + intra group evaluation (30%) + inter group evaluation (30%) + self-evaluation (10%), which can not only exercise students' teamwork and communication skills, but also cultivate their creative thinking and legal literacy. When assessing group tasks, it is important to incorporate self-evaluation, evaluations within the group, and evaluations between groups to accurately recognize each student's contribution to the team's success and the benefits of working together.

4.3 Periodic Assessment of the BOPPPS Teaching Model

In the final stage of the BOPPPS teaching model, students answer evaluation questions on the economic law unit, while teachers review crucial economic law concepts and check student mind map assignments. Utilizing the ChaoXing teaching platform, we have developed an extensive question bank on economic law, containing more than 1000 questions, covering various question formats such as individual-choice, multiple-choice, and case analysis questions. Since 2020, the ChaoXing platform has been crucial in aiding the seamless transition to blended learning through its facilitation of self-assessment and online mid-term exams for every unit in Economic Law. Simultaneously, students are tasked with unit-specific mind map assignments, where each crucial unit mandates that students either manually sketch or digitally design a mind map to aid in their comprehension of pivotal and challenging concepts.

4.4 Knowledge Assessment of the BOPPPS Teaching Mode

For the course of Economic Law, knowledge assessment usually refers to the closed-book exam at the end of the term. Based on the specified learning goals and criteria for the

Economic Law course, we implemented a standardized closed-book examination at the end of the term. The examination has a variety of question types including single-choice (10 points), multiple-choice (20 points), short-answer (10 points), and comprehensive questions (60 points). The short answer questions do not provide students with a specific scope, only a few nouns that may be tested, in order to broaden the students' distinction. We have reformed the format and content of the final exam to address the issue of focusing solely on memorization ability. To enhance students' practical skills, we have adjusted the balance between theoretical and practical questions in our assessments.

5 The Implementation Effect and Reflection of Continuous Assessment and Evaluation in the Course of Economic Law

5.1 Implementation Effect

After one year of practical application, Fig. 1 displays the results comparison showcasing the impact of the reform in the Economic Law course. The article uses students from Grade 21 Class 1–2 of Accounting as the experimental group and students from Grade 20 Class 1–2 of Accounting as the control group. Analysis of Fig. 1 reveals a noticeable enhancement in students' overall academic achievement following the implementation of the assessment and evaluation reform, as compared to both pre-and post-reform periods.

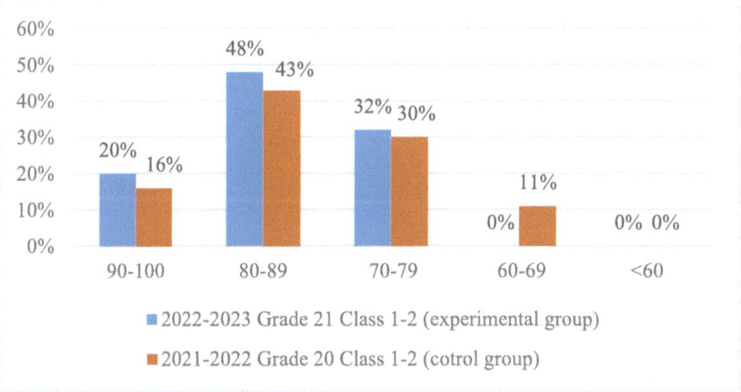

Fig. 1. Comparison and analysis of assessment and evaluation reform

This article compares the teaching classes before and after the teaching reform, with Accounting Grade 20 Class 9–10 as the control group and Accounting Grade 21 Class 9–10 as the experimental group. The control group adopts traditional teaching mode, while the experimental group adopts continuous evaluation teaching mode to observe the teaching effect. The teaching effect is observed. SPSS 20.0 statistical software was used for analysis, and the expression of count data was [n (%)]. The results were analyzed using χ^2 Inspection. If $P < 0.05$, it is a significant difference with statistical significance.

The results showed that the teaching effectiveness evaluation of the experimental group students was significantly better than that of the control group ($P < 0.05$), and

Table 3. Comparison of teaching effectiveness evaluations between two groups of students [n (%)]

Group	Number of people	Continuous assessment reform teaching satisfaction	Beneficial for pre class preparation and post class review	Beneficial for interaction between teachers and students	Facilitating communication and exchange among students
experimental group	90	83 (92.3)	87 (96.7)	84 (93.3)	84 (93.3)
control group	90	74 (82.2)	69 (76.7)	76 (84.4)	66 (73.3)
χ^2 value		13.852	12.296	14.316	10.438
P value		0.004	0.002	0.001	0.003

Table 4. Comparison of evaluations of teaching abilities between two groups of students [n (%)]

Group	Number of people	Significant improvement in self-learning ability	Enhanced communication between teachers and students	Increased student motivation and initiative	Improved ability to analyze and solve problems
experimental group	90	82 (91.1)	86 (95.6)	88 (97.8)	82 (91.1)
control group	90	58 (64.4)	68 (75.6)	66 (73.3)	58 (64.4)
χ^2 value		8.082	10.328	12.678	8.490
P value		0.002	0.003	0.002	0.003

the evaluation of teaching ability by the experimental group students was significantly better than that of the control group ($P < 0.05$). It can be seen that in the teaching of economic law courses, the application of continuous evaluation has a significant impact on the evaluation of teaching effectiveness and learning ability by students. As shown in Table 3 and 4.

5.2 Pedagogical Implication

During our investigation and practical application of evaluating the Economic Law course, we found that: (1) It became apparent that the diverse student body in the course greatly values standardized scoring methods for their grades. This necessitates thorough discussion and meticulous implementation of evaluation indicators and scoring criteria.

(2) The reform of teaching evaluation requires the construction of rich course resources, which is closely related to the informatization construction of courses and the reform of blended learning models. (3) To successfully implement continuous evaluation in teaching, educators must meticulously plan instructional procedures, coordinate

classroom and extracurricular engagements, and foster extensive collaboration among teachers within the curriculum group for sustained academic advancement.

6 Summary and Conclusion

We utilize the BOPPPS teaching framework within the context of Outcomes-Based Education (OBE) to analyze the implementation and continuous assessment in the Economic Law curriculum on the ChaoXing online teaching platform. Our objective is to offer valuable practical experience to further the advancement of evaluating undergraduate course education in the current age. Our study on continuous assessment for undergraduate students revealed that it has a considerable impact on reshaping students' learning attitudes and enhancing their academic achievements. Implementation of continuous assessment methodologies has the potential to transition students from traditional teacher-centered learning to more creative and spontaneous learning experiences.

Acknowledgment. The research is funded by Key Research Base of Humanities and Social Sciences in Universities of Guangdong Province: Research Base for Digital Transformation of Manufacturing Enterprises (2023WZJD012) and Guangdong Province Undergraduate Online Open Course Guidance Committee: Research on the Construction of Online Open Courses Based on OBE - Taking the Economic Law (CPA Perspective) of OBE Concept + BOPPPS Teaching Model as an Example (2022ZXKC592).

References

1. Farrukh, N., Mahgoub, S.: Student-centered role-based case study model to improve learning in decision support system. Int. J. Mod. Educ. Comput. Sci. (IJMECS) **6**(10), 16–22 (2014)
2. Chen, Y., Yin, J.: Apply student-centered learning on computer education. Int. J. Educ. Manage. Eng. (IJEME) **1**(4), 67–71 (2011)
3. Belay, S.: The practice of continuous assessment in primary schools: the case of Chagni, Ethiopia. J. Educ. Pract. **7**, 24–30 (2016)
4. Niko, A.J.: Educational Assessment of Students. 4th ed. Ohio: Merrill Prentice Hall (2004)
5. Yigzaw, A.: High school English teachers' and students' perceptions, attitudes and actual practices of continuous assessment. Acad. J. **8**(16), 1489–1498 (2013)
6. Gao, L.B.: The concept of continuing evaluation and its function. J. South China Normal Univ. (Soc. Sci. Ed.) (6), 102–106+113–160 (2004). (in Chinese)
7. Cole, J.S., Spence, S.W.T.: Using continuous assessment to promote student engagement in a large class. Eur. J. Eng. Educ. **37**(5), 508–525 (2012)
8. Kangaiammal, A., Silambannan, R., Senthamarai, C., Srinath, M.V.: Student learning ability assessment using rough set and data mining approaches. Int. J. Mod. Educ. Comput. Sci. (IJMECS) **5**(5), 1–11 (2013)
9. Zhao, J.: The application of continuous evaluation in primary school mathematics classroom. Shanghai Normal University (Shanghai) (2018). (in Chinese)
10. Niu, K.: Preliminary exploration of learning continuous evaluation for college students. Dev. Eval. High. Educ. **31**(04), 6–13 (2015). (in Chinese)
11. Mohammad I., Samiullah, Aysha A. Effect of continuous assessment techniques on students' performance at elementary level. Bullet. Educ. Res. **4**(39), 91–100 (2017)
12. Wei, W.: Research on Intelligent Course Patrol System for Classroom Teaching Continuous Evaluation. East China Normal University, Shanghai (2016). (in Chinese)

The Application and Research of Human-Robot Collaborative in Chinese Language Teaching for International Students

Yimeng Mao[1], Haifeng Yang[2](✉), and Wenjie Fan[2]

[1] School of Law and Humanities and Society, Wuhan University of Technology, Wuhan 430065, China

[2] Faculty of Letters, Arts and Sciences, Waseda University, Waseda, Japan

yanghaifengtl@163.com

Abstract. In recent years, with the continuous advancement of technological sophistication and deeper policy orientations, an increasing number of technologies have been applied in the field of education. At the same time, the strengthening of China's national power has attracted more and more foreign students to study in China, leading to rapid development in the field of international Chinese education. The human-robot collaborative mode has attracted attention in teaching classrooms, but research on its application in Chinese language classrooms for international students remains scarce. This paper establishes a new teaching model for the application of human-robot collaboration in Chinese language classrooms for international students. Using a quasi-experimental research method, it compares the changes in listening, speaking, reading, writing, and translation abilities of international students after adopting the human-robot collaborative teaching method, thereby providing empirical evidence for the feasibility of the model. The experiment demonstrates that the human-robot collaborative teaching method significantly improves the basic language skills of listening, speaking, reading, and writing for international students, but there is no significant improvement in the short term for translation, which represents a higher-level language skill. Therefore, this model is beneficial for beginner-level international students learning Chinese, and teachers can promote its application in teaching practice to enhance teaching and learning efficiency.

Keywords: Human-robot collaborative in teaching and learning · International Chinese language education · Artificial intelligence education

1 Introduction

In January 2023, the Ministry of Industry and Information Technology and seventeen other departments issued a notice regarding the implementation plan for the "Robot + " application action. The plan proposes to enhance robot education guidance, improve the teaching content and practical environment of robot courses at all levels of institutions, and promote the integration and application of 5G, artificial intelligence, intelligent

Z. Hu et al. (Eds.): ICCSEEA 2024, LNDECT 242, pp. 634–642, 2025.
https://doi.org/10.1007/978-3-031-84228-3_55

voice, machine vision, big data, digital twins, and other technologies with robot technology. The 2024 Government Work Report also proposed deepening the research and application of big data, artificial intelligence, and launching the "Artificial Intelligence + " action to build internationally competitive digital industry clusters. With more and more universities and primary and secondary school classrooms introducing the human-robot collaborative teaching model, it has greatly stimulated students' interest in learning and improved classroom efficiency to a certain extent. Human-robot collaborative teaching and learning have become important ways of intelligent education practice, and the development direction of educational artificial intelligence has also been internationally recognized [1].

In recent years, with China's economic rise and increased international status, international Chinese education has encountered unprecedented development opportunities, and more and more foreign students are receiving Chinese language education. The human-robot collaborative teaching model, as a teaching method that integrates human intelligence with computer technology, provides new ideas and approaches to enhance the effectiveness and efficiency of international Chinese language teaching. Against this backdrop, how to introduce relevant technologies into Chinese language teaching classrooms to enrich teaching methods and effectively use advanced technology to enhance the interest of international students in Chinese language teaching has become an urgent issue to address. This paper will establish an application model of human-robot collaboration in Chinese language teaching classrooms for international students, proposing ideas and solutions for addressing the difficulties in teaching Chinese to international students and utilizing the human-robot collaborative model to solve related problems.

2 Literature Review

2.1 Research on Teaching Chinese to International Students

On the Chinese language teaching classes for international students, due to factors such as cultural differences, differences in teaching philosophy, deviations in teaching methods, etc., international students are prone to difficulties in classroom enthusiasm, classroom performance, Chinese language learning, and other aspects. In recent years, in research related to teaching Chinese as a foreign language, issues such as distinguishing degree complements, distinguishing synonymous words with similar elements, and other key and difficult problems urgently need to be addressed [2, 3]. At the same time, in the process of teaching Chinese to international students, not only do international students face some difficulties, but some Chinese teachers also have problems such as outdated teaching philosophies, monotonous classroom teaching methods, and poor adaptability [4]. The traditional Chinese teaching model is difficult to completely cover the difficulties in the classroom, so it is necessary to seek new models and methods.

2.2 Research on Human-Robot Collaborative Teaching

Human-robot collaborative teaching refers to a novel instructional approach where teachers, students, and teaching software or intelligent systems cooperate and mutually support each other throughout the teaching process to achieve shared educational goals. This collaborative model signifies a significant advancement in the digital transformation of education and signals a new direction for future learning trends.

In related theoretical research, scholars have identified the main structures that have a significant impact on the adoption of blended learning in higher education and proposed a new technology acceptance model suitable for digital educational tools [5]. Based on a negotiated adaptive learning system, through the negotiation mechanism between learners and intelligent systems, students are assisted in self-regulated learning, clarifying when seeking help is encouraged [6]. The human-machine collaborative model has brought about educational reforms, and future teaching models will be more closely integrated with technology.

The integration of digital technology and innovative methods has had a significant impact on the education system, making the work of learners and teachers easier. In practical research, scholars have proposed a method for automatically assessing and grading students' UML diagrams, which can effectively save time [7]. Li Haifeng et al. explored the development of human-machine collaborative deep exploratory learning systems and smart teaching assistants based on ChatGPT and QQ systems to bring convenience to learners [8]. Additionally, some researchers have conducted studies on the application of chatbots in the field of education [9]. The introduction of related technologies can also cater to special groups in teaching. For example, augmented reality software has great potential in improving the teaching process of educational institutions by creating virtual spaces to minimize barriers for hearing-impaired students learning programming logic [10].

Establishing a Peer Personalized Learning Recommendation System (RiPPLE) under the human-machine collaborative mode can recommend learning peers for learners based on their knowledge states, thereby improving learning efficiency [11]. The realization of human-machine collaborative learning depends mainly on creating intelligent learning environments and revealing the principles and mechanisms of learning occurrence to provide favorable learning support for learners [12]. When learners take the lead, intelligent systems mainly help enhance intelligence by increasing transparency, accuracy, and effectiveness, aiming at human-centered artificial intelligence [13].

Despite the widespread application of the human-robot collaborative model, research on its integration with Chinese language classrooms for international students remains scarce. In the process of revamping Chinese language instruction for international students, the adoption of a human-robot collaborative teaching model and leveraging technology to support teaching can not only invigorate the classroom environment and boost student engagement but also foster the enhancement of learners' language and cognitive skills. Furthermore, it can aid teachers in refining their instructional approaches and enable students to deepen their understanding of the Chinese language and culture.

3 Application Model of Human-Robot Collaborative in Chinese Language Classrooms for International Students

In traditional international Chinese language classes, teaching always focuses on the textbook, with single content and without sufficient practicability. At the same time, there is no enough teacher-student interaction to inspire students' enthusiasm. It's difficult for teachers to grasp the Chinese Proficiency of each students. So teachers only can evaluate their teaching effect relying on the final examination, which is lagging to figure out students' learning ability (Fig. 1).

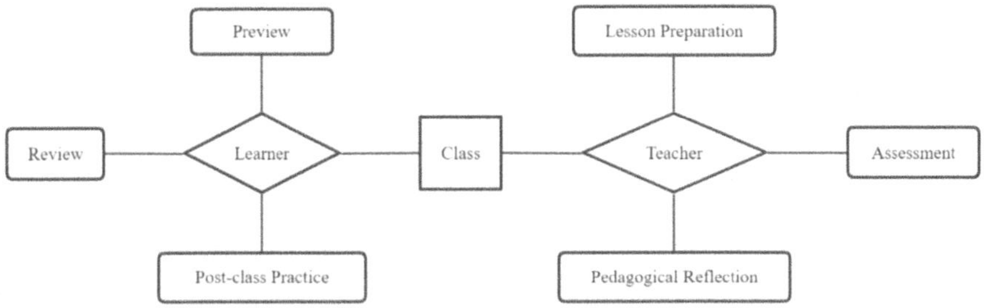

Fig. 1. Traditional teaching mode

Introducing relevant intelligent systems or teaching software into international student classrooms can effectively change simple teaching modes and help achieve corresponding teaching objectives. This study is based on the literature review and learning mentioned above, and constructs an application model of human-robot collaborative in Chinese language teaching classrooms for international students. For second language learners, listening, speaking, reading, writing, and translation are basic skills that need to be mastered. This model aims to improve these skills of international students as the fundamental goal, starting from both learners and educators, to improve the basic work of the learning and teaching stages, establish a teaching database, and ultimately achieve intelligent learning and teaching (Fig. 2).

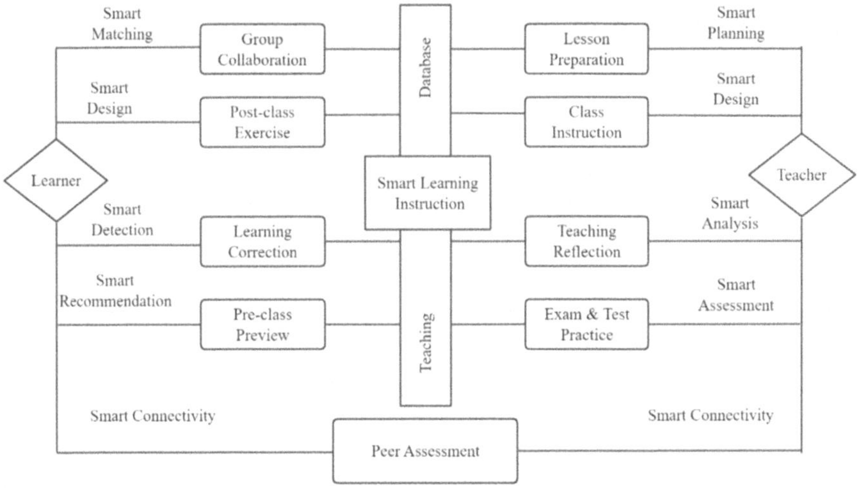

Fig. 2. New applications of human-robot collaborative model on Chinese classes for international students.

3.1 Intelligent Learning

The intelligent learning process for learners encompasses intelligently providing preview content for teaching materials, detecting learning errors post-class, designing after-class exercises tailored to specific individuals, and intelligently grouping learners based on their personality traits, learning habits, and proficiency in Chinese language. This approach fosters group communication both during and after class. The integration of human-robot collaboration enables the delivery of personalized learning materials that cater to individual progress, interests, and learning preferences. The seamless transmission of information facilitates real-time feedback, enabling prompt error correction and reinforcement of learning priorities. Additionally, human-robot collaborative learning can be conducted at any time and place, allowing students to schedule their study time and progress effectively, thereby enhancing the flexibility and convenience of the learning process.

3.2 Intelligent Teaching

The intelligent teaching process of educators includes intelligently planning classroom content based on students' Chinese proficiency; Integrate various multimedia resources, intelligently design teaching ideas, and improve the fun and attractiveness of learning; Intelligent analysis of the key and difficult points in students' after-school exercises, promoting teaching reflection; Set homework and exam content based on classroom teaching content and student intelligent data feedback. Human-robot collaborative teaching can efficiently collect learning data from students, help teachers understand their learning situation through data analysis and mining, guide teaching practice, and optimize teaching effectiveness.

3.3 Database System

(Fig. 3)

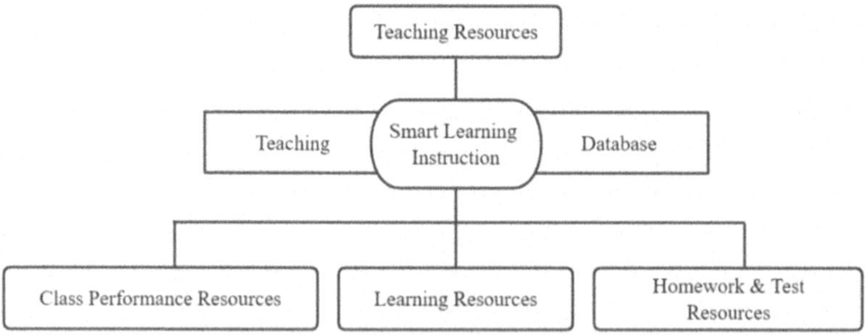

Fig. 3. Intelligent database

Database system is an important data foundation for achieving personalized learning, intelligent learning, and human-robot collaborative learning. Using MySQL and

Excel to establish databases related to learner classroom performance, learning resource databases, homework detection databases, and teaching resource databases, data analysis can achieve data sharing and storage of teaching resources and student information, and also facilitate the sharing and updating of related resources.

3.4 Evaluation Mechanism

Introducing evaluation mechanisms into human-robot collaborative learning mode helps to facilitate mutual feedback between teachers and students, promote interaction and cooperation, stimulate students' learning motivation, adjust learning status in a timely manner, and improve the teaching mode of teachers.

4 Experimental Methods

4.1 Experimental Design

4.1.1 Research Hypotheses

Based on the human-robot collaborative model applied in international Chinese language teaching, this experiment hypothesizes that such collaboration can enhance the Chinese language proficiency of international students. The language skills of international students, including listening, speaking, reading, writing, and translating, serve as crucial criteria for evaluating learning outcomes. This experiment seeks to investigate whether the implementation of a new human-robot collaborative model in Chinese language instruction for international students will result in improved performance in listening, speaking, reading, writing, and translation compared to traditional teaching methods.

4.1.2 Experimental Methods

This experiment utilizes a quasi-experimental research design, involving a total of 120 foreign students with basic Chinese proficiency at Wuhan University of Technology. The participants were randomly assigned to two equal groups. The independent variable in this study is the teaching mode, with the experimental group receiving instruction through a new human-robot collaborative teaching approach, while the control group follows a traditional teaching method. The dependent variables are the test scores of participants in listening, speaking, reading, writing, and translation. The study assesses the participants' scores before and after the experiment and analyzes the differences between the two groups. The experimental period spans from September 2023 to January 2024.

4.1.3 Experimental Process

Throughout the experiment, the teaching tasks and content for both the experimental group and the control group were kept entirely consistent. The instruction was delivered by the same teacher, with strict control over the learning environment and other potential confounding variables. The test tasks assessed the listening, speaking, reading, writing, and translation abilities of the participants in the experimental group. The content of each test remained constant for every participant, although the specific content differed between the pre-test and post-test. Each test had a maximum score of 100 points.

4.2 Experimental Results

The study employed independent sample t-tests to determine the impact of using the human-computer collaborative teaching mode on the listening, speaking, reading, writing, and translation abilities of international students.

The formula for the independent samples t-test statistic is:

$$t = \frac{\overline{x_1} - \overline{x_2}}{s\sqrt{\frac{1}{n_1} + \frac{1}{n_2}}} \tag{1}$$

$\overline{x_1}$ and $\overline{x_2}$ are the mean values of the two sample groups respectively, s is the pooled standard deviation of the two sample groups, n_1 and n_2 are the sample sizes of the two sample groups respectively.

The results revealed differences at the 0.05 significance level in the listening, reading, and writing scores of international students using the human-computer collaborative teaching mode compared to those using the traditional teaching mode. Further comparison of their means showed that international students' listening, reading, and writing scores were higher after using the human-computer collaborative teaching mode than after using the traditional teaching mode, as detailed in Table 1.

From this, it can be concluded that the human-computer collaborative teaching model set up in this study has certain improvement effects on learners' basic language skills training, except for speaking. However, its effect on advanced language skills improvement is not significant.

Table 1. Independent sample t-test results of experimental data

Measurement dimension	Group	Mean value	Standard deviation	T statistical value	P value
Listening	Control group	82.43	6.80	7.520	0.00**
	Experimental group	89.09	6.84		
Speaking	Control group	75.37	5.04	-1.226	0.22
	Experimental group	74.60	4.61		
Reading	Control group	79.37	5.04	13.156	0.00**
	Experimental group	87.60	4.60		
Writing	Control group	75.63	8.73	6.431	0.00**
	Experimental group	82.34	7.31		
Translation	Control group	78.37	5.04	-1.226	0.22
	Experimental group	77.60	4.61		

Note: * $p < 0.05$, ** $p < 0.01$, *** $p < 0.001$

5 Conclusion and Suggestions

This study establishes a new model of human-robot collaborative teaching for the application in Chinese language classrooms for international students and experimentally demonstrates its practical effectiveness. This model not only helps improve teachers' teaching efficiency but also effectively enriches students' learning outcomes. Addressing two different subjects, the following suggestions are proposed.

5.1 For Learners

Firstly, utilize artificial intelligence technology to set personalized learning paths for students based on their language proficiency, learning styles, and interests, customizing course content and learning progress accordingly. Utilize the functions of voice recognition and correction to provide personalized voice correction suggestions.

Use information from student databases to provide intelligent and targeted practice assistance for students. Provide access to learning resources and encourage students to engage in autonomous learning outside the classroom. Establish a platform for peer assessment between teachers and students to provide real-time feedback and promote interaction between both parties.

5.2 For Educators

Firstly, teachers need to master relevant technological tools, including online education platforms and intelligent teaching assistance software, to enhance teaching efficiency. Secondly, provide personalized teaching strategies based on information from student databases, and flexibly adjust teaching content and methods accordingly.

Leverage the advantages of human-robot collaborative teaching, encourage cooperation and interaction among students, and create a positive learning atmosphere.

Acknowledgment. This thesis is one of the achievements of the National Social Science General Project (21BYY138), the Ministry of Education 's Higher Education Division 's Industry-University Cooperation Collaborative Education Project (221006642125730), the World Chinese Teaching Society 's Global Chinese Education Theme Academic Activity Plan (SH23Y38) and the 2023 Hubei undergraduate university provincial teaching reform research project 'Research and Practice on the Deep Integration of Robotic Information Terminals and Chinese International Education and Teaching' (2023111).

References

1. Pedro, F., Subosa, M., Rivas, A., Valverde, P. : Artificial intelligence in education: Challenges and opportunities for sustainable development. UNESCO (2019)
2. Liu, W.: Research on the design of teaching Chinese as a foreign language for the expressions "x de duo", "x de hen", "x de bu de liao". Guangxi Normal University (2024). (in Chinese)
3. Hou, L.: Analysis and research on the differentiation of confusing words "chansheng", "fasheng", "chuxian" and its impact on teaching Chinese as a foreign language. Guangxi Normal University, 2024 (in Chinese)

4. Tong, X., Huang, H., Bai, Y.: Exploration of strategies for dealing with classroom behavioral problems of international students in Chinese teaching. Educ. Res. **42**(5), 82–88 (2019). (in Chinese)
5. Porkodi, S., Tabash, B.K.H.: A comprehensive meta-analysis of blended learning adoption and technological acceptance in higher education. Int. J. Mod. Educ. Comput. Sci. (IJMECS), **16**(1), 47–71 (2024)
6. Chou, C.Y., Lai, K.R., Chao, P.Y., Tseng, S.F., Liao, T.Y.: A negotiation-based adaptive learning system for regulating help-seeking behaviors. Comput. Educ.. Educ. **126**, 115–128 (2018)
7. Jebli, R., El Bouhdidi, J., Chkouri, M.Y.: A proposed algorithm for assessing and grading automatically student UML diagrams. Int. J. Mod. Educ. Comput. Sci. (IJMECS) **16**(1), 37–46 (2024)
8. Li, H.F., Wang, W.: In-depth exploration of human-robot collaborative inquiry teaching mode-taking the human-computer collaborative inquiry learning system based on ChatGPT and QQ as an example. Open Educ. Res. **29**(6), 69–81 (2023). (in Chinese)
9. Bensah, L.E., Tahiru, F., Ankora, C., Noble, A.K.: A bibliometric analysis on chatbot application in education. Int. J. Educ. Manage. Eng. (IJEME) **13**(6), 1–13 (2023)
10. Segura, M., Osorio, R., Zavala, A.: Extended reality model for accessibility in learning for deaf and hearing students (programming logic case). Int. J. Mod. Educ. Comput. Sci. (IJMECS) **15**(4), 1–17 (2023)
11. Potts, B.A., Khosravi, H., Reidsema, C., Bakharia, A., Belonogoff, M., Fleming, M.: Reciprocal peer recommendation for learning purposes. In: The 8th International Conference on Learning Analytics and Knowledge, pp. 226–235 (2018)
12. du Benedict, B.: Escape from the skinner box: the case for contemporary intelligent learning environments. Br. J. Edu. Technol. **50**(6), 2902–2919 (2019)
13. Yang, S., Ogata, H., Matsui, T., Chen, N.S.: Human-centered artificial intelligence in education: seeing the invisible through the visible. Comput. Educ. Artif. Intell. **2**, 100008 (2021)

Methods of Presenting the Structure of the Syllabus of a Training Discipline in a Game Form

Oleksii Beskrovnyi[1], Lesia Kozubtsova[1], Igor Kozubtsov[2](✉), Valerii Lishchyna[2], and Lyudmila Tarshilova[3]

[1] Kruty Heroes Military Institute of Telecommunications and Information Technology, Kyiv 01011, Ukraine
olbezkrovnyy@i.ua, lesia.kozubtsova@viti.edu.ua

[2] Lutsk National Technical University, Lutsk 43018, Ukraine
kozubtsov@gmail.com

[3] NJSC «West Kazakhstan Agrarian and Technical University named after Zhangir khan», Uralsk 090009, Kazakhstan
tarshilova@mail.ru

Abstract. One of the modern requirements of higher education is to ensure that students have the right to freely choose educational components from the list of variable academic disciplines. This is in theory. In practice, it is difficult for students to find logical, not to mention interdisciplinary, connections with the needs of other educational components offered by the department as variable components of the educational component. This is one part of the overall problem. In the future, it is difficult for a teacher in the traditional form of a silobus of an academic discipline to explain to such students the rules of learning, to convey its logical content of structure. The author's research article is aimed at solving these interrelated partial tasks of the general problem. To achieve the purpose of the study, the following tasks were solved: the current state of research and publications in this area was analyzed; theoretical and practical basis for developing a game-based outline was developed. The main part of the study considers the purpose of the syllabus (silhouette) in the traditional form of presentation, and the peculiarities of its writing. An experimental express survey of students' opinions on the sufficient level of information to explain the need to choose this discipline and to evaluate the discipline was conducted. The result confirmed the hypothesis about the need to further improve the structure and content of the curriculum syllabus. At the same time, the students noted a significant improvement in the silhouette sample presented by the lecturer in a game form to improve mental perception. The assumption was experimentally confirmed and students demonstrated an increased interest in the discipline. Thus, the proposed approach to the formation of silhouettes can be announced as a teaching method. The article describes in detail the approach to solving the scientific and applied problem of increasing students' interest in choosing an academic discipline through understanding the structure, logic and necessity of studying academic disciplines. This is achieved by presenting students with an algorithm for reporting stages in a non-standard game form. The scientific result expands the scientific boundaries of pedagogical sciences in the field of teaching methods for higher education students through

Z. Hu et al. (Eds.): ICCSEEA 2024, LNDECT 242, pp. 643–655, 2025.
https://doi.org/10.1007/978-3-031-84228-3_56

the use of simulation as a game-based pedagogical technology. Game-based pedagogical technologies have significant potential in teaching modern students. The proposed solution is fully ready for use by the teacher in practice, but leaves it up to each teacher to adapt the algorithm to a specific discipline. The theoretical results obtained in the course of the research form the basis for its further study and improvement as a teaching methodology. It is expected that the study will have academic and practical value beyond the higher education institutions surveyed.

Keywords: motivation · syllabus · academic discipline · game technology

1 Introduction

1.1 Problem Statement

One of the modern requirements of higher education is to ensure that students have the right to freely choose educational components from the list of variable disciplines.

In practice, unlike in theory, it is difficult for students to find a logical, not to mention interdisciplinary, connection with the needs of other educational components offered by the department as variable components of the educational component.

In the future, it is difficult for a teacher in the traditional form of a silobus of an academic discipline to explain to such students the rules of learning, to convey its logical content of structure. In addition, the discipline consists of semester certification, which includes module tests that are provided for in the curriculum and work program of the discipline. Each type of reporting contains its own criteria for assessing student performance, as well as thresholds for the minimum and maximum level of student knowledge assessment. Module control and certification are not marked in the schedule, so students cannot predict them.

Therefore, not only the authors of the study believe that every teacher faces the need to solve a practical problem: how to increase students' interest not only in the discipline they are studying, but also in the logic of its presentation and interconnection with other disciplines. This is a lifelong problem that has not been fully resolved and is left to each teacher to solve independently.

1.2 Literature Review. Analysis of Recent Research and Publications

The syllabus is an educational-methodical program of the discipline, which includes a description of the subject, goals and objectives, a summary of the content, topics and duration of each lesson, tasks for independent work, consultation time, teacher requirements, assessment criteria and a list of basic and additional literature.

Syllabus – as defined by Cambridge Dictionary and Thesaurus (Cambridge University Press) is a plan showing the subjects or books to be studied in a particular course, especially a course that leads to an examination [1].

For example, in the methodological recommendations [2], proposed an approach to the development of syllabuses in educational institutions of the Republic of Kazakhstan. It has distinctive features from the recommendations for its development adopted in higher education institutions in Ukraine [3].

In the study of J. Parks and M.B. Harris suggest that "a syllabus can serve as a model of professional thinking and writing for students" [4]. They also believe that effective learning requires a complex interaction of skills, such as time management, prioritizing tasks, using technology, etc., and that the curriculum can help develop these skills.

According to J.M. Slattery and J.F. Carlson, a syllabus is nothing more than "a contract between instructors and their students designed to answer students' questions about the course and to inform them of what will happen if they do not take the course" [5].

In [6], D.V. Habanek emphasizes the importance of the syllabus as "a means of expressing responsibility and commitment."

In the book [7] the authors offer their vision on the development of a course curriculum focused on students' learning activities. The book is addressed to teachers of higher education, methodologists, employees of the system of professional development and specialists in the field of university and higher education, university students.

From the point of view of our study, the use of game technology in education is interesting, which has attracted the attention of a significant number of researchers.

Gamification is an educational technology that is rapidly developing, with a huge potential for a positive impact on the effectiveness of the educational process [8, p. 135]. The phenomenon of the concept of gamification in higher education allows it to be considered as: the concept of applying game mechanics and game design methods in a non-game context to engage and motivate people [9, p. 25]; the possibility to consider the formal and informal space of the learner [10]; building learning on the basis of game methods and thus strategically improve learning and education [11].

A separate group of studies are the results of scientists who consider the phenomenon of gamification as a possible pedagogical technology, namely: learning technology [12]; innovative educational technology [13]; technology of professionally-oriented learning [14].

1.3 Highlighting Aspects that Are Understudied

Thus, the traditional scheme of presenting academic disciplines, as evidenced by practice, has outlived its usefulness, and therefore it is necessary to search for new approaches on the example of gamification as a new way to solve the scientific and practical problem of ensuring increased awareness and interest, students to the academic discipline.

This study is a logical summary of the scientific search for the improvement of methods of teaching applicants for higher education through the application of game technologies, which have become one of the trends of modern higher education [15].

1.4 Purpose of the Article

The purpose of the article is to consider an approach to solve the scientific and applied problem of increasing the interest of students (higher education applicants) in understanding the structure, logic and necessity in the study of academic disciplines by increasing their awareness of the reporting stages in a fun game form.

1.5 Research Objectives (Goals)

To achieve this purpose, the following objectives are set: 1. Analyze recent research and publications on the topic of the problem. 2. Explain my own research results.

2 Research Methods

2.1 Research Tools

The main means of scientific and theoretical research. A set of scientific methods that are comprehensively substantiated and consolidated into a single system. To solve the tasks we used methods of theoretical research, namely: analysis and synthesis of the scientific literature on the research problem; method of analytical and comparative analysis for assessing the novelty of the study; synthesis and generalization - to justify the methodological and methodological foundations of research; generalization - formulation of conclusions and recommendations for further research.

2.2 Reliability and Accuracy of Results Reliability and Accuracy of Results

Reliability of the results of the study is ensured by the correctness of the use of mathematical apparatus and research methods.

2.3 Methodological Basis of the Study

The object of scientific and theoretical research is not just a single phenomenon, a specific situation, but a whole class of similar phenomena and situations, their totality. The methodological basis of the research is the game learning activity (A. Novikov, P. Kapterev, K. Kapp, K. Tseas, N. Katsioulas and T. Kalandaridis).

3 Research Results

3.1 Theoretical Foundations for Problem Solving

The problem of students' low interest in courses offered by university departments is not a new one.

Young teachers try to solve this problem with the traditional approach. The first session with students begins with the erroneous praise of the department's academic discipline, its importance and dominance over all others. In doing so, as a rule, they miss the main point of their story: the logic of the structure, types of reporting of the academic discipline, the relationship with other disciplines. As a result of the tremendous work done by the teacher, it is still possible to state low interest and motivation of students to study especially secondary academic disciplines.

Long observation of the learning activities of higher education applicants (the period covers approximately 2009–2021) and subsequent interviewing of applicants in an informal setting allowed us to outline the essence of the problem in more detail. It has nothing to do with applicants' unwillingness to study. As trivial as it turned out, applicants not only do not understand the need to study the discipline offered by the lecturer, but also the logic, structure, and reporting. In addition, the construction of a typical study week leads to a violation of the sequence of statements in the module of types of training sessions. In our opinion, solving this problem will make it possible to develop a unified approach not only to the development of the content, but also the logical structure of the academic discipline, stages, strategies for assessing the progress of higher education applicants.

Possibly introducing the "syllabus" into the modern practice of higher education, thus indirectly trying to solve the described problem. For this purpose, an original explanation of the difference between the "sillabus», the working thematic plan, and the educational-methodical complex in the work [16] is offered. As practice shows, the mere presence of the syllabus does not solve the problem under consideration to the end. In this regard, consider what the approach in the application of the syllabus is.

At the beginning of each semester for each discipline of the department on the website of the institution of higher education, on the page of the department, placed syllabus of educational components, which are freely available to applicants for higher education. According to those who have implemented it the presence of syllabuses will successfully optimize the joint activities of the teacher and the student, including the department of distance learning, while increasing the level of mastering of the course at the expense of accessibility.

In our experience, many faculty members find syllabuses time-consuming to annually update the teaching and learning complex. At the same time, in their opinion, not enough attention was paid to the qualitative change of the educational component of the discipline itself. Which is to be expected.

Therefore, this study aims to try to develop a solution to the outlined problem, by substantiating the development of a method for presenting the reporting stages of the educational discipline to the students. The method should be very simple and provide a presentation in an accessible and cognitive form of the structural and logical scheme of the reporting stages of the discipline studied.

The developed method is based on the method of didactic game. Game methods due to their phenomenal uniqueness have gained widespread adoption due to the high efficiency of teaching in comparison with traditional methods. Great merit in the development of the methodology of game learning activities in an accessible form of presentation belongs to A. Novikov [17].

According to P. Kapterev, games should be recognized as an essential aid to systematic learning; learning and games are not enemies - they are friends, to whom nature itself indicated to go the same way and mutually support each other. Therefore, these methods develop learning and cognitive activities.

The goal of learning in the didactic game is achieved through the solution of various game tasks [18]. Learning the rules of the game, students are disciplined, quickly learn difficult at first glance, the basic concepts and definitions of discipline. Practice proves that didactic games activate students' attention and interest in learning. This is necessary not only for cadets in the study of difficult to understand, military-technical disciplines, but also students. Game moments of the lesson relieve stress and fatigue of students.

As well seen from Table 1, the most appropriate for the formation of knowledge, along with practical exercises, teaching others (mutual learning) and independent work are didactic games [19].

Table 1. Comparative characteristics of different teaching methods

	Teaching methods	Solved tasks						
			form		develop			
			knowledge	skills	thinking	memory	language	gaining experience
Practical	Working with a book (reading)	10%	+	+	+	+	+	
	Educational discussions	50%	++	–	++	+	++	++
	Didactic games	**70%**	++	–	++	+	++	++
	Practical training	75%	+	++	++	+	–	++
	Teaching others is the application of what has been learned	90%	++	–	++	+	++	++

Note: ++ – solve very well; + – solve partially; – – solve poorly.

One of the important conditions, in our opinion, is the strict fulfillment of the requirement for the use of a didactic game, namely that the learning objective does not turn simply into a game [17].

For this purpose, it is necessary for the teacher to manage the process of the game. With the correct and timely comments of the teacher will be the process of managing the learning and cognitive activity of students, so it will be combined with a positive motivational and emotional background of the game.

3.2 Research Methodology

Methodology of the study. In order to find out whether the presentation of silhouettes in the game form (GFS) is really more interesting and understandable in contrast to the traditional form (TFS), an express survey of respondents from among students and cadets was conducted in the form of a questionnaire. The survey was conducted in a relaxed atmosphere. Respondents could choose several answers in the survey. A total of 98 respondents were interviewed. (see Table 2).

Table 2. Quantitative distribution of respondents

Respondents	Educational institution	Specialty	Number of respondents
Cadets	Kruty Heroes Military Institute of Telecommunications and Information Technology (MITIT)	172 Telecommunications and radio engineering, 122 Computer science, 125 Cybersecurity	21
Students	Lutsk National Technical University, (LNTU)	121 Software engineering, 122 Computer science	49
Students	NJSC «West Kazakhstan Agrarian and Technical University named after Zhangir khan», Uralsk, Kazakhstan (WKATU)	7M06100 Information Systems and Technologies	29
Total respondents:			98

Criteria, indicators and parameters used for evaluation. The list of questions (questionnaires) offered to respondents is presented in Table 3. Respondents had the opportunity to choose several answers. Respondents put a " + " sign against the corresponding field. The count was made by simple summation.

Table 3. List of questions (questionnaire)

No	Questions to the respondents	Answer			
		TFS		GFS	
		yes	no	yes	no
1	Does the syllabus contain clear guidelines for assessment (games) in the course of studying the discipline?	+	-	+	-
2	Would you choose an elementary discipline in the form of a silhouette?	+	-	+	-

The distribution of respondents' answers is presented in Table 4.

Table 4. Results of the survey. Distribution of respondents' answers

Questions	Distribution of respondents' answers											
	MITIT				LNTU				WKATU			
	TFS		GFS		TFS		GFS		TFS		GFS	
	yes	no	yes	no	yes	no	yes	no	yes	no	yes	no
1	14	7	18	3	15	34	42	7	7	22	22	7
2	5	16	17	4	5	44	45	4	1	28	17	12

For example, the results of an ex-post survey of students' opinions show an increase in the interest of students (students and surgeons) in choosing an academic discipline (variable) and clearer learning priorities (see Table 4).

3.3 Game Approach to the Presentation of the Logical Structure and Reporting Stages of the Academic Discipline

Let us consider in more detail the approach to composing the logical structure and reporting stages of an academic discipline and presenting them to applicants in a virtual-cognitive game method.

Let us consider the stages of planning and a summary of their main procedures.

The work begins with the first (preparatory) stage, which is carried out before the beginning of the academic semester with the study of a new academic discipline.

The lecturer, who is assigned to the academic discipline of the department, is obliged to make a list of stages of reporting on the academic discipline: numbers of topics of modular controls; control works; laboratory works; practical works (which provide for 100% reporting); other calculation and graphic works; term papers.

The list is made in chronological order, as they are in the academic discipline. All works should be evaluated in accordance with the rating system.

The second stage selects the evaluation criteria for each reportable stage of the discipline. Each reporting topic is assigned an evaluation criterion RD in conditional points. It must correspond to the rating system of evaluation.

At the third stage on the basis of the initial data of the first and second stages an algorithm of study of the academic discipline with the marks of the reporting stages of the academic discipline is made, as shown in Fig. 1. Each content module of the study discipline in the algorithm ends with the block of summarizing and decision making by the lecturer. The assessment block of the study module represents the assessment criteria for making a decision on the admission of an applicant for higher education to the study of the subsequent block.

After gaining the basic minimum of knowledge ($\min \leq RD \leq \max$) the learner is allowed to study the topic further. If a student gets a grade lower than the boundary acceptable ($0 \leq RD \leq \min$) for the assessed module, in this case the lecturer will have every reason to make a decision on the admission to further study of the new module. In exceptional cases, he may be allowed to continue studying the topics of the discipline, on condition of compulsory retake the module before the next session. Similarly to the game learning technology, for example, suggestions in [20].

Thus, the algorithm should correspond to the structural and logical content of the academic discipline (Fig. 1).

Accountable stages of each content block according to the sillabus are combined by lines of direct and feedback.

The fourth stage (implementation). At the first class, the lecturer brings to the students with comments the structural and logical scheme of the academic discipline, its place and role in the further educational process.

Game representation of academic discipline, like any game, has certain rules of the game, obliging all participants of the game (in this case students) to adhere to a certain sequence, analysis and choice of algorithms of response depending on learning conditions, tactical situation deployed on the virtual educational battlefield.

The significance of the game cannot be overestimated. The phenomenon of the game lies in the fact that being entertainment, recreation, it is able to grow into learning, into creativity, into a model of the type of human relations and manifestations in work.

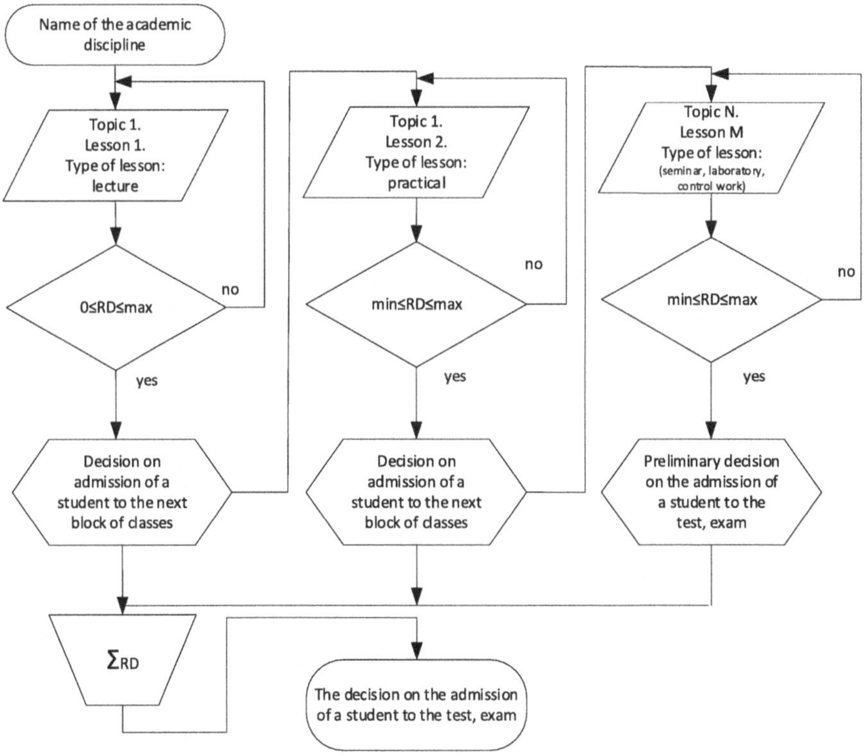

Fig. 1. Example of the structural and logical content of a conditional educational discipline

In the present study it is proposed to consider the game from the perspective of the teaching method. Only then will the method of the pedagogical game assist the applicants in acquiring new knowledge. Unlike games in general, the controlled pedagogical game has an essential feature - a clearly formed purpose of teaching students and the corresponding pedagogical results, which can be substantiated, allocated in an explicit form and characterized by educational and cognitive orientation.

Taking into account the above stated, game-based learning can be represented as a future competition of higher education applicant with lecturer to accumulate assets (knowledge) and be in readiness not only to successfully pass the stages of semester control, but also all sorts of current questions on the topic asked by the lecturer. The functional model is shown in Fig. 2.

3.4 Discussion of Research Results

In the proceedings of scientific and technical conference [20] the author outlined the experience of teaching electrical engineering disciplines by the method of virtual computer game. Unfortunately, the proposed methods were not massively used in military higher education due to the conservative nature of the education system. However, the ideas suggested in [20] were ahead of their time, so they were used unofficially in private practice; further development of the method took place in the first version of this problem [21]. Partially proposed ideas became the prototype of modern syllabuses and a tool that regulates students' learning activities [22], and as a means of organizing students' independent work [23]. The experience of modeling a syllabus, presented in the publication [24], is aimed at students' independent work.

We agree with the authors' opinion that in order to improve teaching and knowledge assimilation it is necessary to use auxiliary technologies [25]. For example, the use of gamification, but in such a way that the method of learning through play does not turn into a pure game [10, 11, 13]. As can be established from the publications [21–24] about the syllabus there is an increasing need not only for it as information about the academic discipline, but also the use of it as a method of teaching applicants for higher education. Thus, the syllabus turns into an auxiliary pedagogical technology of teaching. As noted by the authors [26], the use of the syllabus is aimed at improving the performance of students on their syllabus. The problem of students in the university is that some students cannot develop an effective study plan, and this can lead to failure of their studies.

4 Summary and Conclusion

To summarize the results of this study: to date, there is a problem of low interest of students in the study of academic disciplines offered by the departments due to the difficulty for young teachers to convey to students the structural and logical scheme of the discipline studied; didactic games have a high efficiency of learning compared to traditional ones, so the method of delivering the key elements of the discipline to the student in the form of a game is proposed, namely: structural and logical scheme; its place in the system of training of specialists of this specialty; strategy for evaluating knowledge, skills and competencies; types of accountability in the form of an algorithm. The proposed set of measures is aimed at increasing awareness and motivation of students.

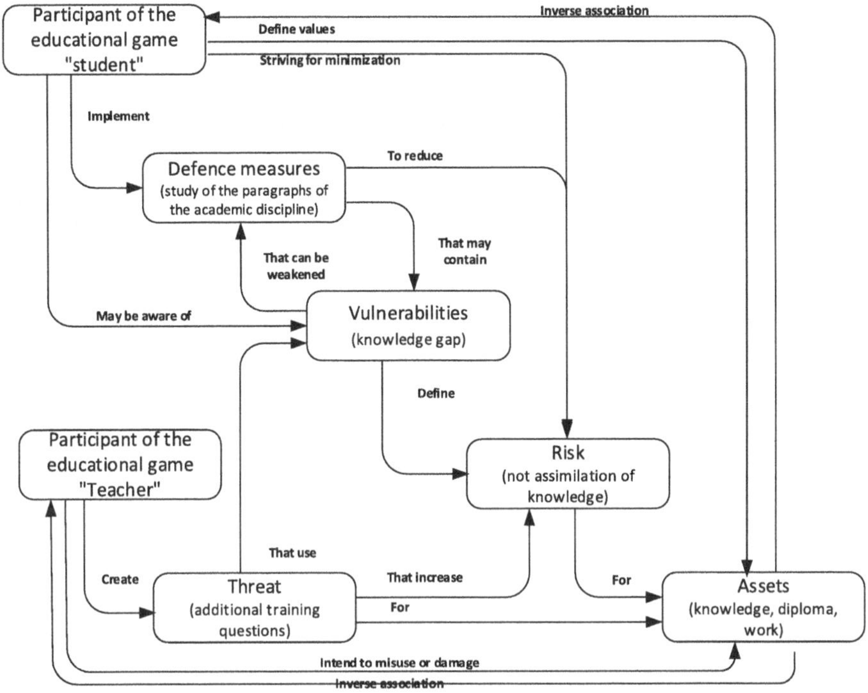

Fig. 2. Functional model of practical implementation of confrontation the teacher-student confrontation during the presentation-study of an educational discipline

Of course, the presented study does not exhaust all aspects of this problem. The authors' developed algorithm of reporting checkpoints on the academic discipline in the form of a didactic game more favorably contributes to all participants in the educational process in achieving the learning objectives. By controlling the process of didactic game, a lecturer naturally activates higher education students' attention and stimulates learning and cognitive activity in studying complex technical disciplines, linking it with a positive motivational and emotional background of the game.

4.1 Expanding the Boundaries of the Scientific Field

The scientific result expands the scientific boundaries of pedagogical sciences in the field of teaching methods of higher education through the use of syllabuses as a game pedagogical technology.

4.2 Scientific Novelty. Scientific Justification

For the first time the hybrid application of syllabus with game pedagogical technology as a method of teaching and motivation of higher education applicants for further study of educational discipline was proposed, which became a distinctive feature from previously known approaches to the development and use of syllabus as an information tool.

4.3 Practical Use

The proposed solution is fully ready for practical use by the lecturer.

4.4 Prospects for Further Research and Study

The theoretical results obtained in the course of this research are the basis for further research and development as a teaching methodology. It is anticipated that the research will bring more academic and practical value.

Acknowledgment. The authors would like to express their gratitude to the organizers of The 7th International Conference on Computer Science, Engineering and Education Applications (ICCSEEA2024), as well as to the general sponsors of the National Technical University of Ukraine "Igor Sikorsky Kyiv Polytechnic Institute", National Aviation University, Lviv Polytechnic National University, Kharkiv National University of Radio Electronics, Kharkiv, Wuhan University of Technology, Polish Operational and Systems Society, International Research Association of Modern Education and Computer Science, for free publication of scientific articles.

References

1. Syllabus. Meaning of syllabus in English. Definition of syllabus from the Cambridge Academic Content Dictionary. Cambridge University Press. https://dictionary.cambridge.org/dictionary/english/syllabus
2. Nurmanbetova, D.N., Nefyodova, L.V.: Methodological recommendations for the development of syllabuses. Third edition, revised and expanded. Astana, L.N. Gumilev ENU. 40 (2011). (in Russian)
3. Procedure for creating and approving work programs (syllabuses) of academic disciplines (educational components) at the Igor Sikorsky KPI. 8 (2021). (in Ukrainian)
4. Parks, J., Harris, M.B.: The purpose of a syllabus. Coll. Teach. **50**(2), 55–61 (2002)
5. Slattery, J.M., Carlson, J.F.: Preparing an effective syllabus: current best practices. Coll. Teach. **54**(4), 159–164 (2005)
6. Habanek, D.V.: An examination of the integrity of the syllabus. Coll. Teach. **53**(2), 62–64 (2005)
7. Kirilyuk, L.G., Krasnova, T.I., Karpievich, E.F.: The curriculum as a guide for students and teachers. Minsk: BSU. 7, 211 (2008)
8. Bugaeva, V.: Gamification as a way of forming active professional behavior of future IT industry specialists. Pedagogy Psychol. **56**, 129–135 (2018)
9. Tseas, K., Katsioulas, N., Kalandaridis, T.: Gamification in higher education. M.S. thesis, Dept. Electrical and Computer Engineering, University of Thessaly. Volos, Greece (2014)
10. Tkachenko, O.: Gamification of education: formal and informal space. Curr. Issues Humanit. **11**, 303–309 (2015). (in Ukrainian)
11. Kapp, K.: The gamification of learning and instruction game-based methods and strategies for training and education. Pfeiffer, San Francisco, USA (2012)
12. Noskov, E.A.: Learning technologies and gamification in educational activities. Yaroslavl Pedagogical Bullet. **6**, 138–142 (2018). (in Russian)
13. Petrenko, S.V.: Gamification as an innovative educational technology. Innov. Educ. **2**(7), 177–185 (2018). (in Ukrainian)

14. Rybka, N.M.: Graization and experience of using computer games in teaching philosophy in technical institutions of higher education. Inf. Technol. Training Tools **67**(5), 213–225 (2018). (in Ukrainian)
15. Volkova, N.P.: Gamification as one of the trends of modern higher education. In: Modern Higher Education: Problems and Prospects: VI All-Ukrainian Scientific and Practical Conference of Students, Postgraduates and Scientists: Abstracts of Reports. Dnipro, pp. 33–35 (2018). (in Ukrainian)
16. Vinnitsky, M.: What is the difference between silabus and RTP and NMK? Kyiv: Kyiv-Mohyla Academy, Profi Summer School+. 18 (2017). (in Ukrainian)
17. Novikov, A.M.: Methodology of gaming activity. Moscow: Egves Publishing House, vol. 48 (2006). (in Russian)
18. Krol, V.M.: Psychology and pedagogy: A textbook for technical universities. Moscow: High School, 319 (2001). (in Russian)
19. Ouahbi, I., Darhmaoui, H., Kaddari, F.: Gamification approach in teaching web programming courses in PHP: Use of KAHOOT application. Int. J. Mod. Educ. Comput. Sci. (IJMECS). **13**(2), 33–39 (2021)
20. Kozubtsov, I.N.: Training of electrotechnical disciplines by the method of virtual computer game. In: Electrotechnologies, Electric Drive and Electrical Equipment of Enterprises: Collection of Scientific Papers of the II All-Russian Scientific and Technical Conference. Ufa: Publishing House of USNTU, vol. 2, pp. 107–110 (2009). (in Russian)
21. Kozubtsov, I.M.: Metod virtualno-cognitive presentation to cadets of the reporting stages of the academic discipline. In: V-A Scientific and Technical Conference "Priority Directions of Development of Telecommunication Systems and Special Purpose Networks. Kiev: Viti NTUU "KPI", pp. 144–147 (2010). (in Ukrainian)
22. Kashtanova, S.N., Kudryavtsev, V.A.: Syllabus as a tool for regulating the educational activity of students. Bullet. Mininsky Univ. **2**(15), 5 (2016). (in Russian)
23. Khaustov, M.M., Bachinsky, R.A., Gordienko, N.A., Poruchikov, V.V., Kolotilov, O.V.: Using Syllabus as a means of organizing student's independent work. In: Innovations in Higher Medical and Pharmaceutical Education of Ukraine (with remote connection of VM(F)NZ of Ukraine via videoconference): Materials of the XVI all-Ukrainian Scientific and Practical Conference with International Participation. Ternopil National Medical University named after I. Y. Gorbachevsky, **142** (2019). (in Ukrainian)
24. Shkunova, A.A., Pleshanov, K.A.: The experience of modeling the syllabus of independent work of students. Bullet. Mininsky Univ. **3**(16), 8 (2016). (in Russian)
25. Adebayo, E.O., Ayorinde, I.T.: Efficacy of assistive technology for improved teaching and learning in computer science. Int. J. Educ. Manage. Eng. (IJEME). **12**(5), 9–17 (2022)
26. Chanamarn, N., Tamee, K.: Enhancing efficient study plan for student with machine learning techniques. Int. J. Mod. Educ. Comput. Sci. (IJMECS). **9**(3), 1–9 (2017)

SMART-Lecture Methodology as Modernization of Blended Training at HigherEducation Establishment in Conditions of Forced Unpredictability

Stanislav Zabara[1], Lesia Kozubtsova[2], Oleksii Beskrovnyi[2], Igor Kozubtsov[3](✉), and Olena Mamonova[4]

[1] Open International University of Human Development "Ukraine", Kyiv 03115, Ukraine
[2] Kruty Heroes Military Institute of Telecommunications and Information Technology, Kyiv 01011, Ukraine
lesia.kozubtsova@viti.edu.ua, olbezkrovnyy@i.ua
[3] Lutsk National Technical University, Lutsk 43018, Ukraine
kozubtsov@gmail.com
[4] Donetsk State University os Internal Affairs, Kropynytskyi 25006, Ukraine
mamonovaddli@ukr.net

Abstract. The subject of the research is the theoretical and practical basis for the use of blended learning in higher education institutions with specific (phenomenal) learning conditions. The object of research is the modernization of lectures to the Smart-lecture format. The relevance of the research topic is confirmed by the urgent need to improve the educational process in higher education institutions with specific conditions for students in wartime to maximize the effective use of short periods of time between air raids. The result of the psychological and pedagogical research is aimed at increasing the level of assimilation of lecture information through the lecturer's successful influence on the areas of visual thinking development. The purpose of the article is to substantiate the content of the concept of Smart-lecture, the methodology of its implementation in a blended learning environment to ensure the safety of all participants in the educational process. The SMART-lecture in the updated format is presented to higher education students as an interactive form of visual education. The solution to the scientific problem is aimed at increasing the level of assimilation of educational information by students through its visualization and creating a safety zone in educational institutions for all participants in the educational process in conditions of war (or force majeure). The author emphasizes the need to use visual means of interactive computer technologies as an integral part of modern education in higher education institutions. It is concluded that the use of such interactive forms of organization of visual study of educational components as SMART-lecture promotes the development of visual learning, activates students in the learning process, which increases the effectiveness of learning. Currently, a scientific and pedagogical experiment is underway to test the effectiveness of the proposed solutions both in an educational institution with specific learning conditions and in ordinary ones. The preliminary scientific result obtained in this work already expands the scientific boundaries of pedagogical science in the field of theory and methods of teaching educational

Z. Hu et al. (Eds.): ICCSEEA 2024, LNDECT 242, pp. 656–669, 2025.
https://doi.org/10.1007/978-3-031-84228-3_57

disciplines of educational components to higher education students in the context of a tendency to reduce motivation, and also clearly contributes to the creation of a safety zone for all participants in the educational process in wartime in educational institutions with specific learning conditions.

Keywords: Smart-lecture · improvement · form · training · teaching · institution of higher education · a student of higher education · lecturer · teacher

1 Introduction

1.1 Problem Statement

The rapid development of computer technologies certainly affects the educational process. Traditional teaching of lectures is obviously not interesting for those seeking higher education, it is necessary to search for new approaches to learning. In lecturing practice, visual methods of training are becoming more and more common, which makes it easier for the teacher to present educational information to students.

It is known that students of higher education in the humanitarian and technical areas perceive the concept of educational material in higher mathematics by means of the traditional model of education in different ways. As a result of different types of perception, students of higher education lose interest in perceiving the educational discipline of the educational component, the effectiveness of the learning result decreases. Based on these circumstances, there is a need to search for approaches to improve the teaching of lectures by visualizing mathematical information, which plays an important role in the perception of information by students of higher education [1]. Since modern students of higher education are increasingly willing to master the digital educational environment [2], it is possible that in the near future lecturers should be ready to conduct lectures using interactive computer technologies, be able to create their own educational and digital environment: either classroom lessons using interactive whiteboards, or distance lessons – using the Internet and platforms for distance lessons.

1.2 Literature Review. Analysis of Recent Research and Publications

Solving the problem of increasing learning efficiency through the visualization of learning information has been the subject of research by many scientists. One of the first publications [3] presented experimental results that confirm the effectiveness of visualizing educational information.

Since the term "visualization" has many meanings, teachers most often mean the process of transforming information into a visually perceptible form: a diagram, graph, figure, scheme, table, etc. However, such an understanding of visualization implies minimal mental and cognitive activity of education seekers, and visual didactic tools perform only an illustrative function [4, p. 113]. Therefore, under the visualization of educational information, we mean the arrangement and structuring of a fairly large volume of information in a visual form that is easily perceived and forms a visual image in the student's memory.

Analysis of further scientific research shows the effectiveness of using interactive SMART boards in the educational process not only for visualization, but to a greater extent as a means of increasing cognitive activity and learning efficiency in computer science classes [5]. Similar confirming results were obtained in the study [6]. Teachers note that the use of interactive SMART boards in lecture practice has significant advantages in terms of motivation, attention and behavior of students of higher education, rather than using other tools. Along with this, there are researchers' opinions [7] on the lack of results that will clearly demonstrate that interactive whiteboards improve academic performance. Interactive whiteboards by themselves, do not improve anything. Only the methodically correct use of visualization tools allows you to influence effectively the educational process of students of higher education.

In their work [8], the authors note that using traditional data visualization methods, it is difficult to visualize abstract facts to obtain valuable information about complex multidimensional data sets. One of the main problems is the high cognitive load during the interpretation of information. Three-dimensional visualization of information in the context of our study is important for justifying the need for such a display of educational information to improve its perception by students of higher education.

1.3 Highlighting Aspects that Are Understudied

Currently, there is a sufficient number of educational and methodological manuals for higher education using visualization tools aimed at developing visual thinking in order to increase the effectiveness of teaching various academic disciplines to students of higher education. These and many other studies confirm that "the informatization of education and the development of electronic subject teaching method fundamentally change the classical theory and methodology of teaching and the methodology of researching scientific problems of subject teaching" [2]. Thus, our study is a logical generalization of scientific research on improving the methods of training cadets.

1.4 Purpose of the Article

The purpose of the article is to consider an approach to solving the scientific and applied problem of increasing the interest of higher education students in studying academic disciplines in conditions of forced unpredictability due to the use of the SMART-lecture methodology.

1.5 Research Objectives (Goals)

To achieve this purpose, the following objectives are set: to analyze the latest research and publications on the topic of the identified problem;;to state your own research results in solving the specified problem.

2 Research Methods

2.1 Research Tools

To solve the tasks defined by the authors, the following methods of theoretical research were used: analysis and generalization of scientific literature by research direction; the method of going from the abstract to the concrete; method of analytical and comparative analysis when evaluating the novelty of research results; generalization when formulating conclusions and recommendations regarding the continuation of further research.

2.2 Reliability and Accuracy of Results

Reliability of the results of the study is ensured by the correctness of using mathematical apparatus and research methods.

2.3 Methodological Basis of the Study

The object of scientific and theoretical research is not just a separate phenomenon, a specific situation, but a whole class of similar phenomena and situations, their totality. The methodological basis of the research is the ideas of L. Vygotsky, P. Halperin, S. Rubinstein (pedagogical psychology of human perception of information).

2.4 Limitations in the Application of the Results

It is important to eliminate any potential biases by focusing on the Ukrainian context, which will provide a more balanced view of the problem. Therefore, the use of the fi ndings in other countries may be considered as one of the possible options at the preliminary analysis stage.

3 Research Results

3.1 Concept of SMART Lecture as a Modern Form of Visual Education

Along with well-known visualization tools, there are also SMART-lectures as a modern interactive form of visual training in higher education institutions. The essence of such a type of initial class as a "SMART-lecture" is based on research in the field of cognitive-visual methods [7] and the selection of some necessary components of visual learning: use of visualization tools; inclusion of specially developed visualized tasks; addition of historical facts, associations in visualized forms; introduction of interactive computer technologies; construction of a visual learning environment.

According to the author [7], all the necessary components of visual learning are successfully combined in such a form of training organization as a SMART-lecture.

The concept of SMART has been used in the field of education since the release of the first interactive SMART Board, when the concept of "SMART-lesson" appeared. SMART translated from English means "clever" and is used in management [9], it is presented in the Table 1.

Table 1. Similarity of constituent concepts in management and lectures

Constituents	Content of components (similarity)	
	Management	SMART lecture
Specific	contains a clearly formed idea	contains a clearly formed proof or algorithm for solving the problem in the form of a closed block diagram, which is opened during the course of the lecture;
Measurable	methods of measurement	the lecturer predicts methods of measurement, memorization of the content of the lecture by students;
Agreed	coordinated with the employee's personal tasks, the company's mission and the client's needs	interactive presentation of information, presented using such information management methods as hyperlinks and the "Drag and Drop" utility;
Realistic	adequate to the current situation, not overstated/understated, opportunities correspond to resources	the lecture is created within the capabilities of the interactive whiteboard software;
Timed	a clear deadline for achieving the goal is set	available to applicants at any time, a video recording of the lecture by means of the interactive whiteboard itself

According to the author [10], the SMART-lecture is an interactive form of organizing visual learning with the following components:

interactive whiteboard is the main technical means of education;

the table is the main means of visualizing the proof of the theorem;

block diagram is a visual representation of the strategy for solving educational problems;

"Drag and Drop" is the main method of operation when solving visualized tasks and deriving formulas;

hyperlinks are the main element of navigation according to the content of the study;

saving video recordings from the board for possible visual repetition of the information presented at any time;

automated control and self-control of knowledge of lecture material – factual or terminological;

formulation of the problem.

Thus, the abbreviation SMART-lecture is formed.

Based on the analysis of Fig. 1, a SMART-lecture can be called a "clever lecture" that promotes effective learning.

The term "SMART-lecture" can be considered in two aspects:

1) as an interactive form of organizing visual learning;

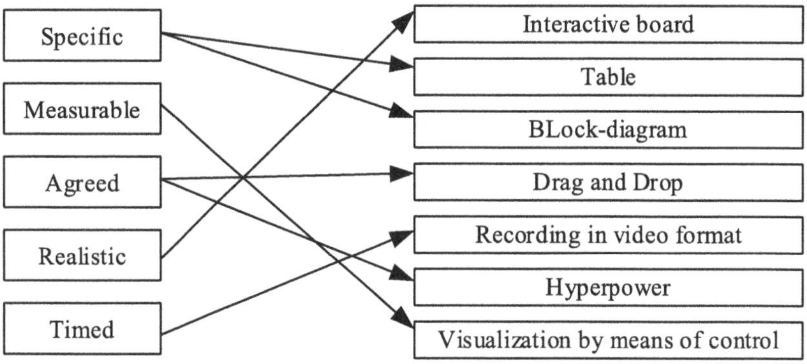

Fig. 1. Relationship between SMART criteria and SMART lecture components

2) as an electronic educational resource (Fig. 2).

A SMART-lecture as an interactive form of organizing visual learning using an interactive whiteboard, and a SMART-lecture as an electronic educational resource is created using interactive whiteboard software using graphic editors, Microsoft Office programs, and browsers. By interactive whiteboard we mean a technical device of interactive computer technologies, represented by a large interactive screen in the form of a white magnetic marker board. To present a SMART-lecture, as an interactive form of organizing visual learning, you need to know the device of the interactive whiteboard and its modes of operation, and to create a SMART-lecture you need to know the software of the interactive whiteboard, be able to use or develop electronic didactic materials in the appropriate software. Since different interactive whiteboards with different sensor technologies have a similar interface, it is enough to master the interface of one software package to create digital didactic materials.

The SMART-lecture technology can be presented in the form of a diagram (Fig. 2).

Selection of Educational Information. The topic of the lecture is set, the lesson objectives are set, and the lesson plan is drawn up. The content of the SMART-lecture must include evidence, historical facts, examples of solved tasks, and questions for summarizing the lecture.

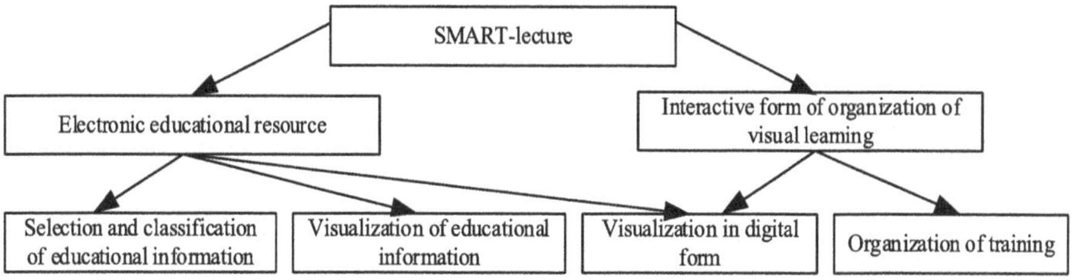

Fig. 2. Model of SMART-lecture construction technology

Visualization of Educational Information. We consider this component of building a SMART-lecture difficult because of the visualization process itself: organizing textual information in such a way that it is visual, easy to read and remember. It is necessary to visualize the prepared tasks on this topic, describe the visual strategy for solving mathematical tasks using block diagrams, think over the visual proof of a theorem or formula.

Visualization of Information in Digital Form. To build a SMART-lecture as an electronic educational resource (EER), you should use SMART Board interactive whiteboard software (SMART Notebook), as well as graphic editors, Microsoft Office programs, and browsers. SMART Notebook software satisfies the principle of an open system for creating EOR, as it is enough to have elementary skills in Microsoft Office programs to use it.

3.2 Methodology of Reading a SMART Lecture

A SMART-lecture is conducted by a leading lecturer in a classroom with connected projection equipment and an interactive whiteboard. Before the start of the lecture, the lecturer or the support staff of the department performs the screen calibration, which is necessary to precisely adjust the touch to the interactive whiteboard. Then it turns on the recording of all actions on the board to create video files. The recorded video file is posted on the portal of the distance learning system.

Before the lecture, students receive the following reference materials: block diagram with empty blocks; a table for proving the theorem with some missing steps; templates for answering questions when summarizing the lecture.

Auxiliary handouts are issued to applicants in advance. Of course, the lecturer makes sure that the applicants note down not only the basic definitions, formulas, and formulation of theorems. The applicants then use the completed handouts in practical, seminar classes.

The results of the lecture by the lecturer are carried out by the lecturer as follows: applicants receive template sheets; an interactive crossword puzzle is included on the board; applicants record their answers in template letters; the lecturer includes the correct answers; applicants test their knowledge.

3.3 Discussion of Research Results and Experience in Conducting SMART-Lectures in Conditions of Uncertainty

In the conditions of war, not only the quality of the lecture plays a key role, but also the lecturer has an effort to focus on creating safe conditions for staying in educational institutions. It should be noted that as a result of the unpredictability of air alarms, there is also a forced reduction of classroom time for lectures. Undoubtedly, this creates conditions for violation of the traditional logic of the educational process. And although the methodological department of the Ministry of Education and Science of Ukraine has developed four models of education organization: in-person, distance, mixed and individual forms (externship or family form) [11, p. 43], they do not completely solve the specified problem.

Scientists from the number of teachers who work in the specified conditions of martial law in institutions of higher education noted the undoubted dominant advantage in the application of a mixed form of education in the organization of the educational process [12]. Studies on the implementation of blended learning, which combines traditional technologies with distance learning, show good results.

In order to ensure balance under such conditions, it is rational to apply in institutions of higher education mixed training of students using the pedagogical technology of the "anticipatory learning method" [13], which has been modernized as the technology of the "inverted" class, the essence of which is the permutation of key components of the educational process [14]. The "inverted" educational process is implemented in the form of successive stages: before the classroom - classroom - after the classroom work, while before the classroom and after the classroom work is implemented in a digital educational and scientific environment. Classroom work is independent work in a digital educational and scientific environment, which continues in the classroom. The transition to a digital educational and scientific environment occurs again at the stage after the classroom work to consolidate the material (Fig. 3).

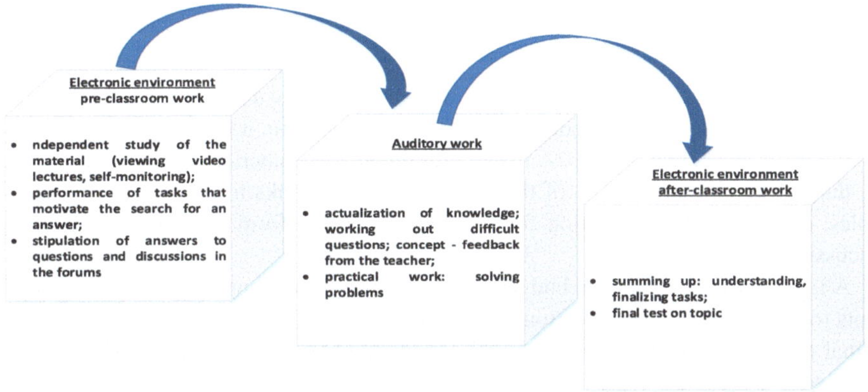

Fig. 3. Inverted educational process in reading a SMART-lecture

The application of the blended learning model in the SMART-lecture of the educational components provides sufficient flexibility and versatility, which are so important in the conditions of war. A mixed form of education is a model built on the basis of integration and mutual complementation of traditional and electronic learning technologies, which involves reducing classroom classes due to the transfer of certain types of educational activities to a digital educational and scientific environment.

Transferring part of the training course to the format of independent distance learning allows you to minimize the hours of work of the lecturer, as well as to create an environment for the implementation of the individual learning trajectory of each student by providing the opportunity to complete part of the course online in a mode convenient for him. For each shortened lesson, the educational activity is redistributed between pre-classroom, classroom and post-classroom work. In the new scheme, the central place is occupied by classroom lessons, around which classroom work is built before and after them (Fig. 3).

The expected effect of the introduction of a SMART-lecture in a mixed format will be calculated based on the results of E. Dale's research on the relationship between the quality of the material learned during training and the forms of its teaching (see Table 2) [15].

Table 2. Cone of experience according to experimental data of E. Dale

Activity	Memorization of information	Acquisition of ability
Reading	up to 10%	Define, list, describe, clarify
Listening to a lecture	up to 20%	
Watching drawings	up to 30%	Perform, Execute, apply
Video watching	up to 50%	
Participation in the discussion	up to 70%	Analyze, create,evaluate, develop
Imitation of real activity	up to 90%	

Based on the data from the Dale cone, it is possible to assume and conclude that due to the use of a mixed form of education and SMART-lecture, it is possible to ensure:

in out-of-class time up to 20% mastery of the lecture material;

during the classroom time of the SMART-lecture thanks to the clarity (drawings, tables, etc.) and its organization for the students in the form of participation in the discussion up to 70%.

A large reserve of the time budget allocated for independent work will allow applicants to study deeply the educational material, prepare for the SMART-lecture and, as a final result, acquire quasi-professional experience [13; 14; 16]. Previously acquired knowledge outside the classroom plays a significant role in the assimilation of new knowledge and skills. Previously acquired knowledge consists of short-term memory and long-term memory. Although it this is a generally accepted learning phenomenon, It is difficult to analyze empirically the influence of prior knowledge on learning [17]. It is confirmed by the results of the conducted research. Academic success of students depends on the time spent on extracurricular activities [18]. And once again about increasing the level of assimilation of knowledge.

3.4 Research Methodology

To find out whether there really is a difference between a traditional lecture (TL) and a SMART-L, a quick survey of respondents from among students and cadets was conducted in the form of a questionnaire. The survey was conducted in a relaxed atmosphere. Respondents could choose several answers in the survey. A total of 110 respondents were interviewed (see Table 3).

Table 3. Quantitative distribution of respondents

Respondents	Educational institution	Specialty	Number of respondents
Cadets	Kruty Heroes Military Institute of Telecommunications and Information Technology (MITIT)	172 Telecommunications and radio engineering, 122 Computer science, 125 Cybersecurity	21
Students	Lutsk National Technical University, (LNTU)	121 Software engineering, 122 Computer science	49
Students	Open International University of human development "Ukraine" (OIUHD)	121 Software engineering, 123 Computer engineering, 125 Cybersecurity	40
Total respondents:			110

Criteria, indicators and parameters used for evaluation. The list of questions (questionnaires) offered to respondents is presented in Table 4. Respondents had the opportunity to choose several answers. Respondents put a " +" sign against the corresponding field. The count was made by simple summation.

Table 4. List of questions (questionnaire)

No	Questions to the respondents	Answer			
		T-L		SMART-L	
		yes	no	yes	no
1	Does the lecture contain a well-formed proof or algorithm for solving a problem in the form of a closed flowchart that opens during the lecture?				
2	Is the information presented interactively?				
3	Is a video recording of the lecture available to students at any time using the interactive whiteboard itself?				

The distribution of respondents' answers is presented in Table 5.

Thus, the results of the survey of students' opinions show 3 out of 5 criteria (Table 1) of the distinctive features of SMART-L compared to T-L.

Table 5. Results of the survey. Distribution of respondents' answers

Questions	Distribution of respondents' answers											
	MITIT				LNTU				OIUHD			
	T-L		SMART-L		T-L		SMART-L		T-L		SMART-L	
	yes	no	yes	no	yes	no	yes	no	yes	no	yes	no
1	4	17	17	4	7	42	42	7	6	34	36	4
2	8	13	19	2	3	46	40	9	2	38	31	9
3	0	21	21	0	0	49	36	13	0	40	29	11

We certainly agree with the opinion of the authors that to improve teaching and learning, it is necessary to use assistive technologies [19]. For example, the use of gamification, but in such a way that the method of learning through the game does not turn into a pure game [20–23].

4 Summary and Conclusion

Using interactive forms of organizing visual training of students in higher education institutions, such as the SMART-lecture, is one of the innovative and effective directions in modern education. Unfortunately, despite the fact that the implementation of SMART-lectures in the educational process in higher education institutions is relevant and in demand, not all lecturers (due to the age limit) are ready to create such lectures due to the large amount of time required for their preparation. However, it should be noted that such expenditure of time is justified, since the SMART-lecture as an electronic educational resource can be used an unlimited number of times, periodically updating it with new tasks. The application of SMART-lectures in combination with the technology of anticipatory learning (flipped class) occurs during independent work, and during classroom work is allocated to tasks, exercises, conducting laboratory and practical research, individual consultations with the teacher. Blended learning allows you to focus on the shortest ways to achieve the planned results, as the usual boundary between independent and classroom work of students is erased. Therefore, the use of a mixed form of training and anticipatory training will allow not only to ensure the implementation of the curriculum for mastering the educational component, but also to maximally ensure the opportunities of the applicants to preserve life and health in conditions of war or other forced contingencies.

4.1 Expanding the Boundaries of the Scientific Field

The scientific result obtained in the work expands the scientific boundaries of pedagogical science in the field of theory and methods of teaching educational disciplines of educational components to students of higher education in conditions of a tendency to decrease motivation. Thus, the expansion takes place in the systemic unity of the philosophy of education and the theory and methodology of professional education.

4.2 Scientific Novelty. Scientific Justification

For the first time, a SMART-lecture was offered in combination with a blended form of education to intensify the training of higher education students and create a safe territory in educational institutions for all participants in the educational process in war (or force majeure) conditions.

4.3 Practical Use

As a result of the damage and destruction of the infrastructure, the premises of higher education institutions, and the loss of educational and research equipment, a flexible universal mechanism is needed to ensure the continuity of the educational process in conditions of war and long-term air strikes. The use of SMART-lectures in combination with a blended form of education allows in practice to overcome new challenges in creating safe conditions for the staying educational institutions of all participants in the educational process, and the most important goal of pedagogical activity in war conditions is the transformation of each educational institution into a security territory. Based on the obtained data, at the end of the experiment, it was found that:

the use of SMART-lectures in the teaching of higher mathematics to students of higher education contributes to a greater extent to meaningful study, rather than formal memorization;

such lectures are interesting for applicants, they are lively;

schematic and tabular presentation of mathematical information contributes to better understanding and memorization, applicants are ready to use such forms of presentation of educational information in other disciplines.

4.4 Prospects for Further Research and Study

The research is expected to bring more academic and practical value. For it, it is necessary to implement systematically the methodology of hybrid use of SMART-lectures in institutions of higher education.

Acknowledgment. The authors would like to express their gratitude to the organizers of The 7th International Conference on Computer Science, Engineering and Education Applications (ICCSEEA2024), as well as to the general sponsors of the National Technical University of Ukraine "Igor Sikorsky Kyiv Polytechnic Institute", National Aviation University, Lviv Polytechnic National University, Kharkiv National University of Radio Electronics, Kharkiv, Wuhan University of Technology, Polish Operational and Systems Society, International Research Association of Modern Education and Computer Science, for free publication of scientific articles.

References

1. Nezhyva, O.: The aspects of smart education in the world. Khazar J. Humanit. Soc. Sci. **24**(3), 62–72 (2021)
2. Pardała, A.: Informatization of mathematics education: didactic opportunities, experience and foreign trends. Inform. Educ. **6**, 49–55 (2019). (in Russian)

3. Shatalov, V.: Fulcrum. Moscow, Pedagogika, **159** (1987). (in Russian)
4. Trukhan, I., Trukhan, D.: Visualization of training information in the learning of mathematics, its importance and the role. Adv. Curr. Natl. Sci. **10**, 113–115 (2013). (in Russian)
5. Ganashok, A.: Interactive whiteboard as a means of increasing the cognitive activity and efficiency of learning in the classroom of informatics. Inf. Technol. Learn. Tools **51**(1), 21–35 (2016)
6. Smith H.J., Higgins S., Wall K., Miller J.: Interactive whiteboards: boon or bandwagon?. a critical review of the literature. J. Comput. Assist. Learn. **21**(2), 91–101 (2005). https://doi.org/10.1111/j.1365-2729.2005.00117.x
7. Dalinger, V.: Theoretical foundations of the cognitive-visual approach to teaching mathematics. Omsk, OSPU **146** (2006). (in Russian)
8. Wijayawardena, A.S.K., Abeysekera, R., Maduranga, M.W.P.: A systematic review of 3D metaphoric information visualization. Int. J. Mod. Educ. Comput. Sci. (IJMECS), **15**(1), 73–93 (2023)
9. Doran, G.T.: There's a S.M.A.R.T. way to write management's goals and objectives. Manage. Rev. **70**(11)(AMA FORUM), 35–36 (1981)
10. Lebedev, V.: Effective training of combinatorics and probability theory. School Technol. **2**, 126–134 (2012). (in Russian)
11. Education of Ukraine under martial law. Information and analytical collection. Kyiv: Institute of educational analytics, 358 (2022). (in Ukrainian)
12. Ducational process under martial law in Ukraine. Materials of the All-Ukrainian scientific and pedagogical advanced training, (Odessa, May 3 – June 13) [C]. Odessa: Helvetika publishing house, 504 (2022). (in Ukrainian)
13. Lysenkova, S.: The method of anticipatory learning: a book for teachers: from work experience. Moscow: Enlightenment, 192 (1988). (in Russian)
14. Bergmann, J., Sams, A.: Flip Your Classroom: Reach Every Student in Every Class Every Day. International Society for Technology in Education, Washington, DC (2012)
15. Dale, E.: Audio-Visual Methods in Teaching. 3rd ed. New York: Holt, Rinehart & Winston (1969)
16. Nechkina, M.: Increasing the effectiveness of a lesson. Communist **2**, 51 (1984)
17. Navandar, S.K., Kiwelekar, A.W., Laddha, M.D.: A performance analysis of the impact of prior-knowledge on computational thinking. Int. J. Mod. Educ. Comput. Sci. (IJMECS) **15**(2), 54–61 (2023). https://doi.org/10.5815/ijmecs.2023.02.05
18. Sharma, N., Appukutti, S., Garg, U., Mukherjee, J., Mishra, S.: Analysis of student's academic performance based on their time spent on extra-curricular activities using machine learning techniques. Int. J. Mod. Educ. Comput. Sci. (IJMECS). **15**(1), 46–57 (2023)
19. Adebayo, E.O., Ayorinde, I.T.: Efficacy of assistive technology for improved teaching and learning in computer science. Int. J. Educ. Manage. Eng. (IJEME). **12**(5), 9–17 (2022)
20. Kapp, K.: The gamification of Learning and Instruction Game-Based Methods and Strategies for Training and Education. Pfeiffer, San Francisco, USA (2012)
21. Tkachenko, O.: Gamification of education: formal and informal space. Topical Issues Humanit. **11**, 303–309 (2015). (in Ukrainian)

22. Petrenko, S.: Gamification as an innovative educational technology. Innov. Educ. **2**(7), 177–185 (2018). (in Ukrainian)
23. Lishchina, V., Kozubtsov, I., Kozubtsova, L.: Choice of pedagogical training technologies as a key problem teacher – subject of the educational process. In: International Scientific and Methodological Conference "Innovative Technologies in Military Education", (Odessa, June 25). Odessa: Military Academy, pp. 225–226 (2021). (in Ukrainian)

Data Analysis Technologies for Enhanced Educational Processes: A Case Study Using the Moodle LMS

Olena Hlazunova, Nataliia Klymenko, Maksym Mokriiev, Maryna Nehrey(✉), and Yevhenii Klymenko

National University of Life and Environmental Sciences of Ukraine, Kyiv 03041, Ukraine
{o-glazunova,nklimenko,m.mokriiev,ye.klymenko}@nubip.edu.ua, marina.nehrey@gmail.com

Abstract. Recent years have seen digital technologies become integrated into traditional educational systems. This integration has facilitated access to a vast amount of educational data, ranging from the aggregation of learning outcomes to the prediction of student behaviour within e-learning courses. Many universities, both in Ukraine and around the world, have adopted the open learning management system (LMS) Moodle for e-learning. In addition to serving as a platform for delivering educational content, Moodle encourages active student engagement through a variety of activities, including quizzes, assignments, and forum-based communication. Through these activities, students create a digital footprint which, when analysed, can be used for management purposes. Modern learning analytics methods enable the identification of behavioural patterns within the data generated by participants in the educational process. These analytics facilitate the construction of individualised learning pathways and the application of newly gained insights to improve the overall quality of the educational process. This paper highlights the range of data analytics technologies applicable to the design of various digital learning environment systems. Particular emphasis is placed on the creation of a system dedicated to the analysis of big data in the context of Moodle distance learning. The paper describes and explains a set of data analysis tools specifically tailored for Moodle LMS courses. Using Moodle's standard analytics capabilities alongside other relevant tools, an experimental study was conducted, exemplified by the analysis of a specific educational portal within a higher education institution. This empirical investigation aimed to demonstrate the practical application of data analytics technologies in the context of a specific e-learning environment, and to highlight the potential benefits and implications for educational institutions embarking on digital transformation.

Keywords: LMC Moodle · BigData · e-learning system · educational data analytics

Z. Hu et al. (Eds.): ICCSEEA 2024, LNDECT 242, pp. 670–682, 2025.
https://doi.org/10.1007/978-3-031-84228-3_58

1 Introduction

Digitalisation is a pervasive global phenomenon that permeates all facets of social existence, including the delivery of educational services. Within higher education institutions, the digital learning environment has become a comprehensive arena that shapes perceptions of professional futures and cultivates productive behavioural strategies [1–5]. This environment typically encompasses both formalised (pre-prepared) and informal (spontaneously generated in the digital space) sources of knowledge, ranging from electronic materials of academic disciplines and individual courses to interactive discussions in forums and personal messages from users.

The proliferation of online educational tools such as Learning Management Systems (LMS) and Learning Content Management Systems (LCMS) has greatly increased the volume of reliable data. A robust learning system requires not only a high quality end result, but also continuous improvement by proactively identifying and addressing potential problems. As with any complex system, learning processes can benefit from analytical methods to monitor and optimise their trajectory.

The emergence of big data in education can be attributed to two main trends in the digital age. Firstly, the digitisation of the recording and storage of institutional data in traditional settings has led to a vast repository of standardised information about students. Second, learning behaviours that were previously difficult to capture in face-to-face classrooms can now be partially elucidated through Learning Management Systems.

In light of these developments, we propose an innovative approach that uses Microsoft Power BI to transform Moodle LMS data into dynamic and interactive visual reports. These reports facilitate informed decision making for all stakeholders involved in the educational process. Managing such a digital environment, along with its technical support and data analysis, is a non-trivial challenge for administrators and information systems managers. Given the access to a wide range of educational data, from the aggregation of learning outcomes to the prediction of student behaviour in e-learning courses, the selection of appropriate tools for the application of different data analysis methods becomes a central concern [6]. The primary objective of this study is to explore methodologies for effectively working with Moodle data using different analytical tools..

2 Literature Review

The basic objective of any digital learning environment in higher education institutions is to facilitate rapid access to information essential for the acquisition of professional and other competences outlined in the educational programme. This digital learning environment is based on a software package designed for information exchange, using publicly available communication channels. This allows users to engage with learning materials regardless of their location, while administrators can analyse these records. Such software, known as learning management systems or e-learning systems (hereafter referred to as LMS), integrates various technological capabilities and tools for remote work, allowing educational institutions to implement various educational process schemes in the digital domain, including comprehensive online learning.

Various digital environments, including open educational platforms (e.g. Coursera, edX, Prometheus, TED, NPTEL, FutureLearn) and distance learning systems (e.g. Moodle, iSpring, Mirapolis, ShareKnowledge, Teachbase, WebTutor), meticulously store digital traces of users' activities. Despite the wealth of research, these capabilities are not widely and effectively used by educators to assess quality and design personalised educational processes. The Moodle system has gained worldwide popularity among educators as a dynamic course creation tool. As of January 2024, statistics show 162,045 active Moodle sites worldwide, hosting 46,572,992 courses in 242 countries, with over 401 million users [7]. Predominant user bases are found in Spain, the United States of America and Germany, while Ukraine boasts 1177 Moodle LMS-based sites.

Statistically, Moodle LMS is mainly used by higher education institutions to deliver the educational process in digital environments. Over time, extensive research has been conducted in several key areas, including the development of e-learning environments [8–10], the identification of behavioural strategies in online learning and their correlation with student characteristics [11–15], the visualisation of data derived from Moodle [16–18], the development and use of tools for analysing Moodle data [19–21], and the establishment of links between students' online activities and their academic success [22, 23].

With the integration of LMS into educational settings, a wealth of data on students' online behaviour has become accessible. Researchers often use this data to predict student performance [24], yielding diverse results that may be influenced by different courses and predictor variables extracted from the LMS. As a result, it becomes difficult to draw generalised conclusions about the mechanisms governing student performance.

In the course of using an LMS, each user leaves a unique 'digital footprint' that includes information about the user, various database records (activities, messages, grades, etc.), individual files and other data. The analysis and formalisation of this information can contribute to the development of a digital footprint model, which provides an optimal approach to the design of a digital learning environment [25–28]. This analysis facilitates an understanding of the popularity of course materials, the timeliness of student submissions, the level of engagement in an e-course, and the identification of students who may be at risk of not completing the course on time.

The analytical exploration of big data in the context of Moodle has attracted the attention of various researchers [21, 29] and developers who are striving to describe and create additional components for in-depth analysis.

3 Results and Discussion

E-learning systems, instantiated through the Moodle Learning Management System (LMS), have specific functionality for monitoring and analysing the processes that take place within them. This functionality is based on the digital footprints of users stored in the database and is applicable at both the individual course level and the overarching LMS level. However, the inherent functionality of Moodle LMS and the existing third-party plug-ins fall short of providing a comprehensive and high quality solution for monitoring, analysing and visualising the work done with the substantial datasets of big data within Moodle LMS.

The proliferation of web-based educational systems has grown exponentially in recent years, generating information of considerable educational value. However, the sheer volume of this data often exceeds manual analysis capabilities. There is a pressing need for tools that can autonomously analyse such large datasets and generate real-time reports. The escalating daily influx of different types of information requires innovative forms of information visualisation to accelerate the processes of understanding, processing and decision making. The primary goal of data visualisation is to transform large amounts of data into a visual format that facilitates the understanding of complex relationships within the data.

Traditional methods of generating analyses that resemble typical databases tend to be time-consuming. To mitigate this, big data processing tools are used to streamline and accelerate the analysis process. As a comprehensive process, data analytics involves the exploration, cleansing, transformation and modelling of large data sets with the aim of uncovering latent patterns, identifying unknown correlations and visually presenting actionable information. Consequently, there is a pertinent need to use educational data mining or Moodle Big Data intelligent analysis tools, as shown in the sequential diagram in Fig. 1.

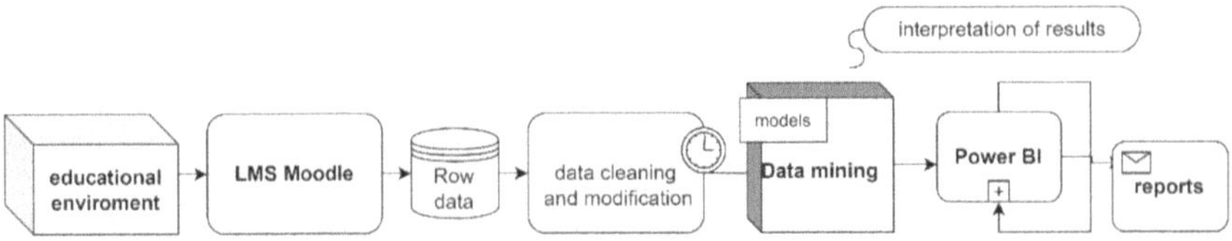

Fig. 1. Educational data mining system

Learning with big data presents inherent challenges, primarily due to the sheer volume of data, which requires significant time and expertise to manage effectively. While many educational institutions are applying machine learning algorithms to LMS data to analyse learner behaviour within specific courses, only a minority are venturing into making predictions for large subsets of courses. Despite these efforts, a comprehensive, large-scale demonstration of the effectiveness of learning analytics in supporting student success remains elusive [30].

The burgeoning growth of digital data from multiple sources has rendered traditional methods of storage, processing and analysis inadequate. Consequently, this limitation has spurred the development of novel approaches for handling, storing, and analysing large datasets [21, 31].

Learning analytics can be performed in two main ways: using the operational data available within the Moodle database, or using transformed and prepared data on an alternative server. The first approach offers various implementations, one of which involves direct access to operational data for real-time acquisition and computation of the required analytical insights. This method ensures that the most up-to-date and comprehensive information is retrieved. In particular, Moodle 4.0 introduced a specialised module known as the 'report builder' to facilitate the extraction of such information, allowing users to construct customised reports from the current dataset (Fig. 2). To avoid overloading the system when viewing reports, there are limits to the number of indicators and reports.

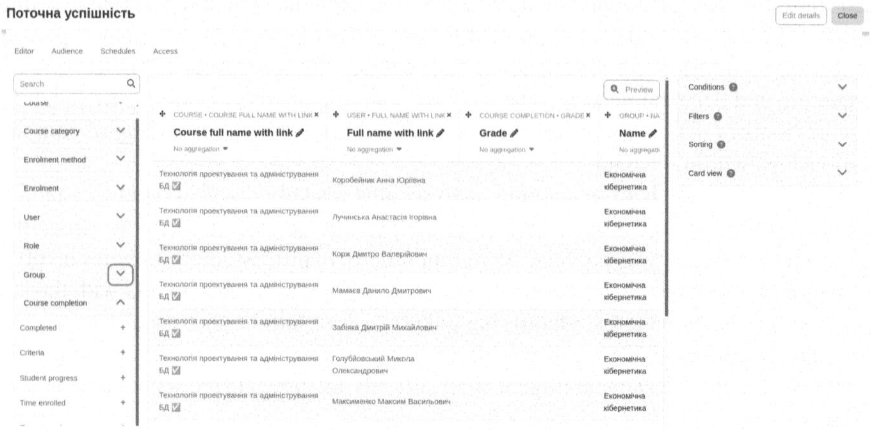

Fig. 2. The report builder in Moodle

Currently, a notable limitation of the built-in report builder is the limited number of indicators available to managers for constructing reports. Despite incremental additions with each Moodle release, the pool of indicators remains inherently limited.

However, there is a solution for instances where specific indicators are required for non-standard reports - the installation of the Configuration Report add-on module. This module allows users to formulate direct queries to the database using the SQL language, thereby enabling the extraction of any information that leaves a digital imprint in the database. The complexity of such reports is limited only by the skill and creativity of the manager creating them. Figure 3 illustrates an example report generated by this module.

In each of these scenarios, the results of the report can be downloaded in a conventional spreadsheet format (e.g. MS Excel). Subsequent analysis can be performed using spreadsheets or specialised statistical and analytical software capable of interfacing with various data formats, including MongoDB, Apache Cassandra, KNIME, R Studio, IBM SPSS, Tableau, Power BI, Xplenty, Adverity and Apache Hadoop. The workflow for such endeavours is illustrated in Fig. 4.

The potential ways of analysing the data collected within the Moodle LMS can be categorised into four distinct groups (Fig. 5):

– Standard system tools;
– Additional plug-ins to the system;
– External tools designed to analyse data directly within the Moodle environment;
– Various multi-functional data analysis programs.

The **standard tools** built into the Moodle LMS include a number of standard reports available in the “Reports” section, as well as several analytical models that are part of the core functionality of the system. For non-technical users, the information in the ‘Event Log’ report may lack clarity and suitability for drawing meaningful conclusions.

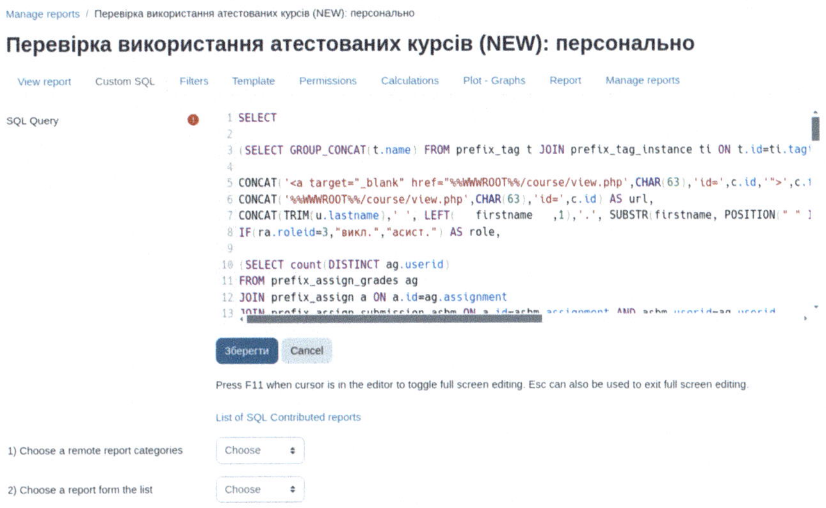

Fig. 3. Report builder in the Configuration Report module

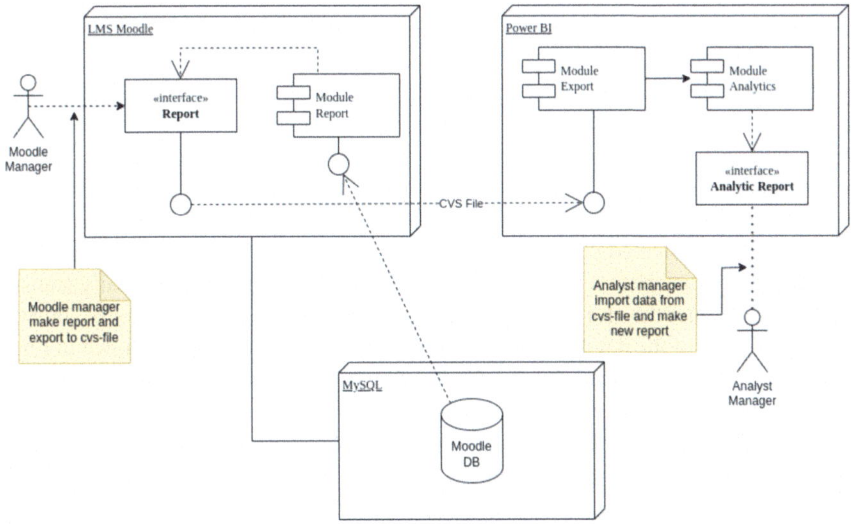

Fig. 4. General scheme for analysing training data

However, third-party data analysis tools often use this data for further analysis. Summary reports can be created based on the Event Log, allowing independent conclusions to be drawn about the most popular course elements among students, exploration of user activity, or engagement with specific course elements. These reports provide a broad overview of the intensity of student use of different course elements.

Additional plugins for the system are used to extend the functionality of Moodle beyond the standard version. The official Moodle Plugins Directory currently contains

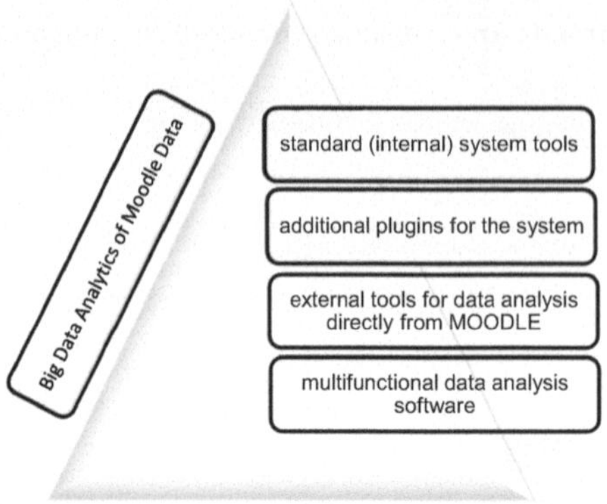

Fig. 5. Classes of LMS MOODLE course data analytics tools

over 2000 applications [32]. Examples of plugins designed for data analysis include SmartKlass, Heatmap and IntelliBoard. This category of tools has notable drawbacks, including the need for administrator installation and the need for regular plugin updates during system upgrades, which can be operationally challenging. In addition, certain machine learning algorithms require significant time and data for training before plug-ins can be used effectively.

While there are inherent disadvantages to working directly with live data for analysis, notably the instantaneous nature of the information and system overload, Moodle developers have addressed the former concern by introducing a basic statistics module. Working on the principles of a data warehouse, this module performs primary processing, adds timestamps and stores the results in designated tables within the Moodle database. Data is collected periodically for statistics, with separate tables divided based on basic time characteristics to ease data retrieval, such as stats_daily for daily statistics collection, stats_monthly for monthly statistics, and stats_weekly for weekly statistics. A partial example of the database for monthly statistics accumulation is shown in Fig. 6.

The creation of a dedicated data warehouse is essential for meaningful analysis of learning information. This involves configuring a system to transform primary data from the live Moodle database into the warehouse. A link to a business intelligence system is then established to allow the generation of a variety of reports, graphs and data panels.

External Tools for Analysing Moodle Data

This category includes standalone solutions designed to analyse Moodle data. Many of these external solutions rely on additional analysis of event log data. Characterised by an intuitive English language interface, these solutions do not incorporate complex, specialised data mining techniques, making them accessible to a broad academic audience. They provide complementary analysis and visualisation of event log data [33].

Various Multifunctional Data Analysis Programs

There are several sets of tools available for analysing data from Moodle:

1. Statistical and mathematical programs (e.g. Statistica, SPSS Statistics, Matlab, Octave, Excel).

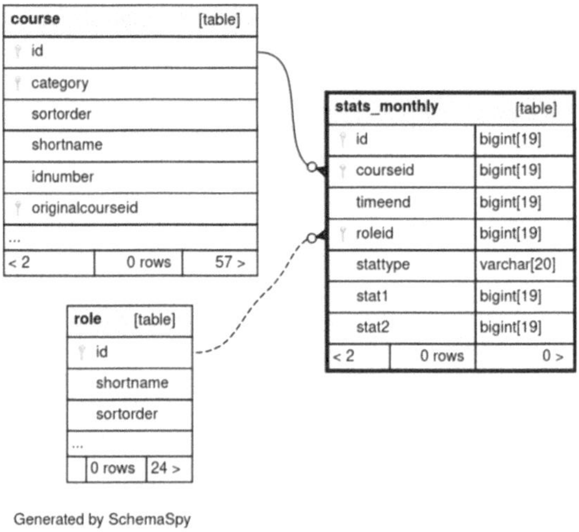

Fig. 6. Data model for comparing e-courses (monthly statistics table)

2. Data mining, business analysis and visualisation software (e.g. Orange, Knime, RapidMiner, Weka, Power BI, Deductor).
5. Programming platforms and languages (e.g. R, Python - the most widely used for data analysis, Java).
6. Database management systems (DBMS) and query languages (e.g. Oracle, MySQL, Microsoft SQL Server) - highlighted by Moodle's use of a relational database.
7. Website analysis tools (using log analysers such as Webalizer, AWStats) - applicable because Moodle functions as an educational website.

The use of these tools typically requires knowledge of statistical methods, data mining algorithms, programming languages, data pre-processing, multi-program workflows and programming skills.

In summarising the review of potential data analysis tools, a brief overview of the advantages and disadvantages of each group is provided. A comparative analysis of the Moodle big data analytics tools is presented in Table 1.

This research presents a system that uses Moodle LMS data for analysis and visualisation through Power BI, a chosen analytics platform for big data analysis. Power BI facilitates the transformation of raw data into clear and specific reports, presented in the form of visualised charts and graphs on desktop, cloud or mobile versions of the platform. Power BI Desktop is a suite of software services, applications, and connectors that work seamlessly together to transform disparate data sources into consistent, visually compelling, and interactive analytics. Users of Power BI Desktop can connect to multiple data sources and combine them into a comprehensive data model. This data model enables the creation of visual elements and collections that can be shared across the organisation in the form of reports. In business intelligence projects, Power BI Desktop is typically used to create reports, while Power BI Service facilitates the distribution of reports to users.

Table 1. Comparative analysis of Moodle big data analysis tools

Moodle Big Data Analytics tools	Benefits	Disadvantages
Standard (internal) system tools (accounts)	Accessible to all teachers and administrators, available by default in Moodle	Insufficient functionality and visibility for decision making
Additional plugins for the system	Implementation of additional features	Typically narrow focus, system administrator installation required
External tools for data analysis directly from Moodle	Focus on learning analytics and working with system data, easy to use	Limited free software, English interface
Multifunctional data analysis software	Broad functionality and capability for qualitative and multidimensional analyses	Require professional IT and mathematics skills

Building on the approach implemented for analysing educational data using the Power BI service, data extracted from the Moodle LMS database is visualised. This allows users to create dashboards of graphs and reports, each showing specific indicators over a defined period of time (Fig. 7).

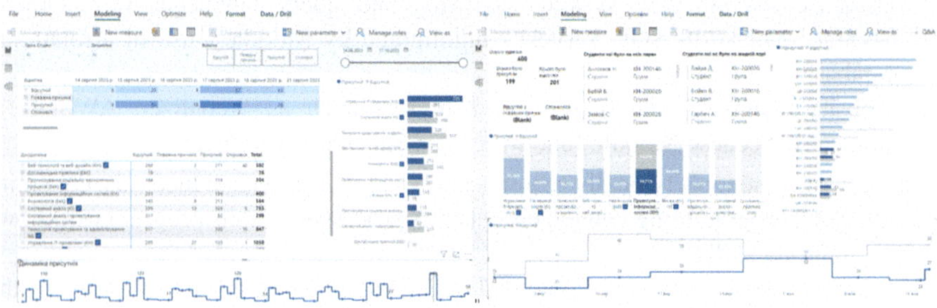

Fig. 7. Dashboard of the educational process indicators monitoring system

We are currently actively monitoring aggregate indicators within the educational process, which includes the dynamic assessment of student activity across e-learning courses and educational components within the curriculum. This involves the establishment of a comprehensive data model and the creation of dashboards with visualisations that detail course participation according to various criteria, ranging from specific disciplines and individual students to overarching aggregate indicators. Our focus also includes a careful analysis of the timeliness with which students submit assignments for review and the subsequent assessment of completed work by teachers. Other parameters under scrutiny include the assessment of the relevance and completeness of users' connections to their electronic course accounts.

A distinctive feature of this system is its ability to be flexibly adapted to the specific nuances of the educational process. This includes the seamless addition of new reports focused on indicators of interest within the educational process, the configuration of real-time dashboards, and the implementation of differentiated access levels for different user groups.

4 Summary and Conclusion

Learning management systems have transformative potential in higher education, providing a substantial reservoir of data that has the capacity to significantly enhance the learning experience for both students and educators. While big data holds great promise for education, it also presents challenges associated with the management of large datasets that require time and expertise. Nevertheless, there has been a remarkable upsurge in the use of big data to improve teaching and learning processes in recent years.

The paradigm of learning with big data has gained traction and offers significant potential within higher education institutions for monitoring, understanding and evaluating educational processes. The influx of digital data from learning environments such as LMSs marks a positive shift towards using this data to inform initiatives aimed at improving learning outcomes and realising high-impact teaching practices. Recognition of this potential leads to the identification of a need for further research, with particular emphasis on the development of systems that utilise data mining tools from e-learning platforms, particularly Moodle. This line of research seeks to address a variety of educational challenges faced by students, teachers, researchers and administrators.

The paper introduces the concept of using Microsoft Power BI as an innovative approach to extracting insights from Moodle LMS data. The envisioned system seeks to empower stakeholders in the educational process by providing dynamic and interactive visual reports. This strategic approach aligns with the need for continuous improvement, proactively identifying issues and improving the trajectory of learning processes.

Educational Data Mining operates at the intersection of computer science, data analysis, education and statistics. It focuses on the development and application of methods and techniques to uncover new models and extract meaningful knowledge from educational data.

In addition, big data holds significant potential within higher education institutions to monitor, understand and evaluate educational processes, thereby guiding efforts to improve learning outcomes and realise high-impact teaching methodologies. Ongoing research is aimed at developing a system that uses data mining tools from e-learning platforms such as Moodle to address various educational challenges faced by students, teachers, researchers and administrators. The proposed system includes the creation of a data warehouse and the integration of a business intelligence system to generate various reports.

In addition to the proposed use of Microsoft Power BI for advanced data visualisation and analysis, future research initiatives should explore the integration of Artificial Intelligence within the Moodle LMS. This intersection between AI and Moodle holds the promise of uncovering deep insights, facilitating adaptive learning pathways, and contributing to the ongoing evolution of educational technologies.

References

1. Nehrey, M., Klymenko, N., Kostenko, I.: Formal and non-formal education of Ukraine: analysis of the current state and the role of digitalization. Lect. Notes Data Eng. Commun. Technol. **181**, 1085–1098 (2023)
2. Volkova, N., Rizun, N., Nehrey, M.V.: Data science: Opportunities to transform education. In: CEUR Workshop Proceedings, vol. 2433, pp. 48–73 (2019). http://ceur-ws.org/
3. Voronenko, I., Klymenko, N., Nahorna, O.: Priority areas of Ukraine's innovative potential in the conditions of digital transformation. Financ. Credit Act. Probl. Theory Pract. **1**(42), 313–321 (2022). https://doi.org/10.55643/fcaptp.1.42.2022.3684
4. Kucherova, H., Honcharenko, Y., Ocheretin, D., Bilska, O.: Fuzzy logic model of usability of websites of higher education institutions in the context of digitalization of educational services. Neuro-Fuzzy Model. Tech. Econ. **10**, 119–135 (2021). https://doi.org/10.33111/nfmte.2021.119
5. Lukianenko, D., Matviychuk, A., Lukianenko, L., Dvornyk, I.: University competitiveness in the knowledge economy: a Kohonen map approach. In: CEUR Workshop Proceedings, vol. 3465, pp. 236–250 (2023). https://ceur-ws.org/Vol-3465/paper23.pdf
6. Fischer, C., Pardos, Z.A., Baker, R.S., et al.: Mining big data in education: affordances and challenges. Rev. Res. Educ. **44**(1), 130–160 (2020). https://doi.org/10.3102/0091732X20903304
7. Moodle Statistics. https://stats.moodle.org/
8. Kuzminska, O.H., Morze, N.V., Osadchyi, V.V.: Digitization of learning environment of higher education institutions: conceptual foundations and practical cases. J. Phys. Conf. Ser. **2611**(1), 012024 (2023)
9. Glazunova, O.G., Voloshyna, T.V.: Hybrid cloud-oriented educational environment for training future IT specialists?. In: CEUR Workshop Proceedings, vol. 1614, pp. 157–167 (2016)
10. Peleshko, D., Rak, T., Izonin, I.: Image superresolution via divergence matrix and automatic detection of crossover. Int. J. Intell. Syst. Appl. (IJISA) **8**(12), 1–8 (2016). https://doi.org/10.5815/ijisa.2016.12.01
11. Akcapınar, G.: Profiling students' approaches to learning through Moodle logs. In: Proceedings of the Multidisciplinary Academic Conference, Prague, pp. 242–248 (2015)
12. Glazunova, O., Morze, N., Golub, B., Voloshyna, T., Parhomenko, O.: Learning style identification system: design and data analysis. In: CEUR Workshop Proceedings, vol. 2732, pp. 793–807 (2020)
13. Kadoic, N., Oreški, D. Analysis of student behavior and success based on logs in Moodle. In: 41st International Convention on Information and Communication Technology, Electronics and Microelectronics (MIPRO), At Opatija, Croatia, pp. 0654–0659 (2018). https://doi.org/10.23919/MIPRO.2018.8400123
14. Zacharis, N.Z.: Classification and regression trees (CART) for predictive modeling in blended learning. IJ Intell. Syst. Appl. **3**(1), 9 (2018). https://doi.org/10.5815/ijisa.2018.03.01
15. Njenga, S.T., Oboko, R.O., Omwenga, E.I.: Use of intelligent agents in collaborative m-learning: case of facilitating group learner interactions. Int. J. Mod. Educ. Comput. Sci. (IJMECS), **9**(10), 18–28 (2017). https://doi.org/10.5815/ijmecs.2017.10.03
16. Gabroveanu, M., Diaconescu, I.: Extracting semantic annotations from moodle data. In: Proceedings of the 2nd East European Conference on Rule-Based Applications, Germany, vol. 428, pp. 1–7 (2008). http://ceur-ws.org/Vol-428/paper1.pdf

17. Khojastehfar, A., Abarbekouh, H., Safai, S., Fatemi, O.: Analysis of teaching and learning in moodle with the help of visualization with power BI: case study of e-learning course of university of Tehran. In: 10th International and the 16th National Conference on E-Learning and E-Teaching (ICeLeT), pp. 1–7 (2023). https://doi.org/10.1109/ICeLeT58996.2023.10139886
18. Poon, L.K.M., Kong, SC., Yau, T.S.H., Wong, M., Ling, M.H.: Learning analytics for monitoring students participation online: visualizing navigational patterns on learning management system. blended learning. New Challenges Innov. Pract. ICBL. Lecture Notes in Computer Science, 10309 (2017). https://doi.org/10.1007/978-3-319-59360-9_15
19. Romero, C., Ventura, S., Garcia, E.: Data mining in course management systems: Moodle case study and tutorial. Comput. Educ. **51**, 368–384 (2008). https://doi.org/10.1016/j.compedu.2007.05.016
20. Aldowah, H., Al-Samarraie, H., Fauzy, W.M.: Educational data mining and learning analytics for 21st century higher education: a review and synthesis. Telemat. Inform. **37**, 13–49 (2019). https://doi.org/10.1016/j.tele.2019.01.007
21. Morze, N., Kuzminska, O., Glazunova, O., et al.: Moodle tools for educational analytics of the use of electronic resources of the university's portal. In: Proceedings of the 1st Symposium on Advances in Educational Technology AET, **20**(2), 444–451 (2022). ISSN 978–989–758–558–6
22. Fenu, G., Marras, M., Meles, M.: A learning analytics tool for usability assessment in moodle environments. J. e-Learn. Knowl. Soc. **13**(3), 23–34 (2017). ISSN:1826–6223, e-ISSN:1971–8829 https://doi.org/10.20368/1971-8829/1388
23. Pal, N., Dahiya, O.: Role of learning management system for evaluating students' progress in learning environment. In: Proceedings of 5th International Conference on Contemporary Computing and Informatics, IC3I 2022: 1800–1806
24. Dogaru, T., Götze, N., Rotelli, D., et al.: Task definition in big sets of heterogeneously structured moodle LMS courses. In: 21. Fachtagung Bildungstechnologien (DELFI), pp. 313–314 (2023). https://doi.org/10.18420/delfi2023-71
25. Farias-Gaytan, S., Aguaded, I., Ramirez-Montoya, M.-S.: Digital transformation and digital literacy in the context of complexity within higher education institutions: a systematic literature review. Humanit. Soc. Sci. Commun. **10**(1), 386 (2023). https://doi.org/10.1057/s41599-023-01875-9
26. Azarova, A., Radomska, L., Azarova, L.: Neuro-fuzzy modelling of economic specialties students' speech-communicative professional competence. Neuro-Fuzzy Model. Tech. Econ. **9**, 3–25 (2020). https://doi.org/10.33111/nfmte.2020.003
27. Izonin, I., Trostianchyn, A., Duriagina, Z., Tkachenko R., Tepla, T., Lotoshynska, N.: The combined use of the wiener polynomial and SVM for material classification task in medical implants production. Int. J. Intell. Syst. Appl. (IJISA), **10**(9), 40–47 (2018). https://doi.org/10.5815/ijisa.2018.09.05
28. Rawat, B., Dwivedi, S.K.: an architecture for recommendation of courses in e-learning system. Int. J. Inf. Technol. Comput. Sci. (IJITCS), **9**(4), 39–47 (2017). https://doi.org/10.5815/ijitcs.2017.04.06
29. Labib, W., Eman Abowardah, D., Abdelsattar, A.: A review of big data's role on higher education. In: Sixth International Conference of Women in Data Science at Prince Sultan University (WiDS PSU), Riyadh, Saudi Arabia, pp. 98–105 (2023). https://doi.org/10.1109/WiDS-PSU57071.2023.00031
30. Ifenthaler, D., Yau, J.YK.: Utilising learning analytics to support study success in higher education: a systematic review. Educ. Tech. Res. Dev. **68**, 1961–1990 (2020). https://doi.org/10.1007/s11423-020-09788-z

31. Khalid, M., Yousaf, M.M.: A comparative analysis of big data frameworks: an adoption perspective. Appl. Sci. **11**, 11033 (2021). https://doi.org/10.3390/app112211033
32. Moodle piugins. https://moodle.org/plugins/?q=
33. Siafis, V., Rangoussi, M.: Visualization of educational data mined from E-learning platforms: a comparative evaluation of tools. In: ICERI–2021, Seville, Spain (2021). https://doi.org/10.21125/iceri.2021

Implementation of the STEM Project "Modeling of Spatial Images of Polyhedra" in the Professional Training of Future Specialists in Engineering and Pedagogical Specialties

Oleksandr Derevyanchuk[1](✉), Vasyl Kovalchuk[2], Serhiy Balovsyak[1], Mickolay Dominikov[1], Hanna Kravchenko[3], Oleksii Pshenychnyi[1], and Andrii Kovalchuk[4]

[1] Yuriy Fedkovych Chernivtsi National University, Chernivtsi 58012, Ukraine
{o.v.derevyanchuk,s.balovsyak,m.dominikov, o.pshenychnyi}@chnu.edu.ua

[2] State Institution «Scientific and Methodological Center for Higher and Pre-Higher Education», Kyiv 03151, Ukraine
v.i_kovalchuk@ukr.net

[3] High State Educational Establishment «Chernivtsi Transport College», Chernivtsi 58000, Ukraine

[4] Oleksandr Dovzhenko Hlukhiv National Pedagogical University, Hlukhiv 41400, Ukraine

Abstract. Modern society requires innovative solutions to solve the complex challenges facing it. The development of science, technology, engineering and mathematics (STEM) is key to ensuring the sustainable development of society and solving its problems. Labor market research shows that there is a demand for skilled professionals in the STEM field, while there is a shortage of personnel in these fields. This is due to rapid technological transformation in all spheres of life. The integration of STEM technologies into university education is becoming an urgent task, which will contribute to the preparation of a competitive generation capable of solving modern challenges. In particular, STEM projects, such as "Modeling of spatial images of polyhedra", develop analytical and creative thinking of students, providing the opportunity to apply theoretical knowledge in practical projects.

In the process of implementing STEM projects, students, using the AutoCad graphic editor, develop the skills of building 3D models of polyhedra using digital technologies.

The author's technology for developing and printing 3D models is presented.

Keywords: STEM-education · STEM-projects · STEM-technologies · digitization of education

Z. Hu et al. (Eds.): ICCSEEA 2024, LNDECT 242, pp. 683–692, 2025.
https://doi.org/10.1007/978-3-031-84228-3_59

1 Introduction

Modern society faces complex challenges that require innovative and technological solutions. The development of science, technology, engineering and mathematics (STEM) becomes an important prerequisite for solving these problems and ensuring the sustainable development of society. In this context, the integration of STEM technologies into the system of university education becomes especially relevant.

Labor market research shows that there is a demand for highly qualified personnel in the field of STEM. According to the study [1], the demand for specialists in the fields of information technology, artificial intelligence, biotechnology and engineering is predicted to grow. At the same time, there is a certain shortage of qualified personnel in these fields. This demand arises from rapid technological transformation in all areas of life, including manufacturing, medicine, transportation, information technology, and many others. When it is necessary to solve complex problems, such as climate change or global security challenges, much attention is paid to the development of STEM technologies. This actualizes the need to integrate STEM technologies into the system of university education. The integration of STEM technologies into university curricula is an important element of the strategy of modern education and meets the challenges of modern society, promotes innovative development and ensures the training of a competitive generation. Knowledge and skills acquired through such education will be key to the success of future generations in various spheres of life [2–9].

Successful integration of STEM technologies into university education requires the implementation of specific projects that contribute to the practical application of theoretical knowledge. One of such projects is "Modeling of spatial images of polyhedra", which has significant potential for professional training of future specialists in engineering and pedagogical specialties.

This project can be organized as an interdisciplinary course that combines knowledge from mathematics, computer science and engineering. Students participating in the project have the opportunity not only to deepen their knowledge of the theory of polyhedra and geometry, but also to learn modern software tools for modeling spatial objects. In particular, students study theoretical knowledge related to polyhedra, their structure and properties. Specialized software tools are used to create three-dimensional models of polyhedra. The implementation of this STEM project contributes to the application of the acquired knowledge in real practical projects. Modeling spatial images requires creativity and innovative thinking from students, which contributes to the development of their analytical and creative abilities.

2 Theoretical Consideration

In the process of modeling polyhedra, we use the graphic editor AutoCad. The task of the project is to build 3D models of regular polyhedra (Fig. 1) on the example of a dodecahedron model.

Completion by students of engineering and pedagogical specialties of the STEM project "Modeling of spatial images of polyhedra" is an important prerequisite for their preparation for the design, implementation and application of computer graphics systems, as well as computer systems for intelligent image processing [10–12].

Fig. 1. Models of polyhedra: (a) tetrahedron; (b) hexahedron; c) octahedron; (d) dodecahedron; (e) icosahedron

The works of S. Balovsyak, O. Derevyanchuk, I. Tereikovskyi, Zh. Hu, Yu. Ushenko and others [13–24]. STEM projects of this field are closely related to the future professional activities of students and are used to solve applied problems. For example, computer image segmentation systems can be effectively used for automatic image processing of various objects (geometric shapes, vehicles, etc.) [13–15]. In addition, during the implementation of such STEM projects, students improve their knowledge of digital image processing methods: filtering [16, 17], segmentation [18], contour selection [14], profile analysis [19], contrast enhancement [20]. During the implementation of the project, it is possible to process images of both modeled polyhedra and photos of objects printed on a 3D printer. Due to digital image processing, it is possible, in particular, to automatically determine the position, size and orientation of polyhedra. In STEM projects, artificial intelligence tools [21, 22] are used to improve the accuracy of image processing, in particular, image recognition methods and fuzzy logic [23], which also helps students acquire the latest digital competencies. In order to evaluate students' success, intellectual data analysis tools are used, in particular, methods of clustering [24] and regression.

The STEM project begins with the design and planning necessary for the development of research skills of students of engineering and pedagogical specialties. The final result of the work on the STEM project is a practical presentation of the results of the activity in the form of a three-dimensional model of a polyhedron, which students print on a 3D printer and present the product of the STEM project.

3 Experimental Consideration

3.1 Technological Design of the Dodecahedron by Means of Three-Dimensional Computer Modeling

To create a frame model of a dodecahedron in the "AutoCad" program, activate the "Polygon" tool. At the prompt of the command line (Enter the number of sides), set – 5 and press <Enter> (Fig. 2). At the request (Enter the center of the polygon), in the command line we set the coordinates – 100,0,0 < Enter >.

The next step will be the answer to the request of the program "Inscribed in a circle, described around a circle", type "In" in the command line and press <Enter> (Fig. 3).

When prompted, enter the radius of the circle, type – 100 and press <Enter> (Fig. 4a). We build another polygon and at the request "Enter the center of the polygon" select

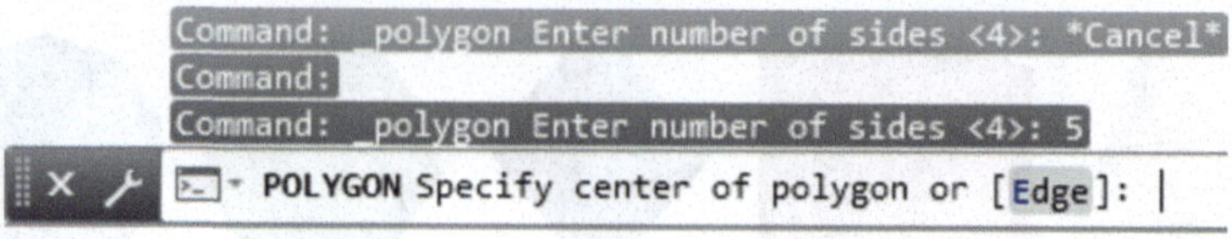

Fig. 2. The program's command line

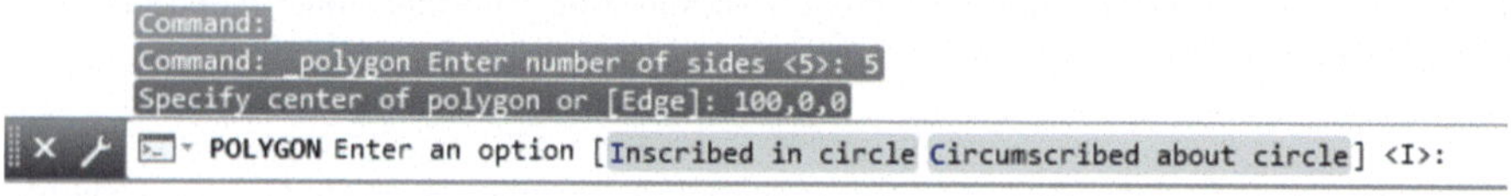

Fig. 3. The program's command line

"Edge", pointing to its edge points of the base (Fig. 4b). The first point with the coordinates of point A, the second with the coordinates of point B and traditionally press <Enter>. The operation is repeated until the pattern of the sweep of the lower part of the dodecahedron is formed (Fig. 4c).

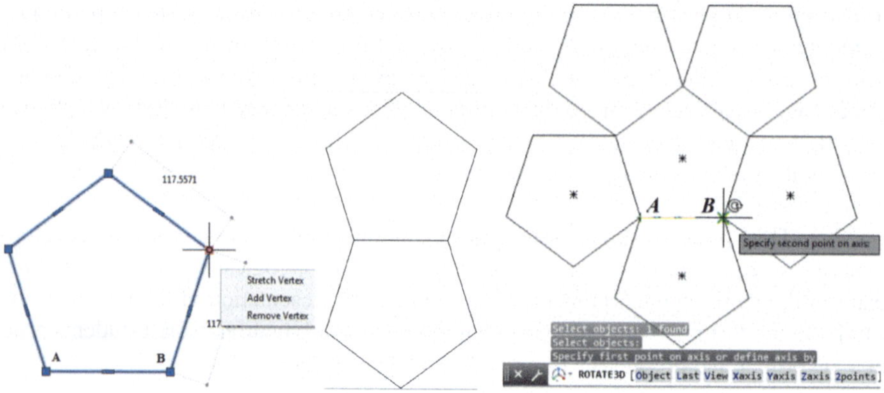

(a) (b) (c)

Fig. 4. The first stages of building a dodecahedron

The received images of pentagons must be rotated to the angle (–63.434948820) relative to the base using the "Rotate 3D" command. For this:

- Activate the "Rotate 3D" command, <Enter>.
- "Select objects": select the contour of the lower polygon.
- The next command is "Specify base point". Place the cursor on the first corner point (A) of the dodecahedron and press <Enter>.
- Move the cursor to the second point (B) – <Enter>.
- Enter the numerical value of the angle (–63.43494882) and press <Enter>.
- We get the following image (Fig. 5a).

- We repeat the given actions in relation to other polygons. As a result, we get half of a dodecahedron. We control the construction process using the "3D Orbit" operation (Fig. 5b).

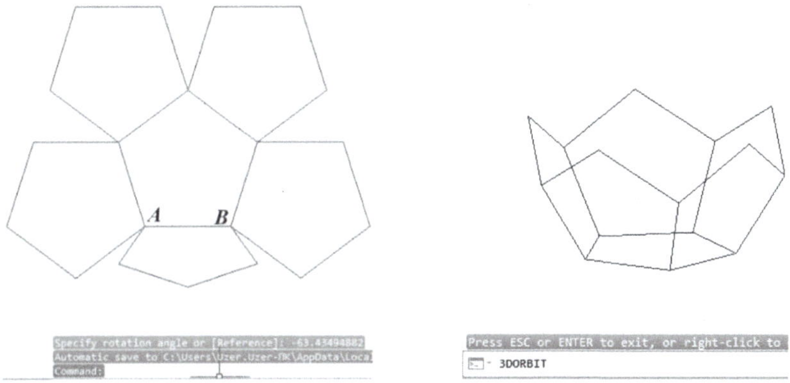

(a) (b)

Fig. 5. Creation and control of graphic constructions

The second half of the dodecahedron is created using the "Copy" and "Align" commands. For this:

- Click on the conditional designation "Copy" in the "3D Modeling" menu, or type "Copy" in the command line and press <Enter>.
- <BUTTON-1> select one of the polygons of the dodecahedron and press <Enter>.
- At the request of the program (Enter the base point), click in any part of the screen. At the second request, move the cursor to a certain distance and click <Enter> (Fig. 6).

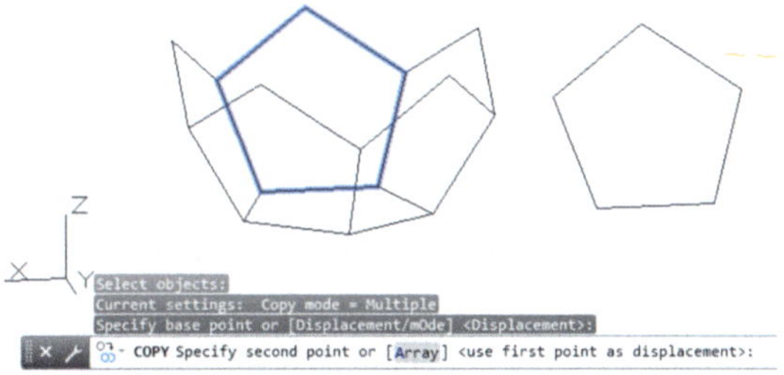

Fig. 6. Creating a copy of the polygon

- The next action is to use the "Align" command. To do this, activate the "Align" command in the "3D Modeling" menu, or type "Align" in the command line.
- At the request (Select objects) - select the copied polygon and press <Enter>.

- At the request "Specify the first starting point", point the cursor at one of the corners of the polygon and click <BUTTON-1>.
- At the request "Specify the first destination point", bring the cursor to the destination of the corner of the dodecahedron and click <BUTTON-1> (Fig. 7a).
- We repeat the previous procedure with the second corner (Fig. 7b).
- Similarly for the third corner. As a result, the copied polygon will take its position (Fig. 8a).

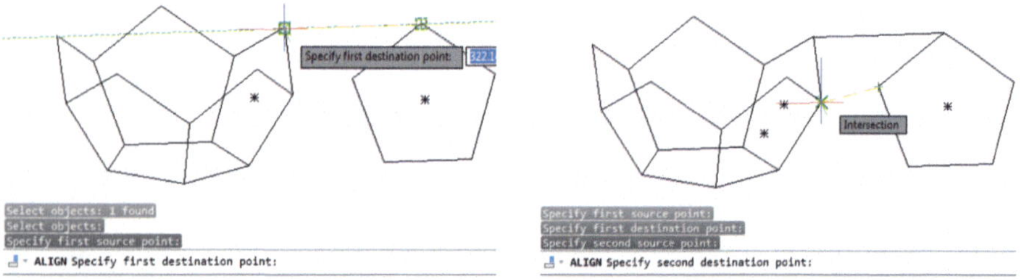

(a) (b)

Fig. 7. Stages of adding sides to the created model

We repeat the "Copy" and "Transfer" process five more times until we get a complete model of the dodecahedron layer. The result will be the creation of a frame model of the dodecahedron layer (Fig. 8b).

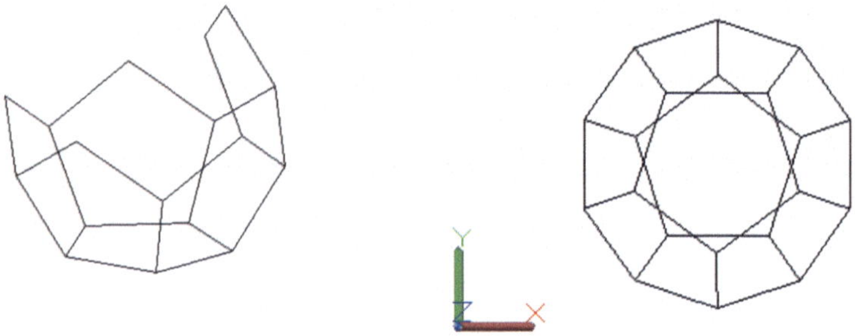

(a) (b)

Fig. 8. Adding sides to the created model and the wireframe model of the dodecahedron layer

After using the "Convert to Surface" command in the "Modify" menu of the AutoCad program, we will go to the planar model of the dodecahedron.

At the request of the program (Select an object), throw a lasso on the created frame (Fig. 9a) and press <Enter>.

After the procedure of zeroing the isolines in the Properties of the planes of the dodecahedron (Fig. 9b), we get a 3D model of the plane dodecahedron (Fig. 9c).

The Object Properties panel is displayed after double-clicking <BUTTON-1> on it. You can also change the object's color there.

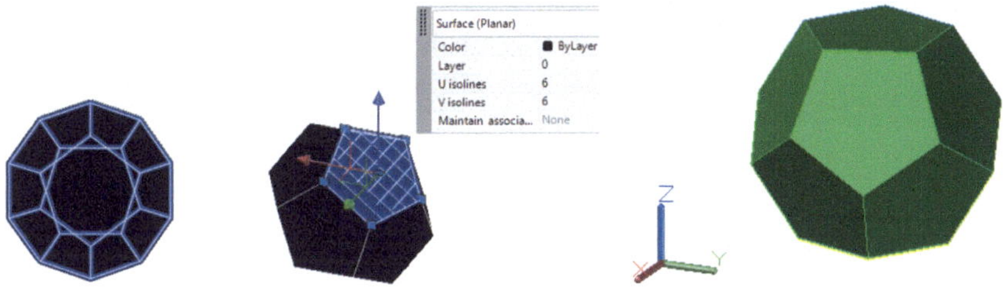

(a) (b) (c)

Fig. 9. Stages of sequential transformation of the frame model of the dodecahedron into a planar model

To create the necessary conditions for printing a model on a 3D printer, it is necessary to transform a planar model into a solid object [25]. For this, we will apply the operation of extruding planes at an angle of 350° (Fig. 10).

In the AutoCad program menu, select the Export Other Formats function. In the "File name" dialog box, the inscription of the name of our file appears. In the "Filer of type" window, select the "Lithography" (*.stl) option.

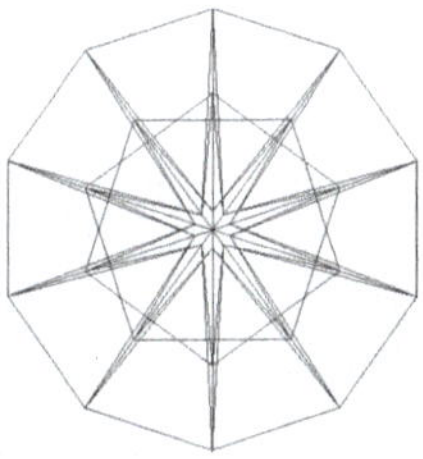

Fig. 10. View of a solid object in a wireframe representation

At the request of the command line "Select Solids or waterproof meshes", click < BUTTON-1> on the solid object we created and press <Enter>. We check the availability of the object we created in STL format.

3.2 Printing a Three-Dimensional Model of a Dodecahedron on a 3D Printer

The production of a part by the 3D printing method includes 3 stages, modeling and preparation for printing, direct printing of the object and mechanical post-processing, if necessary [25–28].

At the first stage, the previously designed model is exported in (.stl,.obj) format for further slicing it into layers in the appropriate software package. Next, the model must be turned and placed on the printer table in such a way as to increase the contact area of the model with the table and minimize the number of supports.

In Fig. 11 shows the window of the Repetier-Host software package with the model ready for printing. Slicing of the model was carried out by the Prusa Slicer plugin, which is part of the Pronterface software package [29]. The parameters and the slicer configuration process are not considered within the framework of this work.

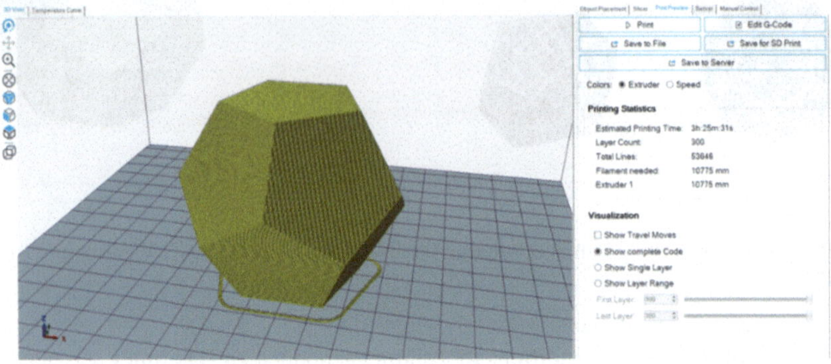

Fig. 11. Repetier-Host program window with cut model ready for printing

After the successful completion of the printing process, if necessary, mechanical post-processing is carried out. Unlike SLA printing, FDM does not require a mandatory post-processing process, and simple mechanical parts, properly placed on the table, can be used immediately after printing. The printing process and the finished printed model of the dodecahedron are shown in Fig. 12.

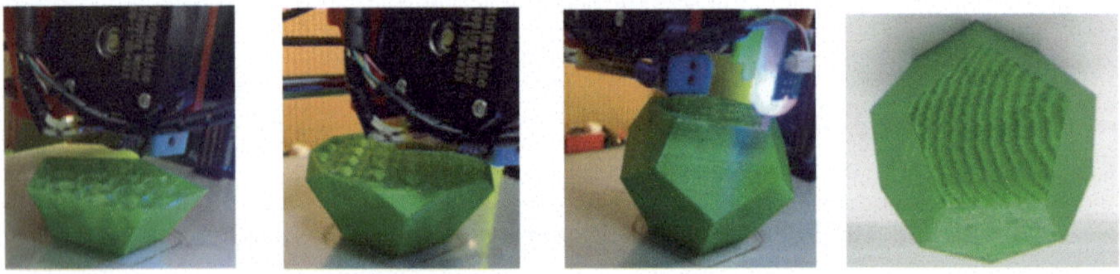

(a) (b) (c) (d)

Fig. 12. The process of printing a dodecahedron

4 Summary and Conclusion

The implementation of the STEM project "Modeling of spatial images of polyhedra" has great potential for training future specialists in engineering and pedagogical specialties. During its implementation, students will develop a wide range of skills and competencies that are important in the modern labor market. They will learn to apply mathematical concepts for the analysis and modeling of complex spatial objects, develop the ability to work with modern technologies, master the skills of working with programs for three-dimensional modeling and visualization, be able to analyze problems and find optimal solutions, learn to set specific goals, develop strategies and creatively approach to problem solving, develop communication skills and creative thinking.

References

1. The Future of Jobs Report 2023. https://www.weforum.org/publications/the-future-of-jobs-report-2023/
2. Li, Y., Xiao, Y., et al.: A systematic review of high impact empirical studies in STEM education. Int. J. STEM Educ. (IJSTEM) **9**(72), 1–18 (2022)
3. Li, Y., Xiao, Y.: Authorship and topic trends in STEM education research. Int. J. STEM Educ. (IJSTEM) **9**(62), 1–7 (2022)
4. Merchant, S., Morimoto, E.T., Khanbilvardi, R.: An integrated STEM learning model for high school in engineering education. In: Integrated Stem Education Conference (ISEC), [Princeton, NJ, 8–8 March 2014], 4–9 (2014)
5. Derevyanchuk, O.V., et al.: Implementation of STEM education in the process of training of future specialists of engineering and pedagogical specialties. Proc. SPIE **12938**, 214–217 (2024)
6. Zhaohao, N.: International comparison of STEM teacher education. Open Access Libr. J. **9**(8) (2022)
7. Voitovych, I., et al.: Training of teachers STEM disciplines to work with students in distance education. inequality, informational warfare, fakes and self-regulation in education and upbringing of youth, Youth Voice J. (I), 103–112 (2023)
8. Kovalchuk, V., Shevchenko, L., Iermak, T., Chekaniuk, K.: Computer modeling as a means of implementing project-based activities in STEM-education. Open J. Soc. Sci. **9**(10), 173–183 (2021)
9. Kovalchuk, V., Maslich, S., Tkachenko, N., Shevchuk, S., Shchypska, T.: Vocational education in the context of modern problems and challenges. J. Curriculum Teach. **11**(8), 329–338 (2022)
10. Davies, E.R.: Computer and Machine Vision: Theory, Algorithms, Practicalities, Elsevier, 912 (2012)
11. Kim, D.E., Hwang, D.: Intelligent Imaging and Analysis, vol. 492. MDPI, Switzerland, Basel (2020)
12. Gonzalez, R., Woods, R.: Digital Image Processing. 4th edidion, Pearson/ Prentice Hall, NY 1022 (2018)
13. Balovsyak, S.V., Derevyanchuk, O.V., Kravchenko, H.O., Kroitor, O.P., Tomash, V.V.: Computer system for increasing the local contrast of railway transport images. In: Proceedings of SPIE, Fifteenth International Conference on Correlation Optics, (12126), 121261E1–7 (2021)
14. Derevyanchuk, O.V., Kravchenko, H.O., Derevianchuk, Y.V., Tomash, V.V.: Recognition images of broken window glass. Proc. SPIE **12938**, 210–213 (2024)
15. Kramar, V.M., Pugantseva, O.V., Derevyanchuk, A.V.: Spatial confinement, self-polarization and exciton-phonon interaction effect on the location of exciton line in lead iodide nanofilms. Low Temp. Phys. **40**(8), 766–770 (2014)
16. Balovsyak, S.V., Odaiska, K.: Automatic determination of the gaussian noise level on digital images by high-pass filtering for regions of interest. Cybern. Syst. Anal. **54**(4), 662–670 (2018)
17. Balovsyak, S.V., et al.: Adaptive oriented filtration of digital images in the spatial domain. In: Proc. SPIE, 11176, Photonics Applications in Astronomy, Communications, Industry, and High-Energy Physics Experiments, vol. 11176 111761A-1–111761A-6 (2019)
18. Tereikovskyi, I., Hu, Z., Chernyshev, D., Tereikovska, L., Korystin, O., Tereikovskyi, O.: The method of semantic image segmentation using neural networks. Int. J. Image Graph. Sign. Process. (IJIGSP), **14**(6), 1–14 (2022)
19. Balovsyak, S.V., Derevyanchuk, O.V., Fodchuk, I.M.: Method of calculation of averaged digital image profiles by envelopes as the conic sections. In: Hu, Z., Petoukhov, S., Dychka, I., He, M. (eds.) Advances in Intelligent Systems and Computing, Springer International Publishing, vol. 754, pp. 204–212 (2019). https://doi.org/10.1007/978-3-319-91008-6_21

20. Balovsyak, S., Odaiska, K., Yakovenko, O., Iakovlieva, I.: Adjusting the Brightness and Contrast parameters of digital video cameras using artificial neural networks. In: Proceedings of SPIE, vol. 12938, pp. 129380I-1–129380I-4 (2024)
21. Lytvyn, V., Lozynska, O., Uhryn, D., Vovk, M., Ushenko, Y., Zhengbing, H.: Information technologies for decision support in industry-specific geographic information systems based on swarm intelligence. Int. J. Mod. Educ. Comput. Sci. (IJMECS) **15**(2), 62–72 (2023)
22. Prokipchuk, O., et al.: Intelligent analysis of Ukrainian-language tweets for public opinion research based on NLP methods and machine learning technology. Int. J. Mod. Educ. Comput. Sci. (IJMECS) **15**(3), 70–93 (2023)
23. Fayek, A.R.: Fuzzy logic and fuzzy hybrid techniques for construction engineering and management. J. Constr. Eng. Manage. **146**(7) (2020)
24. Balovsyak, S., Derevyanchuk, O., Kravchenko, H., Ushenko, Y., Hu, Z.: Clustering students according to their academic achievement using fuzzy logic. Int. J. Mod. Educ. Comput. Sci. (IJMECS) **15**(6), 31–43 (2023)
25. Lukianchuk, I., Tulashvili, Y., Podolyak, V., Horbariuk, R., Kovalchuk, V., Bazyl, S.: Didactic principles of education students 3D-printing. IJCSNS Int. J. Comput. Sci. Netw. Secur. **22**(7), 443–450 (2022)
26. Pandya, H.: 3D printing technology: fundamentals and application. Studera Press, 416 (2021)
27. Soloman, S.: Additive manufacturing technology – 3D printing and design – the 4th industrial revolution. Independently Published, 417 (2020)
28. Kazmer, D.O., Colon, A.: Injection printing: additive molding via shell material extrusion and filling. Addit. Manuf. **36**, 101469 (2020)
29. Ishengoma, F.R., Mtaho, A.B.: 3D printing: developing countries perspectives. Int. J. Comput. Appl.Comput. Appl. **104**(11), 30–34 (2014)

Advantages and Disadvantages of Distance Education and Blended Learning as a Compromise Form of Education

Anna Serednytska(✉)

Lviv Polytechnic National University, 12 Bandera Street, Lviv 790013, Ukraine
anna.y.serednytska@lpnu.ua

Abstract. Information technologies have changed the life of society and the educational process. They made distance education possible. Distance education is cost-effective for universities and students and environmentally-friendly. It provides access to higher education for broad groups of the population, regardless of place of residence, level of prosperity, state of health, and employment. The student may choose any university, pace, schedule, place, time of study. Curriculums become more modern, learning – more interactive, and information becomes more diverse and visualized. However, distance education has its drawbacks. It is difficult for students to concentrate, understand the material and learn without the teacher's supervision. They may feel social isolation, lose communication skills. Blended learning combines the advantages of in-person and distance education. Students attend classes and work remotely in a Virtual Learning Environment (VLE). The National University Lviv Polytechnic uses the VLE based on the Moodle platform. At the VLE students study theoretical courses, perform tasks, solve tests and write exams.

Keywords: Distance Education · Blended Learning · Virtual Learning Environment

1 Introduction

The societal life has undergone significant changes as a result of the emergence of Internet and the introduction of information technologies into the educational process. As a result of changes in the field of information and communication technologies, a new form of education has emerged – distance education. Distance education can be defined as conducting the educational process at a distance, by employing interactive communication between a teacher and a student by means of modern information and communication technologies: multimedia, video, audio, etc. [1]. Although the student communicates with a teacher or a tutor during such education, he also masters a significant amount of educational materials located in the virtual learning environment of the educational institution independently [29]. The term "distance education" was coined in 1982 in USA [22]. Distance education is very popular and develops rapidly, taking on different forms. As a new type of education, distance education has a number of favorable features for both the educational institution and students.

Z. Hu et al. (Eds.): ICCSEEA 2024, LNDECT 242, pp. 693–701, 2025.
https://doi.org/10.1007/978-3-031-84228-3_60

2 Advantages of Distance Education

Distance education is very cost-effective. It allows universities to increase the number of students significantly and simultaneously reduce the costs of organizing the educational process, as expenses for renting premises, providing technical equipment, and utility payments are reduced [4, 12]. Distance learning is environmentally-friendly, because it reduces the amount of natural resources needed for studying and reaching the university. The presentation of the material becomes visual, as information is presented in the form of schemes, diagrams, graphs, roadmaps, spider diagrams, causal chains, mind maps [2]. The structure of the curriculum, as well as teaching methods are becoming more flexible, and the information becomes more up-to-date [18, 24]. Students work with electronic libraries, study various types of educational materials online. Distance learning makes it possible to use a wide variety of information and computer technologies. Thanks to modern technologies, change, improvement and update of educational materials on online resources can happen instantly.

The learning process itself acquires more interactive features. A new learning model appears - group communication and group activity of students [18, 24], the amount of feedback between the teacher and the student increase [20]. The amount of time allocated for independent student work is increases as well [16].

Distance education has many advantages for students too. First of all, it allows them to save money, which makes education accessible to all students [19], irrespective of their level of financial support [27]. Distance education ensures social equality, because it provides an opportunity to get an education, regardless of geographical location, health and social status [7, 12, 24]. There is no need to spend money to reach the university or on accommodation. Distance education "can help to improve the income of vulnerable groups and promote social equality… distance education institutions should promote the idea that distance higher education can improve the income growth of low-income groups, and increase the financial support for low-income groups to access distance higher education." [27]. Distance education is available for students with health problems or people with disabilities [19]. There is no need to buy educational materials and textbooks. Study materials are easily available and the student simply has to download them.

Another favorable feature of distance learning lies in the fact that it is less time-consuming. There is no need to travel from a remote province or even another country to a university [6, 10, 24]. The choice of an educational institution, place and time of study is no longer restricted [19, 24]. A possibility arises to receive an education abroad without spending a lot of money on education services, travel and accommodation. A student can receive an education from home, which is essential for young parents, or combine employment with education [10]. A student can study in comfortable home conditions or anywhere at a convenient time [8]. There is no need to follow the dress code, as one can study in comfortable clothes.

Distance education is favorable for introverts, as there is no need to adapt to the environment, to overcome a psychological barrier. Sometimes distance education is the only way to avoid bullying [11]. Distance education saves time, because students do not distract each other and learning takes place as quickly and efficiently as possible. Therefore, distance education is the optimal choice for those students who cannot concentrate on their studies, since they are excessively distracted by social activity and communication with peers [11].

Learning takes place at a pace convenient for the student [10]. Online recordings of lectures can be listened to again until the information becomes clear. In addition, the issue of classes missed due to illness or for other reasons loses its relevance.

Students can choose courses and programs that interest them, can easily communicate with the teacher, ask questions, receive recommendations, discuss the topic with the teacher or other students. They acquire the skills of systematic self-education, which are relevant in modern society [7].

3 Disadvantages of Distance Education

Distance education has become especially relevant during the pandemic. However, after it ended, a tendency to restore in-person education began to appear in Ukraine. The return to the traditional format is taking place despite hostilities, frequent air alarms, missile attacks, etc. This is probably because distance education, despite a number of advantages, also has significant disadvantages.

During distance education it is difficult to concentrate and perceive the material [23], because direct contact between the teacher and the student is lost [26]. "Technical means of group communication activity between the teacher and student create an artificial and inferior, in the traditional sense, communicative space" [15]. In addition to this, if students perceive the material only in the form of recorded video lectures, there is a lack of direct contact with the teacher, which can lead to mistakes or misunderstanding of the material [11, 19]. In addition, not all students have sufficient self-discipline and motivation for independent study [9, 19]. Some may lose the desire to finish the work, as some students are used to being constantly encouraged and motivated by the teacher.

The lack of a fixed schedule of classes traditionally considered an advantage of this type of distance learning. However, a flexible schedule, lack of clearly defined days and hours for studying one or another subject can negatively affect the learning process. Firstly, not all students know how to plan their time. Secondly, not everyone has sufficient motivation for independent learning [19]. I. Mala claims that "he opportunity to study at a convenient time can turn not into systematic study, but into a constant habit of postponing this type of activity". [19]. Some students feel the need for external supervision, otherwise they cannot avoid procrastination. Constant contact with the Internet during distance learning can distract from tasks and prevent concentration [11, 19].

Lack of physical control by the teacher can cause cheating. Students do not always have high moral responsibility and perform tasks personally. Unscrupulous students sometimes replace themselves with an expert on an exam, during testing, or writing a test paper. Since it is not always possible to establish who exactly performs the tasks online, some institutions conduct the examinations in-person [12].

One of the most significant problems of distance education is social isolation [30]. Some students become unsociable as a result of distance education. A person who constantly works alone loses communication skills, the ability to work in a group or does not acquire them at all. In a traditional school, such students go beyond their comfort zone and learn to communicate and interact with people. Distance education does not give them such an opportunity [11, 15]. It is very difficult or even impossible to hold social events online that contribute to deepening communication between students [11]. Communication in chats is limited and is significantly different from real communication. If a student communicates only in chats, he mostly writes short messages. As a result, he does not learn to fully express his thoughts, lead a discussion and make arguments, speak in front of an audience [12, 15, 19].

The constant use of technical means to communicate and study educational material leads to deterioration of vision, nervousness and fatigue [28]. Excessive virtual communication can negatively affect a person's psyche, especially in adolescence, lead to a lack of life experience [5], inability to communicate and work in a team.

A student may have technical problems with the Internet or equipment during classes, exams, testing, etc. Students from remote regions suffer from this issue above all, despite the fact that distance education is generally convenient for them. A significant problem is the lack of equipment or its low quality, as well as low digital literacy [26, 30].

4 Blended Learning as a Compromise Form of Education

In our opinion, the optimal form of distance education may be the form of blended learning [3, 21]. Blended learning can help to combine the advantages of distance and in-person education and eliminate the disadvantages of distance education in its pure state [14]. During blended learning, students attend classes, listen to lectures, participate in practical classes, perform laboratory works and work remotely with educational materials in a virtual environment.

One of such virtual environments is Moodle, a system for creating distance courses and supporting distance education [13, 20]. According to Moodle Statistics [17], the platform is used by educational institutions in 239 countries around the world as a support system for distance, part-time and in-person education. Moodle users exceed 404 million as of February 2024.

The Virtual Learning Environment of Lviv Polytechnic National University, based on the Moodle platform, is a powerful tool that allows students to acquire new knowledge, repeat the material they have previously studied, and pass training tests. The teacher can upload theoretical materials, explanations and instructions for each topic, dictionaries, lists of literature, video and audio materials, presentations, etc. into the system. The teacher can also upload tasks, practice tests, independent and control works for checking and evaluating knowledge. Students can work with the scientific and educational materials posted in it independently of the teacher. They choose the time, place and pace of tasks that is convenient for everyone, which is typical for distance learning. Self-study in the Virtual Learning Environment develops responsibility, independence, self-organization and purposefulness, the ability to plan one's time to complete certain tasks. I. Mala calls such training asynchronous. In her opinion, "the use of this mode has significant educational potential, because it provides an opportunity to achieve an important systemic task of modern education - not only to provide the future specialist with a system of knowledge, but also to form the ability for self-education throughout life" [12].

During asynchronous training, students study theoretical material and work with practical tasks. They have the opportunity to communicate with the teacher, ask him questions, receive consultations both with the help of the Virtual Learning Environment, in particular on its forum, and in-person, during lectures or practical classes. Teachers can monitor the learning activities of students in the Virtual Environment. Tests and exams are checked by the environment and automatically graded in the journal, and thus the results of the work become instantly available to students and the teacher. In addition, the teacher can keep a journal in the Virtual Learning Environment, which allows students to constantly be aware of their progress.

Thus, the Virtual learning environment of Lviv Polytechnic National University is an effective supplement to the in-person learning process, as it provides constant access to learning materials, the opportunity to repeat the material and train practical skills.

5 Results of the Empirical Research

To conduct the experiment, we involved a social group of 158 Lviv Polytechnic National University students, of whom 52.5% were under 18 years old, 47.5% were under 24 years old; 74.1% are studying, 24.7% are both studying and working. The survey was held in classrooms of the higher education institution in February 2024, the survey time was from 8-30 to 13-00. The students belonged to the specialties 061 journalism, 053 psychology, 231 social work, and 122 computer sciences. The survey was voluntary and anonymous.

Students consider the opportunity to obtain education for those population groups for whom face-to-face education is not available to be the primary advantage of distance education: distance education can be obtained by those who have health problems or are working. Students consider cost-effectiveness a significant advantage of distance education, as well as the opportunity to save money on transportation, accommodation, and nutrition. However, views on whether distance education allows low-income citizens to get an education are divided. There was also an equivocal response to the lack of necessity to buy paper textbooks, perhaps because students download materials from the Internet more often than purchase textbooks.

Students consider the use of modern educational technologies (visual representation of educational material, the use of audio and video technologies) a significant advantage. Students find it beneficial that distance education accustoms them to independent self-education, which is difficult to do without in the modern society, and also provides a convenient pace of learning.

Students consider psychological factors to be of less importance, in particular, the lack of bullying from the other students. The interviewees highly appreciate the importance of communication, and due to this they consider the opportunity to focus exclusively on learning and not communication to be a negative rather than a positive aspect of distance education. Students did not pay attention to such a statement as the environmental friendliness of distance education. Perhaps this is due to the fact that they do not think enough about environmental problems (see Fig. 1).

The results of the survey on the disadvantages of distance education show a greater discrepancy in the attitude of students to the proposed statements. It is obvious that the occurrence of certain problems during distance education is very individual and depends on a student's personal traits. However, most statements have a high level of

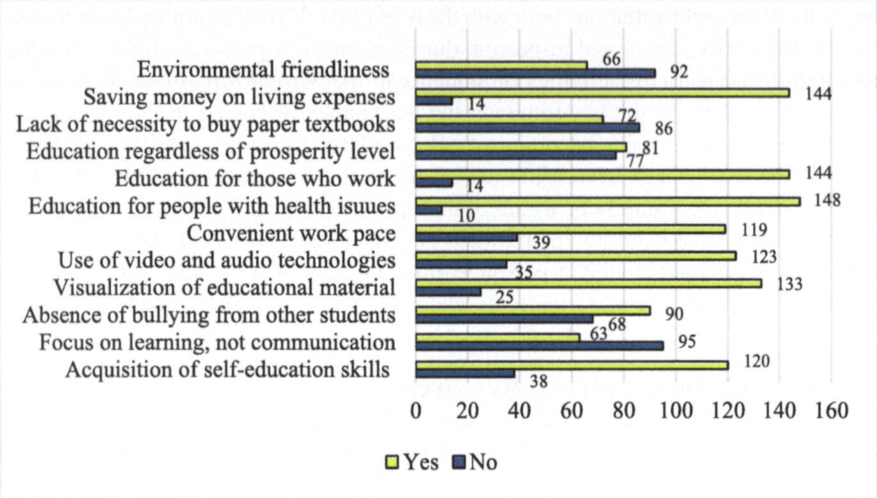

Fig. 1. Advantages of distance education.

positive responses (96-100 responses). Students consider the problems related to the technical support of the educational process, such as poor Internet, lack of equipment or inability to work with it, to be the most important. Students paid considerable attention to the problem of social isolation, which leads to the loss of the ability to work in a team and communicate with people. Loss of direct contact with the teacher leads to misunderstanding of the material, inability to concentrate on the material. Due to the lack of constant supervision by the teacher, students often lose motivation to work and develop a tendency to procrastinate. Lack of teacher control can also cause dishonesty, cheating, which negatively affects the quality of knowledge. Statements related to the deterioration of health as a result of distance education, such as vision problems from constant work with technology; fatigue, nervousness from constant work with technological devices, received significant support from the responders (see Fig. 2).

The majority of surveyed students (44.9%) have a negative attitude towards distance education. The attitude of 28.5% is positive, and 26.6% see both positive and negative sides in distance education.

Students have a positive attitude to face-to-face education. 63.9% perceive it exclusively positively, 21.5% – ambivalently, 14.6% – negatively.

Students prefer blended learning, which combines lessons at the university and work in a virtual learning environment. 78.5% of students have a positive attitude to such education, 14.6% see both advantages and disadvantages in the blended form of education, only 7% have a negative attitude.

The results of the survey make it possible to state that the surveyed students prefer the blended form of education, as it combines the advantages of the face-to-face form of education with the educational opportunities provided by modern information and communication technologies.

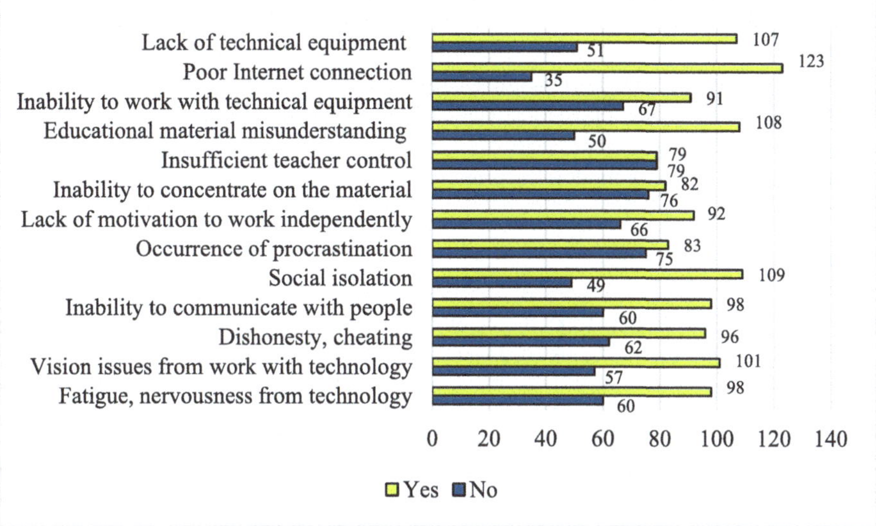

Fig. 2. Disadvantages of distance education.

6 Conclusions

The life of society has changed after the advent of the Internet and the development of information technologies. A new form of education has appeared – distance education, which has many advantages. Among them, 158 surveyed students consider the following to be the most important: the opportunity to get an education for those with health issues; the opportunity to get an education for those who work; saving money on transportation, accommodation, nutrition; visualization of educational material; use of audio and video technologies; development of self-education skills; comfortable learning pace; absence of bullying; the opportunity for the low-income citizens to get an education; no need to buy paper textbooks; environmental friendliness.

However, distance education has significant disadvantages: issues due to poor Internet; misunderstanding of the material without contact with the teacher; social isolation; lack of technical equipment; vision problems from constant work with technology; inability to work in a team and communicate with people; fatigue and nervousness from constant work with technology; dishonesty, cheating; lack of motivation to work independently; problems due to inability to work with technology; occurrence of procrastination; inability to focus on the material; insufficient teacher control. Because of this, the majority of surveyed students (44.9%) have a negative attitude to distance education, 28.5% have a positive attitude, and 26.6% see both positive and negative aspects of distance education.

Students of the National Lviv Polytechnic University have a much better attitude to face-to-face education. 63.9% perceive it exclusively positively, 21.5% – ambivalently, 14.6% – negatively.

In our opinion, a compromise between face-to-face and distance education can be blended learning, which combines lessons at the university and work in the Moodle virtual learning environment. It is to this form of education that students show the best attitude. 78.5% of students have a positive attitude to blended learning, 14.6% see both advantages and disadvantages in blended learning, only 7% have a negative attitude. The results of the survey make it possible to assert that the blended form of education makes it possible to combine the advantages of face-to-face education and use the educational opportunities provided by modern information and communication technologies.

References

1. Antoshchuk, S,V., Hravit, V.O.: Basics of distance learning organization in postgraduate pedagogical education: textbook NIKO publishing house, Sumy (2015). (in Ukrainian)
2. Bezuhlyi, D.: Methods of visual presentation of educational information. Physico-Math. Educ. **2**(3), 7–15 (2014). (in Ukrainian)
3. Clark, T., Barbour, M.: Online, Blended, and Distance Education in Schools: Building Successful Programs. Routledge, New York (2023). https://doi.org/10.4324/9781003446194
4. Herasymenko, I.V.: The use of distance learning technologies in the preparation of future computer science bachelors. Inf. Technol. Teach. Aids **41**(3), 232–247 (2014). (in Ukrainian)
5. Hetta, V., Yermak, S., Dzhevaha, H., et al.: Distance education: didactics, methodology, organization. Chernihiv (2017). (in Ukrainian)
6. Ifenthaler, D.: A systems perspective on data and analytics for distance education. Distance Educ. **43**(2), 333–341 (2022). https://doi.org/10.1080/01587919.2022.2064828
7. Kryshtanovych, S., Chorna-Klymovets, I., Semeriak, I., Mordous, I., Zainchkivska, I.: Modern technologies for the development of distance education. Int. J. Comput. Sci. Netw. Secur. **22**(9), 103–108 (2022). https://doi.org/10.22937/IJCSNS.2022.22.9.16
8. Kucherenko, N.: Distance education as a challenge of modern university education: philosophical and legal dimension. Vysnik Natl. Lviv Polytech. Univ. **906**, 34–40 (2018). (in Ukrainian)
9. Kurkina, S., Zhuravlov, V.: Distance education as a component of educational reforms in Ukraine. Pedagogical Sci. Theory Hist. Innov. Technol. **5**(109), 60–71 (2021). (in Ukrainian)
10. Littlefield, J.: 10 Reasons to Choose Online Education to Earn Your Degree. https://www.thoughtco.com/reasons-to-choose-online-education-1098006. Accessed 08 Feb 2024
11. Littlefield, J.: Pros and Cons of Earning Your High School Diploma On-line. https://www.thoughtco.com/pros-and-cons-of-online-diplomas-1098439. Accessed 08 Feb 2024
12. Mala, I.B.: Distance education as an effective tool of management education. Vcheni zapysky Universytetu «KROK» **2**(66), 132–151 (2022). (in Ukrainian)
13. Maldonado, F., González, J., Abril, A.: B-learning and Moodle as a strategy in university education. Rev. conrado **16**(76), 215–220 (2020). WOS:000579112600030
14. McCarthy, S., Palmer, E.: Defining an effective approach to blended learning in higher education: a systematic review. Australas. J. Educ. Technol. **39**(2), 98–114 (2023). https://doi.org/10.14742/ajet.8489
15. Melnyk, Y.V., Bohdanova, N.V.: Peculiarities of communication links in modern higher education. development of professional competencies of civil servants: communicative aspect. In: 2016 Kyiv, National Academy of State Administration, Materials of the Annual Scientific and Practical Conference with International Participation, p. 460 (2016). (in Ukrainian)
16. Melnyk, Y.V., Borodenko, N.D., Bohdanova, N.V.: Some organizational issues of distance education in universities. In: 2017, V International scientific and practical Conference «Theory and Practice of Using the Moodle Learning Management System» (2017). (in Ukrainian)
17. Moodle Homepage. https://stats.moodle.org. Accessed 08 Feb 2024

18. Moore, M., Kearsley, G.: Distance Education: A Systems View. Wadsworth Publishing Company, Belmont (1996)
19. Nenko, Y., Kybalna, N., Snisarenko, Y.: The COVID-19 distance learning: insight from Ukrainian students. Rev. brasileira de educado do campo-Brasilian J. Rural Educ. **5**, 1–19 (2020). https://doi.org/10.20873/uft.rbec.e8925
20. Ohiienko, O.I.: Distance pedagogical education: foreign and domestic experience. methodological guidelines. Natl. Acad. Pedagogical Sci. Ukraine (2012). (in Ukrainian)
21. Praveen, P., Jebakumar, A.A.: Conscientious study of blended learning and cardinal tools. J. Educ. Teach. Trainers **13**(4), 7–12 (2022). https://doi.org/10.47750/jett.2022.13.04.002
22. Rumble, G.: The Planning and Management of Distance Education. 1st edn. Routledge, London (1986). https://doi.org/10.4324/9780429288661
23. Sydorenko, T.M., Petkun, S.M., Novak, O.M.: Distance education: problems and prospects. Innov. Pedagogy **35**, 45–48 (2021). (in Ukrainian)
24. Tsilmak, O., Revenko, N., Kryvun, N., Fedorenko, O., Tsisaruk, I.: Prospects for the development of distance education in Ukraine: methodological aspect. Eduweb-revista de tecnologia de informacion y comunicacion en educacion **17**(1), 157–166 (2023). https://doi.org/10.46502/issn.1856-7576/2023.17.01.15
25. Vermisli, S., Cevik, E., Cevik C.: The effect of perceived stress and digital literacy on student satisfaction with distance education. Rev. da escola de enfernagen da USP **56** (2022). https://doi.org/10.1590/1980-220X-REEUSP-2021-0488en
26. Vyshnivskyi, V.V., Hnidenko, M.P., Haidur, H.I., Ilin, O.O.: Organization of distance education. Creation of electronic training courses and electronic tests: textbook. State University of Telecommunications, Kyiv (2014). (in Ukrainian)
27. Wang, L., Li, F.: Who gets the highest return to distance higher education? Int. Rev. Res. Open Distrib. Learn. **24**(3), 164–184 (2023). WOS:001104862300010
28. Wu, J.-H., Tennyson, R.D., Hsia, T.-L.: A study of student satisfaction in a blended e-learning system environment. Comput. Educ. **55**(1), 155–164 (2010). https://doi.org/10.1016/j.compedu.2009.12.012
29. Yaroshenko, T.O.: Distance learning in the system of higher education: modern trends. Inzhenerni ta osvitni tekhnolohii **7**(4), 8–21 (2019). https://doi.org/10.30929/2307-9770.2019.07.04.01 (in Ukrainian)
30. Yilmaz, R.K., Karadag, N.: Effects of emergency distance education on teacher training process in turkey: instructors' opinions. Educ. Sci. **13**(9), 1–24 (2023). https://doi.org/10.3390/educsci13090920

Reform and Exploration of the Course "Materials Chemical Synthesis and Characterization Experiments" Based on Outcome-Based Education

Yuan Zhang and Hu Yang(✉)

School of Materials Science and Engineering, Wuhan University of Technology, Wuhan 430070, China
983811939@qq.com

Abstract. In the context of the new engineering discipline, based on the Outcome-based Education orientation, this article takes "Materials Chemistry Synthesis and Characterization Experiment" as an example to analyze the main problems in the course, puts forward improvement measures, and establish a more reasonable course evaluation system by enriching teaching forms and updating experimental content. Finally, a course system of "teaching promotes learning, learning promotes evaluation, and evaluation promotes teaching" has been formed, achieving good teaching results.

Keywords: Experimental teaching · Outcome-based Education · curriculum reform · curriculum system

1 Introduction

The reform of engineering education has always been a key issue of global concern. The rapid growth of engineering graduates has always been a core concern in the field of engineering education on how to ensure the quality of their learning. Whether students have sufficient professional ability, such as problem-solving ability, life-long learning, communication skills, leadership skills, and teamwork skills, is also the top priority for many companies when recruiting engineering graduates. In this context, the Outcome-based education (OBE) system has emerged as a panacea for high-quality engineering education [1–3].

OBE is a new educational model that is student-centered, out-put oriented, and emphasizes continuous improvement. This concept was proposed by William G. Spady in 1994 [4]. Student-centered learning [5, 6] has three basic elements: curriculum objectives, curriculum teaching, and curriculum evaluation. The course objectives describe the expected goals that students must achieve at the end of the course. Course teaching is a way to help students achieve the expected goals. Finally, course evaluation measures students' performance and serves as a basis for continuous improvement. Setting

Z. Hu et al. (Eds.): ICCSEEA 2024, LNDECT 242, pp. 702–714, 2025.
https://doi.org/10.1007/978-3-031-84228-3_61

course objectives in advance can help students understand their learning goals and facilitate their learning. Teachers can also have diverse teaching methods and design various student-centered learning activities to stimulate their initiative and creativity. The role of teachers has also shifted from being a single knowledge professor to guiding students to explore knowledge independently.

In recent years, many scholars have conducted curriculum research and reform based on the OBE concept, achieving good results [7–13]. The Department of Mechanical Engineering at University Malaysia Pahang has carried out educational model reforms based on the OBE concept, improving the quality of teaching [7]. Ningbo University of Technology in China has implemented a sustainable teaching model in the field of computer science and technology based on the concept of OBE and the TSEM (Teaching, Learning, Evaluation, and Management) framework, and has achieved good results [12]. Yang Hu and others from Wuhan University of Technology have reformed the material training system based on output orientation, and the training effect has significantly improved after the reform [13].

The Materials Chemistry major is a characteristic major of our school and has passed the certification of Engineering Education in 2022. The course of "Materials Chemistry Synthesis and Characterization Experiments" is a core mandatory course for the Materials Chemistry major, supporting multiple graduation requirements and indicators, and playing an important role in cultivating top-notch talents in the Materials Chemistry major. This article analyzes the main problems in the current experimental curriculum system through investigation and research. Based on the OBE concept, a new "Materials Chemical Synthesis and Characterization Experiment" curriculum system is constructed and put into practice to address these problems.

2 Problems Existing in the Course of "Materials Chemical Synthesis and Characterization Experiments"

(1) The correlation between course teaching objectives and graduation indicators is not strong, and the supporting relationship is not clear.

    In recent years, with the rapid development of technology, the country urgently needs a group of outstanding talents who can lead the development of new materials, which also puts forward higher requirements for the cultivation of materials chemistry professionals. The previous course teaching objectives were no longer able to support the graduation target points well, and due to insufficient comprehensive and in-depth understanding of the graduation target points, the alignment between the course teaching objectives and the graduation target points was not high.

(2) Lack of updates in teaching content, unable to meet the needs of talent cultivation.

    Based on the concept of integrating science and education, experimental courses have added experimental projects such as quantum sensitized solar cells, preparation of nanoribbons, and gas sensing performance testing in 2015, 2016, and 2017, respectively. However, in the past five years, the teaching content has not been updated, and the development of materials science is advancing rapidly. The demand for interdisciplinary studies is strong, and the existing experimental content can no longer meet the needs of professional talent cultivation [14].

(3) The teaching method is single, and the approach to achieving it does not focus on achieving the course objectives and supporting the relevant graduation indicators.

At present, the teaching methods of experimental courses mainly include experimental preview, offline teaching by teachers, independent experiments by students, and teacher guidance. Students mechanically carry out experiments step by step according to the teacher's explanations and experimental lecture notes, lacking innovation and enthusiasm. The teaching method is mainly traditional nanny style teaching, and the teaching process focuses on the teacher's "teaching", while neglecting the student's "learning". The student's innovative and hands-on abilities cannot be cultivated.

(4) The curriculum evaluation system is not perfect enough, the assessment and evaluation are blurred, and there is a lack of effective assessment methods.

Course evaluation, as the core part of internal evaluation, is an important basis for continuous improvement of courses and plays a crucial role in improving teaching quality and enhancing the support of experimental teaching for graduation indicators [15, 16]. The evaluation of the course "Materials Chemical Synthesis and Characterization Experiments" mainly relies on the students' paper experimental report scores as the evaluation basis, and the evaluation method of the course is single, which cannot well match the teaching objectives of the course. The experimental report correction is subjective and prone to mutual learning and plagiarism among students, lacking authenticity and comprehensiveness. It cannot reflect the true learning level of students, accurately evaluate whether the course objectives have been achieved, and provide effective basis for continuous improvement of the course.

(5) Lack of a complete teaching system.

The OBE philosophy practices a student-centered and continuous improvement approach. At present, due to the weak links in various aspects of experimental teaching, a complete and mutually supportive teaching system that can be continuously improved and circulated cannot be formed.

3 Curriculum Reform of "Materials Chemical Synthesis and Characterization Experiments" Based on OBE Concept

Our school's "Materials Chemistry Synthesis and Characterization Experiment" course has a total of 120 h and is a supporting experimental course for professional courses such as "Materials Synthesis and Processing" and "Principles and Applications of Materials Chemistry". This article is based on the OBE concept to reform the course of "Materials Chemical Synthesis and Characterization Experiments", setting new teaching objectives, updating experimental teaching content and textbooks, enriching teaching methods, and improving the assessment system.

(1) Course teaching objectives

This article conducts visits and research to analyze industry demands, and determines the training goals for materials chemistry professionals based on these demands. Based on the talent cultivation goals and graduation index point decomposition of the Materials Chemistry major, course teaching objectives that are more in line with industry needs and graduation index points have been determined. Table 1 shows the supporting relationship between course teaching objectives and graduation index points.

Table 1. Support of Course Teaching Objectives for Professional Graduation Requirements

Course objectives	Graduation indicators supported by course objectives
1. Master the methods of chemical synthesis, preparation and processing, composition, structure and performance characterization of materials, and be able to select technical routes and design experimental plans based on the requirements of experimental objectives for material structure and performance	Be able to select technical routes and design experimental plans based on the structural and performance requirements of new materials in practical applications
2. Able to select the correct method and equipment to construct an experimental system according to the experimental plan, carry out chemical synthesis, preparation, and processing of materials, characterize the composition, structure, and properties of the materials prepared in the experiment, and collect experimental data during the experiment to improve students' experimental skills and hands-on abilities	Ability to construct experimental systems, implement experimental plans, and collect experimental data based on experimental plans
3. Able to apply professional knowledge and techniques in materials chemistry, materials science, and engineering to summarize, analyze, and interpret experimental data, obtain reasonable and effective experimental conclusions through information synthesis, write experimental reports, deepen students' understanding of the interrelationships between material chemical synthesis, preparation technology, material composition, structure, performance, etc., and improve their hands-on ability and problem-solving ability	Be able to apply professional knowledge and technology to analyze and interpret experimental data, draw reasonable and effective experimental conclusions through information synthesis, and write papers or reports
4. During the experiment, clearly express one's problem-solving solutions, understand different opinions and communicate effectively, share experimental information with group members, and work together to complete the experiment	Having team awareness and collaboration skills, able to effectively communicate with team members, understand the importance of the team, share information with other members, and work collaboratively

(continued)

Table 1. (*continued*)

Course objectives	Graduation indicators supported by course objectives
5. During the experiment, be able to independently complete the tasks assigned by the experimental group and be competent in assuming responsibilities within the experimental group	Capable of independently completing assigned tasks in a multidisciplinary team and capable of assuming responsibilities within the team

(2) Teaching content

The modern materials research industry is developing rapidly, with various new materials and preparation and characterization methods emerging one after another [17–19]. Timely updating of experimental content to keep up with technological development in experimental teaching is necessary to cultivate innovative talents needed by the industry. Therefore, based on the OBE education philosophy (output oriented) and curriculum teaching objectives, combined with the current trend of new material development, this study has updated the experimental teaching content. The updated experimental projects are shown in Table 2. The entire experimental course consists of 9 experimental projects, including 2 design research experimental projects and 7 comprehensive experimental projects, covering the fields of nanomaterials, optoelectronic materials, gas sensitive materials, etc., including various chemical synthesis methods and performance characterization methods.

Compared with the original experimental content, some repetitive experimental hours have been reduced, such as "Electrochemical Performance Analysis of Inorganic Oxide Thin Films" and "Comprehensive Analysis of Lithium Battery Materials and Devices Preparation and Related Performance", which are both studies on the electrochemical performance of materials, and the content is repetitive. Added virtual simulation experiments such as "Preparation and Performance Analysis of Semiconductor Chips", "Melting Preparation of High Temperature Alloys", and "Preparation, Adsorption, and Optical Performance Analysis of Graphene Oxide". Semiconductor chip technology is currently a hot and difficult issue in scientific research, and the industry urgently needs talents related to semiconductor chips. However, semiconductor chip technology experiments are difficult to complete in the laboratory. Therefore, this course adds a project on "Preparation and Performance Analysis of Semiconductor Chips". Through virtual simulation experiments, students can have a close and in-depth understanding of the preparation and performance analysis of semiconductor chips. High temperature alloy materials are key materials for the core components of rocket engines, combustion chambers, and turbopumps. Graphene oxide materials have broad application potential in various fields such as medicine and optoelectronics, and are currently a hot topic in scientific research. Through these three experiments, students can understand and master the preparation and characterization methods of new materials, and stimulate their interest in learning.

At the same time as updating the teaching content, the experimental supporting textbooks have been changed from traditional listing of multiple experimental projects to summarizing and classifying materials, preparation methods, and testing methods. Students can independently choose appropriate preparation and testing methods based on the experimental content to design experimental plans. For example, in the preparation experiment of graphene oxide, there are multiple preparation routes for graphene oxide, and students can independently choose the preparation route to complete the experiment.

Table 2. Updated Experimental Projects

Serial Number	Entry name	Project	Project duration (h)
1	Preparation of ultrafine powder and analysis of semiconductor properties	Design research oriented	16.5
2	Preparation of graphene oxide and testing of its adsorption and optical properties	Design research oriented	13.5
3	Solid phase synthesis, polarization, and piezoelectric performance analysis of dielectric ceramics	Integrative	16.5
4	Preparation and structural analysis of organic-inorganic molecular assembly materials	Integrative	13.5
5	Preparation of Nanobelts and Comprehensive Analysis of Gas Sensitivity Performance	Integrative	13.5
6	Melting preparation, structural characterization, and performance testing of high-temperature alloys	Integrative	13.5
7	Comprehensive Analysis of Lithium Battery Materials and Devices Preparation and Related Performance	Integrative	13.5
8	Preparation and Optoelectronic Performance Analysis of Quantum Dot Sensitized Solar Cells	Integrative	13.5
9	Virtual simulation experiment of functional material preparation	Integrative	6

In addition, cultivating students' awareness of experimental safety and enhancing their experimental literacy play a crucial role in enhancing their hands-on ability. The traditional experimental teaching mainly aims to teach students relevant professional knowledge and skills, neglecting the cultivation of experimental norms, experimental literacy, and laboratory safety emergency capabilities, resulting in poor experimental habits, insufficient consideration of experimental safety issues in experimental plans, weak emergency response capabilities for sudden safety incidents during the experimental process, and insufficient hands-on ability. And what enterprises need are high-quality all-round engineering talents. Therefore, we introduce the 6S (Seiri, Seiton, Seiso, Seiketsu, Shitsuke, Safety) management concept of enterprises into undergraduate experimental teaching. While improving laboratory management level, we cultivate students' safety awareness, enhance their professional experimental literacy, and enhance their hands-on ability.

(3) Teaching methods

With the rapid development of information technology and the continuous improvement of online teaching platform functions, the blended teaching method of online and offline has been widely applied [20]. This course uses the assistant engineering course teaching platform to change the original offline "cramming" teaching mode to an experimental teaching method of "combining online and offline", thereby improving teaching effectiveness and enhancing students' self-learning ability. The teaching of this course is divided into three stages, and the specific implementation process is shown in Fig. 1.

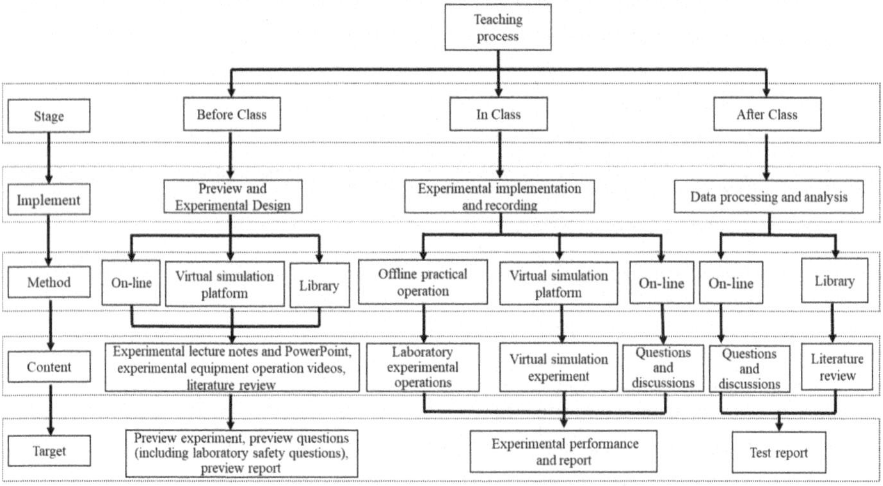

Fig. 1. Teaching flowchart

Before class, the main task is for students to complete preview and experimental design. Before the experiment begins, the teacher will build online learning resources (teaching materials, videos, and preview questions, etc.). Students can preview the experimental content in advance and complete corresponding questions (including laboratory safety questions) in the system. At the same time, relevant literature should be consulted to complete the design and implementation plans of each experiment. The experiment can only be conducted after the experimental plan is approved by the supervising teacher and the experimental preview report is completed. Students can preview online, and teachers can monitor their preview progress in real-time. At the same time, students can also learn independently on the platform anytime and anywhere, greatly improving learning efficiency and effectiveness.

In class, the main focus is on experimental implementation and recording (offline experiments, virtual simulation platform experiments): using the undergraduate experiment check-in mini program to record the attendance of students in real time. The experiment is conducted in groups of 2–3 people, and each group of students is responsible for division of labor and cooperation, while also independently completing the corresponding work. Teachers supervise and guide students in class. Before the experiment, the teacher will post on the online platform any detailed problems or issues that students may encounter during the experiment that require them to focus on. Students can follow up on the problems and express their own opinions, and discuss and interact with the teacher and classmates on the platform. The platform also has a like function, which adds fun to classroom discussions and motivates students. The virtual simulation platform can use virtual simulation technology for device modeling, optical path simulation, micro particle motion demonstration, etc., to visually display the internal structure of the device that cannot be touched and the fleeting micro experimental phenomena to students. By using "practice mode" and "exam mode", students can fully carry out experimental exploration and stimulate their learning interest through different simulation scenarios [21].

After class, students mainly calculate and analyze the data measured in the experiment, and independently complete thinking questions by consulting reference materials. At this stage, students can also ask questions online and discuss and answer them together. This solves the problem of teachers needing to find time and space to answer questions in the past, and many questions may be common to most students. Teachers answer them in written form on the platform, which can be seen by all students, greatly improving teaching efficiency.

(4) Evaluation system

Experimental class is different from theoretical class. It is a course that focuses on cultivating students' hands-on ability, innovation ability, problem-solving ability, teamwork ability, and coordination and communication ability. It should focus more on process assessment. The difficulty of process assessment lies in how to quantify it. Therefore, combining online and offline teaching modes, detailed formulation of observation points and assessment objectives for course teaching has been carried out, process observation points have been refined, and the corresponding relationship between observation points and teaching objectives has been clarified. Assessment standards have been quantified, providing a basis for course evaluation. Table 3 shows the process assessment observation points and assessment objectives.

Table 3. Observation points and assessment objectives

Process	Observation	Assessment objectives	Score ratio(%)
Preview and Experimental Design	Online platform preview of experimental content	The level of understanding of the experimental purpose and basic principles; Consider social, environmental, safety, health, cost, and other aspects in the design of the experimental plan Supporting Goals 1 and 4	5
	Preview questions		10
	Preview report		15
Experimental implementation and process recording	Attend experiments on time	Emphasize the examination of basic operating standards and skills. Test students' learning attitude, rigor, unity and cooperation, etc Supporting Goals 2 and 5	3
	Familiarity with the experimental process		5
	The level of standardization of basic operations		5
	Emergency response capability for unexpected events		5
	The completeness of the original experimental records		5
	The spirit of unity and cooperation among classmate		5
	Organize experimental supplies as required after the experiment is completed		2
Result evaluation and discussion	Is the analysis result using original recorded data	Test students' ability to process and analyze experimental data. Comprehensive application ability of professional knowledge. Ability to seek truth from facts and critical thinking Supporting Goal 3	10
	Is the calculation result correct and the analysis reasonable		10

(*continued*)

Table 3. *(continued)*

Process	Observation	Assessment objectives	Score ratio(%)
	Is there any analysis, comparison, or judgment between the various contents of the comprehensive experiment		10
Other	Understand the relevant standards/manuals/specifications and other reference materials involved in the experiment. Reflection questions	Assess students' ability to access information and deepen their understanding of relevant standards/manuals/norms Comprehensive ability Supporting Goals 3 and 4	10

In addition, compared to the previous simple evaluation of student learning through experimental reports, the online teaching platform system will record the learning time of each student, the completion of preview questions, and the interaction between students and teachers during and after class. It can also draw portraits based on each student's learning situation, providing a more direct, authentic and effective basis for course evaluation, and providing a basis for continuous improvement of subsequent courses. Enable continuous improvement measures to be implemented effectively.

4 The Effectiveness of Curriculum Reform

This study was conducted in the School of Materials Science and Engineering at Wuhan University of Technology. Taking the Materials Chemistry major at a certain level as an example, the teaching effect after the curriculum reform was analyzed. In order to ensure that students meet the training requirements, an analysis is conducted on the degree of achievement of their abilities (see Table 4). From Table 4, it can be seen that the knowledge, abilities, and qualities of students have been cultivated after the curriculum reform, and the training goals have been achieved, resulting in an increase in student satisfaction.

Table 4. Course achievement analysis

Course objectives	Evaluation basis	Evaluation value					
		Full mark E_1	Conversion ratio F_1	Conversion points A_1 $=(E_1 \times F_1)$	Sample average scoreG_1	Sample conversion scoreB_1 $=(G_1 \times F_1)$	Attainment value C $\sum B_1/\sum A_1$
1. Master the methods of chemical synthesis, preparation and processing, composition, structure and performance characterization of materials, and be able to select technical routes and design experimental plans based on the requirements of experimental objectives for material structure and performance.	Design research experiments (preview and experimental design), comprehensive experiments (preview and experimental design)	30	1	30	27.765	27.765	C_1=0.926
2. Capable of selecting the correct method and equipment to construct an experimental system according to the experimental plan, conducting chemical synthesis, preparation, and processing of materials, characterizing the composition, structure, and properties of the materials prepared in the experiment, and collecting experimental data during the experiment to improve students' experimental skills and hands-on abilities	Design research experiments (experimental implementation and process recording), comprehensive experiments (experimental implementation and process recording)	30	1	30	27.046	27.046	C_2=0.902
3. Being able to apply professional knowledge and techniques in materials chemistry, materials science, and engineering to summarize, analyze, and interpret experimental data, draw reasonable and effective experimental conclusions through information synthesis, write experimental reports, deepen students' understanding of the interrelationships between material chemistry synthesis, preparation technology, material composition, structure, performance, etc., and improve their hands-on ability and problem-solving ability.	Design research experiments (result evaluation and discussion, others), comprehensive experiments (result evaluation and discussion, others)	40	1	40	33.613	33.613	C_3=0.840
4. During the experiment, clearly express one's problem-solving solutions, understand different opinions and communicate effectively, share experimental information with group members, and work together to complete the experiment.	Comprehensive experiments (preview and experimental design, result evaluation and discussion, others)	70	0.7	49	60.218	42.153	C_4=0.860
5. During the experiment, be able to independently complete the tasks assigned by the experimental group and be competent in assuming responsibilities within the experimental group.	Comprehensive experiment (experimental implementation and process recording)	30	0.7	21	26.808	18.766	C_5=0.894
Sample information (Number of students)	Excellent(above 90 points): 11; good(80-89points): 23；average(70-79points): 0；pass (60-69 points): 0；fail(below 60 points): 0.						
Attainment	$D=\frac{C_1+C_2+C_3+\cdots+C_n}{n}\times 100$ =88.44						

5 Conclusion

With the development of technology, society has put forward higher requirements for the talent training system of materials chemistry. In order to cultivate talents that meet the goals of the professional training program, the experimental course of materials chemistry must be reformed from multiple aspects such as course teaching objectives, teaching methods, teaching content, and evaluation methods, in order to improve the experimental course system. Therefore, based on the needs of talent cultivation, this article has revised the curriculum teaching objectives, enriched teaching methods, updated the curriculum teaching content, and improved the curriculum evaluation system. After the implementation of this curriculum system, the degree of achievement of the curriculum objectives has increased, and the support of the curriculum objectives for graduation indicators is stronger.

Acknowledgments. This work has been supported by the teaching reform and research project of Wuhan University of Technology (Grant No. W2022005 and W2022061).

References

1. Eldeeb, R., Nisha, S.: Outcome based education (OBE) - trend review. IOSR J. Res. Meth. Educ. **2**(1), 9–11 (2013)
2. Zamir, M.Z., Abid, M.I., Fazal, M.R., et al.: Switching to outcome-based education (OBE) system. IEEE Trans. Educ. **4**(65), 695–702 (2022)
3. Zeynal, H., Zakaria, Z., Anisseh, M., et al.: Strategic implementation of outcome-based education system in Buein-Zahra Technical University of Iran. In: Proceedings of the 2017 IEEE 9TH International Conference on Engineering Education (IEEE ICEED 2017), pp. 122–127 (2017)
4. William, G.S.: Outcome-based Education Critical Issues and Answers. American Association of school administrators, New York (1994)
5. Felder, R.M., Brent, R.: Designing and teaching courses to satisfy the ABET engineering criteria. J. Eng. Educ. **92**, 7–25 (2003)
6. Gamboa, R., Namasivayam, S., Al-Atabi, M., et al.: Quantitative measurement of students PO attainment for Taylor's university engineering programmes. Proc. Soc. Behav. Sci. **103**, 753–762 (2013)
7. Noor, M.M., Rahman, M.M., Rejab, M.R.M.: Education reform model at faculty of mechanical engineering, University Malaysia Pahang. Int. J. Recent Trends Eng. **5**(1), 166–171 (2009)
8. Chan, M.K., Wang, C.C., Arbai, A.A.: Development of dynamic OBE model to quantify student performance. Comput. Appl. Eng. Educ. **5**(30), 1293–1306 (2022)
9. Rahimullah, B.N.S., Mansor, et al.: Analysis of graduates performance based on programme educational objective assessment for an electrical engineering degree. Asian J. Univ. Educ. **16**(3), 303–309:1823–7797 (2020)
10. Zhang, Z., Cheng, L., Li, L., et al.: Curriculum reform of "signal and system based on OBE teaching concept. J. Shenyang Norm. Univ. (Natl. Sci. Ed.) **37**(3), 284–288 (2019). (in Chinese)
11. Li, L., Yang, L., Wei, M., et al.: Exploration of teaching reform in bioengineering majors based on OBE education concept. J. Zhejiang Univ. Technol. **4**, 452–458 (2016). (in Chinese)

12. Zheng, W., Wen, S., Lian, B., et al.: Research on a sustainable teaching model based on the OBE concept and the TSEM framework. Sustainability **5656**(15), 5656 (2023)
13. Yang, H., Cao, Q.: Research and practice of outcome-oriented innovative talent training system of material specialty. Adv. Comput. Sci. Eng. Educ. VI **181**, 972–983 (2023)
14. Gu, S., Zhu, Z., Lei, L., et al.: Exploration and practice of the integration of science and education in the cultivation of innovative talents in materials chemistry undergraduate majors. Educ. Teach. Forum **46**, 128–129 (2019). (in Chinese)
15. Zhao, J., Xu, Y., Pan, H., et al.: Effectiveness of clinical classroom teaching evaluation system. Basic Clin. Med. **42**(3), 529–532 (2022). (in Chinese)
16. Wang, S.: Evidence-based evaluation for instruction quality: taking the instruction quality assessment of mathematics in the USA as an example. J. China Exam. **2**, 73–80 (2022). (in Chinese)
17. Qian, J., Nam, P.N., Yutaka, O., et al.: Introducing self-organized maps (SOM) as a visualization tool for materials research and education. Results Mater. **4**, 100020 (2019)
18. Elistratkin, M.Y., Lesovik, V.S., Zagorodnjuk, L.H., et al.: New point of view on materials development. IOP Conf. Ser. Mater. Sci. Eng. **327**(3), 032020 (2018)
19. Wibawa, B., Syakdiyah, H., Kardipah, S.: Technology development in thermochemistry materials. J. Phys. Conf. Ser. **1869**(1), 012023 (2021)
20. Lin, J.: The transformation and upgrading of traditional engineering majors under the fourth wave of industrial revolution. Res. High. Educ. Eng. **4**, 1–10 (2018). (in Chinese)
21. Yang, S., Jin, W., Zhou, J.: Design and implementation of photolithography etching virtual simulation experiment. Educ. Teach. Forum **17**, 41–44 (2023). (in Chinese)

Developing Emotional Intelligence of Future Teachers

Anatolii Pashko[1] and Iryna Pinchuk[2](✉)

[1] Department of Theoretical Cybernetics, Taras Shevchenko National University of Kyiv, Kyiv 01601, Ukraine

[2] Educational and Scientific Institute of Pedagogy and Psychology, Oleksandr Dovzhenko Hlukhiv National Pedagogical University, Hlukhiv 41400, Ukraine

pinchuk@gnpu.edu.ua

Abstract. Today, the issue of emotional intelligence of Ukrainians is particularly important, especially the problem of developing the emotional intelligence of future teachers. Teachers will prepare students to withstand any emotional stress in the future, so they need to have a high level of emotional intelligence themselves. In the study, we first tested future teachers' understanding of emotional intelligence and later applied strategies to develop each component of the studied concept in future teachers. The following components of emotional intelligence of future teachers were considered in the research: self-awareness, self-regulation, social awareness, social skills. We used several techniques to improve each of the components of emotional intelligence of university students. Three levels of emotional intelligence were defined: high, average and low. In order to determine the levels of emotional intelligence development, we used the method of D. Mayer, P. Selovey and D. Caruso - MSCEIT, V2.0. The final stage of the study was to test the effectiveness of the proposed strategies in developing emotional intelligence in university students by means of the Mann-Whitney U-test. The results show that the strategies used are effective.

1 Introduction

The problem of developing the emotional intelligence (EI) of future teachers has always been important, as it affects the mental health and development of students. This aspect is especially important today in Ukraine. Nowadays the issue of preserving a person's psychological and spiritual balance has become extremely relevant because in the context of aggression, when the lives and well-being of citizens are under threat, the mental state of the population becomes a problem that is important not only for psychologists but also for the entire community. Therefore, the issues of the relationship between teacher-student communication and EI have become a priority both for undergraduates and lecturers. Today, all of us – university professors – should be psychologists and psychotherapists who, through our work and communication with students, help young people maintain their psychological health and activate the structures of EI. This happens through the development of emotional awareness, the ability to recognize and express emotions, and to effectively manage stress and trauma.

Z. Hu et al. (Eds.): ICCSEEA 2024, LNDECT 242, pp. 715–727, 2025.
https://doi.org/10.1007/978-3-031-84228-3_62

2 Literature Review

Interest in the development of EI in future professionals has increased in recent years. The paper joins a growing body of literature that employs both theoretical and practical issues of future specialists' emotional intelligence research.

The OECD 2030 Learning Compass [1] suggests key types of skills, necessary to modern employee to meet complex requirements, which include self-regulation, social and emotional skills, empathy, self-efficacy. Social and emotional skills are recognised as necessary for workers whose jobs require complex social interactions that cannot be replaced by technology. Such occupations include the teaching profession. In order to adapt to technological advances, workers therefore need to develop social skills, including persuasion and negotiation [2].

In their seminal paper American scientists J. Mayer, P. Salovey and D. Caruso [3] reveal the notion of EI. According to their research the concept refers specifically to the cooperative combination of intelligence and emotion. Moreover, they suggest it as a kind of intelligence including the social, practical, and personal ones. The scientists described the nature of EI and developed the model, which consist of four branches and proposed the means of its measurement during study.

More recent evidence of M. Andrew [4] research shows that there are four components of EI: self-awareness, self-regulation, social awareness, and social skills. She highlights three main steps of improving EI. They are recognizing own emotions including naming them, asking for feedback and reading literature with complex characters. F. Cassano and etc. [5] had made a survey among nursing and midwifery students that proved the relationship between their EI, sociodemographic and academic variables. The authors believe EI is significantly related to gender because women showed higher scores on the EI factors. In addition, they show EI is closely related to a person's individual characteristics, is different from technical and professional skills, is also a predictor of professional success. R. Gilar-Corbi, T. Pozo-Rico, B. Sánchez and J.-L. Castejón [6] searched the improving EI by means of randomized experimental study of a business-oriented EI training program for senior managers.

In their systematic review N. Jiménez-Picón and co-workers [7] consider EI is an important trait and skill for healthcare specialists. Mindfulness meditation has been shown to be productive in enhancing the well-being of professionals, leading to improved mental health, self-care, and professional satisfaction. In the research the latest evidence on the relationship between mindfulness and EI among healthcare professionals and students were identified. These studies have found a positive relationship between mindfulness and EI, specifically the ability to regulate emotions. In addition, mindfulness is negatively associated with emotional exhausting. The authors proved mindfulness-based training interventions are useful in promoting emotional balance, awareness, acceptance, recognition, expressive suppression, and reducing emotional exhaustion.

Many attempts of researching this problem have been made by Ukrainian scientists. In particular, the Ukrainian scientists [8–11, 13, 14] established that the modern labor market needs workers with a set of important skills, among which EI is one of the key competences. They propose using active teaching methods ("openwork saw," "pyramid," "brainstorming," etc.) with the aim of developing EI of future vocational education specialists.

The researchers from Ukraine O. Banit and O. Merzliakova [9] study improving the resilience skills of teachers. The authors use the multidimensional model of overcoming stress and finding internal stability "BasicPh" of Prof. Mooli Lahad to identify the key resilience strategies of Ukrainian teachers in difficult conditions. The results of the study showed that the most influential way to overcome stress for teachers is their cognitive settings and the worldview of teachers.

Much work on the potential of emotional intelligence developing has been carried out; however there are still some critical issues, noted in the papers of scientists [10–15].

3 Aim

Having analysed the above mentioned works and teaching practice at Ukrainian universities, we can say that the problem of developing EI of future teachers is urgent nowadays. So, the purpose of the paper is to research the EI of undergraduates (future teachers) in Ukrainian universities. In order to achieve the main goal of the research, we should solve the objectives: a) to learn about the understanding of the concept of EI by university students; b) to implement strategies to improve each of the components of EI of future teachers; c) check the effectiveness of strategies of building EI by means of mathematical statistics methods.

4 Results

Based on the results of the analyzed literature [3, 5, 8], we can conclude that EI is the ability to understand, manage and express one's own emotions, as well as to effectively communicate and interact with other people, taking into account their emotional states. This includes the ability to recognize one's own emotions, understand their origins and impact on behavior, as well as the ability to empathize with others, perceive their emotions, and respond appropriately to them. Emotional intelligence is an important component of successful social adjustment, leadership, interpersonal relationships, and overall well-being. People with a high level of emotional intelligence usually cope better with stress, conflict and difficult situations, and have more successful relationships with other people.

Most scientists [3–8, 11–15] define the following main components of emotional intelligence: self-awareness, self-regulation (emotional management), social awareness (empathy), and social skills (communication). So, based on their views, we distinguish four levels of emotional intelligence.

According to the psychologist [16], people with high level of EI don't wait for someone to improve their mood and bring them pleasure. They satisfy their own emotional needs on their own: they cheer, comfort, and calm themselves. For example, if you recognize the emotional state of a colleague, you will try to help and solve the problems (or rejoice for him / her). The other person, looking at you, can also make a rough guess about your state and understand what you can talk about and what should be discussed later.

5 Methods

First, we conducted a survey of university students to determine their understanding of the concept of EI. The results of the survey are shown in Table 1.

Table 1. The answers of future teachers on questions about emotional intelligence before experimental work

Definition of emotional intelligence	Relative scale (Number of answers of the undergraduates)	Absolute Scale (%)
A) EI is the ability to recognize and understand one's own emotions and their influence on thoughts and actions	43	33,09
B) EI is the ability to effectively control one's emotions, including the ability to calm oneself in stressful situations and manage one's mood	46	35,38
C) EI is the ability to empathize with others, understand their feelings and experiences, and feel their joy or pain	9	6,92
D) EI is the ability to effectively express one's emotions and ideas in communication with other people, and to listen to and understand their emotional signals	24	18,46
E) EI is the ability to understand, use, and manage your own emotions in positive ways to relieve stress, communicate effectively, empathize with others, overcome challenges and defuse conflict	8	6,15

After that, there was a series of lectures and practical material on the understanding of the components and the filling of EI. We then conducted another survey to find out the effectiveness of the lectures on theoretical material. The results of the second survey are shown in Table 2.

After the second survey, we used the Mann-Whitney U-Test to evaluate the difference between two samples of university students. We consider the hypotheses: H_0 – the difference is not significant; H_1 – the difference is significant.

This was done by combining the samples and ranking them as follows (Table 3).

Table 2. The answers of future teachers on questions about emotional intelligence after experimental work

Definition of emotional intelligence	Relative scale (Number of answers of the undergraduates)	Absolute Scale (%)
A) EI is the ability to recognize and understand one's own emotions and their influence on thoughts and actions	16	12,31
B) EI is the ability to effectively control one's emotions, including the ability to calm oneself in stressful situations and manage one's mood	19	14,62
C) EI is the ability to empathize with others, understand their feelings and experiences, and feel their joy or pain	28	21,54
D) EI is the ability to effectively express one's emotions and ideas in communication with other people, and to listen to and understand their emotional signals	36	27,69
E) EI is the ability to understand, use, and manage your own emotions in positive ways to relieve stress, communicate effectively, empathize with others, overcome challenges and defuse conflict	31	23,84

$$U_{em} = n_1 n_2 + \frac{n_x(n_x + 1)}{2} - T_x = 5 * 5 + \frac{5 * 6}{2} - 28 = 12 \quad (1)$$

$n_1 = n_2 = 5$ – sample size;

$n_x = 5$ – sample size with maximum rank sum;

$T_x = 28$ – maximum rank sum;

Critical values of the Mann-Whitney test for the level of significance $\alpha = 0.05$

$$U_{kr1} = 17, U_{kr2} = 38$$

$U_{em} < U_{kr1} < U_{kr2}$ – we accept hypothesis H_1 – the difference between the samples is significant.

As we can see, the difference of understanding the concept "emotional intelligence" by the future teachers before and after lectures is significant (Fig. 1).

Table 3. Calculation results

Sample number	Value	Rank	Ranks of the first sample	Ranks of the second sample
1	8	1	1	
1	9	2	2	
2	16	3		3
2	19	4		4
1	24	5	5	
2	28	6		6
2	31	7		7
2	36	8		8
1	43	9	9	
1	46	10	10	
Total			**27**	**28**

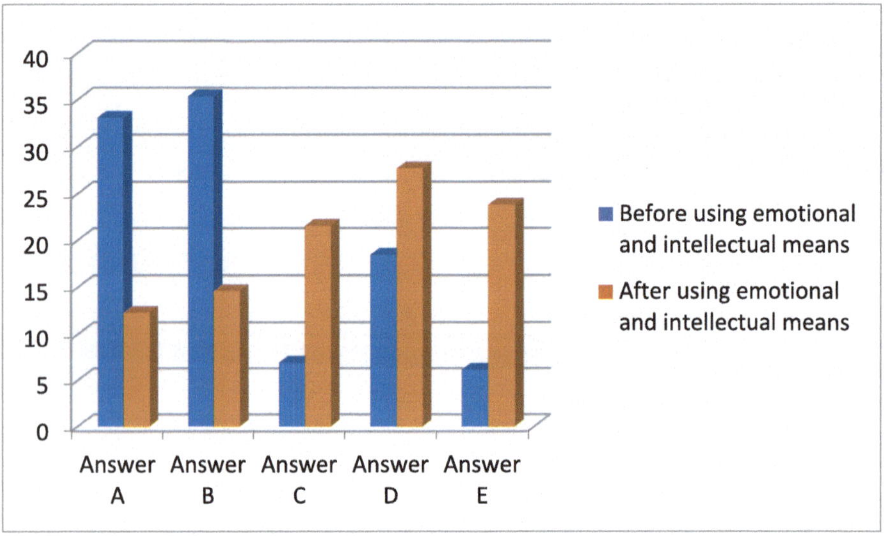

Fig. 1. The results of ascertainment stage of the experiment.

After that it was conducted the formative stage of the experiment, during which we practised different strategies with the aim of developing all the components of students' EI.

We used several techniques (Table 4) to improve each of the components of EI.

In the research, we practiced the Myers-Briggs Type Indicator, the Big Five personality traits for personality self-assessment and the Gallup StrengthsFinder for strengths assessment to gain insight into a student's personality, preferences and strengths. The strategy of challenging assumptions and beliefs made it possible to examine whether the beliefs, biases and assumptions were based on evidence or whether they were limiting growth. Being open to challenging and updating beliefs can expand *self-awareness*. In particular, mindfulness practice was aimed at becoming more aware of one's thoughts, emotions and bodily sensations in the present moment. Reflective journaling helped students gain insight into patterns and triggers that influence their actions and emotions. The

Table 4. The components of future teachers' emotional intelligence

The components of emotional intelligence	The meaning of the components	The strategies of improving the components
Self-awareness	Self-awareness is the ability to recognize and understand one's own emotions and their influence on thoughts and actions. Accordingly, we suggest cognition and understanding of oneself at the first level: what emotions are inherent in a person and how they manifest themselves in different situations	• Mindfulness Practice • Reflective Journaling • Seeking Feedback • Mindful listening • Self-assessment tools • Challenge assumptions and beliefs • Practice self-compassion • Therapy or coaching
Self-regulation	Emotional management is the ability to effectively control one's emotions, including the ability to calm oneself in stressful situations and manage one's mood. Therefore the second level of EI manifestation is the ability to manage reactions and emotions. This is a set of tools that help a person adjust the emotional state	• Mindfulness Meditation • Develop a routine • Set clear goals • Practice delayed gratification • Develop coping strategies • Create action intentions • Practice self-reflection • Seek social support • Develop resilience
Social awareness	Empathy is the ability to empathize with others, understand their feelings and experiences, and feel their joy or pain. So, understanding others is the third level of emotional intelligence that helps to understand how other people behave / look when they are sad, happy, anxious, etc	• Practice active listening • Developing empathy • Observing non-verbal cues • Ask for feedback • Participating in group activities • Reflect on social interactions

(*continued*)

Table 4. (*continued*)

The components of emotional intelligence	The meaning of the components	The strategies of improving the components
Social skills	Social skills (productive communication) are the ability to effectively express one's emotions and ideas in communication with other people, and to listen to and understand their emotional signals. In accordance, we propose mutual regulation as the fourth level of EI, including conflict resolution	• Initiating conversations and developing small talk skills • Making eye contact and using positive body language • Expressing appreciation and gratitude • Practice assertiveness • Join social groups or clubs • Observe social role models • Offer support • Collaborate on problem solving • Build trust • Practice flexibility • Celebrate successes together • Seek professional support

strategy of seeking feedback provided valuable insights into strengths, weaknesses and blind spots that students may not be aware of. Practicing active listening during conversations with others taught them to pay attention to both verbal and non-verbal cues, and to try to understand the perspectives and emotions of the people they're interacting with. Practicing self-compassion helped students to treat themselves with kindness and understanding, especially in times of difficulty or failure, to acknowledge mistakes without harsh self-criticism, and to focus on learning and growth. In addition, for the most challenging situations, we used therapy or coaching with some students, providing objective feedback, guidance and tools to increase self-awareness. By incorporating these strategies into their daily lives, students can cultivate a deeper understanding of themselves, leading to personal growth, improved relationships and greater overall well-being.

Improving *self-regulation* – the ability to effectively manage and control one's thoughts, emotions and behaviours – is crucial for personal and professional success. During the research, we introduced mindfulness meditation to cultivate awareness of one's thoughts, emotions and bodily sensations. Mindfulness helps students observe their impulses without acting on them immediately, giving them the space to choose how to respond. Establishing a daily routine helped students to regulate their behaviour and manage their time more effectively. Consistent routines reduce decision fatigue and provide structure, making it easier to stay focused and disciplined. In addition, setting clear goals allowed learners to define specific, achievable goals and break them down into manageable steps that provided direction and motivation, helping them to stay on track and regulate their efforts to achieve them. Another useful strategy – the practice of delayed gratification – trained students to delay immediate rewards in favour of long-term goals. This involved resisting temptations and impulses that might distract them from their goals. Techniques such as deep breathing, progressive muscle relaxation or taking short breaks helped students regulate their emotions and maintain composure in difficult situations. Students regularly monitored their own thoughts, emotions and behaviours to identify patterns and triggers that lead to impulsive or undesirable actions through self-monitoring. Specific plans for how students will respond to potential obstacles or temptations were formulated using the strategy of creating action intentions. The practice of self-reflection increased self-awareness and helped students to make more informed decisions in the future. The strategy of seeking social support allowed them to surround themselves with supportive friends, family members or mentors who could encourage them and hold them accountable to their goals. Sharing challenges and progress with others can provide valuable perspective and motivation to stay disciplined. Future teachers cultivated resilience by adopting a growth mindset and reframing setbacks as opportunities for learning and growth. Resilient individuals are better able to bounce back from adversity and stay focused on their goals despite obstacles.

Improving *social awareness* is crucial to building strong interpersonal relationships and effective communication. We practised active listening by giving full attention to the speaker, maintaining eye contact and focusing on understanding their message without interrupting, paying attention to both verbal and non-verbal cues to understand the speaker's emotions and intentions. Cultivating students' empathy was implemented by trying to understand other people's feelings, thoughts and experiences and validating others' emotions without judgement. Observing non-verbal communication signals such as facial expressions, body language, tone of voice and gestures helped to convey more information than words alone and to understand others' emotions and intentions more accurately. Seeking feedback from trusted friends, colleagues or mentors about their social interactions and communication style, using constructive criticism, was used as an opportunity for self-improvement and to adjust students' behaviour based on others' perceptions. Future teachers participated in group activities or team projects to practice collaboration, cooperation and interpersonal skills, paying attention to group dynamics and social cues in order to manage interactions effectively and build rapport with others. Reflecting on social interactions and evaluating their own communication effectiveness enabled them to understand how words and actions affect others and to identify areas for improvement in social awareness skills.

Improving *social skills* is essential for building and maintaining meaningful relationships, advancing in your career and navigating social situations with confidence. Mutual regulation refers to the ability of individuals to co-regulate their emotions, behaviours and interactions in ways that promote positive relationships and mutual understanding. Improving mutual regulation involves both self-regulation and the ability to help others regulate their emotions and behaviours. Eye contact conveys trust, attentiveness and sincerity, so it is important to practise making appropriate eye contact during conversations. We have worked with students to pay attention to body language, such as posture, facial expressions and gestures, to signal approachability and friendliness, as this can have a significant impact on how others perceive them. Expressing appreciation for others' contributions, kindness or assistance can go a long way in strengthening relationships and fostering goodwill. Participating in social groups, clubs or organisations that match students' interests or hobbies provides opportunities to meet like-minded people and practice social skills in a supportive environment. Paying attention to people who excel in social situations and observing their behaviour, communication style and interpersonal skills made it possible to emulate positive social role models and incorporate their techniques into one's own interactions. Offering support and encouragement to others when they are experiencing difficulties helped to regulate their emotions or behaviour to help them cope with challenges and manage stressful situations. Working together to solve problems helped them to work with others to identify and address problems or conflicts in a constructive way, allowing for mutual input, shared responsibility and the development of effective solutions. Students built and maintained trust in their relationships by demonstrating reliability, consistency and integrity, which allowed for open communication, vulnerability and emotional support. Practicing flexibility and adaptability in interactions with others and adjusting the approach based on students' needs and preferences promoted mutual regulation by allowing for compromise and collaboration. Acknowledging and celebrating successes and achievements together as a team helped university students to develop a sense of camaraderie and strengthened the bond between individuals. Seeking professional support enabled students to overcome challenges that persisted despite their efforts. Some of them considered seeking support from a therapist, counsellor or mediator who could provide guidance and assistance in improving relationship dynamics.

By implementing these strategies, future teachers improved the mutual regulation of their relationships and created a supportive and harmonious environment in which individuals could thrive and grow together. Furthermore, by incorporating these strategies into their daily lives and consistently practising their social skills, they were able to improve their ability to connect with others, build relationships and navigate social interactions with confidence.

EI is acquired gradually. First, undergraduates learnt about themselves and their emotions, and only then about their environment. Accordingly, we distinguished 3 levels of EI: high, average, low. In order to determine the levels of emotional intelligence development, we implemented the method of D. Mayer, P. Selovey and D. Caruso - MSCEIT, V2.0 (The Mayer Salovey Caruso Emotional Intelligence Test) is a standardised test

Table 5. The levels of future teachers' emotional intelligence before and after experimental work

Levels	High		Average		Low	
Relative scale (%)	Before the experiment	After the experiment	Before the experiment	After the experiment	Before the experiment	After the experiment
Self-awareness	18,46	54,62	63,08	39,23	18,46	6,15
Self-regulation	9,23	47,69	58,46	40,77	32,31	11,54
Social awareness	8,46	24,62	56,15	51,54	35,39	23,84
Social skills	16,15	36,15	50,00	45,38	33,85	18,47

to measure emotional intelligence. The test consists of tasks with correct and incorrect answers. The results of the test can be seen in Table 5.

Similarly, we used the Mann-Whitney U-Test to check for the difference between two samples of students for each level of EI.

6 Discussion

The use of virtual assistants and bots to detect empathy is an interesting and promising direction in the development of artificial intelligence. Such systems can detect emotional signals in users' speech and interact with them on a more empathetic level. We see prospects for further research in this area. Today, research in this direction is beginning.

We have studied some methods that can be used to develop EI. In particular, virtual assistants can analyse the tone of voice, speed of speech, intonation and other parameters to detect emotional signals in the user's speech. For example, the system can respond to a raised tone of voice or certain keywords that indicate stress or excitement. Virtual assistants can also analyse users' text messages to detect emotional signals. They can use natural language processing (NLP) techniques to understand context, identify emotionally charged words and phrases, and analyse sentiment. Based on detected emotional signals, virtual assistants can adapt their responses and reactions. For example, if the system detects signs of stress or irritation in the user's speech, it can offer support, a calm response or help. Some virtual assistants have the ability to express their own emotions using visual elements such as faces, voice effects or emoticons. This allows you to create a more empathetic and engaging communication with the user. Using virtual assistants and bots to recognise empathy can improve the quality of communication and interaction with users. They can provide support, understanding and empathy that can help create a positive technology experience. However, it is also important to ensure that they are used ethically and with user privacy and security in mind.

The virtual assistant can have an animated face that can express different emotions such as happiness, sadness, anger, surprise, etc. For example, the face might smile for positive messages or change to a sad expression for negative ones. A virtual assistant may be able to change the tone of voice or intonation to express different emotions. For example, the voice might be more lively and energetic when expressing joy, or more calm and gentle when expressing sympathy. It can use emoticons to express different emotions. For example, when chatting, the virtual assistant can send hearts or smiley faces to express support or happiness. A virtual assistant can also move or animate to express different emotions. For example, it can clap its hands to show happiness or raise its eyebrows to show surprise. In addition to visual elements, the virtual assistant can also use appropriate words and phrases to express different emotions. For example, it can use the positive word "great" or express understanding when communicating about negative aspects. These elements can help create a more empathetic and cohesive user experience, improving the experience of interacting with a virtual assistant or bot.

7 Conclusion

Our research has led us to conclude on the importance of developing EI in future teachers. The strategies used are a valuable tool to enhance the EI of students in higher education institutions. According to the aim of the study, we carried out some tasks.

1. We learned about the understanding of the concept of EI by university students. After lectures and practical material, they understood the components and meaning of EI.
2. The strategies to improve each of the components of EI of future teachers were implemented during the study. It was found that the most effective of them are "seeking feedback", "challenging assumptions and beliefs" and "therapy or coaching" to develop self-awareness; "setting clear goals", "practicing delayed gratification", "practicing self-reflection" and "developing resilience" to improve self-regulation; "practicing active listening", "asking for feedback" and "participating in group activities" to develop social awareness; "initiating conversations and developing small talk skills", "practicing assertiveness", "observing social role models", "practicing flexibility", "seeking professional support" to improve social skills.
3. At the final stage of the research, we tested the effectiveness of the EI building strategies by using the Mann-Whitney U-test to evaluate the difference between two samples of university students.

We see the prospects for further research in the study of the peculiarities of the implementation of virtual assistants with the aim of developing the EI of future teachers in the course of university studies.

References

1. OECD. Education 2030: The Future of Education and Skills. Position paper (2018). http://www.oecd.org/education/2030/E2030%20Position%20Paper%20(05.04.2018).pdf. (in English)
2. The Standard of Higher Education in Specialty 013 Primary education in the field of knowledge 01 Education/Pedagogy for the first (bachelor) level of higher education, approved by Order No. 357 of the Ministry of Education and Science of Ukraine dated 03/23/2021. https://mon.gov.ua/storage/app/media/vishcha-osvita/zatverdzeni%20standarty/2021/07/28/013-Pochatk.osvita-bakalavr.28.07.pdf. (in Ukrainian)
3. Mayer, J.D., Salovey, P., Caruso, D.R.: Emotional intelligence: theory, findings, and implications. Psychol. Inquiry, **15**(3), 197–215 (2004). https://doi.org/10.1207/s15327965pli1503_02. (in English)
4. Andrew, M.: How to Improve Your Emotional Intelligence – Professional Development [B]. Harvard DCE. (n.d.). https://professional.dce.harvard.edu/blog/how-to-improve-your-emotional-intelligence/ (in English). Accessed 18 June 2022
5. Cassano, F., Tamburrano, A., Mellucci, C., Galletti, C., Damiani, G., Laurenti, P.: Evaluation of emotional intelligence among master's degree students in nursing and midwifery: a cross-sectional survey. Int. J. Environ. Res. Public Health **17**, 6347 (2020). https://www.mdpi.com/1660-4601/17/17/6347. (in English)
6. Gilar-Corbi, R., Pozo-Rico, T., Sánchez, B., Castejón, J.-L.: Can emotional intelligence be improved? A randomized experimental study of a business-oriented EI training program for senior managers. Plos One, **14**(10) (2019). https://doi.org/10.1371/journal.pone.0224254

https://journals.plos.org/plosone/article?id=https://doi.org/10.1371/journal.pone.0224254 (in English)
7. Jiménez-Picón, N., Romero-Martín, M., Ponce-Blandón, J.A., Ramirez-Baena, L., Palomo-Lara, J.C., Gómez-Salgado, J.: The relationship between mindfulness and emotional intelligence as a protective factor for healthcare professionals: systematic review. Int. J. Environ. Res. Public Health **18**(10), 5491 (2021). https://www.mdpi.com/1660-4601/18/10/5491 (in English)
8. Kovalchuk, V., Prylepa, I., Chubrei, O., Marynchenko, I., Opanasenko, V., Marynchenko, Y.: Development of emotional intelligence of future teachers of professional training. Int. J. Early Childhood Spec. Educ. (INT-JECSE), **14**(1), 39–51 (2022). https://doi.org/10.9756/INT-JECSE/V14I1.221006 https://archer.chnu.edu.ua/bitstream/handle/123456789/4206/Development%20of%20Emotional.pdf?sequence=1&isAllowed=y (in English)
9. Banit, O., Merzliakova, O.: Identification the key strategies of resilience of Ukrainian teachers in the conditions of war. Eur. Humanit. Stud. State Soc. **1**, 4–23 (2023). file:///C:/Users/%D0%AF/Downloads/Identification_the_key_strategies_of_res.pdf (in English)
10. Pashko, A.O., Pinchuk, I.O.: Intellectual methods of estimation of intending primary school teachers' foreign language communicative competence. In: IEEE International Conference on Advanced Trends in Information Theory (ATIT), pp. 439–444. Kyiv, Ukraine (2019). (in English)
11. Pashko, A., Pinchuk, I.: Methods of classifying foreign language communicative competence using the example of intending primary school teachers. In: Babichev, S., Lytvynenko, V., Wójcik, W., Vyshemyrskaya, S. (eds.) Lecture Notes in Computational Intelligence and Decision Making. ISDMCI. Advances in Intelligent Systems and Computing, vol. 1246. Springer, Cham (2020). https://ieeexplore.ieee.org/document/9030429 (in English)
12. Machia, L.V., Agnew, C.R., Arriaga, X.B.: Interdependence, Interaction, and Close Relationship, pp. 303–306. Cambridge University Press, Cambridge, UK (2020). https://assets.cambridge.org/97811084/80963/excerpt/9781108480963_excerpt.pdf (in English)
13. Bolotnikova, I., Kucherhan, Y., Vyshnyk, O., Shyrobokov, Y., Dzykovych, O.: Psychological well-being of a pedagogue in the conditions of war. RevistaEduweb, **17**(2), 149–160 (2023). https://doi.org/10.46502/issn.1856-7576/2023.17.02.13 (in English)
14. Asieieva, Y., Trynchuk, O., Furman, A., Melnichuk, I., Koval, K.: Emotional intelligence peculiarities of civil servants of Ukraine. Amazonia Investiga **10**(48), 129–137 (2021). https://doi.org/10.34069/AI/2021.48.12.14 (in English)
15. Boyatzis, R.E.: The behavioral level of emotional intelligence and its measurement. Front. Psychol. **9**, 01438 (2018). https://doi.org/10.3389/fpsyg.2018.01438 https://www.frontiersin.org/journals/psychology/articles/https://doi.org/10.3389/fpsyg.2018.01438/full (in English)
16. O'Connor, P.J., Hill, A., Kaya, M., Martin, B.: The measurement of emotional intelligence: a critical review of the literature and recommendations for researchers and practitioners. Front. Psychol. Sec. Emotion Sci. **10** (2019). https://doi.org/10.3389/fpsyg.2019.01116 (in English)

Author Index

Z. Hu et al. (Eds.): ICCSEEA 2024, LNDECT 242, pp. 729–731, 2025.
https://doi.org/10.1007/978-3-031-84228-3

The manufacturer's authorised representative in the EU is Springer Nature Customer Service Centre GmbH, Europaplatz 3, 69115 Heidelberg, Germany. If you have any concerns regarding our products, please contact ProductSafety@springernature.com

Printed and bound by CPI Group (UK) Ltd, Croydon, CR0 4YY

01/10/2026

02225940-0005